4th International Conference on Energy and Environment

(ICEE 2013)

IOP Conference Series: Earth and Environmental Science Volume 16

Putrajaya, Malaysia
5 – 6 March 2013

ISBN: 978-1-63439-957-9
ISSN: 1755-1307

Printed from e-media with permission by:

Curran Associates, Inc.
57 Morehouse Lane
Red Hook, NY 12571

Some format issues inherent in the e-media version may also appear in this print version.

Printed by Curran Associates, Inc. (2015)

Institute of Physics
Dirac House, Temple Back
Bristol BS1 6BE UK

Phone: 44 1 17 929 7481
Fax: 44 1 17 920 0979

techtracking@iop.org

Additional copies of this publication are available from:

Curran Associates, Inc.
57 Morehouse Lane
Red Hook, NY 12571 USA
Phone: 845-758-0400
Fax: 845-758-2634
Email: curran@proceedings.com
Web: www.proceedings.com

TABLE OF CONTENTS

4th International Conference on Energy and Environment 2013 (ICEE 2013) IOP Publishing
IOP Conf. Series: Earth and Environmental Science **16** (2013) 011002 doi:10.1088/1755-1315/16/1/011002

KEYNOTE SPEAKER I

Speaker: Y.Bhg. Datuk Seri Ir. Azman Mohd
Title: Sustainable Development: TNB's Experience

BIOGRAPHY

Y.Bhg. Datuk Seri Ir Azman Mohd was appointed as the President / Chief Executive Officer of Tenaga Nasional Berhad (TNB) on 1st July 2012. He is actively involved in International Conference on Electricity Distribution (CIRED) Malaysia and was appointed as Chairman from 2008 to 2010. He is now in the process establishing Institute of Asset Management (IAM UK) Malaysian Chapter. He held several key positions in TNB for the past 33 years. He started his career in TNB as an electrical engineer at District Office and gradually positioned himself to higher posts at various offices over the years, holding positions such as District Manager, State General Manager and General Manager of Strategic Management & Organization Development at the Head Quarters before he was appointed as Senior General Manager TNB (Operational Region 2) in 2006. He obtained his Diploma in Engineering from the England Newark Technical College, United Kingdom in 1976 and degree in Engineering from the University of Liverpool in 1979. He then obtained his Master in Business Administration from University of Malaya in 1996.

SYNOPSIS

In the past few decades, Malaysia's rapid economic growth, increasing population and urbanization has resulted in the increase of energy consumption. This development will require the Electricity Supply Industry to adopt strategies relating to sustainable development, energy security and environment. To address the environmental and sustainability challenge, TNB has embarked and planned for various green energy initiatives in its long term strategic policy and plan.

Sustainable utilisation of energy is being given increasing attention in Malaysia where in 2009, National Green Technology Policy was introduced to ensure sustainable development and to conserve the environment for future generation. In the Tenth Malaysia Plan (2011-2015), National Renewable Energy Policy and Action Plan were established.

TNB has established and adopted business strategies and practices that take into consideration the environmental impact. TNB's green initiatives include development of low carbon generation technologies (e.g. solar hybrid, biomass project, mini-hydros and biogas). In support of Government initiatives, TNB also signs Renewable Energy Power Purchase Agreements (REPPAs) with Renewable Energy developers under Small Renewable Energy Programme and Feed-in-Tariff (FiT) mechanism. In addition, Demand Side Management (DSM) and Energy Efficiency (EE) options were also being implemented including time-of-use tariff, continuous communication & awareness programmes, efficiency enhancement at power plants & substations, EE in buildings etc. TNB also conducts carbon footprint study across the whole value chain of its business.

IOP Conf. Series: Earth and Environmental Science **16** (2013) 011002 doi:10.1088/1755-1315/16/1/011002

KEYNOTE SPEAKER II

Speaker: Prof. Dr.-Ing. Isabel Kuperjans

Title: Municipal Climate Protection Concepts:

A Way Towards a Smart, CO_2 Emission Free City

BIOGRAPHY

Prof. Dr.-Ing. Isabel Kuperjans is currently with Institut NOWUM-Energy in Germany. She had her Diplom-Ingenieur in Mechanical Engineering Energy Techniques at RWTH Aachen University in 1994. In 2000, she graduated from her PhD on analysis and assessment of industrial energy systems. From 2001 to 2006, she did her post doctorate in Chair of Technical Thermodynamics, RWTH Aachen University and was coordinator of the research-group "Energy Systems". From 2006 to 2010, she worked as planning engineer at Carpus+Partner AG, Aachen, where she designed energy systems for buildings. She became a full professor at FH Aachen, Campus Juelich, department of energy engineering in 2010 and in 2012 the head of FH Aachen research institute NOWUM-Energy.

SYNOPSIS

Due to climate protection issues, the use of fossil energy and the emission of CO2 should be extremely reduced. The paper describes a way of greenhouse-gas (GHG) emission reduction by development and implementation of municipal climate protection concepts. A systematic approach is described, which is currently pursuit in Germany: In a first step, city councils get funding for the development of a climate protection concept. This concept describes the actual CO2-emissions of the city and the potential and specific measures for GHG emission reduction. After setting up GHG emission reduction goals and the decision for implementation of the concept through the city council, the cities get funding for a so called "climate protection manager" who is - amongst other things - responsible for public relations and education projects concerning municipal climate protection. Further, they can apply for funding for the implementation of some of the developed measures. This approach ensures the integration of most parts of the population in the process of climate protection.

IOP Conf. Series: Earth and Environmental Science **16** (2013) 011002 doi:10.1088/1755-1315/16/1/011002

KEYNOTE SPEAKER III

Speaker: Prof. S.R.S. Prabaharan

Title: Supercapacitors: A Novel Power Storage Option for Sustainable Energy Storage

BIOGRAPHY

Dr. Prabaharan is currently attached to VIT University, Chennai campus, India as a Professor of Electronics and communication engineering. His immediate past appointment was with Manipal International University where he served as a professor. Prior to that, he was a full Professor of Electronics and the head of Power and Energy Research Division in the University of Nottingham Malaysia campus. His esteemed factors include serving various expert committees in Malaysia for instance Multimedia Development Corporation and office of Prime Minister Department (UNIK) and FP7 EU committees. He was the technology consultant to United States Army research project under the Southern University, Baton Rouge, USA. He was the recipient of the prestigious Erasmus Mundus, EU visiting scholar award and held a visiting professorship in France. In 2007, he received a UK-Singapore visiting material scientist award. Recently, he has served as an invited Visiting Professor at IMRAM, Tohoku University, Japan. His expertise centered on the Nanotechnology aspects of advanced batteries (Li-Ion and Li-Air), Super-capacitors for power applications, and RF wireless energy harvesting and storage. He has delivered over 25 invited/keynote talks in International scientific conferences/symposia and organized international conferences (ACSSI-8 and Sym-k-ICMAT2007) as chair. Hither to, he has published over 75 high impact journal papers, 150 conference proceedings and a few international granted patents (USA and Japan). Dr. Prabaharan is known for his notable research in areas of Li- Ion Batteries, Supercapacitors and advanced Power solutions for electronic gadgets. His current research interests include Memristors, Li/Air semi fuel cells, Supercapacitors and advanced power solutions for niche electronic gadgets.

SYNOPSIS

Supercapacitors or Electrochemical Capacitors are intriguing power storage media which store enormous power unlike advanced batteries. Supercapacitors are regarded as a special class of capacitor component exhibits different capacitive characteristics compared to dielectric counterpart. The former operates based on typical double layer charge storage across electrode/electrolyte interface whose separation distance (d) between the layer of positive and negative charge accumulation is approximately 1 nm while the area (A) of the electrode is addressed through porous carbon electrodes having a surface areas > 1500 m^2/g. Thus, both area (A) and the distance (d) between electrodes are addressed simultaneously which enables a huge capacitance to be realised in a relatively small area and thickness. Hence, supercapacitors have found a place of pride in both electrical and electronics systems in which the power is a paramount. Concerning miniaturised smart communication devices, portable power is an utmost priority. That is, designers of portable and miniaturised electronics are challenged to create devices that do everything while running endlessly on a single battery charge. These challenges are addressed effectively by adding a supercapacitors as power delivery component along with batteries (latter delivers energy rather than burst of power).

With devices now sporting large vibrant touch-sensitive displays, multicore CPU and graphics processors, and an assortment of wireless modems for high speed communications anywhere on the planet, high battery capacity has become indispensable. Nonetheless, with the limited capacity and cycles to deliver, batteries alone cannot sustain to this demand. One of the blessing avenues to meet such goals is to look at battery pack as a combination of battery and supercapacitor as a hybrid power pack to meet such demands. These hybrid power packs find tremendous applications ranging from power hunger hand-held electrical gadgets (power tools etc), missile launching, microwave communications and hybrid and/or zero-emission vehicles. The talk highlights various advancements achieved hither to in these power sources and as to how the nanotechnology aspects help enhancing the functionalities of these devices. An advanced hybrid solution by marrying batteries and supercapacitors as hybrid power bank for applications demanding high burst of power will also be discussed. With a proof of concept, the speaker will show as to how a hybrid combination of lithium- -ion batteries and supercapacitor helps the battery's run time as well as life span of lithium – ion batteries taking into some real world electronic gadgets such as a toy remote helicopters. This talk covers speaker's contribution for the development of supercapacitors and hybrid packs over 16 years in the field of advanced portable power sources.

IOP Conf. Series: Earth and Environmental Science **16** (2013) 011002 doi:10.1088/1755-1315/16/1/011002

KEYNOTE SPEAKER IV

Speaker: Dr. Sandra Esteves

Title : Anaerobic Digestion Process - Optimisation and Innovations for Increased Energy Generation and Sustainability

BIOGRAPHY

Dr Sandra Esteves received her Ph.D in 'Monitoring and Control of Biological Textile wastewater Treatment using Artificial Neural Networks (ANN)' at University of Glamorgan in 2002. She launched the MSc Renewable Energy and Resource Management (in 2004) and was the Project Manager for the EU-ESF grant on HE Skills in Renewable Energy. Her research has been funded by EU-FP6, FP7, IEE, EU-KEF, EU-ERDF, WRAP, Welsh Government, Carbon Trust, EPSRC and EPSRC/NERC. She directs the Wales Centre of Excellence for Anaerobic Digestion. She is currently a co-investigator on the EU-FP7-SME ORION Organic waste management by a small-scale Innovative automated system of anaerobic digestion (21 European partners, from both academic and industrial organizations). Her research interests include anaerobic digestion of wastewaters, solid wastes and crops and system monitoring and control, bio-hydrogen and bioplastics production, microbial fuel cells, biogas clean-up and upgrading, organic resources recovery, digestate treatment and dewatering as well as whole system environmental and economic analysis.

SYNOPSIS

Anaerobic Digestion (AD) is a versatile biochemical process that is able to degrade a multitude of organic substrates, from a variety of municipal, industrial and agriculture residues and wastes to energy crops. Resultant from the AD process is methane, which can be used to produce renewable electricity, heat or utilised as a vehicle fuel; and also a digestate, which should be low in easily digestible organic content and can contain valuable nutrients. AD is an environmentally attractive technology and benefits typically include waste treatment, pollution reduction, renewable energy generation and improvement of agricultural practices by recycling of plant nutrients.

The session will provide examples of the latest technical advancements related to process design, operation, monitoring and control that are allowing greater energy generation, improved integration with other technologies and the production of valuable and sustainable products. Some of these advancements and improvements do not only relate to R&D scale operations but they have started in some cases to be demonstrated at full-scale.

4th International Conference on Energy and Environment 2013 (ICEE 2013) IOP Publishing
IOP Conf. Series: Earth and Environmental Science **16** (2013) 011002 doi:10.1088/1755-1315/16/1/011002

KEYNOTE SPEAKER V

Speaker: Prof. Tosaka Hiroyuki

Title : Evaluation of Water Environment / Resources / Disasters through Integration of Technologies

BIOGRAPHY

Prof. Tosaka Hiroyuki did his Bachelor and Master degree in University of Tokyo from the Department of Resources Development Engineering. He worked with Japan Oil Engineering, Co., Ltd. after his Master until 1986. In 1986, he became a Professor in University of Tokyo.

SYNOPSIS:

Our livelihood sphere is a part of the great hydrologic circulation system. It gives precipitation on the land area, forms terrestrial biosphere environment and water resources, and brings about water-related disasters repeatedly. The countermeasures for increasing amenity and safety in residential area should be designed place by place through quantitative evaluation with up-to-date technologies. I will pick up the following topics in the presentation.

(1)Approach to water environment and water-related disasters

The activity of a Malaysia-Japan cooperative project for reducing water-related damages by landslides and floods in Malaysia is introduced. The project was initiated in 2011 as a 5-year SATREPS (Science and Technology Research Partnership for Sustainable Development) PROJECT in association with Malaysia and Japan. The target of the project is to reduce the casualties and socio-economic damages by landslides and inundations in Malaysia, with the use of up-to-date technologies such as satellite information/analysis, field monitoring, and numerical simulation.

The Malaysian team consists of Universiti Sains Malaysia (USM), Multimedia University (MMU), Universiti Tenaga Nasional (UNITEN) and related national research institutes. The Japanese side consists of The University of Tokyo (UT), the Center for Environmental Remote Sensing, Chiba University (CEReS), National Institute of Earth Science and Disaster Prevention (NIED), International Centre for Water Hazard and Risk Management (ICHARM), Vision Tech Inc. (VTI), and other universities of Japan. The project has five groups consisting of researchers of both countries as G1: Remote Sensing Group, G2: Flood/Inundation Analysis Group, G3: Landslide Analysis Group, G4: Database Construction Group, and G5: EWS (Early Warning System) Construction Group.

(2) An integrated simulation technology for terrestrial environment/disasters

An integrated simulation technology is introduced, in which hydrologic fluids and heat circulation including the near-surface atmospheric layer, the surface and subsurface regions of a watershed are treated seamlessly.

IOP Conf. Series: Earth and Environmental Science 16 (2013) 011003 doi:10.1088/1755-1315/16/1/011003

ACKNOWLEDGEMENTS

CHAIRMAN

Prof. Dr. Mohd. Zamri Yusoff

DEPUTY CHAIRMAN

Prof. Ir. Dr. Abdul Halim Shamsuddin

SECRETARIES

Dr. Noor Miza Muhamad Razali

Ms. Hidayah Basri

TECHNICAL PROGRAMME

Prof. Dr. Chandan Kumar Chakrabarty

Prof.Dr. Ibrahim Bin Ahmad
Assoc. Prof Dr. Ir. Lariyah Mohd Sidek
Assoc. Prof Dr. Ir. Norashidah
Assoc. Prof. Dr. Jagadeesh
Assoc. Prof. Dr. Nasri A.Hamid
Assoc. Prof. Dr. Ong Hang See
Dr. Adlansyah Bin Abd Rahman
Dr. Fairuz Bin Abdullah
Dr Goh Su Mei
Dr. Meenaloshini Satgunam
Nagaletchumi Balasubramaniam
Nor Iza Bte. Mohd Rawi
Nur Badariah Bte. Ahmad Mustafa
Nurul Asyikin Bte Mohamed Radzi

PUBLICITY, MARKETING & SPONSORSHIP

Dr. Agileswari K. Ramasamy

Assoc. Prof. Dr. Shahida Begum
Assoc. Prof. Ir. Dr. Tiong Sieh Kiong
Dr. Ir. Chong Kok Hen
Dr. Ir. Johnny Koh Siaw Paw
Norma Bte. Hassan
Jamilah Zakariah
Suryati Bte. Mohamed
Shahibah Bte Bonyamin

LOGISTICS

Assoc. Prof. Dr. Abu Bakar Ghazali

Dr. Mohd Aminur Rashid
Dr. Kong Sih Ying
Aznita Bte. Abdul Aziz
Sazali Bin Muskiman
Intan Nor Zuliana Binti Baharuddin
Rasyikin Bte. Roslan

REGISTRATION

Dr. Fathoni Usman

Dr. Salmia Binti Beddu
Dr. Wong Leong Sing
Ean Lee Woen
Shangari K. Raveendran
Noor Zaimah Mohd Mokhtar
Constance Linda Gnanasegaran
Dr. Grace Pua Fei Ling

FINANCE

Dr. Norazizah Bte Mohd Aripin

Azlina Bte. Abdullah
Suhailah Bte. Leman
Nur Asma Hussain

IOP Conf. Series: Earth and Environmental Science 16 (2013) 011003 doi:10.1088/1755-1315/16/1/011003

ACKNOWLEDGEMENTS

IT SYSTEM & WEBPAGE

Dr. Goh Chin Hock

Aiman Bin Ismail
Zulkifli Bin Ishak

FOOD & BEVERAGES

Dr. Yap Boon Kar

Mohd Azrin Bin Mohd Azau

OPENING CEREMONY EXECUTION & PROTOCOL

Dr. Khairul Salleh Bin Mohamed Sahari

Datin Noor Zarina Musa

Mohd Zafri Bin Baharuddin
Nora Bte. Annuar

CONFERENCE EXECUTION

Prof. Dr. T.M Indra Mahlia

Anas Bin Muhamad Pauzi
Nur Irmawati Bt Om
Sarveswaren a/l Karunanithi
Siti Sarah Ain Bte Fadhil

Support team from: Facility Development & Management Department (FDM), IT & Multimedia Services (ITMS), Finance Department, College of Information Technology (COIT), Corporate Communications Department (CCD)

INTERNATIONAL ADVISORY COMMITTEE

Prof. Ali Sayyigh (World Renewable Energy Network)
Prof. Kaoru Takara (Kyoto, Japan)
Prof. Felix Wu (Hong Kong, RPC)
Prof. K L Lo (Scotland, UK)
Prof. Anthony Griffiths (Cardiff Uni., UK)
Prof. Peter Jimack (Leeds Uni., UK)
Prof. Supitcha Chanyotha (Chulalongkorn Uni., Thai.)
Prof. Abdul Rashid Abdul Aziz (UTP, M'sia)
Assoc. Prof. Amir Al-Falahi (Maple Centre, Canada)
Prof. Kamaruzaman Sofian (UKM, M'sia)
Prof. Pierre Julien (CSU, USA)
Prof. Tosaka Hiroyuki (Tokyo Uni., Japan)
Prof. David J Nagel (Uni. Washington, US)
Prof. Ooi Boon-Teck (McGill, Canada)
Prof. Simon Beecham (Uni. of South Aust.)
Prof. Isabel Kuperjans (FH Aachen, Germany)
Prof. Jean-Bernard Saulnier (Lab. d'Etudes Thermiques, Fr.)
Prof. Farid Nasir Ani (UTM, M'sia)
Assoc. Prof. Dr. Hashem Mohd Ali (Jordan)
Assoc. Prof. Talal F. Yusaf (Uni. of Southern Queensland, Aust.)

TECHNICAL PAPER REVIEWERS

Prof. Dr. Chandan Kumar Chakrabarty
Assoc. Prof. Dr. Ir. Lariyah Mohd Sidek
Assoc. Prof. Dr. Jagadeesh
Assoc. Prof. Dr. Ng Khai Ching
Assoc. Prof. Dr. Saifuddin Hj. M. Normanbay
Dr. Adlansyah Bin Abd Rahman
Dr. Ahmed Hussien Birima
Dr. Cheah Cheng Lai
Dr. Fathoni Usman
Dr. Meenaloshini Satgunam
Ir. Mokhtar Bin Majid
Nurul Asyikin Bte Mohamed Radzi
Prof. Dr. T.M Indra Mahlia
Assoc. Prof. Dr. Au Mau Teng
Assoc. Prof. Dr. Nasri A.Hamid
Assoc. Prof. Dr. Ong Hang See
Dr. Ir. Johnny Koh Siaw Paw
Dr. Agileswari K. Ramasamy
Dr. Azrul Bin Mohd Ariffin
Dr. Fairuz Abdullah
Dr. Leong Yow Peng
Ir. Kumaran Palanisamy
Ir. Zakaria Bin Che Muda

Peer review statement

All papers published in this volume of *IOP Conference Series: Earth and Environmental Science* have been peer reviewed through processes administered by the proceedings Editors. Reviews were conducted by expert referees to the professional and scientific standards expected of a proceedings journal published by IOP Publishing.

IOP Conf. Series: Earth and Environmental Science **16** (2013) 012001 doi:10.1088/1755-1315/16/1/012001

A Preliminary Analysis of Solar Irradiance Measurements at TNB Solar Research Centre for Optimal Orientation of Fixed Solar Panels installed in Selangor Malaysia

A M Hashim[1,3] M A M Ali[2], B Ahmad[1],R M Shafie[1], R Rusli[1], M A Aziz[1], J Hassan[1] and M Z C Wanik[1]
[1]Green Technology Section, TNB Research Sdn Bhd, 1 Lorong Ayer Itam, 43000 Kajang, Selangor, Malaysia
[2]Malaysian Green Technology Corporation, Bandar Baru Bangi, Selangor, Malaysia

E-mail: azlan.hashim@tnbr.com.my

Abstract. The well established rule for orienting fixed solar devices is to face south for places in the northern hemisphere and northwards for the southern hemisphere. However for regions near the equator such as in Selangor Malaysia, the position of the sun at solar noon is always near zenith both to the north and south depending on location and month of year. This paper reports an analysis of global solar radiation data taken at TNB Solar Research Centre, Malaysia. The solar radiation is measured using both shaded and exposed pyranometers together with a pyrheliometer which is mounted on a sun-tracker. The analysis on the solar measurements show that a near regular solar irradiation pattern had occurred often enough during the year to recommend an optimum azimuth orientation of installing the fixed solar panels tilted facing towards east. Even though all the solar measurements were done at a single location in TNBR Solar Research Centre at Bangi, for locations near the equator with similar weather pattern, the recommended azimuth direction of installing fixed solar panels and collectors tilted eastward will also be generally valid.

1. Introduction

The state of Selangor with the nation's capital Kuala Lumpur in federal territory is the most populous region in Malaysia. Presently Malaysia's largest solar farm is also located just south of Selangor at Pajam in the state of Negri Sembilan, hence it is quite important to know the local solar irradiation characteristics in order to optimize the performance of solar energy systems installed in this mid-west coastal region of peninsular Malaysia. The Solar Research Centre at TNB Research (TNBR) located near Bangi in Selangor has been monitoring solar irradiance using precise and accurate solar radiometers at the TNBR Solar House for more than a year. Taking this location to be representative for most parts of coastal Selangor state, this paper report our significant preliminary findings.

2. Factors Affecting Solar Radiation

For places above or below the Earth's tropics, the sun at solar noon is always to the south of zenith in the northern hemisphere and always to the north in the southern hemisphere. So it is obvious that these are the recommended optimum direction to install stationary solar devices at such places.

[3] To whom any correspondence should be addressed.

IOP Conf. Series: Earth and Environmental Science **16** (2013) 012001 doi:10.1088/1755-1315/16/1/012001

However for regions near the equator such as Malaysia, the position of sun at solar noon is always near zenith both to the north and south depending on location and month of year. This will cause the performance of stationary solar panels installed here to be less sensitive to a north or southern oriented azimuth direction and the effect of regular weather patterns which are normally characteristic in equatorial climates becomes more significant in determining the optimal solar panel orientation that will receive the largest amount of incident solar radiation.

The general local weather pattern in this southern part of Selangor area is usually clear skies in the morning. As the ground temperature rises, it tends to produce atmospheric convection currents which push the moist air near ground to the cold air at higher altitudes. This often results in a quite rapid cloud formation that usually occurs near noon and the weather is always likely to remain cloudy in the afternoon and into the evening with frequent thunder storms by the convectional rains in late afternoon during the wet season. The air humidity is high and the often recurring convectional rains accumulate a large amount of total annual rainfall. This near regular local weather occurrences consequently results in a frequently repeated solar irradiation pattern in this area that affects the optimal orientation of fixed solar panels and collectors [1].

3. Solar Radiation Measurements

The new TNB Solar Research Centre was just established in year 2011, so we only manage to monitor and collect slightly more than one year of solar radiation data. The instruments used for measurements of horizontal global and diffuse solar radiation are two Kipp & Zonen pyranometers model CMP-11. The two new pyranometers were mounted and leveled on a Kipp & Zonen Solys-2 sun-tracker with a single shading ball. Mounted on one of the two arms of the sun-tracker is a Kipp & Zonen pyheliometer model CHP-1 to measure direct solar radiation. These solar radiometers were connected to the Kipp & Zonen LogBox for data-logging but presently, all the solar radiation measurements are taken using a Datataker DT-80 data-logger at our solar monitoring tower. The pyranometers and sun tracker are properly levelled and all the solar radiometers are maintained according to the recommended procedures. The solar radiometers desiccants are replaced every 2-3 months upon indicator colour change and the sun tracker alignment is periodically checked and corrected when necessary. The sampling rate of the Kipp & Zonen LogBox was set to record data every one minute while the Datataker DT-80 data-logger now takes one set of readings every five seconds.

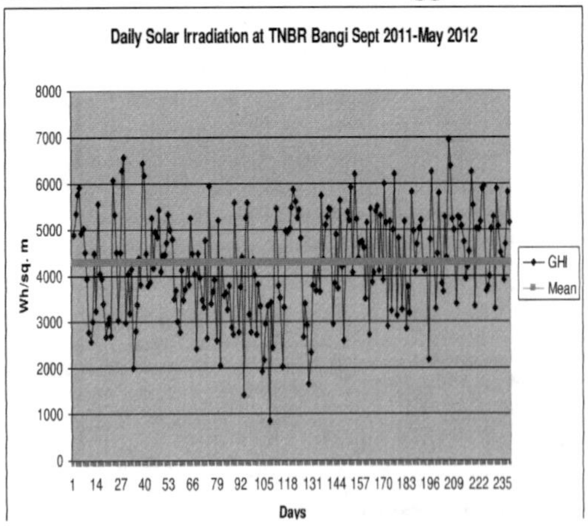

Figure 1. Daily global horizontal solar irradiation (GHI) at TNBR in Bangi, Malaysia.

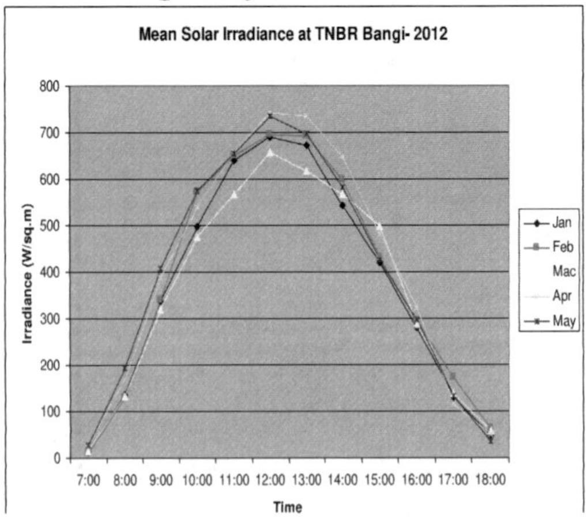

Figure 2. Mean hourly global horizontal solar irradiance at TNBR Bangi, Malaysia.

4. Results and discussion

The solar radiation that is incident on stationary solar panels mainly consists of both direct and diffuse solar radiation which is called the global solar irradiance. The global horizontal irradiance (GHI) is the amount of solar radiation (direct radiation + diffuse radiation) that falls on a horizontal surface [2]. Pyranometer readings that measure the solar GHI were compiled from June 2011 to May 2012 and analyzed. First, the one minute interval GHI readings were summed up for each daylight hour in the day. Then each hourly GHI of the day was averaged for the whole month. Some results of the daily GHI are plotted in Figure 1 and a graph of the mean hourly GHI for the early months of year 2012 is shown in Figure 2. Notice the unpredictable chaotic pattern of the total daily solar irradiation with large variations between each consecutive daily measurement, and that there is hardly any sequence of days with near consistent solar radiation.

Table 1. Mean daily global horizontal solar irradiation data at TNBR in Bangi, Malaysia.

Month	Mean Daily GHI (Wh/m^2.day)			
	Total	Morning	Afternoon	Difference
January	4404.66	2546.69	1857.97	688.71
February	4646.12	2697.80	1948.32	749.48
March	4330.91	2390.75	1940.15	450.60
April	4871.61	2692.19	2179.42	512.77
May	4761.77	2710.54	2051.23	659.31
June	3953.19	2108.85	1844.35	264.50
July	4198.16	2149.08	2049.07	100.01
August	4739.48	2713.69	2025.80	687.89
September	4272.11	2404.35	1867.76	536.59
October	4357.74	2565.77	1791.98	773.79
November	3765.48	2319.28	1446.20	873.08
December	3536.49	2021.04	1515.45	505.59
Average	**4319.81**	**2443.34**	**1876.47**	**566.86**

The daily GHI measurements were then integrated from sunrise to solar noon (morning GHI) and from solar noon to sunset (afternoon GHI) for each day and the monthly average is tabulated in Table 1. During the one year monitoring period, the highest mean daily GHI at TNBR in Bangi is 4871.61 Wh/m^2 for the month of April 2012 and the lowest is 3536.48 Wh/m^2 for the month of December 2011 which is in the wet North-East monsoon season. A monthly plot of the mean daily morning GHI, mean daily afternoon GHI and mean daily total for each month is shown in Figure 3.

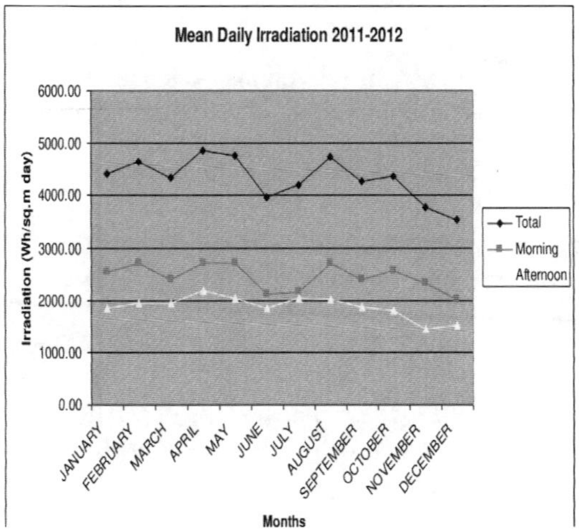

Figure 3. Mean daily global horizontal solar irradiation at TNBR Bangi, Malaysia.

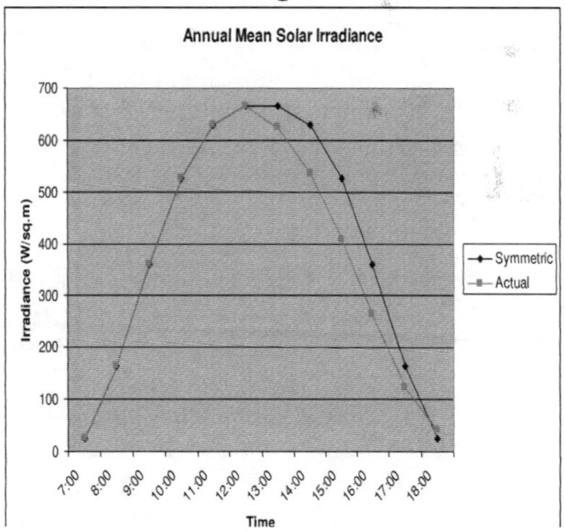

Figure 4. Mean daily global horizontal Irradiance at TNBR in Bangi for June 2011 – May 2012.

If there is no air or the atmosphere is perfectly clear, the amount of daily global horizontal irradiation (GHI) should be symmetrical about solar noon i.e. the morning GHI from sunrise to solar noon should be equal to the afternoon GHI from solar noon to sunset in perfectly clear air, because both the transit time taken and angular motion of the sun is almost the same (see Figure 4). However, due to atmospheric dust scattering, haze and clouds the two amounts of solar energy (morning GHI and afternoon GHI) are obviously not the same. This can be easily seen in Table 1 where the mean monthly morning GHI is clearly higher than in the afternoon when there are more clouds in the sky, for all the months during the measurement period. The mean daily difference for the year is 566.86 Wh/m^2.day that is 13.16% more solar radiation falls in the morning half of the day. This adds up to a significant amount of 206,903.9 Wh/m^2 total for the annual period.

The mean afternoon GHI was found to be quite constant about 2000 Wh/m^2.day for that year with the exception of the wet months of November and December during the North-east monsoon period where the monthly solar irradiance is lowest (see Figure 3). Hence the variation of the total daily solar irradiation is mainly due to the morning GHI that follows almost the same curve pattern as the total daily GHI.

From the acquired results above it is quite obvious that during the one year period of solar radiation monitoring at TNB Solar Research Centre, the amount of solar irradiation in the morning is considerably more than in the afternoon when the sky is usually cloudy. Although these are only preliminary results for just one year, the difference is significant enough that for the proposed absorption cooling system to be implemented at TNBR Solar House [3], we had highly recommended that the flat-plate solar thermal collectors should be installed facing towards east.

5. Conclusion

This paper reports an analysis of global solar radiation data taken at TNB Solar Research Centre, Malaysia. The objective of the analysis is to determine the optimal solar panel azimuth direction that will receive the largest amount of incident solar radiation. The radiation is measured using both shaded and exposed pyranometers together with a pyrheliometer which is mounted on a sun-tracker. The analysis on the solar measurements show that a near regular solar irradiation pattern had occurred often enough during the year to recommend an optimum azimuth orientation of installing the fixed solar panels tilted facing towards east. Even though all the solar measurements were done at a single location in TNBR Solar Research Centre at Bangi which was taken to be representative for the state of Selangor, this does not mean that the results obtained could be applied to places at the immediate surrounding areas in southern Selangor only. For locations near the equator with similar weather pattern, the recommended azimuth direction of installing fixed solar panels and collectors tilted eastward will also be generally valid.

Acknowledgments

The authors would like to acknowledge and express our appreciation to Tenaga Nasional Berhad (TNB), the Managing Director and staff of TNB Research Sdn. Bhd. (TNBR) for their support and contribution towards our Seeding Fund project (TNBR/SF30/2011).

References

[1] Hashim A 2012 Significant Effects of Local Weather Pattern on the Optimum Orientation of Stationary Solar Panels installed in Equatorial Regions *Proc. Electric power supply industry conference (Bali, Indonesia, 15-19 October 2012)*

[2] Duffie J A & Beckman W A 1974 *Solar Energy Thermal Processes* (New York: Wiley–Interscience)

[3] Hashim A 2012 *Development of a Cooling System Using Solar Desiccant Cooling and Earth Tubes – Phase 1: Seeding Fund Green Technology Project Report* (TNB Research Resource Centre, Kajang, Malaysia)

Experimental investigation on the stability and density of TiO_2, Al_2O_3, SiO_2 and $TiSiO_4$

Z Said[a,b*], M Sajid H[a], A Kamyar[a] and R Saidur[a,b]
[a] Department of Mechanical Engineering
Faculty of Engineering, University of Malaya, 50603 Kuala Lumpur, Malaysia
[b] UM Power Energy Dedicated Advanced Centre (UMPEDAC), Level 4,Wisma R & D, University of Malaya, 50603 Kuala Lumpur, Malaysia

Email: zafar@siswa.um.edu.my / zaffar_14@yahoo.com

Abstract. At present, literature data on density of nanofluids are still scarce and controversial. In this paper, density data for nanofluids formed by water, as a base fluid, and solid nanoparticles of two different materials are presented. Density was measured by using a Density Meter DA-130N from Kyoto Electronics. The obtain result is compared with previous models and found incompatible for temperature variations.

1. Introduction

Researchers have always been thoroughly looking for ways to optimize the energy use in various devices. This is achievable through different tools. Nanofluids, defined as the mixture of a base fluid and nano-sized (1-100 nm) solid particles, have been recently introduced as one of the novel solutions in this issue. Because of their enhanced capability in terms of heat transfer, these fluids could be implemented as cooling media in different applications. However, existence of solid particles leads to interesting characteristics in the fundamental thermo-physical properties of nanofluids. Thermal conductivity, viscosity, density and stability have been investigated throughout the recent years by many researchers. . For the first time, Masuda et al. [1] reported a nearly 30% increase in the thermal conductivity of Al_2O_3/water nanofluids. In other early studies different increment in the thermal conductivity were reported like 40% increase [2] for copper and 15% increment [3] for alumina nanoparticles. The mechanism of thermal conductivity enhancement has been discussed by some studies like [4-6]. Brownian motion and thermophoresis have been stated to be the main factors behind the heat transfer behavior of nanofluids.

Viscosity of nanofluids has become of importance since it can majorly affect the performance of the system. Rheological behavior of nanofluid flow has been studied recently by many researchers. Reports indicate that viscosity generally increases due to the existence of nanoparticles [7, 8]. Effective parameters on the viscosity like temperature, particle size and volumetric concentration have been investigated by some researchers [9-11]. Nevertheless, study on the density of nanofluids still seems to be confined to few researches. Sommers and Yerkes [12] investigated the density of alumina propanol nanofluid. Density characterization in this study was done by two different methods.

---

[1] zaffar_14@yahoo.com

Nanofluids with oxide nanoparticles are interesting because they can be produced more easily than metallic or carbon nanotubes nanoparticles, and they are relatively cheap In this study, four different types of water based nanofluids such as, TiO_2, SiO_2, Al_2O_3 and $TiSiO_4$ has been characterized.

2. Experimental Investigation

2.1. Materials
Some Deionised water was used a base fluid. The nanoparticles (TiO_2, P25$\geq$99.5% trace metals basis), (SiO_2, 99.5% trace metals basis), ($TiSiO_4$, 99.8%) and (Al_2O_3, 99.8% trace metals basis) were purchased from Sigma-Aldrich. Field emission scanning electron microscope (FE-SEM) was used to obtain the morphological characterization of the nanoparticles with SIGMA Zeiss instrument (Carl Zeiss SMT Ltd., UK). SEM pictures of the respective nanoparticles are shown in figure 1, where the actual dimensions of the nanoparticles can be determined to be 20-40nm, 20-30nm, and 12-120nm respectively. Clear images of Al_2O_3 and SiO_2 could not be obtained by SEM.

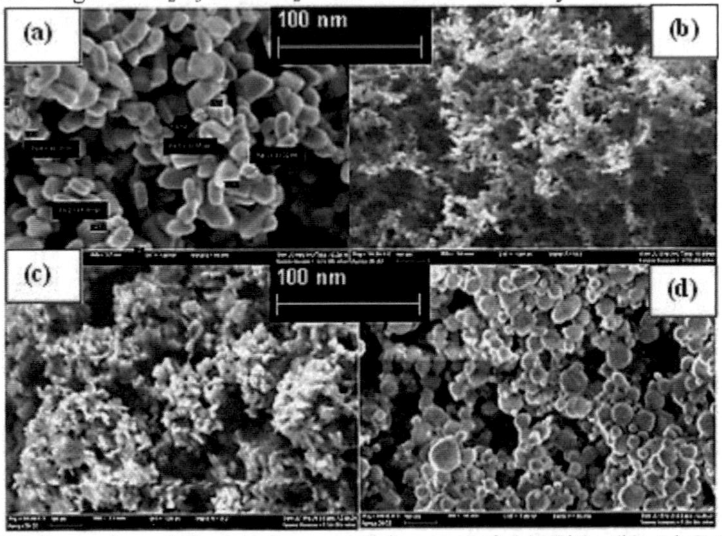

Figure 1. SEM (scanning electron microscope) images of (a) TiO_2 (b) Al_2O_3(c) SiO_2 and (d) $TiSiO_4$ nanoparticles.

Figure 2 given below shows the images of nanoparticles mixed in deionized water, which was obtained using Libra 120 transmission electron microscope at University Malaya. The images obtained show that the particles are indeed very small, whereas some of the particles match the manufacturer's specified size, while some exceeds the specified size (shown in the title boxes).

2.2. Nanofluid preparation
For this investigation, Nanofluids were prepared by dispersing nanoparticles in distilled water using two step method. Nanoparticles were dispersed into the base fluid using ultra sonic agitation. We used 30 mins of sonication time for preparing the samples. Different volume concentrations of nanoparticles (0.05%, 0.5%, 1%, 3% and 5%) were used to prepare distilled water based nanofluids mechanically. The density and stability of the above mentioned nanofluids will be further investigated with different surfactants such as, PEG, HTAB, SDS and PEG.

2.3. Stability characterization of nanofluids
A Zetasizer 3000HSa (Malvern), based on the Dynamic Light Scattering (DLS), was used to analyses the average dimension of the nanoparticles in solution.The Zeta potential of nanoparticles was measured by the Zetasizer 3000 HSa (Malvern).

IOP Conf. Series: Earth and Environmental Science **16** (2013) 012002 doi:10.1088/1755-1315/16/1/012002

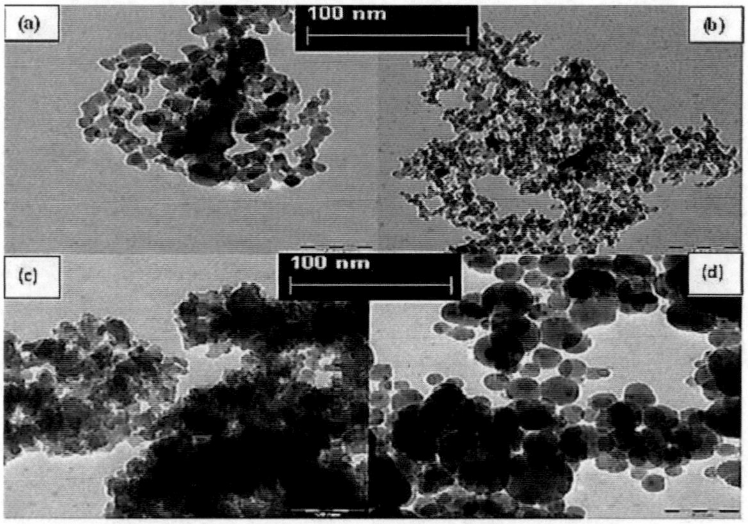

Figure 2 TEM images of (a) TiO2 (b) Al2O3(c) SiO2and (d) TiSiO4 nanoparticles.

2.4. Density of nanofluids

Density is a very important property of a fluid and it affects the friction factor, pump loss, Reynolds number, etc. Pak and Chao [14] considered only the concentration variation at constant temperature (298K) and proposed a formula to measure density of nanofluids. Vajjah et al.[15] has also investigated on density of Al_2O_3, ZnO_2 and Sb_2O_4:SnO_2 EG-W based nanofluids. They compared their result with the proposed formula of Pak and Chao [14] and proposed a correction factor for the relation for their investigated nanofluids. The proposed correction factor also does not reflect the temperature affect. Khanafer and Vafai [16] determined the density of the nanofluids at various temperatures and reported a correlation counting both concentration and temperature. But the correlation is proposed only for the Al_2O_3 water based nanofluid. As the scope of this investigation is mainly to find out the temperature effect on the nanofluids of different concentrations the proposed correlation of Khanafer and Vafai [16] as represented as Equation (1) is used for our work. The Density Meter DA-130N from Kyoto Electronics was used to measure the density of the nanofluids.

$$\rho_{eff} = 1001.064 + 2738.6191\varphi_p - 0.2095T \quad (1)$$

where, $0 \le \varphi_p \le 0.04, 5 \le T(°C) \le 40$

3. Results and Discussion

Equation (1 was compared with our experimental result and represented in the

Figure 1. It is clearly visible from the figure that the equation is only valid for up to 1%vol. Since the above mentioned equation was reported for only Al_2O_3 other nanofluids' results cannot be compared.

Density variation with concentration at constant temperature (25°C) is analysed for the validation of experimental procedure and similar result obtained as claimed by the model shown in **Figure 2**.

4. Conclusion

Thermo-physical property of nanofluids was evaluated in this work. The physical properties at varying temperatures are significant for engineering calculations. The present model of density doesn't satisfy the experimental results. The model was claimed valid for Al_2O_3 for up to 4% vol. concentration but it failed when temperature variation is considered. However, the Pak and Chow model shows an excellent agreement with our experimental value at 25°C. Density variations with temperature at lower concentration (< 1%) are noticeable and significant, thus further investigation should be carried out. New model to meet the experimental results is under study.

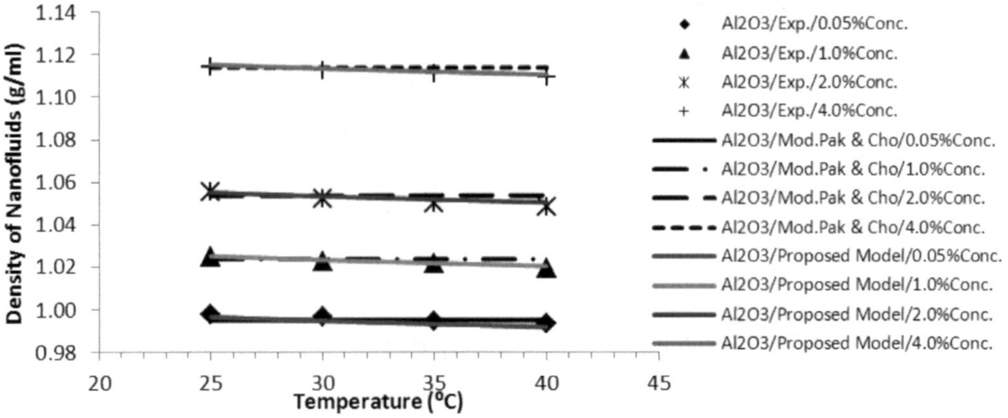

Figure 1. Density vs. Temperature Graph for Al_2O_3 Water based nanofluid at different temperature.

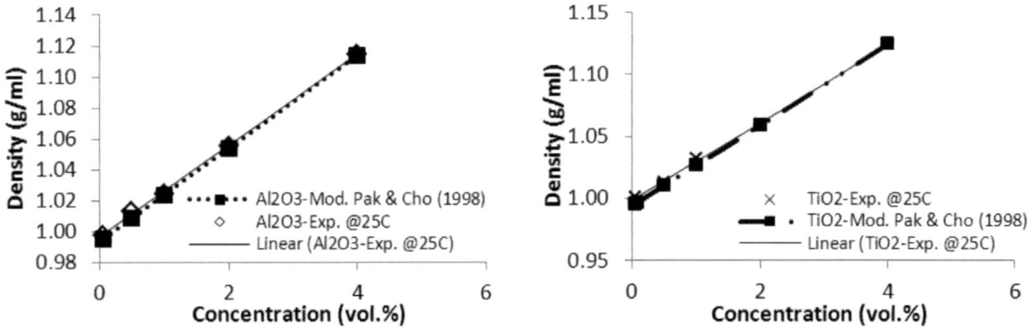

Figure 2. Density vs. Concentration graph for the water based nanofluids

Acknowledgement
The authors would like to acknowledge the financial support from the Ministry of higher education (MOHE), project no: FP019-2011A.

References

[1] Masuda H, Ebata A, Teramae K and Hishinuma N 1993 *Netsu Bussei* **7** 227
[2] Eastman J, Choi S, Li S, Yu W and Thompson L 2001 *Appl. Phys. Lett.* **78** 718
[3] Lee S, Choi S U S, Li S and Eastman J 1999 *J. Heat Transfer* **121** 280
[4] Pak B C and Cho Y I 1998 *Exp. Heat Transfer An Int.J.* **11** 151
[5] Williams W, Buongiorno J and Hu L W 2008 *J. Heat Transfer* **130** 042412
[6] Buongiorno J 2006 *J. Heat Transfer* **128** 240
[7] Murshed S, Leong K and Yang C 2008 *Int. J. Therm. Sci.* **47** 560
[8] Chen L, Xie H, Li Y and Yu W 2008 *Thermochimica Acta* **477** 21
[9] Chen H, Ding Y and Tan C 2007 *New J. Phys.* **9** 367
[10] He Y, Jin Y, Chen H, Ding Y, Cang D and Lu H 2007 *Int. J. Heat Mass Transfer* **50** 2272
[11] Chevalier J, Tillement O and Ayela F 2007 *Appl.Phys. Lett.* **91** 233103
[12] Sommers A D and Yerkes K L 2010 *J. Nanopart. Res.* **12** 1003

IOP Conf. Series: Earth and Environmental Science **16** (2013) 012003 doi:10.1088/1755-1315/16/1/012003

Performance Analysis of AODV Routing Protocol for Wireless Sensor Network based Smart Metering

Hasan Farooq[1] and Low Tang Jung[2]

[1,2]Department of Computer & Information Sciences (CIS) Universiti Teknologi PETRONAS

[1]hasanfarooq7@gmail.com, [2]lowtanjung@petronas.com.my

Abstract. Today no one can deny the need for Smart Grid and it is being considered as of utmost importance to upgrade outdated electric infrastructure to cope with the ever increasing electric load demand. Wireless Sensor Network (WSN) is considered a promising candidate for internetworking of smart meters with the gateway using mesh topology. This paper investigates the performance of AODV routing protocol for WSN based smart metering deployment. Three case studies are presented to analyze its performance based on four metrics of (i) Packet Delivery Ratio, (ii) Average Energy Consumption of Nodes (iii) Average End-End Delay and (iv) Normalized Routing Load.

1. Introduction

This century has witnessed exponential increase in electricity consumption and its unchecked use has brought us to the verge of a power crisis. To combat this, new plans are being devised and we are witnessing a strong push towards the use of renewable energy and smart electric grid deployment.

One of the key components of smart grid is periodic updates on energy consumption of electricity which allows real time tracking of energy demand for optimizing electric load distribution and load forecasting in electrical power. Neighborhood Area Networks in smart power grids comprises of outdoor metering end-points communicating with a gateway. Factors such as large number of metering nodes and harsh outdoor environment possess great challenge for last mile communication. Smart meter reading collection exhibits a many-to-one scenario where all metering nodes communicate with the gateway (sink) so the nodes close to sink will suffer large energy depletion and traffic congestion as they will be involved in most of multi-hop transmissions [1]. Smart Meters can be powered by rechargeable batteries connected to mini-Solar films or in-line powered. In the former case, energy consumption of the routing protocol cannot be neglected.

Recently Wireless Sensor Networks have provided low cost solution for economical, energy efficient and reliable communication between electric utilities and customer meters [2]. The goal of the WSN in smart grid is to provide complete connectivity to all the metering nodes in the network with high reliability and self healing capability. Electric meters equipped with wireless communication capability form an Ad-Hoc self organizing mesh network sending their own and relaying other meters' readings to data collector points thus requiring minimal deployment effort and network management.

Ad-Hoc On-demand Distance Vector (AODV) [3] is a reactive routing protocol for mesh networks. It is adopted as standard routing protocol for Zigbee specification used in Home Area Networks (HANs) [4]. In AODV, routes are only formed when needed. To find new route to destination, source node broadcasts route requests to which destination node replies back (or Intermediate nodes) if they have the route to destination. Every node maintains table of routes with one entry per destination. AODV uses sequence numbers for loop free routing and route maintenance. AODV uses timers to delete expired routes in routing tables. AODV possesses self repair capability by utilizing active neighbor list. AODV is designed for highly dynamic mobile networks with un-predictable topology change in WSN due to fading effects and interference. It is a good choice especially for event driven or periodic data driven WSN applications like Smart Power Grid Metering.

2. Literature review

Most of the existing Research work [4], [5] are concerned with the performance analysis of AODV for Home Area Networks (Indoor appliances sensors communicating with Smart Meter) and very few studies have evaluated it for outdoor Smart Metering Application. S.Ullo et al. [6] did performance analysis of Zigbee based Wireless Sensor Network for smart grid last mile communication and concluded that end-to-end latency and congestion increases with number of nodes. However the study didn't investigate energy consumption of the nodes as well as the effect of meter reading collection frequency and HELLO packet frequency. We believed these factors play an important role in network performance. So a detailed simulation study is required which can outline the performance of AODV for Neighborhood Area Networks.

3. Simulation

We conducted the simulation using Network Simulator NS2. IEEE 802.15.4 Low rate Wireless Personal Area Networks (LR-WPANs) was chosen for Physical and MAC layer as it is similar to 802.15.4g standard for Neighborhood Area Networks. We have simulated four topologies with the number of Smart Meters ranging from 6 to 48 (figure 1). The topologies follow regular grid pattern which is observed in most of the modern housing societies. Simulation parameters are given in table 1.

Table 1. Network Simulation Parameters

Simulation Parameters	
Routing Protocol	AODV
MAC layer/ PHY layer	802.15.4
Channel type	Wireless Channel
Propagation model	Two Ray Ground
Traffic Type	CBR
CBR Packet Size	100 Bytes
Interface Queue Type	Queue/DropTail/PriQueue
Antenna Model	Omni Antenna
Simulation Time	3600 Sec

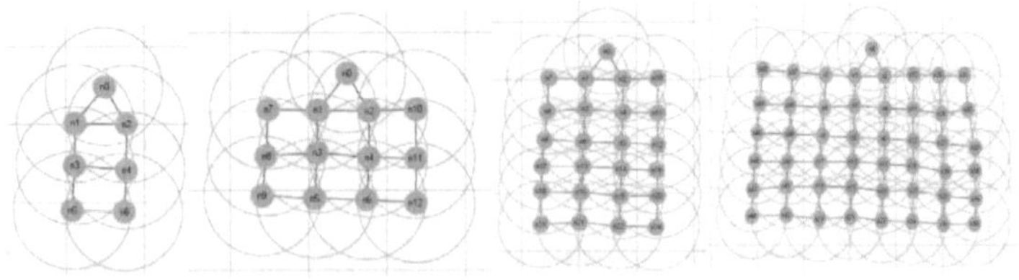

Figure 1. (Left to Right) Topology I, Topology II, Topology III, Topology IV

3.1. Case Study 1

In this case study, we varied number of smart meters and observed their effect on Packet Delivery Ratio. Each meter sent 100 bytes data to the gateway with 30 seconds interval for 1 hour.

The results given in Table 2 and figure 2 indicate that Packet Delivery Ratio continue to decrease with the increase in number of smart meters. Average energy consumption of nodes increases steadily with maximum consumption is observed in Topology IV and II. Average end-end delay ranges from

0.3 sec to 1.8 sec with maximum latency observed in Topology III. Normalized routing load increases exponentially with the increase in the number of smart meters. Bottleneck nodes are the one whose energy is depleted the most and in all scenarios most of them are the ones close to gateway and in the center which will be involved in most of the multi hop transmissions.

Table 2. Case Study 1 Results

Topology	I	II	III	IV
No. of Smart Meters	6.00	12.00	24.00	48.00
No. of Data Packets Sent	720.00	1440.00	2880.00	5760.00
No. of Data Packets Received	412.00	358.00	210.00	451.00
No. of AODV Routing Packets	3579.00	11307.00	26898.00	64249.00
PDR	57.22	24.86	7.29	7.83
Avg. End-End Delay (ms)	343.55	601.63	1732.10	975.77
Normalized Routing Load	8.69	31.58	128.09	142.46
Avg. Energy Consumption (J)	1.23	1.54	1.40	1.64
Maximum Energy Consumption (J)	1.80	2.04	2.32	2.73
Bottelneck Nodes	1, 2, 3	3,6,4	4,6,2	32,4,25,2

Table 3. Case Study 2 Results

Data Packets Interval (sec)	0.50	1.00	30.00	60.00	1800.00
No. of Smart Meters	6.00	6.00	6.00	6.00	6.00
No. of Data Packets Sent	32106.00	21594.00	720.00	360.00	12.00
No. of Data Packets Received	4296.00	4412.00	412.00	229.00	10.00
No. of AODV Routing Packets	5285.00	6943.00	3579.00	1809.00	62.00
PDR	13.38	20.43	57.22	63.61	83.33
Avg. End-End Delay (ms)	301.44	54.26	343.55	433.76	410.26
Normalized Routing Load	1.23	1.57	8.69	7.90	6.20
Avg. Energy Consumption (J)	7.75	5.42	1.23	0.79	0.05
Maximum Energy Consumption (J)	10.00	7.88	1.80	1.07	0.06
Bottelneck Nodes	4,2,1	1,2,4	1, 2, 3	1,2,3	4,3,6

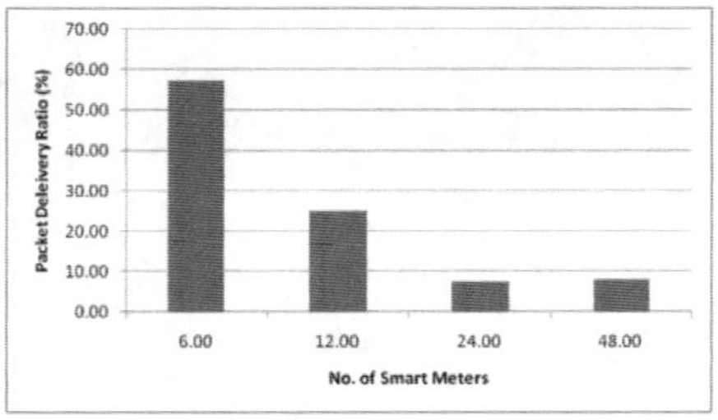

Figure 2. PDR

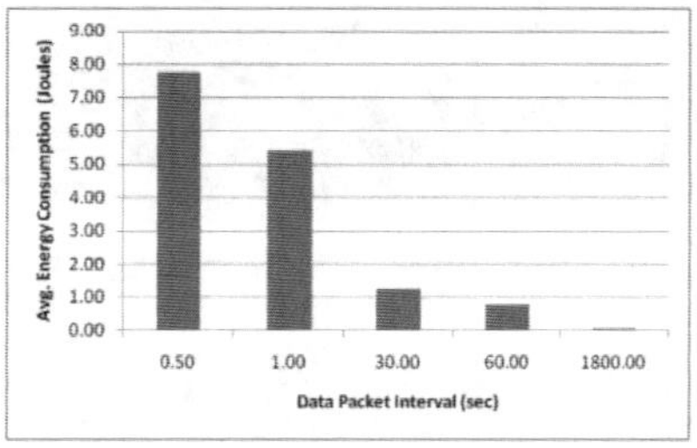

Figure 3. Avg. Energy Consumption

3.2. Case Study 2

In this Case study, Smart Meters Readings Interval was varied to observe its effects on the Performance Metrics. The Topology used is shown in fig.1 with 6 Smart Meters sending their readings to the Gateway for 60 minutes duration. Results given in Table 3 indicate that at Interval of 0.5 to 1 sec, PDF observed is in the range of 20% and increases with the increase in the Data Packet Interval. Avg. end-end delay remains in the range of 0.3 to 0.4 sec except at 1 sec Interval where 0.05 sec latency is observed. Normalized routing load remains in the range of 9%. Average energy consumption suffers heavily at small data packets interval. Analysis of Smart Meter Readings Interval (figure 3) suggests that average energy consumption of the nodes is very high at small data packets interval and so suitable value should be set by the application running on top of AODV to get satisfactory results.

Table 4. Case Study 3 Results for 6 Nodes

HELLO Packets Interval (sec)	1.00	5.00	10.00	30.00	45.00	60.00
No. of Smart Meters	6.00	6.00	6.00	6.00	6.00	6.00
No. of Data Packets Sent	720.00	720.00	720.00	720.00	720.00	720.00
No. of Data Packets Received	181.00	488.00	186.00	78.00	464.00	487.00
No. AODV Routing packets	33809.00	9611.00	7331.00	8595.00	4667.00	4396.00
No. of HELLO Packets Sent	29381.00	5894.00	2929.00	984.00	658.00	499.00
PDR	25.14	67.78	25.83	10.83	64.44	67.64
Avg. End-End Delay (ms)	459.79	507.23	300.18	1173.94	666.15	613.57
Normalized Routing Load	186.79	19.69	39.41	110.19	10.06	9.03
Avg. Energy Consumption (J)	4.40	3.26	1.25	1.59	2.13	2.07
Maximum Energy Consumption (J)	6.61	4.11	1.67	1.86	2.84	2.47
Bottelneck Nodes	4,2,1	2,3,1	4,1,5	4,6,3	2,5,1	4,3,1

Table 5. Case Study 2 Results for 12 Nodes

HELLO Packets Interval (sec)	1.00	5.00	10.00	30.00	45.00	60.00
No. of Smart Meters	12.00	12.00	12.00	12.00	12.00	12.00
No. of Data Packets Sent	1440.00	1440.00	1440.00	1440.00	1440.00	1440.00
No. of Data Packets Received	282.00	541.00	364.00	228.00	143.00	146.00
No. AODV Routing packets	64609.00	20065.00	15709.00	12741.00	13803.00	13885.00
No. of HELLO Packets Sent	54228.00	10927.00	5454.00	1821.00	1218.00	918.00
PDR	19.58	37.57	25.28	15.83	9.93	10.14
Avg. End-End Delay (ms)	680.47	637.06	962.66	836.71	849.20	877.12
Normalized Routing Load	229.11	37.09	43.16	55.88	96.52	95.10
Avg. Energy Consumption (J)	5.66	2.67	1.83	1.34	1.25	1.42
Maximum Energy Consumption (J)	10.00	4.51	2.83	1.93	1.84	2.69
Bottelneck Nodes	2,3,6	1,2,3	2,3,1	3,1,4	2,4,3	6,5,3

3.3. Case Study 3

In this case study, we investigated the effect of HELLO packets frequency on all the Performance Metrics. Two Topologies (I and II) were simulated with 6 and 12 number of smart meters. Simulation was run for 60 minutes and data packet interval was kept at 30 sec.

The results of Topology I are given in Table 4. PDR is largely affected by the HELLO packets Interval. Maximum PDR attained is 67.78% at 5 sec HELLO Interval. Avg. End-End delay is in the range of 0.5 sec to 1.2 sec with low latency observed at small HELLO Packets Intervals. Normalized routing load suffers heavily at small HELLO packets interval around 187% and around 110% when data Packets Interval and HELLO Interval is set at 30 sec (figure 4). Similarly avg. energy consumption of nodes is very high at small HELLO packets interval and decreases steadily with decrease in HELLO packet interval (figure 5).

The Results of Topology II, given in Table 5, indicate maximum PDR of 37.57% is attained at 5 sec HELLO interval. Avg. end-end delay is in the range of 0.68 sec to 0.96 sec with low latency observed at small HELLO Intervals. Maximum normalized routing load of 229.11% is observed at 1 sec HELLO interval respectively. Avg. energy consumption is maximum at small HELLO intervals where Node 2 dies due to maximum energy depletion. Default value of 1 sec, originally designed for mobile nodes, is inefficient for static nodes as in our case leading to maximum energy consumption and routing load.

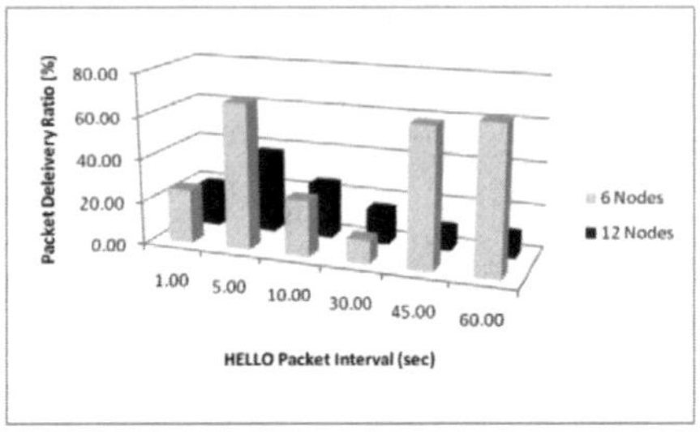

Figure 4. PDR vs HELLO Interval

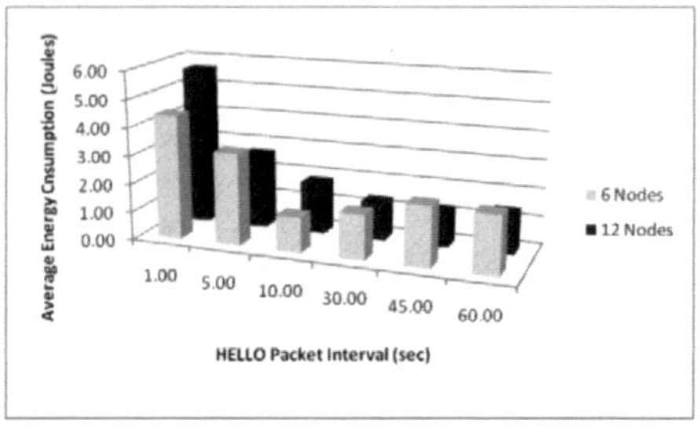

Figure 5. Avg. Energy Consumption vs HELLO Interval

4. Conclusion & future work

In this paper, Performance of AODV routing protocol is investigated for Wireless Sensor Networks based Smart Metering scenario. Simulation results suggest that AODV originally designed for mobile nodes need some modifications for use in Static WSNs. Energy of nodes should be taken into account during routing decision to enhance performance of AODV for such scenarios. HELLO Interval should be increased from 1 to 5 seconds to achieve optimum energy consumption and reliable packet reception in Smart Meters Network. Similarly, Data Packet Interval of 30 minutes should be selected to achieve more than 80% of successful packet delivery ratio. For future work, we want to analyze AODV Performance on Smart Meters Test bed.

References

[1] Saputro N, Akkaya K and Uludag S 2012, *Computer Networks*, **56**(1)
[2] Gungor V C, Lu B and Hancke G P 2010 *IEEE Transactions on Industrial Electronics*, **57**(10)
[3] Perkins C, Royer E B and Das S 2003 *IETF Network Working Group* RFC 3561
[4] Fadlullah Z M, Fouda M M, Kato N, Takeuchi A, Iwasaki N and Nozaki Y 2011 *Communications Magazine, IEEE* **49**(4) 60.
[5] Erol-Kantarci M and Mouftah H T 2011 *Electronics, Communications and Photonics Conference (SIECPC), Saudi International* 1.
[6] Ullo S, Vaccaro A and Velotto G 2010 *MELECON 2010, 15th IEEE Mediterranean Electrotechnical Conference* 443

IOP Conf. Series: Earth and Environmental Science **16** (2013) 012004 doi:10.1088/1755-1315/16/1/012004

Potential of size reduction of flat-plate solar collectors when applying MWCNT nanofluid

M Faizal[1, 2]*, R Saidur[2, 3] and S Mekhilef[4]
[1]Engineering Division, ADP, Taylor's University Lakeside Campus, 47500 Selangor, Malaysia
[2]Department of Mechanical Engineering, University of Malaya, 50603 Kuala Lumpur, Malaysia
[3]UM Power Energy Dedicated Advanced Centre (UMPEDAC), Level 4, Wisma R&D, University of Malaya, 59990 Kuala Lumpur, Malaysia
[4]Department of Electrical Engineering, University of Malaya, 50603 Kuala Lumpur, Malaysia

E-mail: mohdfaizalfauzan@yahoo.com

Abstract. Flat-plate solar collector is the most popular type of collector for hot water system to replace gas or electric heater. Solar thermal energy source is clean and infinite to replace fossil fuel source that is declining and harmful to the environment. However, current solar technology is still expensive, low in efficiency and takes up a lot of space. One effective way to increase the efficiency is by applying high conductivity fluid as nanofluid. This paper analyzes the potential of size reduction of solar collector when MWCNT nanofluid is used as absorbing medium. The analysis is based on different mass flow rate, nanoparticles mass fraction, and presence of surfactant in the fluid. For the same output temperature, it can be observed that the collector's size can be reduced up to 37% of its original size when applying MWCNT nanofluid as the working fluid and thus can reduce the overall cost of the system..

1. Introduction

In Malaysia, most houses are using Electric Water Heater for shower due to the cheap price and easy installation of it. However, Malaysia has a good potential for solar hot water system. Located on the equatorial, Malaysia's climate is hot and humid throughout the year with monthly solar radiation is approximately around 400-600 MJ/m^2 [1]. Solar energy source is sustainable, free, clean and infinite. However, currently solar heater is still expensive, low in efficiency and bulky. One of the effective methods to increase the efficiency is to replace the working fluid with nanofluids. Researches on thermal efficiency by applying nanofluids on flat-plate solar collector have been made in the past few years [2-11]. Experimental investigation conducted by Yousefi, et al. [4] on the effect of Al_2O_3 based nanofluid shown the increase of 28.3% efficiency of flat-plate solar collectors. Lenert and Wang [12] presented a model and experimental study of concentrated solar power application using carbon-coated cobalt (C-Co) nanoparticles and Therminol VP-1 base fluid and concluded that the efficiency is more than 35% with nanofluid and the efficiency will increase with increasing nanofluid height. Lu, et al. [13] shown that the application of Copper Oxide (CuO) nanoparticles in evacuated tubular solar collector will significantly enhance the thermal performance of evaporator and evaporating heat transfer coefficient increased by 30% compared to water as working fluid. 5% improvement in efficiency was found out by Otanicar, et al. [14] by using diversity of nanoparticles with water as base fluid for micro-solar-thermal collector. Shin and Banerjee [15] applied novel nanomaterials in molten salts base fluid for concentrated solar power coupled with thermal storage and experienced an enhancement in operational efficiencies. They also concluded that the cost of electricity will be reduced. Taylor, et al. [10] used graphite based nanofluid in high flux solar collectors that resulting in 10% increase in efficiency. Because of higher thermal conductivity and efficiency of nanofluids, smaller and compact design of solar thermal collector has become possible without affecting the desired output. Smaller size collector can reduced the material usage, cost and energy required in manufacturing [16]. Some studies were made on the potential of size reduction of various engineering applications by using nanofluids. Work had been done by Saidur and Lai [17] in vehicle's weight reduction, Kulkarni, et al. [18] in building heat exchanger's heat transfer area, Leong, et al. [19] on the reduction of air frontal area of a car radiator and Leong, et al. [16] on the size reduction of shell and tube recovery exchanger. None of the studies focus on the size reduction of flat-plate solar thermal collector. Therefore, this study will focus on the potential of size reduction of flat-plate solar collectors when applying MWCNT nanofluid.

2. Collector size calculation

Flat-plate collectors can collect both direct and diffuse solar radiation. To predict and model the collector performance, information on the solar energy absorbed by the collector absorber plate is needed. The solar energy incident radiation on a tilted surface consists of beam, diffuse and ground-reflected radiation [20]. Size reduction calculation is carried out from experimental data of Yousefi, et al. [3]. The data from Yousefi, et al. [3] were selected at 12 noon and the fixed values were taken from Foster, et al. [21] to perform the calculation and analysis.

The important of all for this analysis is the heat-conducting fluid. The fluid will pass through pipes attached to the absorber plate. The fluid will absorb heat from the plate and as it flows through the pipes and increased its temperature to be carried for useful application. The useful heat gain and collectors efficiency can be calculated and compared between the conventional working fluid and proposed nanofluids. After the thermal efficiency of solar collector been determined, the potential of reduction of the size of collector's area can be estimated by:

$$A_c = \frac{\dot{m} C_p (T_{out} - T_{in})}{I_T \eta} \quad (1)$$

3. Results and discussion

Figure 1 compiled the results of size reduction and area of collector when applying MWCNT nanofluid. By maintaining the experimental data from 2 m^2 solar collector area, the area calculated from equation (1) are 1.44 m^2, 1.44 m^2, 1.26 m^2 and 1.38 m^2 for 0.2 wt% MWCNT nanofluid with surfactant, 1L/s, 2L/s and 3L/s 0.4 wt% nanofluid respectively.

The output temperature deceases with increasing mass flow rate due to shorter time for heat to be transferred. However, the efficiency increased with increasing mass flow rate. The highest possible size reduction is observed to be at flow rate 2L/s. Even though the efficiency of 3L/s MWCNT is higher than 2L/s, the difference is not high enough and making the potential of size reduction is higher for 2L/s mass flow rate. When applying 0.2 wt% MWCNT nanofluid the output temperature and efficiency is lower than water because of the increased thermal conductivity of applying nanofluid is small compared to the increased of thermal boundary layer of nanofluid thus decresing the heat transfer as discussed by Ding et al. (2006) and Yousefi et al. (2012b). Therefore, the size reduction is not possible. However, for 0.4 wt % of nanoparticles, the output temperature and efficiency is higher than water. In this case the increased in absorbed energy for higher percentage of nanoparticles in the fluid can compensate the decreasing heat loss. When the nanofluid was prepared at 0.2 wt % with Triton X-100 as surfactant, the efficiency increased and became higher than water. This is due to increase of absorptance of the fluid, improve in heat transfer and better stability and dispersion of nanofluid when applying surfactant (Ghadimi et al., 2011) and (Murshed et al., 2005).

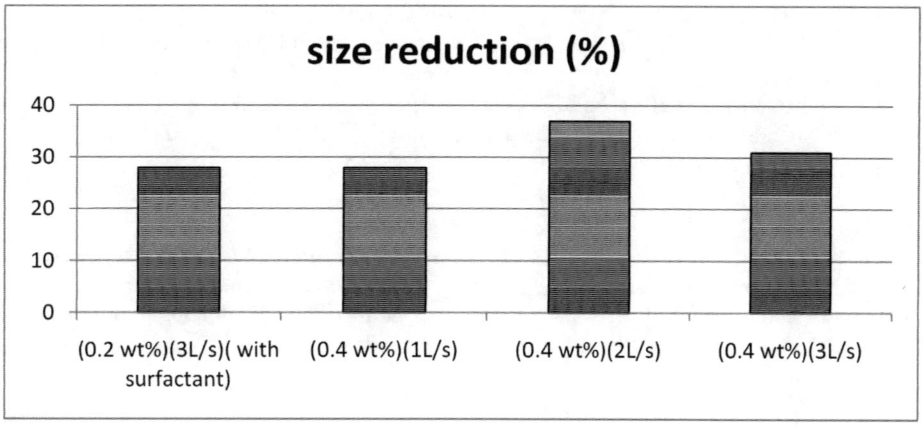

Figure 1. Comparative size reduction and area of solar collector with various working fluid types

4. Conclusion

The thermal efficiency of using the MWCNT nanofluid as absorbing medium in flat-plate solar collector is analyzed based on different mass flow rate, nanoparticles mass fraction, and presence of surfactant. The thermal efficiency of using 0.2 wt% MWCNT nanofluid without surfactant decreased and increased with the presence of surfactant. However, for 0.4 wt% MWCNT nanofluid, the energy efficiency increased even without surfactant. For increasing mass flow rate, the energy efficiency increased while the output temperature decreased. When applying the same output temperature of nanofluid as with water at the same flow rate, it can be observed that the collector's size can be reduced up to 37% of its original size. Therefore, design of a compact and smaller solar collector is possible by using MWCNT nanofluid as working fluid without sacrificing the performance of the system that will directly reduce the material cost and energy required for solar collector's manufacturing process.

References

[1] Mekhilef S, Safari A, Mustaffa W E S, Saidur R, Omar R and Younis M A A 2012 *Renewable and Sustainable Energy Reviews* **16** 386
[2] Yousefi T, Shojaeizadeh E, Veysi F and Zinadini S 2012 *Solar Energy* **86** 771
[3] Yousefi T, Veisy F, Shojaeizadeh E and Zinadini S 2012 *Experimental Thermal and Fluid Science* **39** 207
[4] Yousefi T, Veysi F, Shojaeizadeh E and Zinadini S 2012 *Renewable Energy* **39** 293
[5] Sani E, Mercatelli L, Barison S, Pagura C, Agresti F, Colla L and Sansoni P 2011 *Solar Energy Materials and Solar Cells.* **95** 2994
[6] Saidur R, Meng T C, Said Z, Hasanuzzaman M and Kamyar A 2012 *International Journal of Heat and Mass Transfer* **55**
[7] Otanicar T P 2009 *Direct absorption solar thermal collectors utilizing liquid-nanoparticle suspensions, Doctor of Philosophy, Arizona State University*
[8] Han D, Meng Z, Wu D, Zhang C and Zhu H 2011 *Nanoscale Research Letters* **6** 1
[9] Mercatelli L, Sani E, Zaccanti G, Martelli F, Ninni P D, Barison S, Pagura C, Agresti F and Jafrancesco D 2011 *Nanoscale Research Letters* **6**
[10] Taylor R A, Phelan P E, Otanicar T P, Adrian R and Prasher R 2011 *Nanoscale Research Letters* **6**
[11] Zhu Q, Cui Y, Mu L and Tang L 2012 *Int J Thermophys*
[12] Lenert A and Wang E N 2012 *Solar Energy* **86** 253

[13] Lu L, Liu Z H and Xiao H S 2011 *Solar Energy* **85** 379
[14] Otanicar T, Phelan P E, Prasher R S, Rosengarten G and Taylor R A 2010 *J Renewable Sustainable Energy* **2**
[15] Shin D and Banerjee D 2011 *International Journal of Heat and Mass Transfer* **54** 1064
[16] Leong K Y, Saidur R, Mahlia T M I and Yau Y H 2012 *Energy Education Science and Technology Part A* **30**1
[17] Saidur R and Lai Y K 2011 *Energy Education Science and Technology Part A* **26** 87
[18] Kulkarni D P, Das D K and Vajjha R S 2009 *Applied Energy* **86** 2566
[19] Leong K Y, Saidur R, Kazi S N and Mamun A H 2010 *Applied Thermal Engineering* **30** 2685
[20] Kalogirou S 2009 *Solar Energy Engineering: Processes and Systems* Academic Press
[21] Foster R, Witcher J, Nelson V, Ghassemi M, Mimbela L E and Ghassemi A 2009 *Solar Energy: Renewable Energy and the Environment*: Taylor & Francis

Simulation of existing gas-fuelled conventional steam power plant using Cycle Tempo

M S Jamel[1], **A Abd Rahman**[2] **and A H Shamsuddin**[2]

[1]Mechanical Engineering Department, College of Engineering-Basra University, Basra, Iraq
[2]Centre for Renewable Energy, Universiti Tenaga Nasional, 43000, Kajang, Selangor, Malaysia.

E-mail: mahmood_shaker2005@yahoo.co.uk

Abstract. Simulation of a 200 MW gas-fuelled conventional steam power plant located in Basra, Iraq was carried out. The thermodynamic performance of the considered power plant is estimated by a system simulation. A flow-sheet computer program, "Cycle-Tempo" is used for the study. The plant components and piping systems were considered and described in detail. The simulation results were verified against data gathered from the log sheet obtained from the station during its operation hours and good results were obtained. Operational factors like the stack exhaust temperature and excess air percentage were studied and discussed, as were environmental factors, such as ambient air temperature and water inlet temperature. In addition, detailed exergy losses were illustrated and describe the temperature profiles for the main plant components. The results prompted many suggestions for improvement of the plant performance.

1. Introduction

Generally, the performance of thermal power plants is evaluated through energetic performance criteria based on the first law of thermodynamics. However, recently, the exergetic performance based on the second law of thermodynamics has been found to be a useful method in the design, evaluation, optimisation and improvement of thermal power plants. Therefore, performing exergetic and energetic analyses together can give a complete depiction of system characteristics. In the literature, there are a number of papers concerning the use of energy- and exergy-based studies for plant performance and for comparing different types of thermal power plants, in addition to identifying plant modifications for better performance [1-5]. In the most cases, the authors studied simulations of coal-fired thermal power plants using classical programming methods, in addition to using some software, such as GateCycle, SteamPro and IPSEpro and others. In this paper, the focus is on the simulation of a gas-fuelled steam power plant using Cycle-Tempo software.

2. Simulation

The selected case study is AL-Hartha Steam power plant located in Basra, Iraq. The superheated steam enters the two-stage single reheat steam turbine at 541 °C/125 bar and 541 °C/36.77 bar, for high and intermediate pressure stages, respectively. Steam enters the low pressure stage with a pressure of 6.934 bar; the condenser pressure is 80 kPa. The simulation process and the most important parameters are described in this section. The parameters can be changed to build and simulate different cases and a flow-sheet computer program, "Cycle-Tempo" (Cycle-Tempo-Release 5), is used for that purpose. It is a well-structured package for steady state thermodynamic modelling and analysis of systems for the production of electricity, heat and refrigeration [6,7]. In order to simulate the existing steam plant, the following assumptions were made:

1. Ambient pressure (Po) and temperature (To) of the reference environment are considered as 1.013 bar and 25 °C, respectively (Iraqi climatic conditions).
2. The relative humidity of the ambient air is taken as 60%.
3. The chemical composition of the reference environment model constitutes (in mole fractions): N_2:0.7565, O_2:0.2030, H_2O:0.0312, CO_2:0.0003 and others: 0.009.
4. The terminal temperature difference of the closed feed water heaters (FWHs) is calculated from existing plant design data for each heater[8].
5. Pressure drop in the pipes and steam generator is assumed equal to that in the reference power plants.
6. Isentropic efficiency of fans and pumps is considered as 80% and 85%, respectively.
7. The power consumption by miscellaneous plant (plant control systems, lighting, HVAC), steam turbine auxiliaries and transformer losses of power plants is considered as 5 MW.
8. Fuel gas temperature is equal to ambient air temperature when entering the combustor.
9. Standard air composition is used for plant air inlet.
10. Gas fuel ultimate analysis on volumetric basis is: N_2: 0.0046, CH_4: 0.729, C_2H_6: 0.2208, C_3H_8: 0.0164, CO_2:0.0264, C_4H_{10}: 0.0025, C_5H_{12}: 0.0003[9].

3. Results and discussion

Figure 1 presents a schematic of the simulated plant cycle for fuel gas firing mode in the existing plant. The simulated cycle is first verified with practical data collected from the actual plant and good results are obtained. The plant performance is first evaluated through: cross-plant energy and exergy efficiencies, plant net power output, combustor firing temperature and stack exhaust temperature. Iraqi environmental conditions selected include an ambient air temperature of 0–60 °C and inlet cooling water temperature of 5–40 °C. Performance factors, such as exergy and energy efficiencies, stack exhaust temperature and combustor firing temperature increase with increasing ambient air temperature because of the decrease of heat losses through pipe lines and other station parts. However, this is not the main reason for the increase in thermal efficiency; the mean reason is that the temperature of the air entering the furnace will increase and thus, the demands of the air reheating process will decrease, saving more steam from being extracted for this purpose. Also, the plant net power output decreases with increasing inlet cooling water temperature. The reason behind this increase in the cooling water flow is the relation between its temperature and enthalpy. As the temperature of the water increases its enthalpy will increase too, causing a decrease in its ability to absorb heat energy from the steam in the condenser. Therefore, for the same heat duty in the condenser, an additional amount of cooling water will be drawn to compensate for the temperature increment. Operational factors such as stack gas temperature and excess air percentage are also studied here. The plant net thermal efficiency is greater at minimum stack gas temperature values. As the temperature of the exhaust gases leaving the station increases, the associated energy loss increases too. For this reason, some of these gases are recycled and mixed with the fresh air entering the furnace to minimise these losses.

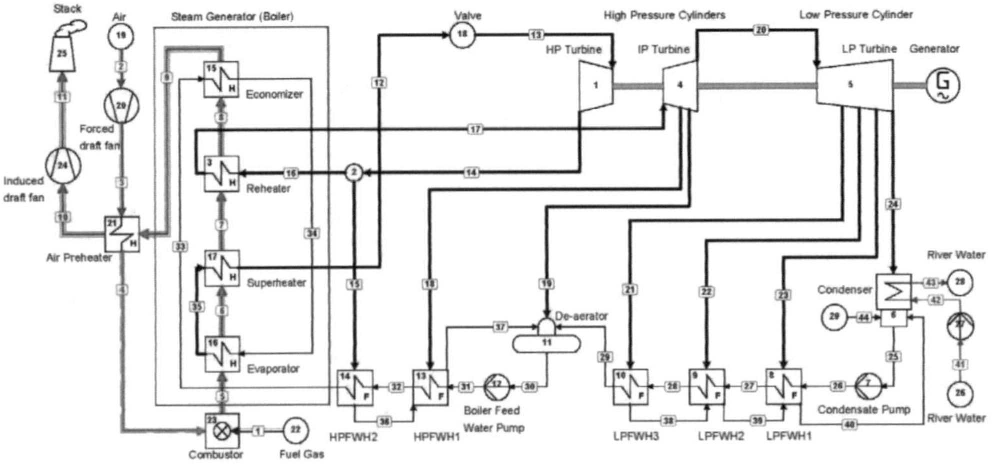

Figure 1. Schematic diagram for 200 MW AL-Hartha Steam power plant.

The change of excess air percentage within the recommended value of 5% to 10% has no positive effect on boiler efficiency; however, a change in excess air percentage of more than 10% will produce a sensible drop in boiler efficiency. Therefore, in order to maintain boiler efficiency at a high value, it is recommended that the excess air percentage must be kept as low as possible, in conjunction with the other parameters that affect boiler efficiency. A detailed exergy analysis showed that the combustor is still the highest relative exergy loss; exergy loss per total exergy input is about 23.4%, as shown in Figure 2. This is due to: the inherent irreversibility in the combustion process, heat loss, incomplete combustion and exhaust losses. Therefore, the combustor requires modification in order to reduce its exergy destructions

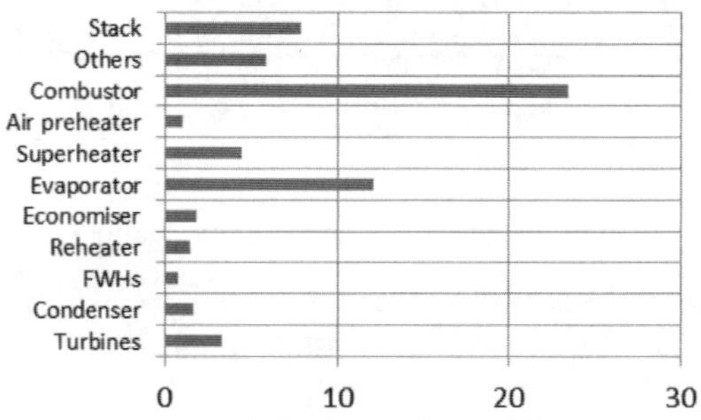

Figure 2. Relative exergy losses in each existing plant component.

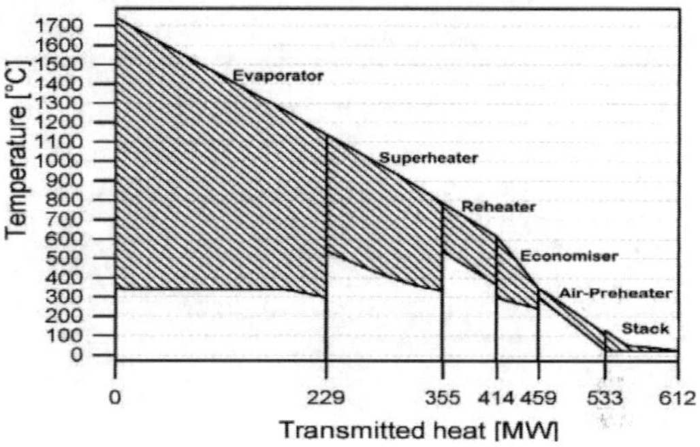

Figure 3. Temperature profiles of the heat exchange in the boiler

and to improve plant performance. Other parts of boiler that have a high percentage of relative exergy losses are the evaporator and superheater with values of 12.08% and 4.43%, respectively. This is because of the large temperature differences during heat transfer between the flue gas and the water or steam. Also, the stack has a high percentage of losses because of high exhaust temperature. The temperature profiles of the heat-exchanging components were studied. The evaporator is the part of the boiler with the highest value of heat transmitted for steam production purposes as shown in Figure 3, and HPFWH6, the high pressure heater for preheating water prior to entering the boiler, has the highest value of heat transmitted.

4. Conclusions

In the present paper, a thermodynamic analysis of a 200 MW gas-fuelled steam power plant has been carried out. The power plant system was simulated and a detailed parametric study undertaken, which involved environmental parameters, such as the temperature of cooling water entering the condenser and the inlet ambient air temperature, as well as some other operational parameters, such as excess air percentage and stack exhaust temperature. It was noted that the excess air percentage should be maintained below 10% and stack exhaust temperature should keep to a minimum. A detailed analysis of exergy losses was made. It was observed that the relative exergy losses in the combustor and evaporator are the highest compared with other parts of the plant. Finally, many recommendations have been suggested for improved plant performance. The present study helped to identify plant site conditions that cause losses of useful energy to take place and also helped to resolve some problems encountered in a 200 MW capacity unit.

Acknowledgements

The authors would like to thank the Energy Technology Section, Delft University of Technology, The Netherlands for licensing a copy of Cycle-Tempo and Mr. Mark Roest for providing technical support. Also, thanks go to the staff of AL-Hartha steam plant for their efforts in providing plant technical data.

References

[1] Aljundi I H 2009 Energy and exergy analysis of a steam power plant in Jordan *Applied Thermal Engineering* **29** 324-8

[2] Bhatt M S and Rajkumar N 1999 Performance enhancement in coal fired thermal power plants. Part II: steam turbines *International Journal of Energy Research* **23** 489-515

[3] Dincer I and Al-Muslim H 2001 Thermodynamic analysis of reheat cycle steam power plants *International Journal of Energy Research* **25** 727-39

[4] Oktay Z 2009 Investigation of coal-fired power plants in Turkey and a case study: Can plant *Applied Thermal Engineering* **29** 550-7

[5] Sengupta S, Datta A and Duttagupta S 2007 Exergy analysis of a coal-based 210 MW thermal power plant *International Journal of Energy Research* **31** 14-28

[6] Woudstra N, Woudstra T, Pirone A et al. 2010 Thermodynamic evaluation of combined cycle plants," *Energy Conversion and Management* **51** 1099-110.

[7] Suresh M V J J, Reddy K S et al. 2010 4-E (Energy, Exergy, Environment, and Economic) analysis of solar thermal aided coal-fired power plants *Energy for Sustainable Development* **14** 267-79

[8] Mitsubishi Heavy Industries Ltd. 1977 Manual for the AL-Hartha PowerStation, Steam Turbine and Auxiliaries **2**

[9] Mitsubishi Heavy Industries Ltd. 1977 Manual for the AL-Hartha PowerStation, Boiler and Auxiliaries **1**

IOP Conf. Series: Earth and Environmental Science **16** (2013) 012006 doi:10.1088/1755-1315/16/1/012006

Study the Growth of Microalgae in Palm Oil Mill Effluent Waste Water

Nabila Selmani, Mohamed E S Mirghani and Md Zahangir Alam
Biotechnology Engineering Department, Kulliyyah of Engineering, International Islamic University Malaysia (IIUM), Gombak, 50728, KL, Malaysia.

E-mail: benlnabila@gmail.com

Abstract. This paper emphasizes mainly on the biomass productivity and lipids content of two microalgae strains known by their high lipids content namely: *Botryoccoccus sudeticus* and *Chlorella vulgaris*. These strains were first screened for the highest biomass and lipids content, then Plackett–Burman design was used to evaluate the significant media for the growth when using POME waste water as culture medium. Results show that *Botryoccocus sudeticus* contains high content of biomass and lipids yield. Moreover, all the three factors have positive effect on the biomass productivity, while using one nutrient factor gives much lower biomass These results can be used further as an insight for optimizing the biomass and the oil productivity of the microalgae.

1. Introduction

Microalgae are prokaryotic or eukaryotic, unicellular or simple multicellular structure, photosynthetic microorganisms which can grow rapidly in suspension and live in difficult conditions [1]. Moreover, these microorganisms are useful for bioremediation applications [2] and [3]; nitrogen fixing biofertilizer [4] and they can provide several types of renewable biofuels.

Biodiesel production from food supplies is reported to need more land for the crop cultivation. However, researchers have estimated that only 2.5% of the whole lands used to produce the crops can be used for algae cultivation, which has 30% of its total mass oil [5]. Many algae species have been identified and reported to contain over than 30% of its weight as oil [6]. Generally, the growth medium must provide with all the nutrients necessary for the algae growth. Those nutrients range from essential nutrients like: nitrogen (N), phosphorus (P) and carbon(C) to facultative nutrients depending on the algae strain and on the final product that we are targeting to produce. The overall nutrients can be summarized in the following formula given by Chisti [5]: $CO_{0.48}H_{1.83}N_{0.11}P_{0.01}$. Besides that, the sources of each essential nutrient also play an important role in term of the final biomass produced and their lipids yield. It was reported in many studies that the combination of glucose with light gave better results in terms of both growth rate and lipids productivity. Moreover, sodium nitrate is considered to be less expensive which will make as a source of interests in the industrial practices [7]. Another essential nutrient is phosphorous; Researchers have found that algae bloom was mainly resulted from the elevation of nitrogen and phosphorus concentration in water, especially phosphorus concentration.

The combination of the microalgae production with the treatment of the waste water might be very promising green technology. The waste water would assist with their organic pollutant as fertilizer during the growth cycle of those valuable microorganisms. Many countries are struggling in finding permanent solution for the discharge of the waste waters especially the ones coming from the industrial sector. Malaysia is facing the same problem with the palm oil industry, which is considered to be one of the greatest sponsors for Malaysian economy. For the last period of time, this industry has well increased to contribute with almost the half of the whole world palm production [8]. Instead of the value added products that it can be recycled from the POME. The aim of this study is to compare the lipids content of two microalgae strains *Chlorella vulgaris* and *Botryoccocus sudeticus* know by their high common occurrence, easy growth characteristics and significant lipids content from one hand. Then, to study the nutrient limiting factor for the biomass and lipids production when using the waste water as culture medium.

2. Material and methods

Two strains of microalgae namely *Chlorella Vulgaris* and *Botryoccocus Sudeticus* are used in this experiment. The first step consists of screening of the potential strain for the high biomass and lipids productivity. Finally Plackett–Burman design used to screen the important variables of media component. 'Table 1' presents the design matrix of the variables under investigation as well as levels of each variable used in the experimental design. The parameters for media components selected for the experiment are: carbon source or glucose, nitrogen source or sodium nitrate and phosphorus source or potassium di-hydrogen phosphate.

Table 1. Plackett–Burman experimental design in coded and actual values with total biomass production

Run	Factor A	Factor B	Factor C	Biomass mg/l
1	+	-	-	403.3
2	-	+	-	386.85
3	-	-	+	471.88
4	+	+	-	898.26
5	+	-	+	528.66
6	-	+	+	508.15
7	+	+	+	1478
8	-	-	-	13.33

Factor A: Glucose, Factor B: Sodium nitrate, Factor C: potassium di-hydrogen phosphate.

3. Results and discussion

Screening of potential strain used for biomass and lipids production: Two strains known by their ability of producing high biomass and lipids content namely *Chlorella vulgaris* and *Botryoccocus sudeticus* were used in order to select the one having the highest biomass and lipids yield, so that it will be used further in our experiment. The results are presented in 'figure 1'.

'Figure 1' shows that both *C vulgaris* and *B sudeticus* have produced more than 1g/l of dried biomass. However, there is a great difference in term of lipids yield. And from those results we can say that for 1.47g/l of biomass from B *sudeticus*, about 38% of lipids can be obtained from, which is making it good candidate for algae biofuel production. In many previous studies, Botryoccucus sudeticus gives higher lipids content ranging from 25-75% of dry weight while for C vulgaris the lipids content can be lower ranging from 14-40% [5]. Increased lipid content also occurred in many microalgae as a response to different culture condition [9]. However, lipid production by microalgae is regulated by environmental factors, and to optimize them, a control is needed [10].

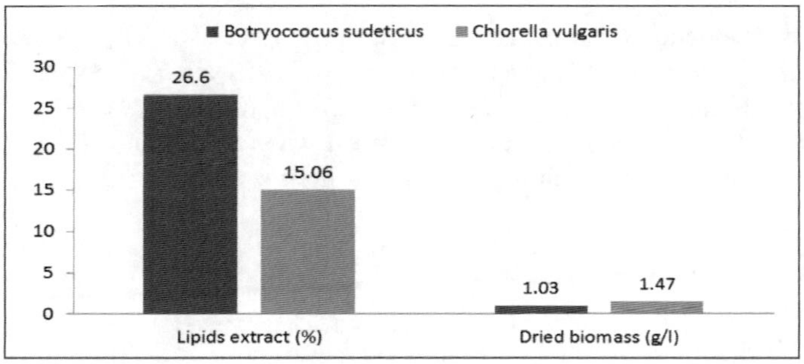

Figure 1. Biomass and lipids extract from *Chlorella* and *Botryoccocus*

3.1. Evaluation of the media components and process parameters for biomass production by Botryoccocus sudeticus using Plackett–Burman design

The main effect of variables on biomass production namely: glucose, sodium nitrate and potassium di-hydrogen phosphate was found to be 510.33, 488.51, 296.31.respectively. The results indicated that all the factors have positive effects on the growth and biomass productivity.

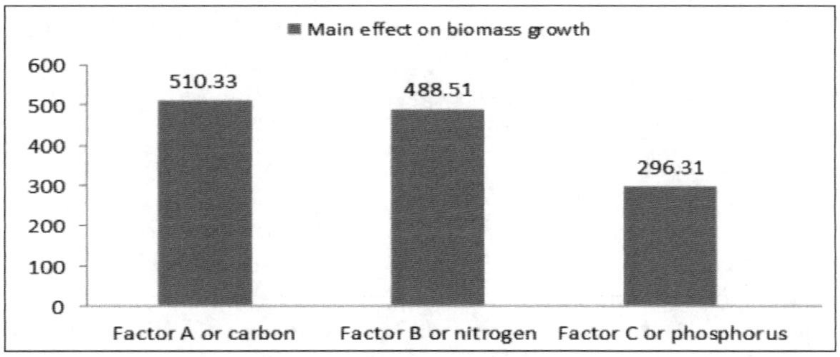

Figure 2. Main effect of the nutrients factors on the biomass production

'Figure 2' shows that the factor A or carbon has more effects on the biomass growth while this effect is decreasing with factor B to be the lowest with factor C or phosphorus. These results are well agreeing with the literature that confirms that carbon, nitrogen and phosphorus are considered to be the main essential nutrients which are very essential for the biomass growth. Chen stated that in order to develop balanced media for optimal microalgae growth, growth medium must be provided with adequate nutrients, most frequently: carbon, nitrogen and phosphorus are the main nutrients causative for biomass production [11]. Organic carbon is essential due to the composition of the half of the microalgae which is made with carbon [12]. Nitrogen is also considered to be the 2nd essential nutrient factor. Most of the microorganisms use nitrogen in form of nitrate, nitrite, ammonia and urea. Those forms can be responsible of some differences in terms of microalgae growth and biomass. Last but not least phosphorus is used to store phospholipids or other molecules related to energy such as ATP, NAD. Furthermore, it is used to produce further biological constituents like coenzymes, proteins, DNA, RNA [13].

4. Conclusions

In this study biomass production from microalgae was carried out using liquid state fermentation with two strains *Chlorella vulgaris* and *Botryoccocus sideticus*. The results have shown that *Botryoccocus sudeticus* offers better lipids prercentage and biomass productivity. It was then used in the second experiment in order to show the essential nutrient needed when using the POMSE as culture medium for the algae growth. Organic carbon, nitrogen and phosphorus have a positive effect on the biomass growth.

Acknowlegment

Research was supported by a research grant approved by the Research Management Centre (RMC), International Islamic University Malaysia (IIUM). The authors are grateful to the RMC and Department of Biotechnology Engineering, IIUM for supporting and providing the laboratories facilities.

References

[1] Li Y, Horsman M, Wu N, Lam C Q, Dubois-Calero N 2008 Biofuels from microalgae *Biotechnology progress* **24** 815-20
[2] Spolaore P, Joannis-Cassan C, Duran E and Isambert A 2006 Commercial applications of microalgae *J Biosci Bioeng.* **101** 87–96
[3] Mallick N 2002 Biotechnological potential of immobilized algae for wastewater N, P and metal removal: a review. *Biometals.* **15** 377–90
[4] Munoz R and Guieysse B 2006 Algal–bacterial processes for the treatment of hazardous contaminants: a review. *Water Res.* **40** 2799–815
[5] Vaishampayan A, Sinha R P, Hader D P, Dey T, Gupta A K and Bhan U 2001 Cyanobacterial biofertilizers in rice agriculture *Bot Rev.* **67** 453–516
[6] Chisti Y 2007 Biodiesel from microalgae *Biotechnology Advances.* **25** 294-306
[7] Guoveia L and Oliveira A C 2009 Microalgae as a raw material for biofuels production *Journal of Industrial Microbiology and Biotechnology* **36** 269-274
[8] Pandian Prabakaran and Ravindran A David. 2012 Influence of different Carbon and Nitrogen sources on growth and CO fixation of microalgae *Advances in Applied Science Research,* **3 (3)** 1714-17.
[9] Jen Chao Liu 2011 Production of fermentable sugars and lipids by microalgae from secondary treated municipal waste water. *A Thesis Submitted to the Faculty of Mississippi State University in Partial Fulfilment of the Requirements for the Degree of Master of Science in Chemical Engineering in the Department of Chemical Engineering, Mississippi State, Mississippi.*
[10] Yanqun Li, Mark Horsman, Bei Wang, Nan Wu and Christopher Q Lan 2008 Effect of nitrogen sources on cell growth and lipids accumulation of green alga Neochloris oleoabundans *Appl Microbiol Biotechnol* DOI 10.1007/s00253-008-1681-1.
[11] M.P.O.B. 2008 *Malaysian Palm Oil Board.* A summary on the performance of the Malaysian oil palm industry. http:// econ.mpob.gov.my/ economy/Performance-130109.htm 1.
[12] Yoo C, Jun S. Y, AHN C Y and OH H M 2010 Selection of microalgae for lipids production under high levels carbon dioxide *Bioresource Technology* **101** S71-74.
[13] Elsey D, Jameson D, Raleigh B and Cooney M J 2007 Fluorescent measurement of microalgae neutral lipids *Journal of microbiological methods* **68** 639-42

IOP Conf. Series: Earth and Environmental Science **16** (2013) 012007 doi:10.1088/1755-1315/16/1/012007

Troubleshooting for the observed problems in processing latex concentrate from natural resource

S Afreen[1, 2], K R Haque[3, 4] and M K Huda[5]

[1] Engineering Officer (Chemical), Department of Engineering, Essential Drugs Co. Ltd, Tejgaon I/A, Dhaka-1208, Bangladesh
[2] Project Member, Essential Latex Processing Plant, Modhupur, Bangladesh
[3] Deputy Manager, Quality Assurance Department, Essential Drugs Co. Ltd, Tejgaon I/A, Dhaka-1208, Bangladesh
[4] Project Manager, Essential Latex Processing Plant,Modhupur, Bangladesh
[5] Managing Director, Essential Drugs Co. Ltd, Tejgaon I/A, Dhaka-1208, Bangladesh and Essential Latex Processing Plant, Modhupur, Bangladesh

E-mail: sadiafreen002@gmail.com

Abstract. Natural latex has special importance in the rubber industry for manufacturing different types of goods like gloves, balloons, male contraceptive and similar thin walled articles. This natural latex is much more sensitive a liquid to handle since it can easily become contaminated and thereby coagulated which makes it unfavourable for centrifuge and getting concentrate from it. Some other related measures also are included in consideration during the processing of concentrate latex from the natural raw latex. The problems that are being faced in a concentrate latex processing plant can be categorized in different groups like, problems related to the latex property, mechanical problems, electrical problems, handling and storage problems, transformation problems, problems related to environmental issues, etc. Among them, the most common and vital problems frequently observed in a concentrate latex processing plant are discussed here with a view to finding the measures for solution which will help to maintain the latex property in any latex processing plant.

1. Introduction

Latex processing industry plays a vital role to produce the raw material for manufacturing various rubber products. Natural rubber offers high tensile strength and resistance to fatigue, from wear such as chipping, cutting or tearing, which carries its acceptance to manufacture various rubber goods like contraceptives, hand gloves and other thin walled materials. Although the process to produce concentrate from natural latex is not intricate but it can easily arise so many complexity if some factors related to the operation is not maintained properly. The most important thing is, in a factory, where processing of concentrate latex is the main concern, it is very much important to take proper care in every stage of the operation, so that no occurrence makes the latex out of order that inhibits to manufacture the desired product. Otherwise, it may result to a huge loss of raw material, energy and money of course. It is therefore recommended to get ready for any unwanted problems in a concentrate latex processing factory as well as have a record of all the troubleshooting to understand the essential steps required to be taken while any obstruction is faced during the production.

2. Process related problems and the possible measures to solve them

2.1. Problems related to latex property

The main parameters of the finished concentrate which are essential to be maintained to ensure the concentrate quality are, Dry Rubber Content (DRC), Mechanical Stability Time (MST), Volatile Fatty Acid (VFA), Potassium Hydroxide No (KOH No.), Ammonia Content.

2.1.1. DRC is found low in the concentrate after centrifuge. DRC of the concentrate is accurately governed by the length of the skim screw, feed rate and the total running time. A slower rate gives a greater separation into cream of higher DRC, likewise short skim screw higher cream DRC. It is normal to run at 60.20% DRC in the concentrate and adjust it later to 60.10%. The skim fraction should be retained in the bowl for as long as possible to remove as much rubber as possible from it [7].

2.1.2. Low ammonia content in the latex after field collection. This may happen when ammonia is not added properly during field collection. To keep the latex free from the possibility of getting coagulated ammonia must be added to field latex during collection. When low ammonia is found, it is required to immediately add ammonia to the latex to save it from contamination.

2.1.3. MST low in latex concentrate. Mechanical stability is the term that defines the stability of the latex concentrate to mechanical shear forces. This is the most important parameter to determine the quality of the final product. For a perfect situation MST of concentrate latex should be within 600-1200 [12]. An amount of MST below 600 indicates that too much DAP may have added at the initial stage before centrifuge the latex which in turns results in low MST value [12]. Another thing is, the mechanical stability of latex may have affected by lowering the temperature. Therefore care should be taken that the sample is not cooled significantly between sampling and testing [7], [12].

2.1.4. High VFA. Another very important parameter for the concentrate latex is VFA or volatile fatty acid. The VFA is recommended to be found less than 0.06 for latex [12]. High VFA is the enemy for both the raw and finished latex. The presence of Mg in the latex normally contributes to increase the bacterial activity and higher VFA. Early ammonia addition is recommended to minimize the VFA increase. In spite of taking this precaution if VFA of the concentrate in the final storage tanks is found higher than the tolerable range, the solution to this problem is to return the latex from the storage into the reception tank, dilute it with water and add ammonia up to 0.4% [12], then re centrifuge it to the desired grade.

2.1.5. Control of KOH Number. It is the measure of the ionic strength of the serum in the presence of ammonia. Thus, if latex is bacterially degraded, the ionic strength increases and also the KOH no. Poorly preserved latex will have a high KOH no but a high KOH no does not necessarily indicate of poorly preserved concentrate. The favourable value of KOH no is 1% [12], if it becomes high it is recommended to dilute the field latex and re centrifuge it.

2.2. Mechanical problems

2.2.1. Field latex overflows into the latex concentrate and skim collection chute. The reason behind this problem is, coagulum completely may have blocked the distributor or the sludge cavity of the centrifuge machine; to avoid this type of problem during production care must be taken on the machine running time. Generally a centrifuge machine can be operated for 3 hours. 10 minutes to reach its operating speed, 2.50 hours to run the field latex into the bowl and the rest 20 minutes for braking and stopping the bowl and letting the motor cool down [7], [12]. It is recommended that the centrifuge machine should not run over its capacity time otherwise it may cause such problem.

2.2.2. Abnormal vibration in centrifuge machine. This type of problem is really dangerous for a centrifuge machine. A centrifuge machine consists of many discs, around 140 discs for 60% concentrate and it generally runs at a speed of 1400 RPM [12]. At this high level of speed, if vibration is observed that means the machine's bowl has not been properly assembled before. There may be some discs are missing, or the discs of two different centrifuge machines may have interchanged, or there may be some damage or wear inside. It is highly recommended that if such vibration is observed the machine should be stopped at once to avoid any unexpected accident or damage of the machine.

2.2.3. Problem related to mechanical agitation. Mechanical agitation or stirring of latex in the storage tank creates friction, shear and turbulence which reduces the stability of latex, it is important that any mechanical stirring must be done slowly, using a slow speed stirrer at 20 to 40 RPM [7], [12].

2.2.4. Pump selection problem. Since the raw material and the finished product both are liquid, it requires pumps to transfer the latex in different zones of the production area. In such case, a right selection of pump is very much important so that no unwanted system loss or hindrance occurs. Generally it is best for latex factories to employ the gravitational method for moving latex, if mechanical pumping is unavoidable, then the latex can be pumped by means of compressed air which is applied at the top of the container/ tanker that is to be emptied. Mechanical pumps have regions in which the liquid is subjected to a high shear or which involve regions of sliding contact, are unsuitable for use with latex, because coagulum may occur at these points. Besides, the latex may be contaminated with lumps of coagulum, causing blockage and seizure of the pump to occur. A further requirement is that the pump should not introduce significant quantities of air into the latex. This condition limits the selection of pumps to only the centrifugal, single screw and diaphragm types [12].

2.3. Handling and storage problem

2.3.1. Ice formation at the nozzle of ammonia cylinder: Ammonia is a case II substance, means it liquefies at standard temperature but increased pressure. As more gas is pulled off, the pressure continually decreases and more gas is able to escape. Because the ammonia is evaporating, the gas cylinder gets very cold. To prevent the valve from freezing and to keep the container at semi-reasonable temperatures, it is better to simply run water over it [4].

2.3.2. No Copper and Magnesium in contact with the latex. The effect of copper or copper material in latex is very severe. Since the presence of copper forms copper amine complex which brings the latex to damage, coagulation occurs and as a result the target of getting concentrate is ceased. Mg is also detrimental in the same way for latex; it reacts with the sugar in the field latex and forms acid which also may be responsible for high VFA. Care should be specially given while handling the latex, that no copper, magnesium or instruments made of these elements are used in latex processing, other wise it will bring contamination.

2.4. Electrical Problem.

Centrifuge machines are driven by electrical motors, the machine should not run above the rated power. The machine should always run at the specified speed and this should be checked periodically using a tachometer [12]. The time requirement for the bowl to reach its operating speed is important and is determined by the efficiency of the linings in centrifugal clutch built into the driving pulley. If the accelerating time increases, then the clutch linings should be examined for wear and replaced if necessary [7], [12].

3. Conclusion

In spite of the advent of synthetic latex, natural latex is still being used as the raw material in different countries like Malaysia, Thailand, Bangladesh, etc. due to its availability from natural resources in those countries.

Almost in every latex processing factory, more or less, the cited problems are observed frequently; sometimes more than one problem is faced at a time. If proper cautions are not taken during the operation these problems may cause severe outcome and sometimes failure of process. It needs further study to search for further obstructions that may arise in this type of process and more research to find their advanced solution. Especially, a more detail study is required on environmental issues. The waste water that comes out from latex processing plant is one of the main concerns of rubber industry. There should be more research to find better solution in this case.

Acknowledgements
Special thanks go to the Poly Era Chemicals, Malaysia, for their consultancy and co operation, and for the information provided by them during latex processing in Essential Latex Processing Plant, Modhupur, Bangladesh. Also thankful to the management body of the Essential Latex Processing Plant and the Essential Drugs Co. Ltd, Dhaka, Bangladesh, for their patronage to do research on this matter. Recognition also goes to Bangladesh Forest Industries Development Corporation (BFIDC) for their co operation during the process operation. Last but not the least, the acknowledgement goes to the workfellows and all other members including workers and technicians of the project "Essential Latex Processing Plant', a concern of Essential Drugs Co. Ltd, Dhaka, Bangladesh; for their contribution and support to help run a process successfully

References

[1] Ngoc H, Duong D T and Thanh N 2007 Waste Abatement and Management in Natural Rubber Processing Sector *School of Environment Resources and Development, Asian Institute of Technology*
[2] Mohammadi M, Man H C, Hasan M A and Yee P L 2012 Treatment of waste water from rubber industry in Malaysia *African J of Biotech.* **9(38)** 6233-43
[3] Tekasakul P and Tekasakul S 2006 Environmental problems related to natural rubber production in Thailand *J.Aerosol Res.* **21(2)** 122-9
[4] Luxem K *Energy/Water Utilization to Prevent Freezing of Ammonia Gas Cylinder Valve*
[5] Sadhan K De and White J R 2001 *Rubber Technologist's Handbook Vol 1, Rapra Technology Limited*
[6] Company Brochure of Thai Rubber latex Corporation (Thailand), Public Co. LTD *http://www.thaitexgroup.com/thai2/main_page/Thaitexgroup.html*
[7] Latex Preservation and Concentration *http://rubberboard.org.in/ManageCultivation.asp?Id=192*
[8] Sharma B K *Industrial Chemistry* Goel Publisher House, India.
[9] Norris R, Shreve and Austin G T *Shreves Chemical Process Industries*, McGraw Hill sed.5
[10] Data and reports provided by the Bangladesh Forest Industries Development Corporation (BFIDC).
[11] Data and Information Provided by Poly Era Chemicals, Malaysia.
[12] Project Report of Essential Latex Processing Plant (ELPP), Bangladesh, 2011, 2012

IOP Conf. Series: Earth and Environmental Science **16** (2013) 012008 doi:10.1088/1755-1315/16/1/012008

Role of Surface to Volume Ratio of Zinc Oxide Arrester Blocks on the Energy Absorption Capability

A N M Karim[1], S Begum[2] and M S J Hashmi[3]

[1] Dept. of Manufacturing and Materials Eng., International Islamic University Malaysia, Jalan Gombak, 53100 Kuala Lumpur, Malaysia
[2]Dept. of Mech. Eng., Universiti Tenaga Nasional Km 7, Jalan Ikram-Uniten, 43000 Kajang, Selangor, Malaysia
[3]School of Mech. and Manuf. Eng., Dublin City University, Glasnevin, Dublin 9, Ireland

E-mail: mustafizul@iium.edu.my

Abstract. The functional life of a zinc oxide arrester block is largely dependent on its energy absorption capability. This capability is an important performance characteristic of a lighting arrester which leads to enhanced reliability of a surge protection system. An arrester block is usually cylindrical in shape with two flat surfaces. Injected energy of transient electrical surge into the arrester body is transformed into heat and dissipated through the surface of the disc body. This study has been conducted to observe whether the higher surface to volume (S/V) ratio of an arrester block enhances the capability of energy absorption or not. The round side or C-surface of the cylindrical disc was ground by diamond wheel to transform into hexagonal shape. By making the modification of the geometrical shape an increase of about 11% in S/V ratio was achieved. ZnO arrester blocks of both shapes were tested for energy. The average energy absorption capability for the hexagonal discs was found to be 483 $J.cm^{-3}$ compared to that of 357 $J.cm^{-3}$ for the discs having the cylindrical shape. Thus, about 35% increase in energy absorption capability is observed for the hexagonal discs. This knowledge can be useful in designing the geometry of the device for improved functional reliability of electrical system.

1. Introduction

Zinc oxide varistors are electronic ceramic devices, the primary function of which is to protect the electrical systems by sensing and limiting transient surges [1]. Application of this metal-oxide varistor technology is very common now-a-days for voltage stabilization or transient surge suppression in electronic circuits and electrical power systems [2-5]. Owing to the improvements of non-ohmic properties and functional reliability, use of ZnO varistors is expanding quite rapidly. Moreover, with the advent of advanced manufacturing technology, varistors having new designs and configurations are expected to have wider application.

In addition to some finishing operations, varistor manufacture basically follows the route of conventional ceramic processing. Compared to the long history of ceramic materials, use of electronic ceramic as transient over-voltage suppression device is a recent development [6]. Performance and reliability of electronic ceramic depends on the grain and the grain boundary phenomena [7- 8]. There are various investigations on particle size and its distribution on physical, mechanical and electrical properties of ZnO varistors [9-11]. The use of binder is very crucial in the processing of electro ceramics. The effect of latex binders is also investigated and it has been observed that binders have significant affect on physical, mechanical and electrical properties of varistor discs [12]. In varistor fabrication the passivation thickness plays an important role. It is applied on the peripheral surface of the discs to provide a collar material and acts as a barrier to heat transfer during energy testing. The influence of passivation thickness on energy absorption capability was investigated and it was observed that the variation of passivation thickness did not have that much influence on the energy absorption capability; however, the mode of failure was changed [13].

1.1. Processing of arrester block

Zinc Oxide varistors produced in the form of cylindrical blocks are often called arrester blocks. The spray dried powder in the form of granulates was compressed into disc-shaped blocks with approximately 55 to 65 percent of their theoretical density. Pressing was performed by uniaxial double action compaction technique. Sintering of the discs was performed by a conventional sintering profile with a peak temperature of about 1120 °C and a total sintering cycle time of about 70 hours. The sintered ceramic body takes the shape of a rigid cylinder with a theoretical density of more than 95 percent. ZnO varistors undergo a liquid phase sintering. During this process, the bismuth oxide melts to form the liquid phase which dissolves, at least in part, the other doping substances and promotes their uniform distribution. Liquid phase also favors the grain growth and dense sintering. Spinel precipitates, on the other hand, inhibit grain growth and help generate a uniform distribution of the ZnO grain size.

Researches in the varistor technology have been primarily aimed at improving the fundamental properties. In this regard much of the work is related to the investigation of material composition, micro-structure, whilst grain and grain boundary phenomenon and defects [14-17] got the maximum attention. Work investigating the effect of Surface to volume ratio on the ZnO varistor properties especially in the context of energy absorption capability has not been reported. This experiment was conducted to check the effect of the surface to volume ratio of the arrester block on the energy absorption capability.

2. Experimental Procedure

2.1. Preparation of samples of hexagonal discs

Hexagonal cross section was achieved by modifying the C surface of the cylindrical discs. Material was removed by using a production-line diamond grinding wheel rotated along a vertical axis. ZnO cylindrical arrester discs were passed under the flat face of the diamond wheel by placing them horizontally on the conveyor belt. A small depth of cut was used to remove material in one pass requiring several passes to complete one of the six flat sides of the hexagonal discs. After one flat plane was obtained, the other two adjacent flat surfaces were developed at angle of 60 degree relative to it. The remaining three flat surfaces were ground one by one by placing the disc on the conveyor belt making contact with the already developed flat surfaces. A cylindrical disc was transformed into a hexagonal shape as shown in Figure 1 by progressively generating all the six flat surfaces.

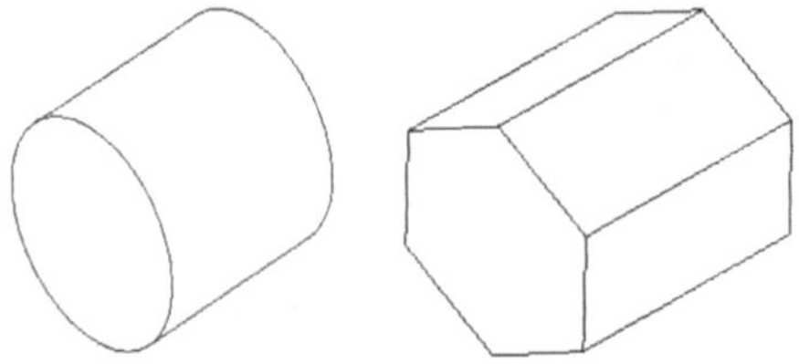

Figure 1. Disc of hexagonal shape as obtained by grinding the cylindrical arrester block.

A cylindrical disc having diameter of 41 mm had a cross-section of 13.2025 cm^2 and the cross-section for a hexagonal disc was reduced to 10.9184 cm^2. Since the disc height of 42 mm remained unchanged, the volumes of the cylindrical and hexagonal discs became 55.45 cm^3 and 45.86 cm^3 respectively with the corresponding values of surface areas of 80.5032 cm^2 and 73.7887 cm^2. As a result the S/V ratio of hexagonal disc is increased to a value 1.609 cm^{-1} compared to the value for the cylindrical disc of 1.452 cm^{-1}. Thus by making the modification of the cylindrical disc into hexagonal shape an increase of about 11% in S/V ratio was achieved for the hexagonal discs.

2.2. Evaluation of Energy absorption capability

Energy absorption capability of a varistor is determined by the maximum energy density injected into the ceramic body up to which it can sustain for a cycle of three shots. The test is initiated with a lower charging voltage so that injected energy remains in the range of about 200 Joule.cm^{-3}. Selection of the lower starting range of energy is attempted to keep the likelihood of failure at the first cycle to a minimum level. Thus the sample size reduces progressively as the disc fails step by step with the increment of charging voltage. The testing procedure was continued until all the discs failed at a certain stage. Energy density of a disc is calculated by using the clamping voltage and the peak current recorded from the cycle of three shots. In Figure 2 the long wave shape for energy testing is presented.

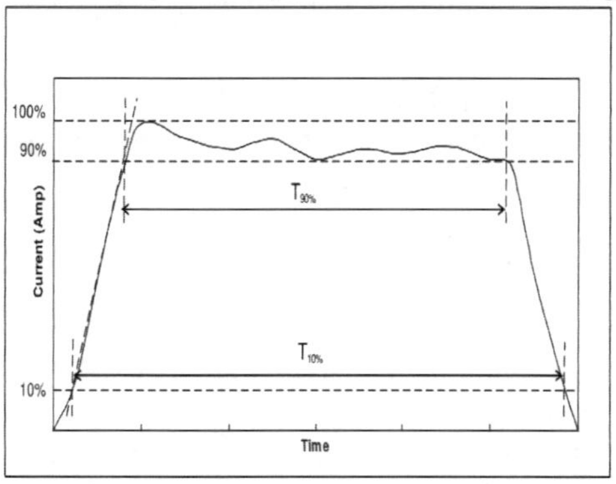

Figure 2. Long wave to measure energy absorption capability

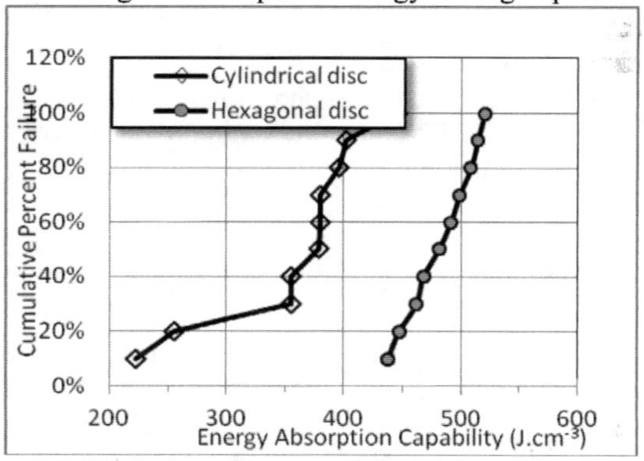

Figure 3. Variation of energy absorption capability with different geometry

3. Results and Discussions

To compare the energy absorption capability a total of 20 arrester blocks (10 cylindrical and 10 hexagonal) was taken. Figure 3 depicts the cumulative percent failure with respect to increasing level of energy. Failure of arrester blocks in the test for energy absorption capability is usually dominated by occurrences of pinhole and flashover. In a very few cases discs failed with crack or fracture alone. The failure mode and the location of damaged marks for each of the discs have been tracked and it has been observed that pinhole accompanied by a flashover is a very common mode of failure. Since the hexagonal discs were prepared by grinding the side surface of the cylindrical discs, variation in energy absorption capability could possibly be linked with the removal of the presumably contaminated surface of the cylindrical discs. But no such link can be validated from the observations of the failure patterns and their distribution.

Out of ten discs for each set of control and hexagonal discs there is no significant difference both in failure mode and the location of the mark or origin of failure. In both the cases, only one disc from each set exhibited failure solely through flash over, 5 and 4 failures by pinholes occur respectively for hexagonal and cylindrical discs. The discs having hexagonal shape have an average of 483 J.cm^{-3} energy absorption capability while the discs having the cylindrical shape yielded an average of only 357 J.cm^{-3}. Thus, about 35% increase in energy absorption capability is observed for the hexagonal discs. In this test the electrical pulses applied is of 2 only millisecond duration and the arrester is considered to be heated adiabatically. But the sample of hexagonal discs having higher surface to volume ratio has exhibited improved energy absorption capability. This phenomenon can be attributable to the effect of increased surface to volume ratio of arrester block.

4. Conclusions

The reliability of a surge protection system is largely dependent on the energy absorption capability. Injected energy from the stroke of transient electrical surge into the arrester body is transformed into heat and dissipated through the surface of the disc body. Higher S/V of the hexagonal disc is found to be highly conducive in heat transfer leading to enhanced life. Hexagonal disc having a S/V ratio of 1.609 cm^{-1} compared to 1.452 cm^{-1} for the cylindrical disc led to an increase of about 11% in S/V ratio. But the average energy absorption capability for the hexagonal discs was found to be 483 $J.cm^{-3}$ compared to that of 357 $J.cm^{-3}$ for the discs having the cylindrical shape which is equivalent to about 35% increase in energy absorption capability. Thus it is evident that the effect of increased surface to volume ratio of arrester block is very significant. Moreover, discs with the hexagonal shape can be produced as a modular unit. By combining hexagonal modular unit of discs in series and in parallel a wide range of surge arrester could be possible to be constructed. This modular concept could tremendously reduce the cost of production as a standard production line could be possible to be made avoiding setups at various points of the lengthy processing line.

Acknowledgements

This work was carried out as part of the DOCERPO project within the BRITE/EURAM II programme (Contract No. BRE2.CT92.0418) and the authors gratefully acknowledge the financial support provided by the Commission of European Communities.

References

[1] Matsuka M 1971 *J. Appl. Phys.* **10** 736-46
[2] Puyane R 1995 *J. Material Science* **55** 268-77
[3] Eda K 1989 *IEEE Electrical Insulation Magazine* **5** 28-41
[4] Gupta Tapan K 1990 *J. Am. Ceram.* Soc. **73(7)** 1817-40
[5] Texas Instruments, Data sheet acquired from Harris corporation, Dallas, Texas 75265, Copyright © 2012. *http://www.ti.com/lit/ds/symlink/cd4094b.pdf*
[6] Craner D C 1991 Overview of Technical, Engineering and Advanced Ceramics *Engineered Materials Handbook* **4** 16-20
[7] Kingery W D 1981 *Advances in ceramics* **1** 1-22
[8] Einzinger R 1980 *Advances in Ceramics* **1** 359-73
[9] Begum S, Karim A N M and Hashmi M S J 2006 *Proceeding of the International Conference of Manufacturing Materials, ICMM2006, Malaysia* 272-77
[10] Lanyi W, Chengxiang L and Guoyi T 2007 *Key Engineering materials, Trans Tech Publications, Switzerland* 739-742
[11] Pillai S C, Kelly J M et al 2003 *J. Mater. Chem.* **13** 2586-90
[12] Begum S, Karim A N M, Hashmi M S J 2008 *J. Am. Ceram. Soc.* **91** 3216-21
[13] Karim A N M, Begum S, Hashmi M S J 2010 *International J. of Mechanics and Materials Engineering (IJMME),* **5** 175-181
[14] Gupta T K and Miller A C 1988 *J. Material Res.* **3(4)** 745-54
[15] Lee J S and Wiederhorn Sheldon M 2004 *J. Am. Ceram. Soc.* **8** 1319-23
[16] Balzer Beate et al 2004 *J. Am. Ceram. Soc.* **8** 1932-38
[17] Karim A N M, Begum S and Hashmi M S J 1997 Processing of Zinc Oxide varistor: Sources of Defects generation and Possible Measures for Their Elimination *Proc. 3d Int. con. on Advanced Methods in Mat. Processing Defects (MPD 3), Ecole Normale Superieure de cahan, France, edited by M Predeleanu and P. Gilormini, Elsevier Science B. V.*

IOP Conf. Series: Earth and Environmental Science **16** (2013) 012009 doi:10.1088/1755-1315/16/1/012009

Preliminary experimental results of Sewage Sludge (SS) Co-digestion with Palm Oil Mill Effluent (POME) for Enhanced Biogas Production in Laboratory Scale Anaerobic Digester

R Sivasankari, P Kumaran, Saifuddin Normanbhay and Abd Halim Shamsuddin

Centre for Renewable Energy, Universiti Tenaga Nasional (UNITEN)
43009 Kajang, Selangor, Malaysia

E-mail: sivasankari@uniten.edu.my

Abstract. An investigation on the feasibility of co-digesting Sewage Sludge with Palm Oil Mill Effluent for enhancing the biogas production and the corresponding effect of the co-digestion substrate ratio on the biogas production has been evaluated. Anaerobic co-digestion of POME with SS was performed at ratios of 100:0, 70:30, 60:40 and 0:100 to find the optimum blend required for enhanced waste digestion and biogas production. Single stage batch digestion was carried out for 12 days in a laboratory scale anaerobic digester. Co-digestion of sludge's at the 70:30 proportion resulted in optimal COD and C: N ratio which subsequently recorded the highest performance with regards to biogas production at 28.1 L's compared to the 1.98 L's of biogas produced from digestion of SS alone. From the results obtained, it is evident that co-digestion of POME and SS is an attractive option to be explored for enhancement of biogas production in anaerobic digesters.

1. Introduction

Biogas is produced naturally when the environment is favourable for anaerobic fermentation or commonly known as anaerobic digestion of organic material [1]. It is generally comprised of 60 – 70% methane (CH_4), 30-40% carbon dioxide (CO_2) and low amount of other trace gases [2]. Sewage wastewater is filtered to separate the solids from the liquid in a modern state of art sewage treatment system. The collected solid sludge is then transferred into an anaerobic digester, to be treated and biogas produced is collected and flared[3]. POME is a thick brownish liquid that contains high amounts of total solids (40,500 mg/L), oil and grease (4,000 mg/L) and COD (50,000 mg/L) [4] [5] that is produced during the processing of Oil Palm. Literature studies have shown that, co-digestions of slurries are known to result in more efficient digestion of organic materials. Better NPK (Nitrogen: Phosphate: Potassium) ratio can also be obtained through co-digestion, which will aid in higher methane production [6]. Promisingly from previous literature reviews done, since both sewage sludge and POME are organic in origin, combined anaerobic digestion seemingly is an ideal solution. Also, improved sludge digestion due to synergistic effects of co-substrate and increased nutrients should result in improved degradation of organic matter. The aim of this work is to investigate the co-digestibility of Sewage sludge and Palm Oil Mill Effluent in laboratory scale single stage batch anaerobic digesters. In this work, the optimum ratio of the sludge for enhanced anaerobic biomethanation process is re-investigated and reported. This paper reports on the enhanced biogas production through anaerobic co-digestion of Sewage sludge and Palm oil mill effluent.

2. Experimental material and methods

The feed sewage sludge was collected from a Sewage Treatment Plant. Palm Oil Mill Effluent (POME) was collected from Seri Ulu Langat Palm Mill, Dengkil. The POME and SS samples were collected and sent for characterization studies upon arrival to the laboratory. The laboratory scale 10 L single stage batch anaerobic digesters were operated under mesophilic conditions at 29°C with a hydraulic retention time of 12 days. Each anaerobic digester was loaded with 5L's of the substrate. Two control systems were fed with 100% POME and 100% SS and its operation is compared to the 2 experimental systems with 60:40 and 70:30 volume ratio of POME: SS. Sample 1 was fed with 60% POME and 40% SS while Sample 2 was fed with 70% POME and 30% SS. Sludge characteristics such as MLVSS, TS, TKN, COD and C: N ratio was measured using APHA 2005 standards at third party laboratory.

Biogas generated was collected in custom fabricated plastic canisters. The volume of gas produced daily is monitored by reading the amount of displaced water.

3. Results and discussion

The results of compositions of the samples are as described in Table 2. It was found that Sample 2 with composition of 70% POME and 30% SS had produced highest quantity of biogas. This was followed by Control 1 (100% POME), Sample 1 (60% POME and 40% SS) and lastly, Control 2 (100% SS). Cumulative production of biogas for all the samples is as detailed in Figure 2.

Table 2. Sludge characterization results

Parameter	Control 1 (POME)	Control 2 (SS)	Sample 1 60:40	Sample 2 70:30
MLVSS (mg/L)	33,000	8,420	22,900	3,000
TS (mg/L)	45,300	63,600	30,900	15,400
COD (mg/L)	103,000	22,800	29,000	50,600
TKN (mg/L)	1,260	1,160	1,200	1,350
C:N Ratio	50.0	2.07	23.3	20.0

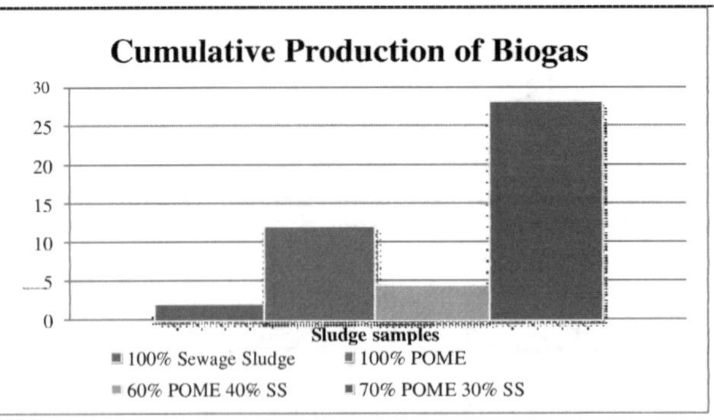

Figure 2. Cumulative production of biogas

The amounts of biogas produced in the respective samples as in figure 2 were 28.1 L's, 11.98 L's, 9.9 L's and 1.98 L's. Control 2 (100% SS) has produced the least amount of biogas as compared to the rest of the test subjects. This finding is concurrent with previous works done, which proved that sewage sludge in combination with high organic content material aids in decomposition of waste and enhanced methane formation [7].

The efficiency of anaerobic digestion for methane production depends on the organic strength of sludge present in the waste. The most important parameters for characterizing these slurries are total solids content (TS) and volatile solids content (VS). Anaerobic digestion typically results in a 50% reduction of organic matter (volatile solids). According to previous works reported, it is derivable that anaerobic digestion of SS as a single substrate is possible, but the amount of biogas produced from this process will be lower than that compared to SS co-digested with other high organic material sludge. The results obtained as in Figure 2 seem to comply with Schmidell's studies which show that a substrate digested individually will produce lesser biogas compared to mixed sludge's [8].

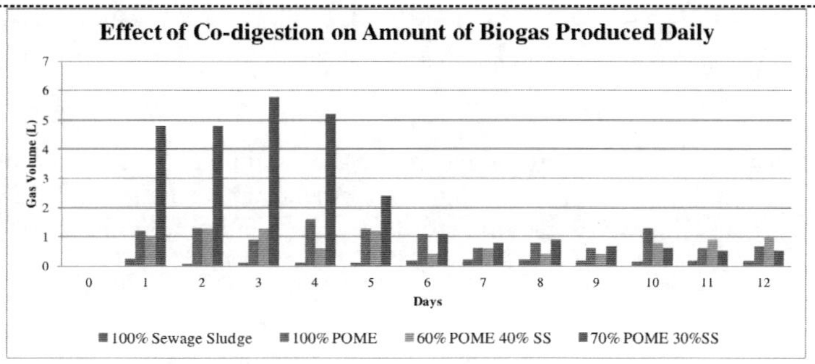

Figure 3. Effect of co-digestion on daily biogas production

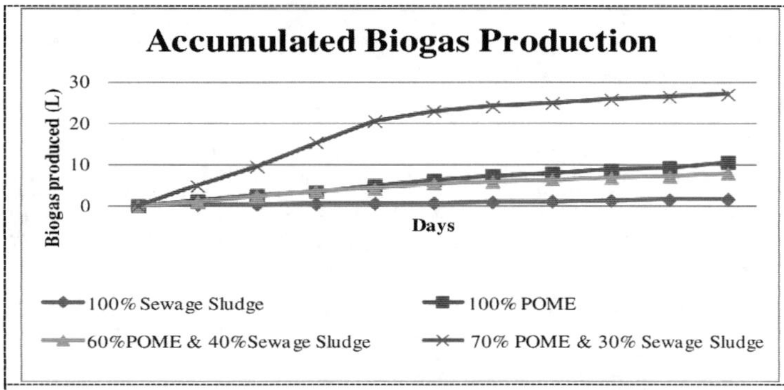

Figure 4. Effect of co-digestion on accumulated biogas production

The biogas production rates for the four experimental samples are shown in Figure 3 and Figure 4. From the Figure 4 it can be seen that Control 2, had produced the least amount of biogas, meanwhile Sample 2, a co-digestion mixture of 70% POME and 30% SS had produced the highest amount of biogas . This can be traced back to the C: N ratio of the sludge. The optimum C: N ratio required for anaerobic digestion is between 20 – 30 [9]. . Sewage sludge, suggests very low C: N ratio, and on the other hand, the C: N ratio of POME is much higher and through co-digesting the sewage sludge and POME, a relatively optimal C: N ratio is achieved, thereby improving the digestion process and correspondingly the by-product methane generation enhancement.

The Chemical Oxygen Demand (COD) is another important parameter which affects the biogas production process. COD value indicates the value of organic compounds present in the water. Recent experimental works has proven that substrates with higher COD value led to better performance in terms of methane production and COD reduction. As visible in Table 2, the COD value of POME is at 103,000 mg/L and COD value of Sewage sludge is a mere 22,800 mg/L. When combined at the ratio of 70:30 (Sample 2) for co-digestion, and optimal rate of 50,600 was achieved. The increased biogas production can be attributed to this particular character of the sludge [10].

Based on the results collected as seen in Figure 4, it can be stated the co-digestion of Sewage Sludge of low organic content with Palm Oil Mill Effluent, which has a higher organic content at the ratio of 30:70 produces increased amount of biogas as compared to its control systems.

4. Conclusion

The objective of this study was to assess the feasibility of co-digesting Sewage Sludge with Palm Oil Mill Effluents. Preliminary findings have concluded that co-digestion on the optimum sludge ratio of 70:30 (POME: SS) has shown increase in biogas production 5 times higher than the samples which were not co-digested. Hence, it can be deduced that mixing SS with POME enhances the decomposition of organic materials leading to higher biogas production. Further research is recommended to evaluate the quality of biogas that is generated from the experiment.

Acknowledgement

We would like to thank AAIBE for funding this research, Indah Water Konsortium (IWK) for providing us sewage sludge samples for this experiment, also thanks to the participation of undergraduate students, Ms.Hephzibah David and Mr. Kuhan Sundrarajoo in assisting with starting up the research process.

References

[1] NNFCC and The Andersons Centre 2011 NNFCC Renewable Fuels and Energy Factsheet: Anaerobic Digestion [Online]. United Kingdom: NNFCC *http://www.nnfcc.co.uk/publications/nnfcc-renewable-fuels-and-energy-factsheet-anaerobic-digestion*

[2] Saija R 2009 Biogas composition and upgrading to Biomethane. *Published PhD thesis. University of Jyväskylä*

[3] Indah Water Konsortium 2012 Sewage Treatment Methods [Online]. *http://www.iwk.com.my/v/knowledge-arena/sewage-treatment-methods*

[4] Loh S K, Noor F M and Ma A N 2009 Characterization of Palm Oil Mill Effluent (POME) for Process Control and Baseline Establishment. Malaysia *MPOB Information Series*

[5] Asian Palm Oil Company Limited 2004 *Waste-to-Energy Project by a Palm Oil Manufacturer. Thailand.*

[6] Fabien M 2003 An introduction to anaerobic digestion of organic wastes *Remade Scotland*

[7] Sharifuddin H A H, Zaharah A R 2004 Utilization of Organic Wastes and Natural Systems in Malaysian Agriculture *Malaysian Palm Oil Board*

[8] Schmidell W and Craveiro A M et al 1986 Anaerobic Digestion of Municipal Solid Wastes *Water Science & Technology* **18** 163–17

[9] Shefali V 2002 Anaerobic digestion of biodegradable organics in municipal solid wastes *Published MSc thesis. Columbia University*

[10] Gamze G D and Goskel N D 2004 Effect of initial COD concentration, nutrient addition, temperature and microbial acclimation on anaerobic treatability of broiler and cattle manure. *Bioresource technology* **93** 109-17

IOP Conf. Series: Earth and Environmental Science **16** (2013) 012010 doi:10.1088/1755-1315/16/1/012010

Evaluating the Mechanism of Oil Price Shocks and Fiscal Policy Responses in the Malaysian Economy

Hussain A Bekhet[1] and Nora Yusma Mohamed Yusoff[2]
[1] *College of Graduate Studies*
[2] *College of Business Management and Accounting*
University Tenaga Nasional (UNITEN)
Muadzam Shah Campus, Pahang, Malaysia

E-mail: profhussain@uniten.edu.my; nora@uniten.edu.my

Abstract: *The paper aims to explore the symmetric impact of oil price shock on economy, to understand its mechanism channel and how fiscal policy response towards it. The Generalized Impulse Response Function and Variance Decomposition under the VAR methodology were employed. The empirical findings suggest that symmetric oil price shock has a positive and direct impact on oil revenue and government expenditure. However, the real GDP is vulnerable in a short-term but not in the long term period. These results would confirm that fiscal policy is the main mechanism channel that mitigates the adverse effects oil price shocks to the economy.*

1. Introduction

The oil price shock is expected to have significant effects on the government budget, supply and demand in domestic markets as well as energy and emissions trends on a global scale [7]. However, the impact of high oil prices on Malaysian economic performance would depend on the exposure of the Malaysian economy to oil, and also to energy elasticity of demand, particularly in terms of domestic consumption even on energy consumption and the extent of the spillover effect of the increase in costs on other products and services.

Thus, the current paper attempts to explore the impact of the symmetric oil price shock on the Malaysian economy and to simulate the effects of oil price shock on government expenditure and revenue in the Malaysian economy. The empirical finding of this study is significant especially to help government and policy planner in giving policy guidelines especially in designing a proper fiscal policy instrument in the macroeconomic level planning. In order to analyze the oil price shocks impacts the GIRF and VDC under VAR model were employed. The VAR model is still the most widely applied empirical approach used to determine the relationship between oil prices and macroeconomic variables [4].

From the previous researches, many researchers have concluded that there is a negative relationship between increases in oil prices and the subsequent economic downturns in the United States [5], [3], [9], [2], [6]. Also, other countries found that co-integration relationship between world oil prices and macroeconomic variables exist in the long run [8], [10]. Earlier, [1] concluded that net-oil exporting countries do not exempt from the negative impacts of high oil prices on their economies. The positive effects of high oil prices would dampen trade and hinder growth through a contractionary effect of trading partners. On the contrary, the response of terms-of-trade to oil price shocks is statistically insignificant. On the other hand, [9] found an asymmetry between the responses of the

GDP and oil-price increases and decreases, concluding that the decreases were not statistically significant. Thus, his results confirmed that the negative correlation between GDP and increases in oil-price was persistent.

2. Data Sources and Methodology

2.1 Data and Variables

The annual data for the (1980-2010) period for oil prices (OILP), real GDP (RGDP), oil revenue (OR), non-oil revenue (NOR), government total expenditure (GOE) and total subsidy (SUB) for Malaysian economy was used. GDP data were collected from the Department of Statistics. GOE, OR, NOR, SUB and OILP data collected from the Economic Planning Unit (www.epu.gov.my). Furthermore, all these variables were measured in constant price (2000 as a base year) and transformed (logarithm base). The logarithm transform procedure is used to ensure that all variables are in stationary [11].

2.2 Model Specification:

The study employed the VAR model to simulate a standard error unit shock on oil price. In order to explain how a VAR is estimated we assume that each equation contains k lags values of Y and X variables. In this case, it can estimate each variable by using the OLS method:

$$Y_{t}= \alpha + \sum_{j=1}^{k} \beta Y_{t-j} + \sum_{j=1}^{k} \gamma X_{t-j} + U_{1t} \qquad (1)$$

$$X_{t}= \alpha' + \sum_{j=1}^{k} \delta Y_{t-j} + \sum_{j=1}^{k} \gamma X_{t-j} + U_{2t} \qquad (2)$$

Where, the $U_t = (U_{1t}, U_{2t})$ is the stochastic error terms for $t = 1, 2.....T$. In addition, U_{1t} and U_{2t} are assumed independent and with zero mean, i.e. E $(U_{1t}) = 0$, k is the lag length criteria, α *and* α' are constant terms, β, γ and δ are the coefficients estimate for independent variables. The above VAR models will be extending to comprise 6 major endogenous economic variables. These are OILP, RGDP, OR, NOR, SUB and GOE.

3. Empirical Results and Findings

First, based on the unit root test results, we fail to reject the null hypothesis of unit roots even at the 10% significance level. However, when we performed the unit root test at first difference, *I(1)*, the results indicate that all variables are *I(1)* since the *P*-value is significant at 1% and 5% level. Second, the result of the LM test for VAR model (at lag order 1) suggests that there is no obvious residual autocorrelation problems exist in the model because all *p*-values are larger than the 0.05 level of significance. Furthermore, the AR Polynomial test confirms that there is no root lies outside the unit circle; hence the VAR models will satisfy the stability condition.

The impact of symmetric OILP shocks that is a standard deviation (S.D) innovation on oil revenue, fiscal tools and real GDP, are presented in the following steps:

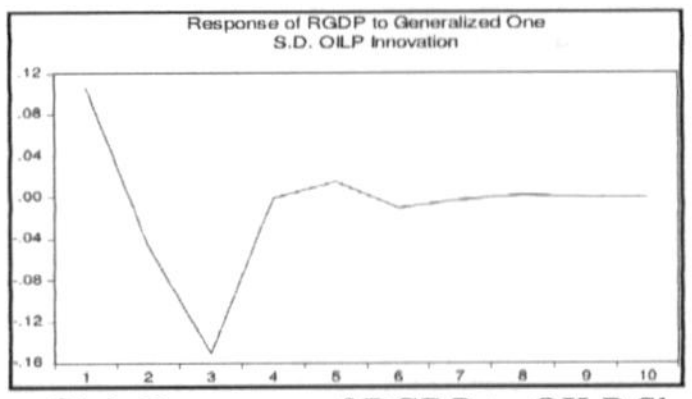

Figure 3(a). Response of RGDP to OILP Shocks.

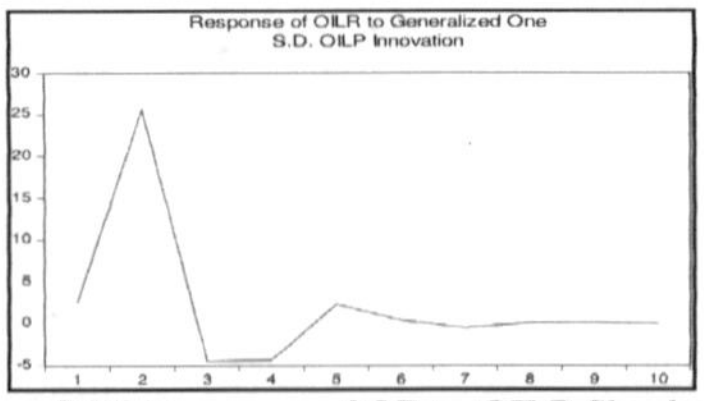

Figure 3(b). Response of OR to OILP Shocks

Third, Figures 3(a) -3(f) show the IRFs for one standard deviation (SD) symmetric OILP shocks to current and future values of endogenous variables over the 10-th period ahead. Figure 3(a) suggests that the OILP shock has an immediate effect, which leads to a decrease in RGDP in the short run, as predicted by theory. The larger negative impact occurs in the 3-rd period which is decreased to negative 0.15%. This is followed by a gradual increase over next periods until the5-th period which are around 2 years.

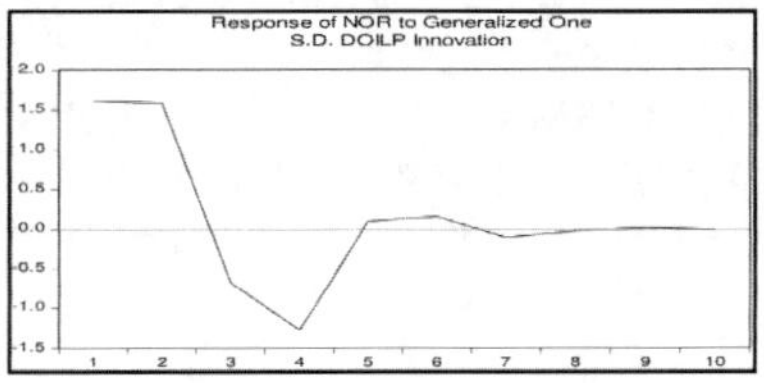

Figure 3(c). Response of NOR to OILP Shocks

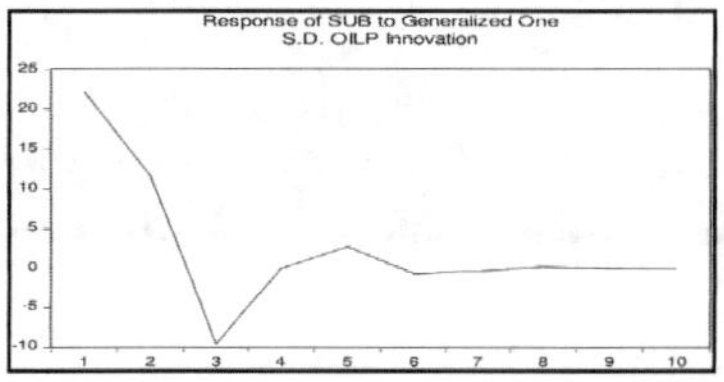

Figure 3(d). Response of SUB to OILP Shocks.

However, the impact on RGDP growth becomes stable or asymptotes to 0 after the 6-th period. This suggests that the impact of OILP shock on the growth rate of GDP is relatively short-lived. It is seen that an OILP shock has instantaneous positive and significant impact on OR until in the 2-nd period and there is substantial volatility in the OR over the 2-nd period (Figure 3(b)). The time path of the impulse response indicates an initial appreciation in the OR, before it decreases after the 2- nd to the 4-th period, finally asymptotes to 0 after the 6-th period.

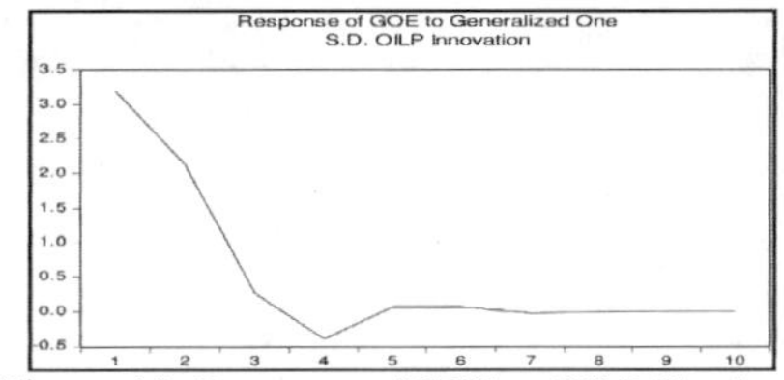

Figure (e): Response of GOE to OILP Shocks

The OR has increased sharply to 26% over the next 2-th period before it falls in the next period. As far as OILP shock is concerned, we confirm that OILP shocks leads to a decrease in NOR until the forth periods (see Figure 3 (c). This situation may be partly explained by the indirect effects of OILP shocks that effects through cost channel. However, SUB and GOE variables show a decreasing effect of OILP shock on a short term trend, with the magnitude effect of SUB larger than GOE (Figure 3 (d) and 3 (e)) respectively. This simply implies that OILP shocks have a strong impact on SUB and GOE variables. A shock of OILP causes government operating expenditure levels to decrease rapidly until the 4-th period. This suggests that the impact of OILP shock on government expenditure is relatively short-lived. Fourth, the accumulated effects of one S.D. of OILP shocks on the endogenous variables for up to the 10-th period found that 1 percent increase in OILP shocks contribute a sharp increase in the **OR** by 22% over the next 10-th period. At the same time, the accumulated response over 10-th period for **NOR** is 1.4%, which is far smaller than the impact on the OR. While the accumulated response for up to the 10-th period is estimated to be positive 5.3% which indicated that 1 percent increase in OILP shocks contributes an increase in the **GOE** by 5.3% over the next 10-th period. On the other hand, the accumulated response for a **SUB** up to the 10-th period is estimated to be positive 26% which indicated that 1 percent increase in OILP shocks contributes a sharp increase in the total subsidy by 26% over the next 10-th period.

Fifth, the *VDC* for the 6 endogenous variables with symmetric OILP shocks are estimated over the10-period horizon for each endogenous variable based on the VAR model, where the decomposition value converging to stable states. In estimating the impact of OILP shocks on other endogenous variables, it can be seen that an **OILP** shock is a considerable source of variation, i.e. contribute substantially to around 74.9% and 32.2% of the volatility in OR growth and GOE, respectively. Also, RGDP and SUB contribute around 27.4% and 22.3%, respectively. Besides the OILP, NOR is also a considerable source of variation for RGDP growth which is contributed around

26%. In terms of per OR, fuel subsidies accounted for 26.45% of oil revenues for the year 2005 and decrease to 16% and 17.22% for the year 2006 and 2007, respectively.

It also shows that fuel subsidy accounted for 0.21%, 0.11% and 0.12% of RGDP for the year 2005, 2006 and 2007, respectively. As Malaysia is an oil exporting country, it is expected that high oil prices in the short term would benefit the Malaysian economy. This is also being supported earlier findings [12] which focused on the 31 OPCs. The studied found that oil revenue is a critical source of fiscal revenue where fiscal oil revenue accounted for more than 25 percent of total fiscal revenue over the (2005-2008) period. In the case of Malaysia, oil revenue has contributed 29.2% up to 36.5% of total revenue for the (2005-2007) period.

4. Conclusion

The results suggest that the impact of symmetric OILP shocks has a direct and positive impact on oil revenue, even if a short-term phenomenon. Also, the macroeconomic impacts of oil price shocks are found to be greater in magnitude for real GDP and total subsidy in the Malaysian economy. Besides, in the long time path the changes of world oil price would have significant impact to GOE through the subsidies channel in the country. Besides that, the findings of the study will help to give clear policy directions especially in designing a better policy instrument system in the macroeconomic level planning, especially for establishing and reviewing existing policies (i.e. fuel subsidy policy, energy and environmental policy). Finally, it is worth pointing out that the results of this paper do no more than suggest that the adoption of expansionary fiscal policy during the oil price shocks can facilitate rapid economic growth. In the sense that, as long as there is a stability and persistence of economic policies within the framework of an appropriate macroeconomic discipline, a higher oil price in a small oil-exporting economy like Malaysia will not necessarily be dissolute simply in terms of higher inflation but will contribute positively to assist in achieving an impressive rate of economic growth in a near term.

References

[1] Abeysinghe T 2001 Estimation of Direct And Indirect Impact Of Oil Prices On Growth *Economic Letters* **73** 147-153
[2] Bernanke B S, Mark G and Mark W 1997 Systematic Monetary Policy and the Effects of Oil Price Shocks *Brookings Papers on Economic Activity* **1** 91-124
[3] Burbidge J and Alan H 1984 Testing for the Effects of Oil-Price Rises Using Vector Auto regressions *International Economic Review* **25** 459-84
[4] Gronwald M 2012 Oil and U.S. Economy *The Energy Journal* **4** 143-160
[5] Hamilton J D 2003 What is an Oil Shock? *Journal of Econometrics* **113(2)** 363-98
[6] Hamilton J D and Herrera A M 2001 Oil Shocks and Aggregate Macroeconomic Behaviour: The Role of Monetary Policy *Discussion Paper 2001-10. University of California, San Diego.*
[7] World Energy Outlook 1999: Looking at Energy Subsidies 1999 *International Energy Agency: Paris, IEA.*
[8] Jones D W, Leiby P N and Paik I K 2002 Oil Price Shocks and the Macroeconomy: What Has Been Learned since 1996? *In: Proceedings of the 25th Annual IAEE International Conference, Aberdeen, Scotland*
[9] Mork K A 1989 Oil and the Macroeconomy When Prices Go Up and Down: An Extension of Hamilton's Results *Journal of Political Economy* **91** 740-44
[10] Rodriguez R and Sanchez M 2004 Oil Price Shocks and Real GDP Growth: Empirical Evidence for some OECD Countries *European Central Bank Working Paper* 362
[11] Sari R and Soytas U 2006 The Relationship between Stock Returns, Crude Oil Prices, Interest Rates, and Output: Evidence from a Developing Economy *Empirical Economics Letters* **5(4)** 205-20
[12] Villafuerte M and Murphy P L 2009 Fiscal Policy in Oil Producing Countries During the Recent Oil Price Cycle *IMF Working Paper* 6.

IOP Conf. Series: Earth and Environmental Science **16** (2013) 012011 doi:10.1088/1755-1315/16/1/012011

Experimental Investigation of A Twin Shaft Micro Gas-Turbine System

Hussain Sadig[1], Shaharin Anwar Sulaiman and Idris Ibrahim

Mechanical Engineering Department, Universiti Teknologi PETRONAS, Bandar Seri Iskandar, 31750 Tronoh, Perak, Malaysia

E-mail: husainsadig@yahoo.com

Abstract. Due to the fast depletion of fossil fuels and its negative impact on the environment, more attention has been concentrated to find new resources, policies and technologies, which meet the global needs with regard to fuel sustainability and emissions. In this paper, as a step to study the effect of burning low calorific value fuels on gas-turbine performance; a 50 kW slightly pressurized non-premixed tubular combustor along with turbocharger based twin shaft micro gas-turbine was designed and fabricated. A series of tests were conducted to characterize the system using LPG fuel. The tests include the analysis of the temperature profile, pressure and combustor efficiency as well as air fuel ratio and speed of the second turbine. The tests showed a stable operation with acceptable efficiency, air fuel ratio, and temperature gradient for the single and twin shaft turbines.

1. Introduction

Gas turbines, which used in power generation, are designed mainly to run on natural gas and kerosene. An attractive feature of gas turbines that it is capable to run with various types of fuels [1]. The main differences in micro gas turbine as compared to industrial size gas turbines, include the way that wasted heat is recuperated, efficiency and capability to run in combined cycles [2]. Due to the challenges which face using fossil fuels, a special attention is required on biomass combined gas turbines.

A pressurized premixed cyclone combustor drive two shaft micro gas turbine have been investigated by Al-ttab and Zainal [3]. CFD modeling was carried out, but it was limited to the design and optimization of a cyclonic combustor. Jaafar et al. [4] designed and fabricated a gas-turbine liquid fuel burner with different radial air swirler angles; the investigation was concerned with the effect of the swirler size on exhaust gas emission. The thermo chemical power group in University of Geonoa developed a new gas-turbine laboratory; a general purpose experimental rig was installed based on a modified 100 kW micro gas turbine [5].

This paper discusses the preliminary experimental results for a low-cost twin shaft gas turbine developed for use with syngas. The tests were conducted using LPG fuel as a benchmark for studying the effect of burning low calorific value gaseous fuel similar to syngas on the gas-turbine performance.

2. Design Approach

The gas-turbines consist of a compressor, combustor, and turbine. The heat input for the gas-turbine is provided by combustors. Gas turbine combustor design falls into three categories; i.e. tubular, annular, and tubo-annular. The amount of heat added to the combustor can be quantified by [6]:

$$Q = \dot{m}_{fuel} CV_{fuel} \tag{1}$$

where $\dot{m}_{fuel}$ and CV_{fuel} are the mass flow rate and the calorific value of the combusted fuel respectively. By knowing the inlet condition, the combustor could be sized. The parameter which facilitate combustion chamber sizing is the reference velocity, U_{ref}, which is the mean velocity at the maximum cross sectional area of the casing in the absence of the liner[7]:

$$U_{ref} = \frac{\dot{m}}{\rho A_{ref}} \tag{2}$$

where $\dot{m}$ is the air mass flow rate, ρ is the air density and A_{ref} is the cross sectional area of the casing. The relationship between the size and pressure loss in the combustion chamber for the optimal cross-sectional area of the casing is given by [4, 7]:

$$A_{ref} = \left[\frac{R}{2}\left[\frac{\dot{m}T^{0.5}}{P}\right]\left(\frac{\Delta P}{q_{ref}}\right)\left(\frac{\Delta P}{P}\right)^{-1}\right]^{\frac{1}{2}} \tag{3}$$

where $\frac{\Delta P}{q_{ref}}$ is the pressure loss factor, and $\frac{\Delta P}{P}$ is the total pressure drop across the combustor. Equation (3) can be used to calculate the reference area for the straight combustors. Table 1 shows some typical values of pressure loss in practical chambers.

Table 1. Combustion chamber pressure loss [4]

Type of chamber	$\left(\Delta P/_{P}\right)$	$\left(\Delta P/_{q_{ref}}\right)$	$\left[\frac{\dot{m}T^{0.5}}{A_{ref}P}\right]$
Tubular	0.07	37	0.0036
Tubo-annular	0.06	28	0.0039
Annular	0.06	20	0.0046

For tubular combustors, the ratio of length to the diameter should be small to reduce the formation of NOx emissions and reduce the weight of the combustor. Furthermore, the increase of liner area results in lower velocities and higher residence time which are good for combustion characteristics.

3. System Descriptions

The small size gas turbine system consists of the following parts:

3.1. Combustion Chamber

The combustion chamber is a tubular combustor with thermal capacity of 50 kW. The consumption rates of LPG fuel and air were calculated using Equation (1). For simplicity and ease of fabrication, a combustor length to diameter ratio of 2 was chosen. In order to reduce the air pressure and increase the flue gas velocity at the combustor exit, the combustor was joined to the turbine by a nozzle. The dimension of the combustor is shown in Table. 2.

Table 2. Combustion chamber geometry specifications

Material	Wall thickness (mm)	Diameter (mm)	Liner/Annulus Ratio	Length/Diameter Ratio	Liner Holes (mm)	Nozzle Height (mm)	LPG Injection Diameter (mm)
S. Steel	4	260	0.7	2	10	100	2

3.2. Turbochargers and Air fuel system

Two turbochargers were selected to for this work. A Nissan H 25 turbocharger was selected and mounted as the high pressure turbine based on its air flow rate and pressure ranges required in this work. For the second turbine where a high torque is needed, the Jasma T70 turbocharger was used. Since both turbochargers rotate with a high rotational speed, a vehicular oil pump driven by one-horse power electric motor was used to provide the turbocharger with the needed lubrication.

The fuel system allows for non-premixing LPG charge, the fuel supplied from LPG cylinders through a pressure regulator and control valve and then injected into the combustor via eight holes distributed in 10 mm diameter stainless steel circular pipe acts as a flame holder. The starter system consisted of a V shape tube connected directly with the compressor inlet; one of the two ends connected to the 1½ hp blower which provided the system with air at the starting stage, the other end was opened to atmosphere.

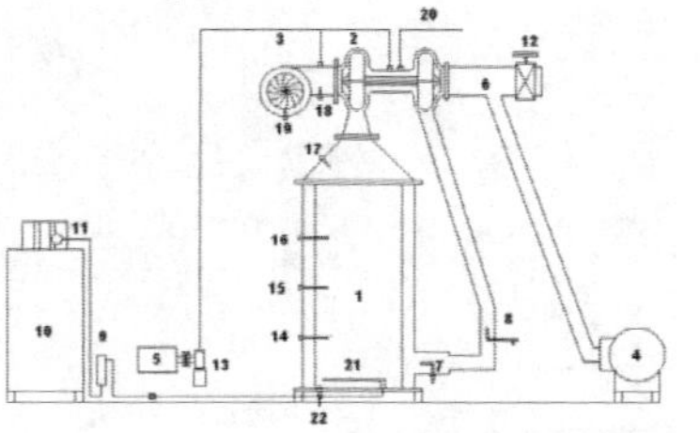

Figure 1. Schematic of the experimental rig (1) Combustor (2) High pressure turbine (3) Low pressure turbine (4) Starting blower (5) Oil pump motor (6) Starting air entrance (7) Pressure gauge (8) Pitot tube (9) LPG rotameter (10) LPG cylinder (11) LPG regulator (12) Entrance air valve (13) Oil pump (14-19) thermocouples (20) Cooling water line (21) Flame holder (22) Ignition plug

4. Test Method

Figure 1 shows a simplified over view of the experimental setup used. Six type K thermocouples distributed in equally axial distance were attached in the combustor to monitor the temperature inside the combustor; the thermocouples were connected with Pico USB TC-08 thermocouple data logger with 0.5°C accuracy to reflect the system temperature. The flow rate of the fuel and the air was measured using 150 mm Concoa flow meter and calibrated Pitot tube respectively. Lutron DT-2259 optical tachometer with an accuracy of 0.1 rpm was used to measure the turbine rotational speed.

In the starter system, LPG and air were injected into the combustor; then the ignition was initiated using a spark plug. For warming up purpose, the system was allowed to run for about 5-10 minutes before the fuel flow rate increasing.

5. Results and Discussions

The single shaft turbine was successfully run with LPG fuel. The temperature and turbine speed increased gradually with increase fuel flow rate. When the temperature in the primary combustion zone reached 1050°C the system started to run, and the compressor could provide the combustor with the needed air. At this stage, the starting blower was stopped.

Shown in Figure 2 is the variation of combustor efficiency at different heat input. The use of the flame tube to control the amount of the combustion and dilution air, affected in improving the combustor efficiency and lead to an excellent temperature gradient inside the combustor. The main problem in this case was the high operation temperature. On the other hand, as shown in Figure 5, removal of the flame tube increase of the amount of admitted air and hence lowered the operation temperature and affected the unsteady combustor temperature gradient, especially in the primary combustion zone which is exposed to the cold air coming from the compressor.

Shown in Figure 3 is the variation of the turbine entrance temperature with heat input. Due to the low amounts of admitted air into the combustor for the flame tube tests, the turbine entrance temperature was relatively higher compared to that one with no flame tube.

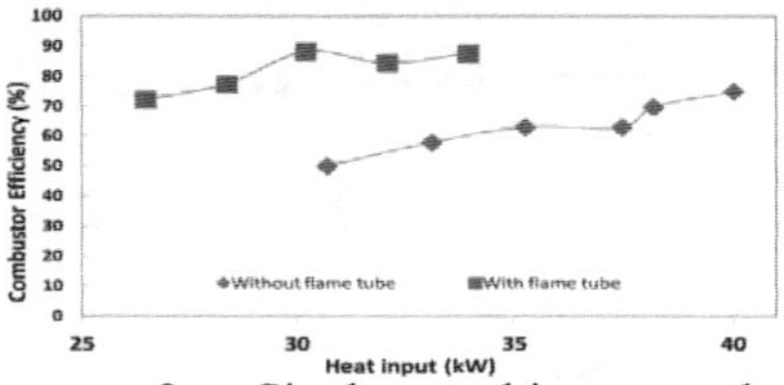

Figure 2. Single turbine combustor efficiency for different heat input

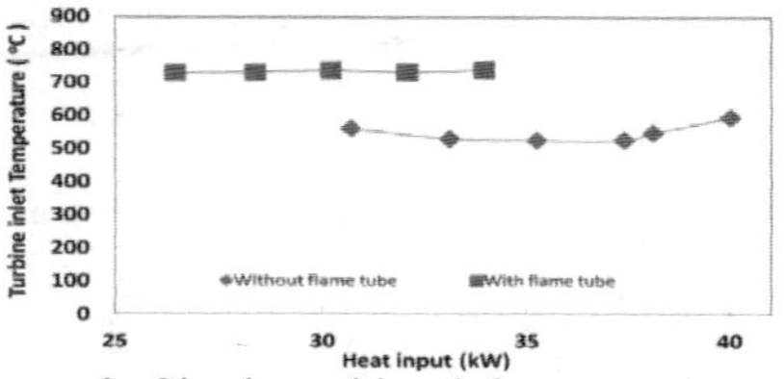

Figure 3. Single turbine inlet gas temperature for different heat input

Generally, as the amount of the burned fuel increased the rotational speeds increased and hence increase the air intake rate into the combustor. The combustor efficiency calculated based on the ratio of the flue gases' thermal power out from the combustor to the combustor thermal power input. As shown in Figures 4 and 5, the tests showed an acceptable air fuel ratio. The excess air range was 120% - 221% which is acceptable for gas turbines.

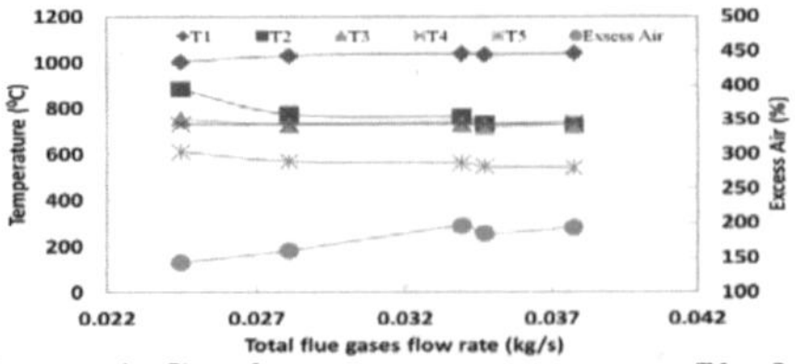

Figure 4. Combustor temperature profile for the single turbine with flame tube

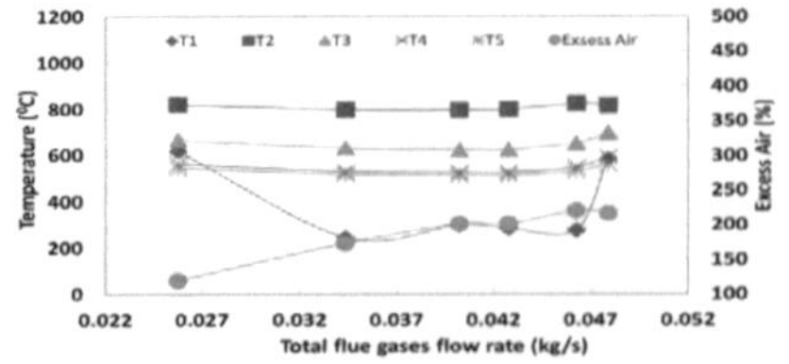

Figure 5. Combustor temperature profile for the single turbine without flame tube

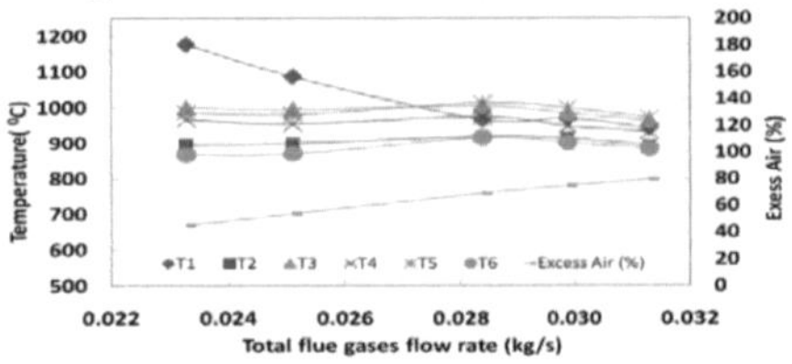

Figure 6. Combustor temperature profile for the twin shaft turbine

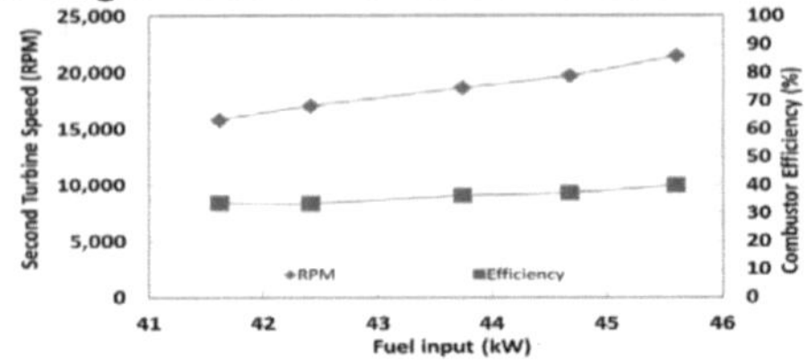

Figure 7. Second turbine speed and combustor efficiency for the twin shafts turbine

For the twin shaft turbine, the compressor could suck the air from atmosphere and inject it into the combustor. For safety reasons, the air was allowed to be sucked through a half inch hole located at the suction line to prevent the high air admission which may affect in flame quenching.

Shown in Figure 6 is the variation of the temperature and excess air with the total flue gases. At the start, the fuel-burning process inside the combustion chamber was at a low air fuel ratio. The first turbine speed was strongly affected by the back pressure of the second turbine. The air-fuel ratio increased gradually by the increase of the first turbine speed, the excess air range was 45%- 217%. As shown in Figure 7, the second turbine speed and combustor efficiency was increased gradually with the fuel increasing. The twin shafts turbine tests showed lower combustor efficiency compared to that one of the single turbines, this is because of the high energy lost by the combustor surface.

6. Conclusions

A 260 mm diameter tubular combustor along with low cost twin shaft micro gas turbine system was designed and fabricated. The system has been characterized with LPG fuel. The test showed a successfully running with an acceptable air fuel ratio, temperature gradient, and efficiency; the combustor efficiency range was 34% - 87% while the turbine entrance temperature was 550°C– 975°C. Due to high heat loss by the combustor surface, the two shaft turbine tests showed a lower combustor efficiency compared with the single turbine.

Acknowledgment

The author would like to acknowledge the support of Universiti Teknologi Petronas in this work, which is internally funded by URIF No. 15/2012

References

[1] Gupta K K, Rehman A and Sarviya R M 2010 *Renewable and Sustainable Energy Reviews* **14** 2946
[2] Ghenai C 2010 *Advances in Mechanical Engineering* **1**
[3] Al-attab K A and Zainal Z A 2011 *Applied Energy* **88** 1084
[4] Jaafar M et al 2005 Development of Low Liquid Fuel Burner *Universiti teknologi Malaysia, Malaysia*
[5] Ferrari M L, Pascenti M, Magistri L and Massardo A F 2009 *Journal of engineering for gas turbines and power* **132**.
[6] Boyce M P 2002 *Gas turbine engineering handbook* Gulf Profisional publishing, 2nd Edition
[7] Murthy J N 1988 *Gas turbine combustor modelling for design* Cranfield University, UK

Energy efficiency monitoring and economic analysis for energy saving potential in UNITEN

M Reyasudin Basir Khan[1], Razali Jidin[1], Jagadeesh Pasupuleti[2], Kang Chin Yew[1] and Sharifah Azwa Shaaya[1]

[1] Center for Automation and Embedded Computing System (CAECS),
[2] Center for Power System Simulation (CPSS),
College of Engineering, Universiti Tenaga Nasional, Jalan IKRAM – UNITEN, 43000 Kajang, Selangor, Malaysia.

E-mail: {Reyasudin, Razali}@uniten.edu.my

Abstract. This paper discusses on energy efficiency survey for typical buildings in Universiti Tenaga Nasional (UNITEN). Undeniably, wastage of energy will cause the increase of operation cost and depletion of fossil fuel resources which contributes to the climate change issue in the world. UNITEN was commenced in the late 1990s and most of the buildings in this university are not equipped with energy management system. Such system is the solution to reduce energy use while maximizing the comfort levels of the occupants. Disregard to the energy management system, the implementation of other energy saving measures is the main objective of this paper. By taking the right measures, the energy wastage in the buildings of this university can be reduced.

1. Introduction

Energy efficiency is an act of using less energy to accomplish the same task. It is something that affects every single individual, from purchasing energy saving household electrical appliances, to the use of efficient industrial equipment. As a result, this will help the end-users to save energy, lower down the operation cost and still be able to generate the same desired output. Most buildings do not have energy efficiency systems installed. Such system can reduce electricity bill by several methods such as demand side management including load shifting. The efficient use of energy will result in conserving the energy and hence will certainly lower down the energy usage and electricity bills. This paper is focusing on survey of loads installed in typical commercial buildings, energy consumption and usage profiling. In this paper, energy consumption of major electrical appliances in UNITEN buildings such as lightings, computers, and air-conditioners will be taken into account.

There are number of energy saving measures in office buildings in recent years such as Kawamoto et al suggest to minimize the power management delay time of office equipment in Japanese offices [1]. In-Ho Yang et al perform an economic analysis of automatic lighting control system installed in office buildings [2]. Study also shows that office lighting system, double glazing windows heating and air- conditioning plays a key role in reducing energy consumption [3- 4].

2. Energy audit

Energy audit conducted at College of Engineering (COE) and Admin Building (BA) to obtain the energy consumption of major electrical equipment and to compare with monthly electricity bill. The major equipment is lightings, air-conditioners and personal computers (PC). The reason this energy audit conducted at COE and BA because these two buildings are among the firsts to be commenced in 1997 and the buildings do not have Building Automation System (BAS) installed, unlike College of Information Technology (COIT) which has BAS. Therefore, further data analysis is conducted at COE and BA to come up with solutions to save electricity usage in these buildings. The energy consumed by electrical equipment and monthly electricity bill calculated by using equation (1), (2) and (3).

$$E = P \times t \quad (1)$$

$$MEB = TE \times \text{Tariff B rating} \quad (2)$$

$$tm = th \times (5\text{days/week}) \times (4\text{ weeks/month}) \quad (3)$$

E: Energy, P: Power, t: operational hour, TE: Total Energy, MEB: Monthly electricity bill
tm: Operational hour per month, th : Operational hour per day, Tariff B: Low voltage commercial tariff.

2.1. College of Engineering (COE)
COE consists of 2 main buildings which are Admin Wing (BN) and Classroom and Lab Wing (BL and BM). BN building is mainly for lecturer's office rooms, BL building is for engineering labs, and BM building is for classrooms. There are 4 floors at BN and 5 floors at BL and BM. On average, COE building consumes about RM75, 000 per month for electricity. The percentage distributions of energy usage by major equipment in COE are 47.8% from air- conditioners, 33.8% from lighting, 17.9% from PC and 0.5% from others.

2.2. Admin Building (BA)
Admin Building consists of classrooms, lecturer's office rooms, administration office rooms, and a café. There are 5 floors in this particular building. On average, Admin Building consumes about RM61, 000 per month for electricity. The percentage distributions of energy usage by major equipment in BA are 55.7% from air- conditioners, 28% from lighting, 15.4% from PC and 0.9% from others.

3. Energy saving opportunities
With reference to the energy audit conducted, air-conditioner consumes most of the energy usage in a building, which is about 50% of the overall energy. The second-highest energy consumption is by lighting (about 30%) and lastly followed by PC (less than 20%). In this section, recommendations on energy saving measures by this major electrical equipment will be proposed. The estimation of energy saving potential will be reviewed, in order for UNITEN Facilities Development and Management (FDM) to take action and implement the possible solutions. The efficient use of energy will result in conserving the energy and hence will certainly lower down the monthly energy usage or electricity bills. The Payback Period and Return on Investment (ROI) for determination of best investment calculated from equation (4) and (5).

$$\text{Payback Period} = \frac{\text{Total Initial Cost}}{\text{Cost Savings}} \quad (4)$$

$$ROI = \frac{\text{Gains} - \text{Investment Costs}}{\text{Investment Costs}} \quad (5)$$

3.1. Lighting

3.1.1. Change the daily operation hour of lightings. Currently at COE, the operation hour of lightings at BN is from 7am until 6pm (11hours) while at BL and BM are from 7am until 8pm (13 hours). At BA, the lightings start to operate from 7am until 8pm (13 hours). Since working hour and classes start at 8am daily, it is recommended that the lightings operate from 8am instead of 7am. By implementing this, UNITEN can save more than RM1, 000 per month. Figure 1 shows cost comparison of different lighting operation hours. From the analysis, the cost savings by implementing this method for COE is RM2, 105 per month and RM1, 292 per month for BA.

3.1.2. Replace current PL-L lamp with MASTER TL5 High Output Eco (Philips). MASTER TL5 uses T5 bulb that can achieve up to 10% energy saving. After 36 months (3 years), the total gain is RM493, 488 (36 x RM13, 708). Therefore, ROI is 49.6% (3 years after installation of T5). Figure 2 shows cash flow after the installation of T5 bulb.

3.2. Personal Computer (PC)

3.2.1. Replace desktops with laptops. From the analysis performed, the average energy consumption, E of a PC are 100kWh per month for a desktop and 4.4kWh per month for a laptop. In conclusion, UNITEN can save up to 95.6% of energy per month in replacing desktop with laptop. By using laptops instead of desktops, UNITEN can cover the replacement cost and starts to gain profit after 4 years as shown in figure 3. After 60 months (5 years), the total gain is RM981, 660. Therefore ROI is 23.3% (5 years after using laptops instead of desktops).

3.2.2. Turn off desktops during weekends. The survey conducted shows 8% of the PC users at COE and 40% at BA leave their desktop turned on unattended during weekends. By turning off desktop during weekends, UNITEN can save up to RM1, 210 (8%) per month at COE and RM3, 692 (40%) per month at BA.

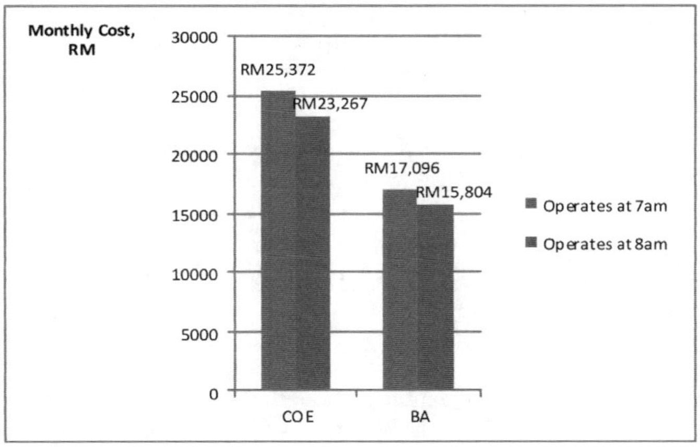

Figure 1. Cost Comparison of different lighting operation hours.

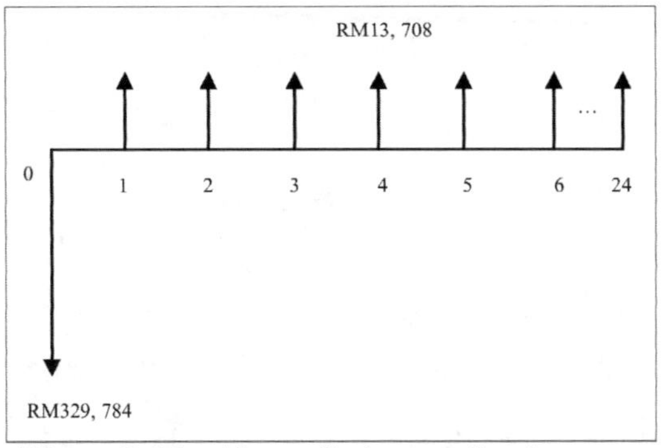

Figure 2. Monthly profits/ gains after the installation of T5.

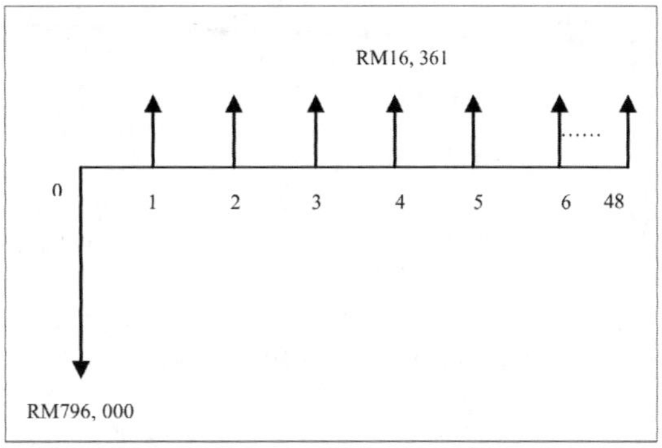

Figure 3. Monthly profits/gains after using laptops at COE.

3.3. Air- conditioner

3.3.1. Aerosol duct sealing. This aerosol duct sealing can reduce duct leakage up to 90%, which can reduce energy used up to 30%. The installation of aerosol duct sealing can save cost RM10, 500 per month. The estimated initial cost for aerosol duct sealing is about RM30, 000 per building. From table 1, it can be concluded that after 3 months of aerosol duct sealing process, UNITEN can gain profit whereby there is a reduction of energy usage by air-conditioner by 30%. Therefore the calculated ROI is 10% (on the 6th month after sealing). The ROI indicates the investment on aerosol duct sealing is the best decision.

Table 1. Expected gain after aerosol duct sealing at COE.

Month	Investment Cost	Cost Savings	Gain/Profit
0	-RM30,000	-	-RM30,000
1	-	RM10,500	-RM19,500
2	-	RM10,500	-RM9,000
3	**-**	RM10,500	***RM1,500*** ⟵
4	-	RM10,500	RM12,000
5	-	RM10,500	RM22,500
6	-	RM10,500	RM33,000

4. Conclusion

In conclusion, the main objectives of this paper are achieved successfully after completing the data analysis of energy usage in College of Engineering and Administration Building. The aim of energy audit to provide methods to reduce energy wastage is also achieved. Moreover, the list of major electrical appliances (lightings, AHU Fan Motors, Terminal Building Pumps, etc.) at COE and BA are recorded for future references. The wattage ratings, operation hour, and computation on energy usage by these appliances are included as well.

Acknowledgments

The author would like to thank UNITEN Faculty Development and Management (FDM) for providing data and Ministry of Higher Education (MOHE) for funding this research (MOHE grant no: 20120236FRGS).

References

[1] Kawamoto K, Shimoda Y, and Mizuno M 2004 Energy saving potential of office equipment power management *Energ. Buildings* 915-23

[2] Yang H and Nam E J 2010 Economic analysis of the daylight-linked lighting control system in office buildings *Sol. Energy* **84** 1513–25

[3] Zografakis N, Karyotakis K and Tsagarakis K P 2012 Implementation conditions for energy saving technologies and practices in office buildings: Part 1. Lighting *Renew. Sust. Energ. Rev.* **16** 4165-74

[4] Tsagaraki K P, Karyotakis K and Zografakis N 2012 Implementation conditions for energy saving technologies and practices in office buildings: Part 2. Double glazing windows, heating and air-conditioning *Renew. Sust. Energ. Rev.* **16** 3986–98

IOP Conf. Series: Earth and Environmental Science **16** (2013) 012013 doi:10.1088/1755-1315/16/1/012013

Building automation: Photovoltaic assisted thermal comfort management system for energy saving

M Reyasudin Basir Khan[1], Razali Jidin[1], Jagadeesh Pasupuleti[2] and Sharifah Azwa Shaaya[1]

[1] Center for Automation and Embedded Computing System (CAECS),
[2] Center for Power System Simulation (CPSS),
College of Engineering, Universiti Tenaga Nasional, Jalan IKRAM- UNITEN, 43000 Kajang, Selangor, Malaysia.

E-mail: {Reyasudin, Razali}@uniten.edu.my

Abstract. Building automation plays an important key role in the means to reduce building energy consumption and to provide comfort for building occupants. It is often that air conditioning system operating features ignored in building automation which can result in thermal discomfort among building occupants. Most automation system for building is expensive and incurs high maintenance cost. Such system also does not support electricity demand side management system such as load shifting. This paper discusses on centralized monitoring system for room temperature and photovoltaic (PV) output for feasibility study of PV assisted air conditioning system in small office buildings. The architecture of the system consists of PV modules and sensor nodes located at each room. Wireless sensor network technology (WSN) been used for data transmission. The data from temperature sensors and PV modules transmitted to the host personal computer (PC) wirelessly using Zigbee modules. Microcontroller based USB data acquisition device used to receive data from sensor nodes and displays the data on PC.

1. Introduction

The position of room inside building contributes to a change of temperature inside the room since the amount of sunlight the room attains in the morning, afternoon or evening. The weather condition either rainy or hot can also affect temperature inside buildings. To overcome this problem, the air conditioning system should be controllable, thus to maintain optimum temperature inside rooms or building. Though such smart building technologies are already available, most of them are very costly to install and maintain. Many of them rendered to useless system after a short period in operation as the high maintenance cost seems as a main factor that hinders it to function. If such a system can leverage on pervasive network technology, office, computers, low cost and not specialized components, more buildings will be able to conserve more energy.

Number of techniques has been proposed in recent years for energy saving in buildings. Study shows that energy saving technologies such as efficient lighting system, double glazing windows, heating and air conditioning contribute to decrement of energy usage in office buildings [1-2]. Evaluation of the air distribution performance in building also results in reduction of energy consumption [3]. Renewable energy also plays an important role in energy saving inside building such as solar assisted air conditioning system been used to reduce the burden by air conditioning system on the electric utilities [4].

2. System components

2.1. USB Data Acquisition Device

This device equipped with Zigbee module (Xbee Pro), temperature sensor (L35DZ) and microcontroller (PIC18F4550). Analog data from temperature sensor converted into digital values through microcontroller which then transmit the data wirelessly using Zigbee module. Influential technology of this device is USB communication protocol which been used for data transfer and power supply, so no external power and no power regulation needed. The device also offers high bandwidth data transfer which is essential for extended data transfer. Figure 1 shows the prototype of the device.

2.2. PV Module Performance Recorder

The important parameters of PV module are voltage, current and power output. This parameters need to be monitored wirelessly from host PC. A device developed using microcontroller (PIC18F452), Zigbee module (Xbee Pro) and Hall Effect current sensor (ACS712). The device capable to measure output voltage, current and power from PV module and transmits the data via wireless communication to the host PC. The device powered by 4 AA batteries. Figure 2 shows the prototype of the device.

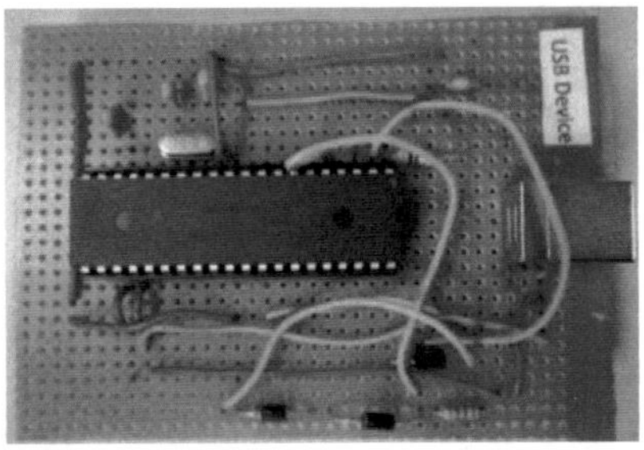

Figure 1. USB Data Acquisition Device.

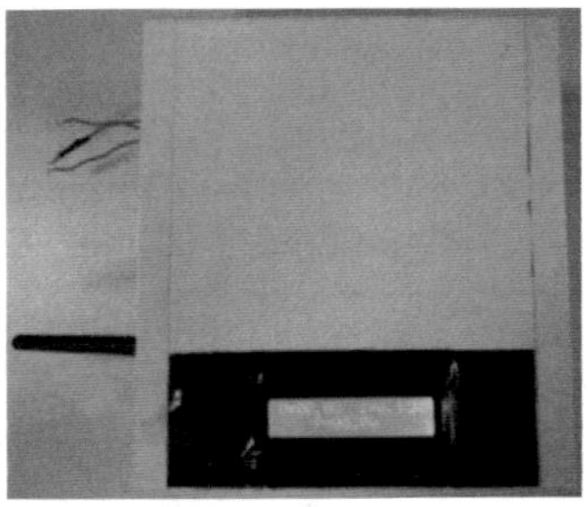

Figure 2. PV Module Performance Recorder.

3. System architecture

The system consists of distributed temperature sensors in each office rooms, PV system and a host PC. The polled data from temperature sensors are for room temperature profiling. The output data from PV modules monitored in order to study the feasibility of PV system for assisting air conditioning system in office buildings. The task of host PC is to monitor all data from the PV system and temperature sensors. Server- client transactions are made via wireless communication protocol. The host PC features output reading from temperature sensors and PV module, real-time graphs and a database system.

3.1. Distributed sensor real time measurement system

Real time temperature for each room needs to be monitored in order to improve the operation of air conditioning system. The system consists of sensors integrated in USB Data Acquisition Device. The device is intelligent to process the temperature values and transmit via wireless communication. The developed device is convenient as it is 'hot pluggable'. Since the case study of this paper is for office buildings, it is assumed that each room possesses a PC. Zigbee modules used in this system is Xbee pro from Digi which capable of data transmission up to 1.5Km outdoor. Since there are many obstacles inside buildings such as doors, roof and walls, few Zigbee modules placed around the buildings in order to hop the data.

3.2. PV system
The PV system comprises of PV modules and PV Performance Recorder. PV Module Performance Recorder installed on each PV modules to record PV output parameters. The data transmitted to host PC via wireless communication using Zigbee module. The architecture for PV system wireless communication is the same as distributed sensors system.

3.3. Software and firmware
Two platforms of software are developed, which consists of embedded system coded in C (PV Module Performance Recorder and USB Data Acquisition Device) and a graphical user interface (GUI) running on host PC in order to establish interaction between user with PV output and room temperatures.

3.4. Network topology
Star, cluster tree and Mesh topology can be implemented using Zigbee technology. Since the proposed system designed for application in small office, which each sensor nodes need to pass the data to coordinator, thus star topology been used.

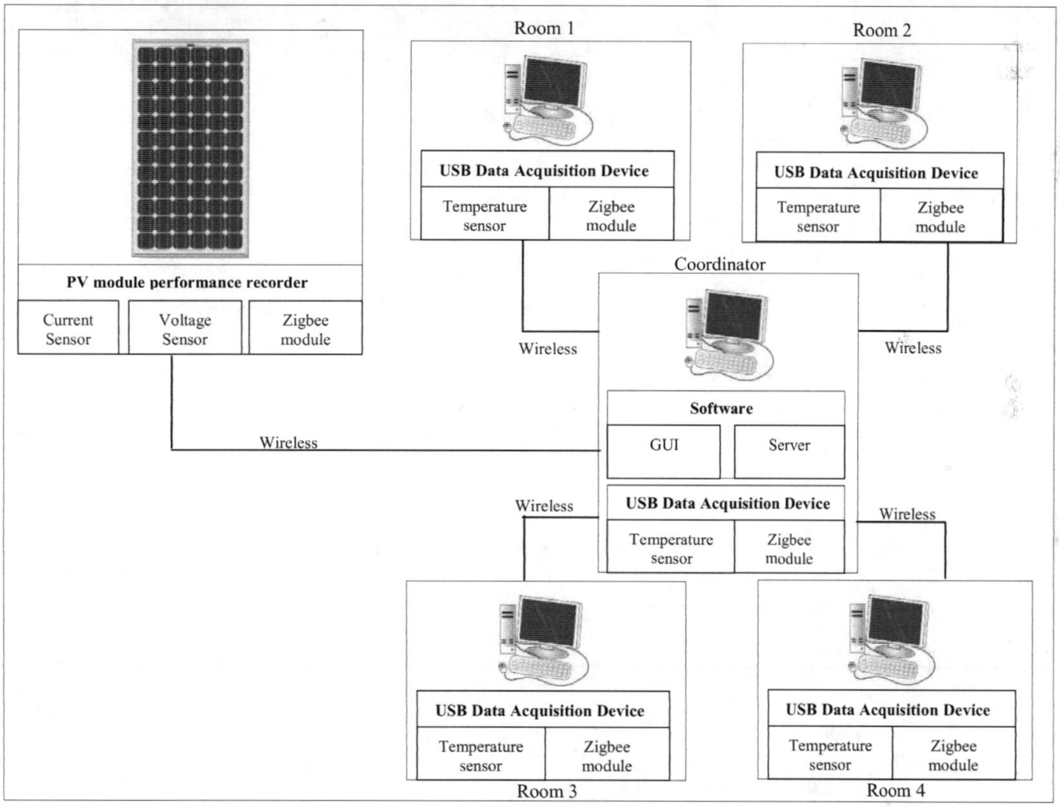

Figure 3. PV module output and room temperature monitoring system block diagram.

4. Testing and implementation
The proof-of-concept system illustrated in Figure 3 consisting of one PV module equipped with PV Performance Recorder device and several remote computers and one host computer equipped with USB Data Acquisition Device. The system operation only use PV module without the complete PV system since this system only monitors output from PV modules. Several phases of testing conducted for the system. The temperature reading from USB Data Acquisition Device from one room monitored

wirelessly from the host PC using a Visual Basic interface. Figure 4 shows the simple GUI developed for monitoring temperature reading inside one room. In order to test PV module wireless communication with the host PC, PV module placed at building rooftop monitored using another Visual Basic interface showed in Figure 5. The interface consists of PV module output current, voltage and power. It also shows variation of the three parameters plotted in graph. GUI shows that both parameters (room temperature and PV module output) can be monitored wirelessly from host PC.

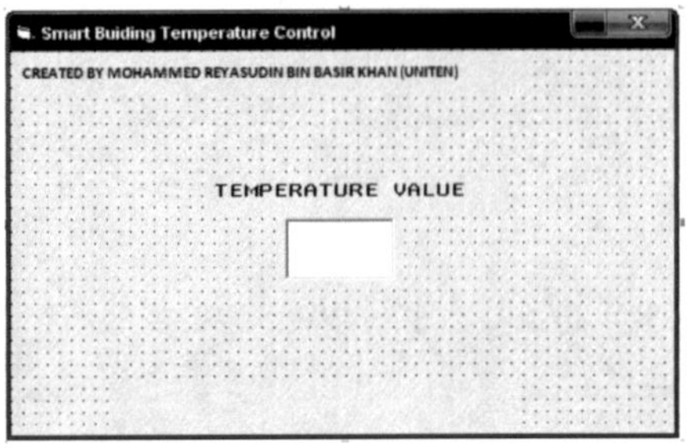

Figure 4. Visual Basic GUI for room temperature monitoring.

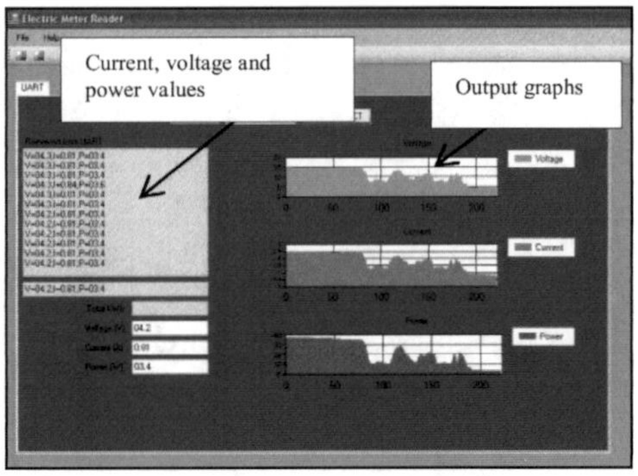

Figure 5. Visual Basic GUI for PV module output monitoring.

5. Conclusion

In this paper, a PV module and room temperature wireless monitoring system introduced. A GUI developed using Microsoft Visual Basic environment to monitor room temperature, current, voltage and power of PV module. The results of the tests and experiments conducted indicate that this system is a promising technology to be implemented in small office building. Currently, the system only capable to monitor output from one PV module and temperature from one room. The complete system still in development. The proposed overall system should be able to utilize energy from PV to support air conditioning system in office building during peak load demand. Monitored room temperature will be used as input parameters for controllable air conditioning system in order to achieve optimum thermal conform in each room.

Acknowledgments

The author would like to thank Ministry of Higher Education (MOHE) for funding this research (MOHE grant no: 20120236FRGS).

References

[1] Zografakis N, Karyotakis K and Tsagarakis K P 2012 Implementation conditions for energy saving technologies and practices in office buildings: Part 1. Lighting *Renew. Sust. Energ. Rev.* **16** 4165–74

[2] Tsagarakis K P, Karyotakis K and Zografakis N 2012 Implementation conditions for energy saving technologies and practices in office buildings: Part 2. Double glazing windows, heating and air-conditioning *Renew. Sust. Energ. Rev.* **16** 3986–98

[3] Gao C F, Lee W L and Chen H 2009 Locating room air-conditioners at floor level for energy saving in residential buildings *Energ. Convers. Manage.* **50** 2009-19

[4] Henning H M 2007 Solar assisted air conditioning of buildings: an overview *Appl. Therm. Eng.* **27** 1734-49

IOP Conf. Series: Earth and Environmental Science **16** (2013) 012014 doi:10.1088/1755-1315/16/1/012014

Preliminary Evaluation of Atomization Characteristics of Improved Biodiesel for Gas Turbine Application

P Kumaran[a], M Gopinathan[a], N M Razali[a], Isabel Kuperjans[b], BHariffin[c] and H Hamdan[c]

[a] Centre for Renewable Energy Universiti Tenaga Nasional (UNITEN), Malaysia
[b]Nowum Energy FH Aachen Campus Julich, 52428 Julich, Germany
[c]TNB Research Sdn Bhd (TNBR), 43009 Kajang, Selangor, Malaysia

E-mail : kumaran@uniten.edu.my

Abstract. Biodiesel is one of the clean burning alternative fuels derived from natural resources and animal fats which is promising fuel for gas turbine application. However, inferior properties of biodiesel such as high viscosity, density and surface tension results in inferior atomization and high emission, hence impedes the fuel compatible for gas turbine application and emits slightly higher emission pollutants due to inferior atomization. This research work focuses on preliminary evaluation of the atomization characteristics of derived from Malaysian waste cooking oil which is the physical properties are subsequently improved by a microwave assisted post treatment scheme. The results shows with improvement in physical properties achieved through the post treatment, biodiesel exhibits significantly better atomization characteristics in terms of spray angle, spray length, sauter mean diameter and shorter evaporation time compared to the biodiesel before improvement and fossil diesel.

1. Introduction

Biodiesel is an alternative fuel produced from renewable resources have recently drawn significant consideration for power generation gas turbine application [1][2]. However, biodiesel or herein after referred as First Generation Biodiesel, (FGB) in this work, has inferior properties such as viscosity, surface tension and density of biodiesel impede the fuel for gas turbine application [3]. Allen [7] claims that, sauter mean diameter (SMD) of biodiesel vary from 5 to 40% higher than diesel which is due to higher viscosity, surface tension and density which results in poor atomization and eventually leads to incomplete combustion. The atomization and evaporation of the fuel spray plays an important role in enabling adequate the mixing of air fuel in combustion chamber which directly affect the engine performance and emissions characteristics [4] [5]. In previously reported work by Gopinathan [6], the improvised biodiesel or called "Second Generation Biodiesel", (SGB) produced by a microwave assisted post treatment scheme (MAPTRES) has significantly improved properties such as viscosity, density and surface tension, favorable to gas turbine application and meets ASTMD2880 fuel specification standard. Correspondingly in this work, the atomization characteristic of SGB has been evaluated in terms of spray angle, spray penetration length, SMD and evaporation time of fuel droplets based on microgasturbine Capstone C30 operation data and compared with FGB and distillate diesel(DD) used as gas turbine fuel.

2. Methodology

2.1 Numerical Evaluation of Atomisation Characteristics & Droplet Evaporation Time

An atomization characteristic is determined via sauter mean diameter (SMD) which is defined as diameter of sphere that has the same volume/surface area ratio as a particle of interest [7]. Numerical evaluation has been done using SMD formula generated by Lefebvre correlation as shown in equation 1. The correlation is adopted to evaluate the SMD because the similar correlation used in works reported by Bolszo [8] to evaluate SMD of diesel fuel in microgasturbine Capstone C30 with air blast atomizer which was experimentally validated using Phase Doppler Anemometer (PDA) [8]. The details of Lefebvre correlation described accordingly and in order to evaluate SMD of SGB, air to liquid ratio (ALR) and relative co-flowing velocity exiting (U_R) value adopted from Bolszo works [10].

$$\frac{D_{32}}{d_o} = 0.48\left(\frac{\sigma}{\rho_A U_R d_0}\right)^{0.4}\left(1+\frac{1}{ALR}\right)^{0.4} + 0.15\left(\frac{\mu^2 L}{\sigma \rho_L d_0}\right)^{0.5}\left(1+\frac{1}{ALR}\right) \quad (1)$$

D_{32} = Sauter Mean Diameter, d_o = Liquid Discharge Opening Diameter (m), ρ_A = Air Density (kg/m^3), U_R = Relative co-flowing velocity (m/s), ALR = Air to liquid ratio, σ = Liquid surface tension (N/m), ρ_L= Liquid density (kg/m^3), μ_L= Dynamic Liquid viscosity (m^2/s)

The fuel properties for the numerical evaluation have been adopted from previously reported work by Gopinathan [6] and the dimensions for the actual injector of microturbine Capstone C30 has been measured at SIRIM laboratory. In this reported work, SGB possess lowest surface tension than FGB and DD which is 15 mN/m, 31 mN/m and 28 mN/m respectively. Besides that, SGB also has kinematic viscosity of 4.5 mm^2/s lower than FGB which is 6.6 mm^2/s. Additionally, slight improvement was observed for density of SGB 0.8756 kg/m^3 compared to FGB is 0.8776 kg/m^3, even density has least affects than viscosity [6]. Overall, SGB has better physical properties than FGB and meets ASTMD2880 and ASTMD6751 standard Bolszo's method was adopted again to evaluate the evaporation time of SGB and compared with diesel fuel. The details of equation 2 given below and effective evaporation constant (λ_{eff}) calculated based on reported data by Bolszo [3],[8].

$$t_e = \frac{D_0^2}{\lambda_{eff}} \quad (2)$$

λ_{eff} = Effective evaporation constant, D_0^2 = Sauter Mean Diameter, t_e = Effective evaporation time

2.2. Experimental Investigation of Spray Angle & Penetration Length

The atomization test rig was designed and developed to investigate the spray angle and penetration length of SGB100, FGB100 and distillate diesel. The equipment comprises of air cylinder, air blast atomizer, observation chamber, fuel and liquid flow meter, pressure regulator and electronic weighing balance. The electronic weighing balance (*Brand : AND ; Model : GF-6000; with super hybrid sensor*) used for fuel mass flow rate measurement and volume measured by flow meter *(Brand : Mcmillan; Model: S-111-5 with accuracy ± 0.5% and flow range 5-500ml)*. Air blast atomizer which is the fuel injector from Capstone C30 has been used to experimentally evaluate the spray angle and spray penetration length of fuel inside observation chamber according to actual engine operation condition at selected load 10, 15, 20 and 25kW at inlet air temperature of 23.8°C to 24.2°C. During atomization test spray images captured using high speed camera to measure the spray length and spray angle.

3. Results and discussion

3.1. Sauter Mean Diameter (SMD) and Droplets Evaporation

The results were presented in Figure 1 and 2 respectively. The patterns of all fuels are identical where SMD decreased gradually as the ALR varies from 0.2 to 0.65. Expectedly, SGB has smaller droplets than FGB that has high viscosity and surface tension which causes the fluid resists agitation, tending to prevent its break up and leading to larger droplets [6]. Results shows SGB have better atomization because lower surface tension, viscosity and density will enhance spray break up and eventually would produce smaller Sauter mean diameter SMD [7]. Droplets evaporation time is another crucial element that that influences the combustion and it's dependent on droplet size of the fuel and SMD directly influence the droplets evaporation time.

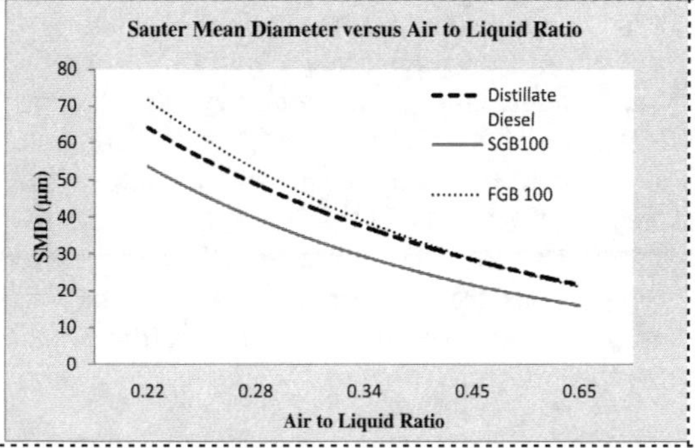

Figure 1. Sauter Mean Diameter versus Air to Liquid Ratio.

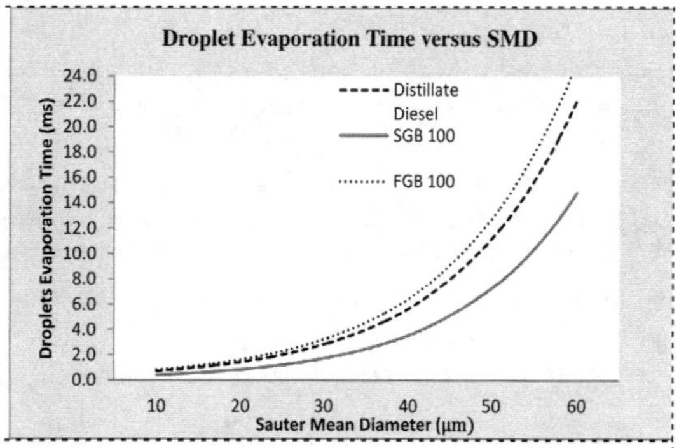

Figure 2. Droplets Evaporation Time versus SMD.

Fuel with higher SMD tends to have longer evaporation time and eventually leads to incomplete combustion which will results in high emission pollutants. According to Bolszo, the ideal residence time to the fuel droplets to be fully vaporize in microgasturbine with diesel fuel is 2ms (milliseconds) with SMD less than 25μm and the reported work indicates that ideal pre-vaporization cannot be achieve in microgasturbine because at full load Capstone C30 required 11ms for the droplets with SMD 50μm to be fully vaporize [8]. Interestingly, Bolszo's method adapted for this numerical analysis has shown that the SGB fuel with lower SMD the evaporation time is reduced significantly compared to diesel. Based on the results, at full load diesel fuel required 11ms for 50μm droplets to be full vaporized whereas SGB only required 7ms. Therefore, SGB has better atomization characteristics than diesel and FGB although the result doesn't meet the ideal evaporation time of 2ms.

3:2. Spray Cone Angle and Spray Length

Spray angle defined as angle formed by the cone of liquid leaving a nozzle orifice where two straight lines wrapped with the maximum outer side of the spray while spray length defined as the greatest distance of the fuel propagates once injected into stagnant air. Nowak [9] reported the key factors determining the efficiency of atomization are droplets size (SMD), spray length, and spray cone angle and evaporation time. The results of spray angle and spray length are presented in Figure 3 and 4.

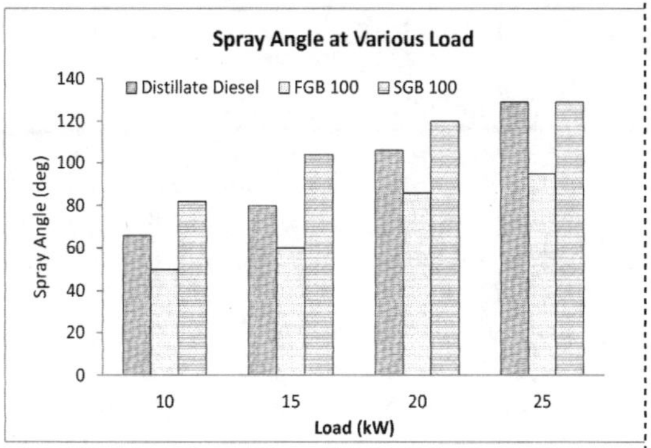

Figure 3. Spray Angle at Various Loads.

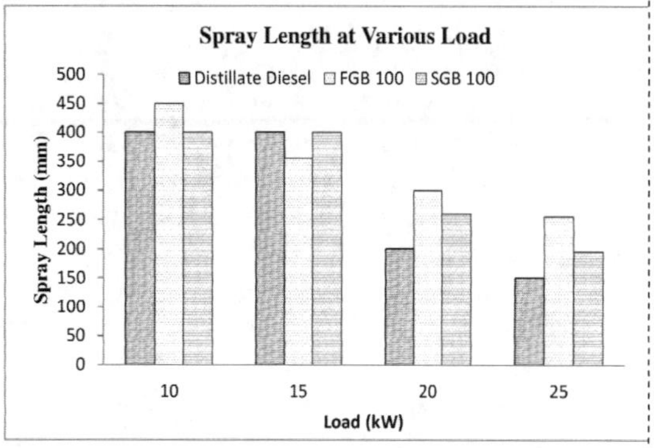

Figure 4. Spray Length at Various Loads.

Interestingly, SGB and distillate diesel has similar spray angle at 25 kW which is apparently due to improvement in physical properties of SGB. Correspondingly, the low surface tension is prone to quicker break up into droplets and wider of dispersion which will cause a relatively larger spray angle, at nozzle exit. FGB has lowest spray angle compared to SGB and distillate. Meanwhile, spray angle for all fuels increase when the load increased because during actual operation in microgasturbine the atomizing air pressure increases when the load is increased. The air pressure increase which will enhance the fuel break up frequency at atomizer tip and produce smaller droplet size (SMD) of the fuels and facilitate evaporation and result in larger spray angle and significant increase of spray coverage. On the other hand, SGB with larger spray angle has shorter spray length while FGB with smaller spray angle has longer spray length since it is inversely proportional to spray angle where larger area of spray dispersion will result in shorter spray length[10]. Based on the results, FGB has longest spray length compared to SGB and distillate diesel. This phenomenon also affirms that FGB with high surface tension and viscosity tend to have lower break frequency than SGB and distillate diesel. Consequently, this will lead to inferior atomization because fuel and also cause longer evaporation for completely vaporize the fuel droplet prior to combustion.

4. Conclusion

Atomization characteristics of improved biodiesel SGB have been evaluated and established that improvement in physical properties such as viscosity, surface tension and density has consequently improvised the atomization characteristics of SGB compared to diesel and FGB. Numerical analysis to determine SMD using Lefebvre equation prove and pre-vaporization of droplets resulting in shorter evaporation time compared to diesel that SGB has smaller SMD than diesel Experimental analysis on spray angle and spray length has shown SGB has wider dispersion angle and shorter spray length due to lower surface tension and viscosity. Therefore, SGB has superior atomization characteristics than diesel and FGB and it has potential to replace distillate diesel for gas turbine application for power generation. Further work recommended to measure SMD with Laser Doppler Anemometer (LDA) for comparison with numerical analysis is to verify the validity of Lebfebrev correlation in SMD analysis.

Acknowledgement

Authors would like to thank TNBR and Tenaga Nasional Berhad for funding this research through its dedicated renewable energy fund and also rendering their lab facilities to perform physical properties test and support of biodiesel research team members for this study.

References

[1] Rahm S, Goldmeer J, Moliere M and Eranki A 2009 Addressing Gas Turbine Fuel Flexibility *POWER-GEN Middle East Conference in Manama, Bahrain*

[2] Hashimoto N, Yasushi Ozawa, Noriyoki Mori, Isao Yuri and Tohru Hisamatsu 2008 Fundamental combustion characteristics of palm methyl ester (PME) as an alternative fuel for gas tubines *Elsevier - Fuel* **87** 3373-78

[3] Bolszo C. D, McDonell V G 2009 Emissions optimization of a biodiesel fired gas turbine *Elsevier- Proc. of Comb. Inst.* **32** 2949-56

[4] Ejim C E, Fleck B A, Amirfazli A 2006 Analytical study for atomization of biodiesel and their blends in a typical injector Surface Tension and viscosity effects *J.Fuel* **86** 1534-44

[5] Yung S L, Hai P L 2010 Study on the spray characteristics of methyl esters from waste cooking oil at elevated temperature *J. Renewable Energy* **35** 1900-07

[6] Gopinathan M, Kumaran P, Ibrahim H, Hamdan 2010 Preliminary Property Evaluation Second Generation Biodiesel for Gas Turbine *SCOReD UNITEN October 2010, Malaysia.*

[7] Allen C A W and Watts K C 2000 Comparative analysis of the atomization characteristics of fifteen biodiesel fuel types *ASAE 0001-2351* **43(2)** 207-11

[8] Bolszo C D 2009 Investigation of atomization mixing and pollutant emissions for a microturbine engine *Research UCI undergraduate research journal*

[9] Nowak D, Dobski T, Slefarski R, Magni F., 2013, Spray Generator for the liquid fuel combustion *http://ufdc.ufl.edu/UF00102023/00466*

[10] Krishna C R 2007 Report on atomization tests for project titled Biodiesel blends in microturbine *Brookhaven National Laboratory*

4th International Conference on Energy and Environment 2013 (ICEE 2013) IOP Publishing
IOP Conf. Series: Earth and Environmental Science **16** (2013) 012015 doi:10.1088/1755-1315/16/1/012015

Exploring EKC, trends of growth patterns and air pollutants concentration level in Malaysia: A Nemerow Index Approach

Hussain A Bekhet[1] and Tahira Yasmin[2]

[1, 2]Graduate Business School, College of Graduate Studies
Universiti Tenaga Nasional (UNITEN), Malaysia

E-mail: profhussain@uniten.edu.my

Abstract. The present study examines an Environmental Kuznets Curve (EKC) hypothesis by analyzing annual data of air pollutants concentartion and per capita GDP as economic indicator over the (1996-2010) period in Malaysia. Nemerow Index Approach (*I*) used to generate a measures of air pollution.The results show that ambient air quality indicators supports the EKC hypothesis which stated that pollution levels increase as a country develops, but begin to decrease as rising incomes pass beyond a turning poin. Also, the *I* result is justifying that most pollutants are showing value less than 1.

1. Introduction

Now-a-days world is facing the challenge of global warming and climatic change issues at large extent. Most economies are looking towards efficient methods of production in order to cut down emissions. The concept of sustainable development arose in the 1970s. Since then there have been discussions in different forums on how to formulate a sustainable development policy which curtails emissions while ensuring continuous progress at the same time.

Many developed countries are adopting new ways and enviornment-friendly technologies to conserve the deteriorating environment. Rapid population growth and consequent urbanization is adversely effecting the quality of the environment as well as the biodiversity of species[1]. The Malaysian economy is characterized by a trend towards increased liberalization, greater openness to world trade, a higher degree of financial integration and greater financial development [2]. Following the conventional trend, changes in economic structure is triggering rapid urbanisation resulting in the advent of a set of new environmental challenges [3]. These emerging environmental challenges are characterised by rise in air pollution caused by land transportation, industrial emissions and open burning [4]. Malaysian government's efforts at protecting the environment are appreciated not only at the local level, but also at regional and international levels.

In Malaysia, the real GDP has grown by an average of 4.5% per annum during (1996-2010) period. This growth rate has helped improve the quality of life for Malaysians and supported widespread advances in education, health, infrastructure, housing and public amenities.Furthermore, there is overall economic improvement and changes in consumption patterns and living standards.

In the early days of abundant resources and minimal development pressures, Malaysia imparted little attention to growing environmental concern but at present, the aim of attaining the status of an industrial country by the year 2020 and the associated rapid economic growth has made it conscious of sustainable development because of rising industrial pollution and the degradation of environment. Depletion of fisheries, air and water pollution, and contamination by industrial wastes have become more serious in

Malaysia in recent years. Out of all the enviornmental problems Malaysia is facing today, air pollution is the most serious issue that has been affecting human health, agricultural crops, forest species and ecosystems.

The aim of this study is to investigate the relatioship between GDP growth rate and ambient air quality, using the concept of Environmental Kuznets Curve (EKC). The relationship yields an inverted U-shaped curve signifying the process of increase inpollution until economic growth reaches a certain point, and then decreases while the growth continues. Many researchers are interested in the relationship between economic growth and environmental quality [5], [6]. Several studies have employed the Environmental Kuznets Curve (EKC) hypothesis which justifies the belief that economic development and environmental degradation can be expressed as inverted U-curve [7], [8], [9]. EKC assumes that environmental pollution increases when per capita income is at a relatively low level, but it will start decline when per capita income reaches a certain point (turning point). Besides, the application of EKC has some implications in choice of variables. [10], [11] discussed the sensitivity of the results to the nations and states, the emission vs ambient concentrations and policy implications. While [12], argued that the choice of methodology and functional form can affect the right application of EKC.

Besides, [13], [14] have analysed the relationship between economic growth and water pollution. Their findings confirmed that there is a simultaneous relationship between pollutants and economic growth in Malaysia. They have explained that it follows the concept of the EKC. Also,the SO_2 and PM_{10} supported EKC in Malaysia. In addition, they provide justification that pollution is one of the determinant of income in Malaysia [15].

The structure of this paper is as follows: section 2 explores the methodology and data sources while result findings are presented in section 3. The paper draws some concluding remarks in Section 4.

2. Data Sources and Methodology

In the paper current available annual data for the economic and environment activities is used. The data for real GDP was collected from DOSM while annual average concentartion of five air pollutants: carbon monoxide (CO), nitrogen dioxide (NO_2), sulfur dioxide (SO_2), particular matter (PM_{10}) and ozone (O_3) was retrieved from annual report of Ministry of Natural Resources and Environment Malaysia. These pollutants indicate the quality of ambient air as concentartion level from 1996-2010.

In order to examine the development-environment relationship in Malaysia, we have considered the standard EKC model [16]. The Nemerow index (I) is introduced as an integrated indicator for overall environmental quality[17], [18]. Specifically, this index reflects both maximum and average values of pollutant concentrations relative to some objective levels (such as the Malaysian Air Quality standards). It is calculated in Equation 1.

$$I = \sqrt{\frac{\text{Max}I_i^2 - \text{Avg}I_i^2}{2}} \quad (1)$$

where; I_i is the pollution index relating the observed concentrationC_i of pollutant i with its objective concentration S_i , $I_i = C_i/S_i$; MaxI_i is the maximum value of all the indices I_i ; AvgI_i is the average of all indices.If $I > 1$then overall environmental quality does not comply with objectives; whereas if $I < 1$, then quality is better than the set objective.In order to indicate the major pollutants, the parameter of K is further introduced in this study. K represents the percentage shares of the polluting substance (See Equation 2).

$$K_i = (I_i/I) * 100 \quad (2)$$

$$I = \sum_{i=1}^{n} I_i \ (i = 1,2,3, \ldots\ldots, n)$$

Certainly, both I values (Equation 1) and K values (Equation 2), are very much determined by the denominator, the set objectives. It is argued that Malaysia and many other developing countries have chosen less stricter environmental objectives than the EU and USA. Therefore, the rationality of the I index and K parameter is arguable. However, Malaysain Ministry of Environment has published the standard values for concentration level.

3. Findings Analysis

The overall results show that only the Nemerow index PM_{10} maintains a level higher than the critical value $I=1$. However the air Nemerow index has been declining in the past 15 years, indicating the air quality improvement in Malaysia. Among the five pollutants, the quality of NO_2, SO_2 and CO is better than the objective level without showing any significant change. Furthmore, the results of the contributory percentage of all five pollutants concentration show that SO_2 has a significant decrease in its contributory percentage from 15% in 1996 to 3% in 2010. This could be attributed to the use of better fuel quality EURO-2M, stricker enforcemnet by the DOE as well as wide usage of natural gas for industrial combustion process and vehicles.

Besides, the result shows the shape of the relationship between per capita GDP and air quality in Malaysia (Appendix A1). The O_3 curve shows an upward trend but after reaching at 19000 RGDP level, it shows declining trend. While the plot of the SO_2 index has a significant downward U-shaped curve trend. Moreover, CO, PM_{10} and Nemerow indices show an inverted U-shaped relationship clearly in terms of turning or critical point. The critical or turning point for Malaysian economy is 17,000 RM, which is quite near to the world mean value (5,000-8,000 USD) [19].

4.Conclusion

The results show that there is consistency between air pollutants and EKC hypothesis. However, SO_2, CO and PM_{10} indices have declined as GDP growth rate is increasing. In fact, the downward and EKC patterns of the ambient air can also be explained by the government's environmental regulations and the citizens' environmental consciousness. The air pollutants support the EKC, as most of them have gone through the turning point, and their environmental qualities are being improved with the economic development. Besides, we have found that policies should be formulated according to the characteristics of each individual pollutant. Also, the findings show that EKC takes various shapes depending on the type of pollutant. Thus, environmental and energy policies need to be 'customized' for each substance, rather than standardization. Figure1 shows the five stages of relationship among GDP and air pollution.We can relate that Malaysain air quality has trend towards improvement to show better policies in order to cut down pollution.

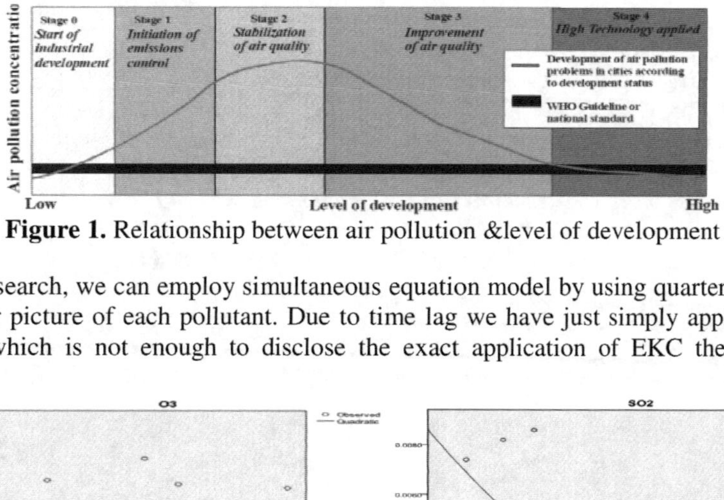

Figure 1. Relationship between air pollution &level of development

For further research, we can employ simultaneous equation model by using quarterly data in order to have more clearer picture of each pollutant. Due to time lag we have just simply applied the Nemerow index approach which is not enough to disclose the exact application of EKC theory in Malaysian economy.

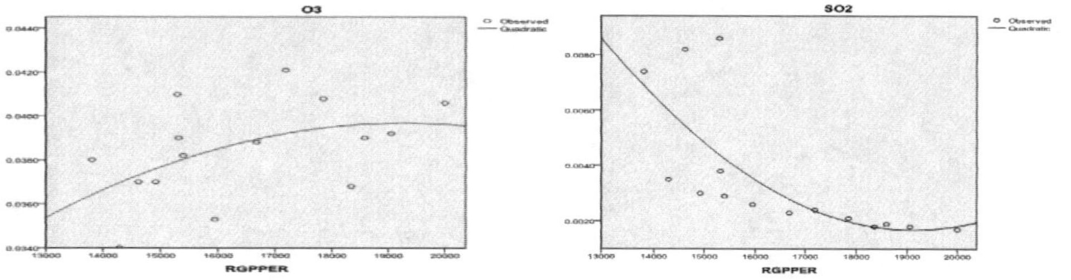

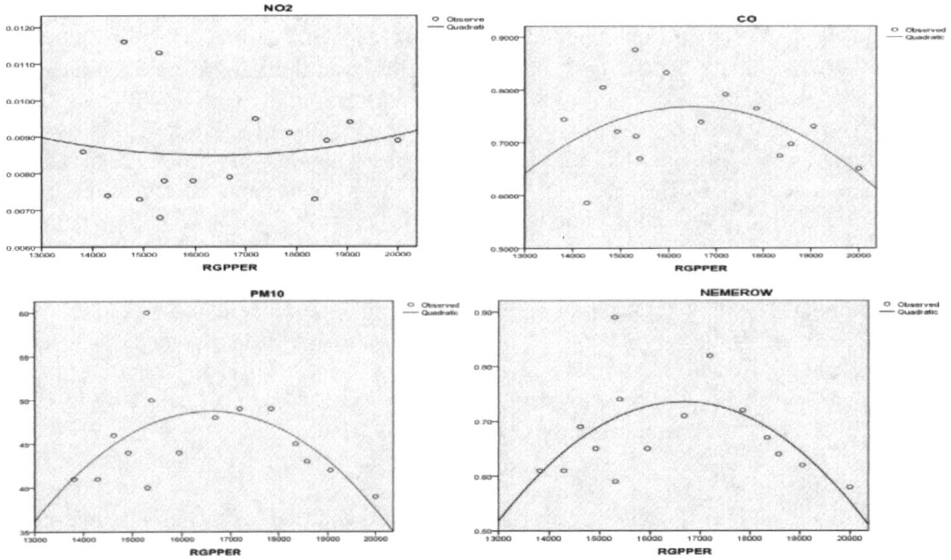

Appendix A1: The relationship between per capita GDP and air quality in Malaysia.

References

[1] Qiao M ,Wang C X, Wang D H and Wang Z J 2006 *J. International Env.* **32** 28–33
[2] Law S H 2008 *J. Asian Economics* **22**161-77
[3] Hezri A A 2006 *J. Natural Res. Forum* **30** 37-50
[4] Afroz R, Hassan M N and Ibrahim N A 2003 *J. Env. Research* **92** 71-77
[5] Lei K P, Chen F P and Wang Z S 2006 *J. Act. Ecol. Sinica* **26** 439-48
[6] Galeotti M 2007 *J. Environment, Development and Sustainability* **9** 427-54
[7] Dinda S 2005 *J. Ecological Economics* **53** 403-13
[8] Orubu C O and Omotor D G 2011 *J. Energy Policy* **39** 4178–88
[9] Zanin L and Marra G 2012 *J. Economic Modelling* **29** 1328–37
[10] Plassmann F and Khanna N 2006 J. *Environment and Development* **15** 22–41
[11] Dasgupta S Laplante B Wang H and Wheeler D 2002 *J. Economic Perspectives***16** 147–168
[12] Harbaugh W T Levinson A Wilson and David M 2002 *J. Eco. And Statistics* **84** 541–51
[13] Borhan H and Ahmad E M 2012 *J. Malaysian Economic Studies* **49** 71-86
[14] Borhan H and Ahmad E M 2012 *J. Social Behavioral Sciences* **42**166 – 174
[15] Borhan H and Ahmad E M 2010 *J.Science, Tech.& Sustainable Dev.* **7**291-308
[16] Stern I D 2004 *J. World Development* **32** 1419–39
[17] Cheng J Shi Z and Zhu Y 2007 *J. Environmental Sciences* **19** 50–54
[18] Dai J Li S Zhang Y Wang R and Yu Y 2008 *J. Env. Monitoring and Assessment* **147** 317–26
[19] Pfaff A Barelli P and Chaudhuri S 2004 *J. Global Environmental Issues* **4** 139-59

Study on Atomization Characteristics for Power Generation Application

Z M Anwar[1], E S Tan[1], R Adnan[3] and Mohd Azree Idris[4]
[1] UNITEN Putrajaya Campus, Jalan IKRAM-UNITEN, 43000 Kajang, Selangor, Malaysia

Email: anwarz171@gmail.com

Abstract: The world is dependent on two common basic needs which is electricity and water supply. Thus, power generation is the pulse of a country's econmony. However, most of today's power is generated from fossil fuel which is non sustainable. This research aims to study the feasibility of biodiesel to substitute diesel fuel for gas turbine application. The objective is to investigate the atomization characteristics using various blend of biodiesel derived from transesterification of palm oil. There are five types of fuels tested which are pure Diesel, B20, B50, B80 and B100 which were experimentally tested in a spray atomizer to evaluate the relationship of the of the fuel blend ratio with atomization characteristics such as spray cone angle, spray tip penetration and Sauter Mean Diameter. Besides that, fuel chemical properties testing were conducted to ensure the fuel chemical properties for tested fuel meet the standard requirements for gas turbine fuel oil. Result shows that the higher blend of biodiesel will give larger SMD, longer spray tip penetration, and a smaller spray cone angle. The SMD was calculated based on a general equation. Meanwhile, spray angle and spray tip was obtained from photos captured and anlayzed using software.

1. Introduction

Power generation depend on fossil fuel to generate electricity from coal, natural gas, diesel and other sources. There are various types of turbine technology used in power generation. Diesel is typically used as a secondary backup fuel for gas turbine as its harmful emission can affect the environment. Thus this study proposes to solve this problem by substituting diesel with biodiesel in a gas turbine. Moreover, biodiesel was successfully used in diesel engine and proved that it can reduce the environmental pollution such as unburnt hydrocarbon (UHC) and particulate matter (PM) emissions [1,3]. Biodiesel influences the performance and emission of the diesel engine. Hence the various blends of biodiesel and diesel were tested in experimental test rig. Biodiesel gives a good impact in environment and performance of the diesel engine [4,5]. In addition, biodiesel is a renewable and sustainbale fuel that is able to handle the depleting of fossil fuel because biodiesel can be derived from various types of sources for the production of biodiesel such as waste cooking oil, vegetable oil, sunflower seed oil and animal fats. The production of biodiesel involves the transesterification process [2]. Palm oil is the most suitable source to generate biodiesel because it has a higher yield and it is one of the most economical oil compared to other vegetable oils. Moreover, palm oil is local resource in Malaysia, so the cost will be reduced and the use of palm oil can boost the economic trasformations of the country. Besides that, palm oil can help to absorb carbon dioxide to reduce greenhouse gas. Therefore, the investigation of atomization characteristic is the main objective of this research because it will affect the performance of gas turbine. The feasibility of using biodiesel in power generation will be evaluated and discussed through experimental investigations in a fuel spray test rig. Spray angle, spray pattern and Sauter Mean Diameter (SMD) are some of the atomization characteristics that will be studies for various types of fuel blends. Five types of fuel will be tested which are diesel, B20, B50, B80 and B100.

2. Research Methodology

Palm oil was used as the raw material to process the biodiesel. Biodiesel was generated from transesterification processes to produce B100 as a pure biodiesel. The subsequent fuel blends were mixed with diesel to generate B20, B50 and B80. Besides that, quality of the fuel will be tested to ensure the fuel meet the minimum requirement of the ASTM standards. The chemical properties of the fuel blends were tested as shown on Table 1. Besides that, the fuel chemical properties will affect the atomization spray because different blends of biodiesel and diesel fuel will contain different amount of fuel chemical properties. Meanwhile, the system of the experimental test rig consists of the control panel, experimental test rig, digital camera, air compressor, pressure tank, solenoid valve and spray gun atomizer. Figure 1 shows the schematic diagram of the atomization test rig. The atomizer will be placed on the top of the experimental test rig to spray thefuel droplets into the rig. Biodiesel fuel is filled inside the pressure tank and the air compressor compresses the air to inject the fuel through the spray gun atomizer. A digital camera is placed in front of the test rig to obtain photos of the fuel droplets and spray patterns. The injector pressure can be adjusted from control panel within the range of 0-5 bars. The experimental test rig is operated by pushing the switch button and then capturing the atomizer spray from the camera. Photos of the spray will be analysed using software to determine the atomization characteristic such as spray angle, spray width and spray length.

Table 1: Important fuel characteristics of Biodiesel and its blend with Diesel.

Fuel Blend	Mixture Viscosity @ 40Celcius (m^2/s)	Density (kg/m^3)
Diesel	3.88×10^{-6}	842
B20	3.71×10^{-6}	852.9
B50	3.95×10^{-6}	860.6
B80	4.27×10^{-6}	869.2
B100	4.55×10^{-6}	874.9

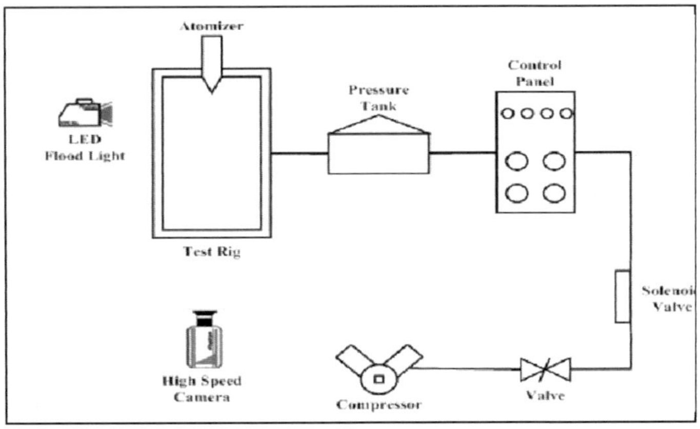

Figure 1: Schematic Diagram of Atomization Test Rig.

3. Results and Discussion

3.1 Sauter Mean Diameter(SMD)

Atomization characteristic consist of spray angle, spray length and Sauter Mean Diameter in which these are the parameters that will affect the performance and emission of the system. Besides that, to ensure the quality of the spray, the variable that has to be taken into consideration is the nozzle type, pressure and fuel properties. Atomization is the process of generating drop or also known as SMD. It begins by forcing liquid fluid through a nozzle or can be explained by aerodynamic force waves on the liquid surface and consequently produces unstable ligaments that eventually disintegrate into droplets on any increase in the relative velocity [6]. Moreover, process of atomization begins with the formation and growth of disturbance waves in the liquid followed by formation of liquid sheet into ligament and fragmentation. Final process is formation of droplets by the further breakup of fragments and it becomes droplets. Phase doppler particle analyser (PDPA) system [7] is one of the most accurate measuring device to measure the SMD. It operates using laser as a detector and measures the droplet size when it hits the laser beam. Due to the complexity of the research, other alternative option is taken to measure the SMD using equation that has to determine the fuel properties such as density, viscosity and surface tension "(1)"[4].

$$\text{SMD} = 2.25\, \sigma^{0.25}\, \mu_L^{0.25}\, m_L^{0.25}\, \Delta P_L^{-0.5}\, \rho_A^{-0.25} \quad (1)$$

where;

σ = surface tension

μ_L = mixture viscosity

m_L = fuel mass flow rate

ρ_A = air density

ΔP_L = liquid fuel injection pressure differential

Figure 2 shows SMD for all the five types of sample fuels. From that figure, it can be seen that B100 has the largest SMD, followed by B80, B50, B20 and D100. It can be seen the trend of the graph is increasing. In addition, this happens because of the fuel chemical properties also affect the atomization characteristics and the higher blend of biodiesel the higher content of density and viscosity. The reason is density can cause a fluid to resist acceleration and viscosity cause the fluids to resist agitation that tending to prevent breakup and leading to larger average droplet size. Thus, the higher content of biodiesel fuels the larger SMD. Another variable that can affect the atomization characteristics is injection pressure whereby higher injection pressure will lead to smaller SMD during the conduct of the experimental testing. On the other hand, injection pressure can be explained through the generation of actuation velocity of the fuel particles which produces smaller droplet size [8].

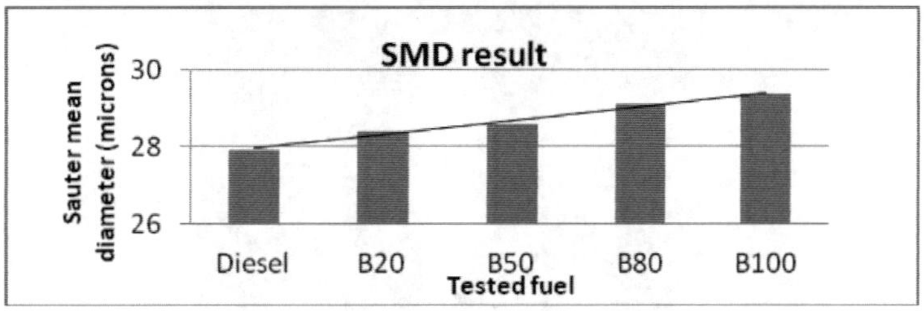

Figure 2: Chart of Sauter Mean Diamater (SMD) for various fuel blends.

3.2 Spray Cone Angle, Spray Tip Penetration and Spray Pattern

Basically, this research is to obtain the atomization characteristic that consists of spray cone angle, spray tip penetration and spray pattern. Previous studies proved that atomization characteristics has an impact on the performance and emission of the diesel engine. Thus the same concept is used for gas turbine and in the early stage of the research this atomization characteristic will be investigatedfor gas turbine. The atomizer spray is captured using DSLR camera for all five types of fuels. Furthermore, a photofit software is used to analyze the atomization characteristics such as spray cone angle and spray tip penetration. Spray cone angle is the angle formed by the cone of liquid leaving a nozzle orifice where two straight lines wrapped with the maximum outer side of the spray [10]. Results show that spray angle is also affected from the fuel chemical properties and injection pressure whereby spray angle decreased when biodiesel blend increase due to the fraction increase but inversely with surface tension because if the surface tension is low, spray droplet are prone to quicker break up and wider dispersion and cause a relatively larger spray droplet [10]. Besides that, higher injection pressure will increase spray angle because there are changes in the flow rate through the nozzle orifice. Hence, SMD will increase together with the spray angle and spray coverage. Spray tip penetration that consist of spray width and spray length were also analysed for all five types of fuels. Spray tip penetration is to determine the size or area of the atomization whereas the length of the spray will be analuzed using the software imposed on the photo. Spray length is defined as the measurement of travel distance of the liquid fuel when it initiate the first spray from the nozzle orifice and spray width is the atomization parameter investigate area of dispersion of the spray [9]. Spray length will become longer when content of the biodiesel blend increase.

4. Conclusion

The atomization characteristics parameters were successfully obtained from the experiment testing. Spray angle, spray tip penetration and spray pattern was determined using software that analyses directly through photographic views captured by DSLR camera. Spotlight is used to get clearer and better image quality of the spray. Meanwhile, sauter mean diameter (SMD) was obtained from equation 1. It consists of fuel chemical properties details such as density, viscosity and surface tension which are required in the equation to enhance the accuracy of the result. There are various factor that effect atomization characteristic such as biodiesel and diesel blend ratio, injection pressure and environmental factor . The higher content of biodiesel blend resulted in larger SMD and longer spray length but spray angle shows a different result whereby it resulted in smaller spray angle. On the other hand, spray angle and spray width become larger when there is an increase in the injection pressure because it tends to break up fuels particle into smaller size. The suggested optimum biodiesel blend is B20 for atomization characteristics. Besides that this blend also have a good impact on the performance and emission. This research could be enhanced in the future by conducting the experiment during different ambient temperatures due to the rapid changes in local weather climate.

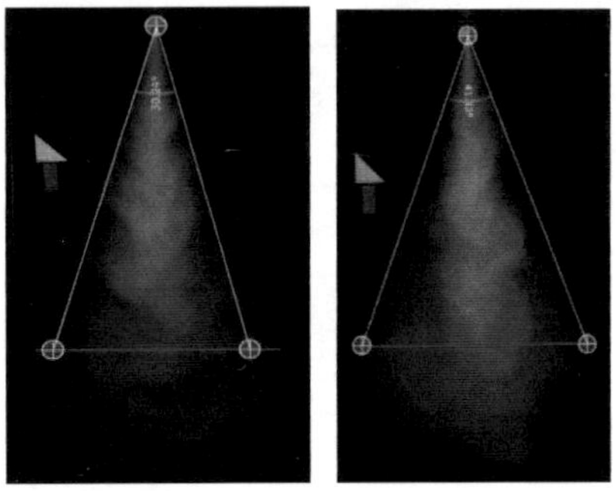

Figure 3: Example of atomization characteristic for B100 (left) and Diesel (right)

4. Acknowledgement

The authors would like to thank Malaysian Ministry of Higher Education (MOHE) for providing financial assistance through FRGS grant. (FRGS/1/2011/TK/UNITEN/03/01). Technical assistance in conducting the performance test provided by the project team members, especially Jan Siong Lim, Ang Jit Yong and Samsul are greatly appreciated.

5. References

[1] Gao Y, Deng J, Li C, Dang F, Liao Z, Wu Z and Li L 2009 *Biotechnology Advances* **27** 616-624
[2] Valentino G, Allocca L, Lannuzzi S and Montanaro A 2011 *Energy* **36** 3924-3932
[3] Som S, Longman D E, Ramirez A I and Aggarwal S K 2010 *Fuel* **89** 4014-24
[4] Park S H, Youn I M, Lee C S 2010 *Fuel Technology* **91** 1753-60
[5] Park S H, Cha J and Lee C S 2012 *Applied Energy* **99** 334-43
[6] Semiao V, Andrade P and Carvalho M G 1996 *Fuel* **75** 1707-14
[7] Chung I P, Strupp C and Karan J 2002 *AIAA journal of prolpusion and power*
[8] Kippax P, Suman J, Virden A and Williams G 2010 *ILASS-Europe 2010, 23rd Annual Conference on Liquid Atomization and Spray Systems , Brno, Czech Republic*
[9] Park S H, Kim H J, Suh H K, Lee C S 2009 *Heat and Fluid Flow* **30** 1183–92
[10] Tan E S, Zulhairi M A 2012 *Proceedings of ASME Turbo Expo 2012: Power for land, Sea and Air, GT 2012-68039*

IOP Conf. Series: Earth and Environmental Science **16** (2013) 012017 doi:10.1088/1755-1315/16/1/012017

Maximum Power Point Tracking With Improved Incremental Conductance Method for Fast Changing Solar Irradiation Level

Tey Kok Soon[1] and Saad Mekhilef[1]

[1]Department of Electrical Engineering, University of Malaya

E-mail: koksoon88@hotmail.com

Abstract. This paper proposed an improved incremental conductance method to track the Maximum Power Point (MPP) for PV Panel under fast changing solar irradiation. When there is increment in solar irradiation level, the conventional incremental conductance method is confused and responses incorrectly. The proposed method response correctly and there is no steady state oscillation compared to the conventional method. Matlab simulation is carried out for both the improved and conventional incremental conductance method under fast changing solar irradiation level. The simulation results showed the system able to track the MPP faster than the conventional method.

1. Introduction

Solar energy has been highly used for electricity generation in the recent past. However, solar energy is an unstable energy source due to the dependence on the surrounding conditions such as the solar irradiation level. Hence, a control system call Maximum Power Point Tracking (MPPT) system has been developed. Varies types of controlled algorithm implemented in the MPPT system to extract the maximum power available from the Photovoltaic (PV) panel.

Some of the MPPT algorithms are Fractional Open-circuit Voltage (FOCV), Fractional Short-Circuit Current (FSCC), Fuzzy Logic, Neural Network, Hill Climbing (P&O), Incremental Conductance and many others [1, 2]. Among all the methods, Hill Climbing or P&O and incremental conductance methods are highly used. These two methods need a DC-DC converter in between the PV panel and load. Then, the MPP is tracked by tuning the duty cycle of the converter. P&O perturb the duty cycle of the converter to find the maximum power of the PV panel [3-5]. After the maximum power is found, the perturbation keeps on going and there is oscillation around the MPP. Meanwhile for Incremental Conductance method [6], the slope of the PV's power against voltage (P-V) curve is used to track the MPP. The PV panel operates at the MPP when the slope of the P-V curve is equal to zero. Incremental conductance method varies the duty cycle of the converter to obtain the zero value in the slope of P-V curve.

In nature, it is rarely to obtain the zero value in the slope of P-V curve for incremental conductance as discussed in [7]. Hence, there is also steady state oscillation in incremental conductance method. Besides, this method is confused by the increment in solar irradiation level and responses incorrectly in the first response. In this paper, an improved incremental conductance algorithm is proposed to overcome the two weaknesses of conventional incremental conductance algorithm which are zero state oscillation and confusion during the increment of solar irradiation level.

2. PV Panel Characteristic

Single diode PV cell as shown in Figure 1 is used in this paper to model the PV panel [8].

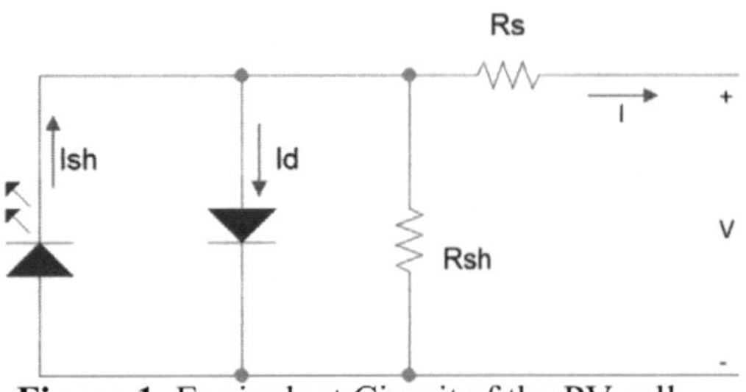

Figure 1. Equivalent Circuit of the PV cell

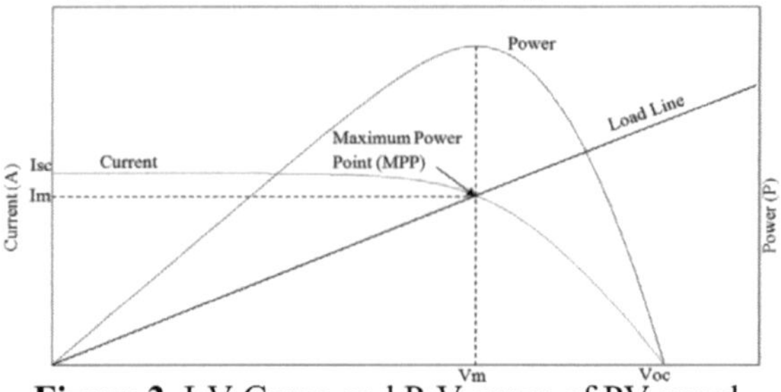

Figure 2. I-V Curve and P-V curve of PV panel

PV panel consists of series or parallel connection of PV cell and the power generated depends on the solar irradiation level. Two important characteristic curves needed in describing the PV characteristics which are current against voltage (I-V) curve and power against voltage (P-V) curve as shown in Figure 2.

Under different level of solar irradiation level, the PV panel produces different level of power. When the PV panel is connected to the load, power is consumed by the load and a load line is imposed onto the I-V curve. PV panel current and voltage is the intersection point of the load line and I-V curve. Then, the PV power is simply the multiplication of current and voltage of the PV panel. MPPT algorithm is used to ensure the PV panel always supplies the maximum power or the load line always crosses MPP in the IV curve.

3. Weakness of Conventional Incremental Conductance Algorithm

In theory, conventional Incremental Conductance method stops the tuning of duty cycle when the MPP is reached. However in real life, it is rarely achieved due to the truncation error of the numerical differentiation and thus there is steady state oscillation.

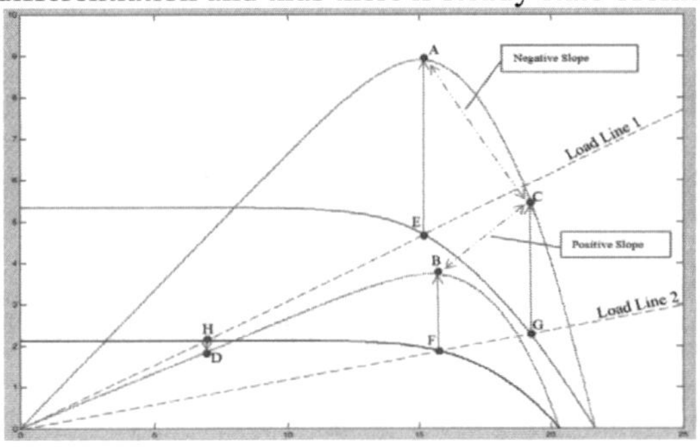

Figure 3. I-V and P-V curve of solar irradiation $1000W/m^2$ and $400W/m^2$

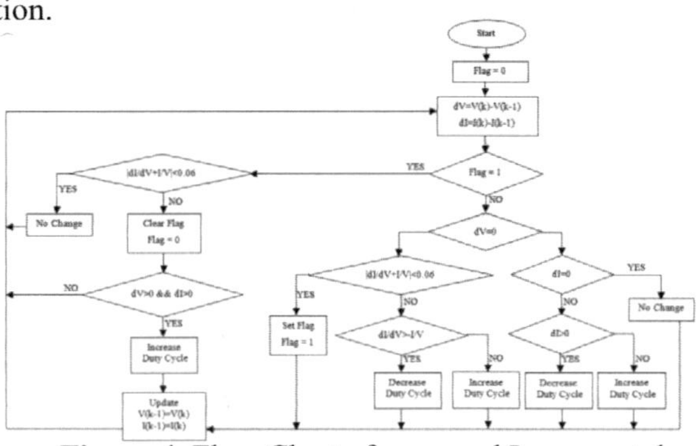

Figure 4. Flow Chart of proposed Incremental conductance Algorithm

Other than that, the conventional method responses incorrectly when the solar irradiation increases. At irradiation level of $400W/m^2$ as shown in Figure 3, the duty cycle of the converter is tuned to obtain Load Line 2 and tracks the MPP (point B). Then, the solar irradiation increases to $1000W/m^2$, but at this moment, the duty cycle of the converter still at Load Line 2, and point G is found by Load Line 2 in I-V curve which corresponds to the power at Point C in the P-V curve. The gradient between point B and C is calculated and it is a positive gradient. In fact the power found by the Load Line 2 is at Point C, the gradient between point C and the MPP (point A) of $1000W/m^2$ P-V curve is in negative value. Hence the gradient value calculated by conventional method just after the increment in solar irradiation is incorrect and thus responses incorrectly in the first step.

4. Proposed Incremental Conductance

There are two main disadvantages in the conventional method, which are the steady state oscillation and also the confusion when the solar irradiation level increases. The shape of the P-V curve is used by incremental conductance method and the shape is affected by the solar irradiation level and the load resistance. Then, the algorithm uses the current and voltage value of PV panel in the calculation. So a well consideration needed on the effect of the solar irradiation and load changes against the current and voltage value of PV panel.

Table 1. Changes in PV voltage and current during the change of solar irradiation and load resistance

		Voltage changes (dV)	Current changes (dI)
Solar Irradiation	**Increase**	Increase	
	Decrease	Decrease	
Load Resistance	**Increase**	Increase	Decrease
	Decrease	Decrease	Increase

Table 1 summarize the responses of PV panel voltage and current value against the changes on the solar irradiation level and load resistance. Refer to Figure 3, when the MPP is tracked at point F, then the solar irradiation suddenly increases, the system operates at point G. So, both of the voltage and current value increased. Conversely, when the MPP point E is tracked, and the solar irradiation suddenly decreased, then the system operates at point H, and both the voltage and current value decreased. These two types of changes are not well considered in the conventional incremental conductance algorithm. Meanwhile, when PV system operates at Load Line 1 and the load resistance increases causes the PV system to operate at Load Line 2. So, the PV voltage increases and current decreases. Vice versa, the PV voltage decreases and current increases when the load resistance decreases. Besides, a permitted error is applied to eliminate the steady state oscillation and equation (1) is used. Figure 4 shows the flow chart of the proposed method.

$$|I + V(dI/dV)| < 0.06 \quad (1)$$

5. Simulation Results

Matlab Simulink used to model the whole MPPT system consists of The PV panel model, SEPIC converter and MPPT controller. The PV panel model designed follows the Kyocera PV panel KC85T module.

The simulation is conducted for both algorithms to investigate the differences between the conventional and proposed incremental conductance algorithm. The duty cycle's step size of the converter is set to be 0.005 and the MPPT controller sampling time is 0.05s. Figure 5(a) shows the simulation results of the conventional incremental conductance method. At the beginning, the solar irradiation is set to be 0.8 kW/m2, then at t=0.2s, MPP is reached and the duty cycle fluctuates between 0.61 and 0.62. PV power also oscillates. At t=0.716s, the solar irradiation is increased to 1.0kW/m2, but the duty cycle still at 0.62, then the power of the PV panel increases. At t=0.75s, the MPPT controller is sampling and it is confused, hence the incorrect duty cycle value changes incorrectly and the power of the PV panel is decreased as shown in Figure 5(a) point A. After that, the algorithm revert back the direction and the PV panel power is increased back again. At t=1.0s, the MPP is reached for solar irradiation of 1.0kW/m2, but there is oscillation during the MPP state.

Figure 5(b) shows the simulation results of the proposed method. Figure 5(b) shows there is no oscillation after the MPP is reached at t=0.2s and t=0.9s. At t=0.716s, the solar irradiation is increased to 1.0kW/m2, and then at t=0.75s, the proposed algorithm detects the increased in solar irradiation and responses correctly as shown in Figure 5(b) (Point C). The proposed method needs only 4 steps to reach the MPP and the conventional method needs 6 steps. Hence, the proposed method is 0.1s faster than the conventional method during the increase in solar irradiation level.

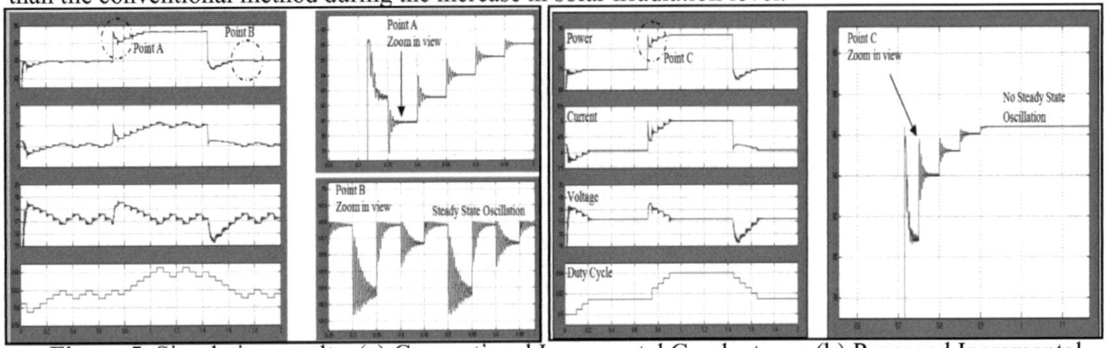

Figure 5. Simulation results: (a) Conventional Incremental Conductance (b) Proposed Incremental Conductance

6. Conclusion

The proposed Incremental Conductance algorithm is used to track the MPP for the PV panel. From the simulation results, the proposed algorithm is able to track the MPP correctly and it is 0.1s faster compared to the conventional algorithm. Then, the proposed method does not show the steady state oscillation and is able to reduce the power losses. As a conclusion, the proposed method reacts faster and correctly against the change in solar irradiation level and it is able to eliminate the steady state oscillation.

Acknowledgement

The financial support of this research by High Impact Research (HIR-MOHE), under the project code UM.C/HIR/MOHE/ENG/17 is greatly appreciated.

References

[1] Esram T and Chapman P L 2007 Comparison of Photovoltaic Array Maximum Power Point Tracking Techniques *IEEE Transactions on Energy Conversion* **22** 439-49

[2] Weidong X, Elnosh A, Khadkikar V and Zeineldin H 2011 Overview of maximum power point tracking technologies for photovoltaic power systems *IECON 2011 - 37th Annual Conference on IEEE Industrial Electronics Society* 3900-05

[3] Weidong X and Dunford W G 2001 A modified adaptive hill climbing MPPT method for photovoltaic power systems *Power Electronics Specialists Conference* **3** 1957-63

[4] Fermia N, Granozio D, Petrone G and Vitelli M 2007 Predictive & Adaptive MPPT Perturb and Observe Method *IEEE Transactions on Aerospace and Electronic Systems* **43** 934-50

[5] Femia N, Petrone G, Spagnuolo G and Vitelli M 2005 Optimization of perturb and observe maximum power point tracking method *IEEE Transactions on Power Electronics* **20** 963-73

[6] Safari A and Mekhilef S 2011 Simulation and Hardware Implementation of Incremental Conductance MPPT With Direct Control Method Using Cuk Converter *IEEE Transactions on Industrial Electronics* **58** 1154-61

[7] Weidong X, Dunford W G, Palmer P R and Capel A 2007 Application of Centered Differentiation and Steepest Descent to Maximum Power Point Tracking *IEEE Transactions on Industrial Electronics* **54** 2539-49

[8] Gonzalez-Longatt F M 2005 Model of Photovoltaic Module in Matlab™ *International Journal of Computer Technology and Electronics Engineering (IJCTEE)* **2**

4th International Conference on Energy and Environment 2013 (ICEE 2013) IOP Publishing
IOP Conf. Series: Earth and Environmental Science **16** (2013) 012018 doi:10.1088/1755-1315/16/1/012018

Impact of Advance Control on Microturbine Generation System Performance

Ahmad Kamil Mat Hussin[1] and Mohd Zamri Che Wanik[2]
Green Technology Section
TNB Research Sdn. Bhd.

E-mail: kamil@tnbr.com.my[1], mzamri@tnbr.com.my[2]

Abstract. Advance control employed in microturbine generation system (MTGS) is expected to improve its performance in responding to grid faults. This paper compares the effect of advance control of MTGS power conversion topology on the performance in riding through the grid faults. The analysis and investigation study through simulation shows there is no significant different on MTGS output performance even advance control is employed for its rectifier.

1. Introduction

Microturbine generation system (MTGS) is a distribution generation (DG) technology which is initially microturbine (MT) was designed for air transport vehicle such as aircraft and helicopter but now found itself as one of the most prospectus technology to be used as DG [1-2]. They can be used to supply a customer's base-load requirements or can be used for standby, peak shaving and cogeneration applications [3].

The typical output power of MTGS is usually between 30 kW and 400 kW for distributed DG application [4] with energy conversion efficiency between 20-30%, but can be increased up to the 60-85% if the waste heat is recovered for combine heat and power system [5]. It also has fuel flexibility where capable of using alternative/optional fuels including natural gas, diesel, ethanol, landfill gas and other bio-mass derived liquids and gases [3].

Power electronic converter is needed in order to connect a MT to the utility or to the grid [6] because the rotational speed of turbine is very high. The most common power converter topology used AC-DC-AC. The high-frequency AC power from MT generator is converted to DC first, then DC component is converted to 50/60 Hz components [2]. Different types of MTGS are available for example in [1] which operates with unidirectional power flow and in [6] which operates with bidirectional power flow.

Converting AC to DC is performed normally by passive rectifier but active rectifier is also utilized. DC to AC conversion is performed using voltage source converter. The selection of using passive or active rectifier depends on the MTGS task. For example, if MTGS is used in shaving peak and only online for certain period, so active rectifier is consider for this task. If MTGS is used for operating continuously and to supply part of based load, in many years major overhaul which is happen only once in a few years so passive diode rectifier is consider [3].

DG connected to electrical grid, in some region has to operate under stringent requirements. The requirement is to make sure DG will not degrade power quality, stability, security and reliability. The example of operating requirements are to lower emission of power quality characteristic parameters, ability to reduce active generation and lastly ability to perform voltage regulation by providing reactive power. During fault condition requirements are such as fault ride-through capability with limitation of short circuit current [1,3].

This paper compares the effect of advance control of MTGS power conversion topology on the performance to ride through the faults. MTGS technology is first reviewed, then the modelling method is described and finally the results of performance study are compared.

2. MTGS Technology and Modelling

There are two primary types of microturbine designs, single shaft and split shaft [4]. The single shaft operates at high speeds of 90,000 to 120,000 rpm where the design is mounted together with compressor and turbine [5]. This type of MTGS is interfaced with the grid through power electronic converter [1]. For the split shaft, the shaft rotates at 3600 rpm where generator and turbine are not mounted together with shaft [2]. The generator usually connected via gearbox, so power electronic converter is not needed for the split shaft design [6].

2.1. Microturbine

MT model is developed based on the gas turbines model that has been suggested in [7]. The model has been used in by several authors and researchers, for example in [1], [3], [5] and [6]. The equation for the system is given by:

$$t_m = f = 1.3\,(w_F - w_{MIN}) + 0.5(\omega_R - \omega_{R_{ref}}) \tag{1}$$

where t_m is mechanical torque, ω_R is the speed of PMSM, ω_R, $\omega_{R_{ref}}$ is speed reference, w_F is the fuel signal for turbine and w_{MIN} is fuel value at no load operation.

2.2. Rectifier

AC component from machine is converted to DC by rectifier. For the active rectifier control, Hall Effect method is used where it provides logic signal back electromotive force (e.m.f.) in controlling power switch [9]. The relationship between Hall Effect and back e.m.f can be represented in Table 1 which is known as "Block Commutation" or "Brushless DC Motor Timing Diagram". Table 1 shows high and low state of power switch or converter device. Usually, this information is provided by manufacturer in datasheet [9].

Table 1. Block Commutation

Hall			***Emf***			***Rectifier***					
a	**b**	**c**	**a**	**b**	**c**	**Q^1**	**Q^2**	**Q^3**	**Q^4**	**Q^5**	**Q^6**
0	0	0	0	0	0	0	0	0	0	0	0
0	0	1	0	-1	+1	0	0	0	1	1	0
0	1	0	-1	+1	0	0	1	1	0	0	0
0	1	1	-1	0	+1	0	1	0	0	1	0
1	0	0	+1	0	-1	1	0	0	0	0	1
1	0	1	+1	-1	0	1	0	0	1	0	0
1	1	0	0	+1	-1	0	0	1	0	0	1
1	1	1	0	-1	+1	0	0	0	0	0	0

2.3. Line Side Converter

In grid connected mode, MTGS operates as the current source [1]. The other objective of the controlling of inverter is to keep the DC-link voltage constant, regardless of the magnitude and direction of the rotor power [6]. Inverter is typically controlled using technique of field control or vector control originates from electrical machines theory. Voltage and current in a-b-c axis is converted to d-q-0 axis. This method reduces the complexity of coupled electrical quantities. Referring to the figure 1, the inverter control algorithm can be derived based on the relationship between converter voltage U_{CON} and grid voltage U_G behind its grid inductor [1].

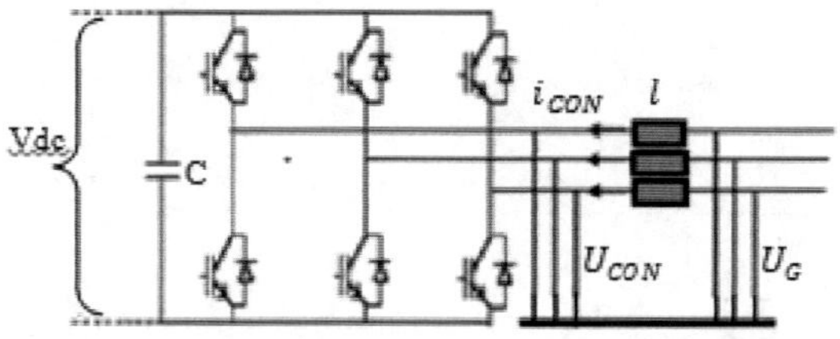

Figure 1. Converter connected to the grid.

From Figure 1 above, the voltage grid in d and q component after mathematical manipulation are [1,3]

$$U_{Gd} = l\frac{di_{cond}}{dt} - \omega l.i_{CONq} + U_{CONd} \quad (2)$$

$$U_{Gq} = l\frac{di_{conq}}{dt} - \omega l.i_{CONd} + U_{CONq} \quad (3)$$

Equations (2) and (3) are the basis of converter controller algorithm. With this controller, active power is controlled through d component and reactive power output is controlled through q component [1]. All these equation of the current controller is implemented in Matlab/Simulink software [8].

3. Simulation Result and Discussion

This section compares the results of the simulation. The comparison is made between the output of 60 kW-MTGS which used passive rectifier and the 60 kW-MTGS which used active rectifier. For both models, LSC converter controlled is set to follow grid code that requires the fault ride through and injection of reactive current for voltage support.

3.1. Under Fault Occurrence Operation

During fault operation, the MTGS must be able to operate without disconnecting the system and at the same time must able to inject the reactive current for supporting. The maximum current that can be injected by converter is 2.0 p.u [10], but in this study 1.5 p.u of rating current is allowed to be injected by LSC to the grid [1].

In Figure 2 and Figure 3, the output of the both MTGS responding to temporary fault at t = 0.5 s until t = 0.65 s are presented. The faults were single-phase to earth fault and three-phase to earth fault. During the fault, the voltage grid dipped about 50 %-60 % for single phase fault and 80 %-90 % for three phase fault. The total nominal current injected to grid during fault for both systems look the same at about 1.5 p.u. of the nominal current injected to the grid for both single phase to earth fault and three phase to earth fault.

DC voltage during the faults rises above nominal value until the fault is cleared. The rising of DC voltage is due to imbalance of power transfer between the rectifier and converter. During the fault the required power cannot be transfer to the grid due to low voltage condition. The power accumulates in DC circuit and causing voltage rise. After fault is cleared, the DC voltage returns to nominal value after power transfer is balanced. From the comparison made, it is seen that the performance of MTGS in riding through grid faults is not influence much on the power conversion topology on the machine side.

4. Conclusion

This paper compares the effect of advance control of MTGS power conversion topology on the performance to ride through the faults. The analysis and investigation study through simulation shows there is not significant different on MTGS output performance even advance control is employed for rectifier.

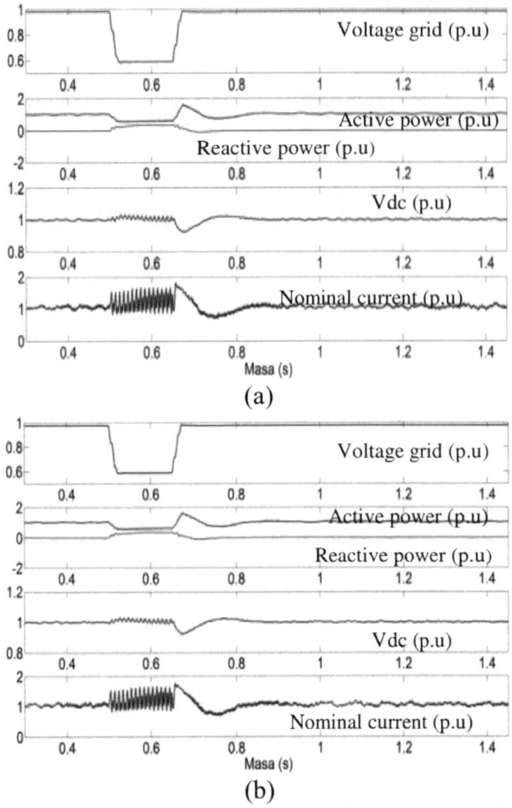

Figure 2. Single-phase to earth fault: (a) MTGS use passive rectifier (b) MTGS use active rectifier.

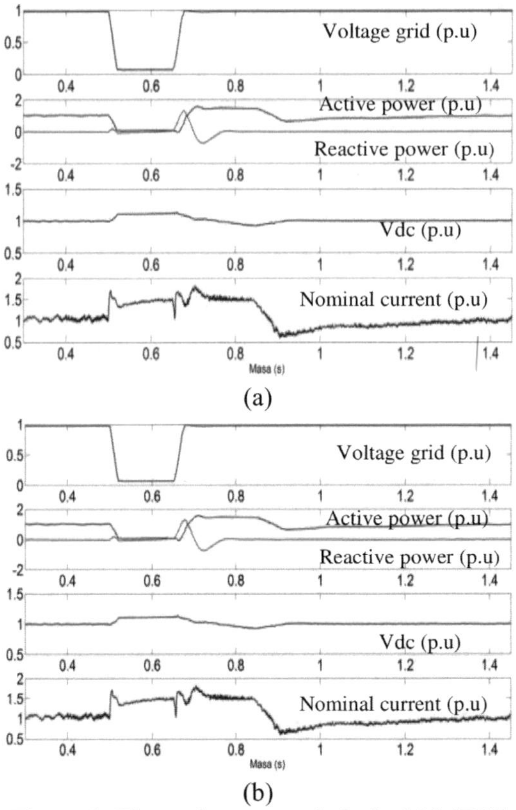

Figure 3. Three-phase to earth fault: (a) MTGS use passive rectifier (b) MTGS use active rectifier.

References

[1] Wanik M Z C 2011 *Dynamic Simulation and Intelligent Management of Distributed Generation* Thesis Doctor of Philosophy Electrical Engineering, University Duisburg-Essen.

[2] Kramer W, Chakraborty S, Kroposki B and Thomas H 2008 *Advanced Power Electronic Interfaces for Distributed Energy Systems Part 1: Systems and Topologies* Golden, Colorado National Renewable Energy Laboratory Tech. Rep. NREL/TP-581-42672 38-45

[3] Wanik M Z C and Erlich I 2009 Simulation of microturbine generation system performance during grid fault under new grid code requirements *IEEE Bucharest, Romania*

[4] Saha A K, Chowdry S, Chowdhury S P and Crossley P A 2008 Modelling and simulation of microturbine in islanded and grid-connected mode as distributed energy resource *General Meeting: IEEE*

[5] Nichols D K and Loving K P 2003 *General Meeting: IEEE* 2314-15

[6] Gaonkar D N, Pillai G N and Patel R N 2008 *Electrical Power Components and Systems* **36** 1031-47

[7] Rowen W I 1983 *ASME Trans. Journal Engineering for Power* **105** 865-69

[8] MATLAB/SIMULINK Sim Power System Documentation [online] *http://www.mathworks.com/help/toolbox/physmod/powersys/ref/brushlessdcmotordrive.html.*

[9] Pmdcorp, Brushless DC Hall Effect Sensor Configuration *http://www.pmdcorp.com/downloads/app_notes/BrushlessSensorConfig .pdf*

[10] Staunton R H and Ozpineci B 2003 *Microturbine power conversion technology review* Oak Ridge National Laboratory Tech. Rep. ORNL/TM-2003/74, Washington. DC.

IOP Conf. Series: Earth and Environmental Science **16** (2013) 012019 doi:10.1088/1755-1315/16/1/012019

Back Propagation Artificial Neural Network and Its Application in Fault Detection of Condenser Failure in Thermo Plant

Firas B. Ismail[1], **Vinesh Thiruchelvam**
School of Engineering, Asia Pacific University, Kuala Lumpur, Malaysia

E-mails: firas@apu.edu.my, Dr.vinesh@apu.edu.my

Abstract. Steam condenser is one of the most important equipment in steam power plants. If the steam condenser trips it may lead to whole unit shutdown, which is economically burdensome. Early condenser trips monitoring is crucial to maintain normal and safe operational conditions. In the present work, artificial intelligent monitoring systems specialized in condenser outages has been proposed and coded within the MATLAB environment. The training and validation of the system has been performed using real operational measurements captured from the control system of selected steam power plant. An integrated plant data preparation scheme for condenser outages with related operational variables has been proposed. Condenser outages under consideration have been detected by developed system before the plant control system"

1. Introduction

Power plant trips always cause a serious problem to the industry. It will bring a huge disaster if it is left unsolved. Usually trips may cause abnormal operation to the plant, such as air in-leakage, tube corrosion, cooling malfunction, inadequate air removal, condenser failures and so on. This will reduce the efficiency of the condenser in cooling exhaust steam [1]. Therefore if these faults could not be resolved in time it may cause the system or plant to shutdown, equipment damage and loss [2] and even disasters such as core meltdown, steam explosion and so on. Due to these trips or faults, a process of fault detection and diagnosis plays an important role in a power plant. The process of fault detection and diagnosis is to monitor the condition of machines operating and to ensure all are functioning at normal and safe levels in the power plant [3]. There are a lot of parameters and complex measuring variables that need to be monitored continuously [2]. Thus, an intelligent real time system for fault diagnosis is required. There are many methods used to process fault diagnosis and play an important role both in the industry and academic sector, and the most common method is Artificial Neural Network (ANN) [2], [4]. This method is still under development phase since there are several other techniques that can be applied in this technology [5]. The BP algorithm uses the gradient of the performance function to find out the weights that maximize performance. Back-propagation is the technique used to determine the gradient of the performance, and this technique performs computations backward through the network [5]. The architecture of BP model includes multiple layers of neurons. Each layer consists of weight matrix, bias vector, input vector and output vector. A three-layer BP model may involve different numbers of neurons in different layers. The network is able to extract higher-order statistics when there are more hidden layers.

[1] To whom any correspondence should be addressed.

2. Methodology

In this work, the feed-forward back-propagation network (BP) is implemented as supervised system, so the input and a target output are provided. Some training parameters need to be considered for implementing the BP network, such as epochs, learning rate (lr), show, goal and root means squared error. The training algorithm that is used in this BP training process is "traingd" The "traingd" (Batch Gradient Descent) is a batch mode and slow training process. The weights and biases are updated according to gradient descent [3-4].The transfer function with the weights and biases input are used to generate output. The most commonly used transfer functions of the BP network are *logsig* (range: 0 to 1), *tansig* (range +1 to -1) and *purelin*. The transfer function that is chosen to be applied in this BP network is *logsig*. It is the most commonly used for the multilayer networks to generate an output in the range 0 to 1. Even though other transfer functions can be used, but since the target output is set into the range from 0 to 1, the transfer function *logsig* is more suitable to be used.

2.1 Selecting of learning rate

During the configuration of the ANN training parameter, the learning rate η must be chosen carefully [4]. Figure 1 illustrates the graph of error against the weight for large and small learning rate. According to [4], the learning rate η should be chosen within $0 < \eta < \frac{2}{\lambda_{max}}$, where λ_{max} is the maximum eigenvalue of the autocorrelation matrix.

$$R_x = xx^T \qquad (1)$$

Where x is the input matrix of the ANN

$$\det(R_x - \lambda I) = 0 \qquad (2)$$

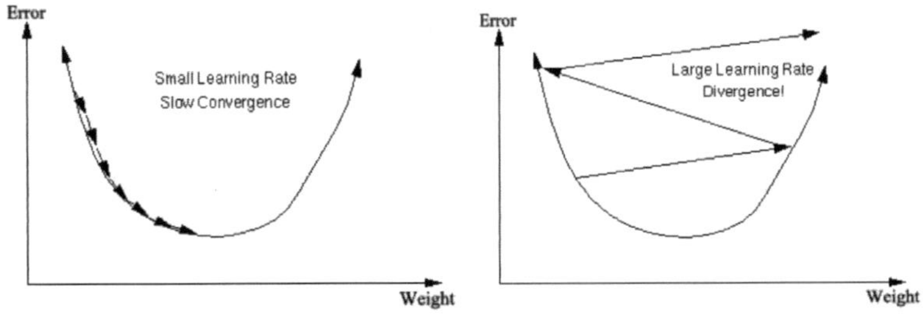

Figure 1. Changes of the Weight with Small and Large η [4]

2.2Selecting the number of hidden layers

The next step is to select the number of hidden layers. The maximum number of hidden layers should not exceed three [4]. The difference between the numbers of hidden layers can be summarized in Table1.

Table 1. Effect of Hidden Layers Number in Neural Networks

Number of Hidden Layer	Result
None	Only capable of representing linear separable functions.
1	Is able to approximate arbitrarily with any systems that contains continuous mapping from one finite space to another.
2	Is able to represent an arbitrary decision boundary to arbitrary accuracy with rational activation functions and can approximate any smooth mapping.

2.3 Selecting the number of hidden neurons

Once the number of hidden layer is selected, the next step is to select the number of neurons in the hidden layer. The hidden neurons in the hidden layer have large influence on the final output. If used too less, it will result in underfitting. Here, the neurons are too few to able to adequately detect the signals in a complicated system. If uses too many, the result would be overfitting. Here, the neural network has too much information processing capacity that the limited amount of information contained in the training set is not enough to train all the neurons [5].There is no proper method of selecting the correct number of neurons in hidden layer, but with the forward selection method, we are able to obtain suitable hidden neuron number systematically. The forward selection method is summarized in figure 2.

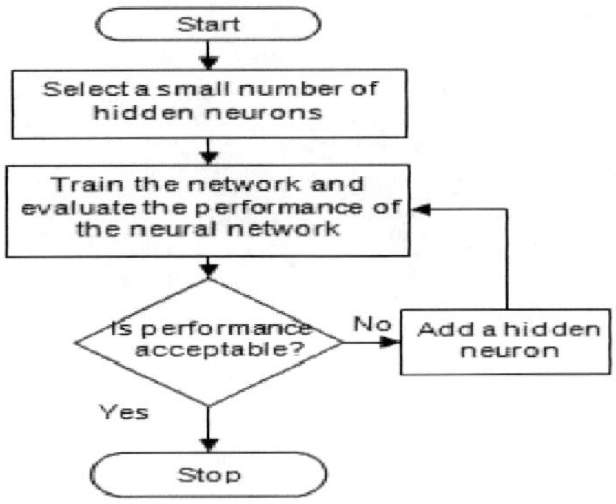

Figure 2. Selecting the Number of Hidden Neurons with Forward Selection Method [5]

2.4 Performance Function: Root Mean Square Error

The Root Mean Square Error (RMS) is the measurement difference between the predicted values and the actual values [5].

$$\text{RMSErrors} = \sqrt{\frac{\sum_{i=1}^{n}(\hat{y}_i - y_i)^2}{n}} \quad (3)$$

Where the $\hat{y}_i$ is the predicted value and y_i is the observed value.

3. Pre Process

Before all the reading can be feed as training data into the ANN system as training data and verifications, they are being pre-processed. The missing values are being treated using interpolation. The trend of parameters with missing values can be represented with a linear straight line. This will show the behaviour of each parameter from normal operation to failure.

After all missing values are being treated; they are being normalized between the ranges of 0 to 1 which is to meet the requirement length size for the ANN inputs. This is done with max-min normalization, where the value is mapped to a value between 0 and 1, where 0 represents the minimum value and 1 represents the maximum value in a range. Max-min normalization is an accurate representation of a value for use in the ANN and is thus used.

$$\text{value}_{\text{normalised}} = \frac{\text{value}-\text{min}}{\text{max}-\text{min}} \quad (4)$$

4. Implementation

There are 1817 input vector with 241 conditions. The ratio for data segmentation process is 0.7(70%):0.3(30%). That's means the training input data is divided into 70% for the training and 30% for validation testing based on 241 conditions. According to Shahin [6], the more data input was used, the smaller the root mean square error. Besides that, it is also used to ensure that the ANN can

IOP Conf. Series: Earth and Environmental Science **16** (2013) 012019 doi:10.1088/1755-1315/16/1/012019

be applicable to actual data other than the training set. At this step, we have to make sure that the verification data have a 100% success rate to tell that the system is able to handle with practical situation. Therefore, all the training input data are arranged into the dimensions of 169×1817. While for the validation testing data is 72×1817 dimensions.

5. Results and Observation

The result of using the Batch Gradient Descent is shown in figure 3. The network is able to train all 169 data successfully and the errors are reduced based on the figure shown. The errors performance decreased smoothly until it reaches the performance goal. In the beginning, the weights are in random condition and the network doesn't "know" which direction to search. Thus, the algorithm is trying to adjust the weight in order to minimize the error. Once the algorithm had found the suitable weights, the errors performance will decrease linearly as shown in figure 3.

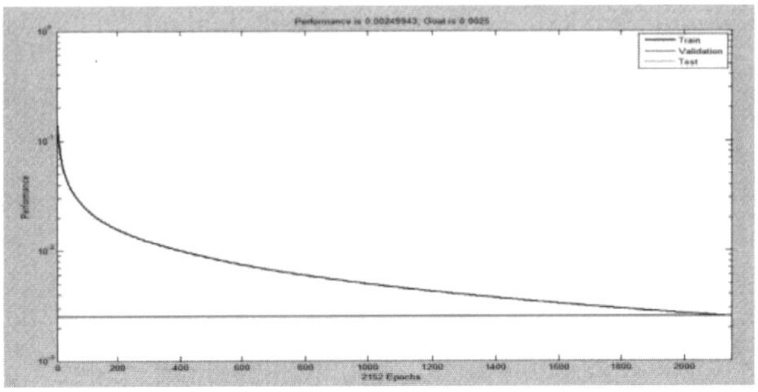

Figure 3. The plot performance of the training

6. Conclusion

BP neural network was introduced into the fault diagnosis of the condenser for coal power plant and a program was constructed. The preparation of the data was explained. "traingd" training algorithm was used in the program along with the data segmentation of 0.7(70%):0.3(30%) and total of 241 conditions. 169 from the total condition for training was successful as the RMS error was lower than 0.05. The result of the training performance had explained the smooth decrease of errors and reaching to the performance goal. After 2152 epochs, the program had reached the performance goal of 0.0025. This proposed fault diagnosis method can be used as a practical application in the power plant system.

References

[1] I. Homepage, 2011. [Online]. Available: http://www.intekflow.com/technical-services/power-plant-condenser-experts/default.html.

[2] Eslamloueyan R 2011 *Applied Soft Computing* **11(1)** 1407-1415

[3] Tan S C, Lim C P, Rao M V C 2007 *Engineering Applications of Artificial Intelligence* **20(2)** 203-213

[4] Nimmo I 1995 *Chemical Engineering Progress* **91(9)**

[5] Zhong C, Ding Y, Fu J 2010 *Intelligent Computation Technology and Automation (ICICTA), 2010 International Conference on IEEE* **2** 414-417

[6] Shahin M A, Maier H R, Jaksa M B 2000 *Evolutionary data division methods for developing artificial neural network models in geotechnical engineering* University of Adelaide, Department of Civil and Environmental Engineering

IOP Conf. Series: Earth and Environmental Science **16** (2013) 012020 doi:10.1088/1755-1315/16/1/012020

Constructed Rain Garden Systems for Stormwater Quality Control under Tropical Climates

Lariyah Mohd Sidek[1], Norshafa Elyza Muha.[2], Nur Asmaliza Md Noor[1] and Hidayah Basri [1]

[1]*Centre for Sustainable Technology and Environment, College of Engineering, Universiti Tenaga Nasional, Jalan IKRAM-UNITEN, 43000 Kajang, Selangor, Malaysia*

E-mail: lariyah@uniten.edu.my

E-mail: norshafaelyza@perlis.uitm.edu.my

Abstract. Malaysia has taken an integrated approach to manage storm water that is increasingly becoming a problem in big cities. Rain gardens are recommended as green technology for a new storm water management in Malaysia. The approach is applied in urban planning and design that integrates the total water cycle management into the development process areas. Rain gardens have been effective in reducing peak discharge and consistently reduce the number of storm water pollutants. This paper will examine some of guidelines, laboratory studies and field monitoring that shows great potential and benefit of rain garden. The preliminary results for rain garden performance were reported in this paper. The findings from this research will open avenues for researchers to advance the knowledge in rain garden systems to achieve the sustainable development in Malaysia.

1. Introduction

Rain garden system systems or rain garden is a best management practice (BMP) introduced by Prince George's County, Maryland in the year 1990s [4]. The initial rain garden system system design consisted of an excavated area backfilled with planting soil underlined by a thin layer of sand and planted with native grass, shrub and tree species. The basic rain garden system design has been modified in City of Alexandria, to include an underdrain within the sand bed to collect the infiltrated water and discharge it to a downstream sewer system [7]. In the landscape area, rain garden normally located at low points so runoff cans easily draining inside the systems. Rain gardens systems efficiently can remove pollutant from stormwater runoff by using the vegetation such as trees, shrubs and grasses or other native plants. Nowadays, rain gardens system has been implemented around the world as a popular technology for stormwater management [13]. Experts have confirmed that that rain gardens have a great potential in reducing peak flow, improve water quality in a natural process and give aesthetically pleasing manner [7].

Todays, Malaysia economy has rapidly changed throughout the year. Urban development is needed to cater the increasing of urban population. Due to the urbanization, the construction of paved area significantly changed the hydrologic and hydraulic characteristics of the catchments. For that

[1] To whom any correspondence should be addressed.

reason, occurrence of flash flood in urban area is becoming more cruel from year to year. This urban stormwater has negative impact to environment and ecological effects. So, a proper stormwater management is needed to overcome these problems. Department of Irrigation and Drainage (DID) Malaysia has introducing New Urban Drainage Manual known as Storm Water Management Manual for Malaysia (Manual Saliran Mesra Alam or MSMA) effective from January 2001. These manual introduced a set of guidelines for new development in Malaysia which adopting natural processes solutions for stormwater treatment. The concept of this guideline is 'source control techniques' in order to minimize the runoff flow rate from the contributing area [14]. The manual apply the concept of Best Management Practices (BMP's) to control stormwater at source with the target to achieve zero development impact contribution and improve the stormwater runoff quality. Rain garden systems are considered as a new concept in Malaysia compared with US where rain gardens was widely used [12].

2. Objective

Generally, this research seeks to have an understanding and knowledge of rain garden. The aim of the proposed study in UNITEN is to improve the quality design criteria of rain garden systems in Malaysia under tropical climate for supporting design guidelines in MSMA Manual. The outcome of the study hopefully can be used to highlight the potential of rain gardens in order to maintain the sustainable urban development in Malaysia. In order to achieve the aims above, three objectives are outlined as follows:

- To investigate the performance of rain garden system facilities using lab scale.
- To analyze the performance of rain garden system in removing heavy metal in different media depth.
- To produce optimum design requirements for rain garden system.

3. Research Methodology

The methodology will be carried out in four (4) stages as follows:

- Literature Review - Literature research will be conducted to gather the information research findings of rain garden system in order to identify the design modification, benefit and potential of rain garden system.
- Monitoring Program - Monitoring program will be carry out for assessing the water quality and also hydraulic performance test for rain garden system.
- Data Analysis and Interpretation - All the data from monitoring program will be examined in order to develop a preliminary design requirements, pollutant removal performance curve and charts.
- Finalize the design requirements, monographs, and charts for rain garden systems. At the final stage, the entire design requirement will be finalized and design protocol will be derived.

4. Rain Garden Performance

Rain gardens system has many potential and unique compared to most the common types of Best Management Practices, mainly wet and dry pond. Besides the landscaping aspect, rain gardens have many positive benefits in improving water quality for higher pollutant removal efficiencies for nitrogen, phosphorus, and trace metals [9]. Study by Hatt (2007) showed that bi-oretention systems can reduce runoffs volumes and concentrations of sediments, heavy metals, phosphorus, hydrocarbons, and pathogenic bacteria [12], [13]. All these pollutants and nutrients were being removed through the combination of natural physical and biological processes according to the type of soil and plants used [13]. Besides that, rain gardens system provides low maintenance as compared to most traditional stormwater BMP's.

Extensive research on assessing the performance of rain garden systems has been conducted in field studies and laboratory work. Previous studies showed that rain garden systems can remove heavy metals with load reduction generally in excess of 90% [13]. Majority of Zinc, Lead were removed from stormwater that flowed through rain garden system system [13]. Study by Davis (2007) showed

IOP Conf. Series: Earth and Environmental Science **16** (2013) 012020 doi:10.1088/1755-1315/16/1/012020

that removal efficiencies of Zn, Cu, Pb and Cd were approximately 94%, 88%, 95% and >95%. For laboratory study, Belinda (2009) has performed a large scale of column study with 125 biofilter columns [14]. All the biofilters were constructed in greenhouse in Melbourne, Australia to test the performance of stormwater biofilters on removal sediments. Result showed that all the biofilters were highly effective in removing TSS greater than 95% consistently over the time [13], [14]. The performance of rain garden system systems in field setting by Paul Matthews (2011) in Auckland, New Zealand also showed reduction of TSS from stormwater that flowed through the rain garden system system. Observation from the study showed that concentration of TSS was reduced significantly by the rain garden system system [8].

5. Application of Rain Garden at Humid Tropic Centre

The application of WSUD on Rain Garden at local scale as pilot study was located at Humid Tropic Centre (HTC), Kuala Lumpur. It is a sub division of Department of Irrigation and Drainage (DID). The site consists of two office buildings, one toilet, one prayer building, one store cabin and two parking areas. Beside rain garden, there are also other components such as wetland, green roof, rainwater harvesting as a part of integrated stormwater management that installed at HTC.

The performance of rain garden on hydraulic conductivity and water quality has been monitored. Water quality samples were collected as grab samples from randomly selected storm event. All the samples were put in a plastic bottle and sent to certified laboratory for chemical testing by using APHA Standard Methods. Besides that, in situ water quality probe are used to measure pH and DO. The hydraulic conductivity of filter media has been monitored in situ by using method as described in Practice Note 1: In situ measurement of hydraulic conductivity. Only top layer of the filter media has been tested due to the fact that, the top layer is the one controlling the hydraulic conductivity of the system as a whole. Then the results are being compared to the baseline hydraulic conductivity before the installation of WSUD by using Horton's Method. Results in Table 1, showed that hydraulic conductivity improved with the installation of rain garden system system at the study area. According to Urban Stormwater Management Manual for Malaysia, the suitability of soil for quality infiltration is based on the minimum infiltration rate of 13 mm/hr. Therefore, based on the results observed, the hydraulic conductivity of rain garden system system installed in the study area is above exceeded the MSMA requirement.

Table 1: Hydraulic Conductivity for Rain garden system system

Testing point	Initial Rate of Infiltration, f_0 (mm/hr)	Constant Rate of Infiltration, f_c (mm/hr)	Shape Factor, k (hr-1)	Horton Equation $f = f_c+(f_0 - f_c)^{e-kt}$
Before installation of bioretention	62.15	3	5.86	$f = 3 + (62.15 - 3)^{e-5.86t}$
3 month after installation of bioretention				
Point 1	211.59	76.36	0.744	$f = 76.36 + (211.59-76.36)^{e-0.744t}$
Point 2	118.80	64.15	0.719	$f = 64.15 + (118.80-64.15)^{e-0.719t}$
Point 3	123.39	30.54	0.961	$f = 30.54 + (123.39-30.54)^{e-0.961t}$
Point 4	272.80	30.55	0.917	$f = 30.55 + (272.80-30.55)^{e-0.917t}$
Point 5	413.67	168.00	0.100	$f = 168.00 + (413.67-168.00)^{e-0.10t}$
Point 6	584.65	158.84	1.543	$f = 91.63 + (296.02-91.63)^{e-0.355t}$

For water quality measurement, the parameters that has been analyzed are Total Suspended Solid (TSS), Dissolved Oxygen (DO), Biochemical Oxygen Demand (BOD), pH and heavy metals such as Zinc (Zn) and Copper (Cu). As can be seen from Figure 1, results showed that the water quality level for DO and BOD was within the range of Water Quality Index Class II. As an average, DO levels were between 5 to 7 mg/L and BOD between 1 to 3 mg/L.

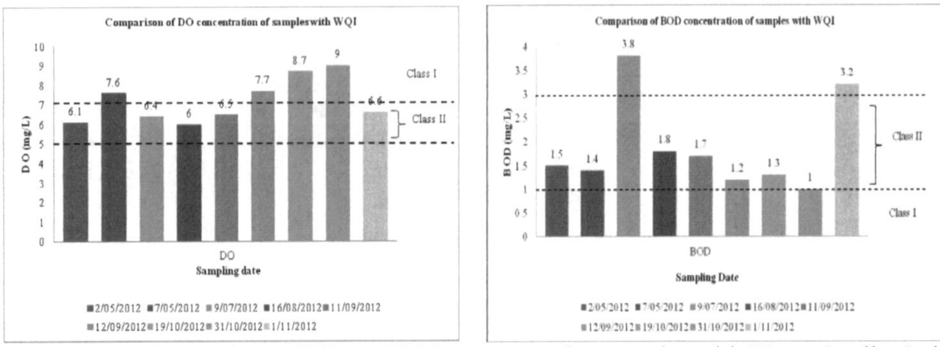

Figure 1: Comparison of DO and BOD concentration samples with Water Quality Index

6. Conclusions

This research is expected to identify the potential of rain garden systems for improvement of stormwater pollutants under Malaysia condition. Furthermore, this research will identify the obstacles to adopt rain garden due to the different rainfall intensity magnitude and make recommendation to overcome the problem in implementing rain garden system.

Acknowledgement

The author would like to acknowledge Humid Tropic Centre, UNITEN and USM for their supporting and collaboration in succeeding this rain garden project. The researcher also wishes to express their gratitude for the financial support from Minister of Higher Education (MOHE) under Long Research Grant Scheme (LRGS).

References

[1] Blecken G T, Zinger Y, Deletić A, Fletcher T D, Viklander M 2009 *Water research* **43(18)** 4590-4598

[2] Bratieres K, Fletcher T D, Deletic A, Zinger Y 2008 *Water Research* **42(14)** 3930-3940

[3] Brown D, Morison P J, Rebekah R 2011 *Lansdcape and Urban Planning* **99** 83-92

[4] Coffman L, Green R, Clar M, Bitte S 1993 *Proc.,20th Annivesary Conference on Water Management in the '90's* 130-133

[5] Collins K A, Llawrence T J, Stander E K, Jontos R J, Kaushal S S, Newcomer T A, Grimm N G, Ekberg M L C 2010 *Ecological Engineering* **36** 1507-1519

[6] Coustumer S L, Fletcher T D, Deletic A, Barraud S, Lewis J F 2009 *Journal of Hydrology* **376** 16-23

[7] Davis A P 2008 *Journal of Hydrologic Engineering* 90-95

[8] Davis A P, Sun X 2007 *Chemosphere* 1601-1609

[9] Hsieh C H, Davis A P 2005 *Journal of Environmental Engineering* **131(11)** 1521-1531

[10] Li H, Davis A P 2008 *Journal of Environmental Engineering* **134(6)** 409-418

[11] Davis A P, Hunt W F, Traver, R G, Clar, M 2009 *Journal of Environmental Engineering* **135(3)** 109-117

[12] Hatt B E, Fletcher T D, Deletic A 2007 *Water research* **41(12)** 2513-2524

[13] Hatt B E, Fletcher T D, Deletic A 2009 *Journal of Hydrology* **365(3)** 310-321

[14] Zakaria N A, Ab Ghani A, Abdullah R, Sidek L M, Kassim A H, Ainan A 2004 MSMA–A New Urban Stormwater Management Manual for Malaysia *In International Conference "ICHE"*

IOP Conf. Series: Earth and Environmental Science **16** (2013) 012021 doi:10.1088/1755-1315/16/1/012021

Energy Efficiency of Biogas Produced from Different Biomass Sources

Shahida Begum[1], **A.H. Nazri**

Centre for Advanced Materials, Department of Mechanical Engineering, Universiti Tenaga Nasional, Kajang, Selangor

E-mail: shahida@uniten.edu.my

Abstract. Malaysia has different sources of biomass like palm oil waste, agricultural waste, cow dung, sewage waste and landfill sites, which can be used to produce biogas and as a source of energy. Depending on the type of biomass, the biogas produced can have different calorific value. At the same time the energy, being used to produce biogas is dependent on transportation distance, means of transportation, conversion techniques and for handling of raw materials and digested residues. An energy systems analysis approach based on literature is applied to calculate the energy efficiency of biogas produced from biomass. Basically, the methodology is comprised of collecting data, proposing locations and estimating the energy input needed to produce biogas and output obtained from the generated biogas. The study showed that palm oil and municipal solid waste is two potential sources of biomass. The energy efficiency of biogas produced from palm oil residues and municipal solid wastes is 1.70 and 3.33 respectively. Municipal solid wastes have the higher energy efficiency due to less transportation distance and electricity consumption. Despite the inherent uncertainties in the calculations, it can be concluded that the energy potential to use biomass for biogas production is a promising alternative.

1. Introduction

Biomass is a name given to any organic matter available on a renewable basis. It can be used to produce biogas in order to reduce dependence on fossil fuels and to achieve environmental benefits. Palm oil wastes, municipal solid wastes, cow dung, sewage waste and landfill sites are the sources of biomass. By anaerobic digestion, the biogas produced can be used for heating, electricity production or as transportation fuel. However, for generating biogas some energy is needed which includes pumping and stirring of biogas reactors, fuels for transporting raw materials and digested residues. The energy efficiency and environmental impact vary with transportation distance, means of transportation, and potential gas production from biomass, conversion techniques and process parameters.

Though Malaysia is endowed with plentiful of natural resources like rubber, palm oil, tin, petroleum and natural gas, but the economic growth based on industrialization, combined with population and urbanization have created an expanding demand for energy[1, 2]. Hence reliable, secure and cost effective energy supply is very much needed. Not only that, efficient utilization of energy sources, diversification of sources and minimization of wastages are also emphasized by government [3]. The two most potential sources of biomass are palm oil residues and municipal solid wastes (MSW). The type and amount of biomass generated from palm oil residues and their heat value is shown in Table 1 and the potential energy from biogas is presented in Table 2 [4-6].

[1] To whom the correspondence may be addressed

Table 1. Biomass generated by palm oil mill

Biomass	Quantity (million tonnes)	Moisture content (%)	Oil content (%)	Heat value (kJ/kg)
FFB	18.25	67	5	18,883
Fibre	11.11	37	5	19,114
Shell	5.55	12	1	20,156
POME	53.16	93	1	17,044

Table 2. Potential Energy from Biogas

Year	Palm oil production (million tonnes)	Biomass (million tonnes)	POME (million m^3)	Biogas (million m^3)
2002	13.3	14.2	45.63	1,278
2007	15.9	18.5	53.55	1,499
2010	16.9	19.6	57.42	1,607

The solid wastes generated in major urban areas are presented in Table 3 [7] which is increasing rapidly due to increase of urban population. The municipal waste is mainly comprised of domestic and industrial waste [8]. It is estimated that the amount of municipal solid waste (MSW) by the year 2020 will be about 9 million tonnes per year and the amount generated per day is 24,650 tonnes [9].

Table 3. Solid waste generated in major urban areas

State	Solid waste (tonne/day)	
	1998	2000
Kuala Lumpur	2350	2800
Penang	1220	1400
Johor Bahru	450	500

2. Calculation Methodology

To calculate the energy efficiency three steps were adopted:

- Searching and collecting data for potential biomass in Malaysia
- Propose a location to produce biogas
- Estimation of energy needed to produce biogas and the output energy

The energy recovery is dependent on the volatile solids, fixed carbon content, inert, calorific value, C/N ratio (carbon/nitrogen ratio) and toxicity.

2.1. Palm oil wastes

The location considered for producing biogas from palm oil mill waste is Serting Hilir Mill. The details are provided in Table 4.

The energy needed for transportation is dependent on the distance of source of palm oil wastes and the lorry specification used for transportation. The capacity of each lorry is considered of 4000kg to calculate energy input for transportation. The mileage of lorry and energy consumed for per litre of oil are 7km/litre and 48 MJ/litre respectively [11]. Hence the energy used for transportation is as follow:

$$\text{Energy use/tonne} = (\text{distance}/7\text{ km per litre}) * (48\text{ MJ/litre}) \div 4\text{ tonne} \tag{1}$$

Energy needed to produce biogas from palm oil waste and energy obtained from biogas is presented in Table 5, considering Johor Bahru as a source of palm oil wastes.

2.2. Municipal Solid Waste (MSW)

It was observed by a study conducted by Japan International Corporation Agency that 45% of Malaysian

Table 4. Palm oil Mill location [10]

Location	Serting Hilir Mill, Negeri Sembilan
Installed capacity	60 T FFB/hour
Effective working days	5 days/week
Effective working hours	20 hours/day
Electricity from grid (MJ/t)	954.47
Methane composition (%)	65
Amount of biogas from POME (m^3)	22,554m^3/day
Calorific value	53,000kcal/m^3

Table 5. Energy ratio for generated biogas from palm oil waste

Energy input (A)	Energy output(B)	Energy ratio
(a)Power supply from grid 954.47 MJ/t (b)Transportation energy: (372km/7 km per litre)*(48 MJ/litre)÷4tonne=637.71 MJ/t Total energy= energy for transportation+ energy from grid= (a) + (b)= 1.59GJ/t	(Amount of biogas per day* calorific value)÷capacity of FEB per day)*methane composition= (22,554m^3/day*221.85MJ/m^3 ÷1200t/day)*0.65= 2.71 GJ/t	B/A= 1.70

MSW consists of food waste [12]. The MSW generation has increased from 5.6 million tonnes to 8.0 million tonnes from year 1997 to 2000 and hence there is an urgent need for a better managed disposal option [13]. In Table 6, the MSW generated in Kuala Lumpur is presented and it will be used as reference to estimate energy efficiency.

Table 6. MSW generated in Kuala Lumpur [14]

year	KL population	Solid waste (T/day)	Amount collected
1998	1,446,803	2,257	1,900
2000	1,787,000	3,070	2,500
2005	2,150,000	3,478	2,922

Table 7. Energy ratio for generated biogas from MSW

Energy input (A)	Energy output(B)	Energy ratio
(a) Electricity used: 13.6MJ/t (b)Transportation energy: 20 MJ/t (c) Handling of raw materials: 0.6 GJ/t Total energy= (a) + (b)+ (c) = 0.634GJ/t	(Amount of biogas per day* calorific value)*methane composition= (200 m^3/t*19.15 MJ/m^3)*0.55= 2.11 GJ/t	B/A= 3.33

Air Hitam Sanitary Landfill was considered as a location to produce biogas from MSW. The potential landfill gas production is estimated to be 1400 million m^3 assuming that one tonne of MSW is able to produce 200 m^3 of landfill gas [15]. The energy input for handling and transportation of raw materials and electricity needed for processing raw materials are 0.6 GJ/tonne, 20MJ/tonne and 13.6 MJ/tonne respectively [16].

Energy needed to produce biogas from MSW and energy obtained from biogas is given in Table 7.

3. Results and Analysis

All the possible energy inputs were considered. The electricity needed to operate the biogas plant, the energy for handling raw materials and transportation were considered as input energy. The energy output was taken from the calorific value of biogas produced. The energy efficiency of palm oil waste and MSW is 1.7 and 3.33 respectively. To utilize palm oil residues higher energy was used from grid. Hence alternate use of energy sources and reduction of transportation distance can enhance energy efficiency with palm oil waste. The transportation cost can be substantially reduced if the biogas plant is located within 20 km of biomass sources.

4. Conclusions

The two main sources of biomass to produce biogas are palm oil waste and MSW. Palm oil residue is the main contributor to produce biogas, while MSW has higher energy potential. The result presented in this investigation is a reasonable estimate based on Malaysian conditions but not the exact values. The energy efficiency of biogas produced from palm oil waste and MSW are 1.7 and 3.33 respectively. However, it can be concluded that whatever is the source of biogas, in terms of energy efficiency it is a suitable energy option.

References

[1] Green J H, Green D J 1996 *Renewable energy systems in Southeast Asia* PennWell Publishing Company 81-122
[2] Yusoff S 2006 *Journal of cleaner production* **14(1)** 87-93
[3] *The Third outline Malaysian perspective plan (OPP 3)* 2001-2010
[4] Gan P Y, Li Z 2008 *Energy Policy* **36(2)** 890-899
[5] Pantzaris T P, Mohd Jaaffar A 2002 *Palmas* **23**
[6] Chow M C, Ho C C 2000 *Journal of Oil Palm Research* **12(1)** 107-116
[7] Agamuthu P 2001 *Solid Waste: Principles and Management with Malaysia Case Study* University Malaya Press
[8] KPPK 2009-3 *http://www.kppk.gov.my* downloaded August 2012
[9] Toh Chin Kang 2008 Study on the Feasibility of Capturing Methane Gas from Existing Landfills in the Country *Final Year Project Thesis* Universiti Tenaga Nasional.
[10] Shirai Y Suzuki S 2002 *Reduction of Methane Released from Palm Oil Mill Lagoon in Malaysia and its Countermeasures*, Ex Corp, Kyushu Institute of Technology and University Putra Malaysia 33–44
[11] Nurul Nazwa Binti Khalil 2007 Potential of Biomass (Agriculture) Energy in Peninsular Malaysia
[12] Cheng Y H, Sang S X, Huang H Z, Liu X J Ouyang J B 2007 *Journal of China University of Mining & Technology* **17(3)** 0403-0408
[13] Fauziah S H, Agamuthu P 2003 *A Comparative Study on Selected Landfills in Selangor, Kuala Lumpur Municipal Solid Waste Management* Institute of Biological Sciences, Faculty of Science, University Malaya
[14] Sivapalan K, Muhd N M Y, Abd H S, Kamaruzzaman S, Rakmi A R 2002 *Proceedings of the Regional Symposium on Environment and Natural Resources* **1** 359-368
[15] NEA *Solid waste collection system* http://app2.nea.gov.sg/topics_waste.aspx downloaded August 2012
[16] Berglund M, Börjesson P 2003 *Energianalys av biogassystem,(Energy systems analysis of biogas systems)* **IMES (No. 44)** EESS Report

Performance Comparison of Voltage Stability Indices for Weak Bus Identification in Power Systems

R.Verayiah[1], A. Mohamed[1], H.Shareef[1], I. Hj. Zainal Abidin[2]

[1] Dept. of Electrical, Electronics and Systems Engineering, Universiti Kebangsaan Malaysia, Selangor, Malaysia

[2] Electrical Power Dept., Universiti Tenaga Nasional, Selangor, Malaysia

E-mail: renuga@uniten.edu.my, azah@eng.ukm.my, shareef@eng.ukm.my, izham@uniten.edu.

Abstract. Voltage collapse event is identified as complex and localized in nature but its effect is extensive once occurred. The vital effect of voltage collapse would be the total system collapse or blackouts which would cost a large loss to utility companies. Eventually, on- line monitoring of power system stability has become an important factor for electric power utilities. The last utmost option to avert voltage collapse incident from occurring is by the implementation of under voltage load shedding scheme. The identification of location for load shedding is the main motivation of the study. The weakest bus in a power system is identified as the location for load shedding. This location is obtained using voltage stability index Ld. The performance and effectiveness of Ld index is compared with Fast Voltage Stability Index (FVSI) and Le Index. The results obtained indicate that Ld Index can be used to identify the weak bus in a power system and consequently for the placement of UVLS relays in a power system network.

1. Introduction

Power systems at present are much more vulnerable to voltage collapse than many years ago because the systems are being stressed due to huge power transfers across the grids which cause transmission lines to operate close to the limits. In addition, generation reserves are minimal and often the reactive power is insufficient to satisfy the load demands [1]. Due to these reasons, power systems become more susceptible to disturbances, outages and instability.

Under voltage load shedding (UVLS) is specifically designed as a safety measure to prevent widespread voltage collapse in the event of severe deficit in local or system wide area reactive power reserves. UVLS has become the preferred strategy by power utilities due to its cost-effective solution to address voltage stability issues. However, before implementing under voltage load shedding, several issues have to be addressed, such as the location of load shed, the quantum to shed at the identified locations and also the timing of when to load shed.

This paper aims to evaluate the performance of various voltage stability indices in identifying the locations for load shedding. The weakest bus in a power network is identified to be the most effective location for load shedding. The various voltage stability indices considered in this study are the stability index, L_d [2], L_e index and FVSI index [3, 4]. Power flow and voltage collapse simulations

[1] To whom any correspondence should be addressed.

were implemented on the IEEE 30 bus test system using the PSS/E software program. The voltage stability indices are then calculated and evaluated at each simulation so as to identify the suitable location for load shedding.

2. Voltage Stability Indices

Typically, line stability indices are formulated based on the power transmission concept in a single line. A single line diagram of an interconnected network is shown in Figure 1.

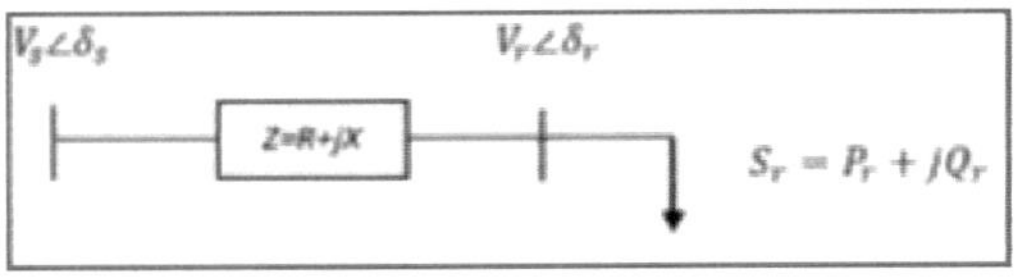

Figure 1. Single Line Diagram of a Two- Bus system [5]

2.2 Voltage Stability Index Formulation for Testing System

There are three types of line voltage stability indices that are used for the analysis. They are Voltage Stability Index Ld, Line Stability Index Le, and Fast Voltage Stability Index (FVSI).

2.2.1 Voltage Stability Index, Ld: The mathematical formulation of this index is derived from voltage equation of the radial distribution system. The developed indicator is given as:

$$L_d = \frac{4\sqrt{((P_{i+1}^2 + Q_{i+1}^2)(r^2 + x^2))}}{V_{i+1}^2} \tag{1}$$

where P_i and Q_i are active and reactive sending end power,P_{i+1} and Q_{i+1} are active and reactive receiving end power,V_i And V_{i+1} are sending end and receiving end voltage and r and x are the line impedance [11]. The line that gives index value closest to 1 will be the most critical line of the system and may lead to wide voltage collapse.

2.2.2 Voltage Stability Index, Le: The equation for this index is given as below:

$$\mathrm{L_e} = \frac{4[V_i^2(P_{i+1}r_i + Q_{i+1}x_i) + (P_{i+1}x_i - Q_{i+1}r_i)^2]}{V_i^2} \tag{2}$$

Similar to the previous index Ld, P_{i+1} and Q_{i+1} are active and reactive receiving end power ,V_i is sending end and receiving end voltage and r and x are the line impedance [11]. It is noted that Le index has to be less than 1 in order to satisfy the stability condition. When the value of the index approaches 1, it indicates that the system in becoming unstable.

2.2.3 Fast Voltage Stability Index, FVSI: This index is proposed by I. Musirin and it is calculated by [12]:

$$FVSI = \frac{4Z^2Q_r}{V_s^2X} \tag{3}$$

As previously defined, Vs and Vr are the sending end and receiving end voltages, Qr is the reactive power at the receiving end while Z is the line impedance and X is the line reactance. FVSI index less than 1 shows that the system is stable while FVSI index value of 0 shows system is collapse.

3. Test System Analysis and Results

The standard model of IEEE 30 Bus test system is used as a test case for the study purpose [5]. The system consists of 6 synchronous generators and 4 transformers. The system has 21 load points totalling 283.4 MW and 126.2 MVAR.

The simulation was carried out and index calculation was performed using MATLAB in order to verify the most sensitive indicator in identifying and locating the weakest bus prior to load shedding. Tables below shows the index value obtained for three different scenarios. Table 1 shows the index value calculated for Base Case. Table 2 shows the results obtained for MW increment at selected load buses, while Table 3 shows the results obtained for MW and MVAR increment at selected load buses. All the indicators range from 0 at most stable to 1 at the most unstable.

From Table 1, it is obvious that all the index values are in the range of most stable. A low index value indicates a high voltage value for a particular bus in the system. Overall, Ld index is found to show a higher value as compared to Le Index and FVSI index.

Table 1. Comparison Study for Index Calculated at Base Case

Case 1	From Bus	To Bus	(Vi, p.u)	FVSI	Ld	Le
	6	7	0.97230	0.01120	0.03738	0.00984
		8	0.97230	0.04568	0.06286	0.04095
		9	0.97230	0.03465	0.05467	0.03321
		10	0.97230	0.04265	0.07527	0.04129
		28	0.97230	0.00536	0.00993	0.00462
	25	26	0.99050	0.05328	0.08164	0.04545
		27	0.99050	0.01528	0.05750	0.01510
	27	28	1.00000	0.10720	0.19067	0.11250
		29	1.00000	0.03640	0.12706	0.04255
		30	1.00000	0.05241	0.21197	0.07541

On the other hand, Table 2 shows the index calculated for a MW increment at loads using PSS/E. The system is stressed as the generators are incapable to cater for the increasing demand at the loads. The system operates until it reaches voltage instability situation. The high index value at certain branches shows that the lines are heavily loaded and can cause cascaded wide area collapse. Bus 26 to Bus 30 exhibits a high index value and correspondingly lower voltage value. Bus 26 to Bus 30 are identified as the weakest point in the whole system.

Results obtained in Table 3 show the index calculated for MW and MVAR increment at loads via PSS/E. Much more severe voltage drop occurs when MVARs are transferred as compared to the drop that occurs when MW are transferred.

Table 2. Comparison Study for Index Calculated with MW Increment at Loads

Case 2	From Bus	To Bus	(Vi, p.u)	FVSI	Ld	Le
	6	7	0.95300	0.01768	0.01881	0.01416
		8	0.95300	0.06476	0.10179	0.05628
		9	0.95300	0.65853	0.66251	0.61723
		10	0.95300	0.18498	0.41243	0.20745
		28	0.95300	0.06489	0.13347	0.05537
	25	26	0.84380	0.48024	1.00000	0.61241
		27	0.84380	0.16388	0.14341	0.09357
	27	28	0.87620	0.03960	1.00000	0.48566
		29	0.87620	0.68881	1.00000	0.87495
		30	0.87620	0.81511	1.00000	1.09513

Table 3. Comparison Study for Index Calculated with MW and MVAR Increment at Loads

Case 3	From Bus	To	(Vi, p.u)	FVSI	Ld	Le

	Bus				
6	7	0.89480	0.00775	0.02725	0.00576
	8	0.89480	0.07389	0.10825	0.05643
	9	0.89480	0.07134	0.29324	0.07476
	10	0.89480	0.05595	0.40746	0.08005
	28	0.89480	0.10624	0.17539	0.07885
25	26	0.70850	1.00000	1.00000	0.51793
	27	0.70850	0.15354	0.14496	0.06247
27	28	0.73620	0.50776	0.99764	0.51114
	29	0.73620	1.00000	1.00000	0.68700
	30	0.73620	1.00000	1.00000	0.81021

The greater voltage drop is caused by the fact that the current is more lagging. The index values found to have reached the unstable value of 1 for Bus 26, 29 and Bus 30. Again, the weakest bus is found to be Bus 30 and all the neighboring bus which are Bus 26 till 29 displays a critical unstable value which is close to 1.

All the results obtained are consistent. In conclusion, Ld Index appears to be the most sensitive index at signifying the weakest bus in a test system, followed by FVSI Index and then the Le Index. Thus Ld index can be used to accurately identify the weak point in a system and hence as an indicator to the best location to perform load shedding.

4. Conclusions

Overall, the performance of the voltage stability index value is compared and analyzed for the purpose of identifying the best location for load shedding. The simulations results run on IEEE 30 Bus Test System show all indices provide the same ranking for the weak buses on the test system. Also, it can be concluded that Ld index appeared to be the most sensitive index in indicating the weak area and lines in the system. Thus, the simulation results demonstrate a strong correlation between Ld Index and the placement of UVLS relays for a power system network.

References

[1] Kundur P, Balu N J, Lauby M G 1994 *Power system stability and control ser. The EPRI power system engineering series.*

[2] Chanda S, Das B 2011 *Power Electronics (IICPE), 2010 India International Conference on* 1-4

[3] Rahman T A, Jasmon G B *Energy Management and Power Delivery, 1995. Proceedings of EMPD'95., 1995 International Conference on* **2** 714-719

[4] Musirin I, Rahman T K A 2002 *Transmission and Distribution Conference and Exhibition 2002: Asia Pacific. IEEE/PES* **2** 1118-1123

[5] *Power System Test Case Archive, 30 Bus Power Flow Test Case* 2012 available at: http://www.ee.washington.edu/research/pstca/pf30/pg_tca30bus.htm

[6] Ekwue A O, Cheng D T Y, Song Y H, Wan H B 1997 *Voltage Collapse (Digest No: 1997/101), IEE Colloquium on* 3-1

[7] Liu Zhuo 2000 *Proceedings of the CSEE* **20** 35-39

[8] Karbalaei F, Soleymani H, Afsharnia S 2010 *IPEC, 2010 Conference Proceedings* 429-432

Optimization of Structural Design for Sustainable Construction of Transmission Tower Based on Topographical Algorithm

Zakaria Che Muda[1], Kamal Nasharuddin Mustapha, Rohayu Che Omar, Fathoni Usman, Md Ashrafu Alam and Sivadass Thiruchelvam[1]

Centre of Forensic Engineering, Universiti Tenaga Nasional, Selangor, Malaysia

Email: MZakaria@uniten.edu.my

Abstract. Optimization of transmission tower structures is traditionally based on either optimization of members sizes with fixed topographical shape or based on structural analysis modelling strategies without taking cognizance of fabrication and constructability issue facing the contractors . This paper look into an integrated optimum design approach strategies whereby size, shape and topology are combined together with the fabrication issues in the construction of the transmission tower. The topographical algorithm is based on changing the inclination degree of the legs of the tower at first with optimum individual members sizing and later rationalized member sizes are performed through member groupings for the ease fabrication and construction of the transmission tower. The optimum weight using topographical algorithm obtained for the transmission tower is 10,924 kg for singular members and 18,430 kg for element grouping at 10° inclination angle.

1. Introduction

Optimization, sustainable design and construction of the transmission tower has become the main concern due to the escalating cost of steel. Designer of transmission tower are having a common aim to develop an efficient way to design a cost effective and optimized transmission tower. It is almost an impossible task to optimize the structures by trial and error method, even by an experience designer, where convergence can be achieve in compliance with the optimum member sizes (size is a variable factor) , topographical factor (structural shape is a variable) in conjunction with fabrication sizes and cost (member grouping is a variable).

Several studies on the structural assessment module of current steel design models for transmission and telecommunication towers have been carried out [1] and [2]. Optimum design module has been studied [3], [4] and [5]. Simultaneous optimization for both sizing and topographical algorithm of the Transmission Tower has been studied by Neves and Fonseca [6] Shape optimization of trusses has been carried out by Saka [7]. A review paper on numerical methods for shape optimization has been given by Vanderplaats [8]. Rajan developed a genetic algorithm that optimizes sizing, shape and topology variables in a truss using a genetic algorithm [9].

The main objective of this paper is to optimize the transmission tower design based on integrated approach combining simultaneously the size factor (member sizes is the variable) and shape factor (tower shape is the variable) in conjunction with member groupings for fabrication cost and ease of

[1] To whom any correspondence should be addressed.

IOP Conf. Series: Earth and Environmental Science **16** (2013) 012023 doi:10.1088/1755-1315/16/1/012023

construction. It is envisaged in future that the design process will be based on an integrated optimum design approach that can both integrate size and topographical algorithm to achieve least fabrication cost in the construction of the transmission tower.

2. Methodology

The selection and development of a topographical algorithm to alter the leg coordinates of the tower is based on the legs inclination angle of the tower. The Visual Basic is used to develop as a pre-processor interface topographical algorithm program. Structural modelling analysis strategy for the Transmission Tower are based mixed truss-beam analysis. This mixed modelling strategy can produce more realistic structural behaviour. In the mixed beam-truss analysis, all the vertical member of the tower are modelled as beam elements whilst all the other the horizontal and incline members inclusive of the cross arms act as a space truss members.

The space algorithm program will extract the .std input file from the STAAD Pro and alter the co-ordinate automatically in accordance with the user define inclination angle of the leg as shown in Figure 1. A series of inclination angle then can be set to obtain the most optimum weight based on the inclination angle and above method of structural analysis.

3. Discussion

Figure 2 shows the process of analysis and optimization of transmission tower in STAAD Pro where the analysis started by setting the new inclination angle of the leg of a transmission tower.

3.1. Loading Trees

A combination of several load which consists wind load, self weight and the load from each condition of the loading trees are considered. There are four combinations of load shown in Figure 3. Firstly, the combination 1 is wind load, self weight and normal condition. Second, the combination 2 is wind load, self weight and ground wire broken. As for the third combination is wind load, self weight, top conductor and middle conductor broken and lastly, it is the combination 4 of wind load, self weight, middle conductor and bottom conductor broken or top conductor and bottom conductor broken. The factor of safety will be included for each load which is 1.0 for self weight, 1.25 for wind load, 2.5 for normal condition and 1.25 for broken condition.

In this case, the application of shape factor in this project is to alter the angle of the leg of the Transmission Tower as shown in Figure 4. Optimization process that arbitrary change the nodal coordinates above the tower legs are likely to lead to violation of electrical clearance requirements [3].

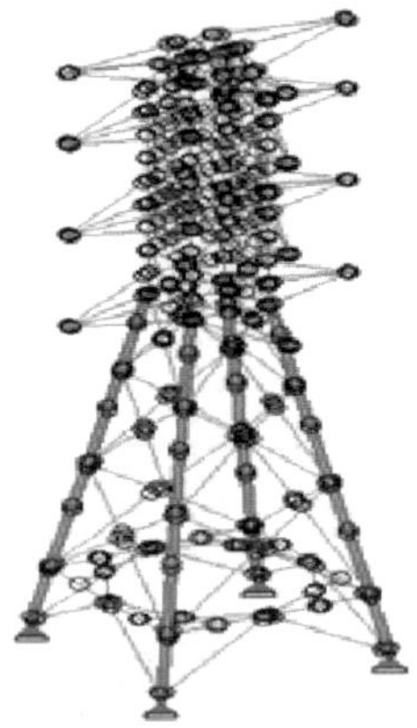

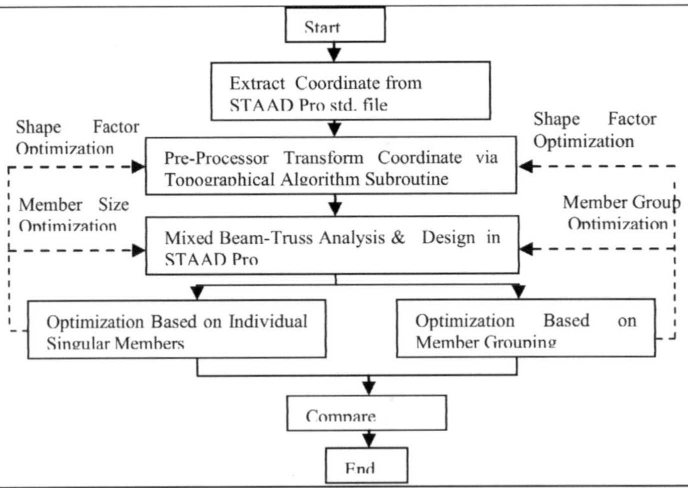

Figure 1. Legs of Tower **Figure 2.** Flow Chart of Topographical Algorithm Interface with

STAAD Pro

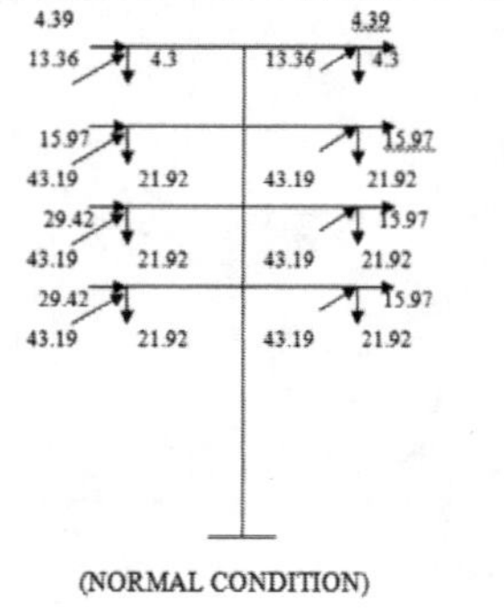

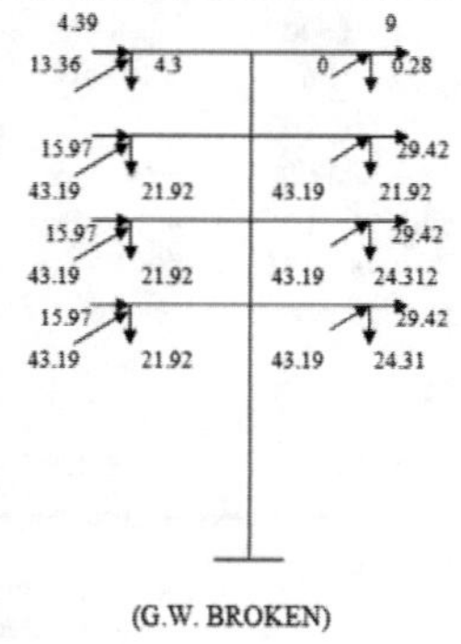

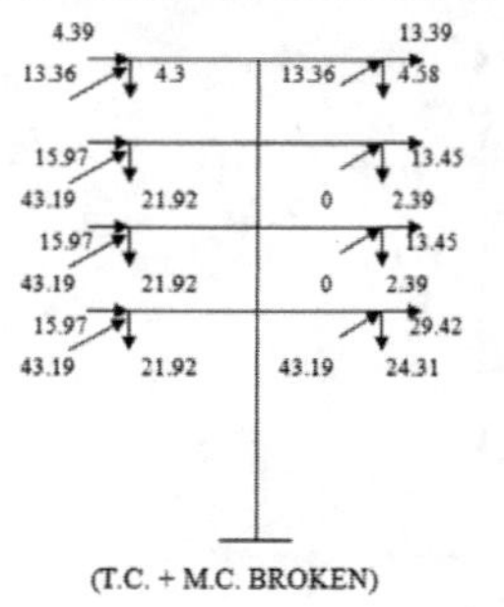

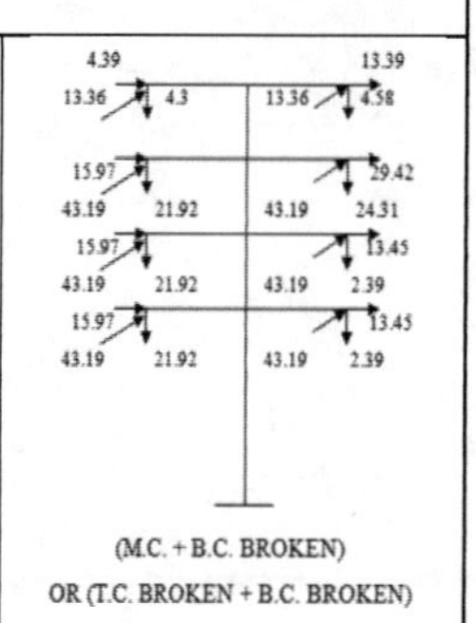

Load Combination 1, Normal Condition (Load in kN)	Load Combination 2, Ground Wire (G.W) Broken Condition	Load Combination 3, Top Conductor (T.C) and Middle Conductor (M.C) Broken Condition	Load Combination 4, Middle Conductor (M.C) and Bottom Conductor (M.C) or Top Conductor (T.C) and Bottom Conductor (M.C) Broken Condition

Figure 3. Loading trees

3.2. Topographical Algorithm in the Structural Analysis.

Mixed beam-truss analysis is a modelling strategy combining beam and truss elements. The main structure will use spatial beam elements. This mixed modelling strategy can produce more realistic structural behaviour with respect to making it as close as possible to the actual built-up structure.

In this paper, the software is tested for 5 different inclination angle of the leg of transmission tower. The inclination angles are 5 degree, 10°, 15 °, 20 ° and 30 °. The weight of the tower at the start of the optimization process mentioned as initial weight is 22,000 kg to 65,000 kg depending on the inclination angle with the optimum weight obtained 10,871 kg to 25,044 kg for singular members and 18.366 kg to 52.425 kg for grouping members. It undergo the optimization design using the topographical algorithm for each singular member weight and later into grouping to come with the rationalize the fabrication cost and ease of construction for the transmission tower.

Generally, optimization based on individual member generates the least weight however practical consideration for fabrication cost and ease of construction requires grouping of members. The selection criteria of member groupings are based on height strata of the members in the tower and its functions as vertical, horizontal or incline members. Thus, the fabrication grouping weight lies in between the initial and the optimize singular members. From the graph shown in Figure 5, the optimum leg tower angle lies from 5° to 15°. Beyond the 15° leg angle, the total weight increases exponentially which is too expensive for the cost of the tower. The optimum weight of the Transmission Tower is about 10°.

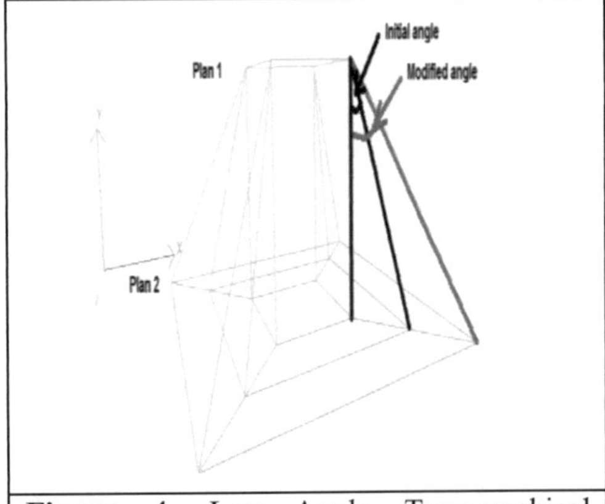

Figure 4. Leg Angle Topographical Algorithm Optimization

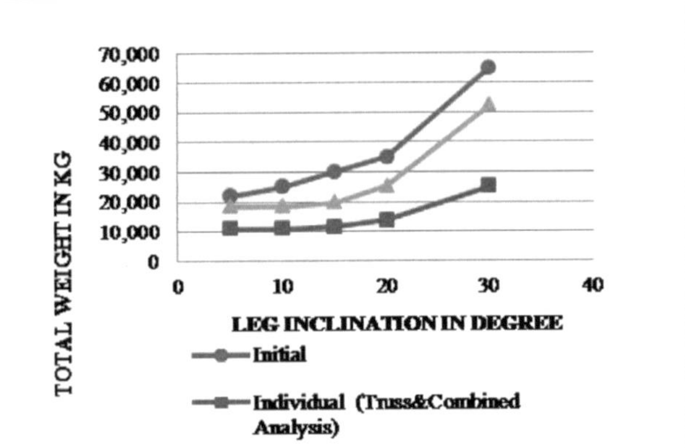

Figure 5. Total Tower Weight versus Leg Inclination Angle

4. Conclusions

The optimum weight using topographical algorithm obtained for this tower is 10,924 kg for singular members and 18,430 kg for element grouping at 10° degree inclination angle after taking due cognizance of foundation loading and deflection. Beyond the 15 degree leg angle, the total weight increases exponentially which is too expensive for the cost of the tower. The topographical algorithm is based on changing the inclination degree of the legs of the tower to obtain the optimum weight in the design and fabrication is viable to be used in the design. The topographical algorithm simultaneously optimized size, shape, and topology variables in the tower and was proven to be successfully integrated in the actual design of the tower. The element groupings can be manipulated within the set criteria to achieve the minimum fabrication cost and ease of construction for the tower.

References

[1] Da Silva J G S, Vellasco P D S, De Andrade S A L, De Oliveira M I R 2005 *Journal of Constructional Steel Research* **61(8)** 1108-1134
[2] Mikitarenko M A, Perelmuter A V 1998 *Journal of Wind Engineering and Industrial Aerodynamics* **74** 1091-1100
[3] Rao G V 1995 *Computers & structures* **57(1)** 81-92
[4] Natarajan K, Santhakumar A R 1995 *Computers & structures* **55(3)** 387-403
[5] Jasim N A, Galeb A C 2002 *Journal of Constructional Steel Research* **58(3)** 413-425
[6] das Nevas F D A, da Fonseca M 2008 Simultaneous Sizing and Shape Optimization of Transmission Towers Using Genetic Algorithm *EngOpt 2008 – International Conference on Engineering Optimization*
[7] Saka M P 1980 *Journal of the Structural Division* **106(5)** 1155-1174
[8] Vanderplaats G N 1981 Numerical Methods for Shape Optimization – an assessement of the state of art *Proc. Int. Symp. Optimum Steel Design*
[9] Rajan S D 1995 *Journal of Structural Engineering* **121(10)** 1480-1487

IOP Conf. Series: Earth and Environmental Science **16** (2013) 012024 doi:10.1088/1755-1315/16/1/012024

Monitoring System for Slope Stability under Rainfall by using MEMS Acceleration Sensor IC tags

S Murakami[1], A Dairaku[1], H Komine[1], N Sakai[2], T Isizawa[2], O Saito[1] and I Maruyama[3]

[1] Ibaraki University, Nakanarusawa 4-12-1, Hitachi, Ibaraki, 316-8155, Japan

[2] National Research Institute for Earth Science and Disaster Prevention, Tennodai 3-1, Tsukuba, Ibaraki, 305-0006, Japan

[3] Fukuyama Consultants Co., Ltd., Hakataeki-higashi 3-6-18, Hakataku, Fukuoka, 812-0013, Japan

E-mail:murakami@mx.ibaraki.ac.jp

Abstract. Real-time warning system for slope failure under rainfall is available to disaster prevention and mitigation. Monitoring of multi-point and wireless measurements is effective because it is difficult to conclude the most dangerous part in a slope. The purpose of this study is to propose a method of monitoring system with multi-point and wireless measurements for a slope stability using MEMS acceleration sensor IC tags. MEMS acceleration sensor IC tag is an acceleration sensor microminiaturized by a technology of Micro Electro Mechanical Systems on board IC tag. Especially, low cost of the sensor will yield to the realization of the system. In order to investigate the applicability of the proposed system, a large-scale model test of artificial slope subjected to rainfall has been performed. MEMS acceleration sensor IC tags has been located on the slope and ground acceleration caused by forced vibration has been measured until the model slope collapses. The experimental results show that the MEMS acceleration sensor IC tag is comfortably available under rainfall, the characteristics of ground accelerations varies with changing the condition of the slope subjected to rainfall, and the proposed method can be applied to a real-time monitoring system for slope failure under rainfall.

1. Introduction

Real-time warning system for slope failure under rainfall is available to disaster prevention and mitigation. Monitoring of multi-point and wireless measurements is effective because it is difficult to conclude the most dangerous part in a slope. This paper proposes a monitoring system by used multi-point and wireless sensor. The sensor is MEMS acceleration sensor IC tag. MEMS acceleration sensor IC tag is an acceleration sensor microminiaturized by a technology of Micro Electro Mechanical Systems on board IC tag as shown Fig.1.

The specifications of the sensor is shown in Table 1.Using 3.6V lithium thionyl chloride battery as C size sell, the data of triaxial acceleration at intervals of 0.05 sec can continue to be sent for approximately 96 hours. The sensor is in a waterproofing case except for a part of an antenna. The data from same MEMS acceleration sensor IC tags is wirelessly received by a personal computer connected with an antenna through an USB cable. Graphs of received time-series data are represented on the display of a personal computer on time. The specification of wireless communication between the sensor and the computer is shown in Table 2. It is necessary to locate where there is no barrier

IOP Conf. Series: Earth and Environmental Science **16** (2013) 012024 doi:10.1088/1755-1315/16/1/012024

Figure 1. MEMS acceleration sensor IC tag

Table 1. The specification of MEMS acceleration sensor IC tag

Title	Specification
Sampling frequency	200Hz
Rated capacity	+ / - 125gal
A/D resolution	16bit
Components of measurement directions	Triaxial directions (X,Y,Z)
Battery Type	C size cell
Mass (included battery and case)	125g

Table 2. The specification of wireless communication

Title	Specification
Type	Zigbee
Standards	IEEE802.15.4
Frequency	2.4GHz
Communication distance	10m – approx. 75m
Transmission rates	250kbps
Power consumption	< 60mW
Association	approx. 6,500

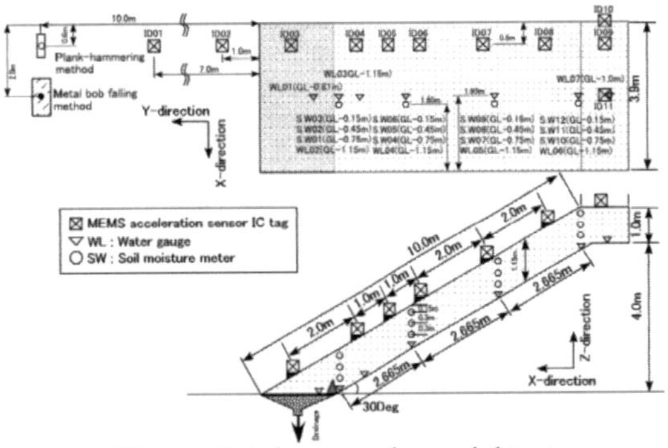

Figure 2.A large-scale model test

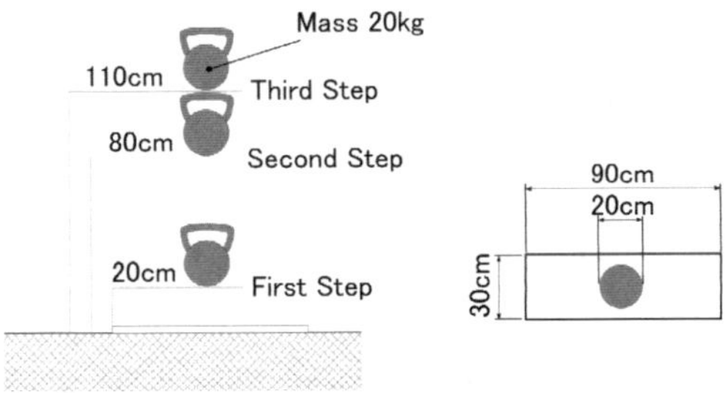

Figure 3. A metal bob falling method

between a transmitter and a receiver because 2.4GHz radio wave has a high straight line. However, the sensor is very useful because a cost of it is low and the multi-point and wireless measurements can be put into practice.

2. Large-scale model test of artificial slope subjected to rainfall

Failure of natural slopes and embankments under rainfall causes due to changing mechanical properties of soils depending on water content. Generally, strength and stiffness of soils decrease with increasing degree of saturation. Therefore, most of monitoring systems for slope failure develop by directly measuring the change of saturation degree of soils and groundwater level in the slope. However it seems that it is difficult to understand the whole condition of slope. The purpose of this study is to apply the MEMS acceleration sensor IC tags to understand variations of slope conditions under rainfall. Ground vibration characteristics of slopes will change the mechanical properties of soils with saturation degree. In order to investigate the possibility and applicability of measuring ground vibrations on a slope subjected to rainfall, a large-scale model test of artificial slope has been performed. On the basis of experimental results, the followings have been confirmed,

1. To measure ground vibrations by forced vibration that cannot affect the stability of slope

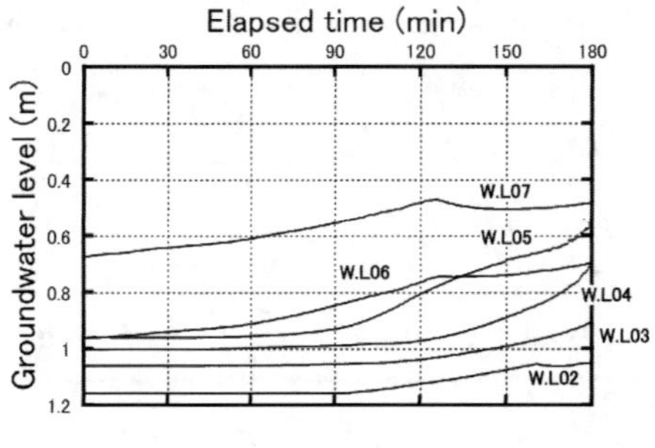

Figure 4. Rise in groundwater level with elapsed time

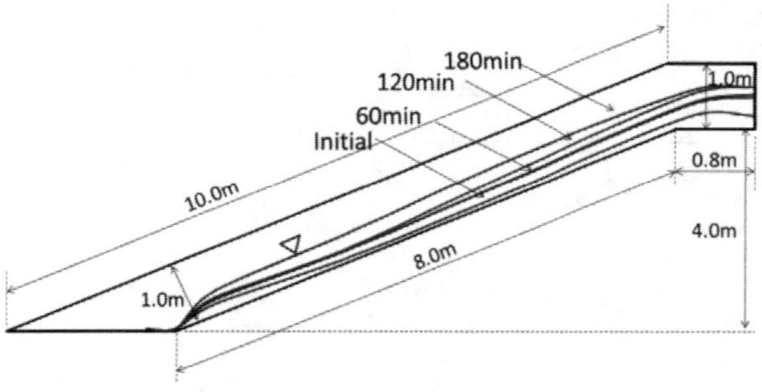

Figure 5. Groundwater level with elapsed time

2. To receive the real-time data in the rainfall.
3. To understand changes of ground vibration characteristics of a slope with elapsed time.

Figure 2 shows a diagram of a large-scale model test of artificial slope. The MEMS acceleration sensor IC tags have been set on the slope as shown in Fig. 2. ID01, 02, 09, and 10 have been located on the level surface. The boxes of the sensors set on the surface, directly. ID03, 04, 05, 06, 07, and 08 have been located on the slope face. Forced vibrations have been applied at the location where is 10m distance from the tow of the slope. Method of forced vibrations is a metal falling bob in shown Fig.3. A size of model slope is 10m in length, 3.9m in width, 1m in depth, and 30 degree in angle. Top of a slope is 0.8m in length. Bottom surface with protuberances is made by mortar and side frames are steel plates. Soil of slope is weathered granite soil so called "Masado" in Japan. Model slope has been constructed by compaction using the soil which initial water content is approximately 7.9%.

In order to make the initial condition of slope, antecedent precipitation and groundwater flow have been applied to the model slope for two week. And then, rainfall with 40mm/h has been applied until slope failure. Figure 4 shows the variations of groundwater levels with elapsed time from starting rainfall to slope failure. Groundwater level increases with elapsed time as shown in Fig. 4. Rise in groundwater level is so fast that the location is upper side of the slope. However, the incremental of rise in groundwater level in middle and lower parts of slope is large after 100min later. Figure 5 shows lines of groundwater level in the slope at 0, 60, 120, and 180 min later. The change of saturation in the slope can be understood in Fig. 5. In order to investigate the change of ground vibrations under rainfall, two kind of forced vibrations have been applied at the elapsed time.

3. Experimental results and discussion

Figure 6 and 7 show the time-series acceleration data due to forced vibrations just before slope failure. No influence of forced vibrations on the deformation and stability of slope have been confirmed by comparing between before and after the conditions of the slope. Figures show that the wireless communication was good in the rain of which intensity was 40mm/h. It is successful to apply the MEMS acceleration sensor IC tags for measuring ground vibrations because the experimental results showed forced ground vibrations did not affect the stability and deformation of the slope and real-time acceleration response has been received in the rainfall.

In order to make sure that characteristics of ground vibrations change under the rainfall, dominant frequency characteristics have been investigated. Dominant frequency has been calculated from the observed time-series records of the triaxial accelerations by using the fast Fourier transform (FFT). Number of acceleration data using FFT is 1,024 and the period is equal to 5.12 seconds. Fourier spectrum values of each forced vibrations have been calculated and the maximum value of frequency in a Fourier spectrum has been obtained. Three forced vibrations in the same conditions have been applied in model test. The dominant frequency has been determined by computing the average of the maximum values of frequency in each Fourier spectrum. Figure 8 shows the variations of dominant frequency. X-direction and Y-direction component of dominant frequency are overall larger with

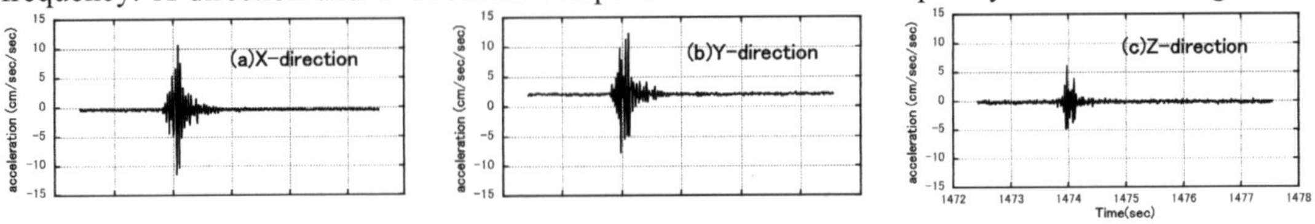

Figure 7. Time-series record by a metal bob falling method

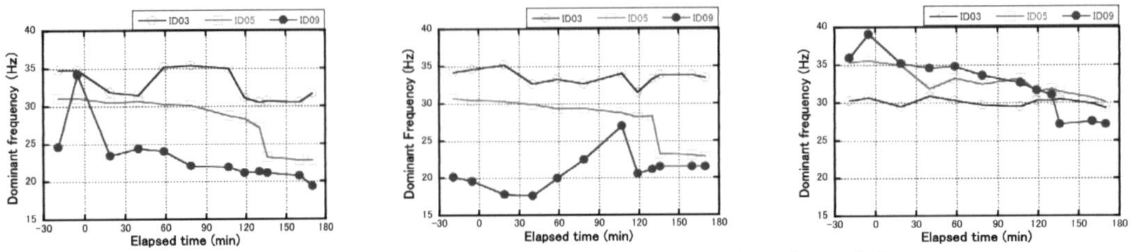

Figure 8. Variations of dominant frequency with elapsed time

crossing to the forced vibration point. And the value of dominant frequency decreases with elapsed time.

The stiffness and strength of soils decrease with increasing saturation degree. As a result, the stability of slope decreases. The saturation degree of soils continues to increase during rainfall because groundwater level in the slope did not reach to a steady-state condition under rainfall and continued to rise during rainfall as shown in Fig. 5. Therefore, the value of dominant frequency decreased because the stability of the slope reduced as the results that the stiffness and strength of soils decreased with increasing saturation degree. The results indicate it is possible to utilize the MEMS acceleration sensor IC tags for monitoring system for slope stability under rainfall.

4. Conclusions

The purpose of this study is to propose a method of monitoring system with multi-point and wireless measurements for a slope stability using MEMS acceleration sensor IC tags. A large-scale model test of artificial slope has been performed. On the basis of experimental results, the followings have been confirmed,

1. To measure ground vibrations by forced vibration that cannot affect the stability of slope
2. To receive the real-time data in the rainfall.
3. To understand changes of ground vibration characteristics of a slope with elapsed time.

The value of dominant frequency decreased because the stability of the slope reduced as the results that the stiffness and strength of soils decreased with increasing saturation degree. The results indicate it is possible to utilize the MEMS acceleration sensor IC tags for real-time monitoring system for slope stability under rainfall.

References

[1] Saito O, Kuwahara Y, Murakami S, Yasurha K 2010 Ambient Network Sensor IC-Tag Application to Geotechnical Practices *J. JGS* **58(5)** 10-13 (in Japanese)

[2] Dairaku A, Murakami S, Komine H, Sakai N, Ishizawa T, Saito O 2012 Investigation of ground vibration characteristics of slope to forced vibration for establishment monitoring method by using MEMS acceleration sensor IC tags *Proc. JSCE Annual Meeting* (in Japanese).

[3] Osaki Y 1994 New Introduction to spectral analysis of ground motion *Kajima Institute Publishing* 37-122 (in Japanese)

IOP Conf. Series: Earth and Environmental Science **16** (2013) 012025 doi:10.1088/1755-1315/16/1/012025

Effect of PID Power System Stabilizer for a Synchronous Machine in Simulink Environment

Tan Qian Yi[1], Gowrishankar Kasilingam, Raman Raguraman
Faculty of Engineering & Computer Technology, AIMST University, Kedah, Malaysia

E-mail: yvonne9011@msn.com

Abstract. This paper presents the use of Proportional-Integral-Derivative (PID) Controller with power system stabilizer (PSS) in a single machine infinite bus system. A PSS is used to generate supplementary damping control signals for an excitation system in order to damp out low frequency oscillations (LFO) of an electric power system. The paper is modelled in the MATLAB Simulink Environment to analyze the performance of a synchronous machine under a wide range of operating conditions. The functional blocks of PID controller with PSS are generated and the simulation studies are conducted based on different test cases to observe the dynamic performance of the power system. Analysis in this paper reveals that the PID-PSS gives better dynamic performance as compared to that of conventional power system stabilizer and also the optimal gain settings of PID PSS obtained at normal operating condition works well to other operating condition without much deterioration of the dynamic responses.

1. Introduction

Considerable research effort has been done for solving the instability problem of power system caused by low frequency oscillation (LFO). To reduce the damping period of oscillations, power system stabilizers (PSS) are used to inject a supplementary control signal at the voltage reference input of the automatic voltage regulator (AVR). It operates by generating an electric torque in phase with the rotor speed. In most cases, the PSS works well in damping oscillations. However, its control has less flexibility because the parameters of PSS are tuned by the original system parameters, which means the control results, are far from ideal if the operating conditions and/or structures of the system is changed. PSS uses auxiliary stabilizing signals to control the excitation system so as to improve power system dynamic performance by the damping of system oscillations. Demello and Concordia [1] analyzed the phenomena of stability of synchronous machines under small perturbations and have provided a set of guidelines for stabilization through supplementary excitation control. Kundur et al. [2] presented a comprehensive approach for conventional tuning of PSS parameters and its effect on the dynamic performance of the power system. This paper presents the use of PID Controller with PSS for a synchronous machine in a MATLAB simulink model environment. The aim is to reduce the damping of low frequency oscillations (LFO) of an electric power system under several operating conditions. The PID controller together with PSS can guarantee to provide a better result to common stabilizer situations with the effectiveness in enhancing and improving the transient ability and dynamic performance of power system [3]. Dynamic responses are also simulated when the power system is subjected to a major disturbance that would result in a significant change in system operating condition. Several simulations have been carried out in order to generate various types of condition using a single machine infinite bus power system.

2. Power System Stabilizer

[1] To whom any correspondence should be addressed.

Damping of low frequency oscillations in interconnected power system is essential for secure and stable operation of the system. Power System Stabilizer (PSS) is the most widely used device for overcoming oscillatory stability problems. PSS provides an electrical damping torque (ΔTm) in phase with the speed deviation (Δω) in order to improve damping of power system oscillations [4]. In this paper an optimal method based on the PID controller is considered to the tuning parameters of the PID type PSS. Power system stability is similar to the stability of any dynamic system, and has fundamental mathematical underpinnings. Power system stability is the ability of an electric power system, for a given initial operating condition, to regain a state of operating equilibrium after being subjected to a physical disturbance, with most system variables bounded so that practically the entire system remains intact. The basic function of PSS is to add damping to the generator rotor oscillations by controlling its excitation using auxiliary stabilizing signal(s). In order to provide damping, the stabilizer will produce electrical torque in phase with rotor speed deviations.

3. PID Controller

Proportional–integral–derivative (PID) controller is a generic control loop feedback mechanism that will correct the error between a measured process variable and the desired input by calculating and give an output of correction that will adjust the process accordingly. A PID controller has the general form: $\mathbf{u(t) = K_p\, e(t) + K_i} \int_0^t e(\tau)\, d\tau + \mathbf{K_d} \frac{de}{dt}$ where K_p is proportional gain, K_i is the integral gain, and K_d is the derivative gain. The calculation (algorithm) involves three separate parameters; the Proportional, the Integral and Derivative values. The weighted sum of these three actions is used to adjust the process for obtaining the desired output.

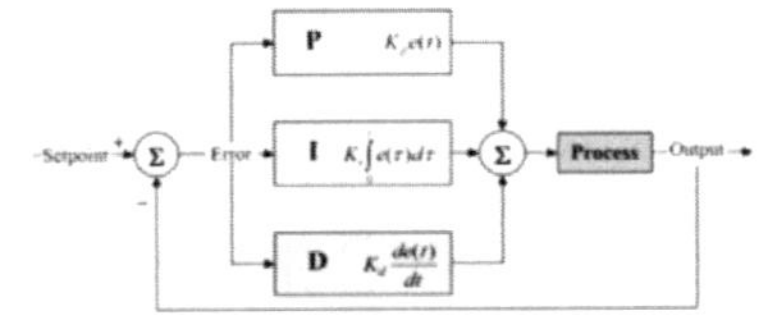

Figure 1. Block Diagram of a PID controller.

4. Modeling of PID Controller with PSS

For the performance analysis of Power System Stabilizer (PSS), a simulation model was developed in Simulink block set using MATLAB software. The effectiveness of PID Controller with PSS is observed for various operating conditions [6]. The functional block diagram of PID with PSS in Simulink environment is shown below. The generator speed deviation Δω is as the input signal of the proposed stabilizer. The objective of using PID controller together with PSS is to provide a better answer to stability problem with the effectiveness much superior to power systems utilizing. The response of the controller can be described in terms of the responsiveness of the controller to an error, the degree to which the controller overshoots the set-point and the degree of system oscillation. The various blocks involved in modeling the synchronous machine are as shown below:

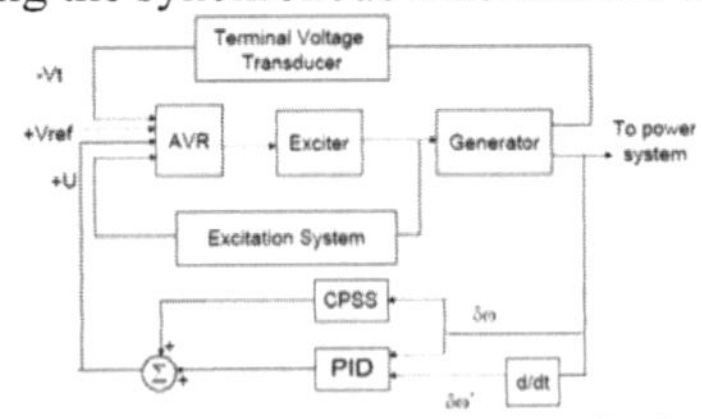

Figure 2. Block diagram of a synchronous generator excitation controller with PID PSS

5. Simulation Results and Discussion

In order to investigate the effect of the synchronous machine on proposed power system stabilizer, numeral simulations were carried out for several scenarios under different operating conditions on the single machine infinite bus power system. In the interest of brevity, we considered a system with the following three typical cases: normal load without vulnerable (fault) condition, normal load subjected a three phase fault condition and heavy load subjected to three phase fault condition.

5.1 Dynamic Performance of the system with PID-PSS

The performance of PID Power System Stabilizer (PID-PSS) was studied in the Simulink environment for different operating conditions. In order to review the significance and robustness of proposed technique, simulation studies are carried out for three phases fault disturbances and fault clearing sequence. The performance of proposed controller under transient and dynamic condition is verified using the PID and PSS. It is recognized that the highest magnitude of power system disturbance is caused by three phase fault. The use of PID PSS is able to track the system operating conditions, and thus, as seen from the results in Figure 3-6, it is able to adjust and provide a uniformly good performance over a wide range of operating conditions and disturbances.

5.1.1 Speed Deviation. Figure 3 shows the dynamic response of speed deviation with different loading conditions. Examining Figure 3, it can be observed that, PID PSS is giving a better response at nominal loading condition, both in terms of peak deviation and settling time. Further, with fault conditions, PID PSS is giving virtually similar response to that of the dynamic response of the system at nominal and also heavy loading conditions. The response with PID PSS is better in terms of settling time which means the system reached steady state faster.

Table 1. Comparison of different cases in PID PSS for speed deviation

Case	Rise Time,Tr (s)	Peak Time,Tp (s)	Settling Time,Ts (s)
Normal Load without Fault	0.2	0.59	2.76
Normal Load with Fault	0.38	0.87	3.75
Heavy Load with Fault	0.41	0.91	3.90

Table 1 shows the comparison results for the speed deviation under different loading condition based on the output obtained in Figure 3. It can be clearly seen that the settling time for normal load without fault condition is the shortest. This is because for normal loading, the speed remains constant when transmission occurs. Besides that, PID with PSS supports the synchronous generator to maintain synchronous speed even at severe fault conditions. It can also provide better damping characteristics than the PSS even at the loading condition. The damping characteristic of PID PSS is insensitive to load change while that of PSS will deteriorate as the load changes.

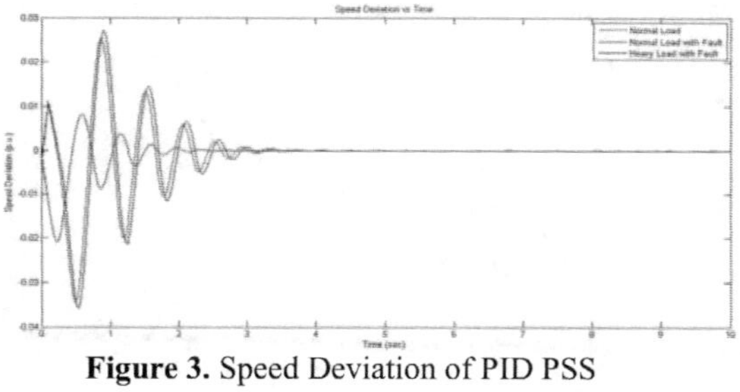

Figure 3. Speed Deviation of PID PSS

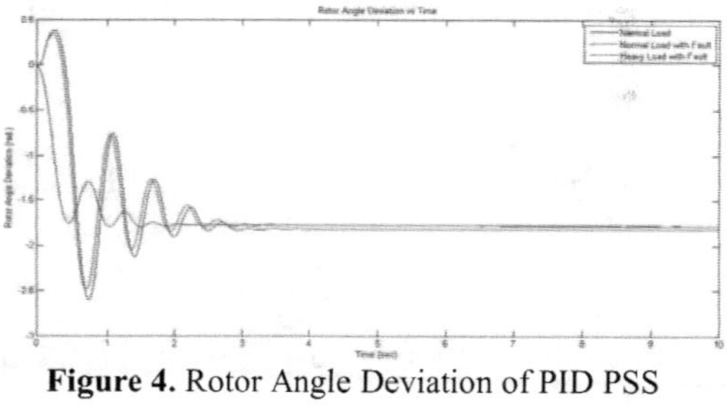

Figure 4. Rotor Angle Deviation of PID PSS

5.1.2 Rotor Angle Deviation. For the rotor angle deviation with different loading condition, the simulation results are shown in Figure 4. It is seen that the acceleration of rotor increases due to fault condition. However, the rotor angle maintained at normal level with the help of PID PSS and also settled down faster compared to PSS. PID PSS improves the rotor angle to the maximum extent by reaching the settling time before 3 seconds.

5.1.3 Load Angle. According to Figure 5, the load angle is maintained at 8 degree for all the three cases after they reached steady state. The settling time for normal load is much shorter than nominal and heavy loads with fault conditions. For this case, it is inferred that PID with PSS supports the synchronous generator to maintain synchronous speed even at severe fault conditions.

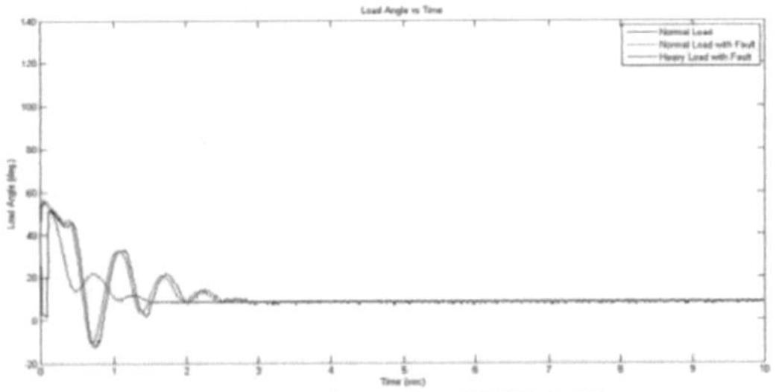

Figure 5. Load Angle of PID PSS

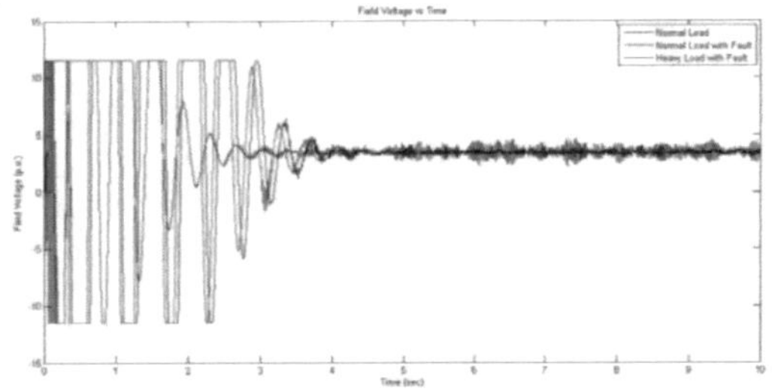

Figure 6. Field Voltage of PID PSS

5.1.4 Field Voltage. Figure 6 illustrates the result of field voltage. It is observed that the normal load produce less overshoot and settle faster compare to the other two cases. By this effect, the field voltage will be stable and in turn it ensures the system stability. We can also conclude that the combination of PID with PSS provides better solution for damping of oscillations. The dynamic performance of the system with PID PSS is better, both in terms of peak deviation and settling time. The performance of the system with PID PSS is also quite robust in wide variations of loading conditions.

6. Conclusions

In this study, the proportional-integral-derivative power system stabilizer (PID-PSS) has been proposed for the enhancement of the dynamic stability of single machine infinite bus. Results from this study indicate that PID PSS gives much better dynamic performance as compare to that of conventional PSS. An excellent improvement in the damping over a wide range of operating conditions has been achieved with PID PSS being used in the excitation control of synchronous machine. Modeling of proposed controller in Simulink environment provides an accurate result when compared to mathematical design approach. Simulation studies described in the paper show that the dynamic performance of the system with PID-PSS is quite robust to wide variations in loading conditions. This results from the fact that, during the faulted period, there will be drastic change in the operating condition. Therefore the proposed PID PSS are relatively simpler than the other controllers to provide better optimal solution and also for practical implementation.

7. Acknowledgements

I would like to express my gratitude to all those who provided me to possibility to complete my paper. Thanks to my dean of faculty and my family members for giving me support and my supervisor for always give me inspirational supervising throughout the research and paper writing process.

References

[1] Demello F P, Concordia C 1969 *IEEE Transactions on power apparatus and systems* **88(4)** 316-328
[2] Kundur P, Klein M, Rogers G J, Zywno M S 1989 *Power Systems, IEEE Transactions on* **4(2)** 614-626
[4] Chi-Jui W, Yuan-Yih H 1988 *Power Systems, IEEE Transactions on* **3(3)** 1059-1064
[5] Kundur P 1994 *Power system stability and control* **4(2)** N. J. Balu, M. G. Lauby (Eds.) New York: McGraw-hill
[6] Gowrishankar K 2012 *International Journal of Automation and Power Engineering* **1(5)** 137-144

Use of Oil Palm Waste as a Renewable Energy Source and Its Impact on Reduction of Air Pollution in Context of Malaysia

Shahida Begum[1], Kumaran P., Jayakumar M.

Centre for Advanced Materials, Department of Mechanical Engineering, College of Engineering, Universiti Tenaga Nasional, Putrajaya, Malaysia

E-mail: Shahida@uniten.edu.my

Abstract. One of the most efficient and effective solutions for sustainable energy supply to supplement the increasing energy demand and reducing environment pollution is renewable energy resources. Malaysia is currently the world's second largest producer and exporter of palm oil and 47% of the world's supply of palm oil is produced by this country. Nearly 80 million tonnes of Fresh Fruit Bunches (FFB) are processed annually in 406 palm oil mills and are generating approximately 54 million tonnes of palm oil mill effluent (POME), known to generate biogas consisting of methane - a Green House Gas (GHG) identifiable to cause global warming. This is 21 times more potent GHG than CO_2. These two major oil palm wastes are a viable renewable energy (RE) source for production of electricity. If the two sources are used in harnessing the renewable energy potential the pollution intensity from usage of non-renewable sources can also be reduced significantly. This study focused on the pollution mitigation potential of biogas as biogas is a renewable energy. Utilization of this renewable source for the production of electricity is believed to reduce GHG emissions to the atmosphere.

1. Introduction

Energy resource is the economic driving force and correspondingly, energy production and utilization is also blamed for the environmental degradation. Malaysia has been observing a relatively high growth rate of energy demand at 9.5% per annum over the past decades and is expected to increase rapidly to 100MTOE (million ton of oil equivalent) to have a developed nation status by 2020 (Asian Pacific Economic Cooperation , APEC 2006). The energy demand is met by both renewable and non-renewable energy sources and the country is abundantly endowed nonrenewable resource such as oil (approximately 3 billion barrels), natural gas (1.61 trillion cubic meters) and coal (776 million tonnes) as well as renewable energy sources like hydro and solar power and biomass (Joanta, 1996). The most usage of primary energy resource goes into electricity production for which the maximum electricity energy demand projections are 40,515 MW for the year 2020 (Yusoff, 2006). On the other hand the gradual deregulation of natural gas prices, with price increase from RM 6.40 per MMBTU to RM 14.31 per MMBTU for power sector in 2008 has often brings natural gas to the downside and it is expected that current gas fields to be depleted by 2027 with new fields of higher carbon dioxide content. In addition, there is a proven declination of oil reserves by 3.0 billion barrels in January 2007, which was at a peak of 4.6 billion barrel in 1996 (Mustapa et al. 2010 and Ralph, 1979). According to (Gan and Li, 2008), Malaysia's total primary energy consumption and Carbon emissions will triple by 2030, since large amount of the energy demand would have to be supported by coal that is a least cost

[1] To whom any correspondence should be addressed.

option. On the other hand, during the COP15, Malaysia has also committed to reduce its carbon intensity by 40% of its 2005 value, which is going to be an uphill task with growing energy intensity of Malaysia and dwindling natural gas resources and increasing coal utilization (Badariah, 2010). One of the solutions would be aggressive and affirmative deployment of available RE resources to ensure supply sustainability and environment conservation. In the recently gazette National RE Act 2011, the share of RE is expected to be 11% and 17% of total electricity generation mix by 2020 and 2030 respectively. This would translate to cumulative CO_2 avoidance of 42.2 and 145.1million tonnes for year 2020 and 2030 respectively.

This paper explores the potential of biomass and biogas generated form Malaysian palm oil industry, as one of the promising alternative energy source and also its impact to mitigate the environment air pollution due to its utilization.

2. Sources of Air Pollution and Air Quality Trend in Malaysia

In achieving industrial country status by the year 2020, Malaysia has undertaken various economic growth activities, which imposed costs in terms of industrial pollution and the degradation of urban environment in several large cities that does not meet national ambient air quality standards. The Air Quality Guidelines due to repeated haze episodes in Southeast Asia are also implemented by Malaysian Government. Suspended Particulate Matter (SPM), nitrogen dioxide (NO_2), CO, O_x, SO_2, and Pb were the major pollutants in several big cities in Malaysia. Mobile, stationary and open burning are the three major sources of air pollution in Malaysia as reported by (Department of environment, 2000). The registration of mobile vehicles such as personal cars, commercial vehicles, and motorcycles had been increased by 26% from year 1996 to 82 % in year 2000 (Department of environment, 2001). Enforcement of the Environmental Quality Act (EQA), the phase-out of leaded gasoline sales and focus on prevention of emissions at the manufacturing and assembly stage had improved the situation. As a result there was a steady decrease in lead levels throughout the western side of peninsular Malaysia.

However, the increase in the sheer number of motor vehicles on the road may offset the result. Stationary sources are related to industry such as power stations produces 9%, industrial fuel burning processes 3% and the rest is domestic fuel burning. In 51 locations of Malaysia, main air pollutants are monitored continuously, while lead concentrations are measured once in every six days at two locations. The air quality trend is then calculated by averaging direct measurement from monitoring sites on a yearly basis and compared with the Malaysian Ambient Air Quality Guidelines.

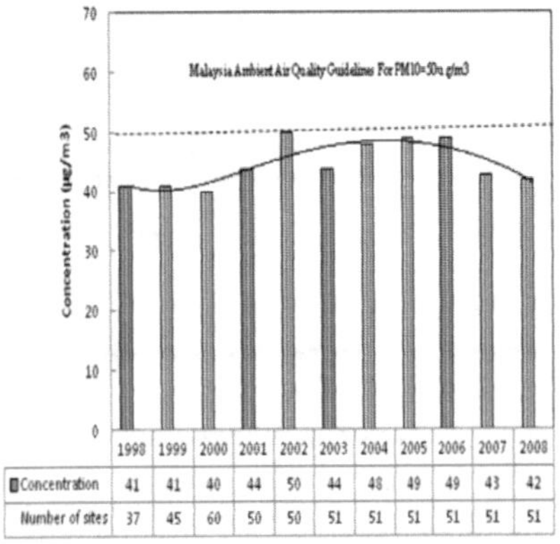

Figure 1. Malaysia: Annual Average Concen. of Particulate Matter (PM_{10}), 1998-2008

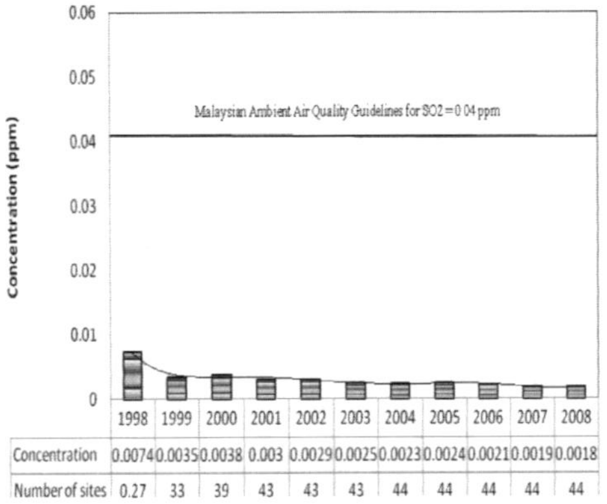

Figure 2. Malaysia: Annual Average Concen. of Sulphur Dioxide (SO_2), 1998-2008

Major air pollutants in different locations of Malaysia such as Klang Valley, West Coast Peninsular Malaysia, East Coast Peninsular Malaysia, Sabah, Labuan and Sarawak are monitored routinely. The six major pollutants as presented in Figures. 1-5 are within the range of Malaysian air quality guidelines. The particulate matter is the predominant pollutant. The concentration of NO_2 has increased by 6% from 2007 to 2008. It remains high in industrial areas and urban areas mainly due to the increase of motor vehicles and combustion process of industrial source. CO emission was increased by 5% from 2007 to 2008. In urban areas the concentration of CO was higher, where the sources of emissions were motor vehicles, which contributes 97% of CO emission (Department of environment, Malaysia, 2008).

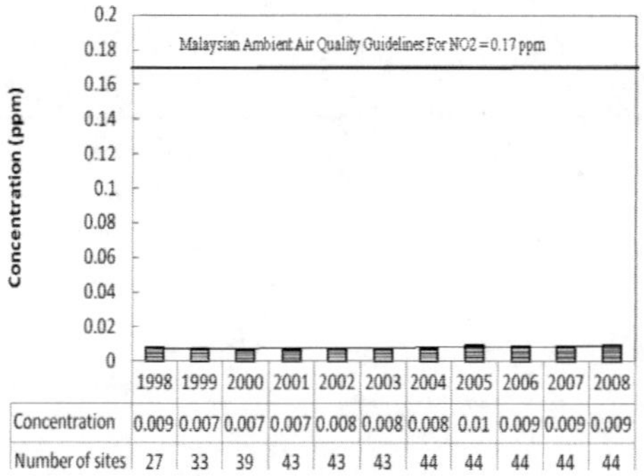

Figure 3. Malaysia: Annual Average Concen. of Nitrogen Dioxide (NO_2), 1998-2008

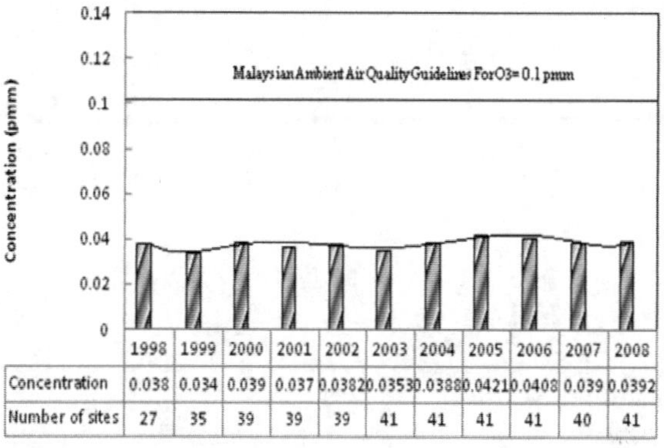

Figure 4. Malaysia: Annual Average Daily Maximum Hour Concen. of Ozone (O_3), 1998-2008

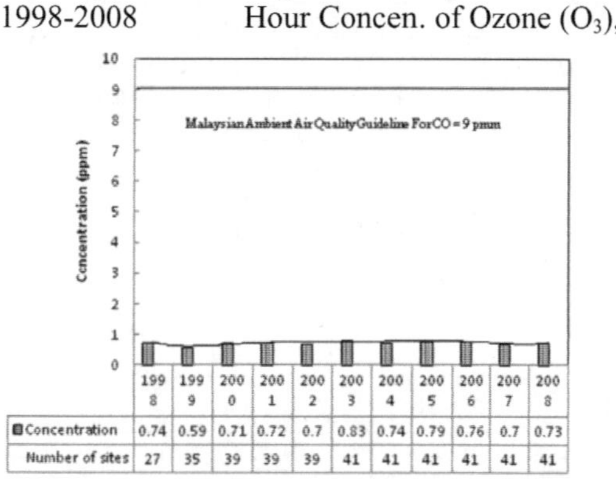

Figure 5. Malaysia: Annual Average Concen. of Carbon Monoxide (CO), 1998-2008

3. Potential Green House Gas Emission Reduction by Biomass & Biogas

Although palm oil waste can be a huge source of energy as data presented in Tables 1but biogas utilization has not been implemented fully in Malaysian Industry. By implementing biogas plant from palm oil wastes GHG gas emission can be reduced significantly as estimated in different project proposals (SSC-CDM-PDD), UNFCCC, version 8 2009).

4. Biomass and Bio-Gas: A Promising Alternative

Malaysia is the world leader in the production and supply of crude palm oil and the plantation industry is projected to grow steadily (Lee, 2009). Consequently, the growth produces more palm oil waste such as fibers, and palm kernel shells. The generated biomass and its heat value are presented in Table 1.

The biogas produced from anaerobic digestion can be captured and utilized as RE to replace fossil fuel/diesel for steam or electricity generation. It has been estimated that a 60 tons FFB/ hr mill, will generate about 12,000 m^3 of gas/day and the energy that can be generated is estimated to be 1.04 MW capacity (Joanta, 1996). The total 261.1MW installed capacity of power from the potential of 406 mills would generate 1.88MWh of electricity (NKEA 2011). It is worth to mention that the CO_2 emissions for the generation of electricity by electric power plants by coal or oil is 1100 g of CO_2 per kW h, whereas, the figure is 600 g of CO_2 per kW h for using gas and biomass use can reduce it dramatically to 16 g of CO_2 per kW h (Yusoff, 2006; Bazmi *et al.*, 2011).

In recent years, a project has been successfully undertaken by Serting Hilir Palm Oil Mill which captures methane and the existing open anaerobic digester tanks in the mill are converted to closed type digester tanks. It is estimated that 647 kW and 1,294 kW of electricity can be generated respectively at phase I and phase II. The emission reduction of methane from open lagoon system can thus be utilized to reduce GHG especially equivalent carbon dioxide. The report by Saad M.F.M., 2010 on this project has shown that annual reduction of 37,251 million tonnes of CO_2 equivalent GHG can be achieved.

Table 1. Biomass generated by palm oil mill

Biomass	Quantity (million tonnes)	Moisture content (%)	Oil content (%)	Heat value (dry) (kJ/kg)
EFB	18.25	67	5	18,883
Fibre	11.11	37	5	19,114
Shell	5.55	12	1	20,156
POME	53.16	93	1	17,044

4 Conclusions

A holistic approach must be undertaken for the promotion of biogas as a renewable energy and for implementation of green energy development. It can be a sustainable source of energy and can play a major role to supplement energy demand, reduce the GHG and mitigate global warming. Affirmative actions need to be taken to reduce environmental effect of nonrenewable energy sources by the GHG gas emissions. Methane generated from effluent of palm oil processing waste is 21 times more potent than CO_2. Hence, it is viable to capture the biogas and utilize to replace non renewable energy to generate electricity which has annual reduction of 37,251 million tonnes of CO_2 equivalent. However, a more detailed study needs to be done to improve on the production of biogas from palm oil waste.

References

[1] APEC energy demand and supply outlook 2006
[2] Badariah A M 2010 Renewable Energy Development in Malaysia *34th APEC Expert Group on New & Renewable Energy Technologies* Kuala Lumpur, Malaysia
[3] Bazmi A A, Zahedi G, Hashim H 2011 *Renewable and Sustainable Energy Reviews* **15** 574–583
[4] UNFCCC 2009 *Clean development mechanism: Simplified project design document for small-scale project activities (SSC-CDM-PDD)* **Version 8**
[5] Department of environment, Malaysia, 2000, 2001, 2008
[6] Gan P Y, Li Z D 2008 *Energy Policy* **36(2)** 890-899
[7] Joanta H G 1996 *Renewable energy systems in Southeast Asia, Malaysia* PennWell Publishing Company 81-122
[8] Lee J S 2009 *6th Biomass Asia Workshop, Hiroshima Japan* 18-20
[9] Saad M F M 2010 Techno Economic Analysis of Biogas production from Oil Palm Waste for the Generation of Electricity *Undergraduate thesis* Universiti Tenaga Nasional
[10] Ralph M R 1979 *Growth in Global energy demand and contribution of alternative supply system, Energy* **4** Pergamon Press Ltd. 881-890
[11] Mustapa, S I, Peng L Y, Hashim A H 2010 Issues and challenges of renewable energy development: a Malaysian experience *In PEA-AIT international conference on energy and sustainable development: issues and strategies*
[12] NKEA 2011 *National Key Economic Area – EPP5 National biogas implementation "Biogas capture and CDM project implementation for Palm Oil Mills"*
[13] Yusoff S 2006 *J. of Cleaner production* **14** 87-93
[14] Zain-Ahmed A 2008 *Contemporary issues in energy and buildings in Malaysia: Focus on R & D and policies, SENVAR+IAESEE: Humanity +Technology* 15-23

IOP Conf. Series: Earth and Environmental Science **16** (2013) 012027 doi:10.1088/1755-1315/16/1/012027

Increase in fault ride through capability of direct drive permanent magnet based wind farm using VSC-HVDC

Hesamaldin Maleki[1], V. K. Ramachandaramurthy, Moein Lak

[1]*Power Quality Research Group, Department of Electrical Engineering, Universiti Tenaga Nasional, 43000, Kajang, Selangor, Malaysia*

E-mail: h.maleki@ieee.org

Abstract. Burning of fossil fuels and green house gasses causes global warming. This has led to governments to explore the use of green energies instead of fossil fuels. The availability of wind has made wind technology a viable alternative for generating electrical power. Hence, many parts of the world, especially Europe are experiencing a growth in wind farms. However, by increasing the number of wind farms connected to the grid, power quality and voltage stability of grid becomes a matter of concern. In this paper, VSC-HVDC control strategy which enables the wind farm to ride-through faults and regulate voltage for fault types is proposed. The results show that the wind turbine output voltage fulfills the E.ON grid code requirements, when subjected to three phase to ground fault. Hence, continues operation of the wind farm is achieved.

1. Introduction

Direct Drive Permanent Magnet wind Turbine (DDPMT) has been introduced to the world since 1991. Many researches have presented the control strategies for this type of wind turbine. In [1], the control system using A Neutral-Point Clamped Converter (NPC) System for Direct-Drive Variable-Speed (DDVS) Wind power unit is presented. The machine-side NPC converter provides torque-speed control of the synchronous generator via vector-control strategy. In [2] and [3], the control of a standalone variable speed wind turbine with a permanent magnet synchronous generator is presented. The results show that the maximum power is tracked via the generator-side converter. On the other hand, the absence of the grid makes the results less reliable. In this paper, continues operation of the wind farm PMSG-based wind turbine connected to the grid is investigated.

1.1. VSC-HVDC for offshore wind farm

Long distance between generation units such as offshore wind farms and load makes HVDC systems more preferable than HVAC systems to transmit bulk power over long distances.

1.2. E.ON grid code fault ride through requirements

The aim of fault ride through (FRT) capability for the wind farms is to increase the reliability of the wind turbines connected to the grid and to make sure that the wind turbines have no diverse effect on the grid operation [4]. There are different FRT curves and their specifications vary from country to country. Likewise Germany's standard, E.ON grid code [5], which is used in this paper to clarify the accuracy of system response.

[1] To whom any correspondence should be addressed.

2. Power system modelling

In this paper, FLT for the wind farm is done via VSC-HVDC system. By implementing two stages 6-pulse Insulated-Gate Bipolar Tyristors (IGBT) at the generator and grid side of the power system, two Voltage Source Converter (VSC) stations are simulated in PSCAD/EMTDC software. Control strategy is shared between stations in order to improve the fault ride through and voltage regulation for the wind farm. Figure 1 shows the power system presented in this paper. Using 20 sets of 2 MW 11kV wind turbines, a 40 MW offshore wind farm is simulated in PSCAD/EMTDC. The power is then transferred via HVDC lines and connected to the grid.

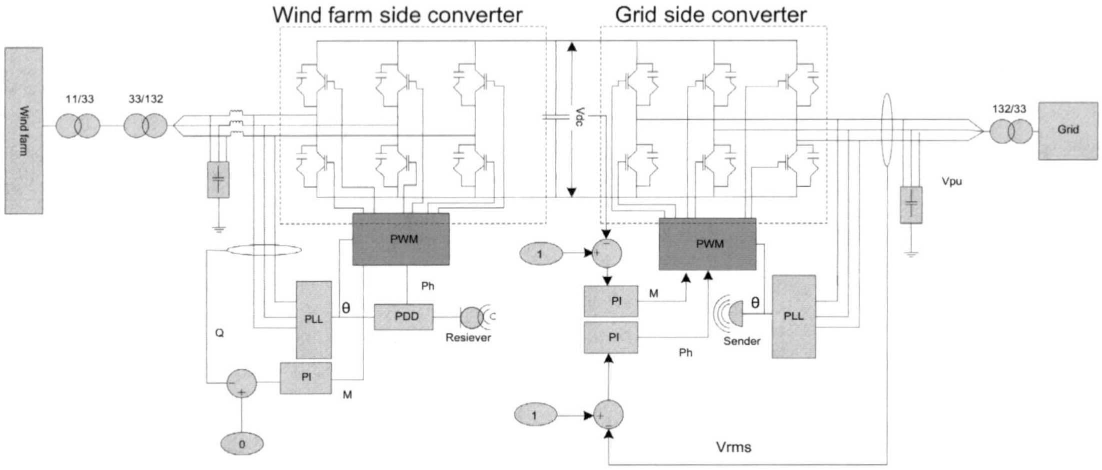

Figure 1. Configuration of the system simulated in PSCAD/EMTDC

Where, Ph1 is the desired phase angle and M1 is the desired voltage magnitude for generator side converter,PH2 is the phase angle andM2 is the voltage magnitude for grid side converter

2.1. Generator side VSC control strategy

Equations 6 and 7 show the principle of this control strategy.

$$P = \frac{U_1 U_2 sin\theta}{X} \tag{1}$$

$$Q = \frac{U_1^2 - U_1 U_2 cos\theta}{X} \tag{2}$$

Where, P and Q are the active and reactive powers between two electrical nodes in AC systems with voltage magnitudes, U1 and U2. θ is the phase-angle difference and X is the line reactance between the two nodes [6]. Equations (1) and (2) show that the active power is mainly controlled by controlling the phase angle difference while, the reactive power is more affected by the magnitude of the voltage. Equation (3) shows the mathematical equations for reactive power control strategy used in this paper.

$$\mathrm{M} = \left(K_p^Q + \frac{K_i^Q}{s}\right)\left[Q_{ref} - Q\right] \tag{3}$$

Where, M is the changes in the voltage magnitude, K_p^Q is the proportional gain and K_i^Q is the integral gain of the PI controller. Q represents the actual reactive power and Q_{ref} presents the reference reactive power. The reactive power is controlled by changing the magnitude of the voltage. A proportional-integral (PI) controller is used to generate the desired magnitude of the voltage (M1). Moreover, equation (4) shows the mathematical equations used to control the real power generated by the wind turbine.

$$\mathrm{Ph} = \left(K_p^p + \frac{K_i^p}{s}\right)[P_{ref} - P] \quad (4)$$

Where, Ph presents the desired phase shift, K_p^p is the proportional gain and K_i^p is the integral gain of the PI controller. P_{ref} is the reference power and P is the actual power of the wind farm [6]. The control strategy is shown in figure 2.

2.2. Grid side converter control system

Control strategy for grid side convertor is presented in figure 3. Grid side converter control system is responsible for controlling the phase and the magnitude of the voltage in the grid side. Voltage magnitude is controlled by controlling the DC voltage. The desired voltage phase is achieved by controlling the phase of the voltage in the grid side. The error signal is fed to the PI controller to determine the magnitude of the control signal (M2). Moreover, the per unit voltage in the grid side is compared with the reference value to determine the voltage phase angle (Ph2) at the grid side.

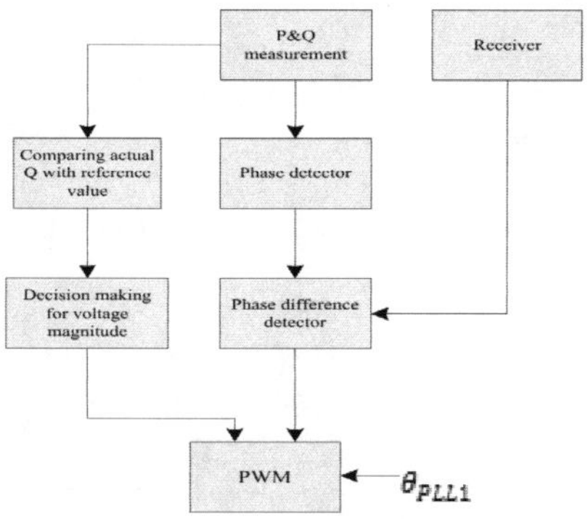

Figure 2. The generator side control strategy

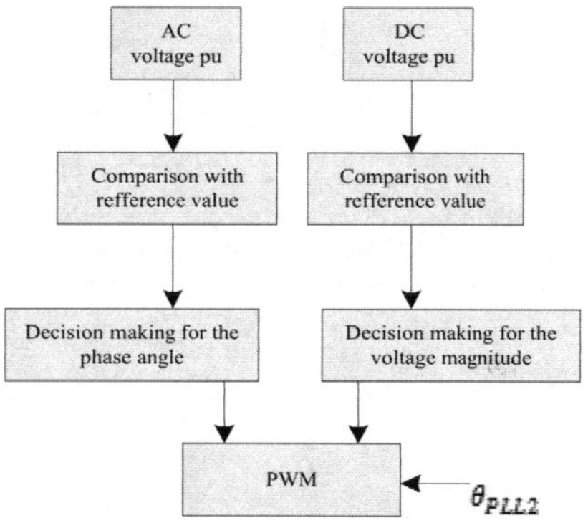

Figure 3. The grid side control strategy

3. Simulation results

A three phase to ground fault is initiated at the grid side to investigate the performance of the VSC-HVDC. At 5s, the fault is applied and is cleared after 150ms. During the fault, the voltage in grid side drops to 0 kV and remains at 0 kV for 150 ms. The DC voltage, which can be controlled to 150kV during the fault, has some oscillation at the beginning of the fault and at the fault clearance. Figure 4 shows the VSC-HVDC response to the three phase fault. The E.ON grid code voltage requirement is also presented in figure 4.d. From the simulation results, it can be seen that the active power is reduced during the fault time and the reactive power is increased and injected to the Point of Common Coupling (PCC) to reduce the voltage sag at the wind farm side.

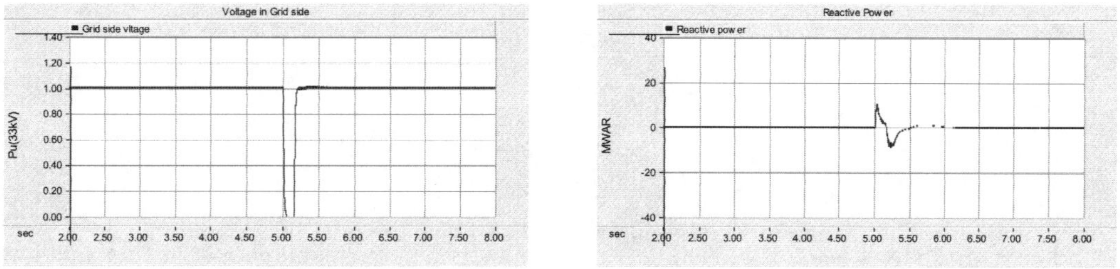

(a) (b)

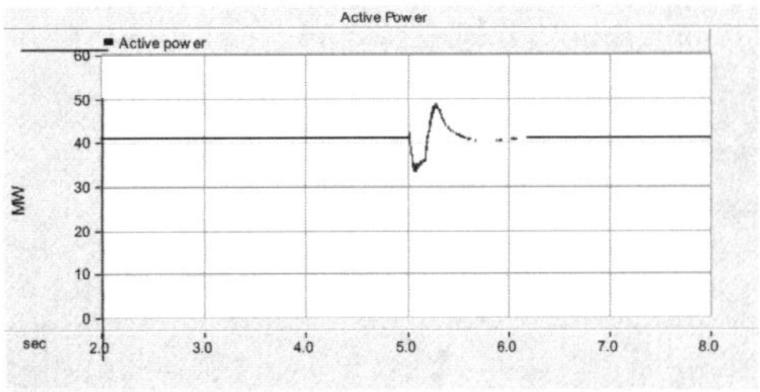

(c)

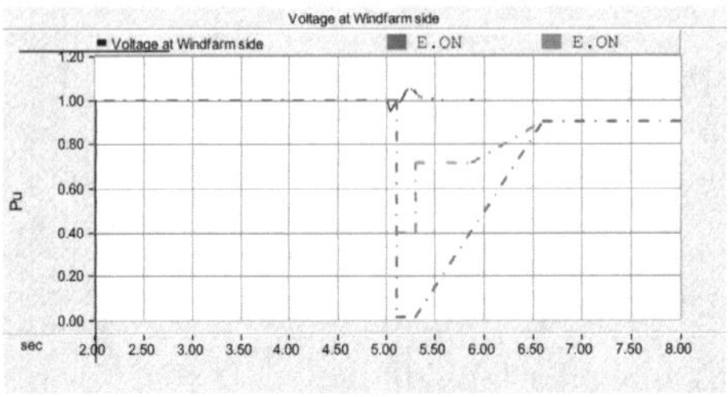

(d)

Figure 4. System response to three phase fault AC voltage at Grid side (b) Reactive power produced by wind farm (c) Active power produced by wind farm (d) AC Voltage at wind farm side

4. Conclusion

In this paper, the proposed control strategy for VSC-HVDC system shows the capacity to increase the FRT of the offshore wind farm DDPM generator based. The results are validated with E.ON standard grid code. Proposed VSC-HVDC control system ensures that the wind farm stays connected during the fault, thus keeping the system stable.

References

[1] Yazdani A, Iravani R 2006 *Energy Conversion, IEEE Transactions on* ***21*(2)** 596-607
[2] Wang X, Vilathgamuwa D M, Tseng K J, Gajanayake C. J. 2009 *Power Electronics and Drive Systems, 2009. PEDS 2009. International Conference on* 752-757
[3] Haque M E, Negnevitsky M, Muttaqi K. M. 2010 *Industry Applications, IEEE Transactions on* ***46*(1)** 331-339
[4] Tsili M, Papathanassiou S 2009 *Renewable power generation, IET* ***3*(3)** 308-332
[5] E.ON Netz Gmbh 2006 *Grid connection regulation of high voltage and extra high voltage*
[6] Zhang L 2010 *Modelling and control of VSC-HVDC links connected to weak ac systems* Doctoral dissertation, KTH

IOP Conf. Series: Earth and Environmental Science **16** (2013) 012028 doi:10.1088/1755-1315/16/1/012028

Probabilistic Evaluation of Wind Power Generation

N.M. Muhamad Razali[1], **Muizzuddin Misbah**
Universiti Tenaga Nasional,
Jalan IKRAM-UNITEN, 43000 Kajang, Selangor, Malaysia

E-mail: noormiza@uniten.edu.my

Abstract. The power supplied by wind turbine generators (WTG) is widely random following the stochastic nature of weather conditions. For planning and decision making purposes, understanding and evaluation of the behaviour and distribution of WTG's output power are crucial. Monte Carlo simulation enables the realization of artificial futures by generating a huge number of sample paths of outcomes to perform this analysis. The paper presents an algorithm developed for a random wind speed generator governed by the probability density function of Weibull distribution and evaluates the WTG's output by using the power curve of wind turbines. The method may facilitate assessment of suitable turbine site as well as generator selection and sizing.

1. Introduction

Wind energy has enjoyed robust growth in recent years, driven by a combination of factors; for example advancements in wind technologies, falling capital costs, abundance of the free resource and a number of government initiatives designed to encourage wind energy. Fuelling much of the government support for wind energy are environmental concerns. Although the impact of wind energy on the environment is not negligible, it is much lower than those of current fossil fuel technologies.

During generation planning activities, it is crucial for planners to take into account as many uncertainties and economic factors as possible that may affect the physical construction, financial obligations and operation of the power plant. Risk management practice is normally based on extensive simulations and analysis specific to the fuel type and role of each generation unit and the local grid system. Wind generators however pose an additional challenge, as the power supplied is widely random following the random nature of weather conditions.

Understanding of wind potential will facilitate estimation of how much energy will be produced. Conventional method for determining suitable wind turbine location is by using "wind power classes" which are dependent on the average wind speed and the wind power density indicating how much energy is available at the site for conversion by a wind turbine.

Wind resources are rarely consistent and vary with time, season, terrain type, height above ground, and from year to year, thus thorough investigation is needed before planning any exploitation [1]. Due to the stochastic nature of the wind resource, a probabilistic approach during investment and operational planning may be much more appropriate than the deterministic approach normally used by electricity utilities [2]. Therefore, knowledge of the wind speed frequency distribution is very important to evaluate the wind turbine potential and economic feasibility of any particular site. This paper presents a methodology applying probabilistic analysis to evaluate the WTG output which can assist in site and turbine selection for wind-type generation planning. Monte-Carlo simulations

[1] To whom any correspondence should be addressed.

IOP Conf. Series: Earth and Environmental Science **16** (2013) 012028 doi:10.1088/1755-1315/16/1/012028

governed by the PDF of the wind distribution at a particular location imitated the uncertainty of average hourly wind speed. Simulations were conducted on MATLAB platform and utilized the in-built Distribution Fitting Tool and Statistical Toolbox. The wind speed data is based on a coastal location in Malaysia.

2. Probability Distribution of Wind Speed

There are several PDFs which can be used to describe the wind speed frequency curve; the Weibull, Rayleigh and Lognormal, as these functions are able to model skewed distributions. The Weibull can be positively or negatively skewed while the Lognormal can only be positively skewed. On the other hand, Rayleigh is a special case of Weibull function with shape parameter of 2. In other words, the Weibull distribution is a generalization of the Rayleigh distribution. Among the three, Weibull is thus more flexible in describing probability distribution of wind speed which varies from location to location. The general type Weibull distribution consists of three parameters, the shape, scale and location parameters. The location parameter is often not considered as it has zero value when the distribution starts at zero in the x-axis, which is the case with wind speed as the lowest recorded speed is usually 0 m/s. Therefore, the two-parameter Weibull distribution has been the most widely used and accepted model in the specialized literature on wind energy [3] and is normally a good fit to the experimental data.

For this study, the wind speed data used to obtain the WTG output are based on the January wind profile at Mersing, Johor, Malaysia [4]. From the historical data at Mersing, it was determined that the Weibull parameters for the month of January are k = 2.776 and λ = 4.900 where k is the Weibull shape parameter and c is the Weibull scale parameter, therefore, this yield the following Weibull PDF:

$$f(x) = \begin{cases} \frac{2.776}{4.9}\left(\frac{x}{4.9}\right)^{1.776} e^{-\left(x/4.9\right)^{2.776}} & x \geq 0 \\ 0 & x < 0 \end{cases} \tag{1}$$

where x is the wind speed. The Weibull shape parameter of Mersing approximate a normal distribution as is falls between $2.6 < k < 3.7$. Noting that the historical data collected at the observation site is at the anemometer hub height, the speed is converted to wind speed at wind turbine generator hub according to the following equation [5]:

$$V_{WTG} = V_r\left(\frac{h_{WTG}}{h_r}\right)^b (1 - 0.01\beta) \tag{2}$$

where,

h_{WTG} is the wind turbine generator hub height in m;
h_r is the reading height for historical wind data in m;
V_r is the historical wind speed data in m/s;
b is the roughness coefficient, and
β is the turbulence coefficient

A very low value of roughness coefficient (e.g. in flat grassy plains) results in almost negligible wind speed difference between h_{WTG} and h_r while a high value (e.g. in cities and forests) results in significant wind speed difference.

3. The Developed Algorithm

The flowchart for the developed algorithm is presented in figure 1. The methodology employs the concept of Monte Carlo simulation which calculates numerous scenarios of a model by repeatedly picking values from a user-predefined probability distribution for the uncertain variables and using those values for the model [6]. As all those scenarios produce associated results in a model, each scenario can have a forecast. Forecasts are events (usually with formulas or functions) that are defined as important outputs of the model. Sufficient number of repetitions or iterations is required to arrive at a statistically viable result.

For each simulated day, the wind turbine's Forced Outage Rate (FOR) of 2% [7] is taken into account. In order to model the generator availability, for each day, a uniform random variable, m with value $0 \leq m \leq 1$ is generated and compared with the threshold of the FOR, $a = 0.02$. If the realization of variable m is less than or equal to a, the wind generator is unavailable as it is subjected to forced outage. On the other hand, if the realization exceeds a, the algorithm generate stochastic wind speed for each hour of the day and convert this speed to generator output using a specified power curve. In this study, the Goldwind 1.5 MW wind turbine model has been used for simulation purposes.

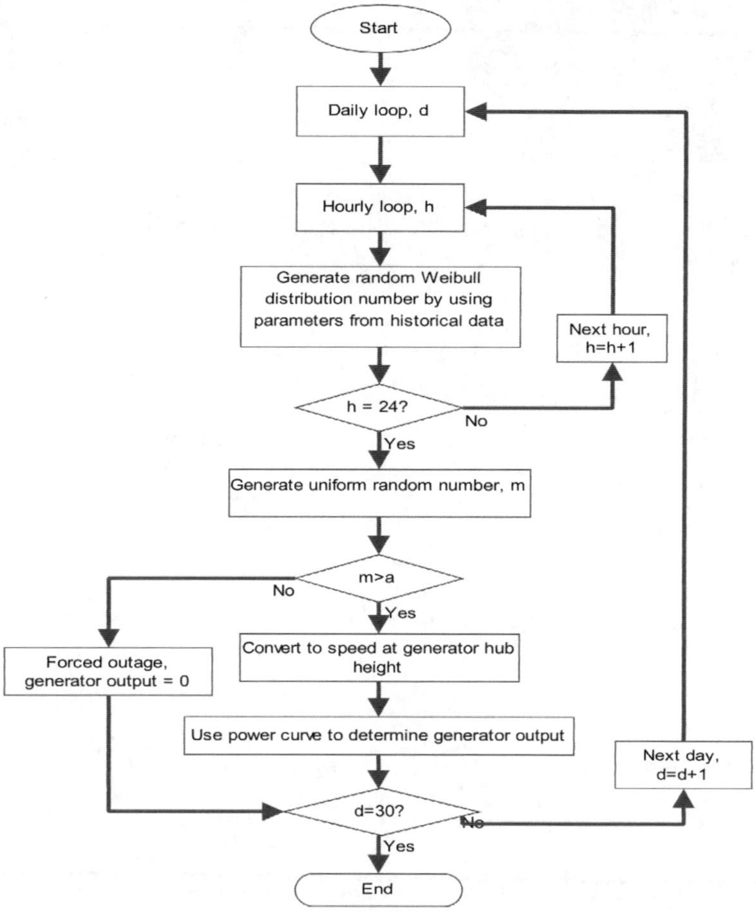

Figure 1. The flowchart for wind turbine output

4. Results and Discussion

Based on the Cumulative Distribution Function (CDF) of the generated wind speed in figure 2, it was found that 92.1% of the time, the wind speed is less than 6.0 m/s which is lower than the rated wind speed of the turbine. From figure 3, it can be observed that 20% of the time, the wind turbine generator (WTG) produces zero output including the time it was put under forced outage. Only 1.11% of the time the WTG produced the maximum output of 1500 kW. However, more than 40% of the time, the WTG produce output more than 80% which is 1200 kW even though the wind speed is low. This is the result from using a low-speed wind turbine with rated wind speed of 6.5 m/s which is efficient in low-wind speed area.

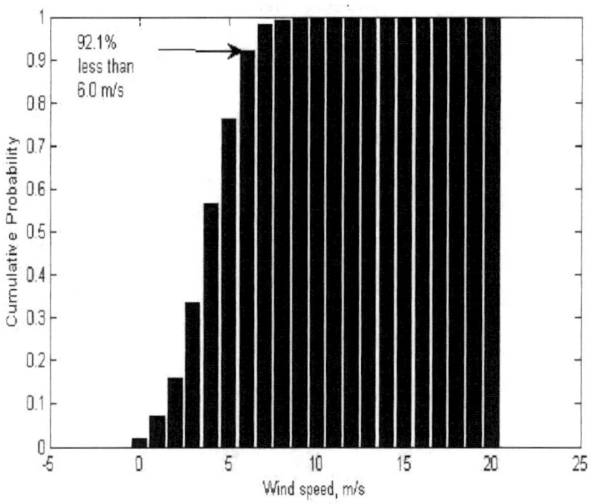

Figure 2. The CDF of generated wind speed at turbine hub height

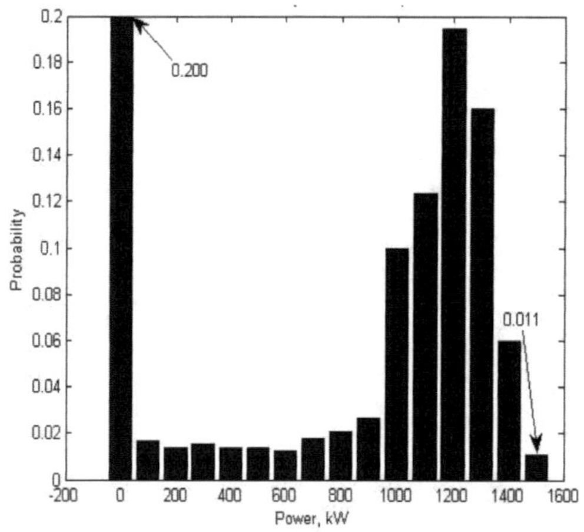

Figure 3. The PMF of average hourly generator output

5. Conclusions

Evaluation of renewable generation technologies which depend on randomly-varying weather conditions such as wind turbine requires stochastic modelling and simulation technique such as the Monte Carlo method in order to capture its inherent behaviour. The results showed that this method is applicable to evaluate the distribution analysis of the wind generator output. The presented method can be used to assess optimal site for WTG as different location would have different wind profile as well as to assist in generator selection and sizing as different types of turbine would have different power curve and rated speed characteristics. The goal of optimizing the design for wind turbine generator in wind farm should be to utilize wind energy as efficiently as possible and to reduce production cost of wind power generation. Another possible application of this analysis is the location-based determination of wind energy tariff and risk management to determine appropriate generation mix and scheduling in a particular system.

References

[1] Bouffard F, Galiana FD 2008 *IEEE Trans. on Power Syst.* ***23*****(2)** 306-316

[2] Muhamad Razali N M, Hashim A H 2009 Cash-flow analysis of a wind turbine operator *Proc. 3rd Int. Conf. on Energy and Environment*

[3] Carta J A, Ramírez P Velázquez S 2009 *Renewable and Sustainable Energy Rev.* **13(5)** 933-955

[4] Sopian K, Othman M, Wirsat A 1995 *Renewable Energy* **6(8)** 1005–1016

[5] Kelleher J, Ringwood J V 2009 *Energy* **34(4)** 401-409

[6] Mitzenmacher M, Upfal E 2005 *Probability and computing: Randomized algorithms and probabilistic analysis* Cambridge University Press

[7] Yichun W, Ming D 2008 *Electric Utility Deregulation and Restructuring and Power Technologies, 2008. DRPT 2008. Third International Conference on* 2487-2491

IOP Conf. Series: Earth and Environmental Science **16** (2013) 012029 doi:10.1088/1755-1315/16/1/012029

Application of Particle Swarm Optimisation in solving Emission Constrained Economic Dispatch

NM Muhamad Razali[1], **YY Teh**
Universiti Tenaga Nasional,
Jalan IKRAM-UNITEN, 43000 Kajang, Selangor, Malaysia

E-mail: noormiza@uniten.edu.my

Abstract. Production of electricity has long been equated with emission of hazardous greenhouse gases and pollutants. One of the ways to reduce this negative impact is via emission constrained economic dispatch (ECED). The ECED is aimed to simultaneously minimize the total cost of generation and the total emission level. In order to overcome the limitations of the widely-used conventional method (CM) such as lack of flexibility and accuracy and the probability of achieving only local minima, meta-heuristic optimisers are gaining popularity. The paper presents the formulation of the ECED problem and proposes the use of Particle Swarm Optimizer (PSO) as a better alternative to the CM. The effectiveness of using PSO to solve the ECED problem as compared to CM is demonstrated and discussed.

1. Introduction

Economic dispatch is a method used to allocate loads to plants or generator units in order to operate for minimum cost while meeting power demand and various operational constraints. Normally, each generator units are arranged in a way that lowest cost generators will be used as much as possible while the generators with higher operating cost comes into production only when the demand increases and the generators with lower generation cost is not capable to handle the power generation. However, the conventional economic dispatch problem does not take into consideration the issue of thermal power stations using coal, oil and gas producing large amounts of greenhouse gases (GHG) and pollutants; altering the global climate and affecting health of living things in the process. At the United Nations Framework Convention on Climate Change (UNFCCC) Conference of Parties – 15 (COP-15) in Copenhagen in 2009, Malaysia has voluntarily pledged to reduce the country's emission intensity of GDP up to 40% by the year 2020 as compared to 2005 levels. Achieving this target is certainly a challenge that requires cooperation and commitment from everyone, including the public and private sectors, and the general public.

In order to reduce GHGs emission, [3] suggested a full range of options available to control emissions which are emission constrained in system operation, fuel switching, switching to lower sulphur coal, energy conservation that lowers SO_2 and NOx emission, purchase or sale of emission allowance, and installing power plant emission control technologies such as cleaning equipments [3]. Those options that require installation of new equipment or modification of the existing systems may involve considerable capital outlay and, hence, they are considered as long-term options. Among all those methods, emission constrained economic dispatch (ECED) is one of the most attractive and

[1] To whom any correspondence should be addressed.

effective short-term options due to it not only control GHGs emission but also help in minimizing fuel cost at the same times.

Conventionally, the economic dispatch (ED) problem is solved by using Lagrange-iteration method with Lagrange multiplier. Reference [4] has shown the incorporation of emission constraint into the fitness function of the economic dispatch problem using this method. The modified objective function is then solved by using Dynamic Programming Based Hydrothermal Coordination program. The simulated result shows that only a slight increase in production cost can cause a large decrease in emission level.

However, this technique may not yield an optimal solution [5]. Besides, the linearization, assumptions and approximations that are used to limit the complexity of the economic dispatch problems causes the solution to be inaccurate [6] and the loss of accuracy by these approximations is not desired. In reality, unit incremental heat rate curves do not exhibit the monotonically increasing shape required by the classical method [7]. The application of intelligent optimizers as well as evolutionary computational techniques is suggested to be used to solve the problems.

This paper demonstrated the use of Particle Swarm Optimization (PSO) to solve the emission constrained economic dispatch (ECED) as compared the solutions of Conventional Method (CM). The method is expected to achieve more accurate results as they are non-gradient dependence. Thus, unlike conventional mathematical methods, they present a higher chance to escape local optima.

2. Formulation of the Emission Constraint Economic Dispatch Problem

2.1 Objective Function

Emission constrained economic dispatch (ECED) problem involves the simultaneous optimization of fuel cost and emission objectives which are conflicting with each other.

i. Fuel Cost

$$\min FC(P_i) = \min \sum_{i=1}^{n} a_i P_i^2 + b_i P_i + c_i \tag{1}$$

where

$FC(P_i)$ = fuel cost of power generation by unit i
a_i, b_i, c_i = fuel cost coefficient of unit i
P_i = power generation of unit i

ii. Emission Cost

$$\min EC(P_i) = \min \sum_{i=1}^{n} d_i P_i^2 + e_i P_i + f_i \tag{2}$$

where

$EC(P_i)$ = emission cost of power generation by unit i
d_i, e_i, f_i = emission coefficient of unit i

The two separate objective functions are combined into a single-objective function to be solved simultaneously by introducing a penalty factor to obtain the solution that gives the minimum total cost.

$$\min TC(P_i) = \sum_{i=1}^{n} FC(P_i) + h_i\big(EC(P_i)\big) \tag{3}$$

$$h_i = \frac{a_i P_{imax}^2 + b_i P_{imax} + c_i}{d_i P_{imax}^2 + e_i P_{imax} + f_i} \tag{4}$$

where
$TC(P_i)$ = total cost of power generation by unit i
h_i = penalty factor of unit i
P_{imax} = maximum power generation by unit i

2.2 Constraints
i. Power Balance Constraint

$$\sum_{i=1}^{n} P_i = P_D \quad (5)$$

where P_D = Power demand
ii. Generator Capacity Constraint

$$P_{imin} \leq P_i \leq P_{imax} \quad (6)$$

where
P_{imin} = minimum power generation allowed for unit i
P_{imax} = maximum power generation allowed for unit i
The problem is solved by using CM and PSO. The PSO algorithm used may be referred to [8].

3. Results and Discussions
In order to determine and compare the effectiveness and accuracy between the solutions of CM and PSO, a test system with a six generator units has been selected. The fuel cost coefficients (a_i,b_i,c_i), the emission coefficient (d_i,e_i,f_i), as well as the generator minimum and maximum capacity limit (P_{imin} and P_{imax}) are given in [8]. Although it is shown in Figure 1 that the total fuel cost for the solutions without emission constraint is slightly lower than the solutions with emission constraint, Figure 2 shows that the ECED formulation has significantly reduced the emission levels of the economic dispatch. These show the effectiveness of the algorithms in achieving its objective. The total generation cost and total emission level generated by the PSO as compared to the CM are presented in Table 1 and Table 2.

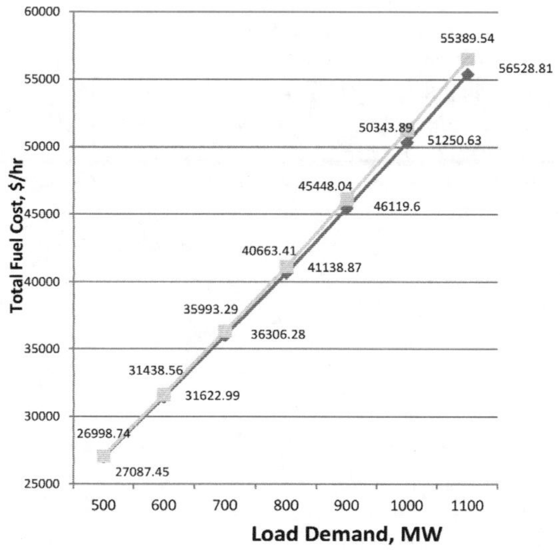

Figure 1. Total fuel cost by PSO

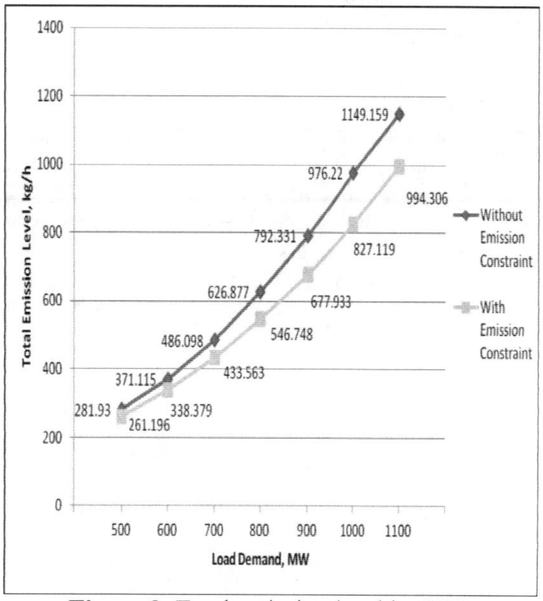

Figure 2. Total emission level by PSO

IOP Conf. Series: Earth and Environmental Science **16** (2013) 012029 doi:10.1088/1755-1315/16/1/012029

Table 1. Total generation costs

Power MW	Total Fuel Cost (FC), ($/hr)	
	CM	PSO
500	27092.5	27091.8
600	31628.7	31627.7
700	36313.9	36312.7
800	41148.3	41146.9
900	46131.8	46130.1
1000	51264.5	51262.5
1100	56546.2	56543.9

Table 2. Total emission levels

Power MW	Total Emission Level (E), (kg/hr)	
	CM	PSO
500	261.634	261.196
600	338.992	338.098
700	434.38	433.563
800	547.796	546.748
900	679.24	677.933
1000	828.72	827.119
1100	996.224	994.306

From the objective function of the ECED (refer equation 3), the emission and fuel costs of each generator are summed up during the evaluation of optimal solution. From Table 1, it is observed that the total generation costs achieved by using PSO are always lower than the CM. Similarly, PSO gives lower total emission level. Hence, it is proven that the PSO is more accurate and effective in solving emission constraint economic dispatch problems.

4. Conclusions

Emission constrained economic dispatch is an effective way to not only minimize generation cost, but also to reduce emission level of GHG. This paper has also shown that the PSO is a better technique to achieve optimal solutions compared to the conventional method employing Lagrange Multiplier. Further improvement can be done by specifying the types of GHGs when solving the problems so that the emission of certain type of GHGs which is considered to be more harmful may be specifically decreased to reduce the damage to our environment. Furthermore, consideration of extra constraints such as transmission losses, power flow and valve-point loading effect may also be considered. By taking into account more factors that can cause the increase of generation cost, more accurate solutions which provide system operator optimal decision making may be achieved.

References

[1] Subramanian S, Ganesan S 2010 *Int. J. of Computer Applications* **8(11)** 39-45
[2] Cheng Y, Xiao W, Lee W J, Yang M A 2009 *North American Power Symp. 2009* **1-5**
[3] Ramanathan R 1994 Emission constrained economic dispatch *IEEE Trans. Power Syst.* **9(4)**
[4] Md. Sayeed Salam 2008 *World Academy of Science, Engineering and Technology* **46** 668-671
[5] Fahmideh-Vojdani A R, Galiana F D 1980 *IEEE Trans. Automatic Control* **25(2)** 213-217
[6] Shebie G B Brittig K 1995 Refined Genetic Algorithm- Economic Dispatch Example *IEEE Trans. Power Syst.* **10(1)** 117-124
[7] Bakirtzis A, Petridis V, Kazarlis S 1994 *IEE Proc. Gener. Trans. Distrib.* **141(4)** 377-382
[8] Zielinski K, Laur R 2007 *Informatica*, **31** 51–59

IOP Conf. Series: Earth and Environmental Science **16** (2013) 012030 doi:10.1088/1755-1315/16/1/012030

Simulation of Breach Outflow for Earthfill Dam

Azwin Zailti Abdul Razad[1], Rahsidi Sabri Muda[1], Lariyah Mohd Sidek[2], Intan Shafilah Abdul Azia[2], Faezah Hanum Mansor[2] and Ruzaimei Yalit[2]

[1]TNB Research Sdn. Bhd, No 1, Lorong Air Hitam, Kawasan Institusi Penyelidikan Bangi 43000 Kajang

[2] Centre of Sustainable Energy, College of Engineering, Universiti Tenaga Nasional

E-mail: azwinzailti.arazad@tnbr.com.my

Abstract. Dams have been built for many reasons such as irrigation, hydropower, flood mitigation, and water supply to support development for the benefit of human. However, the huge amount of water stored behind the dam can seriously pose adverse impacts to the downstream community should it be released due to unwanted dam break event. To minimise the potential loss of lives and property damages, a workable Emergency Response Plan is required to be developed. As part of a responsible dam owner and operator, TNB initiated a study on dam breach modelling for Cameron Highlands Hydroelectric Scheme to simulate the potential dam breach for Jor Dam. Prediction of dam breach parameters using the empirical equations of Froehlich and Macdonal-Langridge-Monopolis formed the basis of the modelling, coupled with MIKE 11 software to obtain the breach outflow due to Probable Maximum Flood (PMF). This paper will therefore discuss the model setup, simulation procedure and comparison of the prediction with existing equations.

1. Introduction

Dams are purposely built for irrigation, power generation, flood mitigation, water supply and recreation. However, substantially huge amount of water body stored behind the standing dam structure could seriously pose severe risks to many. Great level of energy stored in the impounded reservoir will cause unbearable impacts should it be released suddenly to the downstream area. It is therefore important to conduct a dam break study to determine the outflow resulted from such unwanted dam break event to prepare for the necessary Emergency Action Plan (EAP) and to quantify the social, economic, and environmental impacts downstream of the dam. The art of dam break modelling lies primarily in the prediction of the outflow hydrograph as a result of dam failure. This can be done via physical model and laboratory experiment and numerical modelling technique. Physical model is not always financially viable hence the numerical modelling is often taken as a better alternative. Numerical modelling techniques estimate the outflow hydrograph via two main methods; physically based methods and predictor equations.

Physically based methods predict the breach outflow using an erosion model based on hydraulics principles, sediment transport and soil mechanics. This is found to be the most difficult prediction method since many of the dam failure cases contain high degree of uncertainty. Research and data

[1] To whom any correspondence should be addressed.

available from sediment transport under rapidly varying flow are limited, which results in the use of approximation of the standard predictor equations. This method relies heavily on compilation of case study based on actual failure scenarios hence more straightforward calculation. Various predictor equations are developed from the compiled cases to assist in calculation of some important parameters. Amongst the many software used for prediction of dam break outflow hydrograph includes DAMBRK, FLDWAV, BREACH, and MIKE. Tingsanchali (Tingsanchali, 1998) reported that based on Meyer-Peter & Mueller formula, BREACH model computed the peak outflow hydrograph as a result of dam breach is 2.8 times the observed peak. MIKE 11 model computed the breach outflow with a peak equal to 0.167 times of the observed peak and a time to peak equal to 1.167 times of the observed value. This study had therefore used the MIKE modelling package to simulate the dam break event and to determine the outflow hydrograph to be routed to the downstream area. This paper will therefore discuss on the prediction of dam breach parameters, setting up the model using MIKE 11 software and simulation of dam break event under PMF Failure scenario for Jor Dam.

2. Study Area

Jor Dam is part of the existing Cameron Highlands Hydroelectric Scheme, comprises of seven power stations and 3 major dams, namely Sultan Abu Bakar dam at Ringlet, Sultan Yussuf Dam at Jor and Mahang Dam at Kuala Woh. The scheme was constructed in late 1950s and began its operation since 1963. The total installed capacity of the scheme is 262MW, aiming for peaking load. Jor Reservoir was created by damming the Jor River through construction of Jor Dam. Jor Dam is 44.8m high situated at 17th miles, Tapah-Cameron Highlands Road in Perak. Jor dam were constructed in the years of 1965-1967. Major inflows feeding into Jor reservoir are Sg Jor, tailrace of Sultan Yusuff Power Station, Sg Sekam and Sg Batang Padang. The dam is earthfill material with central clay core with total volume of 680, 000 m^3 of earthfill. The 15 km long Menglang tunnel conveys the water from Jor reservoir to the underground Sultan Idris II (Woh) power station, some 420 m below its head pond. Any excess of water from the reservoir will be channelled downstream via siphon bell-mouthed spillway and 8.84m diameter of concrete-lined tunnel. Main purpose of the reservoir is for power generation, which is operated by TNB. Land use in 2008 indicates 99% of the Batang Padang districts are village, cultivated plantation on farmlands and forest.

Table 1. Details of Jor Dam.

Dam Type	Earthfill
Crest Level (m)	EL 496.83
Dam Height (m)	44.8
Crest Length of Dam (m)	210 m
Gross Storage (million m^3)	3.85
Surface area at FSL (km^2)	0.32 km ^{2} at EL 493.45m
Catchment (km^2)	393.9

3. Methodology

The process of obtaining outflow hydrograph using MIKE 11 software requires the prediction of dam breach parameters and setting up model that represents the actual layout on the ground. Those two processes are detailed out in the latter.

3.1 Prediction of breach parameters

Breaching mechanism of a dam is described by dam breach parameters and represented by breach width, b, breach height, h, side slope, s and time of failure, t_f. Dam breach can be specified by trapezoidal, rectangular, or triangular shape. Linear breaching and erosion-based breaching are both relevant but the latter contains high degree of uncertainty. The trapezoidal formation with linear formation mechanism is adopted for the purpose of dam break modeling, based on the assumption that the dams breach linearly with the time. Froehlich and MacDonald and Langridge-Monopolis (MDLM) are the most suitable predictor equations (Wahl, 1998). The process of determining the breach parameters is in the form of loop. It starts off with the assumption that a dam will breach under an overtopping condition due to Probable Maximum Flow (PMF), whereby the water level was set at the maximum crest level of the dam. Simulation predicted that Jor Dam will not fail under overtopping condition, as the maximum water level achieved was EL 496.82 m, less than the dam crest level. The average breach width was calculated using Froehlich is 44.8 m, but the adopted breach width was taken as 210 m to fit the dam geometry. The side slope, s for was taken as 1V:1H for an earthfill dam (Froehlich, 1995b). Failure time, t_f for Jor dam was calculated as 2.376 hrs (MDLM) and 0.244 hours (Froehlich).

3.2 Model Setup

A set of four sub-files was created to complete the dam break model set up, added with two time series file namely dam breach parameters and PMF inflow. The four sub-files required were network, cross section, boundary conditions and hydrodynamic conditions. Network file describe the river channel with the associated dam structure to replicate the actual scenario, named as Sg Batang Padang. Dam structures were described the based on dam height, crest length, and crest height whilst Spillway structure was described using the spillway coefficient, height, width and level. Points were inserted along the channels and defined as chainage, by assigning specific cross section which was described in the cross section file. Storage curve was used to define the reservoir whilst the specific geometry of the spillway described the cross sections of spillway chainages. Three boundary conditions were specified; PMF Inflow time series for Jor Dam inserted at chainage 0 of Sg Batang Padang and defined as 'Inflow' boundary type, 'Dambreak' time series which will compute the progress of breach with regards to width, height and side slope and 'Q-h' or stage – level curve, inserted at the final chainage Sg Batang Padang. For hydrodynamic file, initial water level need was defined at Full Supply Level (FSL) to allow a smooth hydrodynamic computation. The river beds were assumed to be of packed clay with the recommended Manning's (M) equals to 30 (MASMA, 2000).

4. Results and Discussion

The outflow hydrograph reached the peak value of 13,006 m^3/s. The outflow hydrograph showing peak discharge at the dam is illustrated in Figure 1.

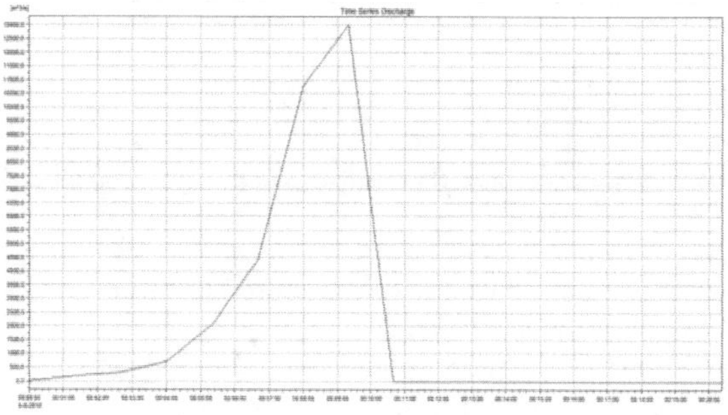

Figure 1. Outflow hydrograph of Jor Dam under PMF failure

Although Macdonald-Langdrige-Monopolis and Froehlich predictor equations are best applicable to earthfill dams, the peak outflow simulated for Jor dam was much higher. However, comparison was also made with other predictor equations, of which deviation from the simulated value ranges from 16% to 69%.

5. Conclusion

Dam could fail due to various causes such piping, overtopping, foundation and many others. In the case of dam break modelling, it is important to predict the breach parameters accurately. The peak outflow discharge due to failure of Jor dam was 13, 006 m^3/s based on time of failure of 0.244 hours, breach width 210m and breach height of 43m. The outflow hydrograph will be coupled with topographical information and flood plain roughness to obtain flood inundation map, which can be modelled using MIKE 21 or MIKE Flood. However, further validation using experimental setup is needed to verify the peak outflow and the breach parameters.

Table 2. Summary of Peak Outflow

References	Peak Flow Equations	Calculated Peak Flow (m^3/s)	Deviation from Simulated Peak Flow (%)
Bureau of Reclamation (1988)	$Q_p=19.1(h_w)^{1.85}$	19545.12	50
MacDonald and Langridge-Monopolis (1984)	$Q_p=3.85(V_w h_W)^{0.411}$	8542.37	-34
Froehlich (1995a)	$Q_p=0.607(V_w^{0.295} h_w^{1.24})$	5277.72	-59
Kirkpatrick (1977)	$Q_p=1.268(h_w+0.3)^{2.5}$	15078.38	16
SCS (1981)	$Q_p=16.6(h_w)^{1.85}$	16986.86	31
Hagen (1982)	$Q_p=0.54(S.h_d)^{0.5}$	6580.47	-49
Singh and Snorrason (1984)	$Q_p=13.4(h_d)^{1.89}$	17851.57	37
Costa (1985)	$Q_p=2.634(S.hd)^{0.44}$	10379.48	-20

Acknowledgement
The author would like to express gratitude and sincere thanks to Tenaga Nasional Berhad for funding of this research project, as well as UNITEN as the collaborator.

References

[1] Che Ros, F 2009 Hydrodynamic Modelling of Emnabkment Dam Breaches: Case Study of Kenyir Dam *ME thesis* Universiti Tenaga Nasional, Malaysia
[2] Department of Irrigation and Drainage Malaysia 2000 *Manual Saliran Mesra Alam Malaysia, Chapter 14: Flow Estimation and Routing* Penerbitan Nasional Berhad, Malaysia
[3] Froehlich D C 1995 *Water Resources Engineering, 1995 ASCE Conference* 887-891
[4] MacDonald T C, Langridge-Monopolis J 1984 *Journal of Hydraulic Engineering* **110(5)** 567-586
[5] Singh V P 1996 *Dam breach modeling technology* **17** Springer
[6] Tingsanchali T, Khan N M 1998 Prediction of flooding due to assumed breaching of Mangla dam *In Proceedings, 3rd International Conference on Hydroscience and Engineering, Bradenburg University of Technology at Cottbus, Cottbus, Germany* **31**
[7] Wahl T L 1998 Prediction of embankment dam breach parameters: a literature review and needs assessment

IOP Conf. Series: Earth and Environmental Science **16** (2013) 012031 doi:10.1088/1755-1315/16/1/012031

Simulation and evaluation of small scale solar power tower performance under Malaysia weather conditions

A M Gamil[1], S I Gilani[1] and H H Al – Kayiem[1]
Department of Mechanical Engineering, Universiti Teknologi PETRONAS (UTP), Perak, Tronoh, Block 17, MALAYSIA

E-mail: jameeel05@gmail.com

Abstract. Solar energy is the most available, clean, and inexpensive source of energy among the other renewable sources of energy. Malaysia is an encouraging location for the development of solar energy systems due to abundant sunshine (10 hours daily with average solar energy received between 1400 and 1900 kWh/m^2). In this paper the design of heliostat field of 3 dual-axis heliostat units located in Ipoh, Malaysia is introduced. A mathematical model was developed to estimate the sun position and calculate the cosine losses in the field. The study includes calculating the incident solar power to a fixed target on the tower by analysing the tower height and ground distance between the heliostat and the tower base. The cosine efficiency was found for each heliostat according to the sun movement. TRNSYS software was used to simulate the cosine efficiencies and field hourly incident solar power input to the fixed target. The results show the heliostat field parameters and the total incident solar input to the receiver.

1. Introduction

All Solar Power Tower (SPT) systems are just commercially employed in certain regions in the world especially in arid areas of mid-latitude zones, such as Spain, Russia, Italy, Germany, Australia, and US [1]. It is usually believed that SPT systems cannot be used in the tropics with relatively high diffuse fraction of global radiation. However, there is no systematic study on this issue [2].

Malaysia is rapidly developing and the growing energy demand requires alternative energy sources to fulfil the demands [3]. Currently most of the solar power used in Malaysia is Photovoltaic (PV) systems on a domestic level only and large scale commercial use is not significant yet [4].

In this paper the solar power tower technology is examined for collecting and redirecting solar power in order to produce thermal energy for latitude 4.34° in Malaysia by designing a small-scale heliostat field then calculating the cosine efficiency for each heliostat and finally simulating the incident solar input to a fixed target mounted on the top of a tower.

2. Model description

2.1. Sun position definition

In order to calculate the hourly heliostat field efficiency, a model for solar position has been utilized. The solar position equations are obtained from Duffie and Beckman [5].

[1] To whom any correspondence should be addressed.

2.2. Heliostat field layout design

Radial staggered layout is the best layout for solar power plants due to its reduced blocking and shadowing losses [6]. The heliostat placement is implemented with the tower centre as the origin using the radial and azimuthal spacing formulas.

2.3. Tower calculations

Several useful mathematical relationships can be obtained from figure 1. These relationships are important in determining the target λ and facing angle $\emptyset$ for each heliostat.

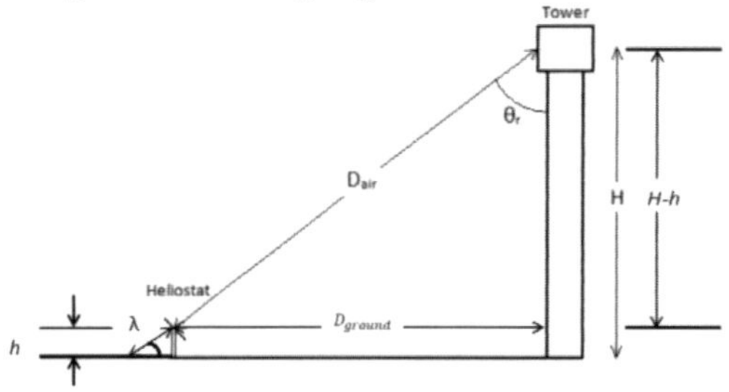

Figure 1. Tower and Heliostat trigonometry relationships.

2.4. Field solar input

The total solar radiation input on the field $Q'_{Solar-total}$ is calculated as in (1):

$$Q'_{Solar-total} = I_{bn} \times A_{total} \tag{1}$$

Where I_{bn} is the direct beam irradiance and A_{total} is the total reflective area of the heliostats.

2.5. Incident solar power on the target

The incident solar power to receiver $Q'_{inc-rec.}$ is calculated by [7] as in (2),

$$Q'_{inc-rec.} = Q'_{Solar-total} \times \eta_{field} \tag{2}$$

Where η_{field} is the field losses.

2.6. Field losses

Heliostat Fields encounter numerous losses, which are referred as field losses. These field losses η_{field} can be calculated as follow [6]:

$$\eta_{field} = \eta_{cosine} \times \eta_{blocking\,\&\,Shadowing} \times \eta_{ref.} \times \eta_{tarcking} \tag{3}$$

Where η_{cosine} is the cosine effect, $\eta_{blocking\,\&\,Shadowing}$ is the blocking and shadowing efficiencies, $\eta_{ref.}$ is the reflective efficiency of heliostat mirror, and $\eta_{tarcking}$ is the tracking efficiency of the heliostat tracking system.

2.6.1. Cosine efficiency

Cosine efficiency has the greatest impact on solar power reflected by heliostat because it depends on the sun position and heliostat location [6]. Cosine efficiency is calculated according to:

$$\eta_{cosine} = \cos(\theta_i) \tag{4}$$

In this study, the average cosine efficiency for a single day (from 9 am to 4 pm) is calculated based on the Azimuth-Elevation (AE) tracking method. The angle of incidence θ_i has been calculated based on the target angle λ , facing angle $\emptyset$, solar elevation angle α_s, and solar azimuth angle γ_s as s [8].

Due to the small number of heliostats, blocking and shadowing efficiency were given the value of 1. The study assumes the use of Sun Tracer™ Dual-Axis Heliostat has a tracking error < 0.1° and 88% reflective efficiency.

3. Analysis

3.1. Heliostat field layout design

Spread-sheet calculations were implemented based on the mathematical model developed in the aforementioned section to determine the field layout. Using the data obtained from the spread-sheet calculations, a simple radial staggered field layout can be plotted as shown in figure 2.

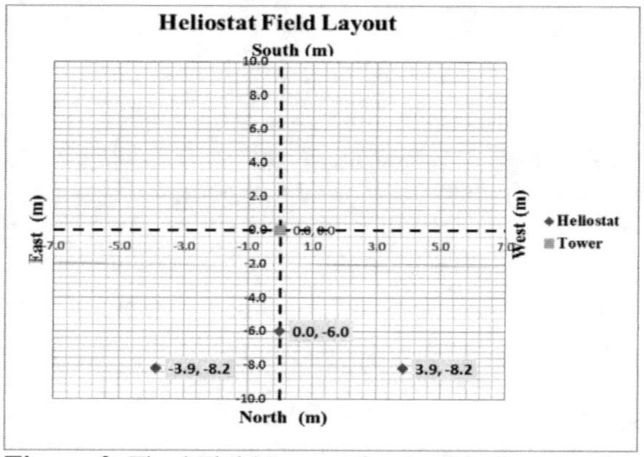

Figure 2. Final Field Layout for the 3 heliostat units.

3.2. Heliostat field modeling with TRNSYS

A TRNSYS model for the proposed heliostat field inputs and parameters has been developed to calculate the incident power to the receiver. All parameters and inputs are inserted according to the data obtained from the spread-sheet calculations for heliostat positioning and field efficiency.

4. Results and discussion

The 3 units of heliostats have been evaluated in terms of cosine efficiency and power incident to receiver as shown in figure 3 and 4. TRNSYS simulation had been carried out for the 3 heliostat units that are located in different positions from the tower as discussed in the previous sections.

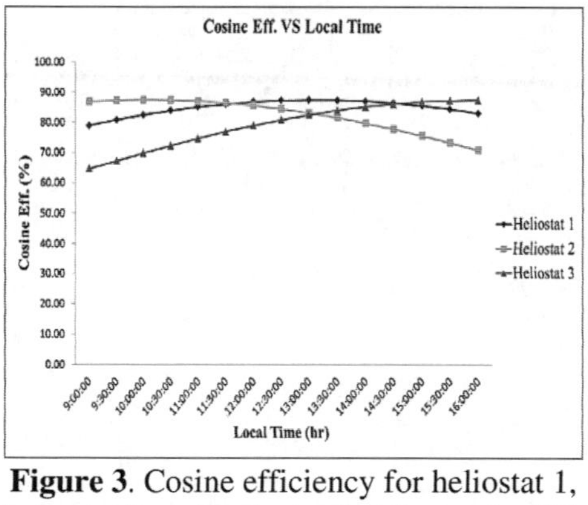

Figure 3. Cosine efficiency for heliostat 1, heliostat 2, and heliostat 3 at September 23.

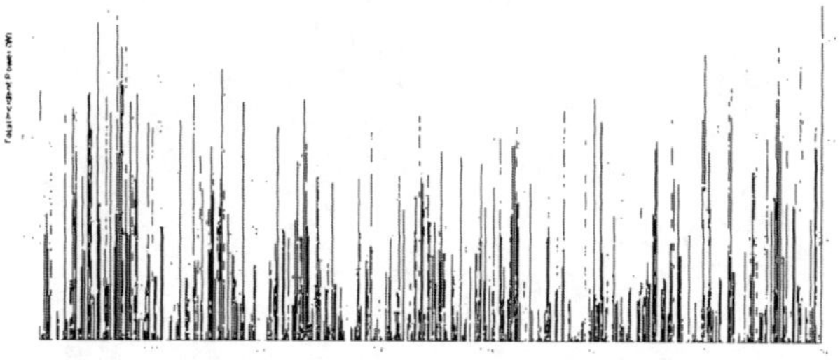

Figure 4. Total incident solar power to receiver from all heliostats throughout the year.

5. Conclusion

The heliostat field design for 3 units of dual-axis heliostat has been carried out, and the incident solar power input to receiver has been simulated for Ipoh, Malaysia. This preliminary layout and specifications shall be used to build a small-scale testing facility for solar power tower evaluation; thus TRNSYS simulations are required to determine the power rating for receiver sizing as next step. The heliostat field parameters were established in this study as shown in table1.

Table 1. Heliostat field parameters.

Heliostat Parameters							
No. of heliostats	Area of single heliostat	Width of heliostat	Length of heliostat	Height of heliostat	Radial spacing	Azimuthal spacing	Height of the tower
3	3.1 m^2	1.65 m	1.9 m	1.5 m	4.4 m	7.7 m	6 m

Heliostat positions						
	Coordinates (m)			Ground distance from the tower (m)	Facing angle (degree)	Target angle (degree)
Heliostat no.	X	Y	Z			
Heliostat 1 coordinates	0	-6	1.5	6	0	37
Heliostat 2 coordinates	3.9	-8.2	1.5	9.1	25.2	26.4
Heliostat 3 coordinates	-3.9	-8.2	1.5	9.1	-25.2	26.4

From TRNSYS simulation, the heliostat field delivers 7.5 kW as a peak value in day 361 (December 27) at solar noon. This amount of power can be increased by increasing the total reflective area of the heliostat field.

References

[1] Anon 2010 Solar Technologies Market Report. U.S. Department of Energy *Energy Efficiency and Renewable Energy*
[2] Janjai S, Laksanaboonsong J and Seesaard T 2011 Potential application of concentrating solar power systems for the generation of electricity in Thailand *Appl. Energy* **88** 4960–7
[3] Ahmad S, Kadir M Z A A and Shafie S 2011 Current perspective of the renewable energy development in Malaysia *Renew. Sust. Energ. Rev.* **15** 897–904
[4] Islam M, Saidur R and Rahim N 2010 Usage of solar energy and its status in Malaysia *Engineering e- Transaction (ISSN 1823-6379)* **5** 6–10
[5] Duffie J A and Beckman W A 2006 Solar Engineering of Thermal Processes 3rd Edition
[6] Stine W B and Geyer M 2001 Power from the Sun
[7] Kistler B 1986 A User's Manual for DELSOL3: A computer code for calculating the optical performance and optimal system design for solar thermal central receiver plants *Energy*
[8] Chong K-K and Tan M H 2012 Comparison Study of Two Different Sun-Tracking Methods in Optical Efficiency of Heliostat Field *Int. J. Photoenergy* **2012** 1–10
[9] Chong K K and Tan M H 2011 Range of motion study for two different sun-tracking methods in the application of heliostat field *Sol. Energy* **85** 1837–50

IOP Conf. Series: Earth and Environmental Science **16** (2013) 012032 doi:10.1088/1755-1315/16/1/012032

High performance silicon lateral PIN photodiode

S. Kalthom Tasirin[1], P. Susthitha Menon[2], Ibrahim Ahmad[1], S. Fazlili Abdullah[1]

[1]Department of Electronics and Communication Engineering, College of Engineering, Universiti Tenaga Nasional, Jalan IKRAM-UNITEN, 43000 Kajang, Selangor, Malaysia
[2]Institute of Microengineering and Nanoelectronics (IMEN), Universiti Kebangsaan Malaysia 43600 Bangi, Selangor, Malaysia

E-mail: sitikalthom@gmail.com, aibrahim@uniten.edu.my, siti@uniten.edu.my, susi@eng.ukm.my

Abstract. Start Photodetectors with high responsivity in Si-related processes are required for high speed optoelectronic applications such as fiber optic data communication systems. In this paper, the performance of a virtual lateral PIN photodiode with intensity response in the 200-1000 nm wavelength range was demonstrated. The results obtained for responsivity, total quantum efficiency and frequency response were 0.62 A/W and 13.1 GHz respectively for design parameters of intrinsic region length of 6 μm, photoabsorption layer thickness of 50 μm, incident optical power of 1 mW/cm^2 and bias voltage of 3 V. As a conclusion, the decrease in device size enabled the realization of a high performance lateral p-i-n photodiode.

1. Introduction

PIN photodiode is the most widely used detector in optical applications because it has high sensitivity at the operating wavelength, high response speed, and low noise. This is because the depletion region thickness (the intrinsic layer) can be tailored to optimize the quantum efficiency, transient response and frequency response [1]. The development of a lateral PIN photodiode based on silicon absorption layer is pursued because of its compatibility in monolithic integration of CMOS-based photonic devices as well as low cost and robustness of the silicon material itself. The lateral PIN photodiodes can be fabricated in this technology [4], giving incident light direct access to the active intrinsic region of the device.

The main properties of a photodiode often considered in analysis are its responsivity and frequency response. In previous research [2], it was reported that a frequency response of 4.38 GHz at 5-V bias for a 7.4 μm intrinsic width of photodetector on silicon-on-insulator (SOI) substrate was achieved. The Si photodiode in standard 0.18 μm CMOS technology exhibited frequency response of 8.7 GHz and responsivity of 0.018 A/W at 11.4-V bias [3]. In another research [4], the silicon nano-photodiodes produced, a frequency response of 5 GHz at 850 nm optical wavelength. Feng et.al [5] produced a PIN modulator on silicon-on-insulator (SOI) substrate, with a frequency response of 12 GHz at 8-V bias voltage. Totsuka et. al [6] reported a responsivity of 0.34 A/W at 590 nm optical wavelength for the dimension of 1.18 mm (width) x 3.8 mm (length) x 5.0 mm (thickness). The active photoabsorption layer of a PIN photodiode must be as thick as 1 μm to obtain high quantum

[1] To whom any corresponding should be addressed.

efficiency and low bias voltages from 2 to 7 V and can be used for optimum operation with low dark current and high speed [7].

In this paper, an investigation on the effect of device size reduction on the frequency response, responsivity and total quantum efficiency of a Si lateral PIN photodiode was explored and the device design is based on the device which was developed in the previous paper [8]-[10]. A two dimensional model of a Si lateral PIN photodiode operating at the 800 nm optical wavelength was developed using an industrial based numerical software.

In this paper, numerical modeling was used to design and characteristize the device. In terms of simulation for the fabrication and device electrical characterization, ATHENA and ATLAS software from Silvaco Int. were used respectively. In this way a more physical insight into the device operation can be obtained. The models in bulk silicon material which were used are the Shockley-Read-Hall (SRH) recombination model, the concentration dependent mobility model, the Auger model and the Fermi-Dirac statistic model [8]-[12].

2. Methodology

The Si lateral PIN photodiode was simulated on silicon substrate (n^+-type) in two dimensional using ATHENA software from Silvaco Int. Then, the n-well was developed using phosphorus diffusion with dopant concentration of 2.02×10^{19} cm^{-3} on the left side of the photodiode with diffusion temperature of 1000 ^0C for 50 seconds. Whilst the p-well was diffused with boron concentration of 8.09×10^{19} cm^{-3} on the other side of the photodiode with temperature of 1000 ^0C for 120 seconds. SiO_2 layer with thickness of 280 nm was deposited on the silicon substrate to act as a passivation layer. The electrode contacts with thickness and length of 500 nm and 6000 nm respectively, were processed by depositing aluminum on the n-well and p-well areas of the silicon photodiode.

The responsivity of a pin-photodiode is given by [10]:

$$\mathcal{R} = \frac{I_T}{I_S}\left(\frac{\lambda}{1.24}\right) \tag{1}$$

where I_s is the source photocurrent, I_T is the cathode current and λ is the optical wavelength.

The frequency response is defined as [10]:

$$f_{-3dB} = 20^{*}\log\left(\frac{I_R}{I_{RO}}\right) \tag{2}$$

where I_R is the real cathode current and I_{Ro} is the real component current.

3. Results and discussion

The developed device structure is shown in Figure 1. The p+ region resides beneath the anode electrode whereas the n+ region resides below the cathode electrode. The area between the p+ and n+ region is the intrinsic (or *i*) region. The device's electrical and optical characteristics were executed using the ATLAS module from Silvaco Int.

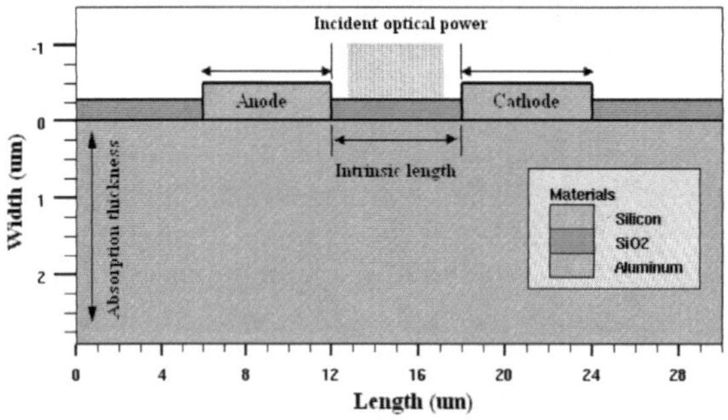

Figure 1. The device structure of Si lateral pin-photodiode

The simulation results for the frequency response of 13.1 GHz is obtained for the structure with 6 intrinsic region, whereas the value drop to 2.27 MHz for 30 μm intrinsic length at optical wavelength of 800 nm as shown in Figure 2 [12]. The result of frequency response for Si lateral PIN photodiode with intrinsic width of 6 μm is increased because of the photoresponse and the speed in the device [13]. The frequency response could be further increased by increasing the and bias voltage and reducing the intrinsic width [12].

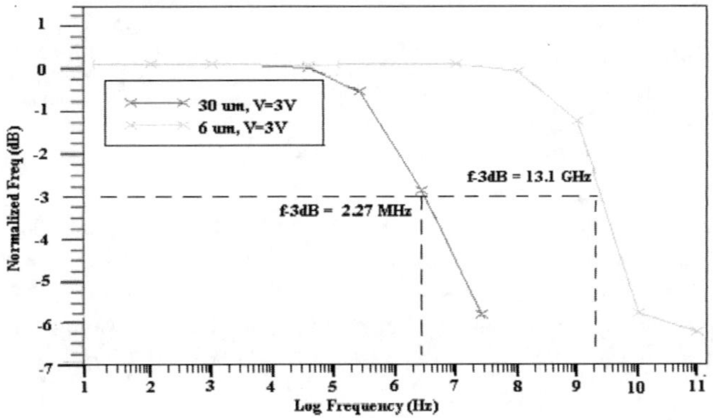

Figure 2. The -3dB frequency response at optical wavelength 800 nm.

Figure 3 shows the responsivity as a function of reverse bias with optical wavelength from 200 nm until 1000 nm. At a small bias of 3 V, the Silicon PIN photodiode for the device with 30 μm and 6 μm intrinsic length produced responsivities of 0.62 A/W and 0.45 A/W at 800 nm, respectively [12]. The result of responsivity for Si lateral PIN photodiode with intrinsic width of 6 μm is increase because of the recombination rate and photoresponse in the device [13]. The responsivity could be further increased by increasing the photoabsorption thickness and bias voltage [12].

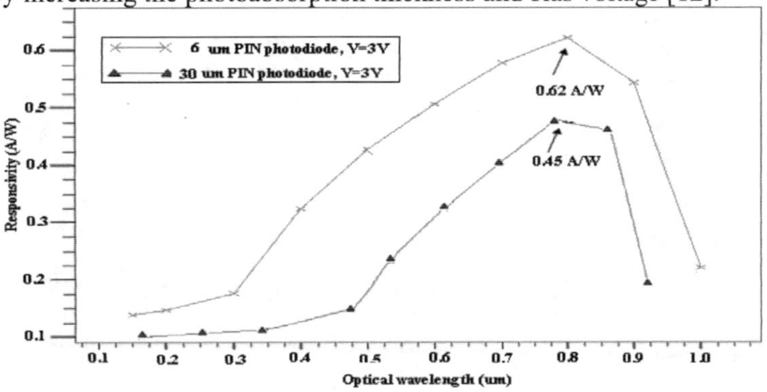

Figure 3. Responsivity for 30 μm and 6 μm intrinsic length of silicon lateral PIN photodiode

4. Conclusion

As a conclusion, the Si lateral PIN photodiode with a high responsivity in the visible spectrum (200 – 1000 nm) is reported. The PIN photodiode device structure is similar to PIN photodiode developed in previous work except for the size reduction of the device. The fabrication of Si lateral PIN photodiode is compatible with complementary metal-oxide semiconductor (CMOS) technology processes, which makes the device reliable and suitable for CMOS integration. In order to make a faster and smaller device in the future, we need to optimize the design of the device including the layer thicknesses, the intrinsic region width and the power intensity.

5. Acknowledgement

The authors would like to thank the Public Service Department of Malaysia (JPA) for their financial support, Photonic Lab, Institute of Microengineering and Nanoelectronics (IMEN), Universiti Kebangsaan Malaysia (UKM) and Pusat Latihan Teknologi Tinggi, Taiping (ADTEC) for the technical and moral support throughout the project. Universiti Kebangsaan Malaysia is also acknowledged for supporting this work under grant Industri-2012-017.

References

[1] Sze S M, et. al. 2007 *Physics of Semiconductor Devices*, A John Wiley & Sons, Inc.

[2] Ramaswamy A, Nunoya N, Johansson L A, Bowers J E, Sysak M N, Tao Yin 2009 *Group IV Photonics,.GFP '09. 6th IEEE International Conference on* **19-21** 9-11

[3] Fang-ping Chou, Ching-Wen Wang, Guan-Yu Chen, Yue-ming Hsin 2010 *OptoeElectronics and Communications Conference (OECC), 2010 15th* 826-827

[4] Fujikata J, Nose K, Ushida J, Nishi K, Kinoshita M, Shimizu T, Ueno T, Okamoto D, Gomyo A, Mizuno M, Tsuchizawa T, Watanabe T, Yamada K, Itabashi S, Ohashi K 2008 *Group IV Photonics 5th IEEE International Conference on* 176-178

[5] Ning-Ning Feng, Shirong Liao, Dazeng Feng, Po Dong, Dawei Zheng, Hong Liang, Roshanak Shafiiha, Guoliang Li, John E. Cunningham, Ashok V. Krishnamoorthy, and Mehdi Asghari, 2010, *Optics Express,* **18**, *799*

[6] Daisuke Totsuka, Takayuki Yanagida, Kentaro Fukuda, Noriaki Kawaguchi, Yutaka Fujimoto, Jan Pejchal, Yuui Yokota, Akira Yoshikawa 2011 *Nuclear Instruments and Methods in Physics Research Section A: Accelerators, Spectrometers, Detectors and Associated Equipment* **659** 399-402

[7] Jang J H, G. Cueva G et. al. 2001 *IEEE Photonics Tech. Letter* **2**

[8] Menon P S, 2005 *Pembangunan diodfoto planar p-i-n silikon (Development of silicon-based p-i-n photodiode)* MSc Thesis Universiti Kebangsaan Malaysia

[9] Menon P S, Shaari S 2005 *Proceedings of the SPIE Symposium on Optics and Photonics: Infrared and Photoelectronic Imagers and Detector Devices, San Diego, USA* **5881** 1-8

[10] Menon P S, Kandiah K, Ehsan A A, Shaari S 2011 *International Journal of Numerical Modelling: Electronic Networks, Devices and Fields* **24(5)** 465-477

[11] Menon P S, Tasirin S K, Ahamd I, Abdullah S F 2012 *World of Applied Sciences Journal (In-press)*

[12] Menon P S, Tasirin S K, Ahamd I, Abdullah S F 2012 *Australian Journal of Basic and Applied Sciences* **6(8)** 275-281

[13] Huang Z, Oh J, Campbell J C 2005 *Appl. Physic Lett.* **17** 1510 – 1512

IOP Conf. Series: Earth and Environmental Science **16** (2013) 012033 doi:10.1088/1755-1315/16/1/012033

Strategies to improve energy efficiency in sewage treatment plants

Mau Teng Au[1], Jagadeesh Pasupuleti[2], Kok Hua Chua[3]

[1] Power Engineering Centre, [2] Center for Power System Simulation, [3] Institute of Energy Policy Research, Universiti Tenaga Nasional, Jalan IKRAM-UNITEN, 43000 Kajang, Selangor, Malaysia.

Email: mtau@uniten.edu.my

Abstract. This paper discusses on strategies to improve energy efficiency in Sewage Treatment Plant (STP). Four types of STP; conventional activated sludge, extended aeration, oxidation ditch, and sequence batch reactor are presented and strategized to reduce energy consumption based on their influent flow. Strategies to reduce energy consumption include the use of energy saving devices, energy efficient motors, automation/control and modification of processes. It is envisaged that 20-30% of energy could be saved from these initiatives.

1. Introduction

Rapid urbanization and increase in population will require wastewater service. It has the potential of increase pollutant loading in municipal wastewater treatment facilities and driving the construction of more treatment facilities and installation of advanced treatment technologies. It will increase energy consumption in treatment facilities unless efforts are in placed to optimize energy consumption in current treatment plants and future treatment plants.

The main objective of sewage treatment is to render sewage become harmless to human and environment before disposal. If untreated sewage is allowed to accumulate, the decomposition of the organic materials can lead to the production of malodor gases. It contains numerous pathogenic or disease causing microorganism and also nutrients that encourage algae growth. Hence it's important to treat sewage to a level that is fulfilling Effluent Standard criteria.

Wastewater utilities in US consume about 2% of the total amount of electricity produced in US [1]. During the next 20 to 30 years, wastewater treatment plant energy consumption in US is expected by 30 to 40 %. In activated sludge treatment, approximately 1100 to 2400 MJ of electricity are required to process each 1000 m^3 of wastewater [2].

According to a report by Electric Power Research Institute EPRI, the main energy consumption is activated sludge aeration (EPRI, 1994). It shows that activated sludge aeration consumed about 55.6 % of total energy usage at a plant.

Wastewater system energy use is dominated by the treatment process itself. Aeration is the most energy intensive part of wastewater treatment, with approximately 70% of energy used in municipal secondary stage wastewater treatment facilities. Sludge conditioning and dewatering processes are also significant energy users in conventional wastewater treatment processes, where about 10% of wastewater sector energy is used.

Preliminary and primary treatment stages are less energy intensive than conventional secondary treatment, while tertiary treatment can be energy intensive as secondary treatment depending on the type and quantity of pollutants being removed and the desired and/or regulated effluent quality.

[1] To whom any correspondence should be addressed.

Most energy usage in waste water treatment is aeration and pumping. Aeration introduces dissolved oxygen into wastewater to support aerobic oxidation and also for nitrogen removal. Pumping is used to move and re-circulate water and solids through the sequence of treatment processes. Sewage aeration at an activated sludge sewage treatment plant accounts for about 30-80% of total plant electricity demand; 67% aeration, 21% pumping, 9% other loads and 3% lighting [3]. Aeration electric demand and energy consumption could be reduced by using fine-pore diffused air system and aeration process control, or lowering the sludge age (mean cell residence time or MCRT) [4].

With certain operating constraints many plants could achieve substantial energy cost savings by treating normal flows in off-peak hours when the cost of electricity is lowest. At plants with excess process capacities, sewerage treatment and sludge management would be minimized during on-peak hours [4]. Normal diurnal sewage flow pattern into a sewage treatment plant closely parallels an electric utility's system demand and energy cost curves. Shifting electrical load from on-peak to off-peak hours or levelizing electricity use throughout the day usually requires temporary storage of the influent waste water either at treatment plant site or within wastewater system, possible sludge storage for batch processing. Costs and benefits of wastewater storage and time of day electricity pricing need further development [4].

Due to oversized equipment, inefficient operations or lack of controls, the amount of air is delivered to the aeration basins is usually much more than required for mixing and biological activity. This excess air represents wasted energy and highly aerated wastewater may lead to sludge setting problems and solids carryover into the plant effluent.

Conversion from coarse bubble or mechanical aeration to a fine-pore system should lower the energy costs for wastewater aeration by at least 25%. Adding a feedback control system that monitors a treatment variable and automatically adjusts the operation of the fine-pore aeration system according to the reading of the measured variable could boost the energy savings to 35% or more. Reducing the sludge age in the aeration tanks at STP has an energy benefit for the activated sludge process.

In [3] and [4] it is reported that energy consumption could be reduced 10% to 30% through the application of proven energy efficiency technologies.

2. Types of sewage treatment plants

2.1. Conventional activated sludge treatment

Conventional activated sludge CAS treatment process consists of screenings, aerated grit chambers, primary settling tank, aeration tank and secondary settling tank and sludge treatment and disposal. It involves the production of activated mass of microbes that digest organic matter aerobically. The aerobic environment in the reactor is achieved by the use of either diffused or mechanical surface aerators. In typical conventional activated sludge process, the aeration time ranges from six to eight hours. After a specific period of time, the mixture of new cells and old cells is passed through settling tank. Some of the settled sludge is returned to the reactor as to maintain the optimum food over microbe's ratio. At mixed liquor temperature of between 18 to 25 ^{O}C, a sludge retention time or sludge age of minimum 3 days is desired.

Currently, activated sludge process may incorporate nitrification, biological nitrate and phosphate removal. These design use reactors placed in series that employ under aerobic, anoxic, ad anaerobic condition.

To maintain high levels of treatment performance, one must focus on the process control such as,

- maintaining dissolved oxygen level in aeration tanks,
- regulating returned activated sludge RAS and
- Controlling waste activated sludge WAS.

The dissolved oxygen concentration in aeration tank should be maintained at 1.5 to 2 mg/l. Higher DO of 2mg/l may improve nitrification, however values above 4 mg/l do not improve treatment performance instead increases aeration cost (Metcalf & Eddy, 2003).

2.2. Extended aeration

Conventional extended aeration system is another form of suspended growth activated sludge process. The aeration is implemented over the longer period than conventional activated sludge process. It operates either as a plug-flow or completely mixed system with high sludge retention time (20 to 30 days). It is possible to have high quality effluent. The typical F/M ratio is 0.04 g substrate/g biomass per day. Generally no primary settling tank is used. It is suitable for small communities. One of the examples of extended aeration design is oxidation ditch.

2.3. Oxidation ditch

OD consists of a ring or oval-shaped channel equipped with mechanical aeration and mixing devices. It allows one directional flow and a relatively long hydraulic retention time. The most common use aeration system is horizontal shaft aerator i,e brush aerator. The mixing and the flow velocity generated allow suspended solids to remain in suspension. As the wastewater leave aeration zone, denitrification process can occur. OD is highly reliable process and simple to operate. It requires less energy as aeration to meet oxygen requirement rather than mixing requirement, It requires more energy than conventional activated sludge process. Normally, the sludge retention time SRT is more than 20 days. Malaysian sewerage design guideline stipulates that the minimum velocity within the channel shall not less than 0.3 m/s.

2.4. Sequence batch reactor

Another type of suspended growth process is sequential batch reactor. Sequential Batch Reactor system is a mixed culture suspended growth activated process which operate in a sequential process of Fill, React (aerate), settle, Draw (Decant) and Idle. For municipal wastewater system at least two basins are used so that one basin is in the fill mode and the other basin goes through the react, settle and decant. The period for each cycle is varies depend on the organic loading. The benefit of SBR is lower space requirement and the treatment efficiency is comparable to conventional activated sludge process. The supporters of SBR system contend that the system is less expensive, more efficient and more controllable than other biological treatment approach particularly for low PE plant. The energy consumption will consist of aeration system, sludge flow and sludge treatment. SBR is a compact facility and with flexible operation with adjustable time for each process mode. With the development of simple inexpensive program logic controllers PLC and availability of level sensors and automatically operated valves, more SBR plants are built for small communities but also in large cities.

3. Strategies to improve energy efficiency in sewage treatment plants

Energy represents a significant cost to sewage treatment as it is used at all stages of treatment process, from the collection of raw sewage to the discharge of treated effluent. The aim of strategies to improve energy efficiency (EE) in Sewage Treatment Plant (STP) is to reduce the amount of energy consumed while maintaining a reasonable amount of reliability to meet effluent quality. For existing plants, EE in STP could be achieved through operational changes, and/or the use of technologies, whereas for new plants, EE strategies could be incorporated at the design stage.

To improve energy efficiency of the sewage treatment plant first we need to analyze in-flow rate and energy consumption of the plant, normally, for high inflow, the electricity consumption is higher. Then identify power profile and energy consumption of major loads using power recorders. The major loads of conventional activated sludge (CAS) plants are blowers, sewage lift pumps and aerators. The major loads of extended aeration (EA) plants are surface aerators and raw sewage pumps. The major loads of oxidation ditch (OD) plants are brush aerators and raw sewage pumps. The major loads of sequence batch reactor (SBR) plants are air blowers and raw sewage pumps.

3.1. Strategies that could be applied to major loads

i. Motors should be operated as close to nameplate voltage as practical because any deviation from nameplate rating affect motor efficiency. In general it is recommended that the motor line drop should not exceed 5% of line voltage.
ii. Right sizing of motors is required based on the connected load to improve the energy efficiency of the motor.
iii. Replacing the standard motors with Energy-efficient motors, which are typically between 2 to 6% more energy efficient than standard motors.
iv. Variable-speed drives are used to vary the speed of a pump to match the flow conditions. The result is a close match of the electrical power input to the pump with the hydraulic power needed to pump the water. The 20% reduction in speed can achieve 50% reduction in energy consumption.
v. Capacitor banks can be installed to improve the power factor of the big motors. Improving power factor is beneficial as it improves voltage, decreases system losses, and decreases power costs where fees for poor power factor are billed.
vi. Installing energy saving device with load detection system in series with big motor. This device continually monitors electrical load at 380 cycles per second, and ensures that optimal

power is being supplied at all times. This automatic feedback corrective system reduces waste and save electrical power.

vii. Pump testing can help where the pump operates on its performance curve. If the over designed condition is encountered then the pump impeller to be trimmed to achieve more efficient operating point. Energy savings result from lowering of pumping capacity to better match system demands and replacing inefficient pumps with high efficiency pumps.

3.2. Other general strategies

i. A common approach is to use tapered aeration to reduce the rate of oxygen supply along the length of a basin. It can be accomplished by placing more diffusers at the inlet to the basin where the organic loading is highest and decreasing the number of the diffusers along the basin's length. Tapered aeration better matches the oxygen demand across the basin by providing more air to the head of the basin where it is needed and less air near the end of the basin where the food-to-microorganisms (F/M) ratio is lower, thereby saving energy.

ii. Intermittent aeration saves energy by reducing the number of hours that an aeration system operates or the aeration system capacity. The methodology involves momentarily stopping air flow to an aeration zone or cycling air flow from zone to zone. The cycle length can be controlled with DO concentration or can be strictly time based. When controlling with the DO concentration, air flow is turned off at a set high level and turned back on based on a lower limit. The cycle length on time-based systems is strictly controlled by a set maximum time.

iii. The oxygen required to maintain biological processes within the aeration basin is proportional to organic and ammonia loading in the influent wastewater. Oxygen demand for aeration, therefore, follows the same diurnal pattern, dipping in the middle of the night and peaking in the morning and evening. Intermittent discharge of ammonia-rich supernatant from sludge dewatering operations can also dramatically increase the oxygen demand in the basin. Conversely, dilution from storm water flow can reduce oxygen needs. In addition to fluctuating oxygen demand of the wastewater itself, the oxygen transfer efficiency in the basin also varies in response to changing air and water temperature and other wastewater characteristics such as concentrations of solids and surfactants. Automated dissolved oxygen control can achieve significant energy saving.

iv. Biogas is a natural complement to operations at many STP's that use anaerobic digestion to stabilize waste water solids. An effective way to use biogas is in a combined heat and power application. All of the biogas is directed to and burned in an engine or turbine that in turn powers an electrical generator. The electricity can be used on STP site. The waste heat is recovered and used to heat the anaerobic digestor to provide heat to the other uses in the plant site.

vii. Processes/systems should be scrutinized for the opportunity to eliminate reduce energy during expensive on-peak rates by shifting them to off-peak rates.

4. Conclusions

Influent flow could be used as a parameter to relate energy efficiency of STP. Energy efficiency studies indicated that there is great potential and numerous energy conservation opportunities in STP that could be implemented which results in 20-30% energy savings. Additionally, there is great potential to harvest renewable energy such as solar, biomass, biogas and minihydro from STP.

References

[1] Batts et al 1993 Demand side management opportunities in wastewater industry *Water Environment Federation, Proceedings of the 66th Annual Conference and Exposition.*

[2] Burton 1998 Saving on Wastewater Treatment Energy *Magazine vol XXIII* **1** 17 – 20 Norwalk CT

[3] Richard Brown 2009 Energy Efficiency and Renewable Energy Technologies in *Waste Water Management Subcommittee on Water Resources and Environment, House Committee on Transportation and Infrastructure*

[4] Lawrence J. Pakenas 1995 Energy Efficiency in Municipal Wastewater Treatment Plants – Technology Assessment *New York State Energy Research and Development Authority*

IOP Conf. Series: Earth and Environmental Science **16** (2013) 012034 doi:10.1088/1755-1315/16/1/012034

Implementing Sustainable Engineering Education through POPBL

D X Lioe[1], G K Subhashini

School of Engineering, Asia Pacific University of Technology and Innovation, Technology Park Malaysia, Bukit Jalil, 57000 Kuala Lumpur

Email: lioe@apu.edu.my

Abstract. This paper presents the implementation of sustainable engineering education to undergraduate student in Asia Pacific University of Technology and Innovation, Malaysia (APU) through Project-Oriented Problem Based Learning (POPBL). Sustainable engineering has already been the paramount term where it is no longer limited to environment, but also to the entire lifetime of the individual engineer. To inculcate every engineering individual with sustainability, education is the way to start off.

1. Introduction

Education is in an important position to lead cultural growth and transformation towards more sustainable futures. Sustainability in education aims to develop students capable of understanding the social, cultural, economic and ecological values, not only at local or national level, but also at international level.Delivery of education should have the impact of inspiring the belief that every individual is able to influence and has the responsibility to make positive change at international level and be a part of mediator in achieving sustainability. In order to do so, education should be in such a way that to focus on what students should know and capable of doing to attain better sustainability. We must have the knowledge on the needs of students to achieve it [1]-[3].

Project-Oriented Problem Based Learning (POPBL) is a new educational model to make sustainable education takes place more efficiently and effectively. The integration of project works into Problem Based Learning (PBL) proves successful in enhancing students' interpersonal skills and abilities towards sustainability, apart from technical qualification upon graduation. Various educational institutions across the world have already implemented or in the progress of implementing POPBL [4]-[5], which the implementation means a structural change to the teaching and learning, and assessment methods.

2. Conceptualization

POPBL gains popularity due to the impact of industry on engineering education where the industry seeks highly on graduates with not only good professional qualification but also graduates with superior soft skills. This is in line with the globalization of society, commerce and culture [6]-[7], where future engineers need to equip themselves with the necessary abilities and competencies to compete and survive in the area of globalization. Implementing POPBL to achieve sustainable

[1] To whom any correspondence should be addressed.

education closes the gap between skills acquired by graduates and the expectation and requirement by the industries. The implementation of POPBL in Malaysia is in accordance with the Outcome Based Education (OBE), which is a student centered learning philosophy to enhance the quality and employability of engineering graduates, adapted by country members of Washington Accord [8].

The educational institutions are constantly facing many challenges trying to achieve sustainability by implementing POPBL. Some of the most spoken about challenges are; resistance by policies, academic staff, students etc. The institution may need to change the policies regarding the teaching and learning activities in order to fully implement POPBL. There should be consensus from briefings and discussions among academic staff for the implementation and transition to be successful. The next step will be the formation of action plan, which is also a critical part in the process of the implementation, where continuous evaluation procedures has to be in action [9].

Students from multicultural background studying together can be a problem to the institution as well as the academic staff, as the diversity of students means a much different personality profile [10]-[11]. Academic staffs need to ensure that the teaching and learning activities takes place for every student. Asia Pacific University of Technology and Innovation (APU) at the moment has students from 144 countries, which translate to the difference in nationality, language, religion and educational background.

2.1. Bringing in the changes

There is a saying of "the only permanent thing in engineering education is change", and this statement clearly shows the engineering education is highly industry driven and reflects the global social changes. Change is needed in engineering education to achieve sustainability in terms of the way it is delivered, but resistance is there for the change. The first hurdle is the top management of an institution. They have to be convinced on the advantages and the importance of sustainable education in order to make changes to the existing institution policies. Policy changes are needed due to the constraint of assessment criteria set. Besides the internal barrier, the changes have to meet with the criteria set by the accreditation body, in the case of Malaysia, is the Engineering Accreditation Council (EAC) of Board of Engineers Malaysia (BEM). The program has to be designed to meet the accreditation requirement and sustainability.

The next barrier will be the resistance from some of the academic staff. Cooperation from all academic staff is not quite possible in the beginning of the process. Generally, academic staff can be divided into four categories [reflect], which are: (1) those who want to change (the pioneers) (2) those interested, but discouraged by logistic, lack of vision, skills, resources or action plan (3) those reluctant to change (4) those against change. Those academic staffs who want to change and those interested are easier to train. Academic staffs whose reluctant or against the change are likely not participating in the training. Besides, senior academic staff may think there is no need for a change because they have done it right all the while and they have dozens of successful examples of their students who achieved and understood the concept of sustainability even without the implementation of POPBL. These groups of academic staffs might need to be pushed by the top management or the institution.

Students will possess opposing voice to the implementation of POPBL too, thinking the workload will be increased [12]. Students need to prepare before coming to the class instead of the conventional way where students come to class to listen to the lecture. Students have to be very independent in completing the project, which some of the students might dislike. Workload in term of hours needed in a particular module in significantly increased. Different students may have different perception on this new concept of sustainable engineering education. With not over assessing students in mind, POPBL has to be properly planned and implemented with care so as not to have students in disadvantage while achieving the objective of POPBL to educate the importance of sustainability for the present and future.

2.2. Multicultural

Students from different culture background seem to take a difference approach at looking at the same problem, and they tend to have different personality which likely to accept a certain teaching style better and affect the academic performance [10],[13]. Teaching the students with multicultural background with single teaching styles might suite only certain group of students, and the rest might feel uneasy with the lecturer. Zaki and Luiz [multi] shows the comparison of engineering students of Saudi and Canada [rosati] using the Myers-Briggs Type Indicator (MBTI) approach, which sorts four major types of personality, Extrovert (E) and Introvert (I), Sensing (S) and Intuition (N), Thinking (T) and Feeling (F), Judging (J) and Perceiving (P). The four major types thus give 16 possible types of personality, and suggest the preference of the individual.

Students from various countries with different culture can have very much different personality which adds to the challenge of providing students a fair teaching style for all. A lecturer would not be able to use different teaching style for each different student. Effort has to be done to inculcate the students to enhance learning in different teaching and learning environment. Team work aspect which is highly important in the implementation of POPBL will be another challenge as personality types do affect team performance [14]-[15]. Students tend to form a group together with friends from the same ethnic or culture rather than good students tend to form groups among themselves and leaving the weaker students working together [16], exactly what happens in APU. Mixing students randomly or based on equally distribution considering gender, ethnic, nationality etc often caused uneasiness to the students. Even though discussions take place among the group members, but the final outcome of a project is often the compilation of individual work rather than a piece of work from the knowledge sharing among group members.

Multicultural scenario in APU comes along with large knowledge gap among the students, which differs the situation here from elite university with diversity of students. Often good students will do most of the parts for the project and leave some part for weak students to complete. This happens as good students not trusting weak students or the weak students just want to be free riders in the team. And as consequences the concept of sustainability does not reach out to all the student has how it was planned out to be. This issue can be addressed by using self assessment and peer assessment [17]-[18]. These assessments are helpful for the lecturer as discussion and collaboration among team members happens not only in the classroom, but also outside lecture hours, where it is not possible for lecturers to be aware of the contribution of each student in their respective group and how the students managed to incorporate the importance of sustainability in their assessment. However these methods may not be reliable if the group members do not reflect the true contribution in the assessment or some might try to sabotage their friends.

3. Conclusion

Achieving sustainable education does not rely on the effort of a single individual; it is the team work and collaboration of several people from different background and levels. It requires for changes in the engineering education to create a better understanding towards achieving sustainability. APU is working on to ensure sustainable engineering education for our multicultural students through POPBL.

References

[1] Australian Government Department of the Environment and Heritage, November 2005

[2] Al-Maati S A, Damaj I 2010 *Transforming Engineering Education: Creating Interdisciplinary Skills for Complex Global Environments, 2010 IEEE* 1-11

[3] Allenby B 2007 *Electronics & the Environment, Proceedings of the 2007 IEEE International Symposium on* 2-56

[4] Lehmann M,Christensen P, Du X, Thrane M 2008 *European Journal of Engineering Education* **33(3)** 283-295

[5] Ruhizan M. Yasin, Saemah Rahman 2011 *Procedia - Social and Behavioral Sciences* **15** 289-293

[6] Downey G L, Lucena J C, Moskal B M, Parkhurst R, Bigley T, Hays C, Jesiek B K, Kelly L,

Miller J, Ruff S, Lehr J L, Nichols-Belo A 2006 *J.Eng. Educ.* **95(2)** 107-121

[7] Rahman R A, Zan M M M, Abidin H Z, Kassim M, Yahaya C K H C K 2010 *Transforming Engineering Education: Creating Interdisciplinary Skills for Complex Global Environments, 2010 IEEE* 1-18

[8] Jaafar M S, Nordin N K, Wagiran R, Aziz A, Noor M J M M, Osman M R, Noorzaei J, Abdulaziz F N A 2008 Assessment Strategy for an Outcome Based Education *International Conference on Engineering Education*

[9] Moesby E 2004 Implementation of POPBL seen from an administration point of view *In Proceedings from International workshop on project organized and problem-based learning–POPBL* Aalborg University Esbjerg

[10] Seddugum Z S, Capretz L F A 2006 *Information and Communication Technologies, 2006. ICTTA'06. 2nd* **2** 3519-3521

[11] Secundo G, Elia G, Margherita A, Passiante G 2012 Student diversity in engineering education: Insights to build a glocal program *In Global Engineering Education Conference (EDUCON), 2012 IEEE* 1-8

[12] Ruiz-Gallardo J-R, Castano S, Gomez-Alday J J, Valdes A 2011 *Teaching and Teacher Education: An International Journal of Research and Studies* **27(3)** 619-627

[13] Felder R M, Felder G N, Dietz E J 2002 *Journal of Engineering Education* **91(1)** 3–17

[14] Bradley J H, Hebert F J 1997 *Journal of Management Development* **16(5)** 337-353

[15] Mazni O, Syed A , Sharifah L, Naimah M H 2010 *Science and Social Research (CSSR), 2010 International Conference on* , 624-628

[16] Seddigi Z S, Overton T L 2003 *Journal of Chemistry Education: Research & Practice* **4(3)** 387-395

[17] Topping K 1998 *Review of Educational Research* ***68*(3)** 249-276

[18] Venkata P K, Jacek U, Tunde Oladiran M, 2010 *World Transactions on Engineering and Technology Education* **8(1)** 56-60

Energy flow models for the estimation of technical losses in distribution network

Mau Teng Au[1] and Chin Hooi Tan[2]

[1] Power Engineering Centre, [2] College of Information Technology, Universiti Tenaga Nasional, Jalan IKRAM-UNITEN, 43000 Kajang, Selangor, Malaysia.

mtau@uniten.edu.my

Abstract. This paper presents energy flow models developed to estimate technical losses in distribution network. Energy flow models applied in this paper is based on input energy and peak demand of distribution network, feeder length and peak demand, transformer loading capacity, and load factor. Two case studies, an urban distribution network and a rural distribution network are used to illustrate application of the energy flow models. Results on technical losses obtained for the two distribution networks are consistent and comparable to network of similar types and characteristics. Hence, the energy flow models are suitable for practical application.

1. Introduction

The amount of energy losses in electrical distribution system is one of the key measures of distribution system performance as it has a direct impact on the utility's bottom line. Energy losses in distribution system can be attributed to technical losses and non-technical losses. Technical losses refers to energy losses resulting from the heating of electrical distribution components such as cables/lines and transformer windings, whereas non-technical losses is associated with the energy unaccounted for due to consumer pilferages, faulty energy meters, incorrect billing, etc. Therefore, technical losses in distribution systems are primarily dependent on network topology, system voltages, and electrical loading. Hence, the accuracy in estimating technical losses in a distribution system would require extensive network and load data for modeling. A variety of approaches to calculate energy losses in distribution system have been reported. Some of the variables commonly used to estimate energy losses in distribution network are load factor, loss factor and load curves/profile.

One of the earliest references on calculating energy loss in distribution system is written by Flaten [1] where percent loading of every distribution components in the distribution system and numerous other related variables, such as power factor, temperature correction, squares correction, load profile correction, etc. are used in the calculation. With numerous input parameters, the approach is seen to be extremely rigorous and requires tremendous effort and resources particularly in dealing with large utility distribution systems. In [2] Grainger presented a simplified approach using load loss factor and diversity factor to calculate energy loss in distribution system. However, the methodology involves a large amount of measured load data as well as network data to establish statistical and physically based models. In reference [3], Shenkman proposed an algorithm for energy loss calculation in network elements using statistical moments obtained by considering statistical daily load curves (DLC) as a set of random variables. Taleksi [4] further improved on the methodology by taking into account element ordering, power calculation with average loads at nodes and applied it to calculate energy loss in radial distribution network. More recent reference on estimating technical losses in distribution system is proposed by Dortolina, et la [5] where energy losses are computed in relative accuracy for specific distribution systems or feeders (known as bottom up approach) and then uses the results to benchmark (i.e estimate technical losses using top down approach) other parts of the network with matching features. The main advantage of the top down/bottom up approach is that it can handle varying degrees of data availability.

It can be seen from the various references [1-5] that the approach taken to estimate technical losses is generally dependent on the amount of data available, distribution system size and the accuracy required for each case. Generally, there is a trade off between amount of data available to estimate technical losses and the required accuracy of its results. With regards to data, most utility companies do maintain records of network data such as medium voltage feeder cable length and types, maximum demand, load profiles, and distribution transformers loading. Based on network information typically available such as those mentioned, this paper presents a methodology to estimate technical losses of utility distribution network using energy flow models associated with each medium and low voltage feeder, and average transformer loading. The approach is applied to estimate technical losses of a distribution system supplying loads to an urban and rural area and found to be efficient and practical. Additionally, the results obtained are within acceptable range which provides an indication of the level of technical losses to be expected from the particular distribution network.

2. Energy losses in network components

2.1. Underground cables and overhead lines

Power losses in underground cables and overhead lines are primarily due to load losses ($P_{Loss}= I^2R$). As the power losses varies with the square of current, the amount of current flowing in the cable will have the highest impact on the energy losses in cables and overhead lines.

Typically, load losses of feeders are calculated under peak demand condition using load flow simulation. The energy losses over a specified period (one month) is then calculated using Loss Factor as follows,

$$\text{Energy Loss}= P_{Loss} \times \text{Loss Factor} \times 24\times 30 \quad (1)$$

Loss Factor is calculated from Load Factor using the expression,

$$\text{Loss Factor} = \alpha\times\text{Load Factor} + \beta\times(\text{Load Factor})^2. \quad (2)$$

Where $0\leq \alpha \leq 0.35$, and $\alpha+\beta= 1.0$.

Input energy to the feeder can be calculated as follows,

$$\text{Input Energy} = \text{Peak Demand} \times \text{Load Factor} \times 24\times 30. \quad (3)$$

$$\text{Technical Losses of the Feeder (\%)} = (\text{Energy Loss}\times 100)/\text{Input Energy} \quad (4)$$

2.2. Power and distribution transformers

Power losses in transformers are contributed by load loss and core loss. Load loss varies with the square of transformer load current, whereas core loss is fixed loss resulting from magnetization and hysteresis loss. The expression to calculate total power loss in a transformer is given as,

$$P_{TX\_Loss} = P_{NL\_Loss} + (CF)^2 (P_{FL\_Loss}) \quad (5)$$

Where CF is the capacity factor defined as the ratio of peak load current over full load current of transformer.

Similarly, following which energy losses in a transformer over a specified period can be calculated using Loss Factor as given in (1) and (2).

3. Energy flow models

In general, the amount of technical losses in distribution systems is influenced by a number of factors such as:

- The geographical size of the network (related to feeder length),
- Quantity of electricity distributed against the capacity of distribution infrastructure, i.e, number of feeders and transformers, and their capacity (capacity factor),

- Ratio of MV network to LV network,
- Load profile (peak demand and load factor),
- Reactive power flow (power factor),
- Dispersion of loads in the network,

For utility distribution systems fed from grid system, energy meters are installed at main in-take substations to record the amount of energy supplied to the whole distribution network. For economic reasons, energy meters are usually not installed at individual outgoing distribution feeders to record its energy flow. Besides the energy meters installed at main in-take substations, other points of energy metering are at bulk customers substations to record energy sold to the respective customers.

In this paper, energy flow models developed to estimate technical losses of the whole distribution network are based on the constraint that only two sources of actual energy and peak demand data are available; that is, the total input energy and peak demand of the whole distribution system, as well as total bulk customers energy and peak demand fed from the distribution system.

Besides the energy and peak demand data set, the energy flow models require realistic distribution network data in order to estimate technical losses in the distribution system.

The network data are:- total numbers of medium voltage (e.g, 33 kV, 22 kV, and 11 kV) feeders, total length of medium voltage feeders according to voltage level, total number of low voltage feeders, and total installed capacity of power and distribution transformers (e.g 33/11 kV and 11/0.4 kV transformers) in the distribution system. In this energy flow model, the average length of low voltage feeders is fixed at 1 km.

The energy flow model then distribute the total input energy and peak demand accordingly to each medium voltage feeder, power and distribution transformer, and low voltage feeder, and calculates the amount of energy (technical) losses at each voltage level and for each components.

4. Case studies

Two case studies are presented in this paper. The first case is a 11/0.4 kV distribution system of an urban area, and the second case is a 33/11/0.4 kV distribution system of a rural area.

4.1. Case I – Urban distribution network

The characteristics of the urban distribution network are:- relatively short feeders, highly load feeders and transformers, and a significant proportion of its energy being sold to customers in bulk at medium voltage. Hence, technical losses of the urban network are expected to be lower.

4.2. Case II – Rural distribution network

The characteristics of the rural distribution network area:- relatively long feeders (two medium voltage level, 33 kV and 11 kV), lightly load feeders and transformers, and a small proportion of its energy being sold to customers in bulk at medium voltage. Hence, technical losses of the rural network are expected to higher.

The network data used to calculate technical losses of the networks are given in Table 1. The results of technical losses calculated based on the energy flow models are given in Table 2.

Technical losses for the rural distribution network are 5.4% compared to 4.97% for the urban network. The higher percentage in technical losses of the rural network is due primarily to technical losses at the medium voltage system (33 kV and 11 kV) where there are long feeders with an additional voltage transformation at 33/11 kV.

5. Conclusions

Energy flow models developed for the estimation of technical losses in distribution network produced results that are consistent with the characteristics of the respective distribution system and comparable with other models. The energy flow models have proven to be practical and efficient, and a useful tool that could be applied to estimate technical losses of distribution system which are in operation as well as for planning future network.

Table 1. Energy, Demand, and Network Data

Network Parameters	**Case I**	**Case II**
Energy Supplied	125,000 MWhr	98,000 MWhr
Peak Demand	250 MW	230 MW
Power Factor	0.9 lagging	0.9 lagging
Load Factor	0.69	0.59
Energy Sold at 33 kV Bulk Cust.	-	0.0
Energy Sold at 11 kV Bulk Cust.	35,000 MWhr	8,000 MWhr
Total Peak Demand of 11 kV Bulk Cust.	61 MW	13.9 MW
Total number of 33 kV Feeders	-	16
Average Length per Feeder (33 kV)	-	30 km
Total Installed Transformer Capacity (33/11kV)	-	800 MVA
Total number of 11 kV Feeders	80	300
Average Length per Feeder (11 kV)	15 km	25 km
Total Installed Transformer Capacity (11/0.4 kV)	600 MVA	500 MVA
Average % Loading of LV Feeders	50 %	40 %

Table 2. Results on Technical Losses

	Case I	Case II
Energy Supplied (Input)	125,000 MWhr	98,000 MWhr
Total Energy Loss in 33 kV Feeders	-	1,341 MWhr
Total Energy Loss in 33/11 kV Transformers	-	233 MWhr
Total Energy Loss in 11 kV Feeders	2,285 MWhr	630 MWhr
Total Energy Loss in 11/0.4 kV Transformers	770 MWhr	460 MWhr
Total Energy Loss in LV Feeders	3,160 MWhr	2,633 MWhr
Total Energy Loss in Distribution System	6,215 MWhr	5,297 MWhr
Percentage of Technical Losses	4.97 %	5.41 %

References

[1] David L Flaten 1988 Distribution system losses calculated by percent loading *IEEE Transactions on Power Systems* Vol. 3 No. 3.

[2] Grainger J and Kendrew T 1989 Evaluation of technical losses on electric distribution systems *CIRED.*

[3] Shenkman A 1990 Energy loss computation by using statistical techniques *IEEE Transactions on Power Delivery* Vol. 5 No. 1.

[4] Taleski R and Rajicic D 1996 Energy summation method for energy loss computation in radial distribution network *IEEE Transactions on Power Systems* Vol. 11 No. 2.

[5] Dortolina C and Nadira R 2005 The loss that is unknown is no loss at all: A top-down/bottom up approach for estimating distribution losses *IEEE Transactions on Power Systems* Vol. 20 No. 2.

Numerical Study of Cofiring Biomass with Coal in Cyclone Combustor.

I. I. Zulkepli[1], H. Hasini, A. M. Ikram , M. Z. Yusoff, A. Abd Rahman, A. H. Shamsuddin.
Centre for Renewable Energy, College of Engineering, Universiti Tenaga Nasional, Putrajaya Campus, Km7, Jalan Kajang-Puchong, 43009 Kajang, Selangor, Malaysia.

E-mail: irieyana@uniten.edu.my

Abstract. This paper investigates the flow and temperature distribution inside a cyclone combustor during combustion of coal and coal blend. The combustion under study is based on the actual cyclone combustor experiment rig used to test the performance of combustion when it be different coal-biomass blend is used. Experiment investigation on 100% coal combustion was also carried out and serves as a basis for the subsequent test for co-firing of different coal or biomass blend. Validation of temperature magnitude along cyclone furnace at 100% coal combustion condition shows good agreement between the measured and CFD results. Subsequent simulation of coal and biomass blend shows very good impact as it gave less error compared to experiment.
Keywords : *Computer Fluid Dynamic (CFD); Empty Fruit Bunch (EFB); cofiring.*

1. Introduction

Global warming is one of the serious consequences of the Greenhouse gas (GHG) emissions which could be attributed to the released of carbon based element into the atmosphere has become global concern since many decades ago. Coal is known to be an important source of and its reserve is estimated to be greater than the other type of fossil fuels [1]. Thus many countries around the world are shifting their fuel utilization by utilizing coal as part of the energy security measure. The benefit of coal is to ignite the combustion inside the combustor and it can be consider as largest reserve and cheapest option at this moment [2]. Coal contributes approximate 40% worldwide electricity generation to be main important role in the energy supply [3]. However, since the coal technologies are available and widespread, demand of the coal increased in the global market and the amount of it become lapse gradually. For the purpose to cover up the coal utilization, the cofiring term were introduced, it's actually mixture coal with biomass to achieve approximately 100% coal at the power plant. Using biomass as few portions to replace pulverised coal represent the greatest potential in existing coal fired power plant. Cost-effective can be save rather than use 100% biomass in the small boiler. Palm residues are abundant in Malaysia due to the fact that Malaysian palm based industry is very large. The palm waste could be divided into three sections which is Empty Fruit Bunch (EFB), Palm Kernel Shell (PKS) and Mesocarp Fiber [4]. Palm residue is suitable as the main biomass element because of the high and comparable heating value.

2. Governing equation

CFD solve the Reynolds Average Navier-Strokes equation in addition to the continuity and energy equations as shown in (1).

$$U_j \frac{\partial U_i}{\partial x_j} = -\frac{1}{P}\frac{\partial P}{\partial x_j} + \frac{\partial}{\partial x_j}\left[V\left(\frac{\partial U_i}{\partial x_j} + \frac{\partial U_j}{\partial x_i}\right) - \overline{u_i u_j}\right] \quad (1)$$

The turbulence is resolved using the k-epsilon (k-ε) turbulent model which consist of turbulence kinetic

[1] To whom any correspondence should be addressed.

energy and dissipation energy equation (ε) are given by (2) and (3) [5]. k-ε standard involve turbulence kinetic energy (k) and (ε) symbolize to dissipation energy equation that used for stress-strain

Relationship [6] [7]

$$\overline{u_i u_j} = \frac{2}{3} k\delta_i - V_t \left(\frac{\partial U_i}{\partial x_j} + \frac{\partial U_j}{\partial x_i}\right) \quad (2)$$

Calculate from

$$V_t = C_\mu \frac{k^2}{\varepsilon} \text{ and } C_\mu = 0.09 \quad (3)$$

V_t is represent transport equation which is solution by k and ε. $C\mu$ instead of constant in (4,5 and 6).

$$U_j \frac{\partial k}{\partial x_j} = \frac{\partial}{\partial x_j}\left[\left(V + \frac{V_t}{\sigma_k}\right)\frac{\partial k}{\partial x_j}\right] + P_k - \varepsilon \quad (4)$$

$$U_j \frac{\partial \varepsilon}{\partial x_j} = \frac{\partial}{\partial x_j}\left[\left(V + \frac{V_t}{\sigma_\varepsilon}\right)\frac{\partial \varepsilon}{\partial x_j}\right] + C_{\varepsilon 1}\frac{\varepsilon}{k}P_k - C_{\varepsilon 2}\frac{\varepsilon^2}{k} \quad (5)$$

$$\text{With}, P_k = -\overline{u_i u_j}\frac{\partial U_i}{\partial x_j} \quad (6)$$

$$\varepsilon = v\overline{\left(\frac{\partial U_i}{\partial x_j}\right)^2}$$

The coefficients values are,

σ_k=1.00, σ_ε = 1.30, $\sigma_{\varepsilon 1}$=1.44, $\sigma_{\varepsilon 2}$ =1.92.

In the numerical setup, the near wall function setup as non-equilibrium and assumes inflexible temperature. Non-premixed combustion includes all the data from the fuels into the inlet at the combustor, and it known as pre-probability density function (Pre-Pdf). Pre-PDF is to import both fuels into combustion process and it has been modelled using non-adiabatic process [5]. A discrete phase model (DPM) used to track trajectories of particles as shown in (7),

$$\frac{du_i^P}{dt} = F_D\left(u_i - u_i^P\right) + g_i(\rho_i - \rho)/\rho_P + F_i/\rho_P \quad (7)$$

3. Model Description

3.1. Cyclone Combustor

The cyclone type combustor is designed with several tangential inlets, located at the primary and secondary chamber [8]. It is able to produce highly turbulence and mixing of air and fuel as injected into the furnace. The mixture environment inside it also determines the firing shape to minimize corrosion on the furnace wall [9]. For the first stage at the primary chamber, fuel has devolatilization process but it will be complete combust in the secondary chamber. Figure 1 is top view of cyclone combustor and the flow inside it. The model developed at Gambit (CFD pre-processor software) before run into Fluent (CFD solver).The full shape and geometry followed the original cyclone combustor from Cardiff University and go through certain modification to be suit with fuel type in Malaysia.

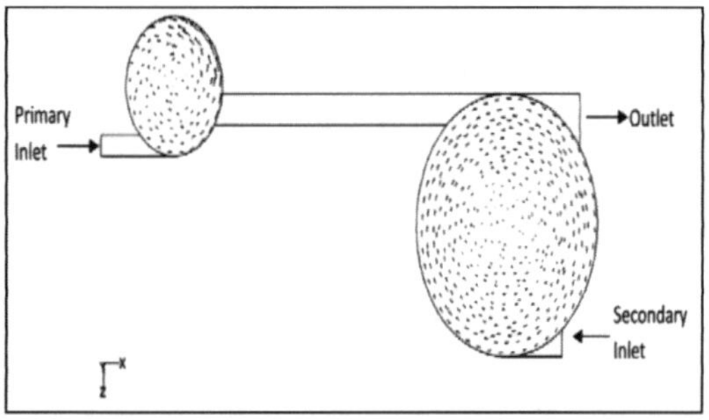

Figure 1. Top view of Cyclone Combustor.

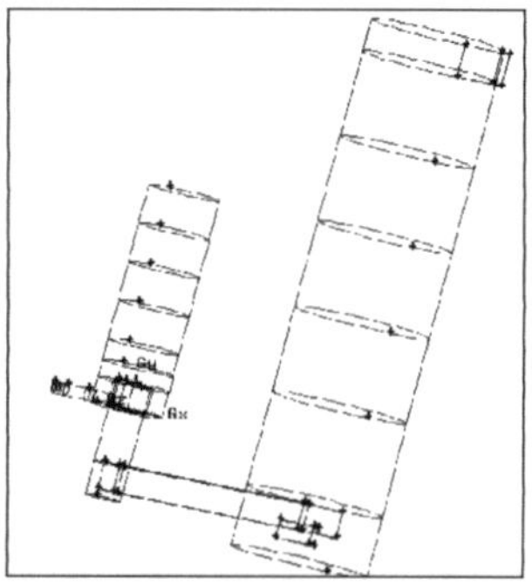

Figure 2. Model Cyclone Combustor in Gambit 3D

3.2.Mesh
Using Hexagonal type of the mesh structure on the geometry can assist it to generate fast and accurate [10]. Geometry selection have very acute angles for each volume and this condition must used

Tetrahedral mesh which is proper for 3D boundary mesh and unstructured geometry containing only triangular faces [10].

3.3 Boundary condition and Model assumption.
For primary chamber temperature is required 800K-1000K lower than the secondary temperature. It is of the combustion started, with devolatization of the fuel than it continues burn completely in the secondary chamber. Both chambers have inlets with different velocity because it includes air that has been divided by 25% and 75% individually with direction normal to boundary. Behavior on the combustion flow and temperature distribution inside the chambers, the model of solver must be noted using pressure based, 3D spaces, absolute velocity formation. Model that has been develop effects the turbulence swirling flow and temperature distribution at every sector inside the combustor. Modeling design is used steady system and activates the energy to see the results at the end of the simulation for the combustion process. Pre-PDF imported in this process to run the simulation until the temperature react and compared with experiment data. The error of the comparison must be lower than 10% for validation which is acceptable for combustion. Error is following the researcher in combustion field.

4. Result and Discussion.
Result at the end of the simulation shows the temperature data has been satisfied with the error percentage below 10% which is acceptable for combustion. The graph in figure 3 shows simulation data compared to the experiment and the model successfully working. The temperature is taken at the secondary chamber is using the thermocouples that installed.

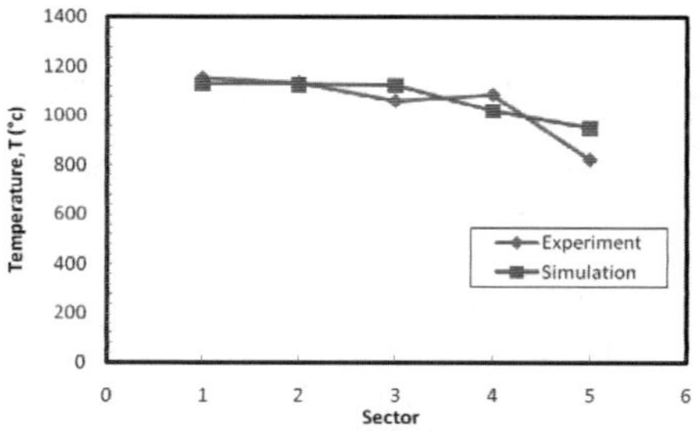

Figure 3. Experiment value versus simulation value of temperature distribution.

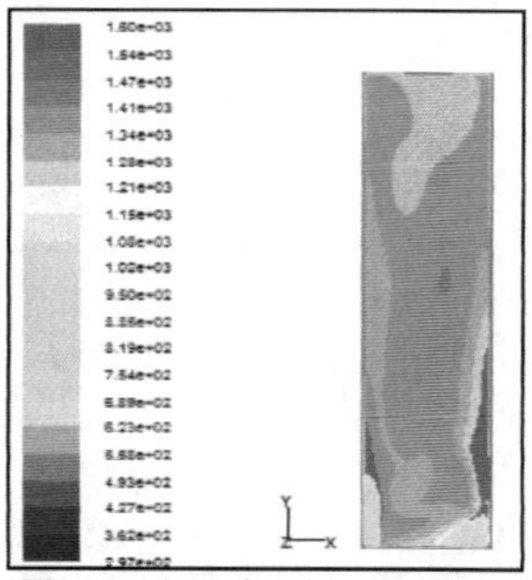

Figure 4. Behaviour flow of the 100% coal combustion.

Simulation on CFD on figure 4 is according to the ultimate analysis of Adaro coal, as shown in table 1. The data followed the ASTM standard and comparison based on the Cardiff University experiment on 100% Adaro coal [11]. Swirling flow of the combustion arrangement makes the bottom of the secondary chamber gain high temperature. It is because complete combustion happens at the region and require more heat to burn all the fuels. Top of it shows low temperature as reason fuels that burn become ash and release to the cyclone exhaust. Continues work has been done by the simulation of combination 97.5% Adaro coal and 2.5% EFB as shown in figure 5 and 95% Adaro coal and 5% EFB as shown in figure 6 which is using the existing model of 100 % Adaro coal as validation. The ultimate analysis for cofiring is based on the calculation of the ultimate analysis data for both fuels. The cooling zones appear at the bottom as the secondary chamber because relative air enters to make the fuel burn rapidly. Highest temperature for figure 5 and figure 6 is above 1600°c it's considered acceptable as combustion with the biomass fuels normally unstable. The contour shows the reactions between coal and biomass as in red in colour (high temperature) appear at the few sections at the bottom secondary combustor and become cool at the top.

Table1. Ultimate Analysis Adaro Coal

Ultimate Analysis	%
C	67.05
H	6.44
N	0.9
S	0.12
O_2	20.41
Gross CV (MJ/kg)	24.65

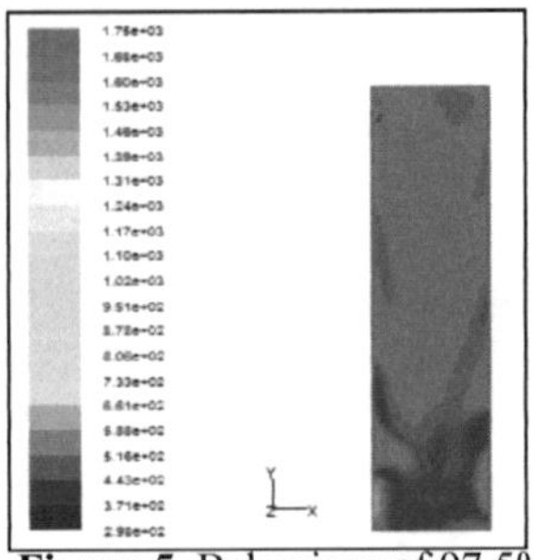

Figure 5. Behaviour of 97.5% Adaro coal + 2.5% EFB.

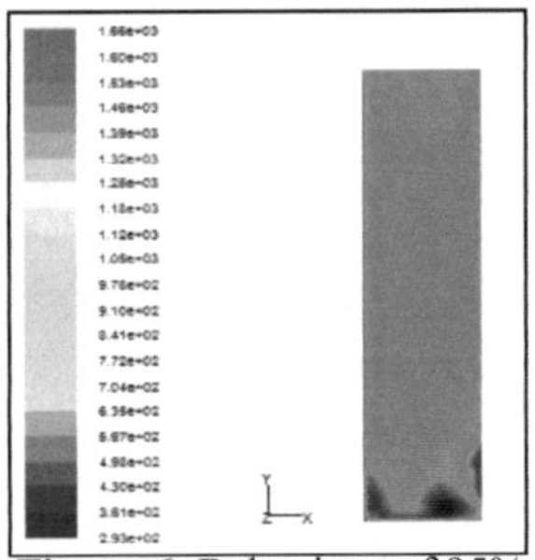

Figure 6. Behaviour of 95% Adaro coal + 5% EFB.

5. Conclusion

The combustion behaviour in cyclone combustor move the fuels and air together in swirling flow and give high temperature at the bottom of the secondary chamber. The fuels go through complete combustion at the bottom of the secondary chamber and leave the ash at the top of it. Model of the 100 % Adaro coal successfully working and can be validation to the cofiring process as it give 6.2% error to the experiment result which is acceptable for the combustion process. Standard (k-ε) turbulence model assist the simulation concept similar with the experimental result.

Acknowledgement

Authors would like to acknowledge for the financial supports given: (1) Long Term Research Grant (01201102 LRGS) from Ministry of Higher Education (MOHE) and Universiti Sains Malaysia (USM)

References

[1] Hashimoto N, Kurose R, Tsuji H, Shirai H 2007 *Energy & fuels* **21(4)** 1950-1958
[2] P.D.O. Energy *Energy Resources* 22 July 2012 [Online] http://www.doe.gov.ph/ER/Coal.htm. [Accessed 13 November 2012]
[3] Pallarés J, Gil A, Cortés C, Herce C 2009 *Fuel Processing Technology* **90(10)** 1207-1213
[4] Shamsuddin H 1989 *RERIC International Energy Journal* 11(2) 25-33
[5] Ikram A, Mostafa N A, Rahman A, Shamsuddin A H 2010 *The 6th International Conference on Combustion Incineration Pyrolysis and Emission Contro* Kuala Lumpur
[6] Bakke A 2002 *Lecture 10 Turbulence Model, Applied Computational Fluid Dynamic* [Online] http://www.bakker.org/dartmouth06/engs150/10-rans.pdf [Accessed 10 October 2012]
[7] Xia J L, Yadigaroglu G, Liu Y S, Schmidli J, Smith B L 1998 *International journal of heat and mass* transfer **41(11)** 1485-1497
[8] Syred C, Fick W, Griffiths A J, Syred N 2004 *Fuel* **83(17)** 2381-2392
[9] Ikram A, Rahman A A, Mostafa N A, Shamsuddin A H 2009 *In Energy and Environment, 2009 ICEE 2009 3rd International Conference on* 482-484
[10] Bakker A *Lecture 7 Meshing, Applied Computational Fluid Dynamic* 2002 [Online] http://www.bakker.org/dartmouth06/engs150/07-mesh.pdf. [Accessed 2012 October 10]
[11] Rahman A 2006 *Studies of Co-firing of Coal with Biomass on a Two Stage Simulator for Utility Boilers* Cardiff University

IOP Conf. Series: Earth and Environmental Science **16** (2013) 012037 doi:10.1088/1755-1315/16/1/012037

Modified Combustion Synthesis and Characterization of ZnO Nanoparticles Using Various Dispersants

N A A Razak[1*], H Misran[1], N N H Shah[1], M A Salim[1], S Z Othman[1]
[1]Center for Micro and Nano Engineering (CMNE), Universiti Tenaga Nasional, Jalan IKRAM-UNITEN Kajang, 43000 Selangor, Malaysia

E-mail: Afiqah@uniten.edu.my

Abstract. Zinc oxide (ZnO) nanoparticles with hexagonal wurtize structure were successfully synthesized *via* simple, cost-effective and environmental friendly modified combustion synthesis route. Three different type of dispersants namely glycerol, palm oil derived fatty alcohol (C8) and fatty ester (C12) were employed to produce the nanoparticles. X-ray diffraction patterns of calcined ZnO nanoparticles indicated the successful formation of ZnO. The crystallite sizes were at ca. 39.42 nm, 27.62 nm and 30.27 nm for ZnO produced using glycerol, fatty alcohol (C8) and fatty ester (C12). The morphology was of spindle-like shape for ZnO produced using glycerol and pseudo-spherical shape for ZnO produced using palm oil derived fatty alcohol (C8) and fatty ester (C12). Energy dispersive X-ray analyses showed the existence of zinc and oxygen peak suggesting successful formation of ZnO using various renewable dispersant.

1. Introduction

Metal oxide nanoparticles received immense interest in research due to their novel and difference in physical and chemical properties as compared to its bulk materials. There are several metal oxides nanoparticles produced for various applications such as copper oxide, nickel oxide and magnesium oxide [1-2]. Among these metal oxides, ZnO nanoparticles received increasing interest due to its various morphology, specific surface area and nanosize grain due to its exceptional properties such as wide band gap of 3.37 eV and a large exciton binding energy of 60 meV. These properties make ZnO nanoparticles a suitable candidate to be incorporated in dye-sensitized solar cells, semiconductor and other industrial applications [3-6]. For example, ZnO nanoparticles are considered as a suitable and cheaper alternative to titanium oxide in solar cells applications because it can replace the role of titanium as effectively [7].

In previous research, various techniques were developed to produce ZnO nanoparticles such as sol-gel method, direct precipitation method, vapour method and microwave-assisted combustion synthesis method [8-13]. However, these methods need to be performed under strict condition which is not favourable for low-cost production. Among them, combustion synthesis method is considerably simple, fast and cost-effective as compared to other methods. This is due to shortened reaction time to produce metal oxide powders and it does not require the used of catalysts to obtain the final product. In this paper, a simple method to prepare ZnO nanoparticles employing new and renewable palm oil derived fatty alcohols and fatty ester using modified combustion synthesis as fast, cost-effective and

* To whom any correspondence should be addressed.

environmental friendly method is reported. Palm oil derived fatty alcohol was reported as a substitute for surfactant to generate carbon template for the formation of silica spheres [14].

2. Materials and methods

2.1. Materials
Zinc nitrate hexahydrate ($Zn(NO_3)_2.6H_2O$) (Sigma-Aldrich, Reagent grade, 98%) was used as the metal precursor and oxidant and three different dispersants which were glycerol (Sigma-Aldrich, 99%), palm oil derived fatty ester (12 carbon chains, 99.7%, Cognis) and palm oil derived fatty alcohol (8 carbon chains, 98%, Emery) were used as received.

2.2. Synthesis of ZnO Nanoparticles
Zinc nitrate hexahydrate ($Zn(NO_3)_2.6H_2O$) act as dual role as metal precursor and oxidant. The ratio of the metal precursor to the dispersant is 2.5:1 and was directly mixed on the heated plate for glycerol and fatty alcohol prior to combustion. The initial temperature was set at 50°C and was raised systematically until the mixture combusted to obtain as-synthesized material. The as-synthesized material is subjected to calcination at 600°C to obtain white ZnO powder. Dry fatty ester was heated using hot plate at initial temperature 50°C until it liquefies. ($Zn(NO_3)_2.6H_2O$) was added in the dissolved fatty ester at 100°C and raised gradually until combustion occured. As-synthesized ZnO particles were calcined at 600°C to obtain ZnO nanoparticles.

2.3. Structural Characterizations
The calcined samples were subjected to characterization by X-ray diffraction (XRD), field emission scanning electron microscopy (FE-SEM) and energy dispersive x-ray (EDX). XRD patterns of the samples were obtained using Shimadzu XRD-6000 with CuKα radiation (30 kV and 20 mA) at 2θ range of 20° to 70°. The morphological feature and elemental analysis of the powder were investigated through FE-SEM using JEOL JSM-7600F field emission scanning electron microscope attached with EDX spectrometer at acceleration voltage of 5.0 kV for FE-SEM and acceleration voltage 20 kV for EDX. The powder being dispersed in ethanol before sonicated for 30 minutes. The homogenous solution was dropped on aluminium coated carbon tape and dried in air. Scherrer equation (1) was applied to the XRD patterns to estimate the crystallite sizes at the highest reflection peak (101) using FWHM method.

$$D = \frac{0.9\lambda}{B\cos\theta} \qquad (1)$$

Where D is crystalline size in nanometer (nm), λ is the wavelength of the radiation (1.54056 Å for CuKα radiation), B is the full width at half-maximum (FWHM-in radian) intensity and θ is the diffraction angle

3. Results and Discussion

3.1. Structural Characteristics of ZnO
The XRD patterns for ZnO nanoparticles using glycerol, fatty alcohol (C8) and fatty ester (C12) are shown in figure 1. All the samples exhibited three prominent peaks of (100), (002) and (101) reflection plane attributable to hexagonal wurtzite structure of ZnO. There are no other peaks observed in the XRD patterns suggesting no other phases formed during the combustion. The results shown in table 1 suggested that the crystallite size was significantly influenced by the type of dispersant used during combustion synthesis.

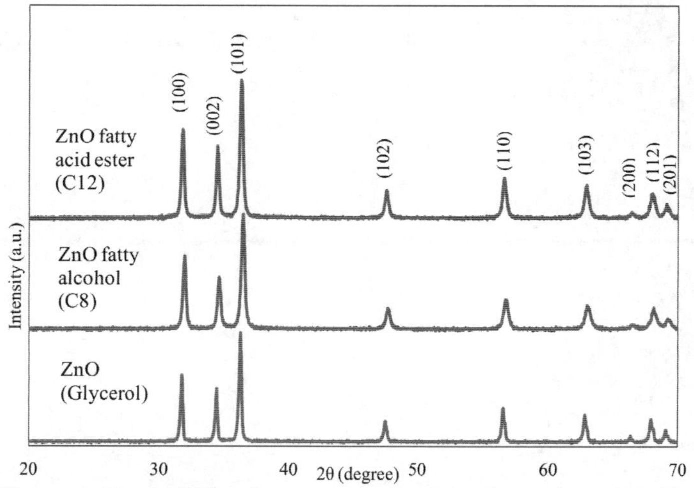

Figure 1. X-ray diffraction patterns of calcined samples of ZnO produced using glycerol, fatty alcohol (C8) and fatty ester (C12).

Table 1. The crystalline size of ZnO nanoparticles

Dispersants	**D (nm)**
Glycerol	39.42
Fatty acid (C8)	27.62
Fatty ester (C12)	30.27

3.2 Morphology and Elemental Analyses

The morphology of calcined samples was investigated using FESEM with acceleration voltage of 5 kV. Figure 2 shows the FESEM image of ZnO nanoparticles using glycerol as dispersant. As shown in figure 2, ZnO nanoparticles produced using glycerol as dispersant exhibited elongated spindle-like shape particles and the particles showed considerable necking due to high calcinations temperature. Figure 3 show the FESEM image of ZnO nanoparticles using C8 fatty alcohol as dispersant. The particles exhibited agglomerated and pseudo-spherical in shape at ca. 100 nm. Figure 4 shows the FESEM image of ZnO nanoparticles using C12 fatty esters as dispersant. It can be observed that the particles were pseudo-spherical in shape at ca. 100 nm with considerable necking due to high temperature. Agglomerations and necking of particles observed in the images were due to the high calcinations temperature. The EDX spectra for all 3 samples exhibited zinc and oxygen peaks suggesting the formation of ZnO.

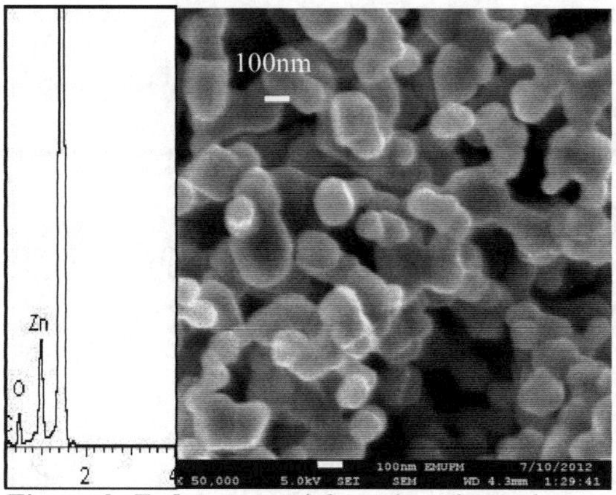

Figure 2. ZnO nanoparticles using glycerol.

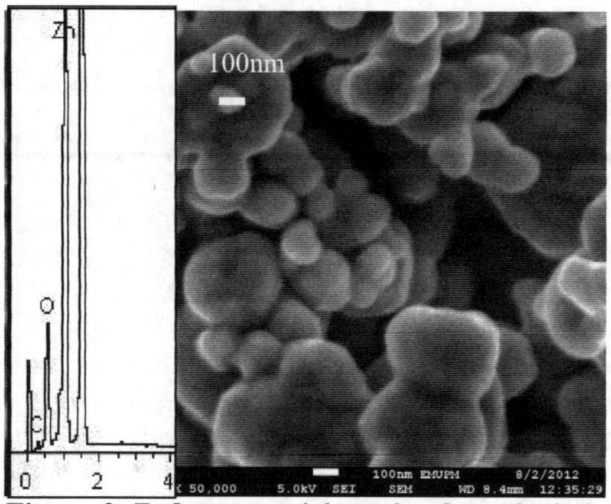

Figure 3. ZnO nanoparticles using fatty alcohol (C8).

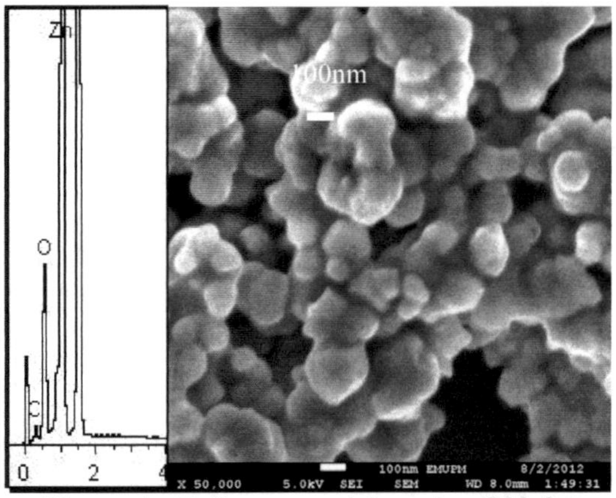

Figure 4. ZnO nanoparticles using C12 fatty acid esters (C12).

4. Conclusions

ZnO nanoparticles with hexagonal wurtzite structures were successfully synthesized *via* combustion synthesis method. New and renewable palm oil derived fatty alcohol (C8), palm oil derived fatty ester (C12) and glycerol successfully played the role of dispersants to produce ZnO. No other phases formed during the combustion and the crystallite sizes were significantly influenced by the dispersants. Thus, it can be concluded that crystallite size produced by combustion synthesis can be reduced and controlled affectively in a simple, quick and cost-effective methods.

Acknowledgment

The authors thank Ministry of Higher Education (MOHE) under FRGS scheme (20110101 FRGS) and Universiti Putra Malaysia (UPM) for facilities.

References

[1] Illy-Cherrey S, Tillement O, Dubois J M, Massicot F, Fort Y, Ghanbaja J and Be´gin-Colin S 2002 *Materials Science and Engineering A* **338** 70
[2] Saravanan M, Dhivakar S and Jayanthi S S 2012 *Material Letters* **67** 128
[3] Bai J, Meng F, Wei C, Zhao Y, Tan H and Liu J 2011 *Ceramics – Silikáty* **55** 20
[4] Ianos R, Lazau I, Pacurariu C and Sfirloaga P *Materials Chemistry and Physics* 2011 **129** 881
[5] Ghotbi M Y 2010 *Journal of Alloys and Compounds* **491** 420
[6] Hana D Y, Yang H Y, Shena C B, Zhou X, Wang F H, 2004 *Powder Technology,* **147** 113
[7] Jang Y J, Simer C and Ohm T, 2006 *Materials Research Bulletin* **41** 67
[8] Yuan D, Wang G S, Xiang Y, Chen Y, Qiao X, Lin G, 2009 *Journal of alloys and compounds* **478** 489
[9] Lin C S, Hwang C C, Lee W H, Tong W Y, 2007 *Material Science & Engineering B* **140** 31
[10] Znaidi L *Materials Science and Engineering, B*, 2010 **174** 18
[11] Chen C, Liu P and Lu C *Chemical Engineering Journal* 2008**144** 509
[12] Hwang C and Wu T 2004 *Materials Science and Engineering, B* 2004 **111** 197
[13] Nehru L C, Swaminathan V and Sanjeeviraja C 2012 *Powder Technology* **226** 29
[14] Misran H, Yarmob M A and Ramesh S 2013 *Ceramics International* **39** 931
[15] Patil K C, Aruna S T, Mimani T, *Current Opinion in Solid State and Materials Science* 2002 **6** 507
[16] Jayalakshmi M, Palaniappa M and Balasubramaniam K 2008 *Int. J. Electrochem. Sci* **3** 96

Modified Combustion Synthesis of ZnO Nanoparticles Using Renewable Fuel

N N H Shah[1, *], H Misran[1], N A A Razak[1], M A Salim[1], S Z Othman[1], A Manap[1]
[1] Center of Micro and Nano Engineering, Universiti Tenaga Nasional, Jalan IKRAM-UNITEN, 43000 Kajang, Selangor, Malaysia.

E-mail: Norliyana@uniten.edu.my

Abstract. Zinc oxide nanoparticles were successfully synthesized using cost effective modified combustion synthesis route employing zinc nitrate hexahydrate as oxidant and palm oil derived C8 fatty alcohol as fuel in various ratios. The calcined zinc oxide exhibited hexagonal phase with wurtzite structure. The crystallite sizes of the samples were ranging from ca. 28 to 40 nm. Therefore, palm oil derived C8 fatty alcohol successfully played the role as low-cost and renewable fuel to synthesize zinc oxide nanoparticles.

1. Introduction

There are vast amount of studies conducted on zinc oxide (ZnO) nanoparticles. This is due to their unique physical, chemical and optical properties making it a potential material for a wide range of applications [1]. ZnO nanoparticles have been applied as electrode in solar cells, catalysts, varistor and in sun protection lotions [2-5]. Various morphology and size of ZnO nanoparticles makes them suitable for different applications. For example, ZnO nanoparticles with high specific surface area enhance photocatalytic ability in the visible light region [6].

ZnO nanoparticles with specific size and morphology can be tailored by choosing the suitable synthesis method to suit desired applications. Various methods have been reported to synthesize ZnO nanoparticles resulting in diverse morphology and particle size. Methods such as pulse laser ablations and thermal decomposition required the use of complicated and expensive equipments [7-8]. In addition, microemulsion and direct precipitation method required strict control of reaction condition such as pH of the precursor solution, temperature and time to obtain the desired nanoparticles [9-10]. Therefore, combustion synthesis is proposed as a simple, fast and economical approach to produce ZnO nanoparticles.

Combustion synthesis involves the redox reaction between an oxidizing reagent usually comprises of desired metal salts and a reducing reagent (organic compound acting as fuel) [4]. Previous studies used various fuels such as urea, glycine and citric acid to produce ZnO nanoparticles using combustion synthesis method resulting various morphologies, sizes and properties [1,4,11]. In this work, palm oil derived fatty alcohol was introduced as a new and renewable fuel to synthesize ZnO nanoparticles. Previously, palm oil derived fatty alcohol was used as organic template to produce mesoporous silica nanoparticles [12]. Palm oil derived fatty alcohol bearing eight carbon chain (C8) with various ratio were used to synthesize ZnO nanoparticles. Then, the morphologies and structural properties of the synthesized samples were investigated.

* To whom correspondence should be addressed.

2. Materials and Methods

2.1. Materials
All materials used in this study were of analytical grade and were used as received. The materials used were zinc nitrate hexahydrate, $Zn(NO_3)_2.6H_2O$, reagent grade 98% from Sigma-Aldrich and C8 fatty alcohol from Emery Oleochemicals (M) Sdn. Bhd.

2.2. Synthesis
Metal precursor was $Zn(NO_3)_2.6H_2O$ while the fuel is palm oil derived C8 fatty alcohol. The $Zn(NO_3)_2.6H_2O$ and C8 fatty alcohol formed almost homogeneous mixtures. The mixture was heated with an initial temperature of 50°C and was increased systematically until combustion occurred at around 200°C producing as-synthesized ZnO powder. The as-synthesized powder was subsequently subjected to calcination at 600°C to obtain the calcined ZnO powder. The calcined samples were named ZnO-1 (ratio 2), ZnO-2 (ratio 2.5), ZnO-3 (ratio 3) and ZnO-4 (ratio 3.5).

2.3. Characterizations
In order to identify the crystalline phase and crystal structure of calcined samples, X-ray diffraction (XRD) analyses were carried out using Shimadzu XRD-6000 diffractometer with CuKα radiation. The crystallite sizes of the samples were determined using Scherrer equation, $d = 0.9\lambda/\beta\cos\theta$ where d is the crystallite size in nm, β is the full width at half maximum (FWHM-in radian) and θ is the Bragg diffraction angle. The crystallite sizes of the calcined ZnO were estimated based on diffraction plane with highest intensity. The morphology of the calcined samples were obtained using field emission scanning electron microscope (FE-SEM, JEOL JSM-7600F) at an acceleration voltage of 5.0 kV. Prior to FE-SEM analyses, a small amount of the samples were dispersed in ethanol and sonicated to form a homogeneous mixture. Then, the homogeneous mixture was dropped onto an aluminium covered carbon taped sample holder and dried in air. The nitrogen adsorption-desorption analyses for specific surface area were completed using Quantachrome Autosorb iQ at liquid nitrogen temperature. Consequently, specific surface area of the samples were estimated by BET method using the adsorption branch of the isotherms.

3. Results and Discussion

3.1. Structural properties
The XRD patterns for the samples are shown in figure 1a). All samples exhibited nine diffraction peaks assigned to the formation of ZnO with hexagonal wurtzite structure. The peaks were at ca. 2θ = 31.9°, 34.6°, 36.4°, 47.7°, 56.7°, 63.0°, 66.2°,68.0° and 69.1° corresponding to (100), (002), (101), (102), (110), (103), (200), (112) and (201) diffraction planes. The narrow and high intensity of diffraction peaks in the XRD patterns suggested high crystallinity of calcined ZnO powder. No other peaks were observed in the XRD patterns, suggesting that no other phases were formed during the synthesis. The crystallite size for ZnO-1 was ca. 28.5 nm while for ZnO-2 was ca. 27.6 nm. ZnO-3 has the highest crystallite size of ca. 40 nm while ZnO-4 was 39.4 nm. However, the crystallite size did not exhibit an increasing or decreasing trend with increased of oxidant-to-fuel ratio. Figure 1b) shows the variation of crystallite size as a function of oxidant-to-fuel ratio. The crystallite size suddenly increased for the ratio of ZnO-2 and ZnO-3 suggesting that it can be further adjusted between these ratios.

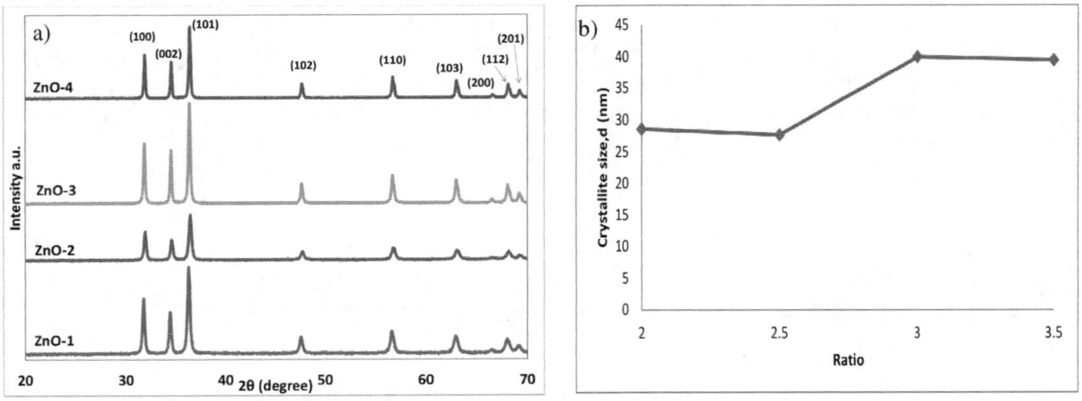

Figure 1. a) XRD patterns for calcined ZnO, b) Crystallite size as a function of oxidant-to-fuel ratio.

3.2. Morphology analysis

The FE-SEM images showed that the particles were agglomerated with irregular shapes and sizes. The agglomeration of particles can be classified as hard (strong bonding) or soft (weak bonds by Van de Waals forces) as reported [12]. However, some particles were observed to be less than ca. 100nm. Particles that are less than 1μm tend to agglomerate due to high energy and surface area [13].

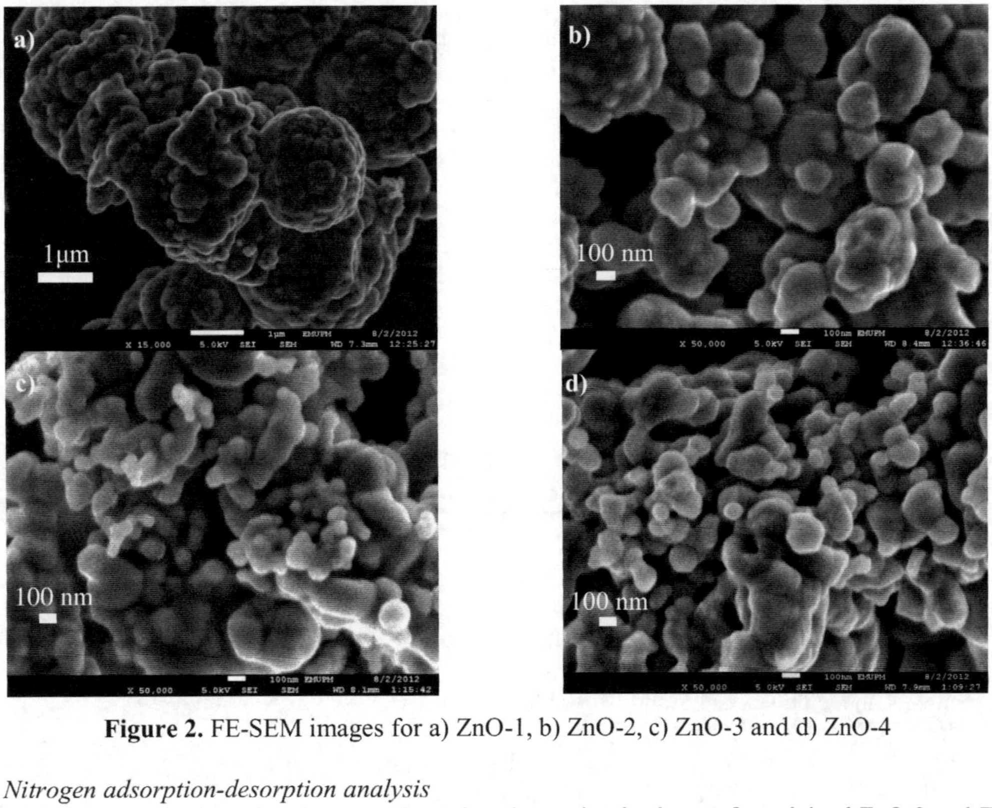

Figure 2. FE-SEM images for a) ZnO-1, b) ZnO-2, c) ZnO-3 and d) ZnO-4

3.3. Nitrogen adsorption-desorption analysis

Figure 3 and figure 4 show the nitrogen adsorption-desorption isotherms for calcined ZnO-2 and ZnO-3, respectively. The samples showed similar isotherm patterns exhibiting Type II normally obtained for non-porous materials. At relative pressure above 0.95, the amount adsorbed was high suggesting that the adsorption occurred on the external surface area of the samples. The inset in figure 3 shows the adsorption branches considered to estimate the specific surface area using BET method. Lower crystallite size for ZnO-2 resulted in higher specific surface area of 3.5 m^2/g. On the other hand, ZnO-3 with highest crystallite size of 40 nm has a specific surface area of 1.8 m^2/g. The low value of specific surface area was probably due to agglomeration of particles as observed in SEM images.

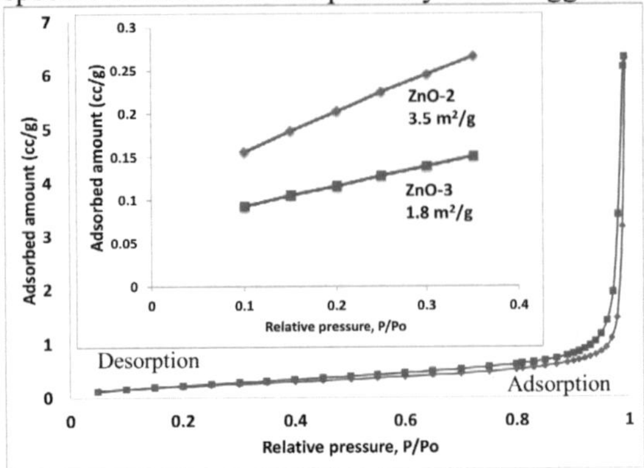

Figure 3. ZnO-2 adsorption-desorption isotherm.

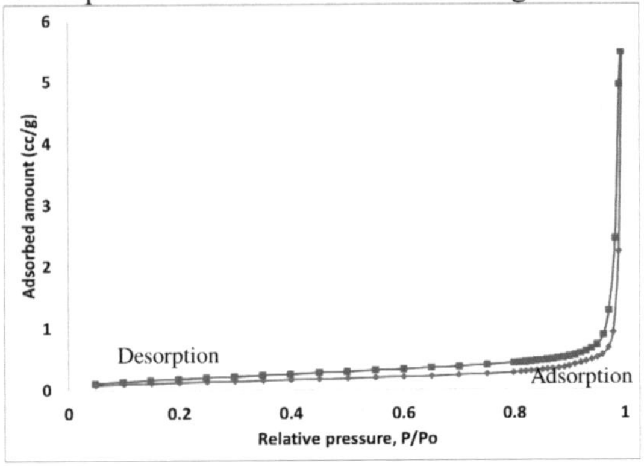

Figure 4. ZnO-3 adsorption-desorption isotherm.

4. Conclusion

ZnO nanoparticles were successfully synthesized *via* modified combustion synthesis employing a new and renewable fuel, palm oil derived C8 fatty alcohol. The calcined samples exhibited ZnO with hexagonal wurtzite structure and the crystallite size range was between 28 to 40 nm. The low value specific surface area of the samples which were 3.5m^2/g and 1.8m^2/g were probably due to agglomeration of particles. Therefore, palm oil derived C8 fatty alcohol successfully played the role as new, renewable and low-cost fuel in producing ZnO nanoparticles.

5. Acknowledgement

The authors thank Ministry of Higher Education (MOHE) under FRGS scheme (FRGS20110101) and Universiti Putra Malaysia for facilities.

References

[1] Kumar V R, Kavitha V T, Wariar P R S, Nair S U K and Koshy J 2011 *J. Physics & Chem. Solids* **72** 290
[2] Rani S, Suri P, Shishodi P K and Mehra R M 2008 *Solar Energy Mat. & Solar Cells* **92** 1639
[3] Jang Y J, Simer C and Ohm T 2006 *Mat. Research Bulletin* **41** 67
[4] Sousa V C, Segadaes A M, Morelli M R and Kiminami R H G A 1999 *Itnl. J. Inorganic Mat.* **1** 235
[5] Strunk J, Kahler K, Xia X and Muhler M 2009 *Surf. Science* **603** 1776
[6] Ahmad M, Hong Z, Ahmed E, Khalid N R, Elhissi A and Ahmad W *Ceramics International*
[7] He C, Sasaki T, Shimizu Y and Koshizaki N 2008 *Appl. Surf. Science* **254** 2196
[8] Bigdeli F, Morsali A and Retailleau P 2010 *Polyhedron* **29** 801
[9] Sui X, Liu Y, Shao C, Liu Y and Xu C 2006 *Chemical Physics Lett.* **424** 340
[10] Chen C C, Liu P and Lu C H 2008 *Chemical Engr. Journal* **144** 509
[11] Tarwal N L, Jadhav P R, Vanalakr S A, Kalagi S S, Pawar R C, Shaikh J S, Mali S S, Dalavi D S, Shinde P S, Patil P S 2011 *Powder Tech.* **208** 185
[12] Misran H, Singh R and Ambar Yarmo M 2008 *Microporous & Mesoporous Mat.* **112** 243
[13] Trommer R, Alves A K and Bergmann C P 2010 *J. Alloys & Compounds* **491** 296

IOP Conf. Series: Earth and Environmental Science **16** (2013) 012039 doi:10.1088/1755-1315/16/1/012039

Numerical Solution of Wave Propagation in Viscoelastic rods (Standard Linear Solid Model)

Abu Bakar Musa[1]
Dept. of Engineering Science, Mathematics and Computing, College of Foundation and General Stidies, University Tenaga Nasional

Email: bakar@uniten.edu.my

Abstract. The studies is about the propagation of waves through a short rod (or slug) of viscoelastic material. The viscoelastic material are modelled as standard linear solids which involve three (3) material parameters and the motion is treated as one-dimensional. In this study, a viscoelastic slug is placed between two semi-infinite elastic rods and a wave initiated in the first rod is transmitted through the slug into the second rod. The objective is to relate the transmitted signal to the material parameters of the slug. We solve the governing system of partial differential equations using Laplace transform. We invert the Laplace transformed solution numerically to obtain the transmitted signal for several viscosity time constants and ratios of acoustic impedances. In inverting the Laplace transformed equations, we used the complex inversion formula because there is a branch cut and infinitely many poles within the Bromwich contour. Finally, we discussed the relationship between the viscosity time constants, ratios of acoustic impedances and the results of the interface velocity discontinuities.

1 Introduction

There are materials for which a suddenly applied and maintained state of uniform stress induces an instantaneous deformation followed by a flow process which may or may not be limited in magnitude as time grows [1]. These materials are said to exhibit both an instantaneous elasticity effect and a creep characteristic. This behavior clearly can not be described by either elasticity or viscosity theories alone as it combines features of each and is called viscoelastic. Viscoelasticity is a generalization of elasticity and viscosity. The ideal linear elastic element is the spring whilst the ideal linear viscous element is the dashpot. Energy is stored in springs as elastic strain energy and energy is dissipated in a dashpot as heat [2]. This paper deals with the transmission of waves through a viscoelastic material which is modeled by a small rod (slug) of the material placed between two semi-infinite elastic rods. The purpose of this paper is to determine the properties of the viscoelastic slug after a wave initiated in the first semi-infinite elastic rod is transmitted through the viscoelastic slug into the semi-infinite elastic rod.

In this problem, we investigate the bahaviours of the longitudinal waves after the waves transmitted through a finite length viscoelastic slug. We model the viscoelastic material as a standard linear solid (Figure 1). Many dynamical systems are so complex that analytical solutions cannot be found and numerical solution is the only answer. Here, we numerically compute the waves transmitted through the slug into the second semi-infinite elastic rod. Results are obtained for several viscosity time constants and several ratios of acoustic impedance

2 Mathematical Model of Wave Propagation in Viscoelastic Material

We model the system where a finite length viscoelastic slug is placed between two semi-infinite elastic rods as an idealization of the experimental work described by H. Kolsky [4] as shown in Figure 2. A velocity discontinuity wave is initiated in the first rod and propagates through the viscoelastic slug moving into the second semi-infinite elastic rod. There are multiple reflections and transmission of waves in the viscoelastic slug. When the waves arrive at the interface between the first rod and the slug, some of the waves are

[1] To whom any correspondence should be addressed.

transmitted and some of them are reflected. The same situation happens when the waves reach the second interface between the slug and the second rod.

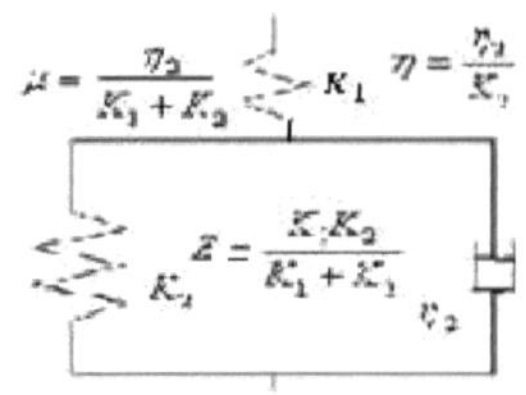

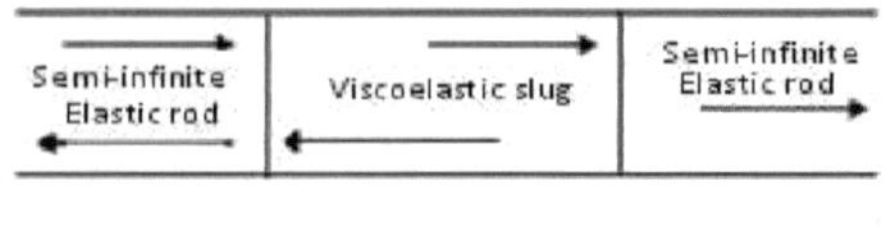

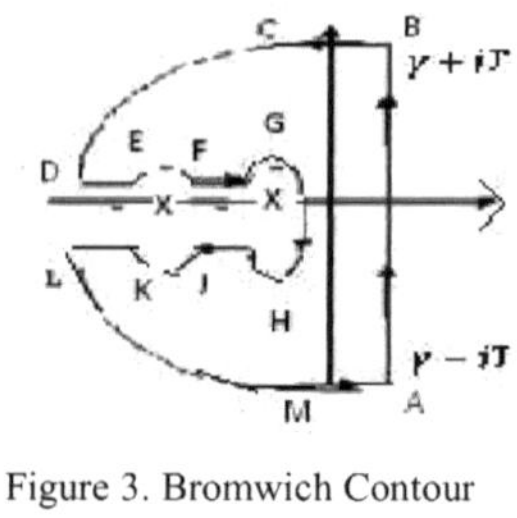

Figure 1. Standard Linear Solid Model Figure 2. Wave Movement in the Rods and Slug Figure 3. Bromwich Contour

3 Governing Equations

Let $\breve{u}$ be the additional displacement in the first rod, and $\breve{v}$ and $\breve{w}$ be the displacements in the slug and in the rod following the wave propagation respectively, where the notation $\breve{}$ indicates dimensional variables. Let $\breve{\sigma}_v$ be the stress in the slug, $\breve{\sigma}_u$ and $\breve{\sigma}_w$ be the stress in the first rod and in the second rod respectively, $\bar{E}$ and E be young's modulus in the slug and in the rod respectively, $\bar{\rho}$ and ρ be the density in the slug and in the rods respectively. We choose the origin of coordinates at the center of the first interface and axis $0\breve{X}$ along the axis of the slug and we assume the wave reaches the viscoelastic slug at time $\breve{t} = 0$. When the wave, initiated in the first rod with constant velocity discontinuity V, reaches the slug at time $\breve{t} = 0$ and $\breve{X} = 0$, we can write the position at time $\breve{t}$ of the cross-section of the first rod which was at location $\breve{X}$ at time $\breve{t} = 0$ as

$$\breve{x}(\breve{X},\breve{t}) = \breve{X} + V\left(\breve{t} - \frac{\breve{X}}{c}\right)H\left(\breve{t} - \frac{\breve{X}}{c}\right) + \breve{u}(\breve{X},\breve{t}) \qquad \text{for} \quad \breve{X} \le 0 \qquad (1)$$

Where $H(t)$ is the Heaviside function and V is a constant. At time $\breve{t}$, the cross-section of the slug which at location $\breve{X}$ at time $\breve{t} = 0$ is $\breve{x}(\breve{X},\breve{t})$ given by

$$\breve{x}(\breve{X},\breve{t}) = \breve{X} + \breve{v}(\breve{X},\breve{t}) \qquad \text{for} \quad 0 \le \breve{X} \le h_s \qquad (2)$$

Where h_s is the length of the slug. We write the position at time $\breve{t}$ of the cross-section of the second rod which at location $\breve{X}$ at time $\breve{t} = 0$ as

$$\breve{x}(\breve{X},\breve{t}) = \breve{X} + \breve{w}(\breve{X},\breve{t}) \qquad \text{for} \quad \breve{X} \ge h_s \qquad (3)$$

The stress in the body is denoted by $\breve{\sigma}(\breve{X},\breve{t})$ and at the cross-section X, the cross-sectional area is A.

$$\left(\breve{\sigma} + \frac{\partial \breve{\sigma}}{\partial \breve{X}}\delta \breve{X}\right)A - \breve{\sigma}A = \rho A\frac{\partial^2 \breve{u}}{\partial \breve{t}^2}\delta \breve{X} \qquad \frac{\partial \breve{\sigma}}{\partial \breve{X}} = \rho\frac{\partial^2 \breve{u}}{\partial \breve{t}^2} \qquad (4)$$

Then the equation of motion [6] in the first rod is $\frac{\partial \breve{\sigma}_u}{\partial \breve{X}} = \rho\frac{\partial^2 \breve{u}}{\partial \breve{t}^2}$ (5)

The equation of motion in the slug is

$\frac{\partial \breve{\sigma}_v}{\partial \breve{X}} = \bar{\rho}\frac{\partial^2 \breve{u}}{\partial \breve{t}^2}$ and the equation of motion in the second rod is $\frac{\partial \breve{\sigma}_w}{\partial \breve{X}} = \rho\frac{\partial^2 \breve{u}}{\partial \breve{t}^2}$ (6)

We model the slug as standard linear solid so that the constitutive equation in the slug is

$$\breve{\sigma}_v + \breve{\bar{\mu}}\frac{\partial \breve{\sigma}_v}{\partial \breve{t}} = \bar{E}\left(\frac{\partial \breve{v}}{\partial \breve{X}} + \breve{\bar{\eta}}\frac{\partial^2 \breve{v}}{\partial \breve{t}\,\partial \breve{X}}\right) \qquad (7)$$

The stress-strain relations in the rods are

$$\breve{\sigma}_u = E\left(-\frac{V}{c}H\left(\breve{t} - \frac{\breve{X}}{c}\right) + \frac{\partial \breve{u}}{\partial \breve{X}}\right) \qquad (8)$$

$$\breve{\sigma}_w = E\frac{\partial \breve{w}}{\partial \breve{X}} \quad \text{Where } c^2 = \frac{E}{\rho},\ \bar{c}^2 = \frac{\bar{E}}{\bar{\rho}},\ z = \frac{\rho c}{\bar{\rho}\bar{c}} \text{ and } \alpha = \frac{c}{\bar{c}} \qquad (9)$$

Now we non-dimensionalize equations (1), (2) and (3), we obtained

$$x = X + \frac{V}{c}(t - X)H(t - X) + \frac{V}{c}u(X,t) \qquad \text{for} \quad X < 0 \qquad (10)$$

$$x = X + \frac{V}{c}\alpha v(X,t) \qquad \text{for} \quad 0 \le X \le 1 \qquad (11)$$

$$x = X + \frac{V}{c} w(X, t) \quad \text{for} \quad X > 1 \tag{12}$$

We then non-dimensionalize (5) – (9), take Laplace transform the equations to obtain and solve the differential equations for the displacement transforms $\hat{u}, \hat{v}, \hat{w}$ in the s domain. We obtain

$$\hat{u}(X,s) = a(s)e^{sX}, \tag{13}$$

$$\hat{v}(X,s) = b(s)\exp\left(\frac{-\alpha sX}{\overline{\beta}(s)}\right) + d(s)\exp\left(\frac{\alpha sX}{\overline{\beta}(s)}\right) \tag{14}$$

$$\hat{w}(X,s) = f(s)e^{-sX} \quad \text{Where} \quad \overline{\beta}^2(s) = \frac{1+\alpha\overline{\eta}s}{1+\alpha\overline{\mu}s} \tag{15}$$

In order to find $a(s), b(s), d(s)$ and $f(s)$, we apply the boundary conditions described
(i)The particle velocity in the first rod is equal to the velocity in the slug at the first interface.($X = 0$)
(ii)The stresses in the first rod and in the slug at $X = 0$ are equal
(iii)The velocities in the slug and in the second rod are equal at the second interface($X = 1$)
(iv) The stresses in the slug and in the second rod at $X = 1$ are equal
We obtain

$$\hat{x} - \hat{X} = -\frac{2}{s^2}\sinh(sX) + \frac{2ze^{sX}\left(e^{\frac{2\alpha s}{\overline{\beta}(s)}}(\overline{\beta}(s)+z) + \overline{\beta}(s) - z\right)}{s^2\left(e^{\frac{2\alpha s}{\overline{\beta}(s)}}(\overline{\beta}(s)+z)^2 - (\overline{\beta}(s)-z)^2\right)} \tag{16}$$

$$\hat{v}(X,s) = \frac{2z\left[\exp\left(\frac{\alpha s}{\overline{\beta}(s)}(2-X)\right)(\overline{\beta}(s)+z) + (\overline{\beta}(s)-z)\exp\left(\frac{\alpha sX}{\overline{\beta}(s)}\right)\right]}{\alpha s^2\left(e^{\frac{2\alpha s}{\overline{\beta}(s)}}(\overline{\beta}(s)+z)^2 - (\overline{\beta}(s)-z)^2\right)} \tag{17}$$

$$\hat{w}(X,s) = \frac{4z\overline{\beta}(s)\exp\left(s\left(\frac{\alpha}{\overline{\beta}(s)}+1-X\right)\right)}{s^2\left(e^{\frac{2\alpha s}{\overline{\beta}(s)}}(\overline{\beta}(s)+z)^2 + (\overline{\beta}(s)-z)^2\right)} \tag{18}$$

4 The Signal Transmitted Through Viscoelastic Slug

Complex inversion formula is applied to invert the Laplace equation (18). Firstly, we apply the Bromwich contour to lay-out the calculation of the complex integrals along the contour and determine the poles and branch points. Secondly, we find all the roots in the real and complex plane until their contribution to the solution is insignificant. After applying complex inversion formula to (18), we obtain

$$w(X,t) = \frac{1}{2\pi i}\int_{\gamma-i\infty}^{\gamma+i\infty} \frac{4z\overline{\beta}(s)\exp\left(s\left(\frac{\alpha}{\overline{\beta}(s)}+1-X\right)\right)}{s^2\left(e^{\frac{2\alpha s}{\overline{\beta}(s)}}(\overline{\beta}(s)+z)^2 + (\overline{\beta}(s)-z)^2\right)}ds \tag{19}$$

We consider a closed curve in the left half of the complex plane and we note that there are branch points of the integrand at $s = -\frac{1}{\alpha\overline{\mu}}$ and $s = -\frac{1}{\alpha\overline{\eta}}$.(see figure 3). Then we make a cut along $-\frac{1}{\alpha\overline{\mu}}$ and $-\frac{1}{\alpha\overline{\eta}}$, and modify the contour in order to avoid crossing the branch cut as shown in figure 3. We let χ be the closed contour ABCDEFGHJKLMA and Γ to be the contour BCD, DE, EF, FG, GH, HJ, JK, KL, LMA. In the closed contour χ, there are a second degree pole at $s = 0$ and poles when the denominator of the integrand (30) is equal to zero. Then it follows from (30), on putting $T = \sqrt{R^2 - \gamma^2}$, that

$$w(X,t) = \lim_{R\to\infty}\frac{1}{2\pi i}\int_{\gamma-iT}^{\gamma+iT} e^{st}\hat{w}(X,s)ds = \lim_{R\to\infty}\left[\frac{1}{2\pi i}\oint_{\chi} e^{st}\hat{w}(X,s)ds - \frac{1}{2\pi i}\int_{\Gamma} e^{st}\hat{w}(X,s)ds\right]$$

$$= \sum residuesinside.\chi - \frac{1}{2\pi i}\int_{\Gamma} e^{st}\hat{w}(X,s)ds = \sum residuesinside.\chi - \int_{BCD} - \int_{DE} - \int_{EF} - \int_{FG} - \int_{GH} - \int_{HJ} - \int_{JK} - \int_{KL} - \int_{LMA} \tag{20}$$

In finding the roots, we used Maple software routine.We found 150 poles in the upper half plane and consequently their complex conjugates. This gave 300 poles in total which gives reasonably accurate results. The residue of the second order pole at $s = 0$ is

$$\lim_{s \to 0} \frac{d}{ds}\left[\frac{4z\bar{\beta}(s)\exp\left(s\left(\frac{\alpha}{\bar{\beta}(s)}+1-X+t\right)\right)}{\left(e^{\frac{2\alpha s}{\bar{\beta}(s)}}(\bar{\beta}(s)+z)^2+(\bar{\beta}(s)-z)^2\right)}\right] = \frac{1}{2z}\left[2z\left(\frac{\alpha(\bar{\eta}-\bar{\mu})}{2}+2-X+t\right)-\alpha(1\_z)^2-\alpha z(\bar{\eta}-\bar{\mu})\right] \tag{21}$$

In calculating the other residues we use $\text{Residue} = \lim_{s \to s_n} \frac{(s-s_n)N(s_n)}{s^2 D(s)}$ (22)

Where $N(s) = 4z\bar{\beta}(s)\exp\left(s\left(\frac{\alpha}{\bar{\beta}(s)}+1-X+t\right)\right)$ and $D(s) = \left(e^{\frac{2\alpha s}{\bar{\beta}(s)}}(\bar{\beta}(s)+z)^2+(\bar{\beta}(s)-z)^2\right)$ and

s_n is the nth. root.

5 Results and Discussions

Having found the required numbers of poles to evaluate the displacement of the wave in the second rod, we obtain the displacement in the time domain

$$w(X,t) = \frac{\alpha(\bar{\eta}-\bar{\mu})}{2}+2-X+t-\frac{\alpha(1\_z)^2}{2z}-\frac{\alpha z(\bar{\eta}-\bar{\mu})}{2z}+\sum residue \quad .inside \quad .\chi \tag{23}$$

The results also show that the first jump at $X=1$ increases as the ratio of acoustic impedances, z increases until the z effective, $z\sqrt{\frac{\bar{\mu}}{\bar{\eta}}}=1$. As the z effective $z\sqrt{\frac{\bar{\mu}}{\bar{\eta}}}$ approaches 1, the graph shows no jumps after the first jump. Then the jumps appear again for values of z effective, $z\sqrt{\frac{\bar{\mu}}{\bar{\eta}}}>1$. The numerical results also show that the subsequent jumps for each ratio of viscosity parameters are getting smaller as time increases. However figures 5 shows that the discontinuities are so small that the velocity graphs appear to be increasing smoothly to a horizontal asymptote at $v=1$. Actually there are jumps according to the viscoelastic discontinuity analysis. Since the numerical values are very small compared to the other jumps, they are not apparent in the results.

For bigger viscosity time constants which are displayed in figures 5 and 6, the velocity curves remain constant in between two discontinuity jumps. However when the viscosity time constants are $\bar{\eta}=5$, $\bar{\eta}=2$ and $\bar{\mu}=1$, the velocity curves show a linear increase between two discontinuity jumps. Furthermore figures 5 and 6 show that the velocity curves are increasing smoothly until the curves settle down as the velocities approaches one.

Another significant finding shown by figures 4, 5 and 6 is that the jumps are bigger for large viscosity time constants. The discussion shows that the effective ratios of acoustic impedance $z*$ and the viscosity time constants $\bar{\eta}$ and $\bar{\mu}$ play a very important role in determining the behaviours of the wave transmitted through the viscoelastic slug.

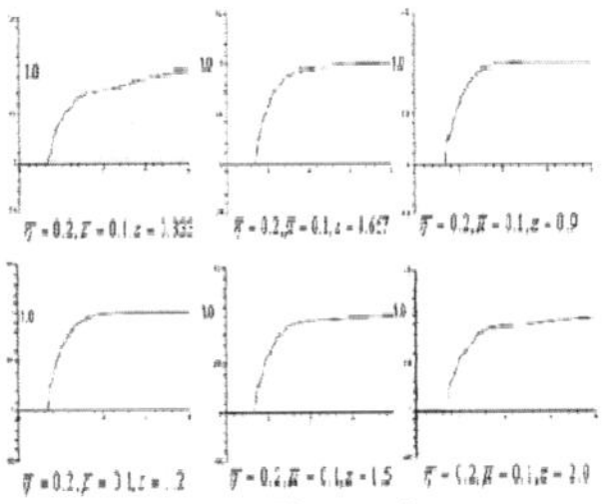

Figure 4

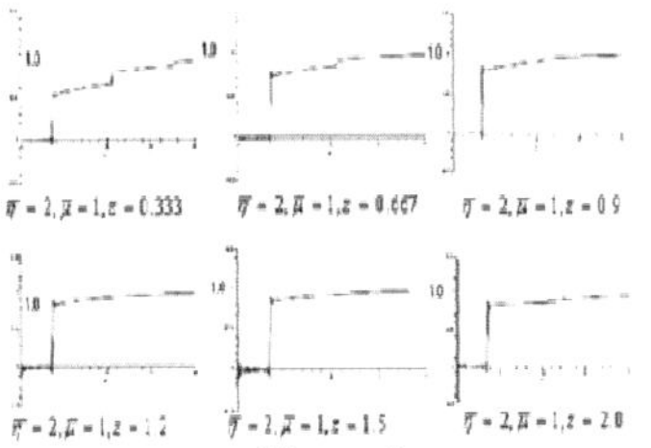

Figure 5

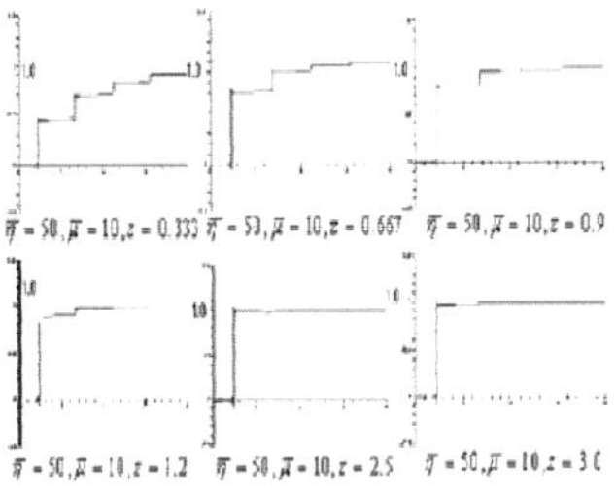

Figure 6

References

[1] Christensen R M 1971 *Theory of Viscoelasticity: An Introduction* Academic Press
[2] Bland D R 1960The Theory Of Linear Viscoelasticity *International Series of Monographs on Pure and Appl. Math.* Pergamon Press
[3] Kolsky H 1963 *Stress Waves in Solids* Dover Publication
[4] Kolsky H 1949 *Proc. Phys. Soc. London* **B62** 676
[5] Spiegel M R 1983 *Advanced Mathematics for Engineers and Scientist* McGraw-Hill Book Co.
[6] Stronge W J 2000 *Impact Mechanics* Cambridge University Press

4th International Conference on Energy and Environment 2013 (ICEE 2013) IOP Publishing
IOP Conf. Series: Earth and Environmental Science **16** (2013) 012040 doi:10.1088/1755-1315/16/1/012040

Fire dynamics simulation of large multi-story buildings Case study: Umm Al-Qura university campus

A F Abdel-Gawad[1] and H A Ghulman[2]

[1]Prof. of Comp. Fluid Mechanics, Umm Al-Qura Univ., Makkah, Saudi Arabia
[2]Dean, College of Eng. and Islamic Archit., Umm Al-Qura Univ., Saudi Arabia

E-mail: afaroukg@yahoo.com

Abstract. The computational fluid dynamics (*CFD*) technique is used to predict the fire dynamics in some main buildings of the campuses of Umm Al-Qura University using the Fire Dynamic Simulator (*FDS*). Important aspects of fire dynamics such as smoke propagation and temperature distribution were investigated. The study contributes in reducing the risks of fires by early prediction of the expected scenarios of fires and associated smoke movement. Hence, early evacuation plans can be established by authorities such as the civil defense. It was found that emergency openings (vents) in the ceiling or side walls that operate in cases of fire, according to appropriate sensors, have a significant role in directing the smoke outside the building. Based on the study, interesting conclusions are drawn and fruitful recommendations/suggestions are introduced. A simple smoke control-scheme is recommended to minimize smoke hazards.

1. Introduction

Big cities attract millions of people every year. People may gather at the same time and place with intensive density. This situation is likelihood to initiate series fires that may lead to heavy losses in human lives. Statistically, most of the deaths in fires are not due to direct fire, but because of suffocation with smoke, fumes and toxic gases. Examples of buildings where smoke-prediction and -control play a significant role include: large worship places, university campuses, shopping centers, large hotels, atrium buildings, large warehouse and industrial buildings, underground structures such as car parks and tunnels, etc. The problem of fire dynamics was investigated by many researchers numerically and experimentally. Men *et al.* [1] presented an approach to the study of gas-phase combustion and convection processes numerically. Kasheef *et al.* [2] conducted a research project to evaluate numerically and experimentally the effectiveness of the current emergency ventilation system (*EVS*) to control fire-smoke spread in two road-tunnels. He and Jiang [3] presented a case study using *FDS* to optimize the location of the Air Sampling-Type Detectors (*ASD*) for Langham Place1 in Hong Kong. Xin *et al.* [4] performed fire simulations of a 7.1-*cm* buoyant turbulent diffusion flame using a mixture fraction-based combustion model. Huo *et al.* [5] used FDS to simulate the typical ventilation in an air-conditioned office. They stated that the diffuser height has the main effect on office temperature. Razdolsky [6] examined some differences and amalgamations between combined effects from multiple fires and "local" explosions on a structural system. Other researchers concentrated on the validation of *FDS*. Zhang et al. [7] assessed the accuracy of *FDS* (V4.07) using the heat flux and flame heights from fires of a single burning item (*SBI*) tests and concluded that *FDS* reasonably predicts the flame heights. Webb [8] performed hot smoke tests (*HST*) according to

[1] To whom any correspondence should be addressed.

Australian Standard AS 4391 -1999 in several buildings such as cinema auditorium and airport concourse to show that fire protection consultants may predict reasonable computational results that are comparable to a full-sized simulation. Smardz [9] assessed the accuracy of *FDS* predictions for smoke spreading from a small compartment into an adjacent larger space (Hotel building), then extracted using powered-ventilation system. Sun *et al.* [10] used experimental data to evaluate the fire-plume properties simulated by *FDS* and the Clark coupled atmosphere-fire model. They indicated that *FDS* produces good agreement with the experimental results even at a relatively coarse horizontal grid size of 4 *m*. Coyle and Novozhilov [11] performed further validation of *FDS* using four smoke filling scenarios reported in the literature. They found that performance of the code varies considerably with complexity of scenario (*e.g.*, geometry). Jahn *et al.* [12] analyzed the sensitivity of *FDS* to input parameters such as fire size and location as well as convection, radiation and combustion conditions. They stated that simulations of fire growth are significantly sensitive to location of fire and its area, flame radiative fraction, and material thermal and ignition properties.

The present study concerns the smoke propagation due to sample fires in two main large buildings of Umm Al-Qura University. A smoke control-scheme is recommended to minimize smoke hazards. Thus, the main objective of the present work is to predict the scenarios of smoke propagation in the university buildings to assist in putting evacuation plans of the students and to suggest simple control solutions to reduce smoke hazards. This objective can be easily adapted to other types of large buildings.

2. General features of the computational modeling

Fire Dynamic Simulator (*FDS*-Ver.5) program [13] is a computational tool for the prediction of fire scenarios and smoke spread that are expected in almost all types of buildings. *FDS* was developed by National Institute of Standards and Technology (*NIST*) [13]. The program prediction depends on the architectural plans of the building in addition to the burning materials. The program is based on the time-dependent solution of the governing equations of flow and combustion due to fire. The building domain is divided into a three-dimensional grid of small cubes (grid cells). The core algorithm of *FDS* is an explicit predictor-corrector scheme, second-order accurate in space and time. Turbulence is treated by means of the Smagorinsky sub-grid scale (*SGS*) model of Large Eddy Simulation (*LES*). More details of the *FDS* program can be found in [14]. There is a big number of attempts for the validation of *FDS*. Some of them are illustrated in the previous section and many others in [15]. Generally, based on these validation investigations, the results of *FDS* can be trusted for almost all fire cases. However, it is important to use a fine mesh to model the problem under-investigation as much as the available computer capacity. For all the present building models, the fire power was set as moderate [14,15]. Smokeview [16] is another program that is used as a post-processing tool to visualize the results of *FDS*. Smokeview is used to represent the results of the present study as will be shown in the coming sections.

3. Governing equations, *LES* simulation and combustion model

FDS solves numerically a form of the Navier-Stokes equations appropriate for low-speed, thermally-driven flow with an emphasis on smoke and heat transport from fires [9]. The basic set of the conservation equations for mass, momentum and energy solved by *FDS* is presented below:

Conservation of mass: $$\frac{\partial \rho}{\partial t} + \nabla.\rho\, u_i = 0 \tag{1}$$

Conservation of momentum: $$\frac{\partial}{\partial t}(\rho u_i) + \nabla.\rho\, u_i u_j + \nabla p = \rho f + \nabla.\tau_{ij} \tag{2}$$

Conservation of energy: $$\frac{\partial}{\partial t}(\rho h) + \nabla.\rho h u_i = \frac{Dp}{Dt} + \dot{q}''' - \nabla.q + \Phi \tag{3}$$

Equation of state for a perfect gas: $$p = \rho RT \tag{4}$$

Where, u_i is velocity in i-direction, i =1, 2, 3, ρ is fluid density, f is summation of external forces, τ_{ij} is shear stresses, p is pressure, h is enthalpy, $\dot{q}'''$ is heat release rate per unit volume (*HRRPUV*), q is the heat transfer, Φ is any heat source, and T is the temperature. Large eddy simulation resolves

IOP Conf. Series: Earth and Environmental Science **16** (2013) 012040 doi:10.1088/1755-1315/16/1/012040

large scales of the flow field and models the smallest scales of the solution by a grid-filtering process. Thus, the filtered term $\overline{u_i u_j}$ is modeled using the sub-grid scale (*SGS*) model of Smagorinsky [17]. *FDS* uses the mixture fraction model as the default combustion model [9]. Mixture fraction is defined as the fraction of gas at a given point in the flow field that originated as fuel. Radiative heat transfer is included in the model via the solution of the radiation transport equation for a non-scattering grey gas.

4. Results and discussions

The first model represents the building of a typical academic department. The building is a part of a complete complex that contains a number of similar departments. It is a three-story building, Fig. 1, with overall plan dimensions of about 41×17 m^2. The offices and facility are concentrated in an area of 18×13 m^2. A central-rectangular hollow-section extends from the first floor to the roof of the third floor with a cross-section of 4.4× 3.2 m^2. The furniture samples that appear in the third floor are placed in the office of the department head. The department council is in the opposite room. The source of fire is a wooden disk that is placed in a small room behind the council room. Figures 1 and 2 show the results after 20 minutes of the fire initiation for *(a)* smoke *(b)* temperature. In the first case, Fig. 1, the smoke propagated in all the three floors of the building. Un-expectedly, smoke moved from the third floor towards the two floors beneath through the central hollow-section. Smoke almost filled all the floors. This means all individuals in all the building are in great risk of smoke hazards. Temperature slightly increased in the ground floor. The second case, Fig. 2, differs from the first one in putting an active outlet vent in the ceiling of the third floor. The vent, 1×1 m^2, is located in the ceiling at the center of the central-rectangular hollow-section. This vent is equipped with a fan to extract smoke out of the building and operates according to the signal of a temperature detector such that the flow speed is 1 *m/s*. The vent sucked out the smoke to the outer surroundings from the third floor. The smoke did not propagate to the lower floors. The temperature of the first two floors remains almost the same without increase.

(a)

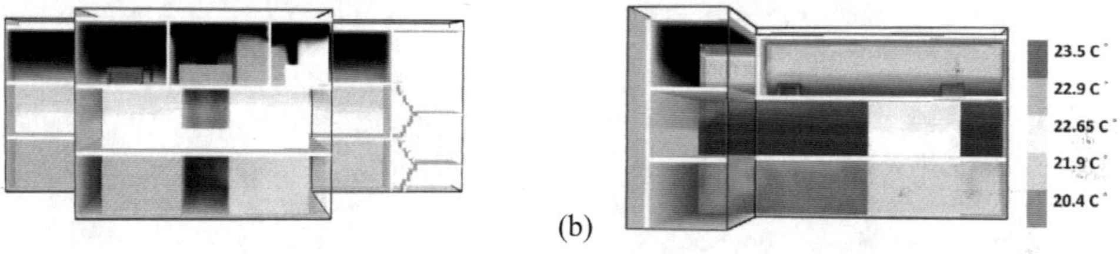

(b)

Figure 1. First model, academic department building without a ceiling opening.

(a)

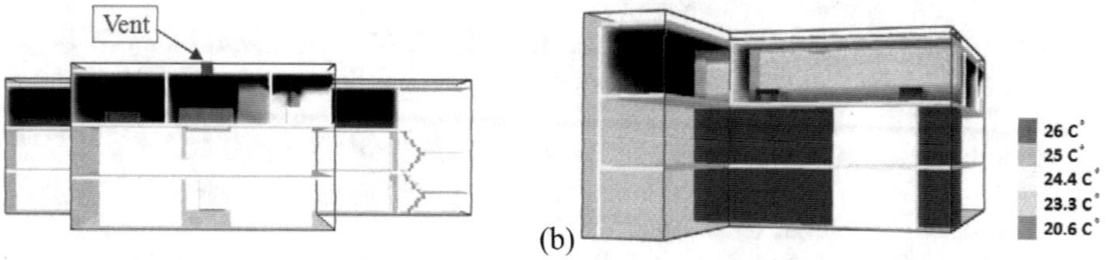

(b)

Figure 2. First model, academic department building with a ceiling opening (vent).

The second model represents another part of the university campus that consists of two buildings; *(a)* the main and *(b)* the auxiliary. A bridge of 40 *m*-long, 10 *m*-wide and 3.5 *m*-height connects the two buildings. The main is a four-story building in the form of a hollow rectangle, Fig. 3a, with outside dimensions of 225×205 m^2. The inner hollow area is about 197×171 m^2. The auxiliary is a two-story building in the form of a hollow rectangle, Fig. 3b, with outside dimensions of 85×225 m^2. The inner hollow area is about 57×185 m^2. Two sources of fire (carpet pile) are supposed to exist simultaneously; one at the main building and the other at the auxiliary. The two sources have the same strength and at the same floor (Second floor). For the main building, after burning time of 30 minutes, the smoke moved down to the ground floor instead of moving up to the higher floors. Due to the large size of the building, smoke did not get out from the openings of the ground floor till 30 minutes of fire. The same scenario was repeated in the case of the auxiliary building except that smoke started to get out of the openings of the ground floor within the thirty minutes of study. Thus, two different evacuation planes should be considered for the two buildings.

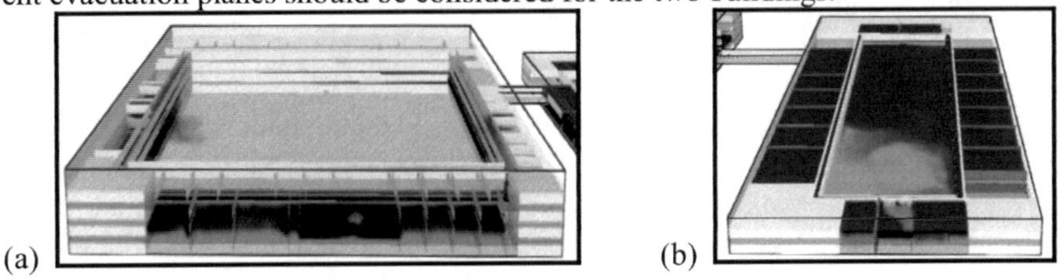

Figure 3. Second model, consisting of two connected buildings.

5. Conclusions/recommendations

Conclusions: (*i*) Emergency openings (vents) in the ceiling or side walls that operate in cases of fire, according to appropriate sensors, have a significant role in directing the smoke outside the building. (*ii*) Smoke propagation depends greatly on the location of the fire according to the building architecture. (*iii*) The increase of the number of openings between different stories increases the possibility of smoke propagation from one story to another. However, the smoke density becomes less. (*iv*) In buildings with complex architecture, smoke may be directed to lower stories instead of moving to the upper ones. This situation complicates both the fire fighting and the evacuation processes.

Recommendations: (*i*) The idea of smart buildings can be adopted by emergency openings (vents). The doors and fans of these openings may be powered by emergency batteries that are charged by solar cells to be independent of the main power supply that to be cut in case of fire. *(ii)* A control system, based on microcontrollers or Programmable logic controller (*PLC*), may be utilized to operate these vents depending on the signals of a matrix of temperature/smoke detectors. (*iii*) Air-Conditioning system must be turned-off or works in opposite direction to suck smoke outside the building. This is a vital point as air conditioning system is a major source of fresh air for fire supply.

References

[1] Men W, Mcgrattan K B, Baum H R 1995 *ASME National Heat Transfer Conference* **2**

[2] Kashef A, Bénichou N, Lougheed G, Debs A 2004 *5th International Conference-Tunnels Fires* London, UK 185-196

[3] He M, Jiang Y 2005 *Vision Fire & Security*

[4] Xin Y, Gore J P, McGrattan K B, Rehm R G, Baum H R 2005 *Comb. and Flame J.* **141** 329-5

[5] Huo Y, Gao1 Y, Chow W 2009 *7th Asia-Pacific Conf. Wind Eng.* Taipei, Taiwan

[6] Razdolsky L 2009 *Proceedings of the World Congress on Engineering* **II** U.K

[7] Zhang J, Delichatsios M, Colobert M 2010 *Fire Technology* **46** 291-306

[8] Webb A 2006 *M.Sc. Thesis*, The Faculty of the Worcester Polytechnic Institute, USA

[9] Smardz P 2006 *M.Sc. Thesis*, University of Ulster

[10] Sun R, Jenkins M A, Krueger S K, Mell W, Charney J J 2006 *Can. J. for Res.* **36** 2894-2908

[11] Coyle P, Novozhilov V 2007 *Int. J. Eng. Performance-Based Fire Codes* **9(1)** 7-30

[12] Jahn W, Rein G, Torero J L 2008 *Fire Safety Science* **9** 1341-52

[13] http://www.fire.nist.gov/fds/

[14] McGrattan K, Hostikka S, Floyd J, Baum H, Rehm R, Mell W, McDermott R 2010 *NIST Special Publication* 1018-5 **1**

[15] McGrattan K, Hostikka S, Floyd J, Klein B 2010 *NIST Special Publication* 1018-5 **3**

[16] Forney G P 2010 *NIST Special Publication* 1017-1,-2,-3

[17] Smagorinsky J 1963 *Monthly Weather Review* **91(3)** 99-164

Municipal Solid Waste Composition Study of Selected Area in Gambang, Pahang

Nadiah Mokhtar[1,3], Wan Faizal Wan Ishak[2,3], Noor Suraya Romali[1,3] and Siti Fatimah Che Osmi[1,3] Mohd Armi Abu Samah[4,3,]

[1]Faculty of Civil Engineering and earth Resources, Universiti Malaysia Pahang, 26300 Gambang, Kuantan, Pahang

[2] Faculty of Science and Technology, Universiti Malaysia Pahang, 26300 Gambang, Kuantan, Pahang

[3] Pusat Penyelidikan Sumber Alam (CERRM) Universiti Malaysia Pahang, 26300 Gambang, Kuantan, Pahang

[4] Faculty of Environmental Studies, 43400 UPM Serdang Selangor

E-mail: nm.nadiah@gmail.com

Abstract. The amount of municipal solid waste (MSW) generated continue to increase in response to rapid growth in population, change in life style and accelerated urbanization and industrialization process. The study on MSW is important in order to determine the composition further seeks an immediate remedy to minimize the waste generated at the early stage. As most of the MSW goes to the landfill or dumping sites, particularly in Malaysia, closure of filled-up landfill may become an alarm clock for an immediate action of proper solid waste management. This research aims to determine the waste composition generated from selected residential area at Gambang, Kuantan, Pahang which represent Old residential area (ORA), Intermediate residential area (IRA) and New residential area (NRA). The study was conducted by segregating and weighing solid waste in the residential area into 6 main components ie., food waste, paper, plastic, glass, metal and others. In a period of four weeks, samples from the residential unit were taken and analyzed. The MSW generation rates were recorded vary from 0.217 to 0.388 kg $person^{-1}day^{-1}$. Food waste has become the major solid waste component generated daily which mounted up to 50%. From this research, the result revealed that the recyclable composition of waste generated by residents have a potential to be reuse, recycle and reduce at the point sources.

1. Introduction

Nowadays, solid waste management becomes the biggest issues to be discussed which may lead to intractable environmental issues. According to Ajadi *et al.*, 2010 has reported that solid waste results from rapid increase in population, massive expansion of the urban areas and the changing lifestyle [1]. According to various report, waste generation in Malaysia has been increasing gradually since year 2000 [2].

The Municipal Solid Waste (MSW) generation rate in Kuantan, the capital of Pahang is continuously rising up every year due to the uncontrollable consumption. Presently, the amount of solid waste produced in Kuantan is about 500 tons daily, consisting of 60% of domestic waste and 40% of industrial and construction waste [3] In Malaysia, the average components of MSW are quite

[1] To whom any correspondence should be addressed.

similar with the largest categories consisting of food waste (45%), plastic (24%), paper (7%), iron (9%) and others (3%) [4].

As disposed of waste in Malaysia is totally end-up in landfill, proper management of solid waste in the point sources need to be emphasized. There were about 230 official dumping site gazetted in Malaysia based on statistic data in 1998. By the year 2002, only 161 disposal sites are actively operated in Peninsular Malaysia [5]. Most landfill in Malaysia were small scales operations with varying levels of design sophistication, while majority of them were poorly manage. The National Strategic Plan for Solid Waste Management was formulated the following service targets to focus plans to improve MSW management and monitor the efficiency of its implementation. **Table 1** illustrated the reduction target for 2020 revised by Malaysia's government where in 2015, closure of dump sites are targeted to achieve 100% with source separation will highly encourage to be done especially in urban area.

Table 1. *Reduction target for 2020 was revised by the Government of Malaysia

Level of Service	Present	2003-2009	2010-2014	2015-2020
Extend collection service	75%	80%	85%	90%
Reduction & Recovery	3-4%	10%	15%	17%*
Closure of dump sites	112 sites	50%	70%	100%
Source Separation (Urban)	None	20%	80%	100%

**National Strategic Plan for Solid Waste Management in Malaysia, 2005*

1.1 Objectives

The main objectives of this study are:

i. To determine the waste composition in three different residential
ii. To analyzed the waste generation rate

2. Methodology

2.1 Overview of the study area

The study was conducted in Kuantan, a capital city of Pahang. As Kuantan consist of several counties, Gambang has been chosen as a study area due to the vast development and expected to face a rapid growth within 10 years due to many development project such as institutional, tourism centre and economic and social hub. The study on MSW composition and management covered 3 residential area at Gambang namely, Taman Tas, Taman Mahkota Aman and Taman Indera Semporna which each of them represent Old Residential area (ORA), Intermediate Resdiential Area (IRA) and New Residential area (NRA), respectively. The selection of the study area was based on the rapid development due to available public accommodation and the population concentration. Although the ages of the residential area is different comprising new (< 5 years), intermediate (5-10 years) and old residential (10-15years), but the rapid growth of accommodation had caused the residential development is skyrocketed at the same level.

2.2 Sample collection, sorting, measurement and recording.

During the data collection at point sources, a few plastic bags were distributed to the sample unit by providing the target sample with two different colour of plastic bag namely black colour of plastic bags for non-biodegradable waste or dry waste; while blue colour of plastic bags for biodegradable or wet waste collection. The waste was segregated into 6 classes which are food waste, plastic, paper, glass, metal and others waste. The waste was weighted and recorded according to its types. The data collection at point sources was administered in four weeks to obtain monthly data of solid waste generation.

3. Result and discussions

The purpose of the analysis is to determine the composition and the average rate (kg/person) of solid waste generated. **Table 2** reveals that in ORA, represented by Taman Tas, food wastes constitutes the largest components of solid waste generated by the people which about 71.74%. This followed by plastics, paper, glass and metal waste with 12.85%, 6.64%, 3.68% and 1.13% respectively. No recorded data was found under other waste category.

Following this is the IRA is represented by Taman Mahkota Aman. Food wastes and plastics constitute the largest components of solid waste generated by the peoples. They constitute 52.6% and 25.67% respectively. Paper and glass waste constitute about 7.54% and 6.1%, respectively. Other wastes such as rags, rubber, grass wastes constitute 5.4% of the total percentage, lastly metal with a minute proportion of 2.69%.

In NRA represented by Taman Indera Semporna, data obtained reveals that 67% of solid waste component constituted by food waste, which still become the largest component generated. This follows by glass and plastics which constitute about 11.98% and 10.8% of the total component. Plastics and metal was become the least produce by residents with 5.5% and 5.3%, respectively, whilst no data recorded for other waste. The analysis of solid waste data reveals that food waste is a major component in solid waste generated in all residential.

Table 2. Composition of solid waste in different residential area

Name of residential area	Food waste %	Plastic %	Paper %	Glass %	Metal %	Others %	Total %
Old Residential area (ORA)(Taman Tas)	71.74	12.85	6.64	3.68	1.13	0	100
Intermediate Residential Area (IRA)(Taman Mahkota Aman)	52.6	25.67	7.54	6.1	2.69	5.4	100
New Residential Area (NRA)(TamanIndera Semporna)	67	10.8	5.15	11.98	5.3	0	100

The higher food waste generate by ORA can be likened to the socio-economic characteristics and household factors in the area. As about 80% of the residents in the ORA are low income earners encourage home-cooked lifestyle. This is very high when compared with IRA and NRA with 53.33% and 78% of the people falling into high income group, respectively (Table 2). **Table 3** also reveals that even though majority of NRA residents were categorize in higher income earners, 67% of food waste were detected to be generated in their house. This may due to the less food provider in the area hence encourage peoples to prepare meal by their own.

In IRA, despite of food waste, the plastics were observed to become the second highest waste component generated by the resident. Plastic waste was composed mainly of packaging, plastic products, hard and flexible plastic household items, PET bottles, jerry can etc [5]. Common plastics component are also come from carrier bags, package wrappers and pouches. Higher percentages of plastics component may due to outside food brought by the resident in this residential area especially during Wednesday night market every week.

In NRA, the proportion of metal/aluminium and glass waste was identified as the highest percentage as compared to other residential. The socio-economic characteristics of the residents of this area can be used to explain the factor. The high status of residents makes it quite possible for them to be able to consume preservative items such as canned foods, milk, beverages, yougart etc. than people who reside in the other residential area.

Table 3. Distribution of income group in different residential area

Name of residential area	Low income	High income	Total
Old Residential area (ORA)(Taman Tas)	80	20	100
Intermediate Residential Area (IRA)(Taman Mahkota Aman)	46.67	53.33	100
New Residential Area (NRA)(Taman Indera Semporna)	22	78	100

Table 4 illustrated the average waste generation rate in ORA, IRA and NRA. The average waste generation rate were recorded varies from 0.217 to 0.388 kg person^{-1}day^{-1}. According to Mohd Shahir Zahari *et al.*, 2010, the national average of waste generated is at 0.5 – 0.8 kg person^{-1}day^{-1}. Thus, the average waste generated by residential in Gambang area were found to be not in range. This may because of Gambang is still a remote area located 20km from the vicinity of Kuantan town.

Table 4. Average waste generation rate in different residential area

Name of residential area	Total Sample	kg person-1day-1	
		Mean	Median
Old Residential area (ORA)(Taman Tas)	10	0.217	0.122
Intermediate Residential Area (IRA)(Taman Mahkota Aman)	15	0.282	0.162
New Residential Area (NRA)(Taman Indera Semporna)	10	0.388	0.236

4. Conclusion

Improper solid waste management may lead to crucial environmental issue. Total dumping of solid waste into the landfill will end up with scarcity of land use area. Separation of solid waste component in the point sources are the best way to identify the component and further reduce the volume by practising composting for organic component and recycling for non-organic component. Both urban and potential rural area should play their roles to reduce waste generation rate toward zero waste to landfill. Socio-economic status gives a greater influence to the waste generation rate to certain area. Demographic factor such as respondent's status, household size, age, education level, occupation and income earners were a greater influence on the type of waste component generated.

References

[1] Ajadi B S, Tunde A M 2010 *J Hum Ecol* **32(2)** 101-108
[2] Dawda Badgie, Mohd Armi Abu Samah, Latifah Abd Manaf, Azizi Muda 2012 *Polish Journal Environmental Studies* **21(3)** 539-547
[3] Mod Shahir Zahari, Wan Mohd Faizal Wan Ishak, Mohd Armi Abu Samah 2010 *International Journal of Engineering Science and Techno*logy **2(5)** 1338-1344
[4] Zamali Tarmudi, Mohd Lazim Abdullah, Abu Osman Md Tap 2009 *Jurnal Teknologi* **51(F)** 1-15
[5] Zaini Zakawi 2011 *Journal of applied Sciences in Environmental Sanitation* **6** 29-38
[6] Sherub Phuntsho, Ichharam Dulal, Dechen Yangden, Ugyen M. Tenzin 2010 *Waste Management and Research* **28** 545-551

IOP Conf. Series: Earth and Environmental Science **16** (2013) 012042 doi:10.1088/1755-1315/16/1/012042

Compact Ultra-Wideband Antenna for Portable Devices

B Kasi[1], L C Ping[2] and C K Chakrabarty[2]

[1] Department of Electrical and Electronics Engineering, Universiti Infrastruktur Kuala Lumpur, Kajang, Selangor 43000, Malaysia
[2]Department of Electronics and Communication Engineering, Universiti Tenaga Nasional, Kajang, Selangor 43000, Malaysia

E-mail: baskaran@iukl.edu.my

Abstract. This article presents a compact ultra-wideband (UWB) antenna for wireless dongle devices. The printed monopole antenna, with the size of 15 mm × 30 mm, has an impedance bandwidth ranges from 2.9 to 13 GHz. The proposed structure comprises of a hexagonal radiator fed by a microstrip line with a modified ground plane. Such a design can be easily integrated with wireless universal serial bus (USB) devices. Having stable radiation patterns and constant gain within the UWB spectrum are significant characteristics of this antenna. The time domain studies on the designed antenna indicate reduced ringing effect for impulse excitation across the band of interest.

1. Introduction

Ultra-wideband (UWB) radio system has been developing in recent years and has become popular in wireless communication applications. This technology offer high data rate, multi-path interference immunity and low power consumption within the allocated bandwidth of 3.1 to 10.6 GHz. Current consumer electronics and mobile devices are using the universal serial bus (USB) wired connectivity technology. However, for terminals using this technology to be truly mobile with higher user data rate, wireless USB are needed. To overcome these limitations, integration of UWB technology into USB device is a potential solution for wireless USB advancements. Therefore, many researchers are working towards the development of USB dongles based on UWB communication specification. Antennas are particularly a challenging issue of wireless USB technology since they are now required to be small and have relative bandwidth due to space constraint of the dongle board. Several UWB antennae for wireless USB applications have been reported recently [1–4].

In this paper, a compact hexagonal monopole antenna for wireless USB dongle applications is presented. The proposed radiator exhibits a compact size of 15 mm × 30 mm and a wide bandwidth covering the UWB spectrum has been obtained. The article is organised in the following sections. Section 2 describes the miniature UWB antenna design. The characteristics of the antenna are presented in Section 3. Section 4 summarises and concludes the study.

2. Antenna Geometry

The geometry of the proposed antenna is depicted in Figure 1. It is designed on a standard RT/duroid® 5880 substrate with a relative permittivity of 2.2 and thickness of 0.787 mm. On the other side of the dielectric substrate, a modified ground plane is printed below the feed line. The hexagonal radiator is

[1] To whom any correspondence should be addressed.

IOP Conf. Series: Earth and Environmental Science **16** (2013) 012042 doi:10.1088/1755-1315/16/1/012042

capable of supporting multiple resonant modes in the presented structure. The optimised antenna parameters are tabulated in Table 1.

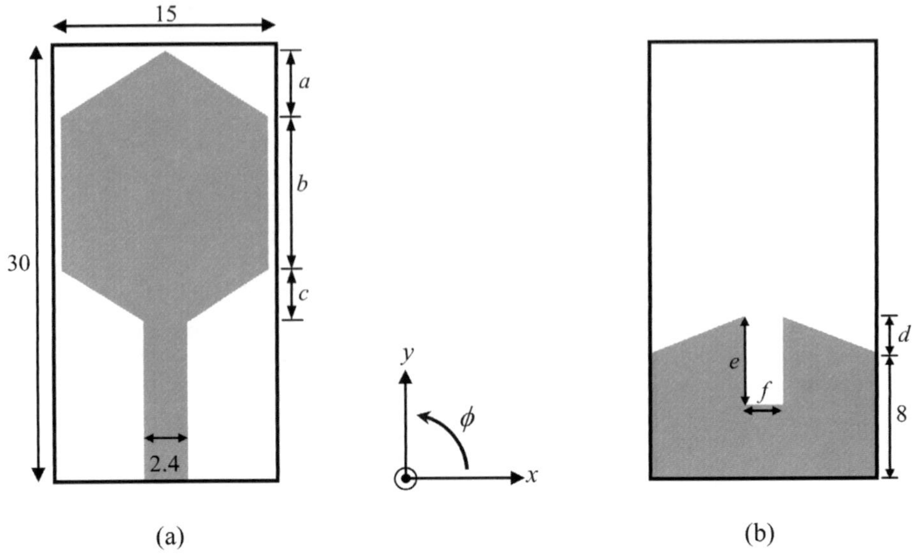

Figure 1. Geometry of the proposed antenna (units in mm). (a) Front view (b) Rear View

Table 1. Antenna parameters.

Parameter	Value (mm)
a	4.8
b	10.1
c	3.0
d	3.0
e	6.5
f	2.8

3. Results and Discussion

All simulations in this article were conducted using the computer simulation technology (CST) Microwave Studio software based on the finite integration technique [5]. The calculated return loss for the optimised antenna is plotted in Figure 2(a). It is noticed that the impedance bandwidth of the antenna satisfies the 10 dB return loss requirement in the frequency band between 2.9 and 12 GHz. Broadside antenna gain at +*z* direction was also simulated, as given in Figure 2(b). The graph shows that the gain performance of the antenna generally increases with the frequency. The proposed antenna radiation patterns in the *E*-plane(*y*–*z* plane) and *H*–plane (*x*–*z* plane) at 3, 6 and 9 GHz are illustrated in Figure 3. Near omnidirectional patterns are observed in the *H*–plane and bi-directional in the other plane.

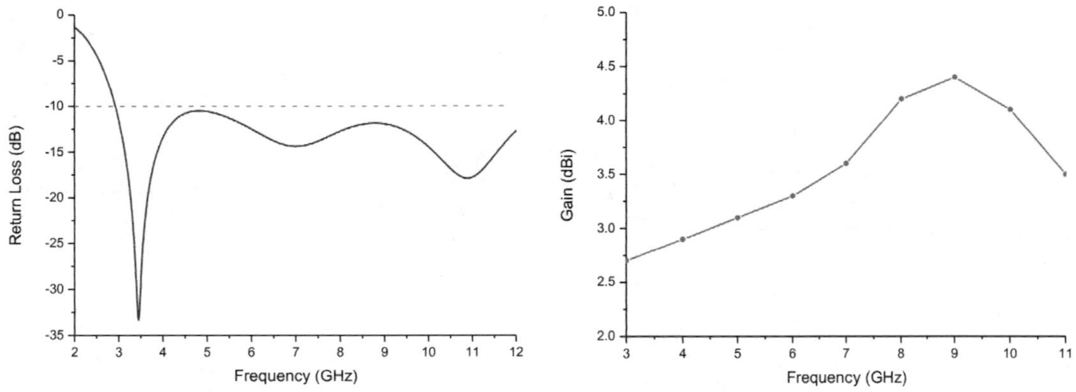

(a) (b)

Figure 2. Input performance of the proposed antenna. (a) Return loss (b) Gain

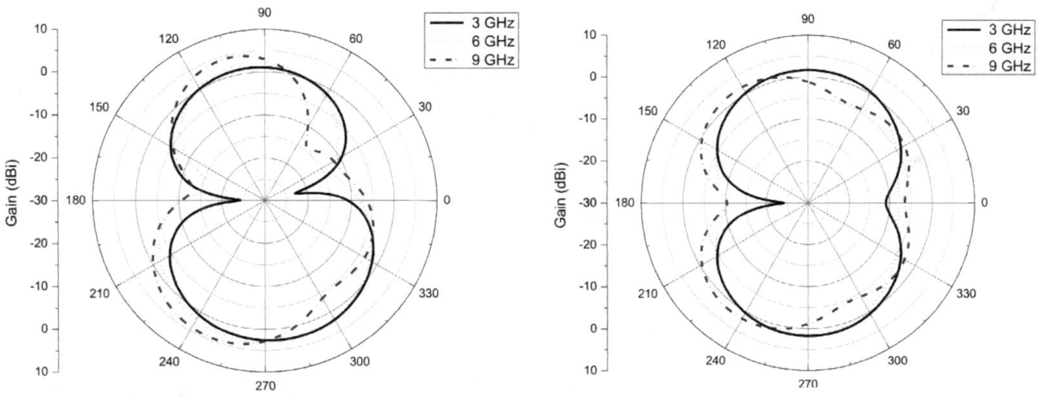

(a) (b)

Figure 3. Antenna radiation patterns at various frequencies. (a) *E*–plane (b) *H*–plane

In designing UWB radio system, it is essential to evaluate the time domain characteristics of the transmitting and receiving antennae. In this work, the system transfer function is investigated with an identical monopole antenna pair. The identical antenna pair are placed 70 mm apart from each other. Figure 4 shows the plot of Gaussian pulse source waveform and the radiated voltage waveform in various directions. It is evident that the received pulses are almost identical with slight ringing effect. As seen in Figure 5(a), the group delay remains almost constant with variations less than 0.5 ns. The simulated transmission coefficient, S_{21}, is plotted in Figure 5(b). It shows a fairly flat magnitude throughout the operating band. These results indicate that proposed antenna is capable of offering good pulse handling capability as required by UWB communication systems

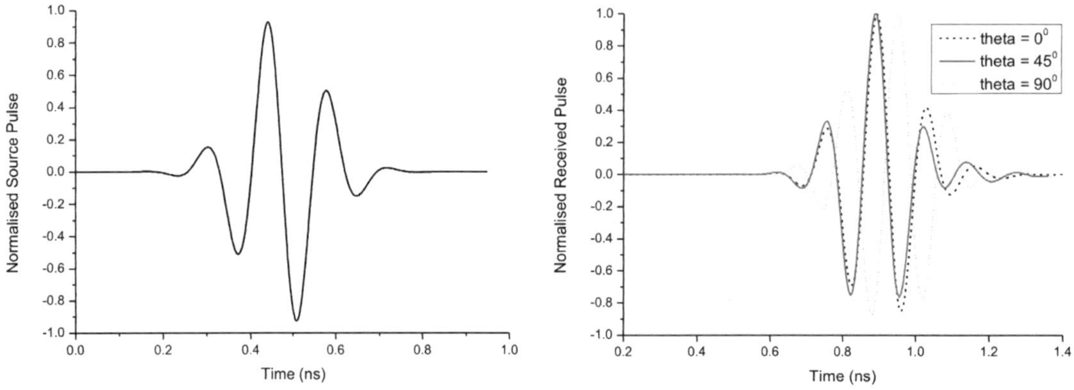

(a) (b)

Figure 4. (a) Normalised input pulse (b) Normalised received signals

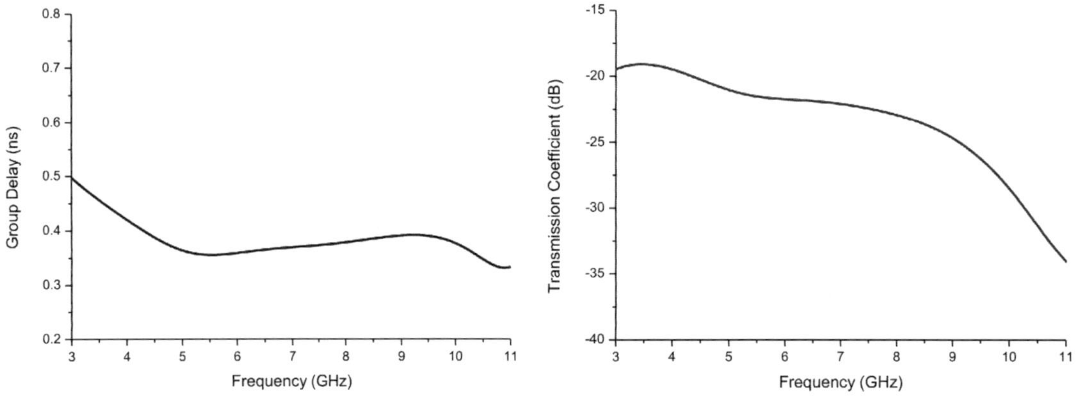

(a) (b)

Figure 5. Proposed UWB antenna (a) Group delay (b) Transfer function

4. Conclusions

A compact UWB antenna for wireless applications is studied in this article. The designed dongle antenna has a small size of 15 × 30 mm^2 while exhibiting ultra-wide impedance bandwidth with reasonable radiation patterns. Furthermore, the antenna features a good system transfer function and constant group delay characteristics. These results make the proposed antenna a promising candidate for emerging portable devices.

References

[1] Su S W, Chou J H and Wong K L 2007 *IEEE Trans. Antennas Propag.* **55** 1180
[2] Shin J, Hong S and Choi J 2008 *Microw. Opt. Technol. Lett.* **50** 1643
[3] Wu Q, Jin R, Geng J and Ding M 2008 *IEEE Trans. Antennas Propag.* **56** 896
[4] Guo L, Wang S, Chen X and Parini C G 2009 *IEEE Antennas Wirel. Propag. Lett.* **8** 554
[5] CST Microwave Studio Electromagnetic Field Simulation Software, Computer Simulation Technology, Darmstadt, Germany.

Energy efficient sensor nodes placement using Territorial Predator Scent Marking Algorithm (TPSMA)

H Z Abidin[1] and N M Din[2]
[1] Faculty of Electrical Engineering, Universiti Teknologi MARA, 40450 Shah Alam, Selangor, Malaysia
[2] College of Engineering, Universiti Tenaga Nasional, Jalan IKRAM-UNITEN, 43000 Kajang, Selangor, Malaysia

E-mail: husnaza@salam.uitm.edu.my

Abstract. The positions of sensor nodes in a Wireless Sensor Network (WSN) must be able to provide maximum coverage with a longer lifetime. This paper proposed a sensor node placement technique that utilizes a new biologically inspired optimization technique that imitates the behavior of territorial predators in marking their territories with their odors known as Territorial Predator Scent Marking Algorithm (TPSMA). The TPSMA deployed in this paper uses the maximum coverage ratio as the objective function. The performance of the proposed technique is then compared with other schemes in terms of uniformity and average energy consumption. Simulation results show that the WSN deployed with the proposed sensor node placement scheme consumes lower energy compared to the other two schemes and is expected to provide longer lifetime.

1. Introduction

An energy efficient Wireless Sensor Network (WSN) with maximum coverage can be achieved through a cost effective deployment mechanism. This can be done by utilizing an effective planning mechanism in arranging the limited number of sensor nodes. Romoozi [1] stated that there is a tradeoff between energy consumption of sensor nodes and network coverage. Closer sensor nodes will reduce the energy consumption but the network coverage will become smaller.

A sensor node placement scheme that utilizes a new biologically inspired optimization technique known as Territorial Predator Scent Marking Algorithm (TPSMA) is introduced in this paper. This is to ensure that WSN deployed with this scheme would be able to reduce the energy usage without jeopardizing the required coverage level. TPSMA imitates the behavior of territorial predators in marking its territories with their odors and in this paper, the sensor nodes placement scheme would mark the area that the sensor nodes would give the maximum coverage ratio. Sensor nodes will then be placed according to the marked area.

The rest of the paper is organized as follows. Section 2 will outline our methodology followed by our simulation model in section 3. Section 4 deals with simulation results and finally section 5 concludes the paper.

2. WSN Sensor Node Placement based on TPSMA

Territorial predator such as tigers and bears can be defined as predators that consistently defend a specific area against animals from other species. The territory is chosen based on certain factor such as food resources. Most territorial predators use scent marking to indicate the boundaries of their territories which are also playing a role in territorial maintenance and as information sites for other members of the population [2]. Scent matching allows a resident animal to distinguish other residents from intruders by recognizing their scent, thereby reducing the need for territorial encounters [3]. The marks may be deposited by urination, defecation, rubbing parts of bodies such as chin and foot, scratching, using glands and vegetation flattening [2].

Territorial predator scent marking behaviour can be adopted in designing the sensor node placement technique where the territory of a sensor node can be scent marked based on the design objectives such as maximum coverage, minimum uniformity, minimum energy consumption and maximum connectivity. This is done based on the scent marking behaviour where normally predator will scent mark the area due to certain factors such as food resources. Sensor node will identify its monitored location based on their marked territories that imitates the scent matching behaviour.

2.1. Assumptions.

- Sensor nodes are homogeneous static nodes and must not exceed the number of potential sensor node locations.
- Number of monitored locations are equal to the number of potential sensor node locations.
- Monitoring area is a 2-dimensional area that is broken up into several monitored locations.
- The center of each monitored location is considered as monitored point.
- Each monitored location can only be equipped with 1 sensor node and may be monitored by more than 1 sensor node.
- At least one path must be there between each sensor node and the base station.

2.2. Sensing Model and Objective Function.

Binary Sensing Model (BSM) is adopted in this study. The covering area of sensor nodes is assumed to be a circular area with a radius of R_s that represents the sensing range. The probability of node $s_j(x_{sj}, y_{sj})$ detects an event occurs on $i(x_i, y_i)$ is [4]:

$$P_{\mathrm{cov}}(s_j, i) = \begin{cases} 1 & d(s_j, i) \le R_s \\ 0 & \mathit{otherwise} \end{cases} \tag{1}$$

where $d(s_j,i)$ is the Euclidean distance between point $i(x_i, y_i)$ and $s_j(x_{sj}, y_{sj})$

i can be considered effectively covered if $P_{cov} > C_{th}$ where C_{th} is the coverage threshold [5]. The objective function is determined by using the following equation which was presented in [5]:

$$\mathit{Objective\ Function},\ f(x) = \mathit{Coverage\ Ratio} = \frac{N_{effective}}{N_{all}} \tag{2}$$

where $N_{effective}$ is the number of locations that are effectively covered and N_{all} is the total number of locations. The following pseudocode describes the proposed TPSMA scheme.

Number of monitored location, $i = 1,2,\ldots,S$

Number of potential location for sensor node, $j = 1,2,\ldots,P$

Step 1:

Check if there is a sensor node located in all j

Step 2:

Compute P_{cov} for all j and i

Compute coverage ratio $f(x)$ for all respective j

Mark territory for sensor nodes in j that has the maximum coverage level to i

(Scent marking by predator on the area that has the highest food resources)

Repeat Step 2 until $i = S$

Step 3:

Monitored locations will be monitored by sensor nodes that are within their marked territory.

(Sensor node identifies its monitored location through scent matching)

Figure 1. Pseudocode for TPSMA based Sensor Nodes Placement

3. Network Model

Simulation parameters are tabulated in table 1and results are then compared with works done by Deyab et al. [6] and Chakrabarty et al. [7] who utilized Integer Linear Programming (ILP). Deyab proposed an algorithm where a monitoring area is divided into few grids and each grid is represented by a critical point. The technique proposed by Deyab specified that sensor nodes and relay nodes could not be placed on the critical point itself. Only specific area is fully covered because the aim of this technique is to reduce the deployment cost. Chakrabarty proposed a sensor node placement technique that uses different costs of sensor nodes. The sensor nodes also have different detection range.

Table 1. Simulation Parameters

Parameter	Value
Number of sensor nodes	10 to 16
Monitoring Area	100m x 100m
Sensing Range, R_s	20m
Number of monitored locations	16
Coverage Threshold, C_{th}	0.9

4. Simulation Results

WSN with smaller uniformity is considered more evenly distributed in the monitoring area hence is expected to have a longer lifetime. Figure 2 shows that the uniformity value reduces as the number of sensor nodes increases because uniformity metric is determined by the communication range and density of sensor nodes [8]. As compared to works proposed by Deyab and Chakrabarty, TPSMA gives the lowest uniformity. Uniformity given by TPSMA is nearly 70% lower than uniformity given by Deyab when the WSN is deployed with 10 sensor nodes and approximately around 77% when it is deployed with Chakrabarty scheme. From this observation, it can be said that, sensor node placement with the proposed TPSMA scheme is expected to provide longer lifetime compared to the other two schemes.

As shown in figure 3, TPSMA has the lowest energy usage while the technique proposed by Deyab consumes the highest energy. The average energy consumption reduced by approximately 12 Joules and 66 Joules compared to techniques proposed by Chakrabarty and Deyab respectively. This graph conforms with the uniformity graph shown in figure 2 where WSN deployed with TPSMA is expected to have the longest lifetime followed by Chakrabarty and Deyab respectively.

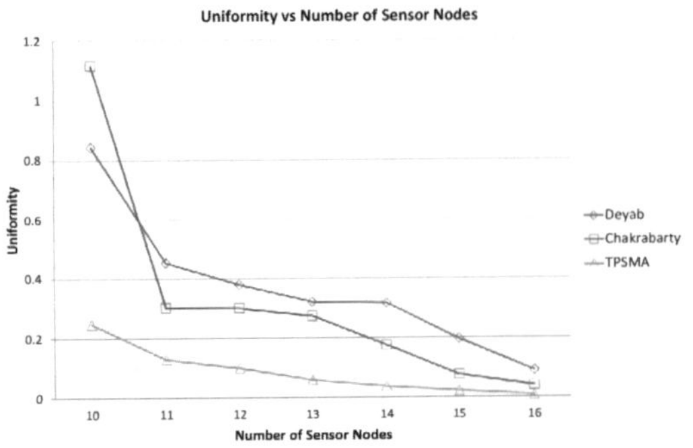

Figure 2. Uniformity vs. Number of Sensor Nodes

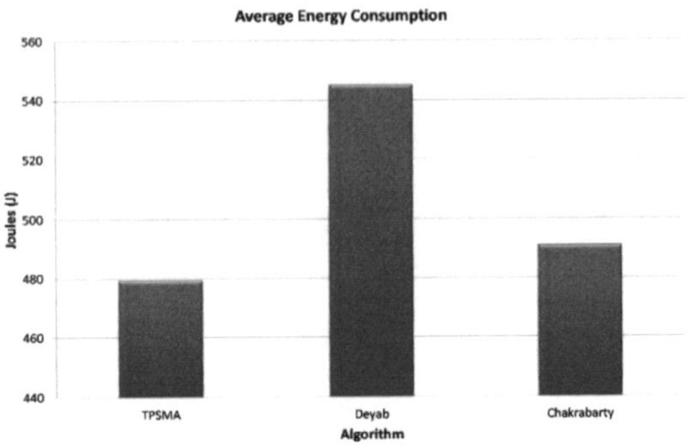

Figure 3. The Averages of Power Consumption

5. Conclusions and Future Work

A sensor node placement scheme based on TPSMA has been proposed in order to ensure a cost effective WSN deployment with minimum energy consumption. Based on the simulation results obtained, it can be concluded that WSN deployed with maximum coverage TPSMA sensor nodes placement scheme is an energy efficient technique that offers considerably lower energy consumption and lower uniformity value compared to the works done by Deyab et al. and Chakrabarty et al. This result implies that, WSN deployed with proposed scheme is expected to be more energy efficient with longer lifetimes. Further work would focus on developing a multi objective TPSMA sensor node placement scheme.

References

[1] Romoozi M, Vahidipour M, and Maghsoodi S 2010 Genetic Algorithm for Energy Efficient and Coverage-Preserved Positioning in Wireless Sensor Networks *Intelligent Computing and Cognitive Informatics (ICICCI)* 22-25

[2] Begg C M, Begg K S, Du Toit J T and Mills M G L 2003 Scent-marking behaviour of the honey badger, Mellivora capensis (Mustelidae), in the southern Kalahari *Animal Behaviour* **66** 917-929

[3] Descovich K A, Lisle A T, Johnston S, Nicolson V and Phillips C J C 2012 Differential responses of captive southern hairy-nosed wombats (Lasiorhinus latifrons) to the presence of faeces from different species and male and female conspecifics *Applied Animal Behaviour Science* **138** 110-117

[4] Salehizadeh S M A, Dirafzoon A, Menhaj M B and Afshar A 2010 Coverage in Wireless Sensor Networks Based on Individual Particle Optimization *International Conference on Networking, Sensing and Control (ICNSC)* 501-506

[5] Zhiming L and Lin L 2009 Sensor node deployment in wireless sensor networks based on improved particle swarm optimization *Applied Superconductivity and Electromagnetic Devices, 2009 (ASEMD 2009)*215-217

[6] Deyab T M, Baroudi U and Selim S Z 2011 Optimal placement of heterogeneous wireless sensor and relay nodes *Wireless Communications and Mobile Computing Conference (IWCMC 2011)* 65-70

[7] Chakrabarty K, Iyengar S S, Qi H, and Cho E 2002 Grid Coverage for Surveillance and Target Location in Distributed Sensor Networks *IEEE Transactions on Computers* **51** 1448-1453

[8] Wang X and Wang S 2010 Hierarchical Deployment Optimization for Wireless Sensor Networks *IEEE Transactions on Mobile Computing* **10** 1028 - 1041

4th International Conference on Energy and Environment 2013 (ICEE 2013) IOP Publishing
IOP Conf. Series: Earth and Environmental Science **16** (2013) 012044 doi:10.1088/1755-1315/16/1/012044

Numerical modelling dam break analysis for water supply project

M S Lariyah[1], M Vikneswaran[1], B Hidayah [1], Z C Muda[1], S Thiruchelvam[1], A K Abd Isham [2], H Rohani [2]

[1] Centre for Sustainable Technology and Environment (CSTEN), College of Engineering, Universiti Tenaga Nasional

[2] HLA Associates Sdn. Bhd.

E-mail: lariyah@uniten.edu.my

Abstract. Dam provides many benefits to the society, but it can also cause extensive damage to downstream area when it fails. Dam failure can cause extensive damage to properties and loss of human life due to short warning time available. In general, dam spillway was designed to drain the maximum discharge from the dam during the Probable Maximum Flood (PMF). The spillway is functioned to prevent the dam from failure due to overtopping, which can lead to the dam failure. Dam failure will result in large volume of water travelling at very high velocity to the downstream area of the dam. It can cause extensive property damage, destruction of important facilities, and significant loss of human life along the way. Due to the potential of high hazard it poses to the downstream area, a dam break analysis is considered very essential. This paper focuses into the dam failure analysis for Kahang Dam by prediction of breach flow hydrographs and generation of inundation map at downstream area. From the PMF scenario simulation, the maximum inflow is 525.12 m3/s and peak discharge from the dam during dam failure is 6188m3/s. The results are able to provide information for preparation of Emergency Response Plan (PMF), in which appropriate steps can be taken by relevant authorities to avoid significant loss of human lives.

1. Introduction

Nowadays, dams are constructed for multiple functions such as supplying drinking water for citizens, process water for industrial and commercial usage, and cooling water for the thermal power stations. Other than their many beneficial uses and value, dams also present risks to property and life due to their potential to fail and cause catastrophic flooding. Dam owners and regulators need carefully analyse and inspect dams to identify potential failure modes and protect against them to mitigate these risks. Since no program for preventing dam failure can ever be certain, and because the potential for loadings exceeding design limits can never be eliminated, another essential part of risk mitigation is simulating potential failures and planning. This prioritization process facilitates is an effective use of financial and human resources to improve public safety and reduce dam failure risk. The analyse area includes two significant urban settlements, namely Kluang township incorporating the old town council area and adjacent villages, as well the Bandar Tenggara township to the east. The remainder area is primarily rural with economic activity concentrated in agriculture which represents the primary economic base for Kluang district.

The proposed Kahang Dam is located across Sg. Kahang, a tributary of Sg. Sembrong. The axis of the dam would be located approximately 1km upstream, of the point where a paved road from the Kluang-Mersing main road to Felda Kahang Timur crosses Sg. Kahang. Kahang Dam is an earthfill embankment dam and the dam core filled with impermeable material, clay. A reinforced concrete outlet structure with control gate will be built across the dam to control releases into the Sg. Kahang downstream.

2. Literature Review

Based on some literature review [1] prediction of the reservoir outflow hydrograph and the routing of that hydrograph through the downstream valley are the two primary tasks in the analysis of a potential dam failure. When populations located close to a dam at risk area, it is important to predict the breach outflow hydrograph and its timing relative to events in the dam failure. A dam break study should cover dam inspection, selection of failure storm, prediction of breach outflow hydrograph and data collection (Coleman, 2002). The breach outflow hydrograph will be utilized to predict the arrival time and flood depth at the reservoir downstream area. Dam breach parameter can be predicted by using empirical and physically based methods. The effective critical breach parameters equations were identified from a numerous researchers [2], [3] and [4]. Equation 1 to Equation 3 shows the common dam breach parameter equation respectively.

$$B_{avg}=2.5h_w+C_b\,;\ t_f=0.015h_w \text{ highly erodible} \quad (1)$$

$$B_{avg}=0.1803K_oV_w^{\,0.32}h_b^{\,0.19}\,;\ t_f=0.00254(V_w)^{0.53}h_b^{\,-0.9} \quad (2)$$

$$V_{er}=0.0261(V_wh_w)^{0.769}\,;\ t_f=0.0179V_{er}^{\,0.364} \quad (3)$$

Where, B_{avg} *is* breach width, t_f *is* failure time, C_b *is* reservoir coefficient and K_o is failure coefficient.

3. Methodology

This dam break analysis involves the field data collection and computer simulation to evaluate the dam failure and the impact of dam break to downstream area. It is particularly more important in this analysis because there are numbers of villages at the downstream area. The field work involves field data collection that is necessary for the estimation of probable maximum precipitation (PMP) and for the estimation of probable maximum flood (PMF). The obtained PMF values will be routed through the reservoir, and checked whether the existing spillway capacity for each reservoirs is adequate for the PMF. If the existing spillway is under designed, the dam will be simulated for its failure. Even if the spillway capacity is adequate for the PMF, it will be assumed that the dam will fail due to poor maintenance, and the impact of its failure to the downstream area be estimated. This paper discussed the first phase of dam break analysis which is prediction of breach outflow hydrograph for all dam failure modes.

The breach parameters include breach width, depth, initiation time and rate of development. These dam break analysis will be carried out using MIKE 11 Dam Break (DB) module, which could generate the outflow hydrograph from a breached dam. MIKE 11 can be linked to hydrologic models, can model reservoirs, perform sophisticated structures, model mechanism of dam failure and can be linked to telemetry and forecasting systems. The flow shall be routed downstream using MIKE 21 in order to determine the maximum water level, discharge along the flood path and the flood wave travel time at different locations of the river downstream of the dam [6], [7].

4. Results and Discussion

For PMF overtopping failure, spillway of the dam was designed to cater PMF (non-overtopping dam). Figure 2 shows the PMF inflow hydrograph for Kahang dam. From MIKE 11 simulation, the reservoir water level under PMF condition was EL 49 m. It is assumed that the initial water level in the Kahang reservoir was at EL 48.5 m start to initiate breach at 52.5 m. The derived failure time of 0.176 hours was estimated using Froelich Equation [3], and the breach was developed until it reached the dam invert level of 37 m with the breach side slopes of 1V:2H and bottom breach width equalled to 107.065 m. Von Thun and Gillette equations was adopted for estimation of breach geometry [4]. The dam peak discharge was simulated to be at 6188m^3/s and illustrated in Figure 3.

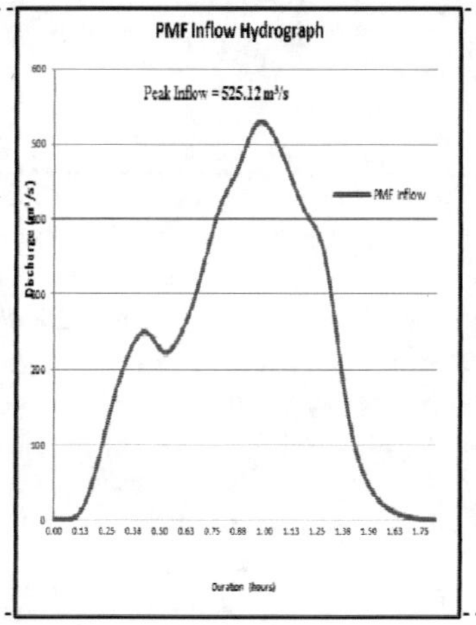

Figure 2. PMF Inflow Hydrograph of 525.13m^3/s for Kahang Dam

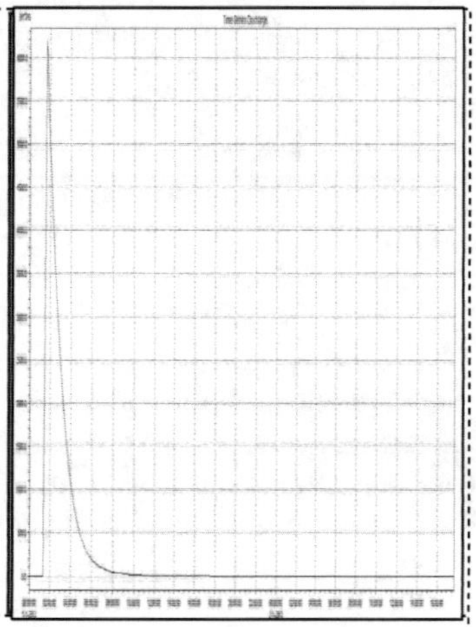

Figure 3. Breach Outflow Hydrograph of 6188m^3/s for Kahang Dam PMF Failure

The integrated dam break study – even when carried out with the latest mathematical models and improved data – falls largely in the realm of predictions based on the set of assumptions made for the study [5]. Therefore, sensitivity analysis was performed in order to estimate the influence of variations made in the model by varying the breach width, breach depth, failure time and breach side slope to determine the sensitivity of the peak breach discharge to these variables [6], [7]. Based on simulated peak discharge, the breach sensitivity analysis for Kahang Dam is summarized in Table 1.

Table 1. Breach Sensitivity Analysis.

Breach Parameters				
Failure time, t_f (hour)	**Side Slope (H:V)**	**Breach Width (m)**	**Breach Depth (m)**	**Q_{peak} (m^3/s)**
0.176	1:2	105.489	15.5	6110.063
1.996	1:2	105.489	15.5	4846.889
0.176	1:1	105.489	15.5	5756.076
0.176	1:3	105.489	15.5	6462.333
0.176	1:2	35.13	15.5	2688.520
0.176	1:2	200	15.5	10211.435
0.176	1:2	105.489	8	1246.185
0.176	1:2	105.489	12	3501.201

The four parameters affected the breach hydrograph's peak. For the same breach side slope, breach depth and breach width, the difference in peak discharge was only 21 % for failure time of 1.996 hours. For the same failure time and breach width, the gentler side slope yielded a slightly higher peak discharge compared to the steeper side slope. The peak value of the breach hydrograph differed by 12% when the breach side slope changed from 1H: 1V to 1H: 3V. For breach width, the difference in breach bottom width had a greater influence on the peak discharge which was in the range of 56 – 67 % for breach bottom width of 35.13 m and 200 m. The higher breach depth also resulted higher peak discharge as compared to shallow breach depth. The difference in peak discharge was 64% when the breach depth changed from 8 m to 12 m. The above results are possible since the Kahang Dam height is 17.5 m with a relatively 460 m crest length, hence the breach area varied drastically with the breach depth and breach bottom width. Since Kahang Dam has a large reservoir area, the failure time was not so critical in influencing the peak outflow discharge. Therefore, the changes of side slope do not given significant different in peak discharge due to short crest length. Thus, breach width was recognized as the most sensitive breach parameters followed by breach depth.

5. Conclusions

The consequences of dam failure will depend on the type of failure scenarios which are Clear Day Failure (CDF) and Probable Maximum Flood Failure (PMF).. In the event of dam failure, there might be structural damage, loss of life, disruption of services and interruption of normal social and economic activities These losses largefly depend on the flood velocity, inundation depth and duration. Depending on these flood parameters, consequences can be classified as minor, moderate and major. To help define these situations the Kahang rivers should be gauged, and caution, warning and danger levels should be determined and shown at the public places of interest.

The generated flood inundation map from MIKE 21 modeling will be used to develop Emergency Action Plan (ERP) for the unexpected dam break event to coordinate the responsibility authorities such as dam owners, police and fire services in the people evacuation plans, other flood warning systems, post-flood actions and hazard classification of affected areas. This dam break study is also important to estimate the property damages and loss of human resulting from a dam breach. This study will enhance the community resilience towards dam failure and stressing the ability to reduce the possible impacts of a disaster as well as to effectively respond a recovery following a disaster.

6. Acknowledgement

The authors would like to take this opportunity to recognize the efforts and assistance from every individual and organizations on all aspects in the preparation of this paper. This study is supported by Ministry of Energy, Green Technology and Water (KeTTHA). The authors also profoundly grateful to ZHL Engineer Sdn Bhd for their ideas, provision of relevant materials during the preparation and development, and providing valuable input to complete this research.

References

[1] Andrews D P, Coleman S E, Webby M G and Melville B W 1999 Noncohesive embankment failure due to overtopping flow *Proceedings, 28th Congress of the International Association for Hydraulic Research* Graz, Austria

[2] MacDonald T C and Langridge-Monopolis J 1984 Breaching characteristics of dam failures *J. Hydraulics Div*. ASCE **(110)5** 567-586

[3] Froehlich D C 1995 Peak Outflow from Breached Embankment Dam *J. of Water Resources Planning Management* **(121)1** 90-97

[4] Von Thun, Lawrence J and Gillette D R 1990 Guidance on Breach Parameters *Unpublished internal document, U. S Bureau of Reclamation* Denver, Colorado

[5] Pillai B R K, Saibal Ghosh and Gupta O P 2012 Integrated Dam Break Analysis – Case Study of Indira Sagar and Omkareshwar Projects *India Water Week 2012 – Water, Energy and Food Security : Call for Solutions* 10-14 April 2012, New Delhi

[6] Sidek L Y, Ros F C, Ibrahim N N N, Razad A Z A and Ahmad M N 2013 Hydrodynamic Dam Breach Modelling of Earthfill Saddle Dam *J. of Energy and Environment* **3(1)**

[7] Sidek L Y, Ros F C , Mohd Noh A, Rahsidi S M, Azwin Zailti A R, Mohd Nor MD, Hidayah B and Intan Shafilah A A 2012 Development of Probable Maximum Flood (PMF) for Sultan Abu Bakar Dam *5th Engineering Conference (EnCon 2012)* 10-12 July 2012, Kuching, Sarawak

Evaluation of green roof as green technology for urban stormwater quantity and quality controls

K H Kok[1], L M Sidek[1], M R Z Abidin[2], H Basri[1] , Z C Muda[1] and S Beddu[1]
[1] Centre for Sustainable Technology and Environment (CSTEN), College of Engineering, Universiti Tenaga Nasional, Jalan IKRAM-UNITEN, 43000 Kajang, Selangor, Malaysia
[2] Humid Tropics Centre Kuala Lumpur, No 2 Jalan Ledang off Jalan Duta 50480 Kuala Lumpur, Malaysia

E-mail: kahhoong_kok@yahoo.com, lariyah@uniten.edu.my

Abstract. Promoting green design, construction, reconstruction and operation of buildings has never been more critical than now due to the ever increasing greenhouse gas emissions and rapid urbanizations that are fuelling climate change more quickly. Driven by environmental needs, Green Building Index (GBI) was founded in Malaysia to drive initiative to lead the property industry towards becoming more environment-friendly. Green roof system is one of the assessment criteria of this rating system which is under category of sustainable site planning and management. An extensive green roof was constructed in Humid Tropics Center (HTC) Kuala Lumpur as one of the components for Stormwater Management Ecohydrology (SME) in order to obtain scientific data of the system. This paper evaluates the performance of extensive green roof at Humid Tropics Center with respect to urban heat island mitigation and stormwater quantity and quality controls. Findings indicate that there was a reduction of around 1.5°C for indoor temperature of the building after installation of green roof. Simulations showed that the peak discharge was reduced up to 24% relative to impervious brown roof. The results show an increment of pH and high concentration of phosphate for the runoff generated from the green roof and the runoff water quality ranged between class I and II under INWQS.

1. Introduction

Depleting natural resources has appealed to sustainable developments. “Green” technology or “green” innovation is the common term for a particular method which can improve quality of natural resources such as water and air by inducing no or minimal pollutants to the nature. Malaysia is a developing country and it is inevitable that more natural resources are sacrificed for purpose of development and commercial activities. One of the prominent depleting natural resource in densely built urban areas is green spaces. It is a norm that green spaces in urban areas are replaced with impervious surfaces like concrete and roadways and phenomenon like Urban Heat Island (UHI) and flash flood are pertinent to excessive impervious surfaces.

Green Building Index (GBI) aims to evaluate the design and performance of local buildings with responses to degree of damages to environment and energy consumption. Green roof is suggested as one of the environmental friendly innovation under category of sustainable site planning and management. In fact, Urban Stormwater Management Manual for Malaysia (MSMA) was firstly introduced in 2001 as a guideline to adopt and design Best Management Practices (BMPs) in controlling stormwater in term of quantity and quality to achieve least impacts of post-development hydrologically [7]. Recently, MSMA second edition was released in 2012 whereby additional BMPs like bioretention was compiled and revised into the manual.

However, green roof was not compiled under MSMA yet though it is one of the BMPs as well. As other stormwater manuals applied at overseas like SUDs (UK), LID (USA), and WSUD (Australia) have been suggesting that green roofs can be used as a practice for stormwater quantity and quality controls, thus there is a urge to study the performances of green roof under local tropical climate. The scientific data obtained can be applied to develop design guideline of green roof system for Malaysia in future.

2. Green Roof Concept

Green roofs are made of a system of manufactured layers deliberately placed over roofing structures which support growing medium and vegetation. Green roof systems can be generally divided into extensive green roof (Eco-roofs) and intensive green roof (Podium Garden & Sky Gardens) [1][6]. Extensive green roof is low-weighted and requires only minimal maintenance. On contrary, intensive green roof is much heavier and requires higher construction cost and constant maintenance. Its substrate layer over 150mm is a common. The basic components of a green roof system regardless of extensive or intensive are vegetation, growth media, drainage layer, filter layer, water proofing layer and protection layer. The type of vegetation selected for the green roof system in HTC is *Zoysia Japonica* (Japan Grass). Figure 1 shows the configuration of green roof components in HTC, while Figure 2 shows the retrofitted green roof system for *surau* located in HTC.

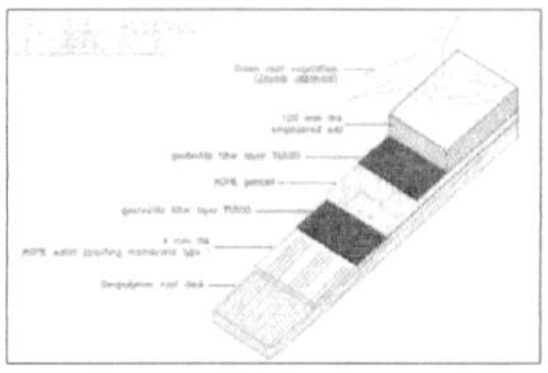

Figure 1. Configuration and details of extensive green roof in HTC.

Figure 2. Extensive green roof for *surau* in HTC

3. Literature Review on Green Roof Performances

Numerous studies and researches conducted overseas have proven that extensive green roof gives positive effects on peak discharge reduction and thermal reduction. Green roofs reduce runoff by delaying the initial time of runoff due to absorption of water in the green roof, reducing the total runoff by retaining part of the rainfall, and spreading the residual runoff over a long time period through a relatively slow release of the excess water that is stored in the substrate layer [2]. Two extensive green roofs installed at North Carolina of America indicated that both of these roofs retained 64% of the total precipitation recorded and each green roof showed reduction in average peak flow by more than 75% [3]. Water storage capacity or stormwater retention ability has close relationship with rainfall intensity [4][5], the ability decreases during heavy storm events.

Green roof contributes to building insulation and energy efficiency by trapping air layer within the plant mass so that the building surface is cooled in summer and warmed in winter [5]. Vegetation of green roof act as a cooling agent by dissipating portion of city heat via evapotranspiration, thus green roof is applied as a way to combat urban heat island effect and to conserve more energy which is initially used to cool the buildings. Green roofs were able to reduce solar energy gained up to 90% when compared with non greened-top buildings, indoor temperature for green roof's buildings was reported to have reduction of 3-4°C as outdoor temperature ranges between 25 °C to 30 °C [4].

Performance of green roof in enhancing water quality is debatable since it sometimes can be a source of nutrients/pollutants to the outflow. Green roofs did remove phosphorus concentration during moderate storms but not the efficient filter for pH, BOD_7, and COD [5]. However, higher concentration of TN and TP were observed in the green roof outflow during the first two years of green roof's life and its media was the origin of the nutrients [3].

4. Methodology

The hydrological performance of green roof in HTC was studied by monitoring the runoff hydrographs that the green roof produced in certain storm events. These runoff hydrographs were named as observed green roof hydrographs. The runoff produced by the green roof was gauged by ultrasonic flow module and a runoff collection tank with V-notch weir. Simulated green roof hydrographs were then plotted using stormwater modeling software (XP-SWMM). Simulation was done both for actual storm events and design storms. For design storm events, the simulation was done based on the rainfall temporal patterns derived for HTC. The specific design storms durations were 10, 30, and 60 minutes with 2 years ARI.

Runoff water samples from the green roof were collected as grab samples for randomly selected storm events. They were sampled into a plastic bottle with capacity of 2 litre and sent to certified laboratory for chemical testing. Besides that, an in-situ water quality probe also installed at the runoff collection tank for

IOP Conf. Series: Earth and Environmental Science **16** (2013) 012045 doi:10.1088/1755-1315/16/1/012045

measuring the physical properties of the runoff such as pH, temperature, and conductivity. Indoor temperature of the green roof building was measured by a digital thermometer which was attached to the inner wall of the building. Differences in maximum temperature recorded before and after installation of the green roof were evaluated.

5. Results and Discussions

Time-Area runoff generation model was used to transform the effective rainfall to runoff hydrograph. The calibrated and verified hydrologic parameters were applied to simulate runoff hydrograph for green roof and brown roof. The peak discharges obtained for green roof were then compared to brown roof in order to determine the ability of the green roof in HTC in reducing the peak discharge. The simulations were conducted for design storms of 2 years ARI with duration of 10 minutes, 30 minutes, and 60 minutes. Local rainfall temporal pattern in HTC was applied to generate the simulation. Figure 3 below indicates the comparison of simulated green roof and brown roof hydrographs for 60 minutes design storm and results for other duration of design storms were summarized in Table 1.

Results show that the simulated green roof hydrographs yielded lower peak discharge than the brown roof in each specified duration of design storm and the reduction ability increased for storms with lower rainfall intensities. Simulation also carried out for actual storm events and the results were compiled in Table 2. Average peak discharge reduction of 23.6% was estimated from the specified storms. Figure 4 shows the comparison of simulated hydrographs for actual storm on 16/8/2012.

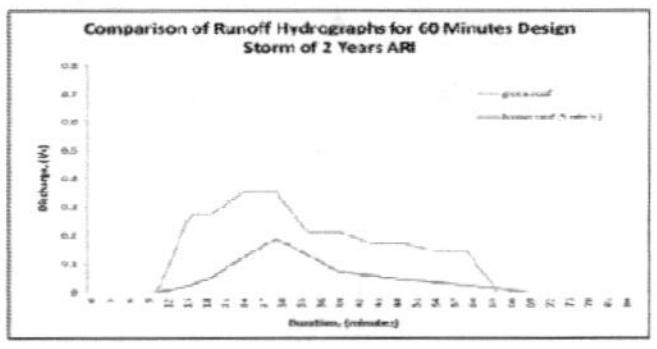

Figure 3. Comparison of simulated green roof and brown roof hydrographs.

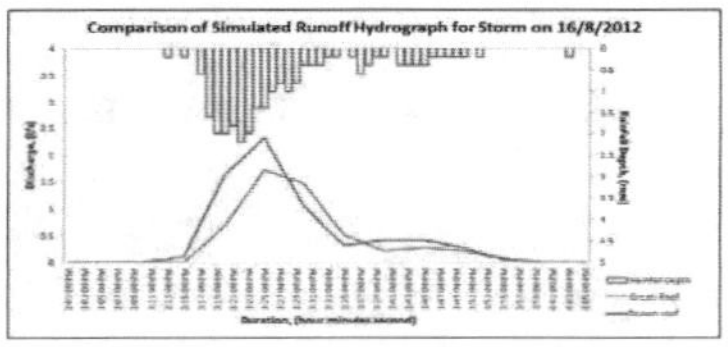

Figure 4. Comparison of simulated hydrographs for actual storm on 16/8/2012.

Table 1. Relationship between intensity with reduction of peak discharge

Duration of Design Storm (minutes)	Rainfall Intensity (mm/hr)	Reduction of Q_{peak} (%)
10	146.6	24.4
30	86.0	40.7
60	59.9	47.3

Table 2. Comparison for simulated hydrographs based on actual storm events

Events	Q_{peak} (*l/s*) (Green Roof)	Q_{peak} (*l/s*) (Brown Roof)	Reduction (%)
6/7/2012	0.881	1.156	23.7
8/7/2012	0.466	0.629	25.9
20/7/2012	0.639	0.840	19.1
16/8/2012	*1.720*	*2.315*	*25.7*

There were total 15 runoff water samples collected from April 2012 to September 2012. The samples were sent to certified laboratory for water quality analysis. Water Quality Index (WQI) of each sample was then calculated based on the results obtained from laboratory and in-situ probe. Table 3 portrays extract of WQI of the samples in several storm events. An average WQI of 92 was estimated out of 15 samples for the runoff generated from the green roof in HTC which indicated that the runoff produced was clean (not polluted).

Although the runoff samples achieved high WQI and considered not polluted, but results told that the green roof in HTC was a source of phosphate (PO_4-P). Average concentration of 2.40 *mg/l* was founded from the samples collected. The substrates (fertilized planting soil) were thought as the origin of this nutrient contained in the runoff. However, the concentration of PO_4 presented in the runoff samples were noticed to be reducing as time lapsed due to uptake by the vegetation. Unestablished vegetation during early stage caused the nutrient leached into the runoff.

Thermal performance of the green roof was monitored by comparing the maximum mean temperature recorded prior and after installation of green roof. Indoor temperature of the green roof building was measured in every interval of 15 minutes daily. In order to succeed the comparison, an assumption was made that the local climate in HTC before installation and after installation of green roof was about the same and fluctuation of surrounding temperature was negligible. Reductions of mean temperature were noticed in both months of January and February after installation of green roof. The amount of the comparison of temperature before and after installation of green roof was illustrated in Figure 5.

Table 3. WQI and SI for runoff samples

Event	TSS	BOD	COD	DO	pH	NH_3-N	WQI	Class
2/5	94	93	70	95	93	89	**89**	**II**
7/5	95	93	80	96	95	95	**93**	**I**
10/5	96	95	80	94	94	91	**92**	**I**
11/5	95	94	75	94	94	90	**90**	**II**
26/6	93	94	75	100	93	84	**90**	**II**
Average	**94**	**95**	**80**	**95**	**94**	**90**	**92**	**I**

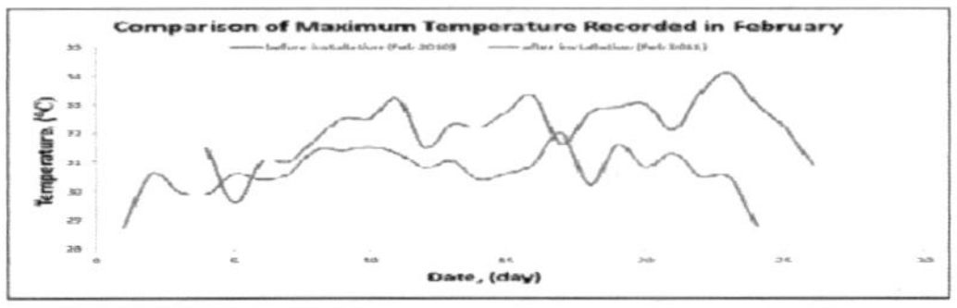

Figure 5. Reduction of room temperature after installation of green roof

6. Conclusion

Findings show that the performances of extensive green roof system in HTC are promising under local tropical climate. Simulations in design storms and actual storm events indicated that it could reduce the peak discharge relative to impervious brown roof. However, its ability reduced for storms with intense rainfall. The water quality of the outflow produced by the green roof was generally good and achieved high WQI. However, it was a source of PO_4 and basic outflow were noticed producing from the green roof during storms. Substrates of the green roof could be the essential factor in affecting the quality of the outflow.Cooler environment was created inside the green roof building since reduction of indoor temperature was noticed after installation of the green roof system.

7. Acknowledgement

The authors wish to acknowledge Humid Tropics Centre, Kuala Lumpur (HTC) and Department of Irrigation and Drainage of Malaysia (DID) for their financial support and collaboration in succeeding this green roof demonstration project. Special acknowledgement to ZHL Engineers Sdn. Bhd for the technical guidance and advices in completing this study.

References

[1] Derek T 2007 Study on Green Roof Application in Hong Kong Final Report 77-81&100-104

[2] Mentens J, Raes D, Hermy M 2006 Green Roof as a Tool For Solving the Rainwater Runoff Problem in The Urbanized 21st Century? *J. of Landscape and Urban Planning* 77 217-226

[3] Moran A, Hunt B, Jennings G 2005 A Field Study of Green Roof Hydrologic and Water Quality Performance *American Society of Agricultural and Biological Engineer* ISSN 0001-2351

[4] Ni J Y 2009 Green Roof Study: Stormwater Quantity, Quality and Thermal Performance *M.S thesis (University of Pittsburgh, USA)*

[5] Villarreal E L and Bengtsson L 2005 Response of a sedum green-roof to individual rain events *Ecological Engineering* 25 1-7

[6] Kah H K, Lariyah M S, Mohamed Roseli Z A, Hidayah B, Salmia B, Shuhairy N and Herdayati K 2012 Green Roof as an Option of Green Technology for Urban Stormwater Management Controls 5th Engineering Conference (EnCon 201) 10-12 July 2012 (Kuching, Sarawak)

[7] Lariyah M S, Mohd Nor M D , Mohamed Roseli Z A, Zulkefli M, Amirah Hanim M P 2011 Application of Water Sensitive Urban Design at Local Scale in Kuala Lumpur 12th International Conference on Urban Drainage (12ICUD) 10- 15 September 2011 (Porto Alegre, Brazil)

IOP Conf. Series: Earth and Environmental Science **16** (2013) 012046 doi:10.1088/1755-1315/16/1/012046

Hydrological assessment for mini hydropower potential at Sungai Pahang @ Temerloh

L M Sidek[1], A Z A Zaki[1], Z Mustaffa[2], M I H Ibrahim, Z C Muda[1], S Thiruchelvam and H Basri[1]

[1] Centre for Sustainable Technology and Environment, College of Engineering, Universiti Tenaga Nasional, Jalan IKRAM-UNITEN, 43000 Kajang, Selangor, Malaysia

[2] ZHL Engineers Sdn Bhd, No 16A, 2nd Floor, Jalan Diplomatik 2/1 Presint Diplomatik 15, 62502 Putrajaya

E-mail: lariyah@uniten.edu.my

Abstract. Sg Pahang at Temerloh was considered for assessment of hydropower potential using hydrological analysis method and hydrological model. The available data related to topography, soil, land use, weather and discharge pertaining to the study catchment were used to characterize the catchment. The characterization was required for water resources hence hydropower assessment. The hydrology of the study catchment was simulated through the model. This hydrological study is required due to the proposed mini hydroelectric power plant at Pulau Temerloh. It is essential to evaluate the existing river flow characteristic and to model the environmental flow assessment of the river. Two rainfalll stations, JPS Temerloh and Pintu Kawalan Paya Kertam Station are selected to develop the Rainfall Intensity Duration frequency (RIDF) Curve to determine the rainfall intensity of the area. Daily river flow were recorded at Sg Pahang at Temerloh and Sg Pahang at Lubok Paku were used to develop the Flow Duration Curve (FDC) to study the characteristic of Sungai Pahang flow. The 7 days low flow with 10 years return period (7Q10 low flow) was obtained using both Gumbel Method and Log Pearson Type III Method. The results from FDC shows that 50% percentage of time the Sg Pahang @ Temerloh is exceeded over a historical period is 400 m^3/s and 50% percentage of time the Sg Pahang @ Lubok Paku is exceeded over a historical period is 650 m3/s. The required environmental flow are set to be 7Q10 low flow which is 64.215 m3/s for Sg Pahang at Temerloh and 79.24 m3/s for Sg Pahang at Lubok Paku. The results show the water resources are abundant and hence boost the mini hydropower potentiality at Sg Pahang.

1. Introduction

Hydropower is a clean, renewable and reliable energy source that serves national environmental and energy policy objectives. It is one of the most important renewable sources for production of electric power due to several obvious reasons. It is renewable in nature, unlike wind, supply variability within shorter time period is less and most importantly green house gas (GHG) emission is least. It is derived from the falling water, either from rivers and streams flowing downhill along the river course due to force of gravity. Harnessing the kinetic energy from the flowing water for driving turbine generates hydropower [1].

This study was conducted as requirement prior to construction of Mini Hydroelectric Power Plant at Pulau Temerloh, Pahang. The planning plays important role in the growth and development of hydropower. The amount of achievable hydropower at any given site is a function of turbine head and the corresponding flow rate. Thus, harnessing hydroenergy requires assessment of the water resource which depends upon the natural processes occurring locally and also the terrain characteristics. Accurate and reliable assessment of water resources leads to successful planning. One of the factors that must be considered when planning the construction of mini hydroelectric power plant is the amount of flow needed to make sure full hydropower electric can be supplied at all time with the consideration to the minimum river flow needed to sustain the ecosystem of aquatic life. In this sense, the study of the flow characteristic of the Sg Pahang is essential especially during dry time where water level is at its minimum. Hydrological methodology relies mostly on the past recorded data of river flow. The flow duration curve provides information about the percentage of time or days that a particular stream flow was exceeded over some historical period. It has long been used as a way of summarizing catchments hydrologic responses, but more recently these curves have been used to validate the output of hydrologic models and compare the observed and modeled hydrologic responses.

Apart from analysis of Flow Duration Curve, low flow analysis was conducted using 7Q10 or 7 days low flow with 10 years return period. This analysis are required for protecting or regulate water quality, serves as general indicator of prevalent drought conditions which is normally cover large area and finding total maximum daily load to assess aquatic life protection. The low flow analysis is able to find minimum quantity of stream flow that are necessary to protect habitat during drought situation, to be considered as the worst case scenario in water quality and lastly to compare the impact of climate change and irrigation on low surface stream flow [2].

2. Methodology

For the purpose of this study, there are several data that must be obtained, which includes the historical rainfall data, historical stream flow data, annual 7 days low flow and the water quality data of Sungai Pahang. Two rainfall stations and two stream flow stations were selected for this study.

The selection criteria of the hydrological stations are to have more than 20 years of historical data. The historical rainfall data will be used to develop Intensity Duration Frequency (IDF) Curve. An IDF Curve is a tool that characterizes an area's design rainfall intensity. A Flow Duration Curve (FDC) will be developed using the stream flow historical data. FDC is a plot that shows the percentage of time that flow in a stream is likely to equal or exceed some specified value of interest. Meanwhile, low flow analysis will be conducted using analytical method i.e. Gumbel's Method and Log Pearson Type III Method and graphical method using Plotting position method.

3. Results and Discussions

The IDF curve for rainfall station located at JPS Temerloh and Pintu Kawalan Paya Kertam was developed for 2, 5, 10, 20, 50 and 100 years ARI as shown in Figure 1. A similar trend was observed, where the rainfall intensity is decreased over time at both stations. Therefore, it can conclude that heavy storm only hit a short time, but minimal storm can go on for long periods. Apart from that, another observation that can can deduced from the IDF curve is that frequency and intensity are related, as rarer rainfall events (greater ARI) tend to have higher rainfall intensity for a given duration.

Flow Duration Curve (FDC) provides information on water availability in the river. The median flow (Q50) is the discharges which is equalled or exceed 50% of the time. The part of the curve with flow below the median flow represents low-flow conditions. The flow duration curves that are constructed for Sg Pahang at Temerloh and Sg Pahang at Lubok Paku will be used to show the percentage of time that flow in the river is likely to equal or exceed some specified value of interest.

The daily river flow value was analysed from as early as 1963 up to 2011 and the maximum flow, Q (m^3/s) recorded was 6563.4 m^3/s while minimum flow was 19.0 m^3/s and mean flow is 542.8 m^3/s. A flow duration curve for Sungai Pahang at Temerloh is shown in Figure 2.

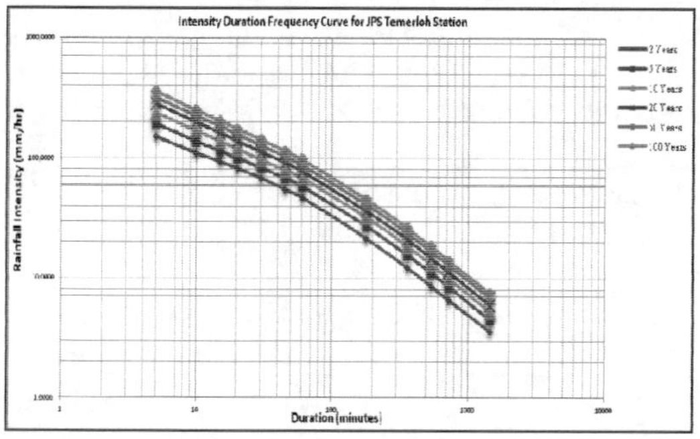

Figure 1. IDF Curve for JPS Temerloh Station

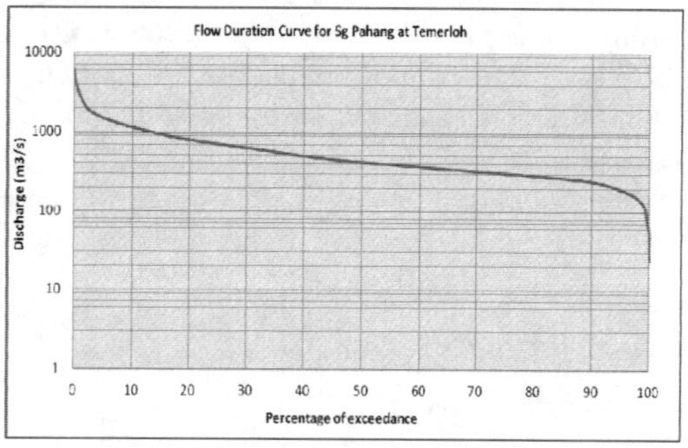

Figure 2. Flow Duration Curve for Sg Pahang at Temerloh

The results show that 50% dependable flow of 400 m^3/s will be exceeded or equalled at Sg Pahang @ Temerloh. A discharge of 200 m^3/s is likely to be available 95% of the time while a discharge of 650 m^3/s is likely to be available 25% of the time. The same graph was plotted for Sungai Pahang at Lubok Paku station, and it is estimated that a 50% dependable flow of 650 m^3/s will be exceeded or equaled for this river. A discharge of 275 m^3/s is likely to be available at 95% of a time and a discharge of 1000 m^3/s is likely to be available 25% of a time.

Table 1 tabulated the 7-days low flow estimates for Sg Pahang at Temerloh and Sg Pahang at Lubok Paku respectively using both Gumbel Method and Log Pearson Type III Method. The average low flow values are obtained by taking the average from Gumbel Graph Plotting and Log Pearson Type III Graph Plotting.

Table 1. 7-days low flow estimates for Sg Pahang at Temerloh and Lubuk Paku

	Sg Pahang at Temerloh			Sg Pahang at Lubok Paku		
Years	Gumbel Method Low Flow (m^3/s)	Log Pearson Type III Method Low Flow (m^3/s)	Average Low Flow (m^3/s)	Gumbel Method Low Flow (m^3/s)	Log Pearson Type III Method Low Flow (m^3/s)	Average Low Flow (m^3/s)
2	137	141.17	139.09	162	169.02	165.51
5	84	83.00	83.50	103	101.49	102.25
10	65	63.43	64.22	81	77.48	79.24
20	53	50.40	51.70	67	62.18	64.59
50	41	38.86	39.93	53	48.42	50.71
100	35	32.79	33.90	46	41.26	43.63

For this study, the 7 day low flow for 10 years return period (7Q10) is used as design criteria to define the environmental low flow condition of the Sg Pahang. Therefore, the values of 7Q10 low flow to be adopted in the environmental low flow assessment for Sg Pahang at Temerloh is 64.2 m^3/s while for Sg Pahang at Lubok Paku is 79.2 m^3/s.

Water quality data for Sg Pahang which includes parameters such as Dissolve Oxygen (DO), Biochemical Oxygen Demand (BOD), Chemical Oxygen Demand (COD), Total Suspended Solid (TSS), pH, NH3NL and Pb are obtained from the Department of Environmental, Malaysia. Table 2 shows the DO, BOD, COD and TSS concentration for Sg Pahang at Temerloh and Sg Pahang at Lubok Paku respectively. Result shows that DO, BOD and COD are classified as Class II according to Interim National Water Quality Standard (INWQS). TSS and pH on the other hand, are categorized as Class III and Class I respectively. The estimated loading was computed by multiplying the value of DO, BOD, COD and TSS with the 7Q10 low flow as shown in Table 2.

Table 2. Environmental Flow Assessments based on DO, BOD, COD and TSS under 7Q10 Low Flow for Sg Pahang at Temerloh and Lubok Paku

	Sg. Pahang at Temerloh				Sg. Pahang at Lubok Paku			
	DO	**BOD**	**COD**	**TSS**	**DO**	**BOD**	**COD**	**TSS**
Value (mg/L)	6.74	2	13	105	6.72	3	16	141
7Q10 Low Flow (m^3/s)	64.215	64.215	64.215	64.215	79.24	79.24	79.24	79.24
Estimated Loading (mg/s)	432,809	128,430	834,795	6,742,575	532,492	237,720	1,267,840	1,117,2840

4. Conclusions

Traditionally the water resources assessment is made using historically observed discharged data at outlet of the catchment. In the present study, the available hydrological data were analysed using various hydrological method to assess water resources and hence hydropower potentiality. The developments of the Flow Duration Curves have enabled the study of flow characteristic of Sungai Pahang both at Temerloh area and Lubok Paku area. In the environmental flow assessment, the 7Q10 low flow value of 64.215 m3/s was chosen as the recommended environmental flow for Sungai Pahang at Temerloh. This low flow value is set as the standard where the river flow must not be lower than this value for the benefits of those who very much depend of the river such as the aquatic life. The result of the hydrological study in Sg Pahang is anticipated to boost the current initiative of mini hydropower production in this region considering untapped vast water resources of Sg Pahang.

5. Acknowledgement

The authors would like to take this opportunity to recognize the efforts and assistance from every individual and organizations on all aspects in the preparation of this paper. This study is supported by Ministry of Energy, Green Technology and Water (KeTTHA).The authors also profoundly grateful to ZHL Engineers Sdn Bhd for their ideas, provision of relevant materials during the preparation and development, and providing valuable input to complete this research.

References

[1] Kusre B C, Baruah D C, Bordoloi P K and Patra S C 2010 Assessment of hydropower potential using GIS and hydrological modeling technique in Kopili River basin in Assam (India) *Applied Energy* **87** 298–309

[2] Arthington A H, King J M, O'keefe J H, Bunn S E, Day J A, Pusey B J, Bludhorn D R and Tharme R 1992 Development of an holistic approach for assessing environmental flow requirement of riverine ecosystem. *In: Piagram, J.J. and Hooper, B.P. (Eds.). Proceedings of an International Seminar and Workshop on Water Allocation for The Environment* Center for Water Policy Research Armidale 69-76

IOP Conf. Series: Earth and Environmental Science **16** (2013) 012047 doi:10.1088/1755-1315/16/1/012047

Effectiveness of the stormwater quality devices to improve water quality at Putrajaya

L M Sidek [1], H Basri[1], M N Md Noh[2], A Ainan[2] and A H Mohd Puad[1]
[1] Centre for Sustainable Technology and Environment, College of Engineering, Universiti Tenaga Nasional, Jalan IKRAM-UNITEN, 43000 Kajang, Selangor, Malaysia
[2] Department of Irrigation and Drainage, Malaysia

E-mail: Lariyah@uniten.edu.my

Abstract. Development of Putrajaya has changed the character of the natural landform by covering the land with impervious surfaces. Houses, office buildings, commercial place and shopping centres have provided places to live and work. The route between buildings is facilitated and encouraged by a complex network of roads and car parks. However, this change from natural landforms and vegetative cover to impervious surfaces has major effect on stormwater which are water quality (non-point source pollution). This paper describes the effectiveness of the stormwater quality devices to improve water quality at selected Putrajaya for demonstration in order to evaluate low cost storm inlet type devices in the Putrajaya Catchment. Five stormwater quality devices were installed and monitored during the study. The devices include Ultra Drain Guard Recycle model, Ultra Curb Guard Plus, Ultra Grate Guard, Absorbent Tarp and Ultra Passive Skimmer. This paper will provide information on the benefits and costs of these devices, including operations and maintenance requirements. Applicability of these devices in gas stations, small convenience stores, residential and small parking lots in the catchment are possible due to their low cost.

1. Introduction

The stormwater treatment devices are designed to reduce the level of one or more constituents of concern in stormwater drainage from a site. In-line quality devices can be categorised as storm drainage insert type devices which are used to catch sediments, filter, and absorb hydrocarbon products from the storm water runoff in the catch basin manhole/sump structure before entering the storm drainage system. These devices can be divided into three categories based on their operating principles: In-line Filtration Devices, Hydrodynamic Separators, and In-drain Filtration Devices. The first types of BMP devices namely In-line filtration devices employ some type of filtration media as the mechanism for removal of stormwater constituents in an in-line device such as StormGate, StormFilter, StormScreen, and Catch Basin StormFilter and etc [1].

Gross pollutants include naturally occurring material, like vegetation, rocks, and soil, as well as litter. High concentrations of O&G can cause toxicity in receiving waters, and most discharge regulations require that there be no discharge of oily wastes that produce sheen on the surface of the receiving water. Sources of oil and grease in storm water include O&G sorbed to trash and other debris; O&G sorbed to particulates; emulsified oils (small drops of oil suspended in storm water); free floating oil; and suspended oil. Research shows that between 83 and 98 percent of total hydrocarbons in storm water runoff are associated with particulate matter, and evidence suggests that a significant portion of these particles are settleable solids, such as sediments [2]. Therefore, storm water Best

Management Practices (BMPs) designed to remove sediments may also remove some oil and grease associated with the sediments. Typically, only free-floating oil concentrations are measured and reported in storm water studies [3].

Sorbents can be divided into three basic categories: natural organic, natural inorganic, and synthetic. Table 1 describes the three materials and their sorbent capacities. This report focuses on synthetic sorbents; however, many BMPs can be fitted with natural organic or inorganic sorbents to help absorb oil and grease. The sorbent capacity for are measured in terms of theirs weight in oil. A study shows that synthetic sorbent can absorb up to 70 times of their weight in oil, compared to organic and inorganic with sorbent capacity up to 15 to 20 times of their weight respectively [4].

This Putrajaya pilot BMP was selected for demonstration in order to evaluate low cost storm inlet type devices in the Putrajaya Catchment. This Pilot Study was conducted to evaluate the performance of selected proprietary devices that can capture gross pollutant, silt, sediments, oil & grease and that can be incorporated into the existing drainage system. This stormwater quality device is considered to be a simple retrofit to improve storm water quality before it enters the storm drainage system. Catch basin insert type units were targeted for this type of stormwater quality device. These devices are presently manufactured and marketed by vendors around the United States and Canada. Most of vendors for these devices were located in Florida and Washington State due to the interest of these municipalities with this type of water quality improvement devices.

The study area located at Putrajaya which is situated 25km south of Kuala Lumpur and 20km north of Kuala Lumpur International Airport (KLIA). Rapid development has been ongoing in Putrajaya catchment since the formation of Putrajaya from year 1995 to date. Post development of Putrajaya has resulted in increased sediment leadings into the Putrajaya Wetland and Lake. Controlling sediment from entering drainage systems is a great challenge, since development of Putrajaya is still ongoing. The selected sites by for the installations of GPTs are Persiaran Utara, Precint 12, Putrajaya Sentral, Precint 7 and Jalan P9C1, Precint 9. Figure 1 shows the actual condition of the Putrajaya Sentral polluted with oil and grease from the buses. Meanwhile, Figure 2 shows typical view of silt and sediments entering the drainage and catch pit from the nearby bareland area at Precint 9. This study will focus on few pollution control devices which are expected to assist in oil & grease and sediment control of the GPTs installed at these 3 selected areas.

Figure 1. Oil & Fuel Spots at Putrajaya Sentral and Precint 9

Figure 2. Silt and sediments entering the drainage and catch pit at Precint 9

2. Methodology

Generally all the collected water samples will be analysed for 6 major parameters which includes DO, BOD, COD, AN, SS and pH to determine the water quality index (WQI). The sieve analysis was also conducted to classify the TSS grain size distribution captured in the study area. The effectiveness of the stormwater quality devices in terms of pollutant removal efficiency and O&M cost has been investigated.

The water sampling was conducted during the rain events. Water sample was collected using 1.5 litre liquid bottles and two bottles of water samples were collected at each sampling points where one plastic bottle for other parameters and another one glass bottle for oil and grease. These samples were sent to certified laboratory for analysis. While the sediment samples were collected by scooping the sediment traps at the pollution control devices installed and were sent to the laboratory for sediment particle size distribution analysis according to BS 1377: Part 2:1990.

3. Results Analysis and Discussion

The results of this paper include the sediment and oil and grease capture rate, as well as operation and maintenance summary and cost assessment of the trapping devices. The basic operations and maintenance (O&M) requirements for sorbent materials include periodic checks to ensure that they have not reached their sorbing capacity or become clogged. The frequency of visits and cleaning of sediment and debris from SQID using sorbent materials depends on the type of SQID and the area in which it is located. In general, all SQIDs using oil sorbent materials should be inspected at least monthly. If the material is placed in an area where it is susceptible to a high oil loading rate, the SQID should be inspected and serviced more frequently.

However, inspections of these devices are depends on the individual unit and vary from monthly to quarterly. For example, the Grate Guard and Drain Guard **should be serviced quarterly to remove accumulated sediment and debris and to check the sorbent**. The Passive Skimmer systems were recommended to be inspected at least once a month to check the position of the skimmer and ensure that it remains on top of the water for maximum oil removal efficiency. The devices were in operation from January 2008 to May 2008. Table 1 and Table 2 shows the results from the monitoring work on the performance of various devices.

Table 1. In-line Quality Devices Sediment Captured Summary

Device	**Average Sediment Captured/ Gallon Filtered (lbs/ 1,000 gallons)**
Putrajaya Ultra Drainguard (Putrajaya Sentral)	2.51
Ultra Grate Guard (Precint 9 – 1 Event)	8.86

Table 2. In-line Quality Devices Oil Adsorption Summary

Device	**Average Oil Captured/Gallon Filtered ((mg/Kg)/1,000 gallons)**
Putrajaya	
Absorbent Tarp (Putrajaya Sentral)	291.13
Passive Skimmer (Precint 9)	326.73
Passive Skimmer (Putrajaya Sentral)	961.69
Sponge under Grate Guard (Putrajaya Sentral)	81.37

There are four main categories are associated with the acquisition and operation of sediment and oil and grease trap, which are: (i) one-time structure cost; (ii) one-time installation cost; (iii) recurring maintenance costs; and (iv) recurring waste disposal cost. The capital costs depend primarily on the model and size of the stormwater quality improvement devices (SQID) purchased. Installation costs vary significantly depending upon whether the SQID is placed in a greenfield or brownfield development. Maintenance costs vary by device, site and practice, while waste disposal costs are location-specific. Table 3 shows a capital and installation cost summary for drain guard, kerb guard, grate guard, oil passive skimmer and absorbent tarp installed at Putrajaya study area.

Table 3. Capital Cost Summary for Stormwater Quality Improvement Devices (SQID)

Device	**Drain Guard**	**Kerb Guard**	**Grate Guard**	**Oil Passive Skimmer**	**Absorbent Trap**
Structure Cost (RM)	630	455	520	530	430
Media	N/A	N/A	350	N/A	N/A
Media Replacement Interval	Every 6 months	More than 2 years	Every 6 months	Every 3 months, based on physical condition	Every 3 months, based on physical condition, applied location
*1st Year Capital Cost (RM)	700	500	580	590	480
Annual Maintenance Cost & Disposal Cost	7600	2300	2300	2300	2300
Total Annual Cost	8300	2800	2880	2890	2780

In general, a typical SQID system can be expected to cost about RM 2,800 or more per year over its lifetime. Again, these costs are preliminary estimation, given their variability by site and application. Higher costs would result for large models, for brownfield retrofits.

4. Conclusions

This study investigates the performance of stormwater quality devices (SQID) in the removal of suspended particles, particularly sediment, Oil & grease and associated pollutants in stormwater. Storm event and dry weather water samples were collected entering and leaving a stormwater quality devices unit and were analysed for Total Suspended Solids (TSS), Heavy Metals and Oil and Grease concentrations. The SQID has shown previously to be an effective stormwater quality device in USA, however the performance of the SQID under tropical climates is not well understood. Therefore, the objective of the study is to investigate the effectiveness of SQID in the removal of sediments and oil & grease.This study suggests a stormwater treatment sequence involving an efficient stormwater quality device, such as the drain guard unit, pre-treatment by a constructed wetland or a bioretention zone can be expected to treat a wide spectrum of pollutants found in stormwater.

At Putrajaya Central and Precint 9, sediment are expected to be contributed from the side of the car park area and sediment from the bare land deposited to the pollution control devices installed at that place respectively. The sediment grain size distribution, mainly consist of Gravelly Sand. It is estimated (using USLE) Soil Erosion Loss at Precint 9 as much as 361.011 t/ha/yr annual. This shows that stormwater quality devices must be installed in the study area in order to remove the sediments in the stormwater based on the estimated soil erosion loss rate. The target is to control the problems at source by installing the SQID or planting vegetation at the bare land slope. The Putrajaya Stormwater Source Control Projects have provided excellent opportunities to demonstrate innovative at-source structural treatments. It is expected that the results of this trial will be valuable for stormwater managers grappling with a multiplicity of pollution issues and limited capital and maintenance budgets.

References

[1] Alsaigh R, Boerma J, Ploof A, and Regenmorter L 1999 Rouge River National Wet Weather Demonstration Project, Nonpoint Work Plan No. URBSW5, Task No. 3, *Evaluation of On-Line Media Filters in the Rouge River Watershed*

[2] Hoffman E J, Latimer J, Mills G, and Quinn J 1982 Petroleum Hydrocarbons in Urban Runoff from a Commerical Land Use Area *J. of the Water Pollution Control Federation* 54(11) 1517- 1525.

[3] Jeffrey N and Barrie M P 2006 Sources of heavy metals and polycyclic aromatic hydrocarbons in urban stormwater runoff *Science of The Total Environment* 359(1-3) 145-155

[4] U.S. EPA 1999b Sorbents Internet site at http://epa.gov/oilspill/sorbents.htm

IOP Conf. Series: Earth and Environmental Science **16** (2013) 012048 doi:10.1088/1755-1315/16/1/012048

Stormwater quality modeling for upscaling MSMA stormwater management ecohydrology

L M Sidek[1], M R Zainal Abidin[2], F Z Esfahani[1] and H Basri[1]
[1]Centre for Sustainable Technology and Environment, Universiti Tenaga Nasional, Jalan IKRAM-UNITEN, 43000 Kajang, Selangor, Malaysia
[2]Director, Humid Tropics Centre Kuala Lumpur, Jalan Duta, Kuala Lumpur, Malaysia

E-mail: Lariyah@uniten.edu.my

Abstract. Langat River is 180 km length has a total catchment area of 2,350 km^2. The Langat River is classified as Class II in the upstream, and as the water flows downstream the quality degrades to Class III and IV. This degradation is caused by the rapid and uncontrolled urbanization in the contributing catchment of the river. The characteristics of the river and its contributing catchment are modeled using Model for Urban Stormwater Improvement Conceptualisation (MUSIC) in order to rehabilitate the river and enhance its Water Quality Index (WQI) to Class II. The historical rainfall data was obtained from DID for the period from 1974 to 2012. This paper presents the effects of various BMP Components on rehabilitating the water quality of the river. In addition, the minimum amount of river flow required for protection of the habitats and the river's ecology has been assessed during the dry seasons. The outcomes of this study suggest the most appropriate Best Management Practices that can be used as solutions for the river's rehabilitation. Simulations and modeling result found out that a configuration of wetlands, bio-retention systems and ponds are capable to reduce pollutants loadings such as TSS, TP and TN by 85.1%, 69.1% and 37.5% respectively

1. Introduction

Langat River is one of the main rivers in the state of Selangor, and starts from Titiwangsa Range in Gunung Nuang and flows into the Strait of Malacca. It has a total length and catchment area of 180 km and 2350 km^2 respectively. The focused study area covers a stretch of 2 km of the river with its respective 20.6 km^2 catchment area, located at Taman Tasik Bangi 3. It is believed that unsustainable human activities in the catchment of the Langat River have resulted in degradation of the river's water quality. Within the catchment there have been several mega projects in the catchment of Langat River, e.g. development of new federal government administrative capital, Putrajaya and the Kuala Lumpur International Airport, KLIA and other mega projects [1]. In addition, this river is not only used as main source of potable water but also is mean to serve other purposes, for example recreational, fishing and irrigation [2] and hence it is vital to have clean water at this river. It is however important to mention that the catchment of this river falls under the semi-residential urban area and due to presence of pollutants in the water of this river, it cannot be used for recreational purposes as well, which consequently impacts its tourism values.

Sg Langat is high in turbidity due to the fact that the river bank is unprotected which causes the erosion of the soil along the river bank. Exposed and unprotected soil along the river bank will be eroded laterally [3]. High turbidity causes a decline in fishery resources and has adverse effect on ecology and aquatic environment of the river. In addition it is said that the river's water quality has deteriorated due to increase of socio-economic activities. It has been found that the previous practices of "end-of-pipe" solutions are not effective in treatment of stormwater and the current popularized practices of urban stormwater management focus on the source control.

The sources of pollutions in the Langat River are 58% from industrial discharge, 28% from domestic sewage from treatment plants, 12% from construction projects and 2% from pig farming. The pollutants in Langat River are point source pollutants, such as manufacturing industries and sewage treatment plants and non-point source pollutants, such as runoff from livestock and pig farms, urban area, agricultural land and solid waste dumpsites. It was found that the typical pollutants of Langat River are organic matters, sediments, Ammonical-Nitrogen, Phosphate-Phosphorus, oil and grease, Phenol, Iron (Fe), Lead (Pb), Copper (Cu) and Cadmium (Cd). Therefore, this study is crucial to investigate balanced ecosystem low flow analysis in order to ensure that a minimum quantity of streamflow exists to protect habitat during a drought situation [3].

2. Study Area

The study area covers a length of 2 km with a respective 20.6 km2 catchment area, located at Taman Tasik Bangi 3, near Universiti Kebangsaan Malaysia, UKM. Figure 1 shows the location of the study area.

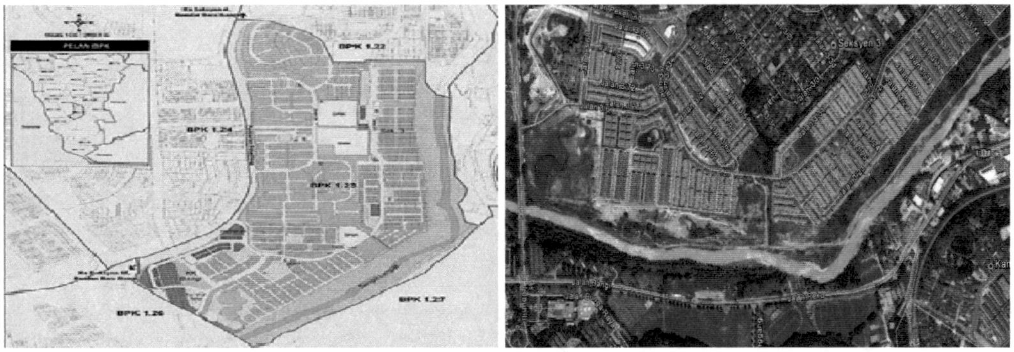

Figure 1. Location of the study area

3. Methodology

The raw data required for this study are collected from DID, and DOE and MPK Bangi. The raw data collected from DID were rainfall data of station 2917001 from 1974 to 2012, for various durations and river flow data for 1 hour and 1, 7 and 30 days were collected for primary analysis and generation of IDF curves, temporal patterns, flow duration curves and low flow analysis. The raw data collected from DOE were the water quality data for better understanding of the river's quality characteristics. This study conducts hydrological analysis and Low flow analysis was also conducted to evaluate the river condition during dry season. The water quality and quantity simulations were carried out using MUSIC software. The simulation result will be used to formulate the best strategies of the up-scaling MSMA stormwater management ecohydrology.

4. Results and Discussion

This paper assesses the quantitative and qualitative behaviour of Langat River and its catchment in order to find solutions for rehabilitating the river. The hydrological analysis show the annual flows and mean annual floods for this river are 35 m3/s and 300 m3 /s respectively. The river's quality current situation and condition was modelled by using MUSIC for qualitative analysis and assessment of effectiveness of various Best Management Practice components. Based on the land use map, the model was set up by dividing the catchment into three sub-catchments, three different Junctions (J1, J2 and J3) and one Receive Node were created for the model. The percentage reduction of each pollutant was simulated by MUSIC model under the two scenarios of with and without BMPs. The assessed pollutants were TSS, TP and TN. Different trials with different components, sequences and sizes were introduced to find the best fitting solution in terms of pollution reduction. The BMP components proposed for this area are 2 wetlands, 2 bio-retention systems and 1 new pond as an additional to the available pond at the study area (see Figure 2). The runoff from the catchment is passed through these systems for filtration and removal of pollution. Table 1 shows the percentage reduction of the pollutants with the proposed BMPS components simulated by MUSIC model.

Table 1. Percentage reduction of the pollutants for three different Junctions used for simulation of Langat River under this study

	Parameter	Unit	Source	Residual Load	(%) Reduction
Junction 1	**Flow**	(ML/yr)	7210	6750	6.50
	Total Suspended Solids	(kg/yr)	1060000	158000	85.10
	Total Phosphorus	(kg/yr)	2440	755	69.10
	Total Nitrogen	(kg/yr)	18700	11700	37.50
	Gross Pollutant	(kg/yr)	174000	0	100.00
Junction 2	**Flow**	(ML/yr)	7230	6760	6.40
	Total Suspended Solids	(kg/yr)	1060000	160000	85.00
	Total Phosphorus	(kg/yr)	2450	760	69.00
	Total Nitrogen	(kg/yr)	18700	11700	37.40
	Gross Pollutant	(kg/yr)	174000	0	100.00
Junction 3	**Flow**	(ML/yr)	24200	23200	4.20
	Total Suspended Solids	(kg/yr)	3680000	1110000	69.80
	Total Phosphorus	(kg/yr)	8370	3730	55.40
	Total Nitrogen	(kg/yr)	63100	46700	25.90
	Gross Pollutant	(kg/yr)	572000	0	100.00

From the modelling result, the reduction of TSS and TP percentages have reached the MSMA targeted percentage while Total Nitrogen percentage is slightly below the target [10]. Figure 3 shows the hydrograph generated by MUSIC and the maximum runoff is 20 m^3/s and the minimum runoff is 0.2 m^3/s. This value 0.2 m^3/s (minimum runoff) is also adequate for the minimum flow required in the dry seasons.

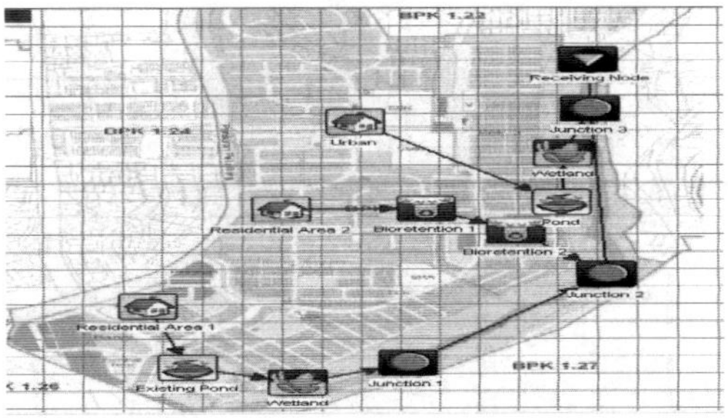

Figure 2. BMPs components in MUSIC model

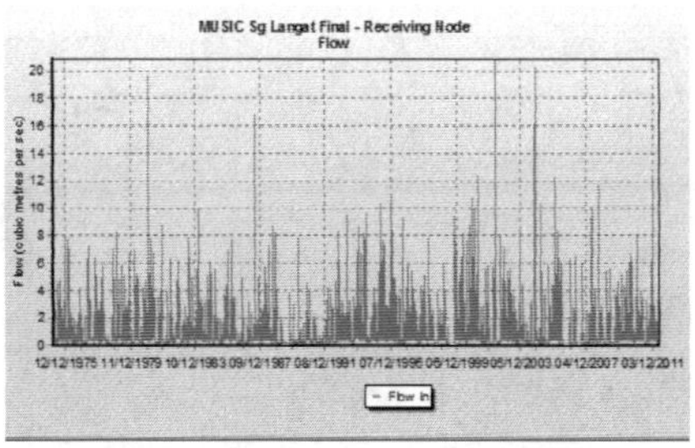

Figure 3. Daily Flow hydrograph generated by MUSIC

It is also important to ensure a healthy river which has the ability to rehabilitate itself and contain a balanced ecosystem. Therefore, a Low Flow Analysis was conducted in this study to ensure a minimum flow in the river to protect the habitats within the river ecosystem process. Gumbel and Log-Pearson method were used for generation of low flow curves and compared with the graphical method. Based on Hydrological Procedure 12 (HP 12) the study area was categorized under Region RC3 and RE3. By comparing the outcome of Low Flow Curves in Figure 4, and Flow Duration Curves in Figure 5, it can be seen that the minimum flow required for the revival of the beneficial ecosystems present in the river is equivalent to 20 years return period. Result from Flow Duration Curve shows that for 100% dependable flow (100% of the time the flow equal or exceeded) the discharge value of 2 m^3/s. This value is well above the minimum required flow in the dry seasons for river water quality protection purposes.

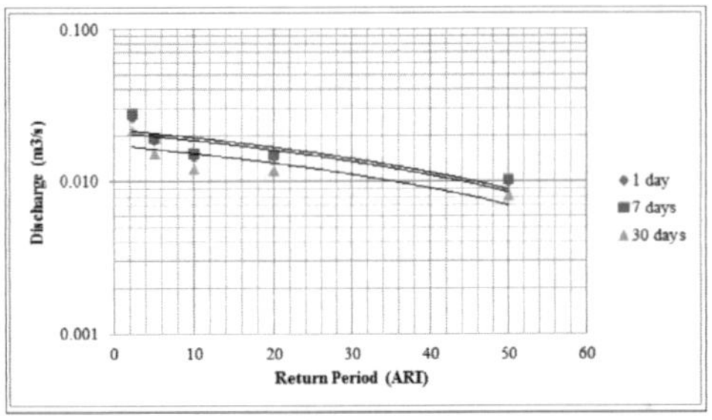

Figure 4. Low Flow Analysis for Sungai Langat at Dengkil (HP 12 Method)

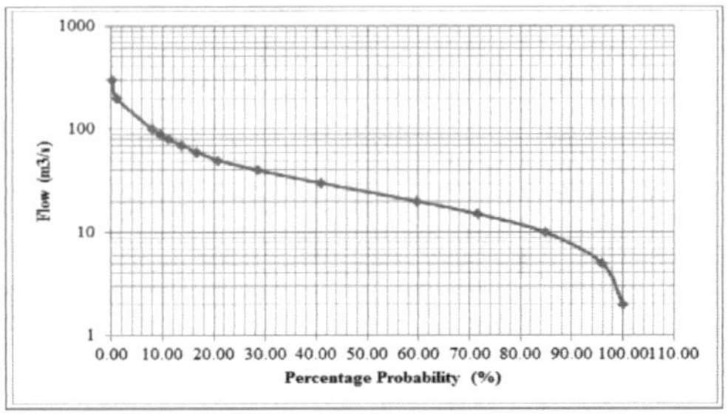

Figure 5. Daily Flow Duration Curve (FDC) at Dengkil Station

5. Conclusion

Langat River's quality simulation was carried out by MUSIC and a number of BMP components were proposed. These components are two wetlands, two bio-retention systems and two ponds. The simulation results indicated that the Flow reduction values vary from 4.2% to 6.5%, TSS reduction ranges from 69.8% to 85.1%, TP reduction ranges from 55.4% to 69.1% and TN reduction ranges from 25.9% to 37.5%. Total Suspended Solid and Total Phosphorus percentage reduction have reached the MSMA targeted range while Total Nitrogen percentage is slightly below the target, due to the fact that Nitrogen removal is slightly difficult. The hydrological analysis shows annual flows and mean annual floods for this river are 35 m^3/s and 300 m^3 /s respectively When assessing the reliability of low flow the Sg Langat river shows 100% dependable flow of 2 m^3/s more than the minimum required for environmental flow protection of the river.

6. Acknowledgement

The authors would like to thank Department of Irrigation and Drainage, Malaysia and Department of Environment, Malaysia for providing the raw data. In addition, the authors would like to thank Universiti Tenaga Nasional for providing the platform to perform this study.

References

[1] Yusof M. A. Hasan N. and Abdullah M P 2002 River water quality in Langat basin, Selangor, Malaysia *Malaysian J. of Environmental Management* **3** 125-142
[2] Toriman M E. 2010 Assessing environmental flow modeling for water resources management: A case of SG.(River) Pelus, Malaysia *J. of Nature and Science* **2(8)** 74-81
[3] Yusof M A. Hasan N. and Abdullah M P 2002 River water quality in Langat basin, Selangor, Malaysia *Malaysian J. of Environmental Management* **3** 125-142

Probable Maximum Precipitation (PMP) over mountainous region of Cameron Highlands- Batang Padang Catchment of Malaysia

L M Sidek[1], M D Mohd Nor [2], P R Rakhecha [2], H Basri[1], W Jayothisa[1], R S Muda[3], M N Ahmad[3] and A Z Abdul Razad[3]

[1] Centre for Sustainable Technology and Environment (CSTEN), UNITEN, Jalan IKRAM-UNITEN, 43000 Kajang, Selangor, MALAYSIA

[2]Centre for Stormwater and Geohazard Management (CSGM), UNITEN, Jalan IKRAM-UNITEN, 43000 Kajang, Selangor, MALAYSIA

[3] TNB Research Sdn Bhd, MALAYSIA

E-mail: lariyah@uniten.edu.my

Abstract. The Cameron Highland Batang Padang (CHBP) catchment situated on the main mountain range of Peninsular Malaysia is of large economical importance where currently a series of three dams (Sultan Abu Bakar, Jor and Mahang) exist in the development of water resources and hydropower. The prediction of the design storm rainfall values for different return periods including PMP values can be useful to review the adequacy of the current spillway capacities of these dams. In this paper estimates of the design storm rainfalls for various return periods and also the PMP values for rainfall stations in the CHBP catchment have been computed for the three different durations of 1, 3 & 5 days. The maximum values for 1 day, 3 days and 5 days PMP values are found to be 730.08mm, 966.17mm and 969.0mm respectively at Station number 4513033 Gunung Brinchang. The PMP values obtained were compared with previous study results undertaken by NAHRIM. However, the highest ratio of 1 day, 3 day and 5 day PMP to highest observed rainfall are found to be 2.30, 1.94 and 1.82 respectively. This shows that the ratio tend to decrease as the duration increase. Finally, the temporal pattern for 1 day, 3day and 5 days have been developed based on observed extreme rainfall at station 4513033 Gunung Brinchang for the generation of Probable Maximum Flood (PMF) in dam break analysis.

1. Introduction

Dams in different parts of the world are built for various purposes such as for fresh water supply, irrigation, hydroelectricity generation, flood control and for recreational activities. These dams however can break due to inadequate spillway design floods which can cause calamities in the downstream area. A survey of dam break data by Lemperiere (1993) showed that about 40 % of all dams break worldwide were because of inadequate spillway capacities. In the design of dams, the spillways are designed for the probable raingauges flood (PMF) derived from the probable maximum precipitation (PMP). Considering that many of the existing dams in Malaysia could be similarly unsafe based on hydrological conditions, a PMP study for Cameron Highlands- Batang Padang (CHBP) was carried out. The catchment has a series of 3 dams known as Sultan Abu Bakar, Jor and Mahang. The present study is devoted primarily to the estimation of the design storm rainfall values for different return periods including the PMP values for 3 different durations of 1, 3, and 5 days storms for rainfall stations in the CHBP catchment and also for the catchments above the 3 dams. The main purpose is to obtain reliable PMP rainfall values for an investigation of the adequacy of the current spillway of the three dams. The PMP and consequent floods were calculated using limited meteorological data then available. The design values were felt to be in need of new PMP estimates in light of more rainfall and hydrological data that are available from the Department of Irrigation and Drainage Malaysia (DID) and TNB organizations. Efforts are made in this study to carefully process these data and to use them as much consistent manner as possible.

2. Study Area

The CHBP catchment and the locations of the three dams is shown in Figure 1. The Telom and Bertam Rivers of Cameron Highlands, Batang Padang River and its tributaries form an integrated drainage system in the central portion of the Peninsular Malaysia that is shared by two states of Pahang and Perak. The Sultan Abu Baker dam (also called Ringlet dam) in the Cameron Highlands utilizes the runoff waters of the hill streams of Telom River and Bertam River originating from the hills of the Main Range and flowing in the Pahang state. The Jor and Mahang dams in Batang Padang utilize the Telom and Bertam waters diverted from the Ringlet dam together with the runoff of the Batang-Padang River and its tributaries. The Sultan Abu Bakar, Jor and Mahang dams are respectively 39.6 m, 44.8 m and 19.2m in the height. The drainage areas above Sultan Abu Bakar dam, Jor dam and Mahang dam are about 110 km2, 275 km2 and 360 km2 respectively. The catchment is mountainous and land surface elevations range from about 500 m to 2130 m above mean sea level making it as excellent potential for the hydropower generation. Table 1 summarizes some of the hydrologic parameters of the 3 dams.

Table 1. Catchment Areas and Height of the three Dams

Dam	Catchments area (km^2)	Dam Height (m)
Sultan Abu Bakar	110	39.6
Jor	275	44.8
Mahang	360	19.2

Figure 1. The CHBP Catchment and the Locations of the Three Dams

3. Rain Gauge Network and Hydrological Data

For the derivation of design storm rainfalls and also PMP depths for a catchment need to use long period actual daily rainfall records supplemented with other meteorological data for several raingauges located in and around the catchment. It has been found that there are about twenty four rainfall raingauges in and around the CHBP catchment whose daily rainfall data are available in the time period from 1947 to 2009. Of these 24 raingauges, 9 raingauges are maintained by DID while the other raingauges are maintained by the TNB.

4. Estimation of PMP by Hershfield Statistical Method

The PMP is defined as the greatest depth of precipitation for a given duration that is physically possible over a given point or a specified area at a certain time of the year (WMO, 1986). Estimates of PMP are required for calculating the PMF which is the design flood for spillways of dams where no risk of dams break can be accepted. A statistical method for estimating the PMP for small areas has been developed by Hershfield (1965) based on a general frequency equation given by Chow (1951).

The method considers the annual raingauges rainfall series of a raingauge or an area and can be used at any place where there is sufficient rainfall data and in particular to make estimates when other meteorological data such as dew point, wind etc are lacking. The general equation for hydrologic frequency analysis

$$X_t = \overline{X_n} + K.\sigma_n \qquad (1)$$

Where X_t is the rainfall for the return period t years, $\overline{X_n}$ and σ_n are the mean, and standard deviation for the series of n annual raingauges rainfall values respectively of a given duration and K is the frequency factor which varies with the different frequency distributions. The frequency factor K is mainly a function of the recurrence interval for a particular distribution. For example, when analyzing the 1- day series of annual raingauges 1-day rainfalls by the Gumbel method (Gumbel, 1958), K is about 3.2 for the 100 year value and about 2.33 for the lognormal distribution.

$$X_{pmp} = \overline{X_n} + K_m.\sigma_n \qquad (2)$$

The calculated the frequency factor K_m for the station by using the equation

$$K_m = \frac{X_1 - X_{n-1}}{\sigma_{n-1}} \qquad (3)$$

Where X_{pmp} is the PMP depth for a given raingauge for a given duration, $\overline{X_{n-1}}$, σ_{n-1} are respectively the mean and standard deviation for this series excluding the maximum value X_1 from the series. Representing graphically K_m as function of the mean ($\overline{X_n}$), the enveloping of the frequency factor K_m (Figure 3(i) and 3(ii)) can be traced thus determining its limit value for the mean of the annual raingauges rainfall series.

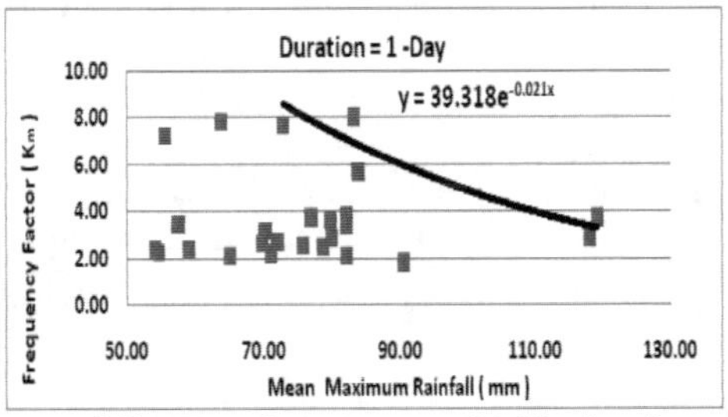

Figure 3(i). Envelope curves for K_m 1 – day

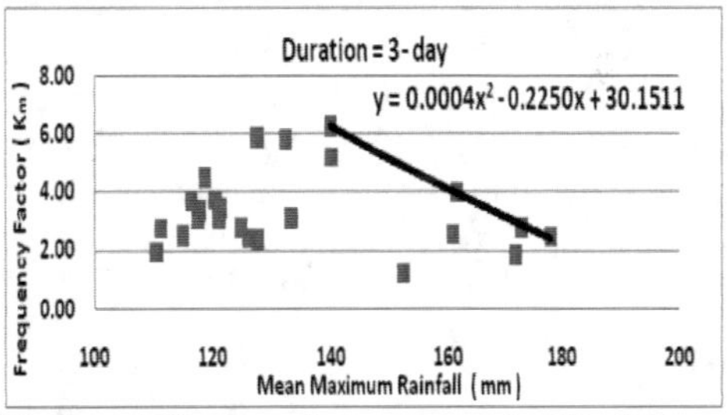

Figure 3(ii). Envelope curves for K_m 3-day

5. Derivation of point PMP values for raingauges in CHBP catchment

The values of $\overline{X_n}$, σ_n, $\overline{X_{n-1}}$, σ_{n-1} and the coefficient of variation (CV) were calculated for all the annual raingauges rainfall series for three different durations of 1, 3 & 5 days pertaining to 24 raingauges in the CHBP catchment. The frequency factors K_m for each of the station for 3 different durations were determined by equation (3). Based on the 24 raingauges, the maximum values of K_m for the 1, 3 & 5days durations were found to be 8.0, 6.3 and 5.9 respectively which corresponds to the return period of more than 10^5 years. It is very interesting to note that in a humid region of Malaysia the K value tends to decrease as the duration of rainfall increase. The maximum value of 8.0 for 1 day duration is obviously an appropriate value for PMP computations for raingauges in the CHBP catchment. The 1-day, 3-day and 5-day PMP estimates for all 24 raingauges were plotted separately on the base maps of the CHBP. Isopleths of PMP were drawn at interval of 50 mm taking into consideration the effect of topography of the catchment. Before drawing the isopleths of PMP, the coefficient of variability (CV) values for all 24 raingauges were plotted separately for 1, 3 & 5 days durations on the maps of the catchment to see whether in certain area the CV value for a station differed too much from nearby raingauges. This needs to be done to smooth out inherent errors indicated by large standard deviations. The PMP values obtained at the dam sites can be transformed using a rainfall-runoff deterministic model into a PMF hydrograph in the dam safety studies. In this process the catchment is taken as a system that has as input the PMP histogram and as output the flow hydrograph.

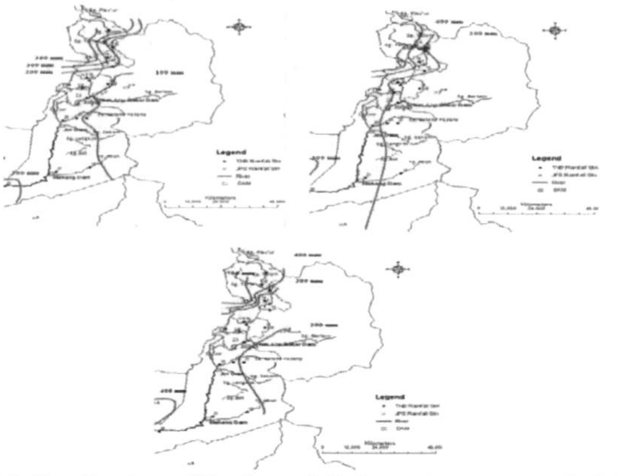

Figure 4. Spatial distribution of 1-,3- and 5-day raingauges rainfall (clockwise).

6. Conclusions

The hydrology analysis revealed that the annual rainfall decreases at much higher elevation thus indicates the orographic effect. This was – 0.570 which is significant thereby indicating that rainfall decreases with elevation. The consistency and reliability of records at Gunung Brinchang station is found to be consistent by double mass curve technique. The maximum values for 1 day, 3 days and 5 days PMP values are found to be 730.08mm, 966.17mm and 969.0mm respectively at Station number 4513033 Gunung Brinchang. The PMP values obtained were compared with previous study results undertaken by NAHRIM. However, the highest ratio of 1 day, 3 day and 5 day PMP to highest observed rainfall are found to be 2.30, 1.94 and 1.82 respectively. This shows that the ratio tend to decrease as the duration increase. Finally, the temporal pattern for 1 day, 3day and 5 days have been developed based on observed extreme rainfall at station 4513033 Gunung Brinchang for the generation of Probable Maximum Flood (PMF) in dam break analysis.

7. Acknowledgement

The authors would like to thank Tenaga Nasional Berhad and the Department of Irrigation and Drainage Malaysia for providing the data used in the study. Special thanks are due to TNBR of Tenaga Nasional Berhad for providing the necessary financial assistance to carry out this study.

References

[1] Chow V 1951 A general formula for hydrologic frequency analysis *Trans. Am. Geophys. Union* **32** 231-237

[2] Desa M N, Noriah A B and Rakhecha P R 2001 Probable maximum precipitation for 24- hr duration over southeast Asian monsoon region- Selangor, Malaysia *Atmos. Res.* **58**, 41-54

[3] Hershfield D M 1965 Method of estimating probable maximum precipitation *J. Am. Water Works Assoc.* **57** 965-972

[4] Koutsoyiannis D 1999 A probabilistic view of Hershfield method for estimating probable maximum precipitation *Water Resour. Res.* **35** 1313-1322

[5] Lockwood T G 1967 Probable maximum 24-hour precipitation over Malaya by statistical methods *Meteorol. Mag.* **96** 11-19.

[6] Myers, V A 1967 Meteorological estimation of extreme precipitation for spillway design floods *WBTM Hydro (U. S. Weather Bureau)* **5** 1-29

[7] Rakhecha P R and Soman M K 1994 Estimation of probable maximum precipitation for a 2-day duration: Part 2- North Indian Region *Theor. Appl. Climatol.* **45** 277-283

Sustainability of hydropower as source of renewable and clean energy

J Luis[1], L M Sidek[2], M N M Desa[3], P Y Julien[4]

[1] Post Graduate Student, UNITEN, Jalan IKRAM-UNITEN, 43000 Kajang, Selangor, MALAYSIA
[2] Centre for Sustainable Technology and Environment, UNITEN, Jalan IKRAM-UNITEN, 43000 Kajang, Selangor, MALAYSIA
[3]Centre for Stormwater and Geohazard Management, UNITEN, Jalan IKRAM-UNITEN, 43000 Kajang, Selangor, MALAYSIA
[4] Department of Civil Engineering, Colorado State University, Fort Collins, USA.

E-mail: jansenla@tnb.com.my, lariyah@uniten.edu.my; mohamednor@uniten.edu.my, pierre@engr.colostate.edu

Abstract. Hydroelectric energy has been in recent times placed as an important future source of renewable and clean energy. The advantage of hydropower as a renewable energy is that it produces negligible amounts of greenhouse gases, it stores large amounts of electricity at low cost and it can be adjusted to meet consumer demand. This noble vision however is becoming more challenging due to rapid urbanization development and increasing human activities surrounding the catchment area. Numerous studies have shown that there are several contributing factors that lead towards the loss of live storage in reservoir, namely geology, ground slopes, climate, drainage density and human activities. Sediment deposition in the reservoir particularly for hydroelectric purposes has several major concerns due to the reduced water storage volume which includes increase in the risk of flooding downstream which directly effects the safety of human population and properties, contributes to economic losses not only in revenue for power generation but also large capital and maintenance cost for reservoir restorations works. In the event of functional loss of capabilities of a hydropower reservoir as a result of sedimentation or siltation could lead to both economical and environmental impact. The objective of this paper is aimed present the importance of hydropower as a source of renewable and clean energy in the national energy mix and the increasing challenges of sustainability.

1. Introduction

All dams are designed to store flood-waters, primarily for the benefit of human beings. The damming of streams and rivers has been an integral part of human civilization from its early history. Reservoirs and dams are constructed by human to deal with the need of water and power. Dams and reservoirs are constructed primarily to function as multipurpose functions including as flood control, drinking water, agricultural water supply, hydro power generation, recreation and others [1, 2, 3].

ICOLD [4] defines large dams as those more than 15 m in height, while including smaller dams up to 10 m height as well, if they are otherwise significant with respect to storage volume, density of population, etc. Large dams generate 19% of the world's total electricity supply. One-third of the countries in the world rely on hydropower for more than half their electricity supply. Half the world's large dams were built exclusively or primarily for irrigation. Some 30-40% of the 271 million hectares irrigated worldwide rely on dams. For hydropower reservoir, the life span design allows for slow deposition of sediments within designated levels above the reservoir bed. However in most cases, during the annual sediment inflow estimates used during the design stage are exceeded half way through the life span or even sooner. This therefore creates serious risk and operational difficulties in term of the functionality of the reservoir to operate as designed.

2. Sustainability of Hydropower

In recent times, the most biggest and important challenge that is faced with the sustainability of the hydro plants are to cope with reservoir sedimentation and loss of storage mainly contributed by uncontrolled deforestation, poor agricultural land management, residential development and infrastructure developments within the vicinity of the once protected forest catchment area. The eroded sediments from the catchment are expected to increase the reservoir sedimentation rates. It is estimated that about 1% of the total storage capacity in the world's reservoirs is lost annually due to sedimentation [6, 7, 8]. Perhaps one of the most sediment affected catchment would be the Cameron Highlands and Batang Padang catchment which contributes to combined total hydropower generation capacity of 262MW.

3. Cameron Highlands and Batang Padang Hydroelectric Scheme

In Malaysia, the Cameron Highlands and Batang Padang Hydroelectric Scheme is located in the central highlands of West Malaysia, at some 160km north of Kuala Lumpur. Ringlet Reservoir is a part of its upper scheme. The upper catchment that contributes to the Ringlet reservoir has an area of 183km2.It utilizes headwater from two major rivers i.e. Sungai Telom and Sungai Bertam for the generation of electricity. Ringlet reservoir originally the reservoir had a total storage capacity of 6.7 million m3 of water, meanwhile Jor Reservoir had a total storage of 3.8 million m3 while Mahang reservoir was designed at 0.4 million m^3.

3.1. Sedimentation Rate

The Cameron Highlands and Batang Padang Scheme appears to have experienced sedimentation problems soon after commissioning. The Batang Padang catchment experiences much less sediment problems and sediment sampling on that catchment started only in the late 1980's. The data suggest that as a result of the large sediment loads carried by the different rivers feeding Ringlet reservoir, the reservoir silted up very quickly. In 1963, at commissioning, the reservoir gross storage was about 6.5 million m^3, of which 4.7 hm^3 were for live storage. The live storage was reduced to 4.2 million m^3 in 1983, 3.2 million m^3 in 1991, and 2.2 million m3 in 1996, 1.7 million m3 in 1998 and to 1.4 million m3 in 1999. The rate of filling increased from 30, 000 m^3/yr in the 1960's to 50, 000 m^3/yr in the late 1970's and early 1980's [11].

3.2. Recent Study and Findings

A study by TNB Research in 2004 concluded that the two (2) major water inflows carrying average sediment inflow of 194,000t/yr or 102,000m3/yr into Ringlet Reservoir. The breakdown of sediment inflows are from Ringlet End (Sg. Ringlet) which contributes approximately 67,000 ton/yr (35,000m3/yr) and Habu End (Sg. Habu, Sg. Bertam and Habu Power Station) with sediments contributes approximately 127,000ton/yr (65,000m3/yr). At present the long term annual capacity loss or sedimentation rate of Ringlet Reservoir in 2008 is estimated at an average of almost 6 fold since commissioning to 139,712m3/yr in 2008 [12].

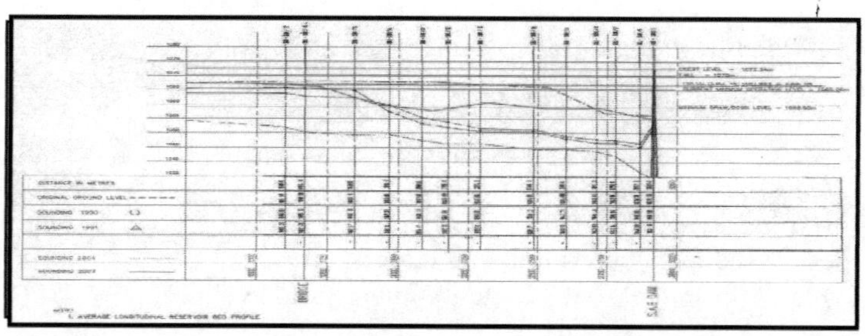

Figure 1. Longitudinal bed profile of the reservoir for different years of depositions as compared to the original bed.

4. Impacts of Sedimentation

A number of studies were conducted in the past to examine the effects of development of the catchment on the hydro scheme. The studies [13] have shown that, the operation and maintenance as well as energy generation of the hydro station of the Cameron Highlands scheme are affected by land development. Among the major impacts directly affected by sedimentation maybe concluded as below:

- Loss of Live Storage which affects the revenue from power generation output;
- Loss of energy production and revenue due to frequent shutdown of intakes and operations of outlet valves which diverts water from reservoir;
- Increase the risk of flooding downstream through controlled spilling especially during monsoon seasons;
- Increased cost of maintenance by carrying out frequent dredging, excavation and disposal of sediments and also replacement of plants mechanical components;
- Reservoir considerably filled with sediments pose a threat for great environmental disaster in case of a dam failure.

The impacts in term of power generation capabilities are also quite significant. Generally as the reservoir storage reduces due to sedimentation so does the power generation as shown in Figure 2. The graph below shows the annual power generation output at Cameron Highlands versus the available live storage since the FY 95/07. It can be observed that in the FY 95/98, the live storage loss due to reservoir sedimentation has gradually reduce the capability of the plants to generate annual output as per the design requirements of 843 GWh. In the FY 97/98, the annual generation output drops to 699GWh below the normal operating limits with a difference of 144 GWh which translates to RM 1.44 million loss of revenue at a basic estimated selling price of RM 0.01/kwh.

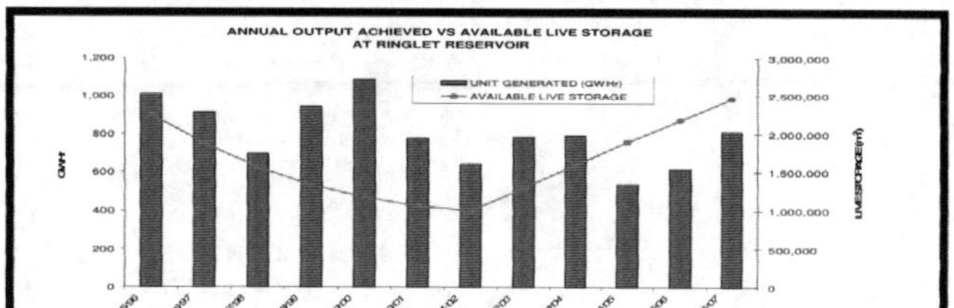

Figure 2. Generation output versus the reservoir live storage

5. Mitigation Measures for Improving Reservoir Sustainability

The most desirable method for controlling reservoir sedimentation is probably by reducing soil loss and erosion from the reservoir watershed. This can be achieved by reducing sediment inflow through soil conservation, watershed management, gross pollutant traps (GPT), vegetative screens and etc.

Some of the most commonly used methods for reducing sediment deposition in reservoirs are as follows:

1. Reducing sediment inflow to the reservoir:
   a. Watershed management and soil conservation
   b. Constructing a check or debris dams upstream to retain relatively coarse sediments
2. Reducing sediment deposition in the reservoir:
   a. Flood flushing through bottom outlets or undersluices
   b. Drawdown flushing to evacuate new and old settled deposits
   c. Reservoir operation policies oriented towards sediment releases.
3. Rehabilitating reservoir storage capacity:
   a. Siphon dredging - easily implemented and very effective for small to medium reservoirs
   b. Reservoir dredging and disposal

The first category, the objective is to reduce sediment deposition in the reservoir by increasing sediment outflow from the reservoir through use of density currents, releasing flows with heavy sediment concentrations during floods events, and/or drawing down and flushing the reservoir during high flows. The second category however is the opposite of the first whereby the reservoir storage capacity is recovered by either dredging it, and/or siphoning the deposited sediments.

6. Conclusions

Hydropower remains an important component of clean energy requirement in the near future. During the 19th Conference of the Electric Power Supply held in Bali recently, most countries in the Asian region as well as globally pledged to increase investment and enhancing clean technology for future sustainable power needs. However the challenges faced with the development and operations of existing plants as shown in the case of Ringlet Reservoir, Cameron Highlands is very real. Ever increasing inflows of sediments annually threatens the sustainability use of hydropower as a source of clean energy. Therefore importance should be made towards including effective sediment mitigation strategies during the development stages. As for the existing hydro power plant, strict law enforcement along with legislation protection towards watershed management should be derived to protect the assets such as dams and reservoir for future power generation.

References

[1] Dickson J C and Gerrard R T 1963 Cameron Highlands Hydro-Electric Scheme, *Proceedings of the Institution of Civil Engineers* 387-424

[2] Choy F K 1987 Cameron Highlands / Batang padang Hydro-Electrci Schemes, Generation Performance Assessment. *Bulletin IEM* February 1987

[3] Choy F K 1989 Cameron Highlands Hydro Electric Schemes – A Study of River Sediment Load and Reservoir Sedimentation *Bulletin IEM* January 1989

[4] ICOLD 1989 Sedimentation control of reservoirs *International Committee of Large Dams*

[5] Mahmood K 1987 Reservoir sedimentation: Impact, extent, and mitigation *World Bank Technical Paper No. 71* World Bank, Washington, DC

[6] Samad E, and Hosien S 2008 Desilting of deposited sediment at the upstream of the Dez reservoir in Iran. *J. of Applied Sciences in Environmental Sanitation* Surabaya, Jakarta 25-35

[7] Yoon Y N 1992 The state and the perspective of the direct sediment removal methods from reservoirs *Int. J. Sediment Res.* **7(20)** 99–115

[8] Annandale G W 2001 Reservoir conservation and sediment management *Water week 2001 Water Bank Group* Washington, DC

[9] Julien P Y 2010 Erosion and Sedimentation. *2nd Ed. Cambridge University Press* 371

[10] Choy F K and Hamzah F B 1997 Cameron Highaldns Hydroelectric Scheme – Performance of Ringlet Rervoir, *Proc of the 19th Congress of ICOLD* Q. 74, R.3, Florence

[11] Jansen L, Soo H T, Lariyah M S , Desa M N and Julien P Y 2011 Soil erosion modeling using RUSLE and GIS at Cameron Highlands, Malaysia *3rd International Conference on Managing Rivers in the 21st Century* Penang, Malaysia

IOP Conf. Series: Earth and Environmental Science **16** (2013) 012051 doi:10.1088/1755-1315/16/1/012051

Enhancement of magnetic flux distribution in a DC superconducting electric motor

N A Hamid[1], **L S Ewe and K M Chin**
Department of Mechanical Engineering, College of Engineering, Universiti Tenaga Nasional, 43000 Kajang, Selangor, MALAYSIA

E-mail: Nasri@uniten.edu.my

Abstract. Most motor designs require an air gap between the rotor and stator to enable the armature to rotate freely. The interaction of magnetic flux from rotor and stator within the air gap will provide the thrust for rotational motion. Thus, the understanding of magnetic flux in the vicinity of the air gap is very important to mathematically calculate the magnetic flux generated in the area. In this work, a finite element analysis was employed to study the behavior of the magnetic flux in view of designing a synchronous DC superconducting electric motor. The analysis provides an ideal magnetic flux distribution within the components of the motor. From the flux plot analysis, it indicates that flux losses are mainly in the forms of leakage and fringe effect. The analysis also shows that the flux density is high at the area around the air gap and the rotor. The high flux density will provide a high force area that enables the rotor to rotate. In contrast, the other parts of the motor body do not show high flux density indicating low distribution of flux. Consequently, a bench top model of a DC superconducting motor was developed where by motor with a 2-pole type winding was chosen. Each field coil was designed with a racetrack-shaped double pancake wound using DI-BSCCO Bi-2223 superconducting tapes. The performance and energy efficiency of the superconducting motor was superior when compared to the conventional motor with similar capacity.

1. Introduction

With the success of fabrication of long length superconducting wire with a very reliable performance, the development of superconducting devices is further intensified [1,2]. Superconducting devices such as magnetic resonance imaging (MRI) and superconducting quantum interference devices (SQUID) have been used extensively. Several others electrical power devices such as motors, generators, fault current limiter, and magnetic storage system had been developed and tested [3-5]. Superconductor required cooling down below the critical temperature, T_c to reach zero resistance. The conventional low temperature superconductors operate at below 20 K, where cooling by liquid helium is required. With the discovery of high-temperature superconductor in 1986 [6], any devices should be able to operate at higher temperatures. Therefore, the used of cheaper refrigerant such as liquid nitrogen is sufficient to cool down the superconducting materials and achieved superconductivity [7].

The special properties of high-temperature superconductor wires such as zero resistivity, high critical current density and high operating temperature, enable the development of high efficiency energy devices. In an electric motor for instance, it estimated that superconducting motor is able to reduce in size and minimize energy losses when compared to the conventional copper-base motor [8]. In developing a superconducting motor, one of the important aspects to be investigated is the magnetic

[1] To whom any correspondence should be addressed.

flux behavior within the components of the motor. By employing the finite element method, the performance and behavior of a system is able to be predicted without the necessity of developing the actual system. In this work, a finite element analysis was employed to study the behavior of the magnetic flux in view of designing a synchronous DC superconducting electric motor. Consequently, a low-power superconducting electric motor is assembled and its performance is compared with a conventional electric motor. All the specifications for both types of motor are similar except for the coils.

2. Experimental procedures

2.1. Finite elements method
In the present work, the finite element analysis method is used to study the behavior of the magnetic flux in a motor system. The scope of the analysis is focused on investigating the behavior of the flux and not its strength. This will help to forecast the behavior of the magnetic flux when subjected to various circuit elements in the motor system.

2.2. Fabrication of motor and performance test
Based on the results from finite element analysis for several preliminary models, a model where the stator is designed with a deep U-shape was chosen. The rotor was mounted at the opening of the U-shape stator together with rotor housing, and the coils were assembled at one of the stator leg. The generated flux produced by the field coils required a magnetic path so that the interaction of opposing magnetic fields at the opening of the U-shape stator resulted in high flux density that provide a high force area, and thus enable the rotor to rotate. DC current was used to provide the input current to the field coils and the rotor. Motor torque measurement was used to gauge the performance of the fabricated motor. Motor torque is one of the motor ratings which is used to indicate motor rotary force produced on its output shaft [9]. During performance tests for both type of motors, a constant current was supplied to the rotor, while the current applied to the coils was varied at fixed intervals as the load mounted on the shaft is increased. The data was taken once the rotor starts to rotate. Testing for the conventional copper coils was repeated with the same procedures except for the cooling procedures.

3. Results and analysis

3.1. Finite elements method
The wireframe of a single coil motor was constructed based on the finite element analysis. The wireframe model which is in the 2-dimension, was converted into a solid 3-dimensional model. All the dimensions and specifications of the model were built with a full scale dimensions. Figure 1 shows the simulation of magnetic flux path in the circuit. Most of the flux is distributed on the desired flux path and followed the designed path. Only small portion of the flux is lost to the surrounding in the forms of flux leakage and fringe effect. The occurrence of leakage is due to the presence of weak flux lines that do not have sufficient energy to flow through the desired path. Apparently, the weaker flux and the flux due to fringe effect at the air gap are dispersed to the surrounding, or are attracted back to the magnetic flux path. Figure 2 shows the simulated result of the flux density contour of the motor system. As shown in the figure, the flux density is highest at the areas around the air-gap and the rotor. The high concentration of the flux at the air-gap is due to the accumulation of the flux generated by the coils and rotor over a small cross sectional area. In contrast, the other parts of the motor body do not show high flux density indicating low distribution of flux. The high flux density at the rotor in the vicinity of air-gap is needed to establish a high force area that enables the rotor to rotate.

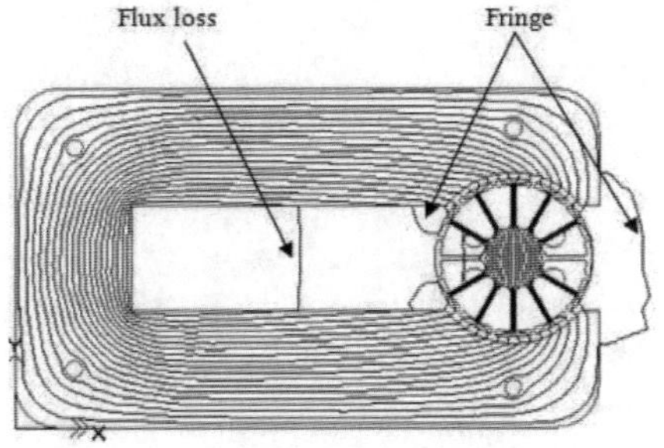

Figure 1. Simulation of the magnetic flux path.

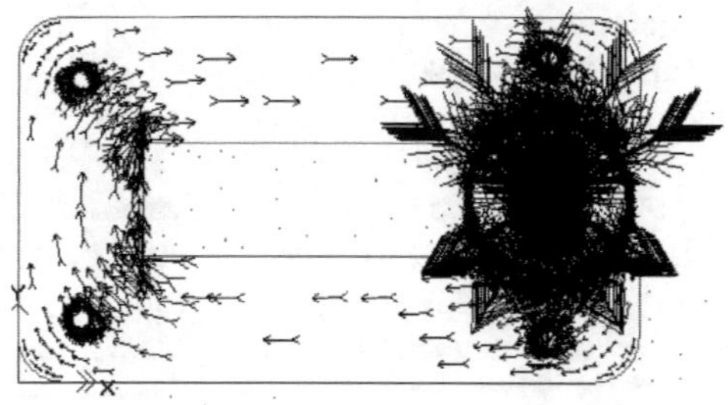

Figure 2. Simulation of the magnetic field density.

3.2. Performance test results

Figure 3 and figure 4 show the graphs of input power against the horizontal torque generated by the shaft for each type of motor. The graphs show that the input power to establish a certain level of torque for the superconducting coils motor is much lower than the conventional copper coils motor. For instance, in order for the motor to generate 0.07 Nm of torque, only about 2.0 W of input power is required by the motor with superconducting coils. In contrast, the motor with copper coils requires 60 times more input power to generate the same amount of torque. This shows that for the same volume of conductor used in the fabrication, motor with superconducting coils performed much better than the motor with copper coils.

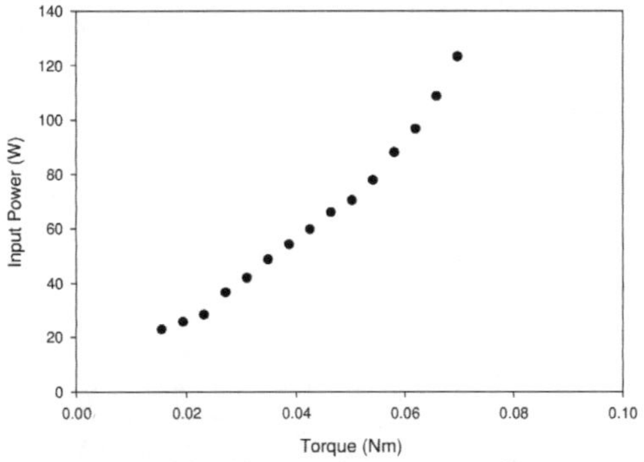

Figure 3. The performance of motor with copper coils.

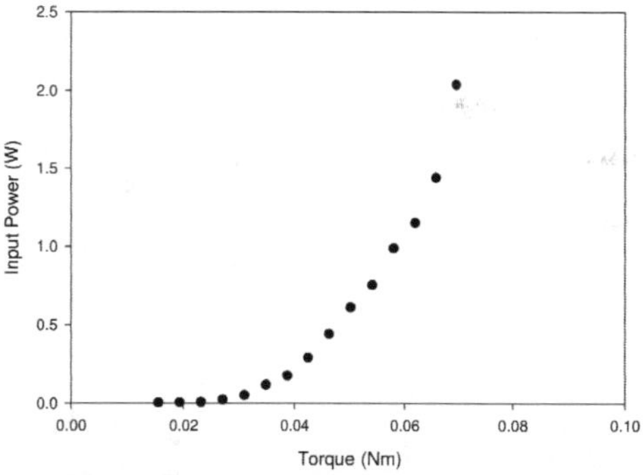

Figure 4. The performance of motor with superconducting coils.

The energy saving in term of percentage of input power for the motor with superconducting coils when compared to the conventional motor with copper coils is significantly higher. But the trend in energy saving is decreasing as torque is increased. Nevertheless, it was reported that with the application of flux collector technology, the energy efficiency of almost 99% could be achieved at all time [10]. The superior performance and better energy efficiency of the electric motor with superconducting coils is attributed to the magnetic properties of the superconducting coils. The Meissner effect contributed to much higher magnetic field concentration along the magnetic flux path in the layered lamination sheets of the stator's core. In addition, the efficiency of the superconducting motor is attributed to lower hysteresis loss inside the superconductor coils [11].

4. Conclusion

The finite element method has proved to be a very useful tool in the designing and fabrication of a bench-top model synchronous DC superconducting electric motor. The model was developed based on the modeling and simulation of the magnetic flux distribution. Successful testing of the superconducting motor proved that the efficiency and performance of motor with superconductor coils is much superior when compared to the conventional copper coils motor with similar capacity. The encouraging results indicate the potential of superconductor applications in machine devices and the possibility of commercialization in the near future.

5. Acknowledgements

This project is supported by the Ministry of Science, Technology and Innovation, Malaysia under the eScience Grant project no. 03-02-03-SF0132.

References

[1] Okazaki T, Hayashi K and Sato K 2006 *SEI Tech. Rev.* **61** 24
[2] Ayai N, Kato T, Fujikami J, Fujino K, Kobayashi S, Ueno E, Yamazaki K, Kikuchi M, Ohkura K, Hayashi K, Sato K and Hata R 2006 *J. of Phys. Series* **43** 47
[3] Chu P C W 1995 *Scientific American* 128
[4] Noe M and Oswald B R 1999 *IEEE Trans. Appl. Supercond.* **9** 1347
[5] Friedman A, Shaked N, Perel E, Sinvani M, Wolfus Y and Yeshurun Y 1999 *Cryogenics* **39** 53
[6] Bednorz J G and Müller K A 1986 *Z. Phys. B: Condens. Matter* **64** 189
[7] McCulloch M D and Dew-Hughes D 1998 *Mater. Sci. and Eng.* **53** 211
[8] Iwakuma M, Tomioka A, Konno M, Hase Y, Satou T, Iijima Y, Saitoh T, Yamada Y, Izumi T and Shiohara Y 2007 *IEEE Trans. Supercond.* **17** 1607
[9] Song M, Yoon Y S, Jang W K, Ko T, Hong G W and Jang I B 1999 *IEEE Trans. Appl. Supercond.* **9** 1241
[10] Takeda T, Togawa H and Oota T 2005 *IHI Engineering Review* 1
[11] Kirtley Jr J L 1999 *IEEE Power Eng. Soc. Sum. Meet.* **2** 1181

IOP Conf. Series: Earth and Environmental Science **16** (2013) 012052 doi:10.1088/1755-1315/16/1/012052

A novel design of DC-AC electrical machine rotary converter for hybrid solar and wind energy applications

K G Mohammed, A Q Ramli and U A U Amirulddin
Department of Electrical Power Engineering, University Tenaga Nasional, Jalan IKRAM-UNITEN, 43009 Kajang, Selangor, Malaysia

E-mail: khalid_alkaisee@yahoo.com

Abstract. This paper proposes the design of a new bi-directional DC-AC rotary converter machine to convert a d.c. voltage to three-phase voltage and vice-versa using a two-stage energy conversion machine. The rotary converter consists of two main stages which are combined into single frame. These two stages are constructed from three main electromagnetic components. The first inner electromagnetic component represents the input stage that enables the DC power generated by solar energy from photo-voltaic cells to be transformed by the second and third components electro-magnetically to produce multi-phase voltages at the output stage. At the same time, extra kinetic energy from wind, which is sufficiently available, can be added to existing torque on the second electromagnetic component. Both of these input energies will add up to the final energy generated at the output terminals. Therefore, the machine will be able to convert solar and wind energies to the output terminals simultaneously. If the solar energy is low, the available wind energy will be able to provide energy to the output terminals and at the same time charges the batteries which are connected as backup system. At this moment, the machine behaves as wind turbine. The energy output from the machine benefits from two energy sources which are solar and wind. At night when the solar energy is not available and also the load is low, the wind energy is able to charge the batteries and at the same time provides output electrical power to the remaining the load. Therefore, the proposed system will have high usage of available renewable energy as compared to separated wind or solar systems. MATLAB codes are used to calculate the required dimensions, the magnetic and electrical circuits parameters to design of the new bi-directional rotary converter machine.

1. Introduction

Finding alternatives for fossil fuel to generate power has become the main motivation towards research on renewable energy in the last few decades. Out of the many renewable energy sources available, extensive research has been focused on harnessing solar energy using photo-voltaic cells [1] and wind energy using wind turbines [2] to generate electrical power.

Studies have shown that solar power is most suited to generate power during the day time whereas wind power is most suited during night time. Hence, extensive research has also been concluded to hybrid solar and wind energy systems for power generation [3].Theses hybrid systems employ the use of power electronic converters to convert the dc voltage from solar PV cells and varying frequency ac voltage from the wind turbine into a constant frequency ac voltage to be supplied to the electrical grid system [4].

This paper aims to introduce a new design of electric machine rotary converter to convert dc to ac voltage and inversely which can be employed on solar and wind energy hybrid systems.

2. Components and Operation of the DC-AC Rotary Converter

The proposed direct current to multi-phase rotary converter machine contains two important stages. The first stage is made up of direct current motor. Naturally it consists of two main parts: stator and rotor. The stationary stator forms the inner fixed part which has wave or lapped windings. These windings are divided into two purposes: main winding, and improving windings. The D.C. motor's speed is determined by the number of main winding poles (2, 4, 6, etc.). The improving windings are devoted to provide a good starting and running performance of the motor.

Generally as in any design, it must be designed based on the demanded specifications of the suggested rotary converter rating, speed, rated input and output voltages. These will be utilized to calculate the engineering dimensions of the machine.

The d.c. stator may have tapered slots for small ratings or rectangular slots for higher ratings with the sizes for stator slots being dependent on the number of conductors per slot. The starting and running windings in the stator windings are always preferred to be wounded in a wave or lap depending on the machine rated power. Figure 1 shows the inner d.c. stator of the rotary converters first stage. The rotor of the first stage consists of permanent magnet pols on the inside part of the middle section of the rotary converter as shown in figure 2. When d.c. voltage is applied to the inner d.c. stator of the rotary converters first stage.

The rotor of the first stage consists of permanent magnet poles placed on the inside part of the middle section of the rotary converter as shown in figure 2. When a d.c. voltage is applied to the inner d.c. stator of the first stage rotary converter, the magnetic field produced by the inner stator will interact with the permanent magnet poles on the inside part of the middle section causing the entire middle section to be rotated.

The second stage of the rotary converter is exactly the same as any conventional synchronous generator. The rotor of the second stage contains the excitation field windings placed on the outer part of the middle section of the rotary converter as shown in figure 2. The stator or armature of the second stage is shown in figure 3. The armature may be comprised of three, six, nine or twelve phases which is made up of a block of laminations mounted in a cast iron or die cast aluminum alloy frame. The armature has tapered slots for small rating machines or rectangular slots in higher rating machines. Armature windings are always preferred to be wounded in lap or progressive coils, which gives a good value of winding factor, reaching more than ninety percent.

The rotor of conventional synchronous generator are driven by an exerted prime –mover. In the case of the second stage of the rotary converter, the excitation field windings are rated by the movement generated from the first stage. This rotation causes the excitation field to induce voltages on the multi-phase outer stator of the second stage to generate ac multi-phase voltages at the output terminals. The three components of the Dc-AC rotary converter are accumulated together as shown in figure 4.

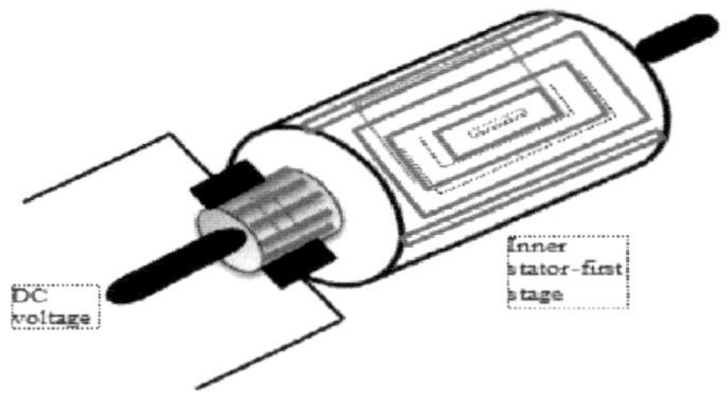

Figure 1. Inner D.C. Stator of the rotary converter's first stage.

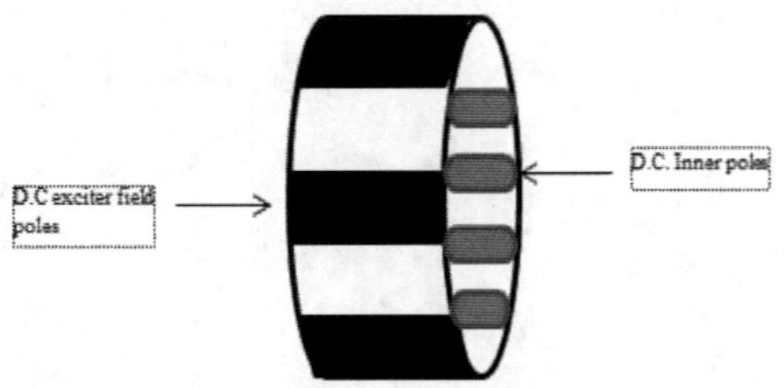

Figure 2. Middle rotating component of the rotary converter consisting of permanent magnet poles of the outer rotor d.c. motor. First stage on the inside part and excitation field windings of the second stage synchronous generator.

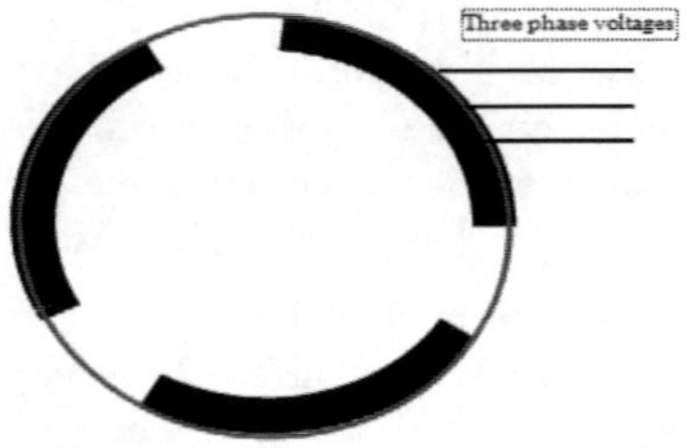

Figure 3. Three phase outer stator of the rotary converter's second stage.

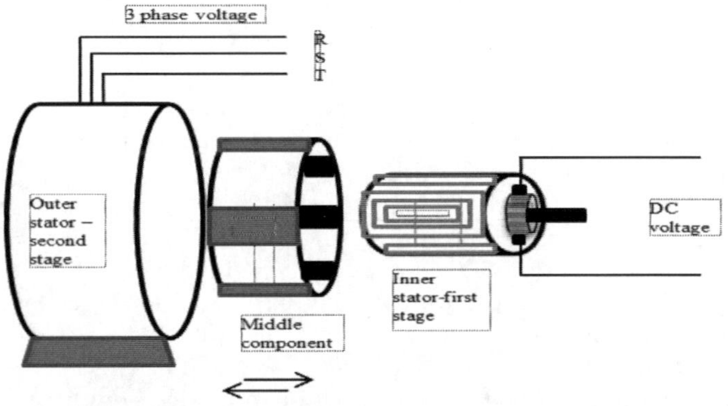

Figure 4. Components of the DC-AC rotary converter.

3. DC-AC Rotary Converter for Hybrid Wind- Solar Energy System

The designed DC-AC rotary converter is proposed in this paper can be employed to harness electrical power for a hybrid wind and solar energy system as shown in figure 6. The d.c. output voltage generated by photo-voltaic cells can b connected as the input to the inner d.c. stator of the rotary converter's first stage. This will then cause the middle rotating component of the rotary converter to rotate and hence generating three- phase output a.c. voltage at the outer stator of the rotary converter which can be connected directly to the grid.

The wind turbine can be coupled to the middle rotating component via gearbox. When the wind conditions are suitable, the wind turbine will be coupled to the middle rotating component causing it to rotate and generating three phase voltages to the grid via the outer stator of the second stage. The rotation by the wind turbine will also generate d.c. voltage at the inner d.c. stator which can be used to charge the back up batteries as shown in figure 5.

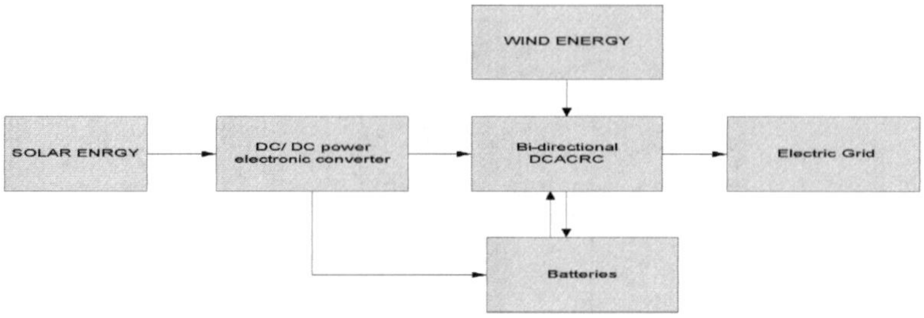

Figure 6. Basic configuration of the proposed hybrid renewable energy system employing the designed Dc-AC rotary converter.

4. Conclusions

The new rotary converter is proposed for hybrid power generation employing stator and wind energy. This can be used as a complement to the power electronic converters because the last type could not be able to violate the solar and wind energies at the same time without need for electric machine, the following merits are relevant to the proposed DC-Ac rotary converter machine:

i. It can be used for generating three, six, or twelve phase.
ii. The output voltages have a high efficiency of balanced multi-phase operation.
iii. It is easy to maintain and re-wind.
iv. It is capable of enduring high ambient temperatures.
v. Its overall noise strength is expected to be similar to normal dc motor.
vi. Its weight and size in within the acceptable range.

References

[1] Khannam W and Yupapin P P 2002 Solar energy conversion using erbium doped fiber pumping technique for feasibility space transmission study *Industrial Technology, 2002. IEEE ICIT'02. 2002 IEEE International Conference* **2** 1241-1244

[2] Narayana M, Putrus G, Jovanovic M and Leung P S 2009 Predictive control of wind turbines by considering wind speed forecasting techniques *Universities Power Engineering Conference (UPEC), 2009 Proceedings of the 44th International* 1-4

[3] Delimustafic D, Islambegovic J, Aksamovic A and Masic S 2011 Model of a hybrid renewable energy system: Control, supervision and energy distribution *Industrial Electronics (ISIE), 2011 IEEE International Symposium* 1081-1086

[4] Mei Q, Zhen-lin X and Wu W Y 2008 A novel multi-port DC-DC converter for hybrid renewable energy distributed generation systems connected to power grid *Industrial Technology, 2008(ICIT 2008) IEEE International Conference* 1-5

IOP Conf. Series: Earth and Environmental Science **16** (2013) 012053 doi:10.1088/1755-1315/16/1/012053

Preliminary evaluation of the effectiveness of moisture removal and energy usage in pretreatment module of waste cooking oil for biodiesel production

K Palanisamy, M K Idlan and N Saifudin
Center for Renewable Energy, Universiti Tenaga Nasional
Km7, Jalan IKRAM-UNITEN, 43000 Kajang
Selangor, Malaysia.

Email: kumaran@uniten.edu.my

Abstract. Waste Cooking Oil (WCO) is a plausible low cost biodiesel feedstock but it exhibits few unfavorable parameters for conversion into biodiesel. One of the parameter is the presence of high moisture content which will inhibit or retard catalyst during the acid esterification or base transesterification causing lower purity and yield of biodiesel. This will effect the post processing and escalate production cost making WCO a not favorable biodiesel feedstock. Therefore, it is important to have an effective moisture removal method to reduce the moisture content below 0.05%wt or 500 ppm in WCO for an efficient biodiesel production. In this work, the effectiveness of moisture removal and the energy usage of a newly develop innovative pretreatment module has been evaluated and reported. Results show that the pretreatment module is able to reduce up to 85% to effectively reduce the moisture content to below 500ppm of the initial moisture content of WCO and only consume 157 Wh/l energy compared to conventional heating that consume 386 Wh/l and only remove 67.6% moisture in 2 hours.

1. Introduction

Biodiesel is predominantly produced from vegetable oils and animal fats using a chemical process known as "transesterification" which involves interchanging of glycerol in the triglyceride (vegetable oil) molecules with alkyl alcohol molecules to form monoalkylesters (biodiesel) in the presence of acid or base catalyst [1]. The base catalyzed transesterification is usually favored for industrial scale biodiesel production for unused vegetable oils are used as feedstock as it is more efficient and faster[2]. In recent years, waste cooking oil (WCO) which is 2-3 times cheaper than edible vegetable oils has been considered as feedstock. However, the use of WCO as biodiesel posses feedstock quality issues and challenges since WCO exhibit high percentage of Free Fatty Acid percentage (%FFA) and high moisture content compared to unused vegetable oils [3].

The favorable level of FFA and moisture content for base catalyzed transesterification are below 1%wt and 0.05%wt respectively [4]. However, WCO generally exhibit FFA more than 2%wt and moisture content beyond 0.05%wt. This will cause formation of soap or saponification reaction process resulting in consumption of catalyst which further will cause emulsion problems during post processing through water washing which is require for removing excess catalyst, alcohol, mono, diglycerol and free glycerol and also reduce the biodiesel yield and purity significantly [5].

The objective of this work is to evaluate the effectiveness of a newly developed innovative pretreatment module for WCO feedstock moisture that hybrids low energy heating and vacuum with a static dehydrator is evaluated. The evaluation of energy consumption has been conducted by measuring the total energy (heating + vacuum) consumed to reduce the moisture content per liter of WCO feedstock and compared with conventional heating over two hour period.

2. Material and Method

WCO samples were collected from local food outlet Mawa De Village, Segamat. Moisture Content is determined using automated Coulometric Karl Fisher titration. The chemical use for moisture content determination is Aquamicron CXU and Aquamicron AXU with purity of 99.99% from Merck. The moisture content of WCO is determined using ASTM D1796 test method as prescribed in ASTM D6751 standard. The testing is conducted at TNBR Qats Sdn. Bhd in Kajang. The energy usage is determined by using single phase Efergy™, energy meter with ±0.1Wh accuracy

A pretreatment module with working volume of 20 liter has been used in the moisture removal experiment where the closed system module is equipped with 1 kW heater in the main tank and preheat tank respectively. The module consisted of two section; filtering section consisting filter size of 5μm, 1μm and 0.5μm and water removal section consisting of nozzle system with 60 numbers of 1.2 mm diameter orifices. The closed system pretreatment is equipped with a vacuum pump and condenser system to reduce the moisture vapor temperature which also has a conical shape static dehydrator section located at the bottom of the tank. For conventional heating method, the experiment is conducted at atmospheric pressure by keeping the heater on with the tank cover and dehydrator removed. The pump recirculation is only deployed during filtering in conventional heating.

2.1. Sample preparation and Measurement of Moisture, FFA and Energy Consumption

A 10 liter WCO sample is filtered using 1mm sieve in order to remove large particulate and preheated up to 100^{0}C before re-filtered in 5.0 μm ,1 μm and 0.5 μm in filtering section to remove smaller particulate at 90°C for about 20 minutes (approximately 5 cycles) without deployment of the vacuum.

The Pretreatment module with nozzle size spray size of 1.2 mm without the presence of dehydrator is first operated for 20 minutes filtering, before the flow is changed to moisture removal section. All measurement is repeated 3 times and the average is calculated for reporting. The energy usage is recorded using the energy meter and the heater temperature is maintained at 100°C. Subsequently, the pretreatment module is equipped with the dehydrator and same procedure is repeated. As for conventional heating the sample is heated up to 100°C in the pretreatment module without closing the top cover (open system). Upon reaching 100°C, 75 ml sample is drawn into sampling bottle and the vacuum pump is deployed first sample was drawn and experiment continued for 120 minutes with samples drawn every 15 minutes for moisture content analysis

3. Result and discussion

Frying activity using triglycerides (cooking oil) at temperatures more than 180 °C promotes hydrolysis results in the fatty acid of triglycerides being disengaged from the glycerol backbone of the triglyceride molecule and become FFA [6]. The increase FFA further attract more water molecules to form hydrogen bond. Commonly, heating is deployed up to the boiling temperature of water at 100°C for removing moisture at atmospheric pressure, which is not energy efficient and also time consuming. In the conventional heating, the bound water/moisture molecules have to overcome the surrounding bulk forces by the triglycerides molecules before gaining enough kinetic energy to liberate out of the oil due to convectional heat transfer and evaporate out through top triglyceride surface. A study by Felizardo et al (2006), proposed the use of vacuum distillation (0.05 bar absolute) to remove moisture from crude WCO at temperature as low as 30-40°C, which consumes less energy for the feedstock heating [7]. However, the disadvantage of this method is the high vacuum instrument consumes high energy and also safety concern in working with vacuum tank. The moisture removed and energy usage from the pretreatment module and for comparison the conventional heating has been summarized in Table 1. The result shows that moisture content for all samples after 2 hours meet the requirement for base catalyzed transesterification which is below than 500 ppm. Sample 3 shows the highest moisture removed which is 87.4% with least energy consumption which is 157 Wh/l. Compared to conventional heating, the energy used is double (386 Wh/l) and only 67.6% of moisture removed. Figure 1 and 2 shows that the moisture content from sample 2 and 3 are already achieve the moisture target (<500 ppm) in 35 minutes while conventional heating required longer time (60 min). Table 1, shows average temperature of only 90°C is required during the whole moisture removal process for sample 2 and 3 because of the pressure in the tank is below the atmospheric pressure which reduces the boiling point of water. This results is in agreement with the work done by Felizardo et al., (2006) but eliminate the need for using pressurize tank where the vacuum pressure in this work is at 0.8 bar absolute [7].

Table 1. Comparison between conventional heating and pretreatment module

Sample no	Initial Moisture (ppm)	Method of moisture removal	Final Moisture (ppm)	Average WCO Temperature (°C)	Amount of Moisture Removed (%)	Energy Consume/ liter (Wh/l)
1	870	Conventional heating	282	100	67.6%	386
2	1243	Pretreatment without dehydrator	305	90	75.5%	157
3	2392	Pretreatment with dehydrator	302	90	87.4%	157

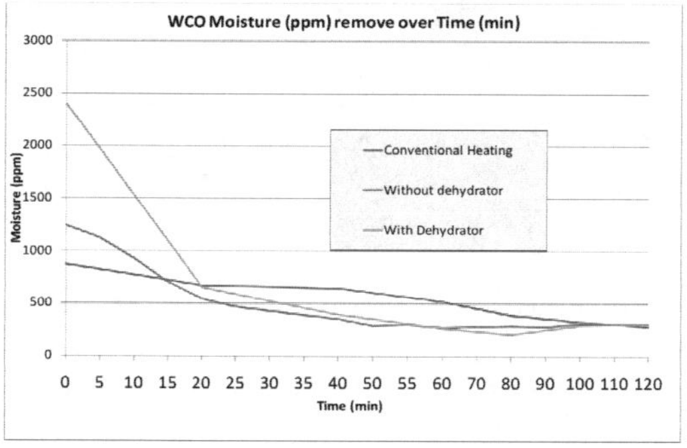

Figure 1. WCO Moisture Content over Time

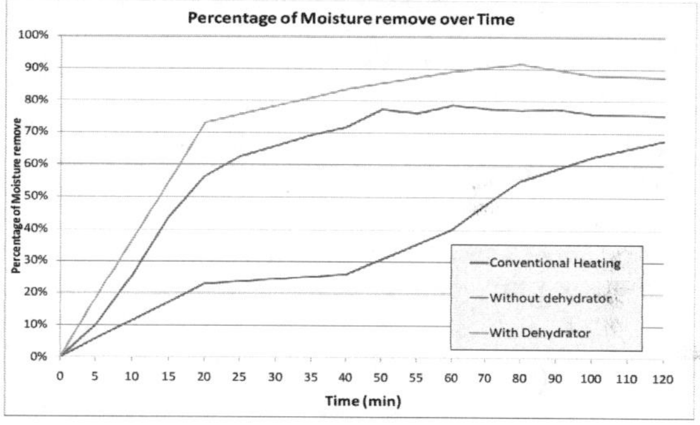

Figure 2. Percentage of Moisture Removal over Time

Another method of moisture removal is by using chemical and desiccants such as magnesium sulphate and silica gel but these methods requires a fine filtering system for the feedstock before entering the base catalyzed transesterification stage [7,8,9,10]. Disadvantages of using chemical and desiccant is the increase of biodiesel production cost due to regeneration limit of the chemical. Figure 1 and 2 have shown that the hybrid use of nozzle spray and static dehydration method enhanced the moisture removal from 75.5% to 87.4% with the same energy consumption. The oil droplet formed after the nozzle spray under vacuum condition enables the water molecules to be easily liberated due to inward heat transfer from the triglycerides molecules to the water molecule which undergoes outward mass transfer to surface of droplet and induced out of the closed tank by the vacuum. Also in the smaller droplet form the influence of hydrogen bond between the water molecule and the triglyceride molecule especially with FFA, di- and mono- glyceride presence in the WCO is less stronger compare to being in the bulk form. Subsequently the droplet of oil that impinges on the dehydrator due to the gravitational force further gets spread into a thin layer over the inclined plane of the conical shaped static dehydrator which further enables remaining moisture trapped within triglyceride to be liberated and also induced out by vacuum. Despite, the starting moisture for the conventional heating being the lowest of 870 ppm the time taken to achieve 282 ppm is two (2) hours compared to the pretreatment module that had higher moisture content, up to 2,392 ppm and need the same two hour to reach 302 ppm. This indicates that the use of static dehydrator is more energy efficient compare to conventional heating better than chemical and desiccant for moisture removal enhancement.

4. Conclusion

The pretreatment module is able to reduce moisture content in WCO below 500ppm and has effectiveness of removing up to 85% of the initial amount of moisture in WCO using 1.2 mm nozzle size with the presence of static dehydrator. The whole process only used 157 Wh/l which is less than half of energy used by conventional heating method. Therefore, the newly develop pretreatment module can be integrated in biodiesel production system that uses WCO as feedstock where it is more energy and cost efficient. Hence, the cost of biodiesel production using WCO as feedstock can be reduced significantly.

5. Acknowledgement

Authors would like to thank TNBR for funding this project and TNBR QATS for rendering laboratories services to perform properties testing for samples.

References

[1] Demirbas A 2009 Progress and Recent Trends in Biodiesel Fuels *Energy Conversion and Management* **50(1)** 14-34
[2] Murugesan A, Umarani C, Subramanian R and Nedunchezhian N 2009 Biodiesel as An Alternative Fuel for Diesel Engines – A review *Renewable and Sustainable Energy Reviews* **13(3)** 653-662
[3] Oilseed : Worldmarket and Trade, United States Department of Agriculture, Foreign Agricultural Service , Office of Global Analysis *Circular Series FOP6-08* June 2008
[4] Leung D Y, Wu X and Leung M K H 2010 A Review on Biodiesel Production using catalyzed Transesterification *Appl Energy* **87(4)** 1083-95
[5] Gui M M, Lee K T and Bhatia S 2008 Feasibility of edible oil vs. Non-edible oil vs. waste edible oil as biodiesel feedstock
[6] Blumenthal M M 1991 A new look at the chemistry and physics of deep frying *Food Technology* **45** 68-71
[7] Felizardo P, Correia M J N, Raposo I, Mendes J F, Berkemeier R, and Bordado J M 2006 Production of biodiesel from waste Frying Oils *Waste Manage 2006* **26** 487-94
[8] Khan N A, and el Dessouky H 2009 Propect of Biodiesel in Pakistan. *Renew Sustain Energy Rev* **13** 1576-83
[9] Canakci M and Van Gerpen J H 1991 Biodiesel Production via acid catalyst *Trans ASAE 1999* **42(5)** 1203-10
[10] Alptekin E and Canakci M 2010 Optimization of pretreatmeant reaction of methyl ester production from chicken fat *Fuel* **89(12)** 4035-4039

IOP Conf. Series: Earth and Environmental Science **16** (2013) 012054 doi:10.1088/1755-1315/16/1/012054

Renewable Decyl-alcohol Templated Synthesis of Si-Cu Core-Shell Nanocomposite

M A Salim[1, 2], H Misran[2], S Z Othman[2], N N H Shah[2], N A A Razak[2], A Manap[2]
[2]Center of Renewable Energy, Universiti Tenaga Nasional, Selangor, Malaysia.

Email: Musdalilah@uniten.edu.my

Abstract. Monodispersed silica spheres with particles size of ca. 450 nm were successfully synthesized using a modified Stöber method. The synthesized monodispersed silica spheres were successfully coated with copper using modified sol-gel method employing nonsurfactant surface modifiers and catalyst. A renewable palm oil based decyl-alcohol (C10) as nonsurfactant surface modifiers and catalyst were used to modify the silica surfaces prior to coating with copper. The X-ray diffraction patterns of Si-Cu core-shell exhibited a broad peak corresponding to amorphous silica networks and monoclinic CuO phase. It was found that samples modified in the presence of 1 ml catalyst exhibited homogeneous deposition. The surface area of core materials (SiO_2) was at ca. 7.04 m^2/g and Si-Cu core-shell was at ca. 8.21 m^2/g. The band gap of samples prepared with and without catalyst was calculated to be ca. 2.45 eV and ca. 3.90 eV respectively based on the UV-vis absorption spectrum of the product.

1. Introduction

Metallic materials have attracted increasing interest for their widespread applications in optical, biological, medical, sensors and optoelectronics due to their superior properties [1-2]. In order to obtain high performance metallic material, it is critical to control the size and morphology of nanoparticles. One significant approach is to design the metallic materials in the presence of support materials such as core-shell nanocomposite. Core-shell nanocomposite are widely studied due to their improved physical and chemical properties as compared to their single material [3]. In recent years, many works were conducted on silica as core and metal as shell for the synthesis of various silica-metal core-shell nanocomposite. Silica was extensively used as core materials because it was cheap and relatively easy to produce [4-5]. In addition, the fabrication of silica-metal core-shell utilized only small amount of metals to cover the surfaces of inexpensive core in order to obtain the desired properties [4-5]. At present, many synthesis routes were explored to synthesize core-shell nanocomposite, such as vapour deposition, plasma-assisted technique, chemical reduction and self-assembly [4-6]. However, in most cases the degree of surface coverage was low and metallic coating was non-uniform [6]. Sol-gel processing is an effective method for preparing core-shell materials since the reactant can be homogeneously mixed at the molecular level [7]. However, conventional sol-gel method employs surfactant as surface modifiers such as triethanolamine, polyetheyleneimine (PEI) and methyltriethoxysilane (MTES) which were toxic and expensive materials. In this study, environmental friendly, cheap and renewable palm oil based fatty alcohol is proposed to substitute the conventional surfactant as nonsurfactant surface modifiers [8]. Si-Cu core-shell nanocomposite will be

[1] To whom any correspondence should be addressed.

synthesized using renewable palm-oil based fatty alcohol (C10) as surface modifiers in the presence of ammonia as catalyst. In addition, the effect of crystallinity and morphology of samples by adding the catalyst during the surface modification process is also reported.

2. Materials and Methods

The silica spheres were prepared according to the method introduced by Misran *et al.* [9]. In a typical synthesis procedure, C10 was added to ethanolic solution in the presence of ammonia to obtain microemulsion solution under vigorous stirring. Then, silica precursor of tetraethyl orthosilicate (TEOS) was added drop-wise to the microemulsion solution. The resulting white colloidal suspension of silica was filtered and washed several time with water and ethanol. The silica powder were dried at 100 °C overnight and calcined at 550 °C to obtained silica spheres. The processing of Si-Cu core-shell nanocomposite was performed following the reported procedure of Xia and Tang [4] with some modifications. Obtained silica spheres were dispersed and sonicated in ethanol and water to prepare the silica sol prior to surface modification. C10 and ammonia were added to the silica sol followed by copper salt under constant stirring for several hours at 80 °C. The solids were collected, washed with ethanol several times and dried at 100 °C. Then, the samples were calcined at 600 °C to obtain Si-Cu core-shell nanocomposite. The final molar ratio of samples was silica : copper : C10 = 1 : 0.32 : 6.4. The calcined samples were denoted as SiCuD-1 (prepared without catalyst) and SiCuD-2 (with 1 ml ammonia). The optical band gap energy of the product can be determined using the equation 1 [10]:

$$(\alpha h\nu)^2 = B(h\nu - E_g) \tag{1}$$

where $h\nu$ is photonic energy, α is the absorption coefficient, B is a constant related to materials, and E_g is the band gap energy.

3. Results and Discussions

The X-ray diffraction patterns of pure silica spheres (SiO_2) as reference (core materials) and Si-Cu core-shell nanocomposite of SiCuD-1 and SiCuD-2 are shown in figure 1. All samples exhibited broad peak centered at ca. $2\theta = 22°$ corresponding to the amorphous structure of silica spheres. SiCuD-1 and SiCuD-2 samples exhibited ten diffraction peaks attributable to monoclinic structure of CuO. The monoclinic copper oxide phase obtained was in good agreement with result reported by Wu *et.al* [11] and standard data (JCPDS 5-0661). The addition of 1 ml ammonia in SiCuD-2 had reduced the intensity of CuO peak with relative crystallinity of ca. 42% as compared to SiCuD-1. These results suggested the crystalline structure of CuO deposited on the silica surfaces were affected by the addition of catalyst.

Nitrogen adsorption isotherms of calcined silica spheres (SiO_2) and Si-Cu core-shell nanocomposite (SiCuD-2) synthesized by modified sol-gel method are shown in figure 2. The adsorption isotherms of both samples exhibited Type II characteristic of nonporous materials where the adsorption occurred on external surface area due to surfaces roughness of nonporous Si-Cu core-shell. The specific surface area of SiO_2 and SiCuD-2 were estimated at ca. 7.04 m^2/g and 8.21 m^2/g respectively using BET method. In addition, the surface area of SiCuD-2 was observed to be higher as compared to pure SiO_2 spheres suggesting that copper nanoparticles deposited on the pure SiO_2 resulted in increased surface roughness that caused higher nitrogen uptake. This will increase the surface area accordingly.

The field-emission scanning electron microscopy (FE-SEM) images and energy dispersive X-ray (EDX) of calcined Si-Cu core-shell nanocomposite prepared with and without catalyst are shown in figure 3 (a-c). SiCuD-1 exhibited spherical morphology at ca. 450 nm interspersed with agglomeration of monoclinic CuO as confirm by XRD and EDX. Homogeneous copper nanoparticles at ca. 50 nm to 100 nm were formed on the silica surfaces and the agglomerations were reduced with the addition of 1 ml catalyst (SiCuD-2). These results were in agreement with previous study by Kim and co-workers that reported the addition of ammonia as catalyst at optimum amount resulted homogeneous copper nanoparticles formed on silica surfaces [12]. The formation of CuO in alkaline condition was aided by ammonia which produced strong nucleophiles that deprotonate hydroxyl ligand in SiOH to generate SiO^-. Then, electrophilic materials like metal ion (Cu^{2+}) can attack and easily bonded with nucleophilic SiO^- group [12]. The EDX analysis of SiCuD-1 (figure 3 (d)) and SiCuD-2 (figure 3 (e)) on the surfaces of Si-Cu core-shell exhibited the presence of copper element. These results suggested that ammonia successfully played the role as catalyst in the surface modification using nonsurfactant fatty alcohol (C10) to aid the homogeneous deposition of copper on silica surfaces.

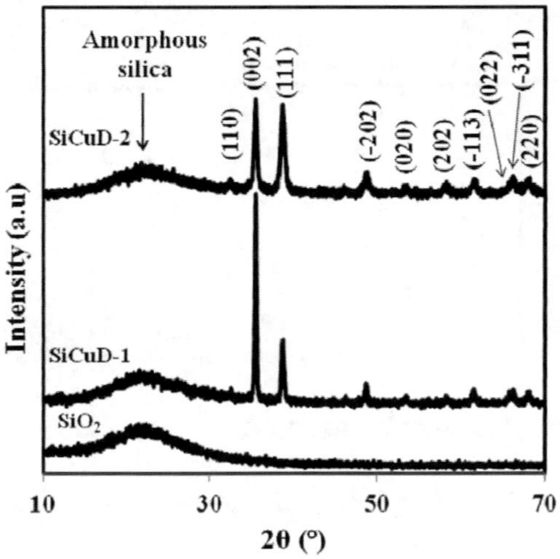

Figure 1. X-ray diffraction patterns of calcined SiO_2, SiCuD-1 and SiCuD-2 core-shell.

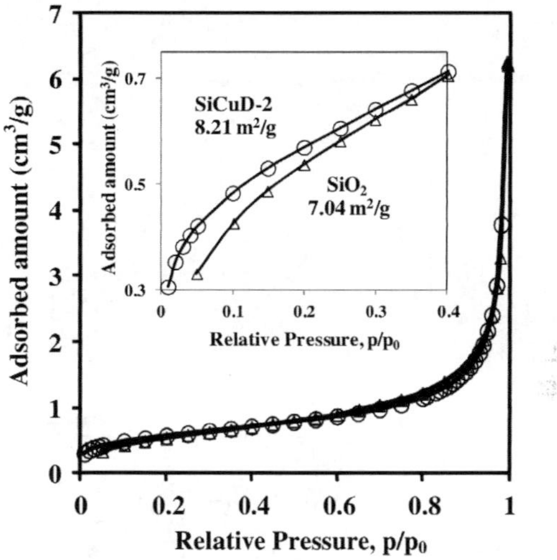

Figure 2. Nitrogen adsorption isotherms of pure SiO_2 and SiCuD-2 core-shell.

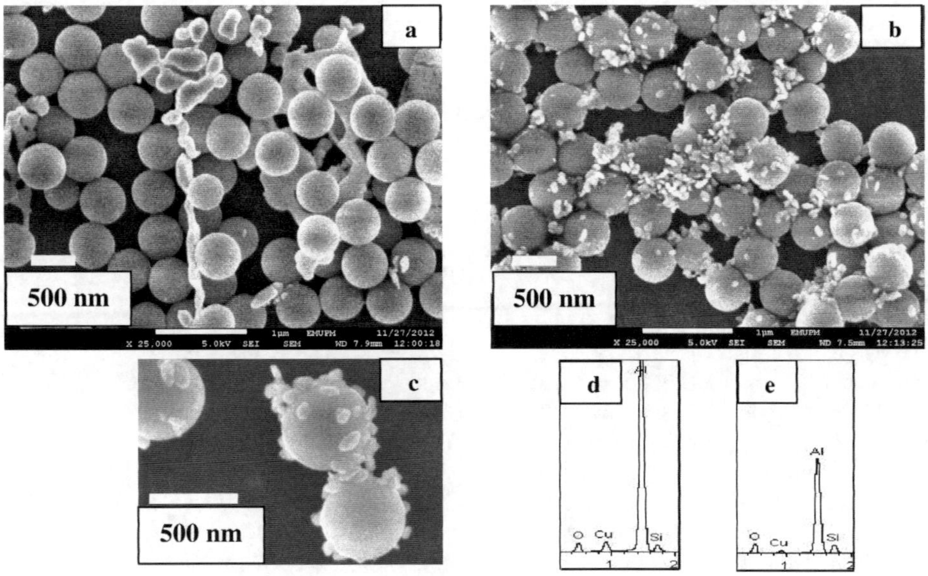

Figure 3. FE-SEM images of Si-Cu core-shell prepared (a) without catalyst (SiCuD-1), (b) with catalyst (SiCuD-2), (c) different magnification of (b); (d) and (e) EDX spectra of (a) and (b), respectively.

The UV-vis spectra (figure 4 (a)) of as-prepared SiCuD-1 and SiCuD-2 dispersed in ethanol exhibited sharp and broad peak centered at ca. 273 nm and ca. 322 nm respectively. The maximum absorption peak was shifted from ca. 273 nm of SiCuD-1 to ca. 322 nm SiCuD-2. The different morphology of Si-Cu core-shell exhibited absorption at different wavelength. These results suggested that the Si-Cu core-shell prepared in this study have possibilities of manipulating light absorption when the size and shape of copper shell on silica were changed. Based on the relation in equation (1), the plot of $(\alpha h\nu)^2$ vs. $h\nu$ are shown in figure 4 (b). The direct band gap energy of both Si-Cu core-shell (SiCuD-1 and SiCuD-2) are estimated to be at ca. 3.90 eV and 2.45 eV correspondingly.

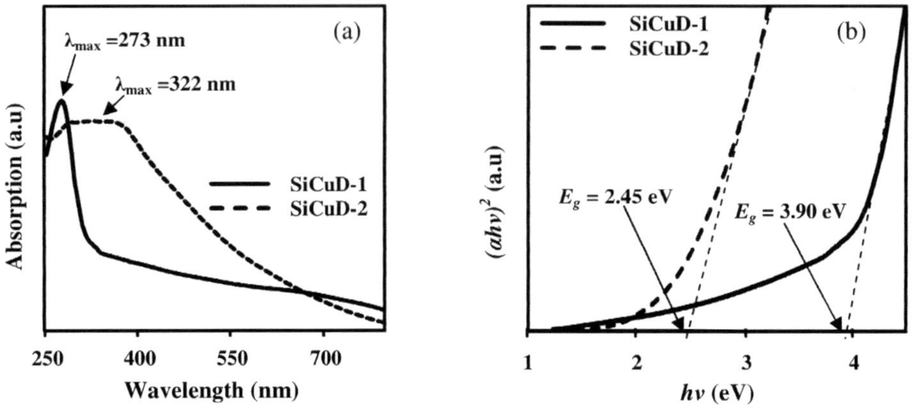

Figure 4. (a) UV-vis spectra and (b) plots of $(\alpha h\nu)^2$ vs. $h\nu$ for SiCuD-1 and SiCuD-2 core-shell.

4. Conclusions

Si-Cu core-shell nanocomposites at ca. 450 nm were successfully synthesized using a simple and quick method. Decyl-alcohol (C10) successfully played the role as surface modifier for depositing copper onto silica core surfaces. However, in the absence of catalyst, the depositions were not controlled resulting in random formation of monoclinic CuO in between the silica spheres. Addition of catalyst resulted in homogeneous deposition of CuO onto the silica surfaces. The optical properties were highly governed by the morphology of deposited copper.

5. Acknowledgment

The authors thank Ministry of Higher Education (MOHE) under FRGS scheme (FRGS20110101) and UPM for facilities.

References

[1] Lu Y, Yin Y, Li Z Y and Xia Y 2002 *Nano Letters* **2** 785-788
[2] Lanke U D and Vedawyas M 1999 *Nucl. Inst. Meth. Phys. Res.* B **155** 97
[3] Jankiewicz B J, Jamiola D, Choma J and Jaroniec M 2012 *Adv. Coll. & Interf. Sci.* **170** 28-47
[4] Xia H L and Tang F Q 2003 *J. Phys. Chem.* B **107** 9175-9178
[5] Jiang Z J and Liu C Y 2003 *J. Phys. Chem.* B **107** 12411-12415
[6] Caruso R A and Antonietti M 2001*Chem. Mater* **13** 3272-3282
[7] Han K, Zhao Z, Xiang Z, Wang C, Zhang J and Yang B 2007 *Mater. Lett.* **61** 363-368
[8] Misran H, Singh R and Yarmo M A 2008 *Micropor. & Mesopor. Mater.* **112** 243-253
[9] Misran H, Yarmo M A and Ramesh S 2012 *Ceram. Inter.* **39** 931-940
[10] Yang C, Su X, Xiao F, Jian J and Wang J 2011 *Sens. Actuat.* B **158** 299-303
[11] Wu H Q, Wei X W, Shao M W, Gu J S and Qu M Z 2002 *Chem. Phys. Lett.* **364** 152-156
[12] Kim Y H, Lee D K, Cha H G, Kim C W, Kang Y C and Kang Y S 2006 *J. Phys. Chem.* B **110** 24923-24928

IOP Conf. Series: Earth and Environmental Science **16** (2013) 012055 doi:10.1088/1755-1315/16/1/012055

Hybrid green permeable pave with hexagonal modular pavement systems

M A Rashid[1], I Abustan[2] and M O Hamzah[2]
[1]Department of Civil Engineering, College of Engineering, Putrajaya Campus, Universiti Tenaga Nasional, Jalan IKRAM-UNITEN, 43000 Kajang, Selangor, MALAYSIA
[2]School of Civil Engineering, Engineering Campus, Universiti Sains Malaysia, 14300 Nibong Tebal, Penang, MALAYSIA

E-mail: aminur@uniten.edu.my

Abstract. Modular permeable pavements are alternatives to the traditional impervious asphalt and concrete pavements. Pervious pore spaces in the surface allow for water to infiltrate into the pavement during rainfall events. As of their ability to allow water to quickly infiltrate through the surface, modular permeable pavements allow for reductions in runoff quantity and peak runoff rates. Even in areas where the underlying soil is not ideal for modular permeable pavements, the installation of under drains has still been shown to reflect these reductions. Modular permeable pavements have been regarded as an effective tool in helping with stormwater control. It also affects the water quality of stormwater runoff. Places using modular permeable pavement has been shown to cause a significant decrease in several heavy metal concentrations as well as suspended solids. Removal rates are dependent upon the material used for the pavers and sub-base material, as well as the surface void space. Most heavy metals are captured in the top layers of the void space fill media. Permeable pavements are now considered an effective BMP for reducing stormwater runoff volume and peak flow. This study examines the extent to which such combined pavement systems are capable of handling load from the vehicles. Experimental investigation were undertaken to quantify the compressive characteristics of the modular. Results shows impressive results of achieving high safety factor for daily life vehicles.

1. Experimental Setup

The designed permeable pavement was constructed in 525 mm x 565 mm Perspex with hexagonal modular which was set up with hole on the bottom plate for water to pass through. Plate 1 shows the overall infiltration rig together designed permeable pavement. A rainfall simulator with 800 evenly spaced 50 mm centre to centre sprays as in Plate 3.2 was set up above the pavement surface to cover an area of 525 mm x 565 mm. Varies condition of simulated rainfall intensities were controlled by a flow meter as shown in Plate 3.3. The water flowing through the pavement was collected from underneath the pavement via a funnel connected only to the central 500 mm x 500 mm area. The infiltration area is limited to 500 mm x 500 mm to prevent boundary effects influencing hydraulic flow within the PVC box. The flat surface at the top of the funnel was screwed on to the bottom plate. There is one infiltrated holes in the middle of the bottom wooden plate (525 mm x 565mm). The diameter of the hole is 12 mm and connected with controlled valve. This valve can be act as different condition of sub base which ranges from fully saturated to highly infiltrate area. In the boundary, 25 drainage holes were drilled on each side of the bottom plate. The pavement is constructed at a zero slope. There are two PVC tank at the bottom of the rig, one is the water inlet storage to supply rainfall simulator and the other is to collect the outflow water. Both of the tank are connected with pipe and can be controlled the amount of water by calibrate the valves.

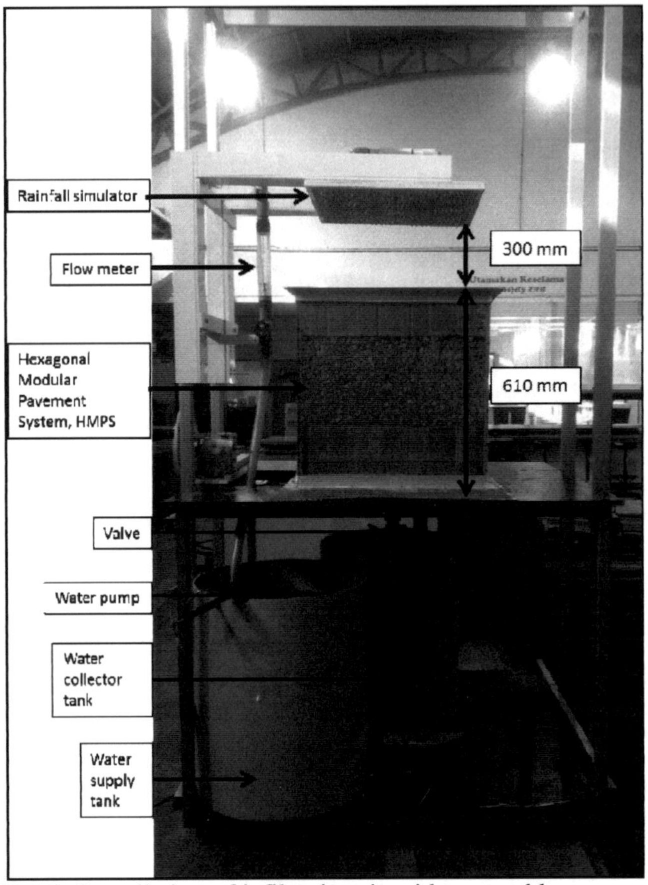

Plate 1. Overall view of infiltration rig with permeable pavement

2. Pavement Structure

This new prototype is a new type of a pavement called Hexagonal Modular Pavement System (HMPS) [1], [2]. The HMPS is a structure of thin-walled hexagonal columns of recycled resins of polyvinyl chloride (PVC) plastic for strength, durability, and green industry benefit. Hexagonal are 10cm diameters, 5mm wall thickness, and 10cm tall. HMPS based on a simple, but impressive technology-a series of rings (hexagonal) arranged side by sides. The modular engineered to withstand significant structural loads and the grid provides stability, flexibility, and continuity for large areas. The point load pressure is transferred from the top of the hexagonal, through the fill material and modular, to the engineered base course [3], [4].

The physical model of the permeable pavement was placed in a vertical rectangular flume as shown in Plate 1 made from 20mm thickness PVC for three sides and 20mm thickness of flexi glass for front view of the model with width and length together were 525mm wide and 565mm length. In this permeable pavement, it has three different layers consists of 110mm thickness of surface layer, 300mm thickness of gravel base and 200mm thickness of sub-base layer. The sub-base layer was constructed by adding two layers of HMPS to the rig in 200mm depth. After completion of the sub-base layer, the gravel-base layer was constructed by adding washed clean cubical aggregate of 15mm to 20mm to the infiltration rig. The surface layer of HMPS consisted of hexagonal columns of PVC with 5mm clean cubical aggregate. After the addition of each layer, the gravel was compacted using Modified Proctor Test with 95% compacted. Porosity for the surface layer was 35%, gravel-base was 40% and sub-surface layer was 92% acted as water storage.

3. Modular Pavement, HMPS

Compression test on hexagonal modular pavement system, (HMPS) have been carried out using Shimadzu Autograph AGS-X Series with loading rate of 0.005mm/min. Maximum stress obtained were 494 N/mm^2, maximum force were 69.7kN with maximum stroke of 3.9mm. Material used in HMPS was high-density polyethylene, HDPE. HDPE is a recycle material that been use widely due to its resistant to many different solvents such as motor oils and fuels. HDPE is highly abrasion-resistant and is unaffected by extremes in pH. HDPE was chosen as it famous in geomembrane for hydraulic applications such as bank reinforcements, geothermal for heat transfer pipe systems, water pipes and corrosion protection for steel pipelines. Figure 1 and Figure 2 shows the results from the compression test on HMPS.

High density polyethylene, HDPE is one of the most chemically inert of all plastics, and therefore is extremely chemical and corrosion resistant. This gives a significant long-term strength advantage over concrete and metal pipe. A study at California State University, Sacramento, proved that even though corrugated polyethylene had a significantly thinner wall than concrete, it was more abrasion resistant. Further, corrugated polyethylene pipe lasted 45 percent longer than concrete under more aggressive conditions.

Polyethylene has demonstrated, through testing and actual application, which it will meet or exceed life service requirements for storm water drainage applications. A service life of more than 70 years is projected in areas where corrugated polyethylene is specified.

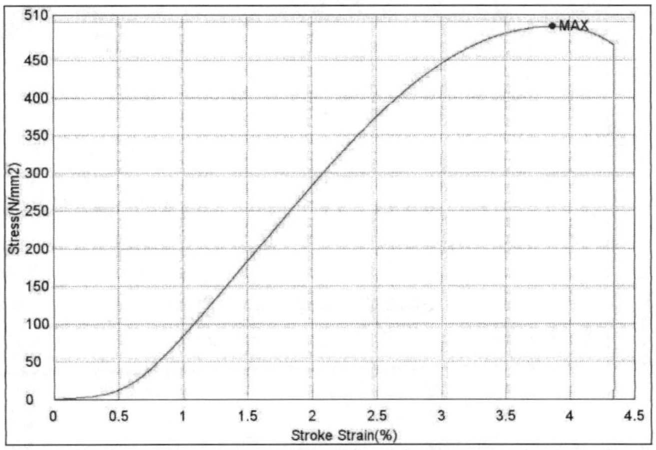

Figure 1. Compression test on HMPS with stress against stroke strain

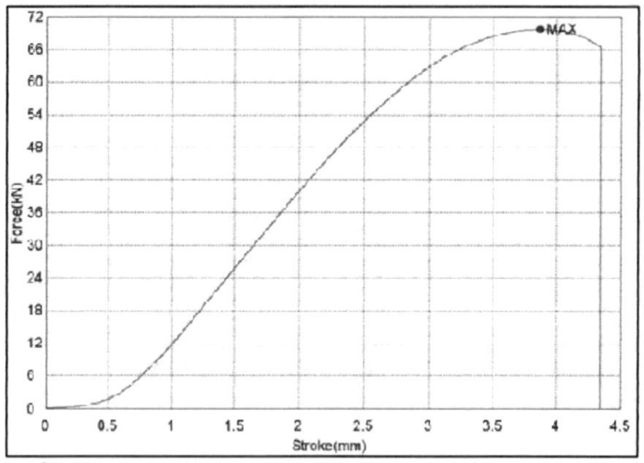

Figure 2. Compression test on HMPS with force against stroke

Here is the summary of laboratory test results on Hexagonal Modular Pavement System, HMPS. Physical modelled unit data made from HDPE was tested with dimension of 5 mm thickness and 100 mm diameter by 100mm high. Lab test data showed that bare rings of 72kN per modular with deflection stopped at 3.9 mm. while sand filled modular with zero deflection has 75kN per modular. Table 1 show HMPS can withstand high load with high safety factor for auto tires, truck tires, F-16 tires and fire truck outriggers. Value of sand filled HMPS with 75kN was use to compare with each items.

Table 1. Examples of usage HMPS in daily life

Item	Maximum load (kN)	Safety factor
Auto tires	275	27
Truck tires	758	10
F-16 tires	2413	3
Fire truck outriggers	558	13

4. Conclusion

This engineered design allows for any street-legal vehicle to park or drive on HMPS surfaces. The point load pressure is transferred from the top of the hexagonal modular to the engineered base coarse. HMPS is 92 percentage void space allowing for more decorative gravel for some of the most attractive paved surface around. This new modular pavement also makes for better runoff coefficients rates. The safety factor of loading criteria exceed over the highest truck tire pressure allowed on public highway.

References

[1] Abustan I, Hamzah M O and Rashid M A 2012 A 3-Dimensional Numerical Study of a Flow within a Permeable Pavement *OIDA International J. of Sustainable Development* **4(2)** 37-44

[2] Abustan I, Hamzah M O and Rashid M A 2012 Review of Permeable Pavement Systems in Malaysia Conditions *OIDA International Journal of Sustainable Development* **4(2)** 27-36

[3] Rashid M A, Abustan I, and Othman M O 2012v Build Green with Permeable Pavement Systems *Awam International Conference on Civil Engineering (AICCE'12.*

[4] Rashid M A , Abustan I, and Othman M O 2012v Infiltration Characteristic Modeling using FLOW3D within a Modular Pavement *International Conference on Advances Sciences and Contemporary Engineering 2012, Procedia Engineering* **50**, 658-667

IOP Conf. Series: Earth and Environmental Science **16** (2013) 012056 doi:10.1088/1755-1315/16/1/012056

Numerical simulation of a 3-D flow within a storage area hexagonal modular pavement systems

M A Rashid[1], I Abustan[2] and M O Hamzah[2]
[1]Department of Civil Engineering, College of Engineering, Putrajaya Campus, Universiti Tenaga Nasional, Jalan IKRAM-UNITEN, 43000 Kajang, Selangor, MALAYSIA
[2]School of Civil Engineering, Engineering Campus, Universiti Sains Malaysia, 14300 Nibong Tebal, Penang, MALAYSIA

E-mail: aminur@uniten.edu.my

Abstract. This numerical study has been performed to predict the flow patterns and characteristics within a storage area Hexagonal Modular Pavement Systems. Throughout the design and planning period for future construction are increasingly integrating computational fluid dynamics (CFD) into the process. A commercially known software, FLOW-3D, is applied to numerically solve the Navier-stokes equations for solution domains which are separated into three regions with overlapping boundaries to efficiently accommodate the grid resolutions, namely the honeycomb shaped modular, gravel and combined honeycomb shaped modular with gravel fill. The filtration of the fluid within the interstices of a permeable pavement is evaluated by integrating the Reynolds Averaged Navier-Stokes equations (RANS) inside the voids rather than making use of the widespread porous media approach. In conclusion, the results from numerical simulation are generally well agreed with the existing data and flow information such as flow patterns at increased flow, discharge rate and pressure is obtained to be used for engineering design purpose. Overall, the potential for FLOW-3D to model various geometries and configurations appears great. It should be noted that CFD should not be considered a complete replacement for physical modelling; however, it can definitely be used as a supplementary tool throughout the pavement design process.

1. The Modeling Unsaturated Flow through Pavements

The unsaturated zone is located above the water table. Within this zone, the pore spaces are partially filled with water whereas the reminders of the voids are taken up by air. Therefore, the volumetric water content is lower than the soil porosity. Due to the fact that water in this zone is held in the soil pores under surface-tension forces, negative pressures or suction pressures are developed. In addition, in this zone both the volumetric water content and the hydraulic conductivity are function of this suction pressure. The soil volumetric water content is held between the soil grains under surface tension forces that are reflected in the radius of curvature of each meniscus. The hydraulic conductivity content increases with increasing the volumetric content. [1]

In most cases, water in pavements is introduced to the pavement system through the process of infiltration into unsaturated pavement layers. Infiltration is a process in which water moves across the atmosphere-soil interface for example water seeps from the pavement ground surface and enters the base, subbase, and subgrade soils. The time rate at which water infiltrates across the atmosphere-soil interface is known as infiltration rate. The total volume of liquid crossing the interface over a given period of time is known as cumulative infiltration [4], [10], [11].

Soil characteristics play an important role in the infiltration rate. Total infiltration of any layer depends upon its porosity, thickness and quantity of water or other liquid present. Soil texture, structure, organic matter, root activity, and other physical properties determine the magnitude of the porosity of a given soil. [5] [6] [15]

2. FLOW-3D background

FLOW-3D has the ability to ignore the air surrounding the flowing water by using the volume of fluid (VOF) method developed [4]. This method also allows the numerical model to create a sharp interface between the water and air without using the fine meshes required by other computational fluid dynamics (CFD) software. Another method, [5] known as fractional area volume obstacle representation (FAVOR) method, is also a trademark of FLOW-3D. It allows the program to use fully structured grids that are very easy to generate throughout the entire flow domain. FLOW-3D utilizes a finite difference solution scheme and also has the ability to calculate solutions using various implicit and explicit solver options. The ability of using multiple and nested meshes as well as the re-run capability available in FLOW-3D are other options that make the numerical model suitable for permeable pavement modelling.

Numerical solutions to the problems based on the full Reynolds Averaged Navier-Stokes equations (RANS) with the Volume of Fluid surface tracking method (RANS/VOF) have indeed been thoroughly tested, and have found their way into engineering practice. Interesting examples of how these issues have been addressed both physically and numerically are, for instance reported in [2], [3], [7], [9], [8], [10], [12], [13], [14].

3. Storage thickness

In order to get a better understanding of the effect of storage layer thickness, the 500 mm long section by 500 mm wide and 620 mm height model was investigate with three difference of thickness with 100 mm, 200mm, and 300 mm as in Figure 1. The analysis results of these models are shown in Figure 2. From the analysis results one can see that the thickness of storage layer, within the range of study, does have effect on drainage, Figure 2 shows volume of fluid shows that 300mm thickness is the highest and have the longest time to fill. Findings indicate differences about 35 sec from one another, 100 mm, 200 mm and 300 mm of storage layer of Hexagonal Modular Pavement System (HMPS). Storage thickness of 200 mm was later chose as part of HMPS for further testing. Actually, this HMPS can have up to one meter thickness of storage layer. It is depend on water table and soil layer of site.

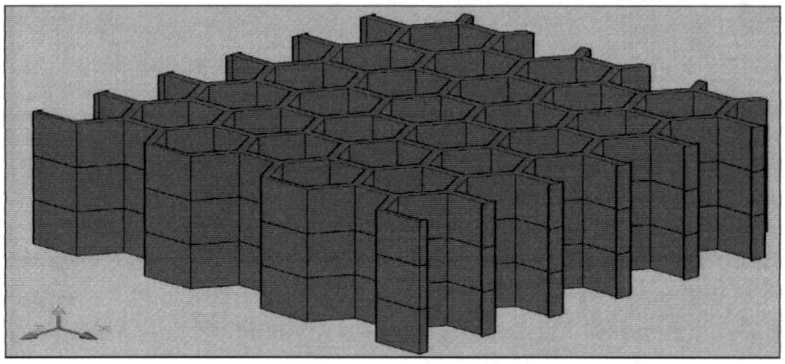

Figure 1. Sub-base pavements of HMPS in FLOW-3D simulation

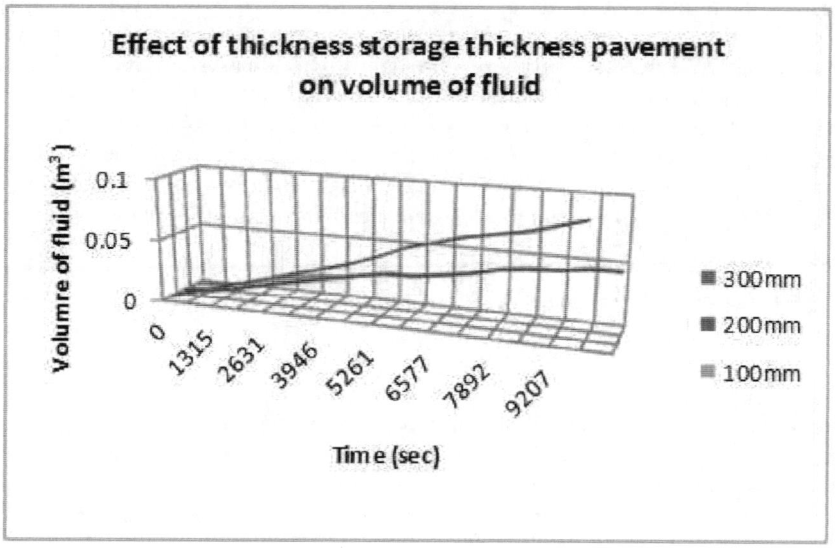

Figure 2. Effect of storage thickness pavement on volume of fluid

FLOW-3D with porous media simulation is user friendly and easy to understand interface, velocity fields are similar however, differences can be observed in the way that the FAVOR method interprets the geometry. Results show that pressures in Figure 3 in reasonably good agreement with one another, with marked differences between the solution of the coarsest mesh and the mesh independent solutions of the two other meshes.

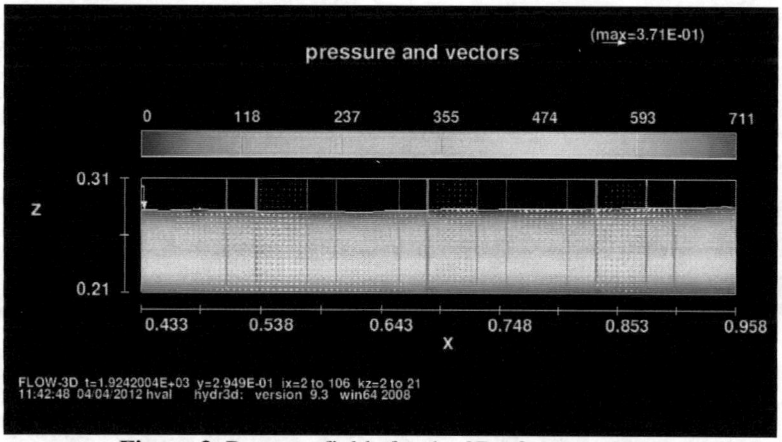

Figure 3. Pressure fields for the 2D of storage layer

4. Conclusion

From the test data and the estimated data, it can be demonstrated that there are many aspects have significant effect on pavement infiltration. To increase the drainage capability, pavements must be designed with materials that have high conductivity. As in second objective of this study, to predict the hydraulic performance of the designed pavement HMPS, a series of models were designed and analysed using the computational fluid dynamic method, FLOW-3D. The analysis results of these models indicate that the use of a permeable base is effective in improving the drainage ability of a pavement. The infiltration performance of HMPS is affected by material permeability. During the drainage process, the pavement works as an integrated hydraulic system and the infiltrated water usually flows through the layer that is easier to be saturated. HMPS base can increase the storage capacity and store a large volume of water to be reuse. To show the effects of rainfall events on the pavement drainage characteristics, a series of experiments were designed and analysed. The analysis results under various rainfall events indicate that rainfall quantity does influence the pavement thickness. The time to fill for a pavement usually depends on the rainfall duration.

References

[1] Chanel P G, and Doering J C 2007 An evaluation of computational fluid dynamics for spillway modelling *Proceedings of the 16th Australasian Fluid Mechanics Conference (AFMC)* 1201-1206

[2] Gracia N, Lara J L and Losada I J 2004 2-D numerical analysis of near-field flow at low-crested breakwaters *Coastal Engineering* **5(10)** 991-1020

[3] Greben J M, Cooper A K, Gledhill I and de Villiers R 2008 Numerical modelling of structures of dolosse and their interaction with waves

[4] Hirt C W, and Nichols B D 1981 Volume of Fluid (VOF) method for the dynamics of free boundaries *J. of Computational Physics* **39(1)** 201-205

[5] Hirt C W, and Sicilian J M 1985 A porosity technique for the definition of obstacles in rectangular cell meshes *Proceedings of the 4th International Conference on Ship Hydro-dynamics National Academy of Sciences* Washington, D.C

[6] Hsu T J, Sakakiyama T and Liu P L F 2002 A numerical model for wave motions and turbulence flows in front of a composite breakwater *Coastal Engineering* **46** 25-50

[7] Hsu T W, Chang J Y, Lan Y J, Lai J W and Ou S H 2008 A parabolic equation for wave propagation over porous structures *Coastal Engineering* **55** 1148-1158

[8] Huang C J, Chang .H H and Hwung H H 2003 Structure permeability effects on the interaction of a solitary wave and a submerged breakwater *Costal Engineering* **49** 1-24

[9] Hur D S and Mizutani N 2003 Numerical estimation of the wave forces acting on a three dimensional body on submerged breakwater *Coastal Engineering* **47** 329-345

[10] Karim M F, Tanimoto K and Hieu P D 2009 Modelling and simulation of wave transformation in porous structures using VOF based two-phase flow model *Applied Mathematical Modeling* **33** 343-360

[11] Lara J L, Garcia N and Losada I J 2006 RANS modelling applied to random wave interaction with submerged permeable structures *Coastal Engineering* **53** 395–417

[12] Lara J L, Losada I J and Guanche R 2008 Wave interaction with low-mound breakwaters using a RANS model *Ocean Engineering* **35** 1388– 1400

[13] Lin P and Karunarathna S A 2007 Numerical study of solitary wave interaction with porous breakwaters *J. of waterway, port, coastal and ocean engineering* 352-363

[14] Rashid M A, Abustan I, and Othman M O 2012 Build Green with Permeable Pavement Systems *Awam International Conference on Civil Engineering (AICCE'12)*

[15] Rashid M A, Abustan I and Othman M O 2012 Infiltration Characteristic Modeling using FLOW3D within a Modular Pavement *International Conference on Advances Sciences and Contemporary Engineering 2012, Procedia Engineering* **50** 658-667

IOP Conf. Series: Earth and Environmental Science **16** (2013) 012057 doi:10.1088/1755-1315/16/1/012057

Potential scour for marine current turbines based on experience of offshore wind turbine

L Chen [a] W H Lam [b 1] A H Shamsuddin [c]
[a] Dept. Civil Engn., Fac. Engn., Univ. Malaya, 50603 Kuala Lumpur, Malaysia.
[b] Dept. Civil Engn., Fac. Engn., Univ. Malaya, 50603 Kuala Lumpur, Malaysia (corresponding author).
[c] Centre for Renewable Energy, Universiti Tenaga Nasional, Jalan IKRAM-UNITEN, 43000 Kajang, Selangor, Malaysia.

E-mail: logchan11@hotmail.com; wlam@um.edu.my; abdhalim@uniten.edu.my

Abstract. The oceans have tremendous untapped natural resources. These sources are capable to make significant contribution to our future energy demands. Marine current energy offers sustainable and renewable alternative to conventional sources. Survival problems of Marine Current Turbines (MCTs) need to be addressed due to the harsh marine environment. The analogous researches in wind turbine have been conducted. Some of the results and knowledge are transferable to marine current energy industry. There still exist some gaps in the state of knowledge. Scour around marine structures have been well recognised as an engineering issue as scour is likely to cause structural instability. This paper aims to review different types of foundation of MCTs and potential scour and scour protection around these foundations based on the experience of offshore wind turbine farm.

1. Introduction

In 1987, the term "sustainable development" has been coined by the United Nations in the Brundtland Report. Sustainable development refers to the development that meets our today generations' needs without compromising those future generations. As the human population increased dramatically in recent years, the energy consumption has been raised up sharply as well. It is not possible to meet future energy demands without risking serious damage to the environment either through continuing to burn the fossil fuels or through placing increasing reliance on nuclear power [1].

In order to achieve sustainable development, exploration of renewable energy becomes highly vital. The oceans have tremendous untapped natural resources. These sources are capable to make significant contribution to our future energy demands. There are several types of ocean sources have been defined as potential sources to generate energy, which include tidal barrage, marine current energy, wave energy, ocean thermal energy and salinity gradient energy [2].

Tidal energy is more predictable compared to wind and wave energy. The tidal sources are easier to be quantified and predicted [3,4]. The marine environment is harsh for the installation and operation of Marine Current Turbines (MCTs). Extreme sea conditions have to be considered for the structural stability of MCTs under marine environment. Scour around marine structures have been well recognised as an engineering issue where scour is likely to cause structural instability. The information regarding the seabed scour around MCTs' foundation is limited in the existing literature. Relevant

[1] Wei-Haur Lam. Associate Professor, Dept. Civil Engn., Fac. Engn., Univ. Malaya, 50603 Kuala Lumpur, Malaysia (corresponding author). Email: wlam@um.edu.my.

research has been done in the field of offshore wind farm. There still exist some gaps in the state of knowledge. This paper reviews the seabed scour concerns around MCTs' foundation and potential scour protections.

2. Different types of foundations of marine current turbines

The support structure of MCTs is of significant importance in the overall marine current energy system. Prior assessment of the support structure of MCTs was carried out before approaching the seabed scour around foundation of MCTs. Based on the current status of MCTs, four basic support structures for MCTs have been described in the following [5]:

1) Gravity structure: This gravity structure is made up of large steel or concrete base column. It is able to resist overturning by its self-weight. The steel component of this gravity structure adds some advantages to itself, such as ease of production, transportation, and installation.
2) Monopile structure: This type of structure is made up of a large-diameter hollow-steel beam. The beam is penetrated approximately 20-30 m into seabed while the seabed conditions are soft. If the rock is harder, pre-drilling, positioning and grouting may be the methods to install this structure.
3) Floating structure: Floating structure provides the optimum solution for the placement of devices in deeper water conditions. This type of structure is made of mounting device and floating vessel which is moored to the seabed using chains, wire or synthetic rope.
4) Tripod/Piled Jacket structure: Each of the corners of the structure's base is anchored to the seabed by using steel piles. The steel piles are driven approximately 10-20 m into the seabed depending primarily on the seabed conditions. This type of structure is well understood since it has been widely used in the oil and gas industry. Compared to other structures, this type of structure has lighter structural loading. All the aforementioned support structures have shown in Fig.1[5].

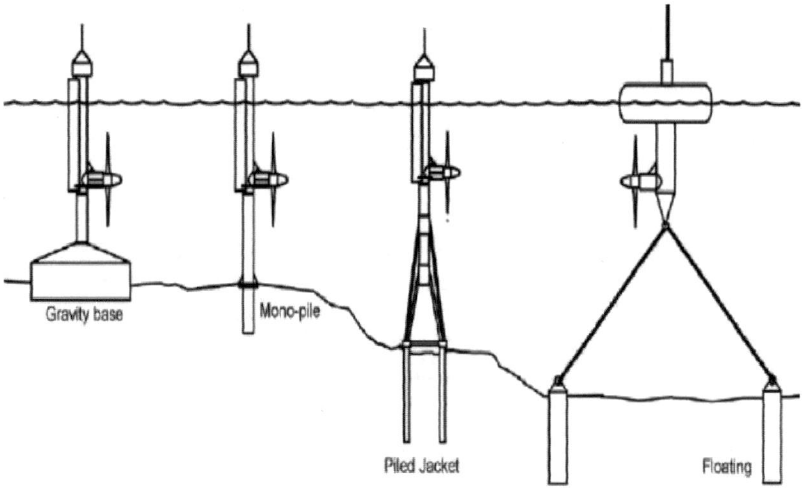

Figure 1. Different types of support structures with horizontal marine current turbines

Horizontal axis and vertical axis MCTs are the two groups of hydrofoil-shaped blades applied in marine current energy devices [6].Horizontal axis MCTs is the most common to-date devices, and they rotate at a horizontal axis which is parallel to the current stream [7]. On the other hand, vertical axis MCTs rotate about a vertical axis which is perpendicular to the current stream [8]. There are many novel prototypes has been invented in the industry to harness the kinetic energy from marine current. Rourke et al. [5] reviewed some of the prototypes in detail which including their dimensions, features and status of development.

3. Potential seabed scour around foundation of marine current turbines

In this section, the local scour around the MCTs foundation is discussed. Scour around marine structures have been well recognised as an engineering issue where scour is likely to cause structural instability. Seabed protection for the foundation of marine structure is required [9-11]. Seabed scour issues have to be taken into account for a sound design of MCTs and their support structures. However, the researches relevant to seabed scour of MCTs foundation are insufficient. The seabed development around MCTs is based on the experience of offshore wind turbine farm.

Whitehouse developed a conceptual model of scour sensitivity which includes full range of marine sediment types [12-13]. According to Whitehouse, as the susceptibility of sediments to erosion reduces, the scour is expected to decrease for both coarser and finer soils. Moreover, muds and clays have more uncertainties regarding the response of scour due to their formation history and degree of compaction. Besides, cyclic loading by waves may result in scour due to pore pressure built up in the soil. Hence, wave-induced liquefaction may occur and it could decrease the strength of soil. [12-13].

In extreme events, the hydraulic force increases whilst the scouring depth of some sediment increases. On the contrary, some sediment waves may result in a decrease of scour depth. In addition, the events either with high energy for erosion occurrence or in long duration increase the rate of scour. Nevertheless, the scour response of the seabed in time-limited event depends on the sensitivity of the seabed to increased shearing force on seabed, and the severity of the event [9]. Due to the intrinsic nature of sea state, sea environment is harsh and extreme events occur frequently, such as hurricane, typhoon, tsunami and storm. The extreme events could bring along extreme wave and strong wind so that the extreme event may causes substantial erosion and scouring around the foundation of MCTs. The scour around the foundation of MCTs may result in instability of the support structure of MCTs.

Whitehouse et al. [13] assessed scour problem at several offshore wind turbine farms and summarised the site characteristics of each farm with description of sediment type and other governing factors such as wave and current conditions. The largest value of scour depth observed was S/D=1.47 (S stands for scour depth, D stands for pile diameter). However, the peak current speeds at these sites are small which not more than 2 m/s. Charlier [14] stated that the marine current sites have a current velocity of 2.5 m/s or more are considered as high energy resources so that seabed around foundation of MCTs may has larger scour depth.

Besides, the type of support structure of MCTs may lead to different impacts on benthic habitats. Byrne and Houlsby [15] reviewed different foundation design for offshore wind farms. The study found that the gravity base foundations require substantial seabed preparation and are susceptible to scouring. The scour protection is installed below the level of the surrounding seabed [5, 15]. On the other hand, the monopile structure does not require seabed preparation [5]. It means monopile structure is less susceptible to scouring during its operational life. Therefore, an appropriate foundation design of MCTs integrated with seabed condition could enhance the structural stability of MCTs.

4. Potential seabed scour protection

Sour protection could be placed around the MCTs' foundation in order to stabilize the support structure of MCTs. Whitehouse [13] reviewed numerous approaches for mitigating scour around foundations. These methods are geotextile containers/sandbags, concrete amour units, concrete block mattress, grout bags/mattresses, gabions/gabion mattresses and flow inhibitors, where rock amour is the most common scour protection in offshore wind farm industry [14]. These scour protection technologies are transferable to MCTs. However, the appropriateness of these sour protections for MCTs has to be identified in further researches.

5. Conclusion

Seabed scour development around offshore wind turbine foundations has been studied extensively. Some of the results and knowledge are transferable to marine current energy industry. However, surrounding environment where MCTs operates is differs from wind marine. MCTs face more challenging problems compare to wind turbine. Some other survival problems MCTs may face during its operational life such as cavitation, high axial force and access of installation and maintenance. The developers of marine current energy have to overcome all the survival problems in order to harness kinetic energy from MCTs and to generate designed electricity.

6. Acknowledgements
The authors wish to extend their gratitude to the Ministry of Higher Education Malaysia for the financial support under the UM/MOHE High Impact Research Grant (H-1600-00-D000047).

References

[1] Fraenke P L 2001 Power from marine current *Proc Instn Mech Engrs Part A:Power and Energy* 216
[2] Robin Pelc and Rod M.Fujita 2002 Renewable energy from the ocean *Marine Policy* 26471–479
[3] Bahaj A S and Myers L E 2003 Fundamentals applicable to the utilisation of marine current turbines for energy production *Renewable Energy* **28** 2205–2211
[4] Watchorn M, Trapp T and Sayigh A A M 2000 Tidal stream renewable offshore power generation *In:World Renewable Energy Congress VI.Oxford:Pergamon* 2664-667
[5] Fergal O. R, Fergal B and Reynolds A 2010 Marine current energy devices:Current status and possible future applications in Ireland *Renewable and Sustainable Energy Reviews* 1026-1036
[6] Bryden I G, Grinsted T and Melville G T 2004 Assessing the potential of a simple tidal channel to deliver uesful energy *Applied Ocean Research* **26(5)** 198-204
[7] Bahaj A S, Batten W M J and McCann G 2007 Experimental verifications of numerical predictions for the hydrodynamic performance of horizontal axis marine current turbines *Renewable Energy* **32(15)** 2479-90
[8] Camporeale S M and Magi V 2000 Stream tube model for analysis of vertical axis variable pitch turbine for marine current energy conversion *Energy coneversion and Management* **41(16)** 1811-27
[9] Hoffmans G J C M and Verheij H J 1997 Scour Manual (Balkema: Netherlands) 224
[10] Whitehouse R J S 1998 Scour at Marine Structures (Thomas Telford: London) 216
[11] Sumer B M and Fredsøe J 2002 The mechanics of scour in the marine environment *World Scientific Pres* Singapore 552
[12] Whitehouse R J S 2006 Scour at coastal structures (Invited lecture) *In:Proceedings Third International Conference on Scour and Erosion* (The Netherland)
[13] Whitehouse R J S, Harris J M, Sutherland J and Rees J 2011 The nature of scour development and scour protection at offshore windfarm foundations *Marine Pollution Bulletin* **62** 73-88
[14] RH C 2003 A "sleeper" awakes:tidal curret power *Renewable and Sustainable Energy Reviews* **6(7)** 515-29
[15] Byrne B W and Houlsby G T 2003 Foundations for offshore wind turbines *Phil.Trans.Roy.Soc.* **361** 2909-2930

IOP Conf. Series: Earth and Environmental Science **16** (2013) 012058 doi:10.1088/1755-1315/16/1/012058

Carbon credit of renewable energy projects in Malaysia

X Lim [a] W H Lam [b 1] and A H Shamsuddin [c]

[a] Research Assistant, Dept. Civil Engn., Fac. Engn., Univ. Malaya, 50603 Kuala Lumpur, Malaysia.
[b] Associate Professor, Dept. Civil Engn., Fac. Engn., Univ. Malaya, 50603 Kuala Lumpur, Malaysia (corresponding author).
[c] Centre for Renewable Energy, Universiti Tenaga Nasional, Jalan IKRAM-UNITEN, 43000 Kajang, Selangor, Malaysia.

E-mail: xinle88@hotmail.com; wlam@um.edu.my; abdhalim@uniten.edu.my

Abstract. The introduction of Clean Development Mechanism (CDM) to Malaysia improves the environment of the country. Besides achieving sustainable development, the carbon credit earned through CDM enhances the financial state of the nation. Both CDM and renewable energy contribute to the society by striving to reduce carbon emission. Most of the CDM projects are related to renewable energy, which recorded 69% out of total CDM projects. This paper presents the energy overview and status of renewable energies in the country. Then, the renewable energy will be related to the CDM.

1. Introduction

The United Nations Framework Convention on Climate Change (UNFCCC) defines climate change as "A change of climate which is attributed directly or indirectly to human activity that alters the composition of the global atmosphere and which is in addition to natural climate variability observed over comparable time periods". Apparently, the UNFCCC believes that climate change is boosted by human activities such as agriculture, industrialization, transportation, deforestation and open burning. Many countries join the UNFCCC in order to reduce average global temperature together. Rises in global temperature lead to climate change.

The effects of greenhouse gases (GHGs) are more outstanding than expected. As a result, the Kyoto Protocol has been adopted as an international agreement under the UNFCCC since 1997 and entered into force on 2005 to reduce the GHGs emission. The Kyoto Protocol has three mechanisms – Clean Development Mechanism (CDM), Joint Implementation (JI) and Emission Trading (ET). Among these three, the CDM is the most acclaimed mechanism in developing countries as it brings profits to them. Furthermore, it involves technology transfer which enhances the environment through sophisticated technology. The CDM involves both the developed and developing countries. It permits the developed country to achieve their commitments by materialising emission reductions in developing country [1]. It is essential in promoting sustainable development via technology transfer and investment. Besides, GHGs those results in climate change can be reduced through this mechanism [2].

---

[1] Wei-Haur Lam. Associate Professor, Dept. Civil Engn., Fac. Engn., Univ. Malaya, 50603 Kuala Lumpur, Malaysia (corresponding author). Email: wlam@um.edu.my.

The Kyoto mechanisms play an important role in realising technology transfer from developed country to developing country [3]. Besides, technology transfer brings economic benefits to the projects that are already economically sound. As a developing country, Malaysia joined the CDM voluntarily as one of the non-Annex I countries. After becoming a member in the UNFCCC on 9 June 1993 and following by the ratification on 13 July 1994, Malaysia became a member in the Protocol officially on 4 September 2002 [4]. Many projects in Malaysia have been successfully registered as CDM projects. This had motivated other corporate sectors in the country such as power manufacturing, waste management, forestry, oil and gas manufacturing, agriculture and transportation sectors to proactively participate in CDM projects applications.

Thereinafter, the Malaysia Government has been highly supportive and instrumental in the CDM participation. A comprehensive administrative committee was set up upon the requirement of the CDM. Besides, the Budget 2008 extends 10 years pioneer status to companies involved in energy conservation, with 3 years tax exemption for income derived from the carbon trading in the country. These are encouraging news to the project developers because the process of application can be done smoothly with the assistance of the group of committee.

2. Demand of Renewable Energy in Malaysia

An estimation done by the International Energy Agency (IEA) demonstrates that, the global energy consumption is foreseen to increase 53% by 2030. [5]. A report by the Ministry of Energy, Green Technology and Water (KeTTHA) in Malaysia shows that, the maximum demand for electricity in Peninsular Malaysia is 15,473 MW in May 2011, which is a 2.7% increment compared to previous year. This projection rate will lead the peak demand to reach 21GW in 2020 and 25 GW in 2030. The electricity consumption growth increased 6.5% in 2011 compared to 2010, which was driven by commercial and domestic sectors with 9.2% and 7.4% rise, respectively. These growth trends illustrate the potential of higher energy demand as the country is striving to grow towards a high-income economy. By 2020, around 10.8 GW of new generation capacity will be required by the country as 7.7 GW of existing capacity will be terminated. By that time, the total installed capacity will rise up 16% if compared to the total installed capacity in 2011 [6].

The electric power consumption in Malaysia is growing substantially from 1970 until now. Over the last 4 decades, the usage of electricity in the country has been largely increased due to the rapid economical growth attributed to industrialization and high density of development [7]. The consumption percentage of coal and coke has been raised from 5.6% to 14.0% out of total commercial energy supply over 7 years from year 2000. This circumstance is like turning the clock back as many proofs had shown that coal burning brings many negative effects to the environment. [8]

The major CO_2 emissions contributor is the energy industry, which accounts for 41 % of entire emissions. It is followed by the transport industry and manufacturing industry, which recorded 23% and 20% of total emissions, respectively [9]. In the Copenhagen forum 2009, Malaysia has pledged to reduce 40% of carbon emissions by 2020 compared to the 2005 levels, subject to assistance from developed countries. This might be a turning point to the local market as environment consideration is not popular among the entrepreneurs, as some of them do not understand the mechanism of global warming and the effects of carbon emissions.

3. Status of Renewable Energies in Malaysia

The Renewable Energy (RE) Act 2011 is an act to provide an establishment and implementation of a special tariff system to encourage renewable energies generation. The SEDA is a statutory body formed under the Sustainable Energy Development Authority Act 2011 to administer and manage the implementation of the Feed-in Tariff (FiT) mechanism which is mandated under the RE Act 2011. The FiT allows electricity generated from indigenous RE (sources from within the country and are not imported from neighbouring countries) to be traded to power utilities such as Tenaga National Berhad (TNB) at a fixed premium price in a certain period. The distribution licensee will pay some amount of incentive to the clean energy power generator depending on the kilowatt hour (kWh) of electricity generated and exported to the national grid. To date, the FiT mechanism accepts electricity generated from four types of RE such as biogas, biomass, small hydro and solar photovoltaic, based on different FiT rates [10].

Most of the renewable energy in Malaysia is generated from biomass as of end of October 2010. Three options of biomass are mass production of electricity by furnaces, fluidised boilers and pyrolysis, combined heat and power schemes for industry such as solution from bagasse (fibres from sugarcane) or paper and pulp as well as biodiesel. Biomass is easily available since the methane and other gases produced from dead organic matters can be found everywhere. However, the burning of biomass contributes to global warming and particulate pollution. It is also difficult in collection and expensive.

Mini hydro is the second largest RE electricity supplier in Malaysia. It performs well in energy efficiency, cost effectiveness and electricity quality assurance. However, it cannot always achieve the required load demand due to the seasonal water flow variations. Solar power is expected to be fourfold more than the world fossil fuel resources. Tropical country like Malaysia is suitable to adapt solar energy due to accessibility of large amount of sunshine and resource. Nevertheless, the cost of solar power is relatively high, which is three to four times of the cost of fossil fuel. Tidal energy is a type of renewable energy that is unfamiliar in Malaysia. However, it is a potential energy source to the country as Malaysia possess high coast per area ratio, which is 14. The implementation of tidal energy needs more supports from the government due to its promising in energy source.

4. Renewable Energy Projects in Clean Development Mechanism

Both renewable energy and CDM are labelled as solution of environmental problems and climate change. Hence, they are in a good match contributing to sustainable development [2]. Both of them lead to sustainable development and contribute positive impacts to the nations.

By August 2012, the CDM had received a total of 4460 CDM projects in the pipelines. Out of the total CDM projects, 69% is renewable energy projects. Most of the CDM projects are related to wind, hydro and biomass energy which recorded 2,523 projects (28%), 2,280 projects (26%) and 906 projects (10%), respectively. Only 1 CDM project is regarding tidal energy which is expected to gain 1,104,000 Certified Emission Reductions (CERs) in 2012. In other word, tidal energy is foreseen to obtain 315 kCERs per year.

Among the projects, there are 108 projects or 2.42% are in Malaysia. Malaysia belongs to the group of non-Annex I. Annex I includes the countries listed under Kyoto Protocol that commit to greenhouse gases (GHGs) emission reduction. They are formed by a group of developed countries such as Denmark, United Kingdom, Australia, United States and etc. As a developing country, Malaysia has no quantitative commitments under Kyoto Protocol. Nevertheless, the country can voluntarily participate in globally reducing emissions of GHGs Besides collaborating with industrialised country investors to develop new industries and technologies, the CDM pushes developing countries like Malaysia to create cleaner environment.

As a commitment to the Kyoto Protocol, the Annex I Parties are committed to reduce quantified emissions to a certain level. The protocol limits the emissions from Annex I countries by an average of 5.2% of the 1990 level over a 5-year commitment period from 2008 to 2012. The emission reductions in developing countries are tradable under CDM. The Annex I countries thus purchase the 'emission reductions' from the non-Annex I countries such as Malaysia. Then, Malaysia can earn extra income for projects that reduce emissions by trading the certified emission reductions (CERs) to Annex I countries, besides contributing to sustainable development.

The Malaysian Energy Centre (PTM) predicts that the yearly potential in Malaysia reaches 18 million CERs in 2010. This is equal to approximately 100 million tonnes CO_2 equivalent from 2006 to 2012. Besides, PTM assumes the price range to be US$ 3-10 per tonne CO_2 equivalent. This is equivalent to a capital flow of US$ 0.3 to 1 billion to Malaysia by earning the carbon credits [11].

Since the ratification of Malaysia in the CDM in 2002, some of the projects' developers in the country started to apply for CDM projects. The Table 1 illustrates the number of projects applied in

CDM from 2002 to 2008. From the 7 CDM projects in year 2004, a total of 552,758 tonnes CO_2 equivalent has been eliminated, which is an encouraging result. Until February 2008, there are 48 PIN and 61 PDD received by the Malaysia Energy Centre (PTM).

Table 1. Number of CDM projects application in Malaysia

Application year	2002	2003	2004	2005	2006	2007	2008
Number of projects	3	2	7	25	21	40	4
Ton of CO_2 equivalent/ year	121,000	407,660	552,758	2.5 million	2.6 million	9.2 million	843, 759

Source: Wan Nadia, 2008 [12]

5. Conclusion

Most of the CDM projects are renewable energy projects. It recorded 69% out of total CDM projects. This means that, renewable energy projects are more viable in getting CDM recognition. Malaysia should diversify the exploitation of renewable energy. The number of CDM projects is still low if compared to countries like China, Mexico, India and Korea.

6. Acknowledgement

The authors wish to extend their gratitude to the Ministry of Higher Education for the financial support under the UM/MHE High Impact Research Grant (H-1600-00-D000047).

References

[1] Nautiyal H and Varun 2012 Progress in renewable energy under clean development mechanism in India *Renewable and Sustainable Energy Reviews* **16** 2913-2919

[2] UNFCCC Background on the UNFCCC: The international response to climate change. (online). http://unfccc.int/essential_background/items/6031.php, access on 30 August 2012

[3] Wohlgemuth N. and Missfeldt F 2000 The Kyoto mechanism and prospects for renewable energy technologies *Solar Energy* **69(4)** 305-314

[4] NRE 2005 Conservation and Environment Management Division. Clean Development Mechanism Process in Malaysia *Putrajaya: Ministry of Natural Resources and Environment*

[5] Mansor S 2008 Keynote address *International energy security forum* Kuala Lumpur

[6] Peter Chin F K 2011 Keynote address: the sustainable energy forum. The Sustainable Energy Forum *Putrajaya: Ministry of Energy, Green Technology and Water*

[7] World Bank. World Development Indicators. (online). http://data.worldbank.org/data-catalog/world-development-indicators, access on 02 March 2012

[8] Ando M et al 1995 Indoor and outdoor air pollution in Tokyo and Beijing supercities *Atmos Environ* **30** 695-702

[9] Borneo Post. Electricity consumption and its impact on the environment is your business. (online). http://www.theborneopost.com/2012/03/31/electricity-consumption-and-its-impact-on-the-environment-is-your-business/, access on 31March 2012

[10] SEDA. FiT Dashboard. Retrieved from Sustainable Energy Development Authority Malaysia. (online). http://www.seda.gov.my/, access on 02 August 2012

[11] Pedersen A 2008 Exploring the clean development mechanism: Malaysian case study *Waste Management & Research* **26** 111-114

[12] Nadia K 2008 The CDM/ Sustainable Energy Market in Malaysia *Copenhagen. Denmark: Pusat Tenaga Malaysia*

Quantitative image analysis of WE43-T6 cracking behavior

A Ahmad[1] and Z Yahya
Department of Mechanical Engineering, Universiti Tenaga Nasional,
Jln. IKRAM-UNITEN, 43000 Kajang, Selangor, Malaysia

E-mail: aazmi@uniten.edu.my

Abstract. Environment-assisted cracking of WE43 cast magnesium (4.2 wt.% Yt, 2.3 wt.% Nd, 0.7% Zr, 0.8% HRE) in the T6 peak-aged condition was induced in ambient air in notched specimens. The mechanism of fracture was studied using electron backscatter diffraction, serial sectioning and in situ observations of crack propagation. The intermetallic (rare earthed-enriched divorced intermetallic retained at grain boundaries and predominantly at triple points) material was found to play a significant role in initiating cracks which leads to failure of this material. Quantitative measurements were required for this project. The populations of the intermetallic and clusters of intermetallic particles were analyzed using image analysis of metallographic images. This is part of the work to generate a theoretical model of the effect of notch geometry on the static fatigue strength of this material.

1. Introduction

Image analysis operations are normally employed to evaluate an image quantitatively and descriptively, so that more information would be gained from it [1]. The operation aims to improve the visual ppearance of images as well as preparing for the measurement of their dimensions. Image morphology involves the study of structure of features and objects in an image which requires that features to be well defined in all factors, such as its brightness, colour, texture etc. By having all these factors optimized, the maximum amount of information may be quantified from the image. Binary morphological processes work on binary images, composed of pixels which are either black or white. The grey-scale morphological processes progress from the binary techniques, in which pixels are defined by their grey-scale level. Erosion and dilation are two of the primary morphological operations which employ masks to remove small, unwanted features like noise spikes and ragged edges [1]. Erosion is often used to remove uniformly pixels that should not be there, like irrelevant pixels representing point noise or scratches which have size that is normally only a single pixel wide. It removes any pixel touching another pixel that is part of the background. This removes a layer of pixels from around the edge of all features and areas, and can cause the dimensions to reduce.

Dilation is used to add pixels to the object, resulting in expansion. Typically, it adds a background pixel that touches another pixel that is already part of a foreground area. The effect is to add a layer of pixels around the edge of all features and areas. The overall dimensions can increase and features may merge. It also fills in small holes within features. Opening and closing are considered as the secondary operations to the erosion and dilation morphological operations. An opening is the combination of an erosion operation followed by a dilation operation. This combination will open up gaps between just-touching features. It works by removing single-pixel object anomalies like small spurs and single-pixel noise spikes, resulting in object contours being smoothed. In contrast to erosion, this operation should retain the original shapes and sizes of objects in the image. Even though erosion shrinks the objects by one pixel, dilation will expand them back to their original shape size. The closing operation employs a dilation followed by an erosion operation. It can close breaks or gaps in objects by closing fills in single-pixel object anomalies, like small holes and gaps. This results in smoothed object contours. This operation should also maintain the shape and size of the object. In order to achieve the necessary effect, both opening and closing operations may be applied many times. In using the erosion and dilation operations, several parameters can be used to adjust them according to the requirement, particularly the neighbour pattern (neighbourhood size) and the number of iterations. These are described below. The neighbourhood size setting is employed in erosion and dilation binary filters, and refers to the minimum number of pixels which must be a different colour to cause a pixel to be changed. For example, in an erosion filter, this number refers to the minimum number of neighbouring pixels that must be white to cause a black pixel to turn white. In the dilation filter, it refers to the minimum number of neighbouring pixels which must be black to cause a white pixel to turn black. The smaller this number, the more aggressive the operation will be. The open/close iterations is the number of consecutive dilations that are performed for a close filter before the erosions are performed (erosions followed by dilations for an open filter). The larger this number, the more aggressive the filter will perform [2].

2. Experimental

Simple image analysis of the raw micrographs can only provide details of the individual particle sizes. The intermetallic tends to appear in clusters, which are readily apparent to the eye (Figure 1). Manual identification and measurement of the size of these clusters is very time consuming. It was therefore decided to use the image analysis techniques of closing (dilation followed by erosion) operations to artificially coalesce the individual particles to form a single feature, which was more readily measurable automatically by the image analysis software. This was done by finding the best convolution setting of intermetallic to represent the realistic crack coalescence mechanism (with the assumption that cracks initiated from the intermetallic cluster). Joining together of the individual intermetallics in a cluster (Figure 1) is chosen as a representation of the crack coalescence mechanism, and was done using the closing operation. This operation fills in single-pixel object anomalies such as small holes and gaps between intermetallic itself. The closing operation was applied multiple times to fill-up larger anomalies and gaps (Figure 1). Besides oxide and pores, the microstructure of WE43-T6 magnesium alloys mainly contains intermetallic clusters of the rare earth-enriched divorced eutectic retained at some grain boundaries and mostly at triple points (Figure 1) and this feature plays a significant role in initiating cracks which leads to failure of this alloy [3].

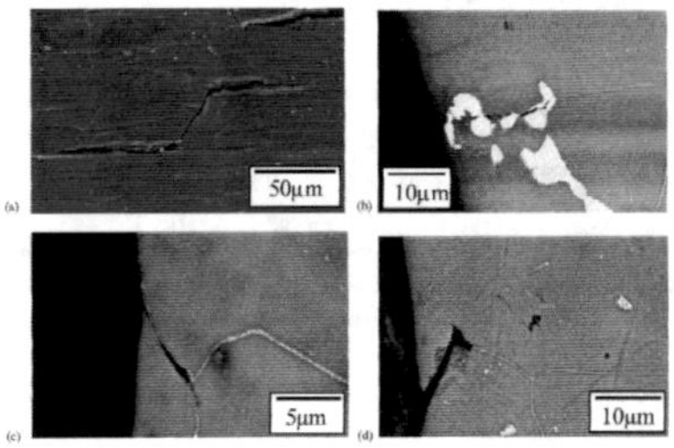

Figure 1. Typical arrested cracks in run-out static fatigue test specimen. (a) Crack coalescence, (b) initiation by fracture of intermetallic, (c) intergranular cracking, (d) transgranular crack arrested at grain boundary ((a) secondary electron, (b–d) backscatter electron images) [3]

3. Results and Discussion

It was observed that the mean gap between intermetallic particles within a cluster was approximately 4μm. Dilation and erosion of thresholded metallographic images were therefore used to coalescence most if not all intermetallic particles with a spacing of equal or less than 4μm (Figure 1**Error! Reference source not found.**). The closing operation was used to simulate the coalescence of the

IOP Conf. Series: Earth and Environmental Science **16** (2013) 012059 doi:10.1088/1755-1315/16/1/012059

features that were separated by less than this distance, and thence to estimate the cluster sizes that could develop from such a coalesence. The best convolution setting had to be determined in this analysis to show that dilation of 4μm would join most if not all gaps. Since the size of the largest clusters was of the order of 200 - 300μm, this level of dilation/erosion was not expected to cause significant error. The operation was therefore expected not to cause growth of the features and the largest features should therefore saturate at single value. At the magnification employed in the analysis of the data (Figure 1), the size of 1 pixel was 0.312μm. The mean gap of 4μm between intermetallic features in the optical microscopy images was therefore 13 pixels. The effects of the dilation and erosion settings (neighbourhood size and number of iterations) on typical intermetallic clusters are shown in Figure 2, in which larger dilations to coalesce more widely separated features were found to saturate.

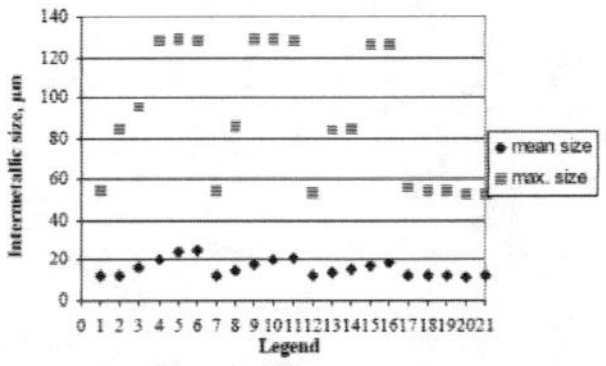

(a)

legend	convolution setting neighbourhood size, close iterations	mean um	max. um
1	no filter	12	55
2	1,1	12	85
3	1,3	16	95
4	1,5	21	129
5	1,7	24	130
6	1,8	25	129
7	2,1	12	54
8	2,3	14	86
9	2,5	18	130
10	2,7	20	129
11	2,8	21	128
12	3,1	12	53
13	3,3	13	85
14	3,5	15	85
15	3,7	17	126
16	3,8	18	126
17	4,1	13	56
18	4,3	12	54
19	4,5	12	54
20	4,7	12	53
21	4,8	12	53

(b)

Figure 2. a) Effect of typical image analysis convolution setting on the size of an intermetallic cluster and b) the legend for the convolution settings. 7 iterations are predicted to be sufficient to coalesce features separated by 4μm.

For example, 5 iterations perform 5 dilations and 5 erosions, each iteration typically increasing the coalescence distance by 0.6μm (*i.e.* 0.312μm x 2). The maximum defect saturates at 5 iterations for neighbourhood 1 and 2 and saturates at 7 for neighbourhood setting 3, which is less aggressive. The setting (2,7) was selected as giving optimum coalescence of features, *i.e.* to coalesce features that are separated by 4μm. Examples for the selected settings of (3,7), (2,5) and (2,7) are shown in Figure 3a, b and c. In Figure 3a, when the setting was (3,7), there were still some parts , separated by less than 4μm, were not coalesced. These are shown by the red circles. For the (2,5) setting (Figure 3b), all the intermetallics separated by less than 4μm were coalesced except for one (shown by the red circle). When the neighbourhood size and number of iterations were set at (2,7), Figure 4c shows that all the intermetallics in the cluster that were separated by equal or less than 4μm were coalesced.

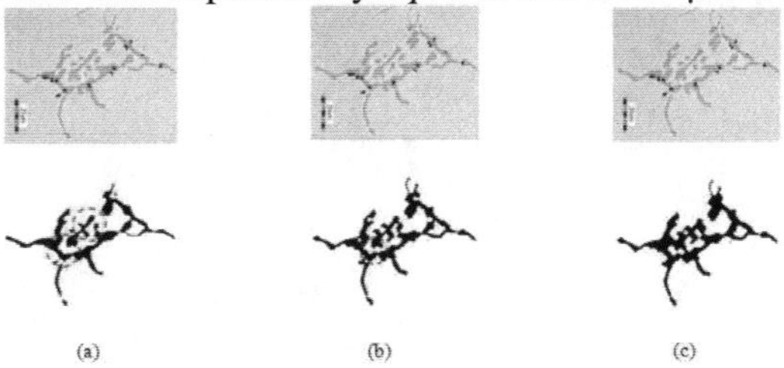

(a) (b) (c)

Figure 3. Image processing of intermetallic clusters of the optical image (left) and thresholded image (right) after applying dilation/erosion convolution filters (neighbourhood, iterations) to coalesce features separated by less than 4μm (a) (3,7) (b) (2, 5) and, (c) (2, 7) settings. Red circle indicates features that were still not coalesced.

Comparison with the mean distances between cracks (Figure 4) shows that they were separated by 4μm. Hence, the observation of saturation of maximum cluster size at the setting (2,7) analyzed above was used to simulate the coalescence of the co-linear crack-like features typically as shown in Figure 4a, based on the assumption that the microcracks were originated from the intermetallic particles within a cluster.

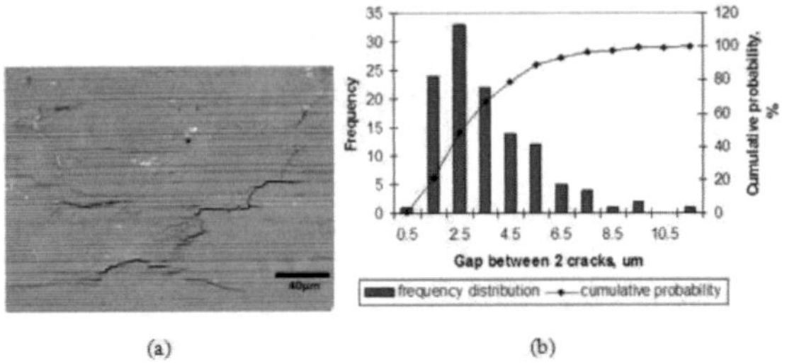

Figure 4. Comparison with mean distances between cracks, (a) typical surface cracks on notch root surface, (b) frequency distribution and cumulative probability of the separation distance between crack tips (data from 119 cracks, stress 279MPa).

4. Conclusion

- Image analysis techniques could be employed to artificially coalesce the individual particles to form a single feature, which is more readily measurable automatically by the image analysis software. This achieved by finding the best convolution setting of the feature to represent the realistic crack coalescence mechanism.
- This method is more practical compared to manual identification and measurement which is very time consuming.
- This analysis shows that the best convolution setting of the intermetallic to represent the realistic crack coalescence mechanism is (2, 7).

References

[1] Baxes G A 1994 *Digital image processing - principles and applications* (John Wiley & sons)
[2] Russ J C 2002 *The Image Processing Handbook 4th ed.* (Boca Raton)
[3] Marrow T J, Ahmad A, Khan I. N, Sim, S M A and Torkamani S 2004 Environment-assisted cracking of cast WE43-T6 magnesium, *Materials Science and Engineering A* 387-389, 419

IOP Conf. Series: Earth and Environmental Science **16** (2013) 012060 doi:10.1088/1755-1315/16/1/012060

Hydraulic Characteristics of a Stepped-slope Floating Breakwater

H M Teh[1] and H Ismail[2]

[1] Offshore Engineering Centre, Universiti Teknologi PETRONAS, Bandar Seri Iskandar, 31750 Tronoh, Perak, MALAYSIA
[2] Coastal & Offshore Engineering Institute, Universiti Teknologi Malaysia City Campus, Jalan Semarak, 54100 Kuala Lumpur, MALAYSIA

Email: [1]heemin.teh@petronas.com.my; [2]hadibah@citycampus.utm.my

Abstract. A stepped-slope floating breakwater is developed to provide wave protection to small ports and harbours. The width of the structure can be enhanced by increasing the number of breakwater units that are placed side-by-side to each other. This produces three types of test model, *i.e.* single-row, double-row and triple-row breakwaters. The test models have been tested in monochromatic waves in a wave flume to determine their hydraulic performance in various wave conditions. The incident and reflected wave profiles in the vicinity of the test models are recorded and analysed by using moving-probe method. The hydraulic performance of the test models are quantified by the coefficients of transmission, reflection and energy loss. The experimental results showed that the stepped-slope floating breakwater is an effective anti-reflection structure and a reasonably good wave attenuator.

1. Introduction

In recent years, the use of floating breakwaters for providing protection from wave disturbance has become prevalent in recreational harbours, marinas and fishing ports that do not require a high level of wave attenuation. For recreational harbours, coastal swimmers and surfers prefer to have acceptable wave condition to suit their sporting activities; and for marinas and fishing harbours, creation of complete still water conditions in the shelter regions may not be a necessity. Due to extensive application potentials in various sectors, floating breakwaters are still being one of the most studied structures in coastal engineering.

Floating breakwaters of various ingenious designs have been developed to cope with a broad range of applications. Breakwaters of different configurations are classified into four types: box, pontoon, mat, and tethered float [1]. Some other floating breakwaters with exclusive features are the Y-frame floating breakwater [2], floating plate breakwater [3], and floating pipe breakwater [4]. The majority of these floating breakwaters suppress the wave energy mainly by reflection, which may, in turn, result in standing waves in front of the structures. The confusing sea states may pose navigation hazard to the small floating vessels in the vicinity of the breakwaters.

Various efforts have been made by different researchers to identify the most optimum floating breakwater design that is capable of providing the desired hydraulic performance, *i.e.* adequate wave attenuation with minimal reflection effect [2,3,4,5].With respect to the geometrical effect of the breakwater, McCartney provided a comprehensive survey on each floating breakwater type [1]. Koftis and Prinos studied the hydraulic performance of box-type, circular-type and trapezoidal-type floating barriers using Reynolds Average Navier-Stokes Equation solver [5]. They concluded that the trapezoidal-type barrier is geometrically more advanced than the other forms of barrier in attenuating wave energy. This is because the trapezoidal barrier provides increased surface area for wave interaction and energy dissipation. This finding was in consensus with the numerical finding by Duclos et al. who simulated vorticity around the trapezoidal barrier with a concave front face [6]. The vortices generated in front of the trapezoidal barrier were found to be more developed than those generated at the rear face. The geometry of the barrier also generated multiple higher harmonic components in the reflected waves resulting in energy dispersion over a large range of angular frequency.

In this research, a trapezoidal barrier with a stepped-slope feature at both front and rear faces of the structure is developed. Figure 1 shows a single row of the stepped-slope floating breakwater model used in the experiment. The stepped slope at the front face of the breakwater is designed to facilitate wave breaking and to minimize the overtopping discharge. The test model has dimensions of 0.80 m length, 0.25 m bottom width and 0.13 m height. The density of the model is 784 kg/m^3 and it generates a draft of 0.08 m in static water. The size of the breakwater model can be enlarged by introducing additional test unit(s) to the primary one with side-by-side connection mode. This produces three types of test model for the stepped-slope floating breakwater, *i.e.* single-row, double-row and triple-row models. The total widths of the respective models, *B*, are 0.25 m, 0.50 m and 0.75 m. Further details of the model set-up are described in the subsequent section.

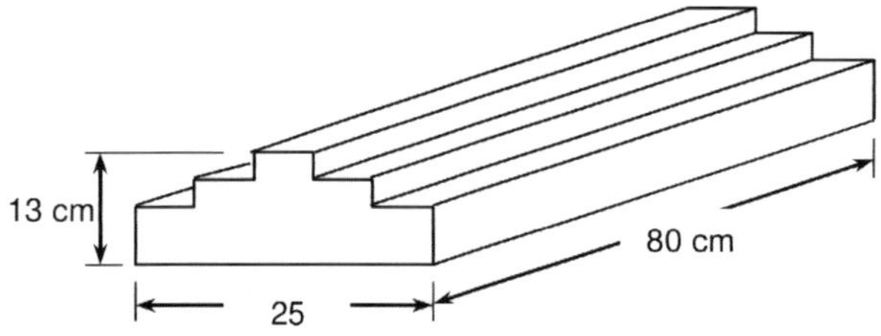

Figure 1. A single-row stepped-slope floating breakwater model.

2. Experimental Programs

The laboratory experiments were conducted at the hydraulic laboratory of Coastal and Offshore Engineering Institute, *Universiti Teknologi Malaysia*, Kuala Lumpur, Malaysia. The models were tested in an 18 m long, 0.95 m wide and 0.9 m high unidirectional wave flume equipped with a piston-type wave generator. At the other end of the flume was a wave absorber so as to reduce the reflected waves in the flume. The test models were made of a composite material (*i.e.* cement, sand and polystyrene) to provide adequate structural durability and buoyancy. Two capacitance-type wave probes were used for the measurement of wave profiles at the seaward and the leeward side of the test models. The seaward probe, which was located at the mid-section of a movable carriage that travelled along the steel rails at the top of the side walls of the flume, was used for measurement of the incident and reflected waves in the flume using the moving-probe method. The transmitted waves were measured by the leeward probe that was placed away from the test models by 3 times the tested water depth. These probes were plugged into a data acquisition system (DAS-800) for data recording. The wave probes adopted in the experiments were well calibrated prior to experiments, on a regular basis.

The three types of model, namely the single-row, double-row and triple-row stepped-slope floating breakwaters were tested in regular waves. Each test model was cross-moored to the bottom of the flume by four nylon ropes such that no initial pre-tension was present in the mooring lines. These breakwater models were subjected to 9 wave periods ranging from 0.9 s to 1.7 s in two water depths, *i.e.* 0.20 m and 0.33 m. For each test, the models were respectively exposed to waves of two different amplitudes. In total, 108 series of tests were conducted to study the hydraulic behavior of the stepped-slope floating breakwater models.

3. Results and Discussion

The hydraulic performance of the breakwater can be expressed in terms of the coefficients of transmission, reflection and energy loss. The transmission coefficient, C_T is the ratio of the transmitted wave height-to-the incident wave height, *e.g.* a lower C_T value indicates the breakwater is an effective wave attenuator. The reflection coefficient, C_R is represented by the ratio of the reflected wave height-to-the incident wave height, *e.g.* a lower C_R value implies the breakwater is an effective anti-reflection structure. Since the energy dissipated at the breakwater involves complicated processes and is difficult to measure experimentally, it is therefore mathematically estimated based on the principle of conservation of energy, giving the energy dissipation coefficient, $C_L = 1 - C_T^{\,2} - C_R^{\,2}$. The C_L value indicates the percentage of the energy dissipated at the breakwater by the incident waves. Hence, a good energy dissipater always yields a high C_L value.

3.1. Wave transmission

In this study, the energy coefficients (*i.e.* C_T, C_R and C_L) are plotted with respect to the relative breakwater width, *B/L*, where *B* and *L* are the breakwater width and wavelength, respectively, as shown in Figure 2. For Figure 2(a), with a relative breakwater draft $D/d = 0.24$ (where D = breakwater draft and d = water depth), it was observed that the C_T of the single-row, double-row and triple-row models decreased with the increase in *B/L*. This implies that the stepped-slope floating breakwaters exhibit higher wave attenuation performance when exposed to shorter period waves. This is sensible as the shorter waves tend to have more intense interactions with the floating structure. On the other hand, wave attenuation efficiency of the breakwater in longer period waves is not much affected by the number of models used. It is interesting to note that the double-row model performed more efficiently than the triple-row model for the tested wave conditions. For instance, the double-row model is capable of reducing the incident wave height by nearly 90% at $B/L \approx 0.4$ whereas the triple-row would require a $B/L \approx 0.7$ for the similar degree of wave attenuation. It is also seen from Figures 2(a) and 2(b) that the wave suppression ability of the breakwater models improves in deeper waters (*i.e.* as *D/d* increases). This is due to the fact that wave energy of the deeper waters, which is well distributed at the upper column of the water, was efficiently dissipated by the stepped-slope feature of the breakwaters. Overall, it can be deduced that the double-row stepped-slope floating breakwater is superior to the single-row and triple-row breakwaters, particularly in deeper waters.

3.2 Wave reflection

The reflectivity of the stepped-slope floating breakwaters is demonstrated in Figures 2(c) and 2(d). It is learnt from the figures that the C_R values of the test models are barely beyond 0.4 (equivalent to a reflection of 16% of the incident wave energy) which is relatively small compared to the amount of waves reflected by the conventional breakwaters. The C_R of the test models do not exhibit a strong correlation with *B/L*, indicating that the reflectivity of the breakwaters is less influenced by the wave period. The variation of C_R grows gradually as *D/d* increases from 0.24 to 0.40. This is attributed to the fact that the C_R values are governed by the effect of wave height more in deeper waters. In addition, it is also noticed that the C_R values are not subjected to the number of breakwater unit used, *i.e.* increasing the number of test models (*i.e.* double- and triple- row models) will not further amplify wave reflection in front of the breakwaters.

3.3 Energy dissipation

The mechanisms of energy dissipation observed in the experiments were (i) wave breaking at the seaward slope of the model, (ii) wave run-up on the seaward slope of the model, (iii) wave overtopping, (iv) wave run-down at the shoreward slope of the model, and (v) vortices formed at the bottom edges of the floating model. The energy loss posed by these hydraulic phenomena is estimated by the coefficient of energy dissipation, C_L. Figures 2(e) and 2(f) demonstrate the energy dissipation by the breakwater models tested in D/d = 0.24 and 0.40, respectively. It is evident that the test models are highly dissipative when exposed to shorter period waves, particularly in deeper waters, among which, the double-row model is the most efficient and viable energy dissipater, resulting in energy loss of almost 95% at *D/d* = 0.24 and almost 82% at *D/d* = 0.40, with both occurring at *B/L* ≈ 0.4. Increasing the number of breakwater units does not seem to boost the dissipative performance of the breakwater considerably. Therefore, it is suggested that the double-row stepped-slope floating breakwater be designed at *B/L* = 0.4 so as to achieve the optimal hydraulic performance.

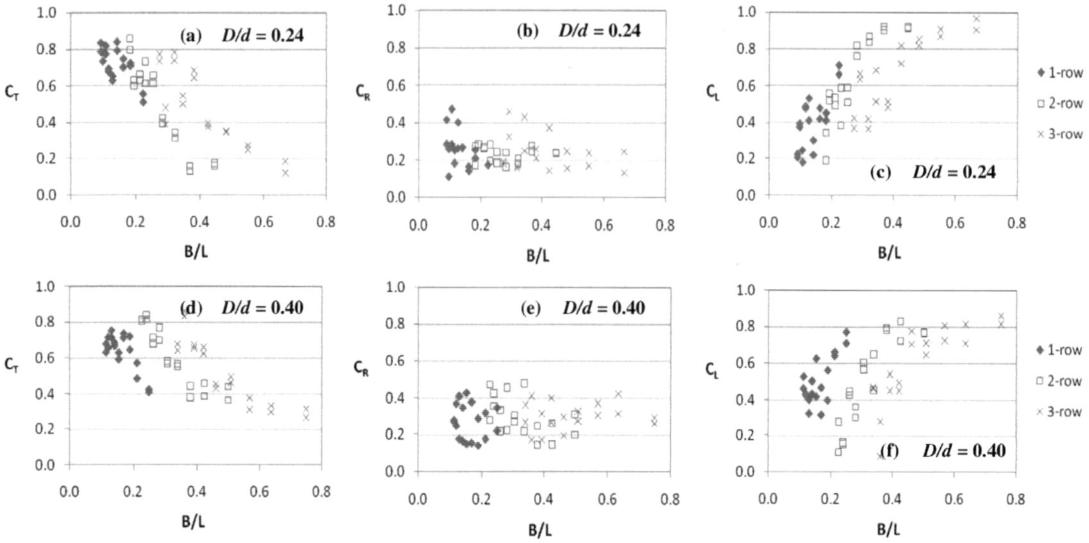

Figure 2. Energy coefficients with respect to the relative breakwater width, *B/L*.

4. Conclusion

Laboratory experiments were conducted to study the hydraulic characteristics of a stepped-slope floating breakwater system in various wave conditions. The experimental results revealed that the hydraulic performance of the breakwater models was strongly influenced by the effects of relative breakwater width. The breakwaters were effective anti-reflection structures with high dissipative ability, particularly when subjected to shorter period waves in deeper waters. Due to the highly energy dissipative ability, the double-row stepped-slope floating breakwater was claimed to be the most hydraulically viable structure compared to the single-row and the triple-row breakwaters. It achieved the optimum hydraulic performance at *B/L* = 0.4, whereby it attained wave attenuation and energy dissipation as high as 95%, and the maximum wave reflection anticipated at this range was about 40%.

References

[1] McCartney B L,1985. *J. Waterway, Port, Coastal and Ocean Engineering.* **111**, 307-17
[2] Mani J S 1991 *J. of Waterway, Port, Coastal and Ocean Engineering* **117** 105-19
[3] Kumar K S V. and Sundaravadivelu R 2001 *Proc. 1st Asia-Pasific Conf. on Offshore System* Kuala Lumpur 159-64
[4] Purusthotham S, Sundar V, and Sundaravadivelu R 2001 *Proc. 1st Asia-Pasific Conf. on Offshore System*, Kuala Lumpur 165-70
[5] Koftis T, and Prinos P 2005 *IASME Transactions* **7(2)** 1180–9
[6] Duclos G, Josset C, Clement A H, Gentaz L, and Colmard C 2004 *J. Waterway, Port, Coastal and Ocean Engineering* **130(3)** 127–33

IOP Conf. Series: Earth and Environmental Science **16** (2013) 012061 doi:10.1088/1755-1315/16/1/012061

Performance of double –pass solar collector with CPC and fins for heat transfer enhancement

Ebrahim M A Alfegi[1], Alhadi A S Abosbaia[1], Khaled M A Mezughi[2] and Kamaruzzaman Sopian[3]
1Department of Mechanical Engineering, Azzawia University, LIBYA
2Department of Chemical Engineering, Azzawia University, LIBYA
[3]Solar Energy Research Institute, Universiti Kebangsaan Malaysia, 43600 Bangi, Selangor, MALAYSIA

E-mail: Alfegi@yahoo.com

Abstract. The temperature of photovoltaic modules increases when it absorbs solar radiation, causing a decrease in efficiency. This undesirable effect can be partially avoided by applying a heat recovery unit with fluid circulation (air or water) with the photovoltaic module. Such unit is called photovoltaic / thermal collector (pv/t) or hybrid (pv/t). In this unit, photovoltaic cells were pasted directly on the flat plate absorber. An experimental study of a solar air heater with photovoltaic cell located at the absorber with fins and compound parabolic collector for heat transfer enhancement and increasing the number of reflection on the cells have been conducted. The performance of the photovoltaic, thermal, and combined pv/t collector over range of operating conditions and the results was discussed. Results at solar irradiance of 500 W/m2 show that the combined pv/t efficiency is increasing from 37.28 % to 81.41 % at mass flow rates various from 0.029 to 0.436 kg/s..

1. Introduction

Solar energy is one of the most important source renewable energy that world needs, the major applications of the solar energy can be classified into two categories; thermal system and photovoltaic system. Normally these systems are used separately, now these systems can be combined to become hybrid photovoltaic thermal system (pv/t). A number of researches and development programs have been carried out to improve the applications of solar energy systems. Several design of photovoltaic thermal solar air collector has been proposed in the past. Among the first, Kern and Russel [6] are the first who give main concept of photovoltaic thermal collector using water or air as the working fluid. Hendrie and Raghuraman [5] have been made a comparative experimental study in (pv/t) collectors with liquid and air as the heat removal fluid (working fluid). Cox and Raghuraman [3] suggested air type photovoltaic thermal system by analysis the effect of various design variables on the performance of the system. Lalovic et al. [7] fabricated photovoltaic thermal collector using amorphous Silicon pv cell and its performance was tested. Garg et al. [4] presented the theoretical study of (pv/t) collector with reflectors; they found that the system is well suited for solar drying applications. Bharagava et al. [2] and Prakash [8] reported the effect of air mass flow rate, air channel depth, and packing factor. Sopain [9] have successfully demonstrated the improved performance of steady state double pass collector over the single pass collector due to efficient cooling of pv cells. Bergene and Lovvik [1] found that the thermal efficiency may increase only by a factor of 0.1 if flow rate increase from 0.001 to 0.075 kg/s. Sopian et al. [10] developed and tested a double pass photovoltaic collector suitable for solar drying applications and they comparison between theoretical and experimental results.

IOP Conf. Series: Earth and Environmental Science **16** (2013) 012061 doi:10.1088/1755-1315/16/1/012061

Tripanagnotopoulos et al. [11] built and tested various photovoltaic thermal collector models with both water and air as the working fluids. Zondag et al. [12] compared the efficiency of seven different design types photovoltaic thermal collectors.

2. Experimental Setup

The design of the double pass photovoltaic thermal collector with (CPC) and fins which considered in this paper is shown in Figure 1.

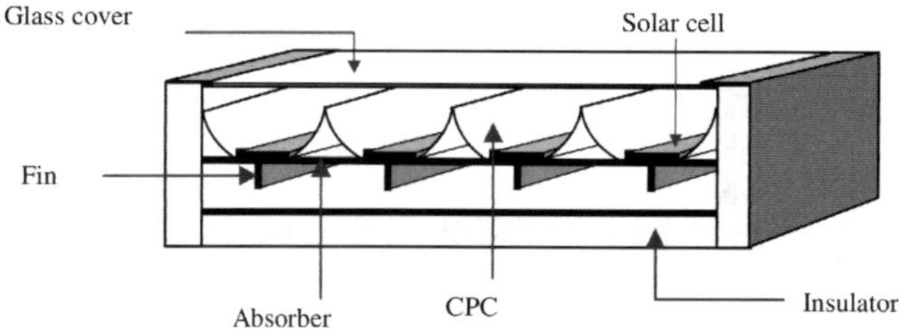

Figure 1. Schematic diagram for double pass (pv/t) collector with (CPC) and fins.

The system have three essential static components which are: a glazing on the top, back plate, absorber plate painted black and containing solar cells, CPC with concentration ratio of 1.86 is used as a reflector and located parallel to the air flow.29 rectangular fins with density of 0.384 fin / cm, each fin 2.5 cm high and thickness of 0.1 cm attached along the length of the back of the absorber plate. 23 tungsten halogen lamps each rated at 500 W used to simulate solar radiation during the test. The intensity of the incoming solar radiation was measured by Eppley pyranometer. Ambient temperature and other temperatures at several positions of the system are measured by K-type thermocouple. The air is supplying to the collector at uniform rate from electrical blower, the air mass flow rate calculated from the air velocity which measured by HTA 4200 anemometer of the van type probe head. All these things connected direct to the data acquisition system to feeds them to the computer. The collector dimension is (0.755 * 1.22) m. the high of the upper channel is 16.5 cm and the lower channel is 12.5 cm. the total area covered by solar cells is 0.38 m2. The sides and bottom of the collector are insulated with 5 cm of glass wool. In our configuration; inlet air enters through the upper channel formed by the glass cover and absorber plate and then through the lower channel formed by absorber plate with fins and the back plate. Air is circulated for one hour prior to the period in which data are taken. Data are sampled using the application program of the data acquisition system named DasyLab. Data include the solar radiation, temperatures, and mass flow rate.

3. Results and Observations

The performance parameters of combined photovoltaic thermal solar collector are obtained in terms of the solar cell efficiency and the thermal efficiency. The thermal efficiency of the double pass collector with CPC and fins is as follows:

$$\eta_{th} = \frac{m \cdot c_p \int (T_0 - T_i)\, dt}{A_c \, CR \int I_{tot}\, dt}$$

The solar cell efficiency is as follows: $\eta_{pv} = \eta_{ref} \, [1 - 0.0045 \, (T_{pav} - T_{ref})]$

η_{ref} is the reference efficiency of the solar cell at $T_{ref} = 25^{o}C$ which is in our study 10 %. The efficiency of the combined photovoltaic thermal collector is defined as the sum of thermal efficiency and electrical efficiency as:

IOP Conf. Series: Earth and Environmental Science **16** (2013) 012061 doi:10.1088/1755-1315/16/1/012061

$$\eta_{pv/t} = \eta_{\text{thermal}} + \eta_{\text{electrical}}$$

The electrical efficiency is as follows:

$$\eta_{elect} = \frac{\int \tau_g \, I_u \, \eta_{pv} \, \alpha_{pv} \, P \, CR \, d \, \rho_R^{n'} \, [1 + (\frac{\rho_{pv} \, \rho_g \, \rho_R^{2n'}}{CR}) \, dt}{CR \int I_{tot} \, dt}$$

$$I_u = \frac{I_{tot}}{CR}$$

The electrical maximum power of the collector increase with the radiation intensity increase as can be seen in Figure 2. The air temperature rise decrease with increasing of mass flow rate as shown in Figure 3. Figure 4 shows the effect of radiation intensity on the temperature at various flow rates. In this Figure it can see that the temperature rise linearly to the solar radiation, so the collector which operate with high solar radiation levels experience high temperature rise. It can see also that at the same solar radiation the collector which operates at low mass flow rate will experience high temperature rise too. The effect of the mass flow on the efficiencies (photovoltaic, thermal, and combined pv/t) of the collector was shown in Figure 6. The experimental results show that, when the collector is operating at high mass floe rate; the efficiencies (photovoltaic, thermal, and combined pv/t) will increase. This is expected when the photovoltaic panel is cooled by the incoming air. As seen in figure 5 the photovoltaic efficiency is 9.72% to 9.78% at solar radiation of 500W/m^2 and inlet temperature of 25°C. The thermal efficiency is 27.55% to 71.68%. The combined pv/t efficiency is 37.28% to 81.41%. We can see that the combined pv/t efficiency is decrease at low flow rate because the mean photovoltaic temperature is high. Therefore, cooling of the photovoltaic cells by increasing the mass flow rate will increase the combined photovoltaic thermal efficiency.

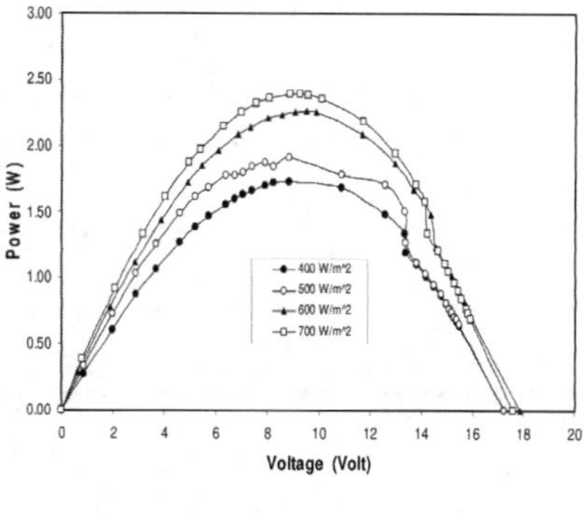

Figure 2. Power against voltage at mass flow rate of 0.029 kg/s and T_i = 25°C (Solar Irradiance (400 – 700 W/m^2).

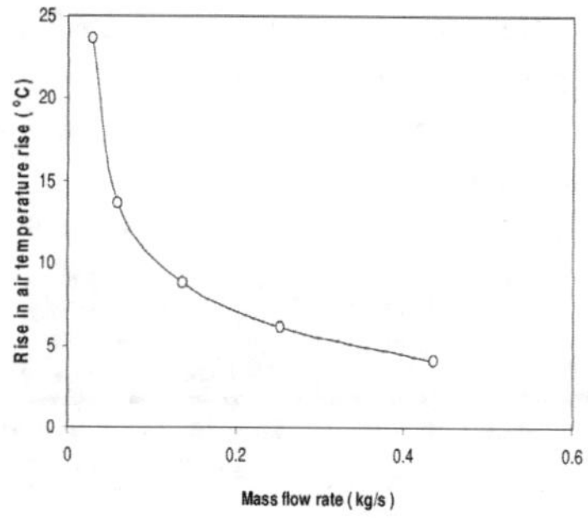

Figure 3. Variation of maximum air temperature rise with air specific mass flow rate at Solar Irradiance 500W/m^2 and Ti = 25°C

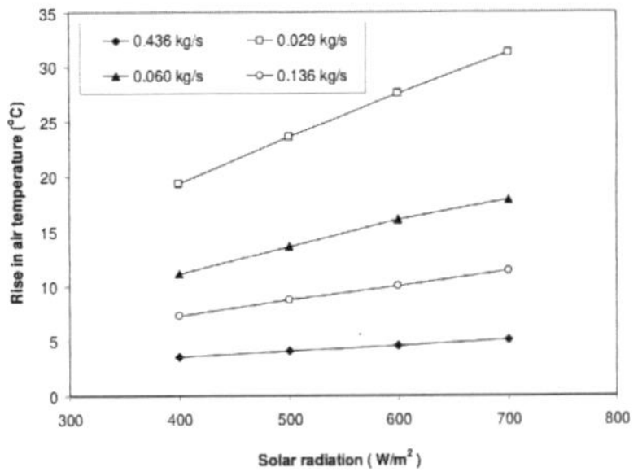

Figure 4. Variation of maximum air temperature rise with air specific mass flow rate at Solar Irradiance 500 W/m^2 and Ti = 25 °C.

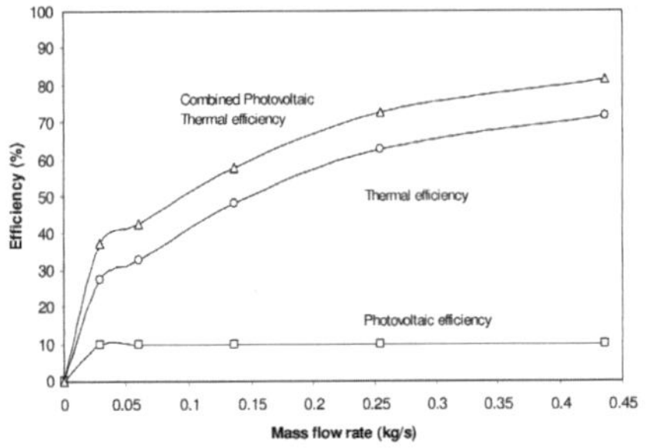

Figure 5. The effect of mass flow rate on efficiencies at 500 W/m^2 and Ti = 25°C.

4. Conclusions

Results show that the electricity production in a PV/T hybrid module decreases with increasing temperature of the air flow. This implies that the air temperature should be kept as low as possible. On the other hand, the system should deliver hot air for other purposes. A trade off between maximizing electricity production and producing hot air of useful temperatures is thus necessary. The simultaneous use of hybrid PV/T, CPC and fins have a potential to significantly increase in power production and reduce the cost of photovoltaic electricity.

References

[1] Bergene T, Lovvik O M 1995 Model calculation on flat plate solar heat collector with integrated solar cells *J. Solar Energy* **55** 453

[2] Bhargava A K, Garg H P and Agarwal R K 1991 Study of a hybrid solar system-solar air heater combined with solar cell *J. Solar Energy* **31(5)** 471

[3] Cox C H and Raghuraman P, 1985 Design considerations for flat-plate photovoltaic/thermal collectors *J. Solar Energy* **35** 227

[4] Garg H P, Agarwal R K and Bhargava A K 1991 The effect of plane booster reflectors on the performance of a solar air heater with solar cells suitable for a solar dryer *Energy Conv. Manage* **35(6)** 543

[5] Hendrie S D and Raghuaraman P 1980 Comparison of theory and experimental for photovoltaic thermal performance *Proc. of 14th IEEE Photovoltaic Specialist conference*, San Diego

[6] Kern Jr. EC and Russell M C 1978 Combined photovoltaic and thermal hybrid collector system *Proc. of 13th IEEE Photovoltaic Specialist* 1153–57

[7] Lalovic B 1986 A hybrid amorphous silicon photovoltaic and thermal solar collector *J. Solar Cells* **19** 131

[8] Prakash J 1994 Transient analysis of a photovoltaic-thermal solar collector for co-generation of electricity and hot air/water *Energy Conv Manage* **35(11)** 967

[9] Sopian K K S, Yigit H T, Liu S Kakac and Veziroglu T N 1995 Thermodynamics and the design Analysis, and improvements of energy systems *ASME Int. Mec. Eng. Congress and Exposition, San Francisco* **35** 89

[10] Sopian K, Yigit K S, Liu H T and Kakac S 2000 Performance of a double pass photovoltaic thermal solar collector suitable for solar drying system *Energy Conv. Manage* **41** 353

[11] Tripanagnostopoulos Y, Nousia Th, Souliotis M and Yianoulis P 2002 Hybrid photovoltaic / thermal solar system *Solar energy* **72(3)** 217

[12] Zondag H A, Vries D W, Van Hendel W G J, Van Zolingen R J C and Van Steenhoven A A 2003 The yield of different combined Pv – thermal collector designs *J. Solar energy* **74** 253

IOP Conf. Series: Earth and Environmental Science **16** (2013) 012062 doi:10.1088/1755-1315/16/1/012062

Impact resistance of sustainable construction material using light weight oil palm shells reinforced geogrid concrete slab

Z C Muda[1], G Malik[3], F Usman[1], S Beddu [1], M A Alam[1], K N Mustapha[1], A H Birima[1], O S Zarroq[2], L M Sidek[1] and M ARashid[1]

[1]Centre of Sustainable Technology and Environment , Universiti Tenaga Nasional, Jalan IKRAM-UNITEN, 43000 Kajang, Selangor, Malaysia
[2]Centre of Advance Computational Engineering , Universiti Tenaga Nasional, Jalan IKRAM-UNITEN, 43000 Kajang, Selangor, Malaysia
[3]Former Undergraduate Student of Universiti Tenaga Nasional , Malaysia

E- mail : mzakaria@uniten.edu.my

Abstract. This paper investigate the performance of lightweight oil palm shells (OPS) concrete slab with geogrid reinforcement of 300mm x 300mm size with 20mm, 30mm and 40 mm thick casted with different geogrid orientation and boundary conditions subjected to low impact projectile test. A self-fabricated drop-weight impact test rig with a steel ball weight of 1.2 kg drop at 1 m height has been used in this research work. The main variables for the study is to find the relationship of the impact resistance the slab thickness, boundary conditions and geogrid reinforcement orientation. Test results indicate that the used of the geogrid reinforcement increased the impact resistance under service (first) limit crack up to 5.9 times and at ultimate limit crack up to 20.1 times against the control sample (without geogrid). A good linear relationship has been established between first and ultimate crack resistance against the slab thickness. The orientation of the geogrid has minor significant to the crack resistance of the OPS concrete slab. OPS geogrid reinforced slab has a good crack resistance properties that can be utilized as a sustainable impact resistance construction materials.

1. Introduction

The search of sustainable green materials for the construction industry has reached its critical time to mitigate the negative impact of the climatic change. It is critical to study the impact strength characteristics and assess its performance for eco-green construction materials for various potential use in the building industry.

There is lack of research investigation been carried out on impact resistance of light weight geogrid reinforced concrete. Impact resistance represents the ability of concrete to withstand repeated blows and absorb energy without adverse effect to cracking and spalling. Impact scenario can also be classified into low velocity impact and high velocity impact. Gopalartnam and Shah [1] measured the performance of impact resistance to analyse different compositions of materials. After detailed observation and interpretation of results, they concluded that the projectile impact effect is based on number of factors such as configurations of the loading, geometry of specimen, compliance of the tests system, rate of loading and the prescribed criteria of failure. Mahmoud et. al [2] in 2010 mentioned that, out of all these testing options, the drop weight impact test which is recommended by the ACI Committee 544 [3] is the simplest.

The review paper on impact resistance on concrete target has been published by Zakaria Che Muda et al [4]. Impact resistance of oil palm shells lightweight concrete slab with bamboo fibers has been studied by Zakaria Che Muda et al. The results indicate that 2% volume fraction of bamboo fibers has an optimal performance for first crack resistance and ultimate crack resistance [5]. Zakaria Che Muda et al has also studied the effect of various amount geogrid content on the impact resistance of the OPS reinforced concrete slab of 20 mm thick [6].

The objective of the paper is to study the relationship of impact resistance of lightweight OPS geogrid reinforced concrete slab with different OPS content against the slab thickness, boundary condition and the geogrid orientation in its main direction and cross direction.

2. Materials

Oil palm shells (OPS) has unique physical properties, which make it a suitable replacement for conventional aggregate. OPS are naturally sized, hard and lighter than the conventional aggregates. The lightness in mass of oil palm shells make them a good substitute in the production of lightweight concrete. In the mix design, OPS to cement ratio by weight of 0.45, 0.50 and 0.60 are used.

Geogrid is made of high strength geosynthetic material which will be used as reinforcement in the lightweight oil palm shell concrete slab. The strength and flexibility of geogrid made it an ideal replacement of conventional steel reinforcement. The geogrid used in the experiment as reinforcement is Type A 160/50 having characteristic short term tensile strength of 160kN/m in the main direction (MD) and 50kN/m in the cross direction (CD). Each slab has one layer of reinforcement.

Ordinary Portland cement complying to ASTM Type I cement are used with a water cement ratio of 0.45 for all experimental design of concrete. The silica fume is supplied by EIKEM. It has the specific gravity of 2.2. The size of particle is 0.15 µm. It is grey in colour and slightly soluble in water. Silica fume used in the design mix is 5% of cement density. The amount of super plasticizer used will be equal to 2 % weight of binder (silica fume and cement). The specific gravity of the superplasticizer used in this experiment is 1.12.

3. Methodology

The potential energy due to the drop body is absorbed as strain energy, generating stresses that causes cracks in the target element. The width, depth,length of the crack developed and its failure mode is associated with the intensity of the energy, the amount of energy absorbed and the properties of concrete. It is assumed that the total computed energy imparted is fully absorbed by the specimens. The relationship of potential energy of a drop-weight projectile and the strain energy dissipated in cracks development is expressed as following formula as proposed by Kankam [7];

$$N*e = R_u * l_c * d_c * w_c \qquad (3.1)$$

Where, N = No. of Blows, e = Energy per blow (Joules), l_c = Total length of all cracks, d_c = Maximum crack depth, w_c = Maximum crack width, R_u = Ultimate crack resistance

Another dimensionless factor "impact crack resistance ratio□ was also defined:

$$C_r = R_u / f_{cu} \qquad (3.2)$$

Where, C_r = Impact crack resistance ratio, f_{cu} = cube compressive strength of the concrete slab.

A total of 72 sample slabs of size 300mm x 300mm with 20 mm,30mm and 40mm thickness were casted and tested with 0.45 OPS (f_{cu}=*28.4 Mpa*), 0.5 OPS (f_{cu}= *27.5 Mpa*) and 0.60 OPS (f_{cu}= *26Mpa*).

The study used a self-fabricated low velocity drop-weight impact test using a steel ball weighing 1.25 kg with drop height of 1000mm impacting the specimen of size 300mm x 300mm with thickness of 20 mm, 30 mm, and 40 mm mounted on the steel rack frame with different boundary conditions. The slab specimen will constitute the three different amount of OPS/C ratio of 0.45, 0.50 and 0.60. The slab specimen will also be subjected to different boundary conditions with geogrid placed with different orientation based on its main direction (MD) and cross direction (CD) facing the supports. At the first crack and ultimate (failure) crack, the total crack length, the crack width and the crack depth measured by filler gauge with its total numbers of blows recorded.

4. Results and Discussion

4.1 Relationship between crack resistance against slab thickness

There is a strong linear relationship between the crack resistance and slab thickness as shown in Figure 1 and Figure 2. The maximum first crack resistance for 0.45 OPS with 40 mm thick slab is 98.6 N/mm^2 and ultimate crack resistance is 633.9 N/mm^2. As the thickness increases, 0.45 OPS with a concrete cube strength of 28.4 Mpa has the highest crack resistance values as compare with 0.5 OPS (f_{cu}= *27.5 Mpa*) and 0.60 OPS (f_{cu}= *26Mpa*).

The first crack resistance for 0.45 OPS with 20mm thick slab using the geogrid, as compare to control (without geogrid) increase up to 3.1 times and its ultimate crack resistance increased by 15.2 times. The first crack resistance for 0.50 OPS with 20mm thick slab as compare to control increase up to 3.8 times and ultimate crack resistance is 16.1 times. The first crack resistance (0.60 OPS with 20mm thick slab) as compare to control increase up to 5.9 times and (0.50 (OPS0.6, 20mm thick slab)ultimate crack resistance is 20.1 times. The highest impact crack resistance ratio obtained is 22.3 for the 0.45 OPS with 40mm thick slab.

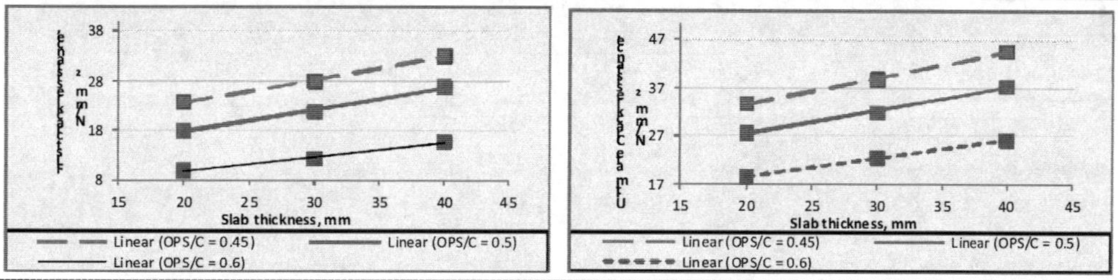

Figure 1. First and Ultimate Crack Resistance against Slab Thickness Control Sample (without geogrid) for Four (4) Support Boundary Condition

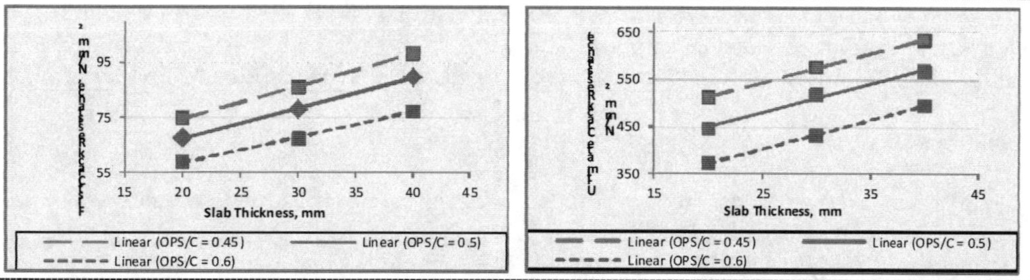

Figure 2. First and Ultimate Crack Resistance against Slab Thickness with Geogrid for Four (4) Support Boundary Condition

4.2 Relationship between crack resistance against boundary conditions and geogrid orientations.

Generally the first and ultimate crack resistance increases with increasing number of supports at the boundary as shown in Figure 3 and Figure 4. The first crack resistance increase from 24.8% to 29.8% f and from 26.8% to 40.6% for ultimate crack for four (4) supports as compare to two (2) supports boundary condition for various OPS/cement ratios.

Comparing the first and ultimate crack resistance of the slab due to the orientation of the geogrid main direction with its cross direction, the increase varies only from 0.8 % to 5.5 %. As such the orientation of the geogrid has minor significant to the crack resistance of the OPS concrete slab.

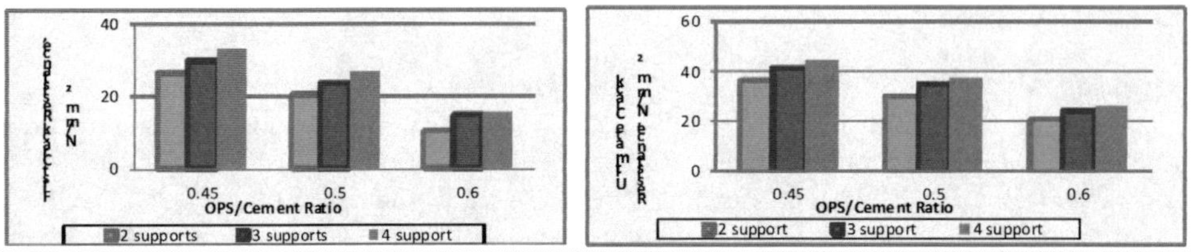

Figure 3. First and Ultimate Crack Resistance for Control Sample without Geogrid against Boundary Condition for 40 mm Thick Slab.

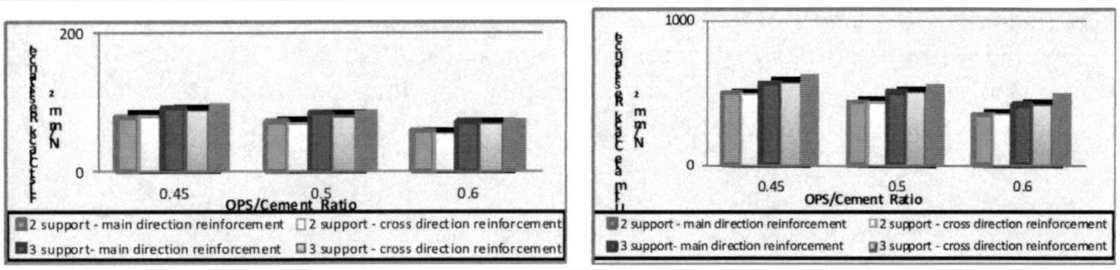

Figure 4. First and Ultimate Crack Resistance against Boundary Condition and Geogrid Orientation for 40 mm Thick Slab

5. Conclusions

The good linear relationship for the first and ultimate crack resistances against the slab thickness for various OPS/ cement ratio are obtained.

The first and ultimate crack resistance increases with increasing number of supports at the boundary up to 29.8% for first crack resistance and up to 40.6% for ultimate crack for four (4) supports as compare to two (2) supports boundary condition for various OPS/cement ratio.

The orientation of the geogrid in its main direction as compare to its cross direction is insignificant on the first and ultimate crack resistance of the OPS concrete slab.

OPS geogrid reinforced slab has a good impact resistance properties that can be utilized as a sustainable construction materials that have a maximum ultimate crack resistance of 633.9 N/mm^2.

References

[1] Gopalartnam V S and Shah S P 1986 Properties of fibre reinforced concrete subjected to impact loading *ACI J. 1986* **83(1)** 117–26

[2] Nili M and Afroughsabet V 2010 Combined effect of silica fume and steel fibers on the impact resistance and mechanical properties of concrete *International J. of Impact Engineering 37*(**8**) 879-886

[3] ACI Committee 544 1996 State-of-the-art report on fiber reinforced concrete *ACI Committee 544 report 544.1R-96* Detroit: American Concrete Institute; 1996

[4] Muda Z C, Ying K S, Salah F A Sharif, Sidek L M and Farhan N S 2013 A Review - Local Failure On Concrete Target Due To Projectile Impact *International J. of Science and Engineering Research (IJSER)* **4(1)**

[5] Muda Z C, Sharif S F, Sidek L M and Farhan N S 2013 Impact Resistance Of Oil Palm Shells Lightweight Concrete Slab With Bamboo Fibers *International J. of Science and Engineering Research (IJSER)* **4(1)**

[6] Muda Z C and Sharif S F *2012* Impact Behavior of Lightweight Oil Palm Shells Concrete Slab by Geogrid Reinforcement *International J. of Scientific and Engineering Research (IJSER)* **3(11)**

[7] Kankam CK 1999 Impact Resistance of palm kernel fibre-reinforced concrete pavement slab *J Ferrocement 1999* **29(4)** 279-86

IOP Conf. Series: Earth and Environmental Science **16** (2013) 012063 doi:10.1088/1755-1315/16/1/012063

Impact resistance performance of green construction material using light weight oil palm shells reinforced bamboo concrete slab

Z C Muda[1], F Usman[1], S Beddu [1], M A Alam[1], S Thiruchelvam[1], L M Sidek[1], H Basri[1] and S Saadi[2]
[1]Centre of Sustainable Technology and Environment, Universiti Tenaga Nasional, Jalan IKRAM-UNITEN, 43000 Kajang, Selangor, Malaysia
[2]Former Undergraduate Student of Universiti Tenaga Nasional

E-mail: zakaria@uniten.edu.my

Abstract. This paper investigate the performance of lightweight oil palm shells (OPS) concrete with varied bamboo reinforcement content for the concrete slab of 300mm x 300mm size reinforced with different thickness subjected to low impact projectile test. A self-fabricated drop-weight impact test rig with a steel ball weight of 1.2 kg drop at 1 m height has been used in this research work. The main variables for the study is to find the relationship of the impact resistance against the amount of bamboo reinforcement and slab thickness. A linear relationship has been established between first and ultimate crack resistance against bamboo diameters and slab thickness by the experiment. The linear relationship has also been established between the service (first) crack and ultimate crack resistance against the bamboo reinforcement diameter for a constant spacing for various slab thickness using 0.45 OPS and 0.6 OPS bamboo reinforced concrete. The increment in bamboo diameter has more effect on the first crack resistance than the ultimate crack resistance. The linear relationship has also been established between the service (first) crack and ultimate crack resistance against the various slab thickness. Increment in slab thickness of the slab has more effect on the crack resistance as compare to the increment in the diameter of the bamboo reinforcement.

1. Introduction

The search of sustainable green materials for the construction industry has reached its critical time to mitigate the negative impact of the climatic change. It is critical to study the impact strength characteristics and assess its performance for eco-green construction materials for various potential use in the building industry.

There is lack of research investigation been carried out on impact resistance of light weight bamboo reinforced concrete. Impact resistance represents the ability of concrete to withstand repeated blows and absorb energy without adverse effect to cracking and spalling. Impact scenario can also be classified into low velocity impact and high velocity impact. According to Dancygier [1], the nuclear structures are designed to resist impact loading caused by projectile or missiles travelling up to 1000 m/s. The response of reinforced concrete structures under impact loading is different from the static loading especially in the case of high velocity impact of rigid projectile. For impact velocities up to 10 ms^{-1}, the failure modes are generally the same as the static failure, except there is increased tendency for local damage or shear failure to occur. The review paper on impact resistance on concrete target has been published by Z Che Muda et al [3]. Impact resistance of oil palm shells lightweight concrete slab with bamboo fibers has been studied by Z Che Muda et al. The results indicate that 2% volume fraction of bamboo fibers has an optimal performance in first crack resistance and ultimate crack resistance regardless of its fiber length with a potential to be used as an impact resistance composite structures in the future [4]. Z Che Muda et al has also studied the impact resistance of the OPS reinforced with geogrid. Type 80/80 geogrid increased the impact resistance under service (first) crack up to 3.2 times and ultimate crack up to 8.3 times than the control specimen [5].

2. Materials

Oil palm shells (OPS) has unique physical properties, which make it a suitable replacement for conventional aggregate. OPS are naturally sized, hard and lighter than the conventional aggregates. The lightness in mass of oil palm shells make them a good substitute in the production of lightweight concrete. In the mix design, OPS to cement ratio by weight of 0.45, 0.50 and 0.60 are used.
Bamboo Reinforcement

Buloh kuning *(Bambusa vulgaris Schrad)* is used to prepare the bamboo reinforcement material. The bamboo are splice and cut into the required diameters of 2.5 mm , 5 mm and 7.5 mm diameter and tied to form a mesh at 50 mm spacing.

Ordinary Portland cement complying to ASTM Type I cement are used with a water cement ratio of 0.45 for all experimental design of concrete. Silica fume used in the design mix is 5% of cement weight to achieve an optimum concrete grade. Two percentage (2%) of super plasticizer is used in the design mix to achieved the desired workability.

3. Methodology

The potential energy due to the drop body is absorbed as strain energy, generating stresses that causes cracks in the target element. The width, depth,length of the crack developed and its failure mode is associated with the intensity of the energy, the amount of energy absorbed and the properties of concrete. It is assumed that the total computed energy imparted is fully absorbed by the specimens. The relationship of potential energy of a drop-weight projectile and the strain energy dissipated in cracks development is expressed as following formula as proposed by Kankam [6];

$$N^*e = R_u{}^* l_c{}^* d_c{}^* w_c \quad (3.1)$$

Where, N = No. of Blows, e = Energy per blow (Joules), l_c = Total length of all cracks, d_c = Maximum crack depth, w_c = Maximum crack width, R_u = Ultimate crack resistance

Another dimensionless factor "impact crack resistance ratio" was also defined:

$$C_r = R_u{}^* f_{cu} \quad (3.2)$$

Where, C_r = Impact crack resistance ratio, f_{cu} = Cube compressive strength

Residual impact strength ratio as defined in Equation 3.3 helps to evaluate the 'post – crack' behavior and use to measure the ductility of the bamboo reinforcement composite imparted in the OPS cement matrix.

$$Residual\ impact\ strength\ ratio\ (Irs) = \frac{Energy\ absorbed\ at\ ultimate\ failure}{Energy\ absorbed\ at\ initiation\ of\ first\ crack} \quad (3.3)$$

A total of 36 sample slabs of size 300mm x 300mm with 30 mm,40mm and 50mm thickness were casted and tested with 0.45 OPS and 0.60 OPS.

The study used a self-fabricated low velocity drop-weight impact test using a steel ball weighing 1.25 kg with drop height of 1000mm impacting the specimen mounted on the steel rack frame. The test sample is 2 way simply supported.

At the first crack and ultimate (failure) crack, the total crack length, the crack width and the crack depth measured by filler gauge with its total numbers of blows recorded.

IOP Conf. Series: Earth and Environmental Science **16** (2013) 012063 doi:10.1088/1755-1315/16/1/012063

4. Results and Discussion

4.1 Relationship between Crack Resistance and Bamboo diameter

There is a strong linear correlation between impact resistance against bamboo diameter as shown in Figure 1 and Figure 2. Generally the first and ultimate crack resistance increases with increasing diameter. For 0.45 OPS, crack resistance increase by up to 21% for first crack and 10% for ultimate crack as the bamboo diameter increases whilst the effect is only marginal for the 50mm thick slab. For 0.60 OPS, crack resistance increase by up to 32% for first crack and 19% for ultimate crack as the bamboo diameter increases.

The increment in bamboo diameter has more effect on the first crack resistance than the ultimate crack resistance.

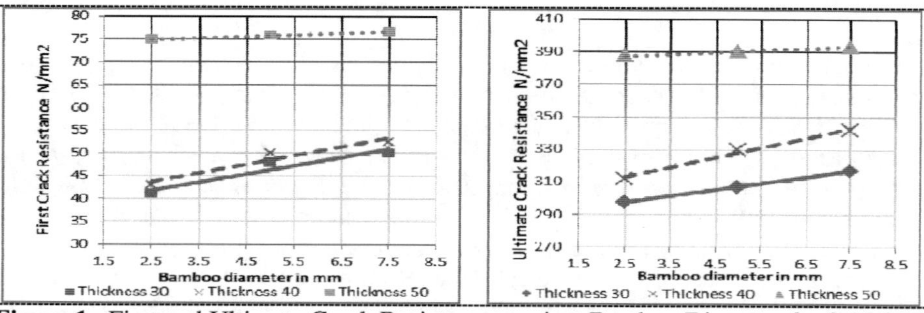

Figure 1. First and Ultimate Crack Resistance against Bamboo Diameter for 0.45 OPS

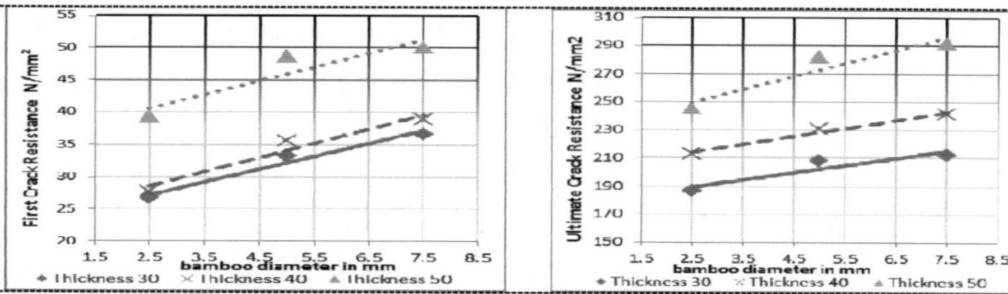

Figure 2. First Crack and Ultimate Crack Resistance against Bamboo Diameter for 0.60 OPS

4.2 Relationship between Crack Resistance and Slab Thickness

There is a linear relationship between the crack resistance and slab thickness. Increase in crack resistance is more substantial. For 0.45 OPS, the first crack resistance increase up to 85% and 31% for ultimate crack resistance as the slab thickness increases. For 0.60 OPS, the first crack resistance increase up to 47% and 37% for ultimate crack resistance. Thickness of the slab has more effect on the crack resistance as compare to the diameter of the bamboo reinforcement.

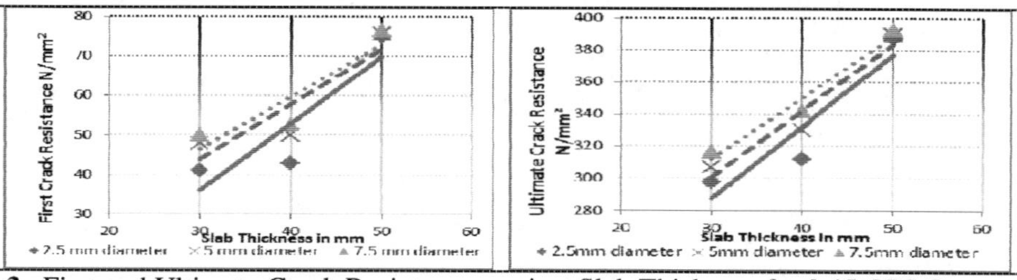

Figure 3. First and Ultimate Crack Resistance against Slab Thickness for 0.45 OPS Content

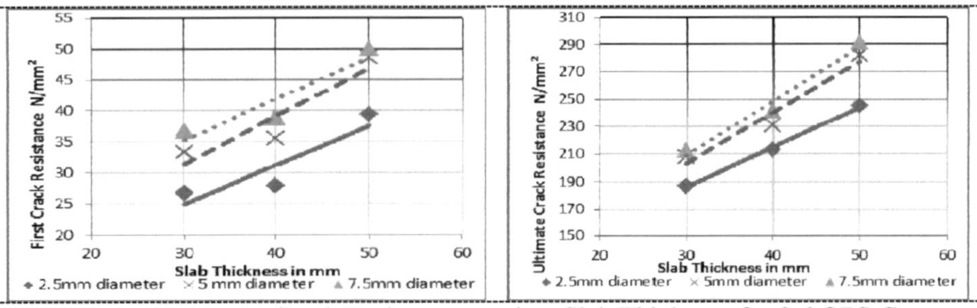

Figure 4. First and Ultimate Crack Resistance against Slab Thickness for 0.6 OPS Content

5. Conclusions

The correlation for the first and ultimate crack resistances are presented and discussed. OPS percentage effect impact resistance, bamboo diameter outcome on the impact resistance and also slab thickness effect on the impact resistance have different contribution on impact resistance.

The linear relationship has also been established between the service (first) crack and ultimate crack resistance against the bamboo reinforcement diameter for a constant spacing for various slab thickness using 0.45 OPS and 0.6 OPS bamboo reinforced concrete. The increment in bamboo diameter has more effect on the first crack resistance than the ultimate crack resistance.

The linear relationship has also been established between the service (first) crack and ultimate crack resistance against the various slab thickness using 0.45 OPS and 0.6 OPS bamboo reinforced concrete. For 0.45 OPS, the first crack resistance increase up to 85% and 31% for ultimate crack resistance as the slab thickness increases.

Increment in slab thickness of the slab has more effect on the crack resistance as compare to the increment in the diameter of the bamboo reinforcement.

References

[1] Dancygier A N 1997 Effect of reinforcement ratio on the resistance of reinforced concrete to hard projectile impact *Nuclear Engineering and Design* **172** 233-245

[2] Muda Z C, Ying K S, Sharif S F, Sidek L M and Farhan N S 2013 A Review - Local Failure On Concrete Target Due To Projectile Impact *International J. of Science and Engineering Research (IJSER)* **4(1)**

[3] Muda Z C, Sharif S F, Sidek L M and Farhan N S 2013 Impact Resistance Of Oil Palm Shells Lightweight Concrete Slab With Bamboo Fibers *International Journal of Science and Engineering Research (IJSER)* **4(1)**

[4] Muda Z C and Sharif S F 2012 Impact Behavior of Lightweight Oil Palm Shells Concrete Slab by Geogrid Reinforcement *International J. of Scientific and Engineering Research (IJSE*R) **3(11)**

[5] Kankam CK 1999 Impact Resistance of palm kernel fibre-reinforced concrete pavement slab *J Ferrocement* **29(4)** 279-86

An experimental investigation of pump as turbine for micro hydro application

N Raman, [1]I Hussein, K Palanisamy and B Foo
[1]Department of Mechanical Engineering, Universiti Tenaga Nasional (UNITEN), Jalan IKRAM-UNITEN, 43000 Kajang, Selangor, Malaysia

E-mail: nraman57@yahoo.com; ibrahim@uniten.edu.my

Abstract. This paper presents the results of an experimental investigation of a centrifugal pump working as turbine (PAT). An end suction centrifugal pump was tested in turbine mode at PAT experimental rig installed in the Mechanical Engineering Laboratory of Universiti Tenaga Nasional. The pump with specific speed of 15.36 (m, m^3/s) was used in the experiment and the performance characteristic of the PAT was determined. The experiment showed that a centrifugal pump can satisfactorily be operated as turbine without any mechanical problems. As compared to pump operation, the pump was found to operate at higher heads and discharge values in turbine mode. The best efficiency point (BEP) in turbine mode was found to be lower than BEP in pump mode. The results obtained were also compared to the work of some previous researchers.

1. Introduction

Combinations of highland areas with huge river networks promise Malaysia with a lot of hydro power potential. To date Malaysia had utilized this potential mainly in the range of large and mini hydropower but very few in the micro hydro range. Micro hydro refers to power generation in the range of 5kW to 100kW. Micro hydro can provide alternative renewable energy sources especially in areas with small rivers or streams flowing. One of the reasons for the micro hydro left undeveloped is because of economic constraints. Although the running cost for micro hydro scale are very low, the initial capital cost are high especially the turbines cost. Reduction in cost is possible by considering pump as turbine (PAT) system for micro hydro installation. The system utilize standardize equipment which are readily available in the market with capability of good performance and reliability in generating electricity. This paper presents the results of an experimental work on PAT. The aim of the study is to have a better understanding of the performance characteristic of PAT. An experimental rig was designed and constructed, and a pump with specific speed of 15.36 (m, m^3/s) was installed and tested. The pump was selected based on the results of reconnaissance study to identify micro hydro potential sites done by Nathan et.al. [1,2], using the available head and flow data obtained from the site survey of Sg. Kering in the Kenaboi region in Negeri Sembilan. The characteristic curve of the PAT was determined and the results obtained were compared with work done by previous researchers.

2. Experimental Investigation of PAT

Figure 1 shows the general layout of the PAT experimental rig installed in the Mechanical Engineering Laboratory at Universiti Tenaga Nasional (UNITEN), and Figure 2 shows the schematic diagram of the rig. A Calgon (HES 40-200) end suction centrifugal pump of specific speed 15.36 (m, m^3/s) was selected to operate in the turbine mode. The maximum head and flow of the pump are 22m and 8.3l/s respectively. The pump operating in turbine mode was tested at the speed, N_t of 1500rpm. Fifty (50) pieces of bulbs, 100W each were used as ballast load for the generated power. A 230V synchronous AC generator of 5KW was directly coupled with the pump operating in turbine mode to generate the electricity. PVC pipe with diameter of 100mm was used for the experimental rig and the water was re-circulated within the polythene tank having capacity of 150 gallon. A bypass pipe was connected from the main pipe to regulate excess flow and pressure. A feed pump having maximum head and flow of 100m and 20l/s respectively was used to supply the water at the desired head and flow rate.

Figure 1. PAT experimental rig

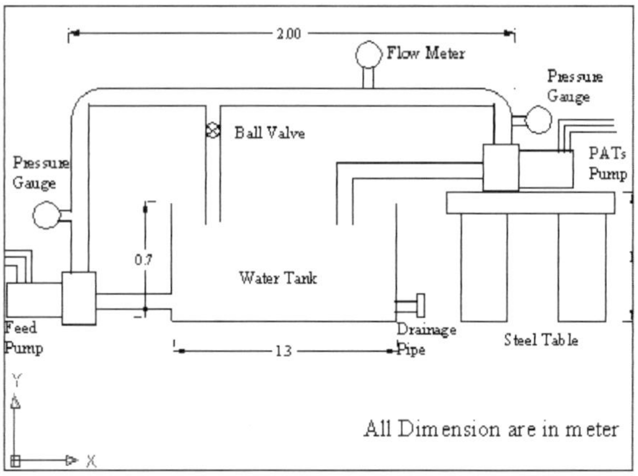

Figure 2. Schematic diagram of PAT rig

The control panel used in the experimental rig consists of PAT pump RPM reading, generator voltage meter, amps meter, feed water amps meter, on/off toggle switches for the bulbs and an inverter. Current readings of the feed pump were taken directly from the inverter. On/off toggle switches were used to increase the ballast load by switching on the 100W bulbs. The inverter was used to regulate the feed pump motor frequency, thus controlling the feed water flow and pressure to the PAT pump. The PAT pump shaft rotation speed was measured using a pulse counter which receives the signal from the rotary encoder installed at the PAT pump coupling.

Based on the measured parameters the theoretical power, actual power and efficiency were determined [3]. The head versus flow curve, efficiency versus flow curve, power output versus flow curve were plotted to analyze the performance of the pump operating in turbine mode. The results were presented in the form of non-dimensional parameters i.e. head coefficient (ψ), discharge coefficient (ϕ) and power coefficient (π). These non-dimensional parameters are expressed as follow:

$$\psi = \frac{gH}{n^2D^2}; \quad \phi = \frac{Q}{nD^3}; \quad \pi = \frac{P}{\rho n^3 D^5}$$

3. Results and Discussion

The experiment shows that the low specific speed end suction centrifugal pump can be used as turbine at various flow rates and head without any operational problem. The dimensionless performance curves are shown in Figures 3, 4 and 5. All data were obtained with PAT constant operation speed of 1500rpm.

Figure 3 shows that the flow increases with head in the turbine mode. Comparison with the pump characteristic curve supplied by CALGON manufacturer, the tested pump was found to operate at higher heads and flow rates in turbine mode than in pump mode. The higher values of flow rate in turbine operation indicate larger handling of water as compared to pump operation.

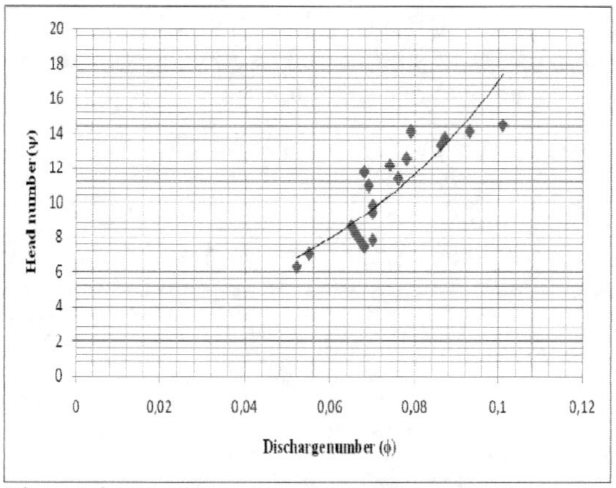

Figure 3. Head number versus discharge number

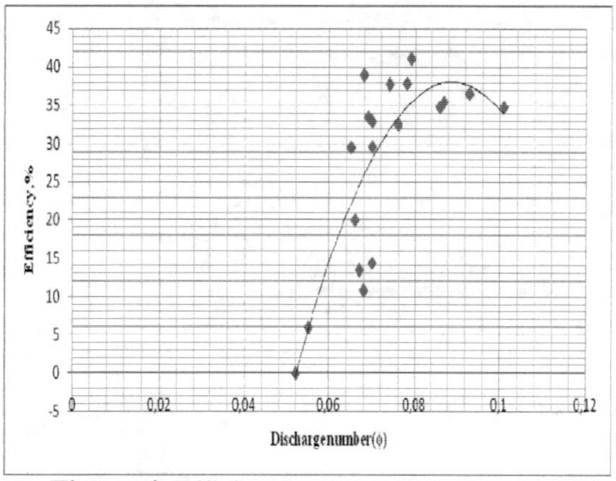

Figure 4. Efficiency versus discharge number

However, there is a limit to the pump operating in turbine mode to generate electricity at the best efficiency point (BEP) of PAT. This is critical in PAT system since the maximum power generation from the available site head and flow is important for a successful PAT system operation. Figure 3 agrees with the characteristics of PAT computed by other researchers [4,5,6,7]; where the head and discharge values at BEP in turbine mode were found to be higher than in pump mode, and the BEP in turbine mode is normally lower than BEP in pump mode. The BEP of the tested pump in turbine mode was only 39.0 % with head and flow of 30m and 13.52 l/s compared to 58% in pump mode. Lower efficiency of above tested PAT indicates higher losses in turbine operation as compared to pump operation. Figure 5 shows that the power generated is directly proportional with the flow rate, however, from Figure 4, the power generated shows lower efficiency with increasing flow rate.

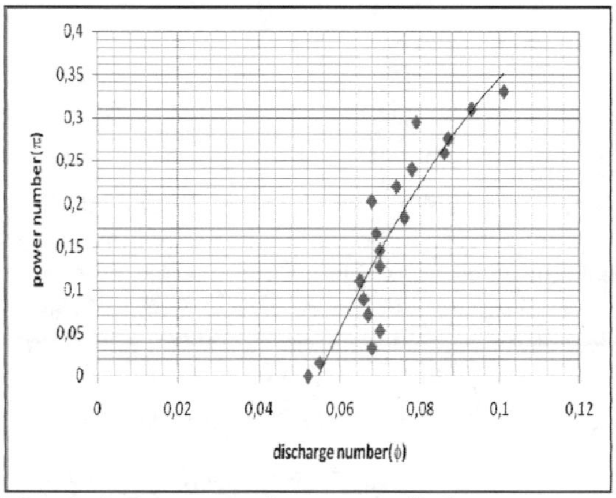

Figure 5. Power number versus discharge of number

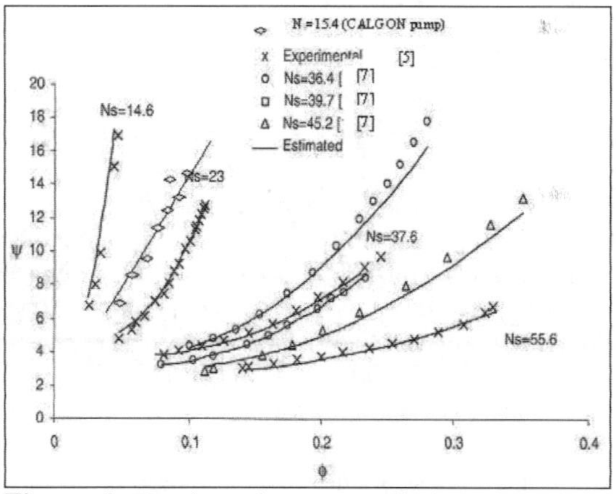

Figure 6. Head number versus discharge number: comparision with other PAT works

The results obtained were compared with experimental results obtained by Nautiyal et.al. [4], Derakhshan et. al. [5,6] and Singh [7]. Nautiyal's [4] experimental work was conducted for Kirloskar (KC 100-65- 315) centrifugal pump with specific speed of 18 (m, m^3/s), maximum head of 32.8m and maximum flow of 14.8l/s at the speed of 1500rpm. The performance characteristic of the Kirloskar pump [4] working in turbine mode was found to be similar to the CALGON centrifugal pump, except for the efficiency which noted higher efficiency compared to 39.0% obtained for CALGON pump. Generally, the efficiencies of pump in turbine mode operation are often lower by several percentages

[4,5,6,7]. Figures 6, 7 and 8 show the characteristic curves of the present work compared to the results obtained by other researchers [5,6,7]. The results obtained showed similar trend although the specific speed for the current work is lower and the efficiency obtained is also slightly lower.

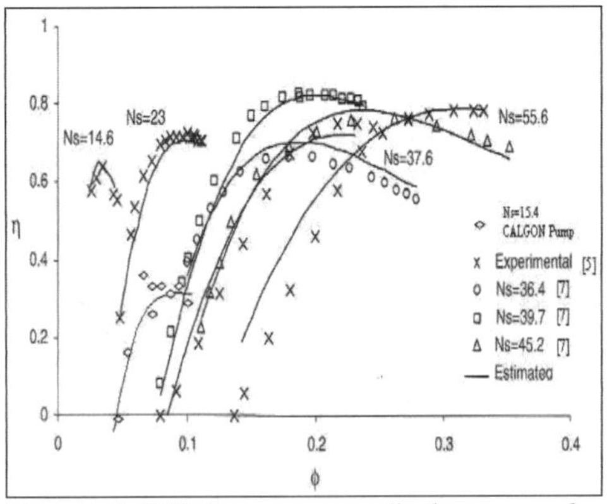

Figure 7. Efficiency versus discharge number: comparision with other PAT works

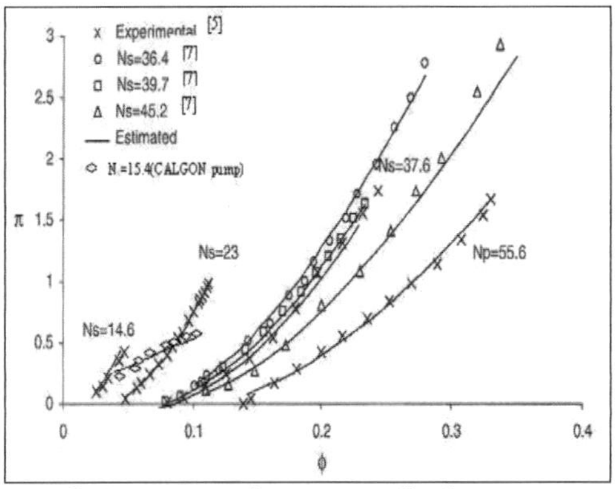

Figure 8. Power number versus discharge number :comparision with other PAT works

4. Conclusion

A CALGON end suction centrifugal pump of specific speed of 15.36 (m, m^3/s) was experimentally tested to study the pump operating in turbine mode characteristic. The experiment work showed that a centrifugal pump can satisfactorily be operated as turbine without any mechanical problems. As compared to pump operation, the pump operates at higher heads and discharge values in turbine mode. The BEP in turbine mode was found to be lower than BEP in pump mode. The BEP of the tested pump in turbine mode was only 39.0 % with head and flow of 30m and 13.52 l/s respectively.

References

[1] Raman N and Hussein I 2010 Reconnaissance Study to Identify Micro Hydro Sites in Malaysia *European J. of Scientific Research* **41(3)** 354-372

[2] Raman N and Hussein I 2009 Micro Hydro Potential at Sg. Sering, Kenaboi Region, Jelebu, Negeri Sembilan *International Seminar on Advances in Renewable Energy (ISARET2009* 27-28 April 2009 Universiti Tenaga Nasional

[3] Layman 1998 Layman's Guidebook on How to Develop a Small Hydro Site (Second Edition : European Small Hydropower Association (ESHA))

[4] Nautiyal H, Varun V, Kumar A and Vanday S Y S 2011 Experimental Investigation of Centrifugal Pump Working as Turbine for Small Hydropower Systems *J. of Energy Science and Technology* **1(1)**

[5] Derakhshan S and Nourbakhsh A 2008 Experimental Study of Characteristic Curves of Centrifugal Pumps Working as Turbines in Different Specific Speeds *J. of Experimental Thermal and Fluid Sciences* **32** 800-807

[6] Derakhshan S and Nourbakhsh A 2008 Theoretical, Numerical and Experimental Investigation of Centrifugal Pump in Reverse Operation *J. of Experimental Thermal and Fluid Sciences* **32** 1620-1627

[7] P. Singh 2005 Optimization of internal hydraulics and of system design for pumps as turbines with field implementation and evaluation P.h.D.M.E. Thesis, Karlsruhe University

IOP Conf. Series: Earth and Environmental Science **16** (2013) 012065 doi:10.1088/1755-1315/16/1/012065

Extraction of natural coagulant from peanut seeds for treatment of turbid water

A H Birima[1]*, H A Hammad[2], M N M Desa[1], Z C Muda[1]
[1]Centre for Storm water and Geohazard Management, University Tenaga Nasional, Malaysia.
[2]Former master student of Universiti Tenaga Nasional , Malaysia

E-mail: ahusseien@uniten.edu.my

Abstract. This study investigates the potential of peanut seeds as an environmental friendly and natural coagulant for the treatment of high turbid water. The peanut seeds have been used after oil extraction; and the active coagulation component was extracted by distilled water and salt solution of different salt concentrations. The salts used were NaCl, KNO_3, KCl , NH_4Cl and $NaNO_3$. Synthetic water with 200 NTU turbidity was used. Peanut extracted with NaCl (PC-NaCl) could effectively remove 92% of the 200 NTU turbidity using only 20 mg/l, while peanut seeds extracted with distilled water (PC-DW) could remove only 31.5% of the same turbidity with the same dosage. The coagulant dosage did not affected by the concentration of the salt solution, however, residual turbidity decreased with increasing the concentration of the salt; and the relationship was found to be a second order polynomial curve with R^2 of 0.9312. The other salts tested were also found to be good solvents to extract the active coagulation component with no much difference from NaCl solution in terms of efficiency.

1. Introduction

The production of potable water from most raw water sources usually entails the use of a flocculation/coagulation stage to remove turbidity in the form of suspended and colloidal material. Aluminum and iron salts are the chemicals most commonly used together with synthetic organic polymers. With aluminum salts, there is always the concern about residuals in the treated water, and Alzheimer's disease and, strong carcinogenic properties, the cost of any imported chemicals can be a serious problem for developing countries. Thus, in recent years, there has been considerable interest in the development of natural coagulants [1].

Natural coagulants have been used for domestic household for centuries in traditional water treatment in tropical rural areas. Some reports describe natural coagulants from Nirmali seed and maize [2], mesquite bean and Cactus latifaria [3], and Moringa Oleifera seeds [4].

The main advantages of using natural plant-based coagulants as water treatment material are apparent; they are cost-effective, unlikely to produce treated water with extreme pH and highly biodegradable. Naturally occurring coagulants are usually presumed safe for human health [5].The objectives of this study were to investigate the potential of peanut seeds in coagulation of the turbid water, as well as to improve the coagulation activity by extracting the active coagulant component by using salt solution.

2. Materials

2.1 Preparation of Peanut Seed Powder

The peanut seeds used in this study were obtained from Sudan. The seed cover was shelled by hand just before extraction. The extraction of the active ingredients was carried out by removing the shell to collect the kernel inside the shell. In order to ensure the efficiency of Peanut seeds extraction, the kernels have been crushed and grinded to medium fine powder by using the domestic blender (Assparo, Model 900) every time when the preparation of Peanut seeds extraction was needed.

2.3 Oil Extraction of Peanut Seeds

Oil extraction from Peanut seed was carried out as in Ali et al. [6] method.

2.4 Preparation of Peanut Seeds Extraction

Peanut seeds extraction was prepared by using a salt solution of NaCl with varied concentrations, and Peanut cake in 5% (w/v) suspension mixed with a domestic blender (Assparo, Model 900) for 10 minutes and left settling for 10 minutes. The suspension was then filtered using a vacuum pump filter with filter paper of 70 μm pore size (Whatman). The five salts used were, NaCl, KNO_3, KCl , NH_4Cl and $NaNO_3$. Different concentrations for each type were tested. This extraction was used to obtain the required dosage of coagulant.

2.5 Preparation of synthetic turbid water

Synthetic turbid water for coagulation tests was as in [7]. This kaolin suspension was used as the stock solution for the preparation of water samples in the coagulation tests. The suspension was diluted using tap water to prepare synthetic turbid water of about 200 NTU

2.6 Coagulation test

Jar tests were carried out by using a jar tester (Phipps and Bird, Model 300) to evaluate coagulation activity at several dose levels of Peanut seeds coagulant. Six 500-ml beakers were filled with 500 ml of synthetic water were placed in the slots of the jar tester. One beaker acts as a control while the other five beakers were fed with different dosages of coagulant. The synthetic waters were agitated at 80 rpm. During this agitation, various amounts of Peanut seeds coagulant were added to each beaker and agitated for 2 min at 80 rpm. The mixing speed was reduced to 40 rpm and was kept for 30 min for slow mixing. After sedimentation for 1 h; five ml of the sample was collected from about 1 cm below the surface of the water and residual turbidity of each coagulated water sample was measured using a turbidimeter (HACH 2100N).

3. Results and Discussion

3.1 Improvement of extraction of the active coagulant component by NaCl solution

Figure 1 shows turbidity removal by coagulation with peanut seeds extracted by distilled water (PC-DW); and that coagulated by peanut seeds extracted by NaCl (PC-NaCl) with different concentrations, namely, 1, 2 and 3 mole/l. The initial turbidity of synthetic water was in 208 – 214 NTU range. The coagulant dosage is expressed in mg/l throughout this study. Figure 1 shows that the residual turbidity obtained by PC-DW was 140 NTU at an optimum dosage of 20 mg/l. On the other hand, the residual turbidity obtained by PC-NaCl of 1, 2 and 3 mole/l, was 129, 57 and 31.5 NTU respectively, and the corresponding optimal dosages were 30, 20 and 20 mg/l. It can be significantly observed that NaCl solution gave lower residual turbidity than distilled water. The efficiency of the 3 mole PC-NaCl in terms of residual turbidity was 4.4 times better than that of PC-DW.

The improvement of turbidity removal implies improvement in coagulation activity. This could be losing-up of the protein associations leading to more soluble and coagulation active species in solution. This mechanism was reported by [7] , in that study, 1.0 mole of NaCl solution was significantly improved the coagulation activity of Moringa ofeifera seeds. This mechanism was also cited in [8] and [9] as increasing protein solubility with the addition of 1:1 salts or the salting-in effect. They found that addition of salt solution enhanced the breaking of protein associations, leading to increased protein solubility.

Based on the above mentioned mechanism and the results of the present study; it could be concluded that the peanut seeds have a high potential of coagulation of turbid water; and that the protein associations inside the peanut seeds are responsible for coagulation activity.

Figure 2 shows residual turbidity and the corresponding percentage of turbidity removal for the synthetic water. Results are represented by residual turbidity at the optimum dosage of each concentration. The optimum dosage for all concentrations was 20 mg/l except for 1 mol/l NaCl, which was 30 mg/l. The initial turbidities of all tests varied between 200 to 217 NTU. It can be observed that the residual turbidity decreases with increasing the concentration of NaCl, this relationship can be represented by second order polynomial curve with a coefficient of determination (R^2) of 0.9312. On the other hand, the percentage of turbidity removal increases by increasing the concentration of NaCl and the relationship also can be represented by second order polynomial curve with $R^2 = 0.9212$.

The residual turbidity achieved by 6 mol/l PC-NaCl was 8.2 times better than that achieved by PC-DW. This result (Increasing coagulation activity by increasing NaCl concentration) is in agreement with [7]. The decrease in coagulation activity at 7.0 mol/l could be due to saturation stage where by the NaCl solution reaches saturation stage at 7 mol/l. The phenomena that solubility of proteins decrease with salt concentration at high salt ionic strength is also well known as well as salting-in [8] [9].

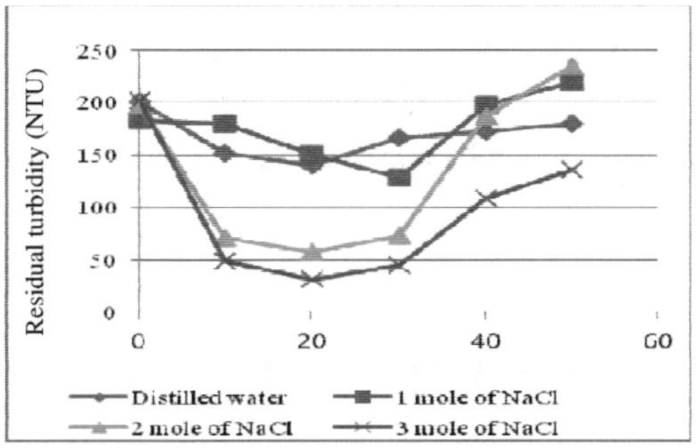

Figure 1. Coagulation of Synthetic Water with Peanut Seeds Coagulant Extracted by Distilled Water and Nacl solution.

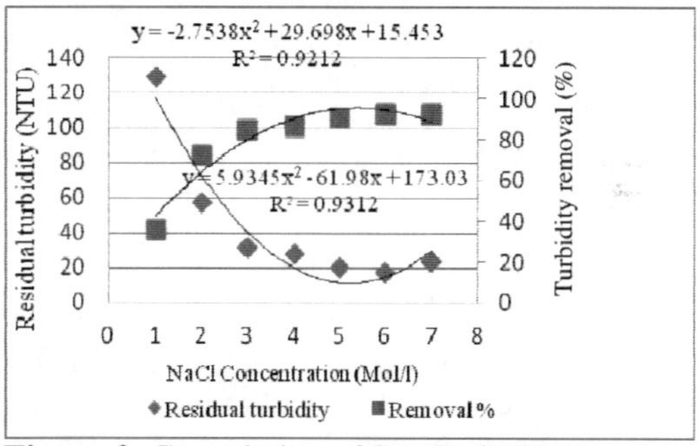

Figure 2. Coagulation of Synthetic Water Coagulated by peanut seeds coagulant extracted by NaCl solution of Different Concentrations. Results are taken at optimum dosage of each test

Figure 3 shows coagulation by peanut seeds extracted by NaCl, KNO_3, KCl, NH_4Cl and $NaNO_3$. The turbidity removal for each concentration was calculated at the optimum dosage. The optimum dosage for most concentrations of the different salts was found to be 20 mg/l, whereas, it was 30 mg/l for both 1 mol/l of NaCl and 1 mol/l of $NaNO_3$. It is observed that for all types of salt, turbidity removal increases with increasing salt concentration. It varies from 22 % at low salt concentration to 93.2% at higher concentrations. This result confirmed the mechanism of salting-in effect that discussed in Figure 1. Turbidity removal up to 3 mol/l is rabidly increases, however, the rate of increment become slowly at higher concentrations, as the salt solution approaches its saturation stage. No significant difference in removal efficiency for NaCl, NH_4Cl and $NaNO_3$ was observed, however, for KNO_3 and KCl the removal efficiency was little bit lower. This could be due to the difference in ionic strength. This result indicates that not only NaCl can be used to extract the ingredient coagulant, but others salts can also be used for the same purpose. However, NaCl seems to be more suitable as it is cheap and available everywhere.

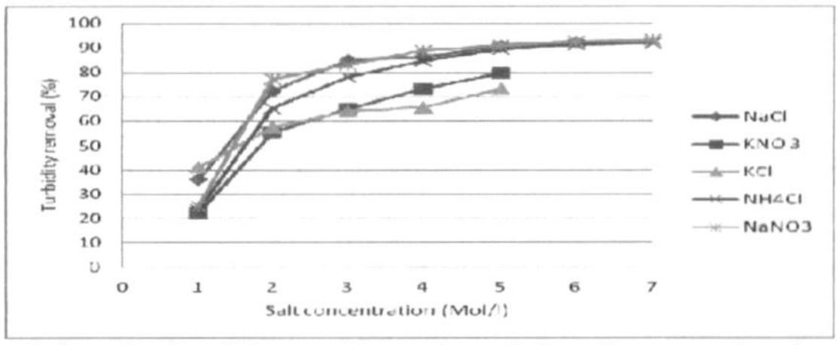

Figure 3. Coagulation of Synthetic Water Coagulated by Peanut Seeds Coagulant Extracted by Different Salts with Different Concentrations. Results are taken at optimum dosage of each test.

4.0 Conclusions

The specific conclusions derived from this study are as follow:

The peanut seeds have a high potential for coagulation of turbid water with an initial turbidity of 200 NTU. Thus, turbidity removal reached up to 93.2 %.

PC-NaCl improved the coagulation activity significantly. PC-NaCl extracted by 6 mol/l NaCl solution could effectively remove 92% of the 200 NTU turbidity using only 20 mg/l, while PC-DW could remove only 31.5 % of the same synthetic water with the same dosage of coagulant. This result implies that the protein associations inside the peanut seeds are responsible for coagulation activity.

The optimum dosage of PC-NaCl for coagulation of the turbid water for all concentrations was 20 mg/l except that of 1 mol/l NaCl solution which was found to be 30 mg/l. Nevertheless, residual turbidity decreased with increasing the concentration of NaCl, and the relationship was found to be a second order polynomial curve with a R^2 of 0.9312.

KNO_3, KCl, NH_4Cl and NaNO3 solutions were also found to be good solvents to extract the active coagulant component for peanut seeds; leading to improvement of coagulation activity; with no much difference from NaCl solution in terms of efficiency.

References

[1] Muyibi S and Alfugara A 2003 Treatment of surface water with Moringa Oleifera seed extract and alum – a comparative study using a pilot scale water treatment plant *Intern. J. Environ. Studies* **60** 617-626

[2] Raghuwanshi N, Singh R, Wallender W and Pruitt W 2002. Estimating evapotranspiration using artificial neural network *J. of Irrigation and Drainage Engineering* **128** 224

[3] Diaz A, Rincon N, Escorithuela A, Fernandez N C and Forster E C F 1999 A preliminary evaluation of turbidity removal by natural coagulants indigenous to Venezuela *Process Biochem.* **35 (3)** 391-395

[4] Ndabigengesere A & Subba Narasiah K 1998 Quality of water treated by coagulation using Moringa oleifera seeds *Water Research* **32** 781-791

[5] Yin C Y 2010 Emerging usage of plant-based coagulants for water and wastewater treatment *Process Biochemistry* **45** 1437-1444

[6] Ali E N, Muyibi S A, Salleh H M, Alam M Z and Salleh M R M 2010 Production of Natural Coagulant from Moringa Oleifera Seed for Application in Treatment of Low Turbidity Water *J. of Water Resource and Protection* **2** 259-266

[7] Okuda T, Baes A U, Nishijima W and Okada M 1999 Improvement of extraction method of coagulation active components from Moringa Oleifera seed *Water Research* **33** 3373-3378

[8] White A, Handler P and Smith E L 1968 Principles of Biochemistry, 4th edn (McGraw-Hill, New York)

[9] Voet D and Voet J G 1990 Biochemistry (John Wiley& Sons, New York)

Private Finance 2 (PF2): Re-inventing the Wheel?

N A W A Zawawi[a], A R Abdul-Aziz[b], M F Khamidi[a], I Othman[a], A Idrus[c], A A. Umar[a],

[a]Universiti Teknologi PETRONAS, Bandar Seri Iskandar, 31750 Tronoh, Malaysia
[b]Universiti Sains Malaysia, Pulau Pinang, 11800 Penang, Malaysia
[c]Universiti Pertahanan Nasional Malaysia, Kem Sg. Besi 57000 KL, Malaysia

E-mail: *amilawa@ petronas.com.my*

Abstract. The Procurement policy of any government is the most influential factor in determining the efficiency of infrastructure and service provision like roads, water supply and energy. The United Kingdom's HM Treasury released its new guidelines on private involvement in infrastructures provision and services towards reforming the popular Private Finance Initiatives (PFI) policy. This new approach, it now refers to as the Private Finance 2 (PF2) is meant to correct the imperfections which have bedeviled the older version-PFI. However, the 'new guidelines' contained nothing really new in the area of private financing and operation of public infrastructures, at best it is akin to 're-inventing the wheel' rather than being 'new'. While dwelling extensively on issues relating to cheaper financing sources, risks transfer, counterpart funding by government and improving public sector procurement skills, this paper argues that some countries in the developing world have long recognised these issues and taken practical steps to correct them.

1. Introduction

Public procurement practices and policies greatly influence the provision of effective and efficient services, hence the constant quest for optimal procurement practices that would help government deliver on its policy objectives. The Private finance initiative (PFI) was one such outcome of the quest for optimal performance in essential public services like electricity, water supply and transportation [1]. The PFI is one form of Public Private Partnerships (PPP), which is a co-operation between the government and private sector firms for the design, financing, building, operation, and maintenance of public infrastructure and services to pre-agreed standards over a duration ranging from 10-60 years in return for a service charge. The PFI procurement strategy evolved in the UK in 1992 in response to the persistent poor quality of government infrastructures and the need to keep government debts at healthy levels. However, over the years a number of other reasons have been advanced for the use of the PFI, according to [2] these include government's aim to deploy private sector management expertise, innovative technologies and operational efficiencies in addition to mobilising private funds to meet the tremendous demands on infrastructure systems. However, structuring of the PFI contract should be such that it meets the approval of the private sector partners and their financiers and sufficient to tip the value for money balance in the project's favour [3]. There is no doubt that the adoption of the PFI has increased efficiency and has helped in expanding utilities coverage to increased number of consumers. But it has also thrown up some disasters which made [4] conclude that the practice as it presently is should be referred to as 'government-private' rather than 'public–private partnerships'. Given the failures that have occurred and criticisms against the practice, many countries in the developing world have taken pro-active steps to re-evaluate the strategy and have made significant structural changes to reflect local realities and cater to public concerns. However, the UK, to which the credit for the PFI has often been attributed, had continued to defend the practice citing the much criticised Arthur Andersen Report, 'Value for Money Drivers in the Private Finance Initiatives', as its main evidence of the strength of the strategy. After 20 years of defending the practice in the face of public disenchantment, failures and buy backs, the UK government through its HM Treasury has recently reviewed and reformed the PFI into what it now terms the 'Private Finance 2' (PF2) [5]. This study argues that though the UK might be credited with the 'invention' of the PFI but cannot take the credit for PF2 as the proposed changes expected to be the hallmark of the PF2 have already become a part of PFI model in some countries long before now.

2. The Road to PF2

If the global financial crisis put any development model on trial, it was the free-market or neoliberal model, which emphasizes a small state, deregulation, private ownership, and low taxes [6]. Since the first sets of PFI projects were commissioned, problems and failures began to be observed in some of them like the Eurotunnel, Railtrack, motorways concessions in Mexico or Urban motorway TEO in Lyon [7]. Researchers, civil society groups, labour/trade unions and the public have analysed and raised concerns with certain aspects of the PFI model. These weaknesses and problems faced by the PFI have been identified a long time ago and reforms advocated by researchers and practitioners including, [8], [9], [10], [11], [12], [13], [14], [15], [16] and [17]. Other evidence of the inevitability of a move to PF2 include [18] and [19]. For over 20 years the UK government had consistently tried to justify its use of the PFI model with reports such as [20].

But other countries did not wait for the UK to take the first step in reforming the PFI, instead countries like Malaysia and Chile have taken local realities into considerations and reformed the PFI practice in their various domain as it affects sources and cost of financing, employee transfer issues, debt to equity ratio and ownership structure. The reforms had resulted in increased affordability, reduced failures and cancellations, improved public support, improved bureaucratic quality, better project selection and generally better partnerships.

However, the UK government has finally admitted [5] that experience in past projects has demonstrated that 'significant' aspects of the model are 'not satisfactory', these include:

1. The PFI procurement process has often been slow and expensive for both the public and the private sector. This has led to increasing costs and has reduced value for money for the taxpayer;
2. PFI contracts have been insufficiently flexible during the operational period, so making alterations to reflect the public sector's service requirements has been difficult;
3. There has been insufficient transparency on the future liabilities created by PFI projects to the taxpayer and on the returns made by investors;
4. Inappropriate risks have been transferred to the private sector resulting in a higher risk premium being charged to the public sector; and
5. Equity investors in PFI projects are perceived to have made windfall gains, and this has led to concerns about the value for money of projects.

In response to the above weaknesses, the government has mapped out a plan of action which includes restructuring in a number of areas including competitive debt finance sources, improved bureaucratic quality, appropriate risk transfer strategies, more flexible service provision, shorter procurement periods and the strengthening of Infrastructure UK with the appropriate skills. And one of the critical tasks in strengthening Infrastructure UK according to the guidelines includes [5],

- [1.] A five year capabilities plan for the whole of Civil Service to identify which skills and capabilities are in deficit and how these gaps will be filled; and
- [2.] The establishment of the Major Projects Leadership Academy to train senior leaders in the delivery of major projects. In future only project leaders who have successfully completed this programme will be able to lead a major government project.

3. The UK arrived late

Despite the overwhelming evidence that a reform was needed, the UK government refused to blink while countries thousands of miles away who adopted the PFI from the UK took steps to reform their model in line with local realities. For example in Chile, the government's response to disenchantment and lack of public support was the concession of about 35% of the investment in the electricity sector to pensioners [21]. While in response to higher cost financing from banks & multi-lateral financial institutions, the Malaysian government employed low cost Employees Provident Funds (EPF) and Pension Trust Fund (PTF) in financing PFI projects [22]. Islamic project finance has also featured prominently in PFI financing according to [23] the Truman Park, Maconda Park Apartments, & East Cameron Gas Company in United States, Utility Power Projects in Saudi Arabia are examples of projects financed using Islamic project finance due to the absence of 'cost of capital' in their transactions. Furthermore in Malaysia, [24] the government created a RM20 billion 'Facilitation Fund' (FF) to bridge the private sector viability gap on PFI with respect to projects with strategic impact and huge economic spill overs. In the area of bureaucratic quality, [25], [26] & [27] had already argued that public sector procurement skills needed to be improved for better PFI results. Noting the importance of bureaucratic quality, the Malaysian government had in 2010 planned the establishment of a world class civil service college to help civil servants increase their knowledge, and competency level [28]. From the foregoing, it is clear that the intent of the UK [5] to establish a Civil service training centre is a concept which had already been implemented in response to skills shortage in the public sector. The new guideline saw the skills of the civil service as important because according to them [5] "a significant proportion of the responses to the call for evidence on PFI reform considered that there was a need to improve public sector procurement skills". After 20 years of defending the PFI model, it is the global financial crisis that finally brought the message home that the PFI model is no longer tenable anymore, anywhere, around the globe.

4. Conclusions

The UK is just conceding that there is a need to reform the PFI after 20 years of defending the model, to take into consideration the original aims of the strategy, which is to provide affordable, effective and efficient public services for citizens. While the credit for the PFI can rightly be attributed to the UK, they should not take credit for the PF2 because all the intended reforms being advocated have already been carried out by other countries for close to a decade. Hence rather than waste resources in trying to structure the PF2, they should emulate the practices in countries like Malaysia, Chile and New Zealand which have successfully reformed their PFI models to suit local realities. The UK should also look in the direction of Islamic Financing for their PF2 debt finance and take advantage of the 'interest-free' Islamic project finance instruments. Therefore any attempt to hype the PF2 as another original strategy from the UK would be tantamount to 're-inventing the wheel'.

5. Acknowledgement

The authors are grateful to MOHE for its financial assistance through its Exploratory Research Grant Scheme (ERGS) and Universiti Teknologi PETRONAS for both the Short Term Internal Research Fund (STIRF) & Graduate Assistance Scheme (G.A)

References

[1] Fewings P 2005 Construction Project Management: An Integrated Approach (Abingdon: Taylor & Francis)

[2] Zhang Q X and Kumaraswamy M M 2001 Hong Kong Experience in Managing BOT Projects J. *of Construction Engineering Management* **127 (2)** 154-162

[3] Froud J 2003 The Private Finance Initiative: Risk, Uncertainty and the State *Accounting, Organizations and Society* **28** 567–589
[4] Hayllar M R 2010 Public-Private Partnerships in Hong Kong: Good Governance – The Essential Missing Ingredient? *Australian J. of Public Administration* **69 (1)** 99-119
[5] Treasury H M 2012 A New Approach To Public Private Partnerships London
[6] Birdsall N and Fukuyama F 2011 The Post-Washington Consensus: Development After the Crisis Foreign Affairs **90 (2)** 45-53
[7] Meunier D and Quinet E 2010 Tips and Pitfalls in PPP Design *Research in Transportation Economics* **30** 126-138
[8] Gaffney D, Pollock A M, Price D, and Shaoul J 1999 PFI in the NHS-Is there an Economic Case*? British Medical J.* **319** 116-119
[9] Abdul-Aziz A R 2001 Unraveling Of Bot Scheme: Malaysia's Indah Water Konsortium *J. of Construction Engineering and Management* **127 (6)** 457-460
[10] Nickson A and Vargas C 2002 The Limitation of Water Regulation: The Failure of the Cochabamba Concession in Bolivia *Bulletin of Latin American Research* **21(1)** 99-120
[11] Pollock A, Shaoul J, and Vickers N 2002 Private Finance and "Value for Money" in NHS Hospitals: A Policy in Search of a Rationale? *British Medical J.* **324** 1205-1209
[12] Hillary J 2004 Profiting From Poverty:Privatisation Consultants *DfID and Public Services*, London
[13] UNISON 2005 PFI: Against the Public Interest.Why a 'licence to print money can also be a recipe for disaster U.K
[14] Mehra N 2005 Flawed, Failed, Abandoned *100 P3s: Canadian and International Evidence. Ontario*
[15] Estache A 2005 PPI partnerships versus PPI divorces in LDCs: (or are we switching from PPPI to PPDI?) Washington D.C.
[16] Estache A, Goicoechea A, and Trujillo L 2007 Infrastructure, Reforms and the Poor in Latin America.
[17] Shaoul J, Stafford A, and Stapleton P 2012 Accountability and Corporate Governance of public private partnerships *Critical Perspectives on Accounting*
[18] House of Commons 2010 Financing PFI projects in the credit crisis and the Treasury's response: Ninth Report of Session 2010–11
[19] HM Treasury 2011 Major Project approval and assurance guidance, London
[20] HM Treasury 2000 Public Private Partnerships :The government's approach, London
[21] Tenenbaum B 1996 Regulation: What the Prime Minister Needs to Know *The Electricity J.* 28-36.
[22] Syuhaida I and Aminah M D 2009 The Provision Of Infrastructure Via Private Finance Initiative *Theoretical and Empirical Researches in Urban Management* 76-86
[23] Alexander A J 2011 Shifting Title And Risk: Islamic Project Finance With Western Partners *Michigan Journal of International Law* **32** 572-612
[24] EPU Tenth Malaysian Plan (2010-2015) Putrajaya, June 10, http://www.epu.gov.my/html/themes/epu/html/RMKE10/rmke10_english.html
[25] Morledge R and Owen K 1998 Critical Success Factors in PFI Projects *14th ARCOM Annual Conference, University of Reading* 565-574
[26] RICS Project Management Forum 2003 *PFI and the Skills of a Manager*, UK.\
[27] Serebrisky T, Azumendi S L, and Andres L A 2011 Institutional Design and Governance of Airport Regulators: The Case of Latin America *J. of Air Transport Management* **17** 207-210
[28] Mohd Najib Bin Tun Abdul Razak. (2010, June) 10th Malaysian Plan (2010-2015). http://www.pmo.gov.my/dokumenattached/speech/files/RMK10_Speech

IOP Conf. Series: Earth and Environmental Science **16** (2013) 012067 doi:10.1088/1755-1315/16/1/012067

Temperature profile and producer gas composition of high temperature air gasification of oil palm fronds

F M Guangul[1], S A Sulaiman[1], A Ramli[2]
[1]Department of Mechanical Engineering, Universiti Teknologi PETRONAS, Bandar Seri Iskandar, 31750 Tronoh, Perak, Malaysia
[2]Department of Fundamental and Applied Sciences, Universiti Teknologi PETRONAS, Bandar Seri Iskandar, 31750 Tronoh, Perak, Malaysia

E-mail: fiseham2002@yahoo.com

Abstract. Environmental pollution and scarcity of reliable energy source are the current pressing global problems which need a sustainable solution. Conversion of biomass to a producer gas through gasification process is one option to alleviate the aforementioned problems. In the current research the temperature profile and composition of the producer gas obtained from the gasification of oil palm fronds by using high temperature air were investigated and compared with unheated air. By preheating the gasifying air at 500°C the process temperature were improved and as a result the concentration of combustible gases and performance of the process were improved. The volumetric percentage of CO, CH4 and H2 were improved from 22.49, 1.98, and 9.67% to 24.98, to 2.48% and 13.58%, respectively. In addition, HHV, carbon conversion efficiency and cold gas efficiency were improver from 4.88 MJ/Nm3, 83.8% and 56.1% to 5.90 MJ/Nm3, 87.3% and 62.4%, respectively.

1. Introduction

Nowadays emissions of green house gases from fossil fuels, which are the main source of energy, have become a serious environmental problem. On the other hand, an ever growing demand for energy source which is in dissonance with a limited supply of fossil fuels has also become another related concern [1-3]. Thus, it is imperative to find other reliable, abundant, and eco-friendly alternative sources of energy. In this respect, biomass is one of the promising sources of energy to alleviate the aforementioned problems [3, 4].

Currently in Malaysia the share of renewable energy from the total electric energy is only 1.2% (332.4 MW), and hence the government has intended to increase to 5.9% by 2030 on its energy diversification plan. In the diversification plan, oil palm wastes are considered as one of the major option among others. In Malaysia the palm oil sector generates 195.6 million tons of biomass wastes every year from plantation and milling activities. These wastes constituted 82.21% of the total energy that could be obtained from the total biomass wastes. The contribution of oil palm fronds (OPF) constituted 46.71% (97 million tons per year) of the total oil palm wastes. Currently, fronds are dumped in the plantation site without significant contribution except for the use of erosion control and soil conservation [3, 4]. Thus, OPF can be used as alternative energy source for the production of producer gas through gasification. Recently, high temperature air gasification of biomass waste has received high attention because of the improvement gained in the quality of producer gas and reduction of tar compared to that of the conventional process [5-8].

Studies showed that when the gasifying air temperature kept high, amount of H_2 was increased. CO and CH_4 concentration showed different trends when different feedstocks were used [5-8]. Regardless of the feedstock and gasifying agent, carbon conversion efficiency, cold gas efficiency and HHV were increased as temperature increased [2, 5-8]. Using high temperature gasification process wood pellets, cypress, hemlock, cedar, rice husk, camphor wood, sawdust, etc., were investigated and significant improvements were obtained on the quality and yield of producer gas, and performance of the process [6-8, 9].

2. Methodology

2.1. Material and experimental setup

The feedstock, oil palm fronds, was collected at Felcra Classic oil palm plantation in the state of Perak, Malaysia. Leaves were shredded and the petioles were chopped to a maximum size of 25 mm using a chopper. The basic characteristics of OPF are shown in table 1. From the experiment CO, CO_2, CH_4 and H_2 of the producer gas were collected and analyzed. In addition, HHV, gas yield, carbon conversion efficiency and cold gas efficiency were determined from the output and input parameters.

Table 1. Basic characteristic of the oil palm fronds

Proximate analysis (% dry basis)	
Volatile matter	83.5
Fixed carbon	15.2
Ash	1.3
Ultimate analysis (% dry basis)	
Carbon	44.58
Hydrogen	4.53
Oxygen	48.80
Nitrogen	0.71
Sulphur	0.07
HHV (MJ/kg)	17.00
Moisture content (wb)	10%

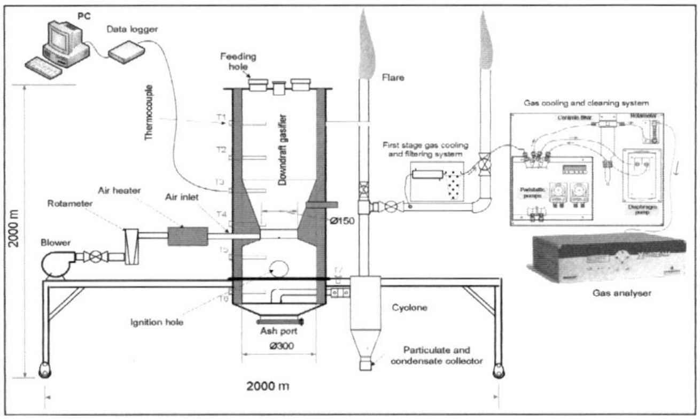

Figure 1. Experimental setup design

2.2. Experimental procedure

Before the gasification process started the thermocouples were fixed in position and 8 kg of OPF feedstock was added. In the case of unheated air to start ignition, scrap paper and ignition lighter were used and ignition was started through the ignition starting hole (figure 1). After ignition the air was supplied by the blower and the flow rate was adjusted to the required volume and was measured using a rotameter. When preheated air was used as a gasifying medium the feedstock self-ignited. The temperature data was recorded automatically using a data logger and computer. For the gas composition analysis sample gas was tapped from the syngas outlet pipe. The char and ash were collected from the ash box and weight measurement was taken.

3. Results and discussion

3.1. Temperature profile of the gasification process

Figures 2 and 3 show the temperature profile of the gasification process recorded using K-type of thermocouples at seven points (Figure 1) for unheated and preheated air against time. When unheated air was used the time taken to consume the feedstock was longer than that of the preheated air. The oxidation zone temperature was exceeded 700°C after 8 min from the gasification starting time, while for preheated air it took less than 5 min. In addition, the pyrolysis zone temperature (T4) was relatively stable and high when preheated air was used as a gasifying agent.

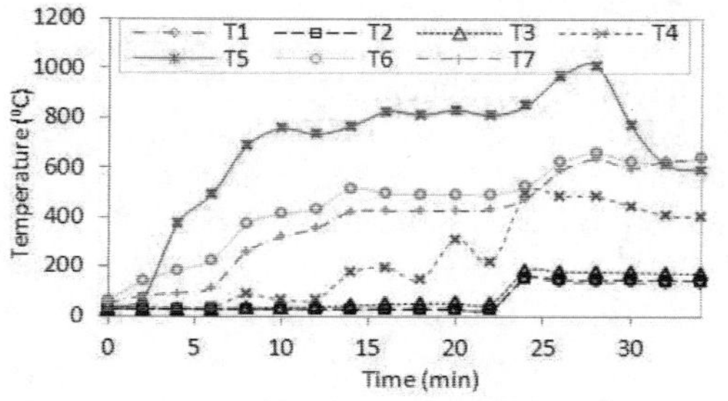

Figure 2. Temperature profile for a gasification process with unheated air

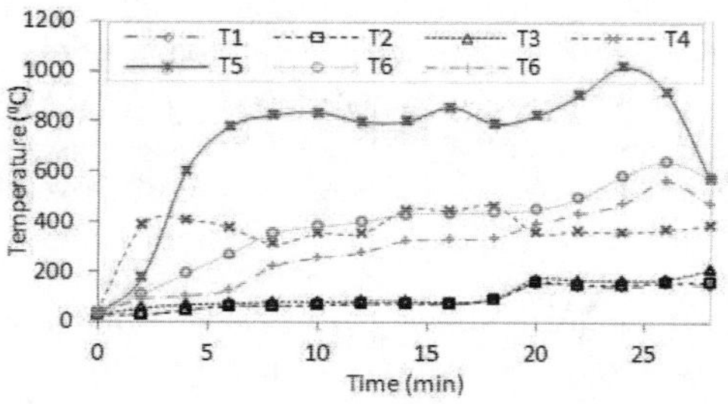

Figure 3. Temperature profile for a gasification process with 500°C preheated air.

The maximum oxidation temperature (above 1000°C) was recorded in both cases close to the end of the processes. Wei [10] had obtained similar temperature profile with the current trend in the co-gasification process of hardwood chips and crude glycerol. The reason could be the accumulation of high char on the grate and the depletion of the pyrolysis gases. When the pyrolysis gas decreased, its cooling effect in the oxidation zone while it passed via the char bed could decrease as most of the reactions between the char and the gas are endothermic. In addition, as the pyrolysis gases decreased high amount of air could have a chance to react with the char instead of reacting with the incoming pyrolysis gases, and more oxidation processes take place which could result in high temperature in the oxidation zone.

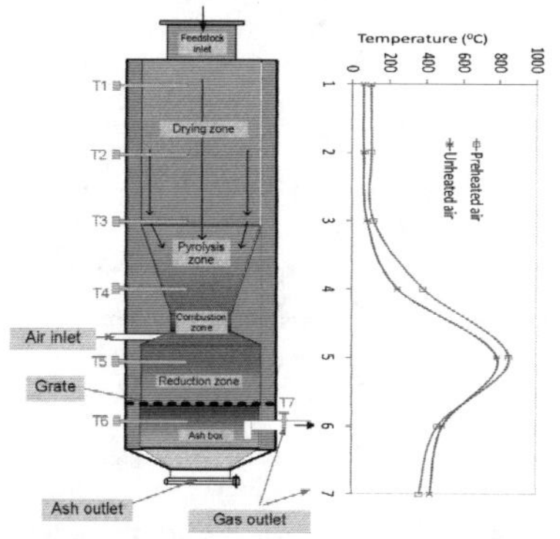

Figure 5. Temperature profile of the gasification process inside the gasifier

Figure 5 shows the average temperature of the gasification process along the height of the gasifier and at the outlet of the producer gas for preheated and unheated air. The temperature profile of the gasification process using preheated air was higher than that of unheated air except in the ash box and at the outlet point of the producer gas. The higher temperature obtained in preheated air gasification was owing to the latent heat obtained from the air. The higher temperature obtained in the ash box and at the outlet point of the gas for unheated air could be due to the higher amount of char and ash accumulated in the ash box compared to that of the preheated air process [11]. Due to the higher amount of char and ash formed in unheated air gasification process, the temperature of the ash box could be high and as a result the drop in temperature of the producer gas which was passing via the ash box could be minimal.

3.2. Effects of preheating the gasifying air on the producer gas composition

Table 2 shows the producer gas composition results obtained from unheated and preheated air gasification processes. Preheating of the gasifying air improved the volume percentage of CO, CH_4 and H_2 from 22.49, 1.98 and 9.67 to 24.98, 2.49 and 13.58, respectively. HHV, carbon conversion efficiency and cold gas efficiency were improved from 4.88 to 5.90 MJ/Nm3, from 83.8 to 87.3%, and from 56.1 to 62.5%, respectively. The amount of char and ash collected in the ash box and on the grate at the end of the gasification process was decreased from 741 g to 675 g when the preheated air was used. The reduction of ash and char was due to the higher reaction temperature created as a result of air preheating which would enhance the conversion of the feedstock [11].

Table 2. Comparison of the Results of OPF Gasification Outputs using Unheated and Preheated Gasifying Air

Parameter	Gasifying medium		Increase (%)
	Unheated air	Preheated air	
CO (vol. %)	22.49	24.98	10
CH_4 (vol. %)	1.98	2.49	20.5
H_2 (vol. %)	9.67	13.58	25.1
Syngas HHV (MJ/Nm^3)	4.88	5.90	17.3
Gas yield (Nm^3/kg)	1.95	1.80	8.0
Carbon conversion efficiency (%)	83.8	87.3	4.0
Cold gas efficiency (%)	56.1	62.5	10.2
Equivalence ratio	0.27	0.21	-28.6

4. Conclusion

By preheating the gasifying air at 500°C the gasification temperature was improved and as a result significant improvements in the concentrations of the combustible gases of the producer gas and performance of the process were obtained. The volumetric concentrations of the combustible gases in the producer gas were improved from 22.49, 1.98, and 9.67% to 24.98, to 2.48% and 13.58%, respectively. The performance evaluation parameters, i.e., HHV, carbon conversion efficiency and cold gas efficiency were improver from 4.88 MJ/Nm3, 83.8% and 56.1% to 5.90 MJ/Nm3, 87.3% and 62.4%, respectively. The amount of char and ash was reduced from 741 g to 675 g.

References

[1] Bernstein L, Bosch P and Canziani O 2007 Synthesis Report Valencia
[2] Kumar A, Eskridge K, Jones D D and Hanna M A 2009 *Bioresource Technology* **100 (6)** 2062-2068
[3] Shuit S, Tan K, Lee K, and Kamaruddin A 2009 *Energy* **34 (9)** 1225-1235
[4] Yusoff S, 2006 *Journal of cleaner production* **14 (1)** 87-93
[5] Doherty W, Reynolds A and Kennedy D 2009 *Biomass and Bioenergy* **33 (9)** 1158-1167
[6] Lucas C, Szewczyk D, Blasiak W and Mochida S 2004 *Biomass and Bioenergy* **27 (6)** 563-575
[7] Yang W, Ponzio A, Lucas C and Blasiak W 2006 *Fuel Processing Technology* **87 (3)** 235-245
[8] Zhou J, Chen Q, Zhao H, Cao X, Mei Q, Luo Z and Cen K 2009 *Biotechnology Advances* **27 (5)** 606-611
[9] Li X, Grace J, Lim C, Watkinson A, Chen H and Kim J 2004 *Biomass and Bioenergy* **26 (2)** 171-193
[10] Wei L, Pordesimo L O, Haryanto A and Wooten J 2011 *Bioresource Technology* **102 (10)** 6266-6272
[11] Moilanen A, Nasrullah M and Kurkela E 2009 *Environmental Progress & Sustainable Energy* **28 (3)** 355-359

Influence of operating conditions on the air gasification of dry refinery sludge in updraft gasifier

R Ahmed[1], C M Sinnathambi[2*]

[1]Department of Chemical Engineering, Universiti Teknologi PETRONAS
[2]Fundamental and Applied Sciences Department, Universiti Teknologi PETRONAS, Bandar Seri Iskandar, 31750 Tronoh, Perak, Malaysia.

E-mail: chandro@petronas.com.my

Abstract. In the present work, details of the equilibrium modeling of dry refinery sludge (DRS) are presented using ASPEN PLUS Simulator in updraft gasifier. Due to lack of available information in the open journal on refinery sludge gasification using updraft gasifier, an evaluate for its optimum conditions on gasification is presented in this paper. For this purpose a Taguchi Orthogonal array design, statistical software is applied to find optimum conditions for DRS gasification. The goal is to identify the most significant process variable in DRS gasification conditions. The process variables include; oxidation zone temperature, equivalent ratio, operating pressure will be simulated and examined. Attention was focused on the effect of optimum operating conditions on the gas composition of H_2 and CO (desirable) and CO_2 (undesirable) in terms of mass fraction. From our results and finding it can be concluded that the syngas (H_2 & CO) yield in term of mass fraction favors high oxidation zone temperature and at atmospheric pressure while CO_2 acid gas favor at a high level of equivalent ratio as well as air flow rate favoring towards complete combustion.

1. Introduction

Air gasification technology is being studied by many researchers for various types of fuels. Gasification is the conversion of hydrocarbons in the presence of air/steam or oxygen to produce syngas which is a mixture of hydrogen and carbon monoxide. Since there is a lack of information available in the open journals about gasification of dried refinery sludge (DRS) using updraft reactor, the goal of this work is to investigate the effect of three important process conditions namely oxidation zone temperature, operating pressure, and equivalence ratio on gas yield in term of mass fraction for of H_2, and CO (desirable) and CO_2 (undesirable). Taguchi orthogonal array design is used here to evaluate the optimum conditions for DRS gasification in updraft gasifier.

2. Taguchi Orthogonal Array Design (TOAD) Methods

TOAD is a classical factorial method of analysis used in processes involving many process variables. For the successful interpretation or prediction, TOAD requires numerous experiment data for optimization purposes. In classical statistical method, if the number of experiments to be carried out is "N" and the number of levels or factors (process variables) is "m", this can be translated by the equation ($N = L^m$). Some of the experiments carried out may prove less significant or redundant. Omitting these less significant experiments or process variables gives huge saving and reduces the time frame or good analytical results and prediction. Selection of the favorable experiments can be obtained by. Using classical statistical methods for this purpose is not an easy task as indicated by Mahamuni & Adewuyi [8]. For our gasification of DRS both TOAD and DOE methods were used.

3. Results and Discussions

Table 1 list the various factors and level performed on a Taguchi orthogonal array design or TOAD simulation. For this simulation study, four factors were chosen (A:A for oxidation zone temperature, B:B for operating pressure , C:C for the flow rate of air and D:D for equivalent ratio) and three levels for each factor were selected using a standard L_9 Orthogonal array based on Kul and Cetinkaya [6], and Mohan and Ahmed [10] model. The description of gasification conditions are shown and highlighted in Table 1.

Table 1. DRS gasification factors and levels.

Factor	Levels		
	1	2	3
A:A Temperature (CO_2)	950	750	500
B:B Pressure (bar)	20	14	1.01
C:C Air flow (L/min)	280	240	180
D:D Equivalent Ratio	0.34	0.24	0.14

Next the statistical analysis of our equilibrium model results was carried out using analysis of variance (ANOVA) and tabulated in Table 2. The purpose of ANOVA is to investigate the factors which significantly affect the response factor which in turn grades the percentages contribution of each process variables mentioned above. This is then applied to the DRS gasification simulation results as per our present study. The Model F-value of for all the factors shows <1 except for factor D:D which is calculated to be 3.79 which implies there is an 8.61% chance that a "Model F-Value" this large value could occur. It should be noted that the ER is affected by air flow rate and feed rate. As a result the large value accounted is a result of combination of two variables in one.

Table 2. ANOVA Analysis for contributions of four factors on Carbon monoxide mass fraction

Factors	DOF	Sum Sq.	F Value	Prob>F	Cont. %
A:A	2	4.033E-003	0.24	0.7927	7.45
B:B	2	0.013	0.91	0.4524	23.23
C:C	2	7.295E-003	0.47	0.6477	13.48
D:D	2	0.030	3.79	0.0861	55.84

3.1Carbon monoxide-mass fraction yield

The percentages contribution of individual factor for CO mass fraction using TOAD method on simulated results from Table 3 is shown in Figure1. From Figure 1, it can be seenthat the highest contribution for CO yield is from ER factor which accounts for 55 % followed by operating pressure factor (23 %). The reaction temperature and air flow factors have smaller effect on the yield for CO mass fraction.

IOP Conf. Series: Earth and Environmental Science **16** (2013) 012068 doi:10.1088/1755-1315/16/1/012068

In accordance to Lijun et al., [7] high reaction temperature favor high conversion rate of carbon. Since there are few species of carbon produced during gasification reaction, it will be wise if we determine the four process variables on an individual basis to understand its sensitivity towards changing process variable and its effect to CO yield. The result from this operation is shown in Figure 2. From the same figure it can be observed operating the gasification process at ER of 0.28 gives the optimum process condition for maximum yield of carbon monoxide formation at high temperature and low pressure. Although Figure 2 shows that increase in reaction temperature and lower operating pressure and air flow rate favours high CO yield, but ER of 0.28 far exceed the other 3 process variable yield on CO.

Our results is in agreement with Konda et al [5] and, Atnaw et al [1], documented in their simulation study of downdraft gasification of oil-palm fronds using ASPEN PLUS that, the mole fraction of CO decrease sharply after 5 bar of the operating pressure. Khadse et al., [4] and (Lijun et al., [1] found that higher oxidation zone temperature maximize CO yield for biomass using updraft gasification.

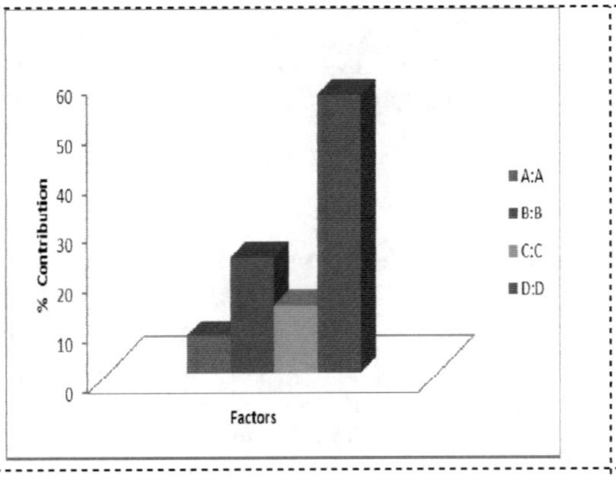

Figure 1. Percentages contribution of factors on CO mass fraction.

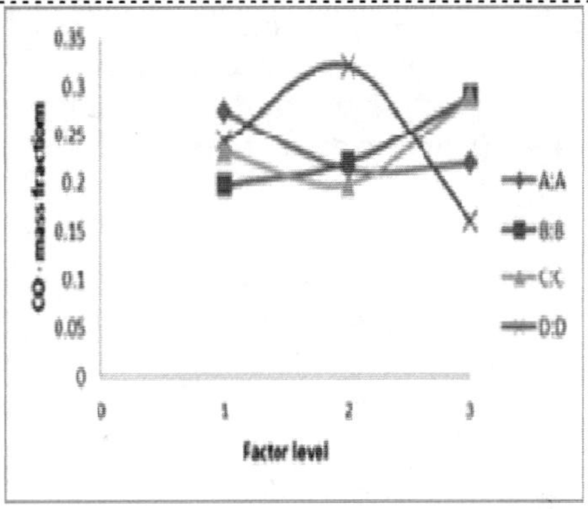

Figure 2. Effect of factor level on CO mass fraction.

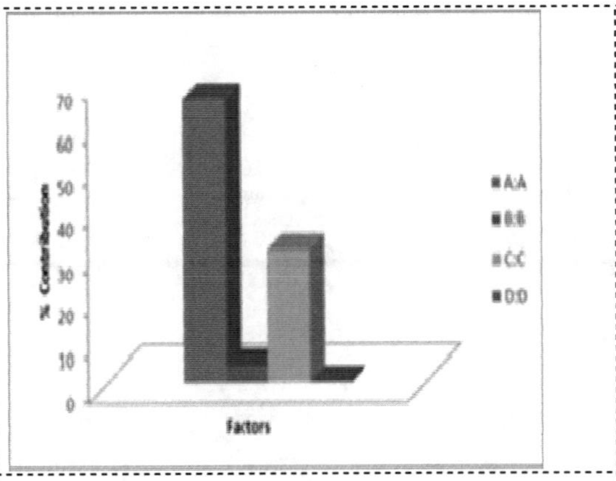

Figure 3. Percentages contribution of factors on H_2 mass fraction.

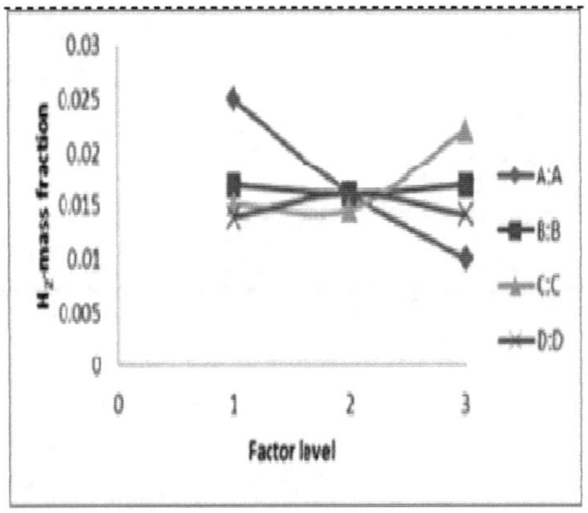

Figure 4. Effect of factor level on H_2 mass fraction.

3.2 Hydrogen-mass fraction yield

The percentages contribution of individual factor for hydrogen mass fraction is described in Figure 3. From the same figure it is can be observed that the oxidation temperature is the main factor which contributes about 67% favoring increase H_2 yield follwed by air flow rate about 34%. Pressure and ER on the other hand has little affect on the H_2 yield. Studies on effect of individual process variables' contribution on H_2 yield was undertaken and presented in Figure 4. From the same figure it can be seen operating at high temperature and low flow rate favors high H_2 yield. The process pressure has little effect on H_2 mass fraction yield. Although lower flow rate also do show some significant increase in H_2 yield, a comprise must be made between flow rate and ER to maximise H_2 yield. This prediction is in agreement with literature as the yeild of CO and H_2 is known to decrease with pressure Higman and Van de Burgt, [3].

4. Conclusion

The TOAD statistical was successfully used as a tool to identify the most significant operating conditions on gasification of dry refinery sludge in updraft gasifier making use of ASPEN PLUS Simulator software. From our results and finding it can be concluded that the syngas yield in term of mass fraction favors at high oxidation zone temperature and at atmospheric pressure while CO_2 acid gas favor at a high level of equivalent ratio as well as air flow rate. This is because the reaction tends towards complete combustion by increasing of actual air flow. It can be concluded that the optimum operating conditions for CO and H_2 formation from DRS gasification is to operate the process system at atmospheric pressure 1.01 bar (114.7) Psi and under an equivalent ratio of 0.28, with an air flow rate of 240-280 l/min and at a moderately high temperature of 750 to 950°C.

References

[1] Atnaw S M, Sulaiman S A and Yusup S 2011 A simulation study of downdraft gasification of oil-palm fronds using Aspen Plus *J. Applied Sci.* **11 (11)** 1913-1920

[2] Atnaw S M and Sulaiman S A 2009 Modeling and simulation study of downdraft gasifier using oil-palm fronds *Proceedings of the 3rd International conference on Energy and Enviroment* Dec. 7-8 Malacca Malaysia 325-330

[3] Higman C and Van Der Burgt M 2008 Gsaification 2nd Edn (Gulf Professional Publishing, Barlington, MA)

[4] Khadse A, Parulekar P, Aghalayam P and Ganesh A 2006 Equilibrium model for Biomass gasification Department of Chemical Engineering, Bombay. India

[5] Konda R E, Sulaiman S A, and Ariwahjoedi B 2012 Synga production from gasification of Oil palm fronds with an Updraft gasifier *J. Applied Sci.* **11 (11)** 1812-5654

[6] Kul M and Cetinkaya U 2010 Application of the Taguchi Orthogonal Array Design to Cr (VI) solvent extraction *Minerals and Metallurgical Processing*

[7] Lijun W, Weller C L, Jones D D and Hanna M A 2008 Review: contemporary issues in thermal gasification of biomass and its application to electricity and fuel production *Biomass Bioenergy* **32** 573-581

[8] Mahamuni N N and Adewuyi Y G 2010 Application of Taguchi Method to Investigate the Effects of Process Parameters on the Transesterification of soybean oil using high frequency ultrasound *Energy Fuels* **24** 2120-2126

[9] Mathieu P and Dubuisson R 2002 Performance analysis of a biomass gasifier *Energy Conversion Manage* **43** 1291-1299

[10] Mohan C and Ahmed R 2012 Optimization of n-hexane reforming using Taguchi Orthogonal array Design *Sains Malaysiana* **41 (7)** 883-890

[11] Nikoo M B and Mahinpey N 2008 Simulation of biomass asification in fluidized bed reactor using ASPEN PLUS *Biomass Bioenergy* **32** 1245-1254

[12] Zainal Z A, Ali R, Lean C H and Seetharamu K N 2000 Prediction of performance of a downdraft gasifier using equilibrium modeling for different biomass materials *Energy Conversion Manage* **43** 1499-151

IOP Conf. Series: Earth and Environmental Science **16** (2013) 012069 doi:10.1088/1755-1315/16/1/012069

Current control loop of 3-phase grid-connected inverter

A F Jabbar and M Mansor
Dept. Of Electrical Power Engineering, Universiti Tenaga Nasional,
Jalan IKRAM-UNITEN, 43000, Kajang, Selangor, Malaysia

E-mail: j.alfattah@gmail.com; muhamadm@uniten.edu.my

Abstract. This paper presents a comparative study of current control loop in 3-phase inverter which is used to control the active and reactive output power. Generally, current control loop, power control loop and phase lock-loop are the conventional parameters that can be found in an inverter system controlled by the conventional linear control type, for instance proportional (P), integral (I) and derivative (D). If the grid remains stable throughout the day, PID control can be use. However variation of magnitude, frequency, voltage dips, transient, and other related power quality issues occur in a 3-phase grid often affects the control loop. This paper aims to provide an overall review on the available current control techniques used in grid connected system.

1. Introduction

Fossil fuels are non-renewable energy resource because they take millions of years to generate. As the amounts of fossil fuel used are more than the ones produced, it is expected that one day all the reserves fuels will be depleted [1]. In respond to this, the Malaysian government has shown interest in green energy developments and aims to ensure 11% of the country's total energy capacity is powered using renewable energy by 2020 [2]. A positive outcome especially to the environment is expected as 42.2 million tons of carbon emission will be eliminated with addition of generating more job opportunities [3]. By the year 2020, it is anticipated that addition 10.8-gigawatts of electricity will be needed in the country [4]. An ongoing research is being conducted on renewable energy sources such as solar, wind and biomass to ensure safe, cheap and reliable energy sources to replace fossil fuel. Among all, production of electrical energy from renewable sources in photovoltaic (PV) and wind-energy has been given wide attention. To avoid expensive cost on storage devices, researchers have considered implementing these renewable energy sources into the grid which is called grid-tie system.

2. Current control techniques

Control scheme such as PID, hysteresis free running and Dead Beat control has been used in the current control loop for the past decade. Current loop controls are robust as their insensitive characteristics towards electrical noise. A current controller operates by generating signals for the pulse width modulation so it produces a switching state according to the desired reference point. Figure 1 shows an example of an inverter using a simple current controller scheme.

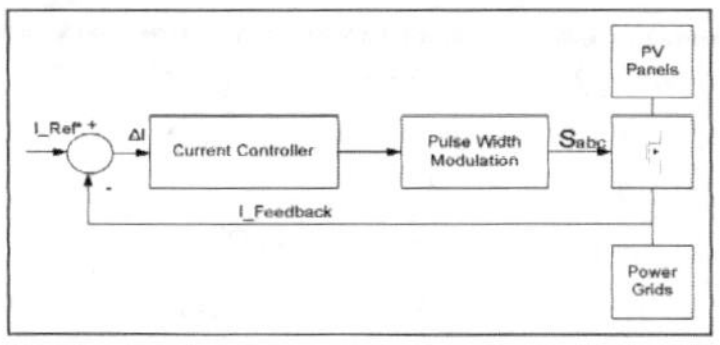

Figure 1. Basic block diagram of current control PWM inverter

The current error ΔI will decrease over time as the output current reaches the desired reference current. Designing an error compensator within the current control must be taken into consideration as it affects the system stability, transient respond and dynamic behavior [5] [6]. Controlling the output current protects the inverter from overload and ensures sufficient power delivering.
Advantages of current control:

- Output unaffected by load parameter changes.
- Peak current and overload protection.
- Extremely good dynamics.

In grid-connected system, current controllers must be carefully design and can be unstable if the modeling does not includes nonlinearities such as switching dead time, delays and semiconductor behavior[6][7]. An adaptive control approach was proposed to overcome the model uncertainties and successful in minimizing power quality deteriorate for harmonic and unbalance output [8].

2.1. Deadbeat current control

Classified as predictive control group because it predicts the future behavior of the output controlled variables [9][10]. This scheme calculates the require reference voltage vector v*(k) within each sampling period k+1 such the load current i(k) will be equal to the reference current i*(k). Figure 2 shows the block diagram of deadbeat current control scheme.

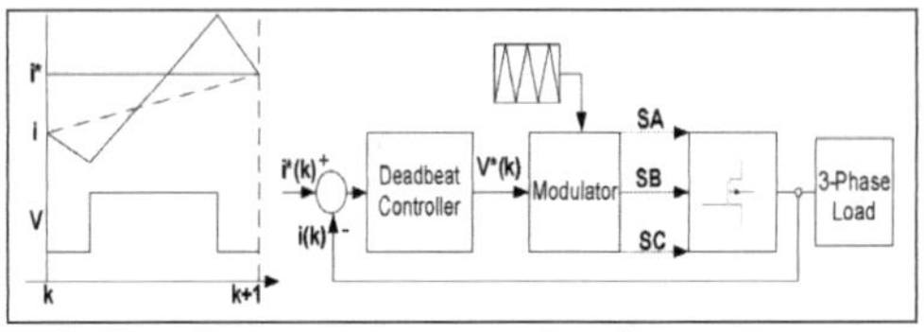

Figure 2. Deadbeat current control and waveform operation

Advantages:

- Capable of having the fastest transient response.
- High precision in current control.

Despite its advantage and superior traits compare to conventional methods. Research has discovered two main practical issues related to deadbeat control [10][11][12]:

- Bandwidth limitation due to the inherent plant delay.
- Sensitivity to plant uncertainties.

Ideally to obtain zero reference error, all the plant definite parameters must be known to achieve zero time delay response. Changes in the plant parameter due to resonance at output filter (LC or LCL) will make the preset controller gain not able to achieve deadbeat response. Figure 3 shows the pole placement of different resistance value from no load to infinite load [11].

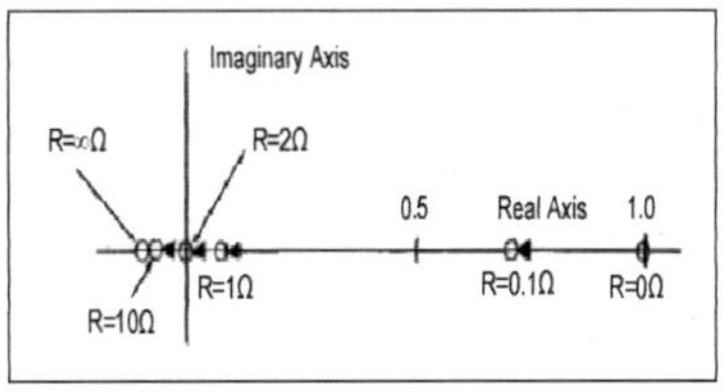

Figure 3. Reference signal waveform

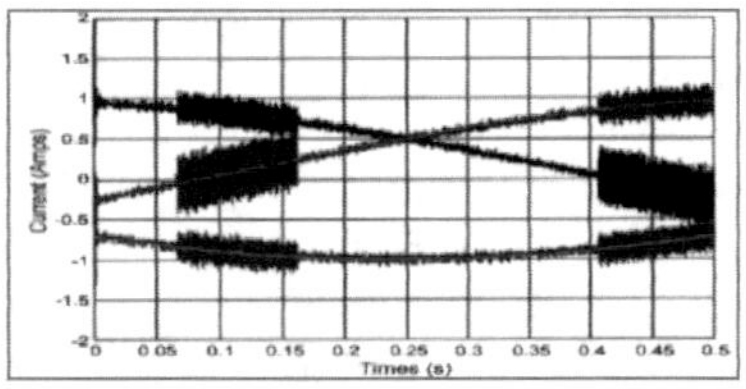

Figure 4. Zero current clamped oscillation [13]

To eliminate plant uncertainties in deadbeat control, an integral compensator was proposed to compensate the inaccurate plant model and variation of DC voltage source [8]. As the output performance is acceptable, the performance of PI controller is inadequate against nonlinearities and dynamics of grid system. Another problem associated to deadbeat control is zero current oscillation, this happens if the measured phase current crosses zero during dead time. Figure 4 shows the waveform of zero current oscillation measured from output current. An improved predictive current regulation (PCR) was introduced [10] [13], this technique further increases the system robustness to load parameter mismatch and was able to reduce zero current clamped oscillation.

2.2. Hysteresis current control

Free-running hysteresis has been used in variable speed ac-drives, active power filters (APF) [10], power factor correction (PFC) [11] and multilevel inverters [12]. It operates by making the output ia, ib, ic close to the desired reference current ia*, ib*, ic* within a hysteresis band as shown in figure 5.

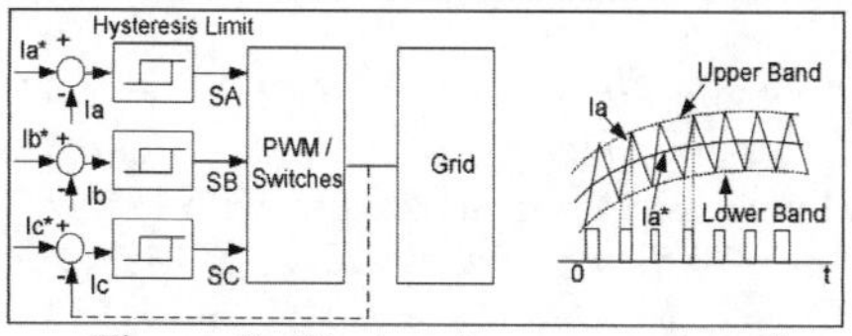

Figure 5. Hysteresis current control

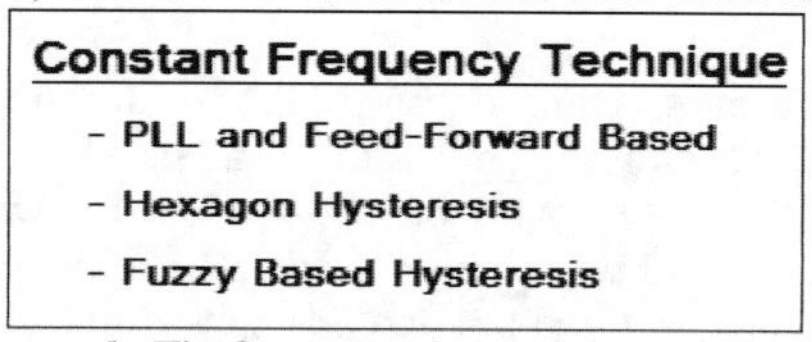

Figure 6. Fix frequency hysteresis current control

The advantages of hysteresis current control:

- Simple implementation.
- Good stability and dynamic performance.
- Independence of load parameter changes.

Having three separate hysteresis control, each signal SA, SB, SC will produce switching states independently to maintain it's current within the hysteresis limit. Due to coupling effect, the performance at zero-crossing deteriorates and has related issue in variable switching frequency [10].

Disadvantages:

- High current ripple
- Injecting high frequency components at output
- High switching losses

Adaptive hysteresis current control was proposed to overcome the variable switching frequency. This technique enables the limit bands to be varied according to output load hence achieving constant switching frequency and lower switching losses [13][14][15]. The types of adaptive hysteresis current control is shown in figure 6 Implementation of hexagon based control was done in synchronous rotating d-q coordinates [16][17]. The entire current error is maintained within a hexagon area like shape which rotates about the reference current as shown in figure 7 Selection of zero-voltage vectors reduces the switching frequency and eliminates the interaction of coupling effect [10][18].

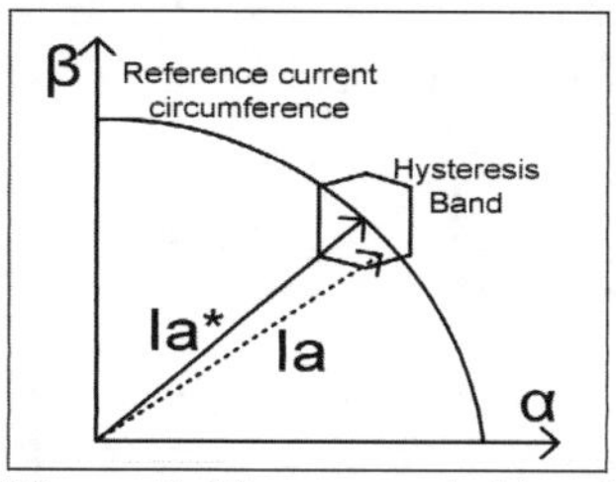

Figure 7. Hexagon switching

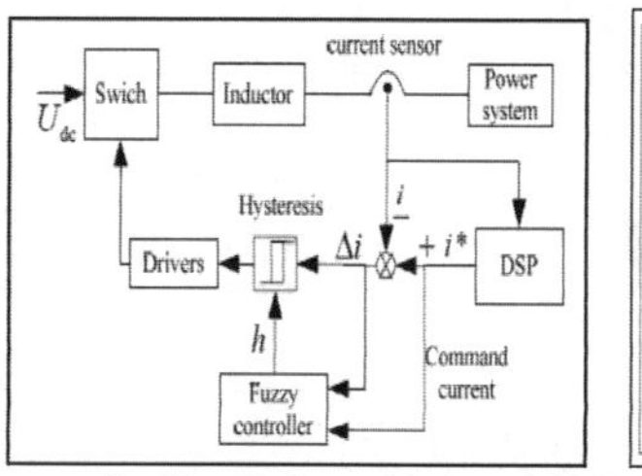

Figure 8. Fuzzy based control [10]

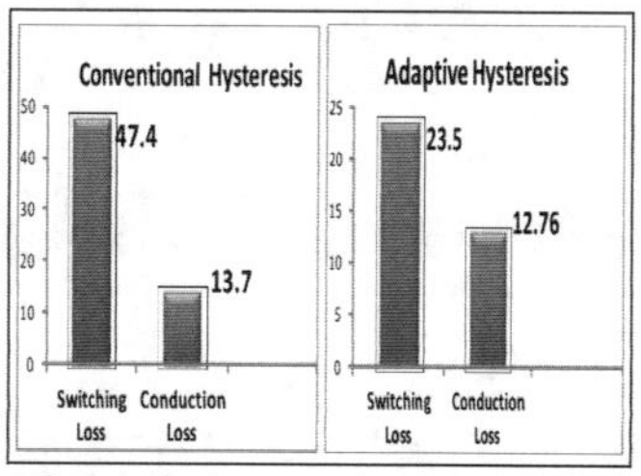

Figure 9. Losses comparison [13]

An intelligent control such as fuzzy based in adaptive hysteresis control is shown in figure 8 [10]. With fuzzy logic controller, transient overshoots and tracking error was improve significantly and increase the robustness of overall system [7]. Comparisons between conventional and adaptive hysteresis current control method has been done on a 1650-watts HERIC "High efficient and reliable inverter concept" [13]. As shown in figure 9 adaptive hysteresis has smaller values in switching and conduction loss. Although overall efficiency between the two methods are small which is 96.2% for conventional and 97.6% for adaptive, this experiment was conducted on a single-phase six switch inverter. This method is suitable for topology that has more switches such as multilevel inverter.

2.3. Comparison study

Comparison of current control loop on a grid connected system was conducted using matlab-simulink. The inverter parameters are shown in table 1.

DC voltage	1200V
Frequency	50Hz
LC Filter	L=7.5mH , C=1.2mF
Load	415V, 45kW, +1kVAR
Current Control Methods	
Conventional PID	SVPWM ,Fs=2000Hz
PID - Decoupling	SVPWM ,Fs=2000Hz
Hysteresis	PWM
Deadbeat	PWM

Table 1. Inverter specification

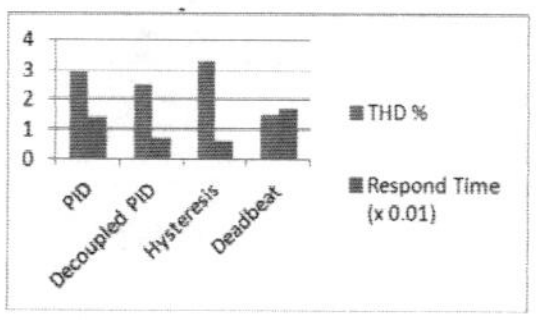

Figure 10. THD and respond time comparisons

A comparison of THD and respond time of the current control technique is shown in figure 10. It is observed that hysteresis current control has the highest THD with 3.3% but opportunely has the fastest respond time with 6ms. Deadbeat control has the smallest THD with 1.57% but has slowest respond time with 17.5ms. For the conventional PID and decoupled PID, an improvement is seen for the decoupled PID with THD of 2.5% compared to 2.98% and respond time 7ms compared to 14ms.

3.0. Conclusion

This paper has presented a general review of current control techniques used in grid connected system with the general problems associated to each methods. Improvement approach was also presented. It is concluded that comprehensive study and analysis on current control loop must be conducted to ensure safe, reliable and optimum performances in renewable system since the performance is largely correlated to the quality of the current control loop. Further improvements to the current loop can be made by using complex computation techniques such as fuzzy and neural network control schemes.

References

[1] http://www.solarpowerwindenergy.org/2011/04/27/how-grid-tie-system-works/, Dec. 2011

[2] http://thestar.com.my/news/story.asp?file=/2012/2/16/nation/20120216103037&sec=nation, Dec, 2012

[3] http://biz.thestar.com.my/news/story.asp?sec=business&file=/2012/10/11/business/20121011155913

[4] The Future of Energy in Malaysia *KETTHA-Malaysia* April 21.2012

[5] Zeng Q 2004 SVPWM-based current controller with grid harmonic compensation for three-phase grid-connected VSI *IEEE Trans.* June. 2004

[6] Kawabata T, Miyashita T and Yamamoto Y 1990 Dead Beat Control of Three Phase PWM Inverter *IEEE Tran.*

[7] Zeng Q 2005 Development of an SVPWM-Based Predictive Current Controller for Three-Phase Grid –Connected VSI *IEEE Trans.*

[8] Le-Huy H, Slimani K and Viarouge P 1991 Analysis and Implementation of a Real-Time Predictive current controller for Permanent-Magnet Synchronous Servo Drives *IEEE Trans.*

[9] Prodanovic M and Green T C 2002. Control of Power Quality in Inverter-Based Distributed Generation *IEEE Trans.* 0-7803-7474-6

[10] Jun L and Dazhi W 2009 Study and Simulation of a Novel Hysteresis Current Control Strategy *IEEE Trans* 978-0-7695-3804-4/09

[11] Karaarslan A 2008 Hysteresis Control of Power Factor Correction with A New Approach of Sampling Techniques *IEEE Transactions* 1-4244-2482-5/08

[12] Shukla A, Ghosh A and Joshi A 2008 Improved Multilevel Hysteresis Current Regulation and Capacitor Voltage Balancing Schemes for Flying Capacitor Multilevel Inverter *IEEE Transactions on Power Electronics* **23**

[13] Vazquez G, Rodriguez P, Ordonez R, Kerekes T and Teodorescu R 2009 Adaptive Hysteresis Band Current Control for Transformerless Single-Phase PV Inverters *IEEE Transactions* 978-1-4244-4649-0/09/

[14] Chun T W and Choi M K 1996 Development of Adaptive Hysteresis Band Current Control Strategy of PWM Inverter with Constant Switching Frequency *IEEE Transactions* 0-7803-3044-7/96

[15] Malesani L, Mattavelli P and Tomasin P 1995 Improve Constant-Frequency Hysteresis Current Control of VSI Inverters with Simple Feed-Forward Bandwidth Prediction *IEEE Transaction* 0-7803-3008-0/95

[16] Chang T Yand Pan T C 1994 A Predictive vector control Algorithm for μ-based induction motor drives using a new space vector controller *IEEE Trans. Ind. Electron.* **41** 97-103

[17] Kazmierkowski M P, Dzieniakowski M A, and Sulkowski W 1991 Novel space vector based current controllers for PWM-inverters *IEEE Trans. Power Electron.* **6** 158-166

[18] Ackva A, Reinold H and Olesinski R 1992 A Simple and Self-Adapting High-Performance Current Scheme for Three Phase Voltage Source Inverter *PECS* Toledo

IOP Conf. Series: Earth and Environmental Science **16** (2013) 012070 doi:10.1088/1755-1315/16/1/012070

Early flame development image comparison of low calorific value syngas and CNG in DI SI gas engine

Ftwi Yohaness Hagos[1], A. Rashid A. Aziz[1] and Shaharin A. Sulaiman[1]

[1]Centre for Automotive Research and Electric Mobility, Universiti Teknologi PETRONAS, Bandar Seri Iskandar, 31750 Tronoh, Perak, Malaysia

E-mail:ftwi@yahoo.com

Abstract. The early flame development stage of syngas and CNG are analysed and compared from the flame images taken over 20° CA from the start of ignition. An imitated syngas with a composition of 19.2% H_2, 29.6% CO, 5.3% CH_4 and balance with nitrogen and carbon dioxide, which resembles the typical product of wood biomass gasification, was used in the study. A CCD camera triggered externally through the signals from the camshaft and crank angle sensors was used in capturing of the images. The engine was accessed through an endoscope access and a self-illumination inside the chamber. The results of the image analysis are further compared with the mass fraction burn curve of both syngas and CNG analysed from the pressure data. The analysis result of the flame image of syngas validates the double rapid burning stage of the mass fraction burn of syngas analysed from in-cylinder pressure data.

1. Introduction

The history of in-cylinder combustion and flow visualization is associated to Rossweiler and Withrow in 1938 when for the first time they modified a spark ignition engine to have an optical window [1]. The combustion behavior of fuel in an internal combustion engine can be studied with the help of flame imaging. This technique is mainly helpful in the study of flame development stage of the combustion. There are different techniques of capturing flame images in the literature. Among these, the most common ones are planar laser-induced fluorescence (PLIF) imaging with the help of laser diagnostic technique and direct visualization techniques like single shot camera or continuous video recording and imaging the chemiluminescence signals of combustion gas radicals. Detail techniques of the visualization techniques are explained elsewhere [2]. The general objective of this study is to explore the employment of syngas in a direct injection spark ignition internal combustion gaseous engine as a potential replacement for fossil derived fuels. A direct visualization technique is preferred that use a self-luminosity assisted by an image intensifier to compare the flame characteristics of syngas and CNG. Further the flame visualization analysis result is compared with the mass fraction burn results if any relationship can be drawn.

2. Methodology

The study was conducted in a four-stroke, single cylinder, direct injection spark ignition research engine setup with a compression ratio of 14:1, schematic diagram, its specification and description about the operation stated elsewhere [3]. The fuel injection pressure was set to be at 18 bar and the engine was run at 1800 rev/min wide open throttle. The injection timing was set to be 120° CA BTDC and the ignition timing at 30° CA BTDC for both fuel types.

2.1. Combustion Analysis

The combustion analysis mainly the mass fraction burn (MFB) was analysed from the pressure readings from the in-cylinder combustion. MFB is a fraction of a chemical energy release as a function of crank angle with a scale of 0 to 1 determined based on burn rate analysis, a method originally developed by Rassweiler and Withrow. The rate of burning process is slower at the time of ignition, reaches its maximum midway and then decreases to close to zero at the charge extinction stage. For a spark-ignition engine, it is commonly represented by the Wiebe Function as:

$$X_b(\theta) = 1 - exp\left[-a\frac{(\theta - \theta_o)^{m+1}}{\Delta\theta}\right] \quad (1)$$

where X_b -MFB, θ_o -Start of combustion, $\Delta\theta$ -Duration of heat release, m -Weibe form factor and a - Weibe efficiency factor [4].

2.2. Flame Imaging

In this study flame image was captured with the help of a Charge-Coupled Device (CCD) camera assisted by an image intensifier to overcome the weak luminescence. The image was taken through the endoscope access installed at 30° from the horizontal in between the two intake valves. The main requirement in the flame imaging was the synchronization of the imaging system with the engine operations like speed and piston position. In the current study, the synchronization was done with a LabVIEW control program along with the cam position, crank angle and TDC signal. Figure 1 shows the setup arrangement of the flame imaging. These all three signals were calibrated and synchronized with the help of a four channel oscilloscope. LabVIEW program was then used to control the external camera triggering position taking the crank angle cam position signals an input from NI-DAQ system.

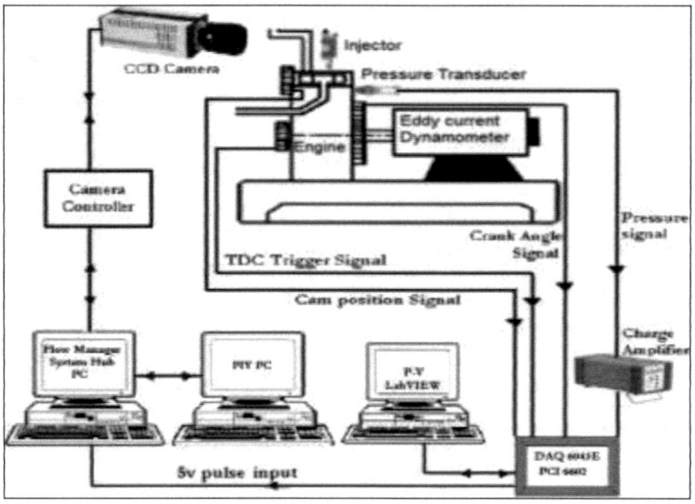

Figure 1. Flame imaging setup

Flame image was captured starting from the ignition up to 20 CA degrees for every 2° CA interval. It is impossible to capture image beyond 10° BTDC as the piston do obstruct the endoscope. A total of 40 flame images were captured at every required crank angle in order to get the average properties of the developing flame for the specific crank angle.

2.3. Fuel Used

A pre-mixed imitated syngas was preferred for this investigation. This was similar in composition with that of syngas used by Papagiannakis et al. [5]. It was supplied in a gas bottle with 160 bar pressure and the net heating value is 7.67 MJ/Nm^3. The CNG used in this test was obtained from local NGV stations pressurized in a bottle with 200 bar. The net heating value was in the range of 38.13-38.96

MJ/kg. Engine warming up and stabilization operation was done with CNG as imitated syngas cost was much higher. Fuel switchover was done once the engine attained stable operation. Since a common injector was used, the engine remained to operate with syngas for more than 5 minutes before formal reading was taken. This pre-condition was to avoid fuel contamination. Besides this, the endoscopic access has a cooling system with a continuous compressed gas at 6 bar flowing. To avoid an image distortion and damage to the endoscope, the engine was shut down after every 10 minute data collection.

3. Result and Discussion

Figure 2 shows the comparison of MFB of low calorific value syngas (producer gas) and CNG at 1800 rev/min analyzed based on Equation (1). MFB of syngas is observed to deviate from the normal s-curve, experiencing a double s-curve with a plateau shape around 60-70% MFB. This is similar to the Double-Weibe function used to model the combustion process in HCCI [6].

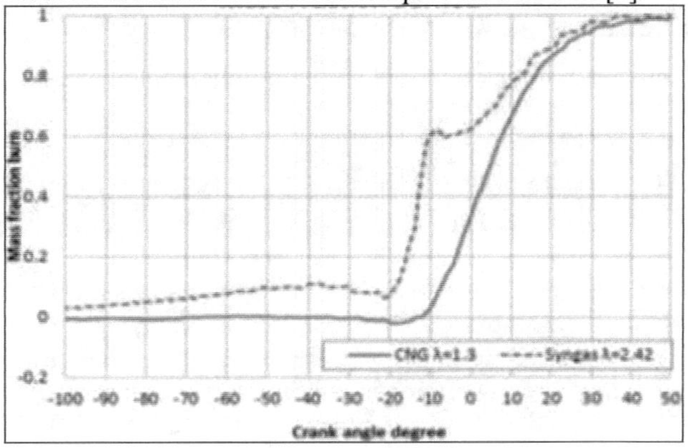

Figure 2. Mass fraction burn

The main purpose of the flame imaging process is to validate the double-Wiebe shape of the MFB of syngas in Figure 2. Though the image capturing should last from the start of ignition up to 10° ATDC to see the full picture of the variation of MFB, there is limitation with the endoscope access of the engine. It gets obstructed in the region from 10° CA BTDC to 10° CA ATDC due to the piston covering the endoscopic access. Therefore, flame is captured from the start of ignition up to 10° CA BTDC.

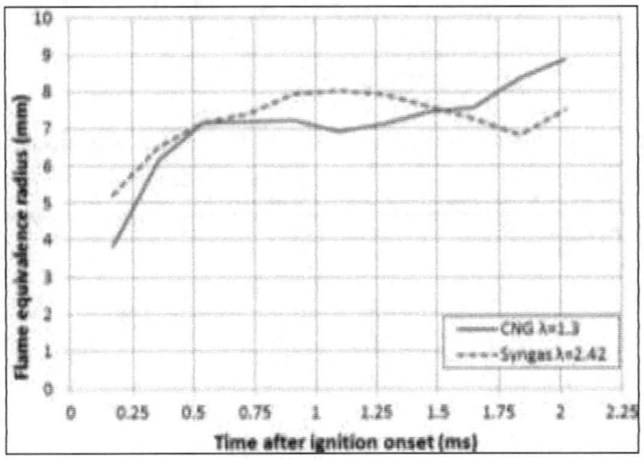

Figure 3. Flame equivalence radius with time for 1800 rev/min

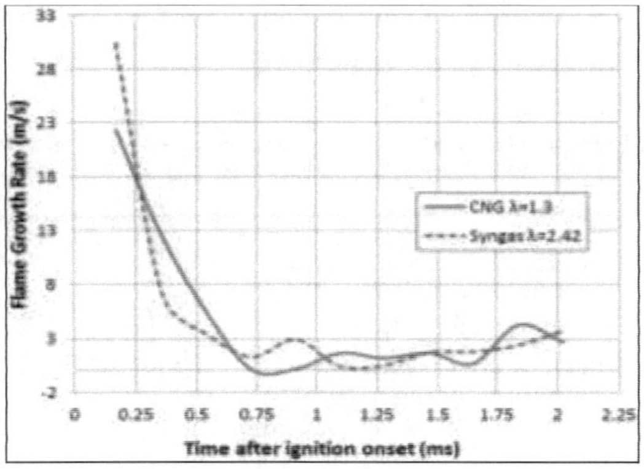

Figure 4. Flame growth rate with time for 1800 rev/min

Images captured by CCD camera are rich in information. Besides, they are accompanied by noises. As a result, the images are pre-processed with an image filtering process before they are subjected to analysis. A detail procedure of the image processing was reported elsewhere [2]. The images are processed for the equivalence radius and flame growth rate. Figure 3and 4 show the profile of the flame equivalence radius and flame growth rate verses time after the start of ignition for both syngas and CNG. The flame equivalence radius of syngas gets its maximum at 1.1 ms after the onset of the ignition and then start declining up to 1.82 ms. After this it again starts rising. For CNG, the flame equivalence radius starts rising at around 1.75 ms. The flame equivalence radius trend indicates that another peak radius for syngas is projected to appear sometime after 2 ms. This matches with the double rapid burning stage of the MFB of syngas, double-Wiebe function in Figure 2. The flame growth rate is observed to start at higher rate and it dropped sharply for both fuels. This is attributed to the shock wave of plasma plasma kernel [7]. Syngas started with higher radius compared to CNG with 30.4 m/s and 22.36 m/s for syngas and CNG, respectively. The minimum value of flame growth rate for syngas is observed at 1.1 ms from the onset. For CNG the minimum value of flame growth rate is observed at 0.75 ms from the onset. The flame growth rate remained constant for both fuels until a rise at around 1.75 ms and 1.82 ms for CNG and syngas, respectively. Both fuels experience different flame-growth rate profiles.

4. Conclusion

The experimental data from a four-stroke, single-cylinder, direct injection spark ignition engine is analyzed for the MFB and compared with the flame image analysis taken over 20 CA degrees for every 2° CA interval. The current fuels under investigation are syngas (a multi-component fuel with five constituent gases in the makeup with different proportions) and CNG. The profile of the equivalence radius of both CNG and syngas observed to match the mass fraction burn curves. While CNG remained in its early flame development stage in this duration, Syngas observed to experience its first rapid-burning stage. Due to the endoscope access limitation in the region from 10° CA BTDC to 10° CA ATDC, the second rapid-burning of syngas and the rapid-burning stage of CNG could not be observed. However, there is indication for such occurrence from the profile of equivalence radius profile as the profile start rising around 1.75 ms for both fuels. Therefore, this conform that the combustion of syngas takes the double-Weibe function of mass fraction burn, experiencing a double rapid burning stage. The reason behind is unknown and needs further investigation. The flame growth rate is also analyzed and presented, both fuels experiencing differently at the early stage.

References

[1] Ferguson, C. R. and A. T. Kirpatrick, *Internal Combustion Engines: Applied Thermosciences*: John Wiley and Sons, Inc., 2001

[2] Anbese, Y. T., "Early Flame Characteristic Study in a Lean CNG DI Combustion," PhD, Mechanical Engineering Department, Universiti Teknologi PETRONAS, 2011

[3] Hagos, F. Y., A. R. A. Aziz, S. A. Sulaiman, and Firmansyah. Combustion Characteristics of Late Injected CNG in a Spark Ignition Engine under Lean Operating Condition. 2012 *J. Applied Sci.* **12** 2368-2375

[4] Heywood, J. B., *Internal Combustion Engine Fundamentals*. New York: McGraw Hill International, 1988

[5] Papagiannakis, R. G., C. D. Rakopoulos, D. T. Hountalas, and E. G. Giakoumis. Study of the Performance and Exhaust Emissions of a Spark-Ignited Engine Operating on Syngas Fuel 2007 *Int. J. Altern. Propul.* **1** 190-215

[6] Yasar, H., H. S. Soyhan, H. Walmsley, B. Head, and C. Sorusbay. Double-Wiebe Function: An Approach for Single-Zone HCCI Engine Modeling. 2008 *Appl. Therm. Eng.* **28** 1284-1290

[7] Anbese, Y. T., A. R. A. Aziz, and Z. A. B. A. Karim. Flame Development Study at Variable Swirl Level Flows in a Stratified CNG DI Combustion Engine using Image Processing Technique 2011 *J. Applied Sci.* **11** 1698-1706

IOP Conf. Series: Earth and Environmental Science **16** (2013) 012071 doi:10.1088/1755-1315/16/1/012071

Characteristics of compressed natural gas jet and jet-wall impingement using the Schlieren imaging technique

M. A. Ismael[1], **M. R. Heikal**[2], **M. B. Baharom**[3]
Center for Automotive Reseach and Electric Mobility, Universiti Teknologi PETRONAS, Bander Seri Iskandar, 31750 Tronoh, Perak, Malaysia

E-mail: xmhadosx@yahoo.com

Abstract. An experimental study was performed to investigate the compressed natural gas jet characteristics and jet-wall impingement using the Schlieren imaging technique and image processing. An injector driver was used to drive the natural gas injector and synchronized with camera triggering. A constant-volume optical chamber was designed to facilitate maximum optical access for the study of the jet macroscopic characteristics and jet-wall impingement at different injection pressures and injectors-wall distances. Measurement of the jet tip penetration and cone angle at different conditions are presented in this paper together with temporal presentation of the jet radial travel along the wall.

1. Introduction

With increasing concerns about the energy crises and the harmful effects of conventional fossil fuel emissions such as those from diesel and gasoline fuels, alternative fuels have become attractive to power internal combustion engines. Compressed natural gas (CNG), and other gaseous fuels have been widely used as alternative fuels in internal combustion engines because of their potential to maintain engine performance and reduce the emissions of hydrocarbon [1]. Many researchers have used CNG in different engine types such as spark ignition engines, port injection lean burn engines, dual-fuel-pilot injection engines, and direct injection engines [2]. Significant research has been carried out on these engines, the most promising of these; the direct injection engine requires further development in order to investigate the injection full potential. There are two ways for direct injection natural gas engines: to start the injection at an early stage of the compression stroke and ignite it by a spark ignition system or to start the injection at a late stage of the compression stroke taking advantages of high compression ratio of diesel engines. Still many problems associated with the second method of direct injection compression ignition (DI CI) engines. For example, the characteristics of natural gas is differs considerably from those of the diesel fuel. Furthermore, previous studies focused on liquid fuel spray characteristics and wall impingement [3]. The results of these studies demonstrated the importance injection strategy as it plays an important role in the diesel fuel spray, atomization, droplet size and evaporation which in turn have a significant effect on the mixture formation inside the engine. Although the gas plays an equally important role in gas engines, its effect on the mixture preparation is different than liquid jets. Many studies on high pressures natural gas jet characteristics have been conducted using different techniques [4]. However, high injection pressures of the gas require high performance injectors and associated with high energy consumption in the production of such high pressures. One of the objectives of this study is to examine the effect of low gas injection pressure on the characteristics of the jet, wall impingement and hence mixture formation.

In the present study, the characteristics of a natural-gas jet (12-20 bar) and its wall-impingement into a constant-volume chamber were extensively investigated under different injection pressures and injector-wall distance using the Schlieren system and the obtain images were processed using Matalb software.

2. Experimental methods

2.1. Experimental test rig

Spray research has been performed in a number of test rigs: special research engines, high pressure, high temperature (bombs), flow rigs and constant volume chambers. a constant volume test rig was considered as the best option for fundamental research as it combine good optical access, large observation areas and good control of scavenging before injection. The chamber was 90mm high and 94mm wide was designed and built to allow for optical analysis of both the jets and wall impingement. The bottom plate of the test rig was from aluminum (piston material). Different methods have been used in previous studies for achieving representative temperature conditions in the rig e.g. some used direct heating of the walls (bottom) of the test rig [5]. While others used an electric heater [6]. In this study, a controlled electric heater was used to heat the bottom plate up to 200°C and a thermocouple was inserted in the plate to measure the wall temperature. The variation of the injector-wall distance, a holder of the CNG injector was mounted in the center of test rig top plate. The injector-wall-distance was changed in the range 10-80mm by moving this top plate up and dawn.

2.2. Experimental Procedure

The experimental conditions are listed in Table 1. The natural gas found in Malaysia which consists mainly of mainly methane, ethane and propane was used in this project. Typical composition of this natural gas found in Malaysia is shown in the Table 2. the CNG injection system consisted of a CNG cylinder (the pressure inside the cylinder was up to 200 bar), two pressure regulators (double stage), pressure gage, pressure regulating valve, and a CNG injector (solenoid-valve injector with maximum pressure of 20bar). A National Instrument driver and a power supply (12V) were used to drive the injector with injection duration10 ms. The gas flow meters was calibrated using the positive displacement method described in [7]. The injection duration (solenoid valve opening) against the mass flow rate with different injection pressures are shown in Figure 1. The images were captured using a Schlieren system which consists of a lamp, two concave mirrors, knife edge and a high speed video camera (Phorton, FASTCAM-APX) operated at speed of 4,000 frames per second with effective pixel size of 640x128. A Nikon 60 mm f/2.8D micro-Nikkor lens was used to accompany the camera.

Table 1. Experimental Condition

Fuel	Compressed Natural Gas
Injection pressure	12-20bar
Injection duration	10 ms
Ambient pressure	1 bar
Wall position	80-40 mm
Wall temperature	300-473 k

Table 2. CNG Composition

Items	Volume Fraction
CH_4	96.160
C_2H_6	1.096
C_3H_8	0.136
N_2	0.001
CO_2	2.540

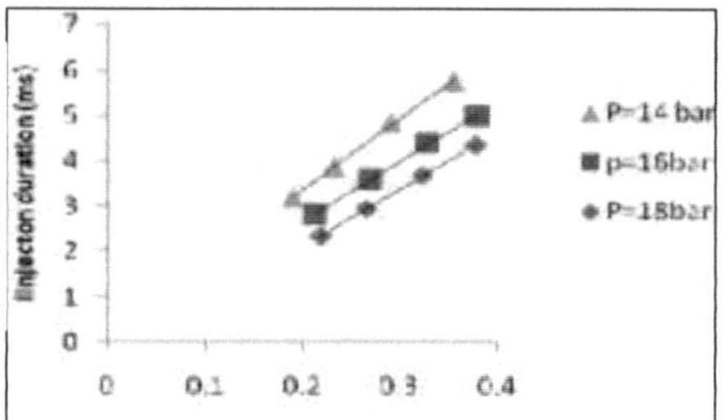

Figure 1. Effect of injection pressure on time-resolved injection rate

2.3. Image Processing

The captured images were processed in order to calculate a number spray characteristic such as the spray penetration, spray angle and impingement foot-print. The spray edges were detected using the canny image detection technique [8]. Several methods were presented in the literature for obtain a spray contour, but no consensus was found on the definition of a spray (cone) angle equivalent spray angle or standard angle. Figure 2 shows how the spray tip and the spray cone angle were defined. The

IOP Conf. Series: Earth and Environmental Science **16** (2013) 012071 doi:10.1088/1755-1315/16/1/012071

spray angles were measured by drawing horizontal lines at 20 mm from the nozzle tip and measuring the angle between the edges of the spray [9]. The spray tip penetration was measured along the centerline axis of the spray as the distance between the injector nozzle tip and the furthest spray point. To convert pixels measurements to SI metric measurements a calibration image of a graduated scale was taken. Calibration of the gas jet dimensions was also achieved by taking an image of a scale attached to the fuel injection vertical plane. The value of the one pixel was found out to be equivalent to 0.27 mm.

3. Results and Discussion

3.1. Effect of Injection Pressure

Figure 3 shows the Schlieren images of the CNG under ambient condition and injection pressure of 18, 16, and 14 bar at different time intervals from 0.25 to 2.75 ms at interval Of 0.25ms. As shown in the spray images, the injection pressure has a large influence on the spray structure as expected. However, for the case of low injection pressure 14 bar, a weak cone shape and lower tip penetration were observed.

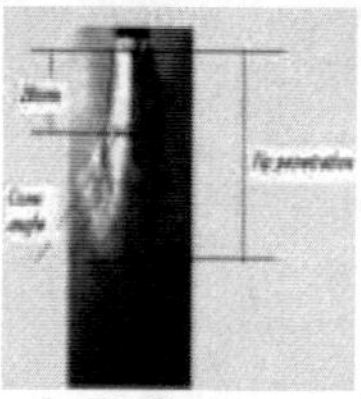

Figure 2. Definitions of jet penetration and jet cone angle

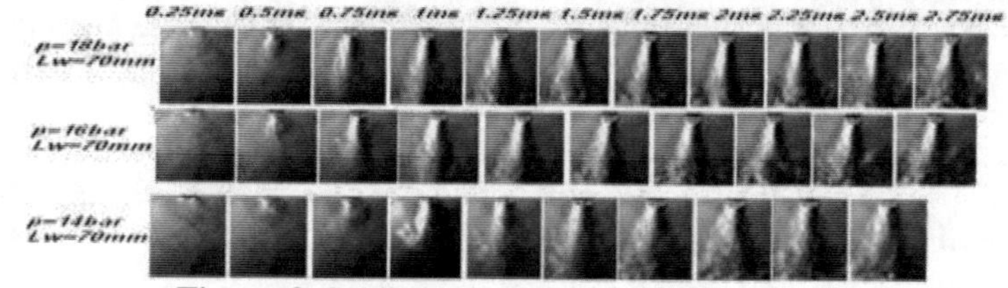

Figure 3. Schlieren images for CNG at 18, 16 and14 bar and injector-wall distance 700mm

With an increase of the injection pressure to 20 bar, a strong vortex in the spray tip plume was found to develop. Figure 5 shows is the spray cone angle measurement against the time after the start of injection. The spray cone angle was measured at a distance of 20mm from the end of injector nozzle was found to increase with increase of injection pressure.

Generally, both the penetration and cone angle (See Figure 4 and 5) have shown an increase with the increase of injection pressure. Figure 6 shows the effect of the wall distance from the injector on the tip penetration. It can be seen that the tip penetration rate decreases with the jet travel as the cone angle and hence the aerodynamic resistance increase and the loss of jet momentum due to air entrainment by the jet. The effect of injection pressure on the development of the impinged jet is shown in Figure 7. The results showed that as the injection pressure increased, both the radial penetration and the height increased due to the effect of the higher momentum of the moment of impingement.

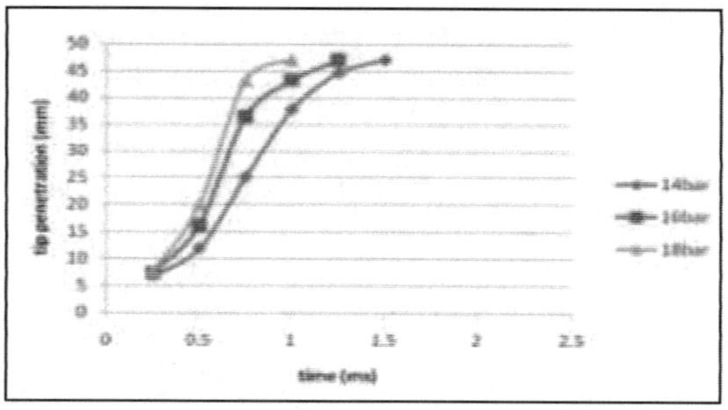

Figure 4. Spray tip penetration of CNG against time after start of injection (P= 18, 16 and 14 bar)

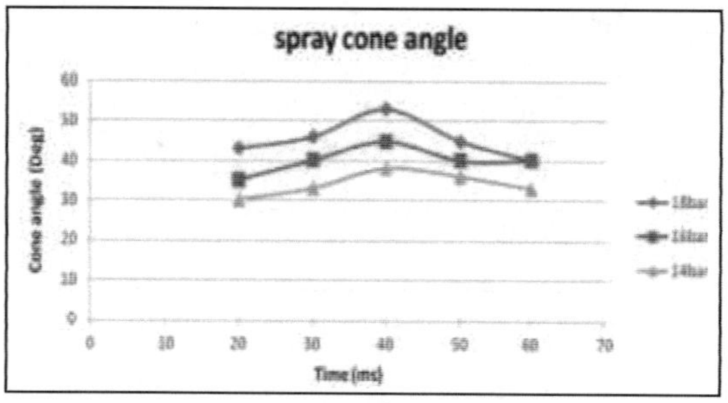

Figure 5.SFpary cone angle of CNG against time after start of injection

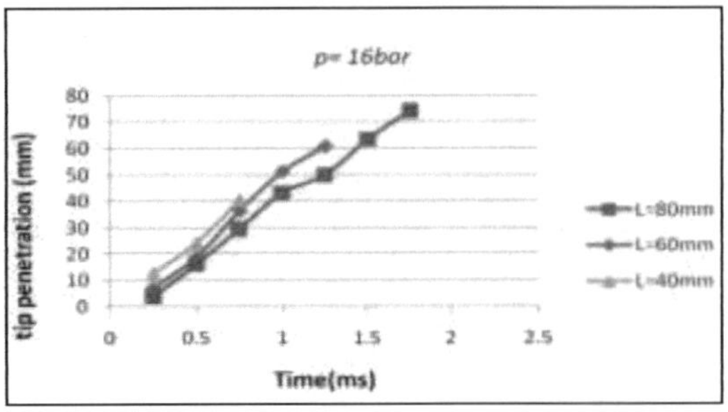

Figure 6. Effect of injector-wall distance on and tip pentaration against time.

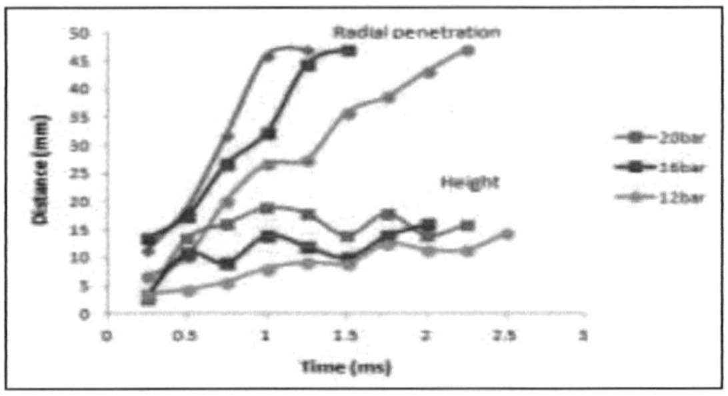

Figure 7. Effect of injection pressure on the radial penetration and the height of the jet impinged on the wall.

4. Conclusion

As expected the Schlieren images of natural gas jet showed, the jet structure is largly influnced by the injection pressure. Generally, the tip penetration, cone angle and radial penetration all increased with the increase of the injection pressure. However, the rate of jet penetration decreased with the jet travel as a result of the increase in the cone angle and hence the aerodynamic resistance as well as the loss of jet momentum due to air entrainment. The radial spraed of the impinged jet on the wall was also found to increase with the increase in injection pressure while the hight of the jet plume on the wall was not significantly affected.

Refferences

[1] S. H. Schneider the Greenhouse Effect : Science and Policy 1989 *Science* **243** 771 - 781

[2] Semin, Rosli A. B. A Technical Review of Compressed Natural Gas as an Alternative Fuel for Internal Combustion Engines 2008 *American J. of Engineering and Applied Since* **1** 302 - 311

[3] Kennaird D., Crua C., Lacoste J., M. R. Heikal, M. R. Gold, and N. S. Jackson In-Cylinder Penetration and Break-Up of Diesel Sprays Using a Common-Rail Injection System 2002 *SAE Technical Paper* **1** 1626

[4] T. Senoo, M. Sasaki, and M. Shioji, Spark-Ignition Stability of Natural-Gas Jets with Impinggement on Cavity Wall 2010 *FISITA World Automotive Congress*

[5] Y. S. Shim, G. M. Choi, and D. J. Kim Numerical Modeling of Hollow-Cone Fuel Atomization, Vaporization and Wall Impingement Processes under High Ambient Temperatures 2008 *International Journal of Automotive Technology* **9** 267-275

[6] M. Mbarawa, B. Edward, and R. Thomas Experiments and Modelling of Natural Gas Combustion Ignited by a Pilot Diesel Fuel Spray 2001 *International Journal of Therma Sciences* **40** 927 - 936

[7] C. Pinho, "The Positive Displacement Method for Calibration of Gas Flow Meters. The Influence Of Gas Compressibility 2012 *Applied Thermal Engineering* **41** 111-115

[8] J. Canny A Computational Approach to Edge Detection 1986 *IEEE Transactions on Pattern Analysis and Machine Intelligence* **8** 679-98

[9] Chengli, Zhang Application Research on Digital Image Technology in the Measurement of Nozzle Spray Cone Angle 2011 IEEE *Instrumentation and Measurement Technology Conference* 1-5

Comparative study of water quality of rivers used for raw water supply & ex-mining lakes in Perak, Malaysia

K U Orji*, N Sapari, K W Yusof, R Asadpour, and E Olisa

Department of Civil Engineering, Universiti Teknologi PETRONAS, Bandar Seri Iskandar, 31750 Tronoh, Perak, Malaysia

E-mail: suunyboy2008@yahoo.com

Abstract. Ex-mining lakes are seldom used as sources of raw water for the treatment of public water supply due to the general view that they are highly polluted. This study examined the water quality of these lakes, compared and contrasted them to the water quality of the rivers used for Perak drinking water supply. Ten water samples were analyzed from different ex-mining lakes. Two water samples were from Kinta and Perak rivers. They were analyzed for physico-chemical properties such as temperature, pH, EC, TDS, SO_4^{2-}, COD, Cl^-, Na^+, Fe, As, and Pb. The results showed that temperature varied from 28.1°C to 34.1°C, pH 6.2 to 9.0, EC 55 to 400 µs/cm, turbidity 5.6 to 74.2 NTU, TDS 36.8 to 268mg/l, Cl^- 0.483 to 3.339mg/l, SO_4^{2-} 0.051 to 15.307mg/l, Na 0.669 to 3.668mg/l, Fe 0 to 0.14mg/l, As 0 to 0.004mg/l, and Pb 0.019 to 0.075mg/l. All the samples were highly turbid, had slightly high concentration of Pb, and had common water quality problem. The ex-mining lakes can also be used to supply water after treatment since these rivers are already being used by the Metropolitan Utilities Corporation for water treatment. The ex-mining pools can be used as alternative sources of drinking water supply to the people of Perak.

1. Introduction

Malaysia is facing an ever growing need for the management of her water resources due to rapid socio-economic development [1]. For the future water needs to be met, the government must make every effort to develop, conserve, utilize and manage all the available potential water sources.

River is the major source of raw water supply. Due to the increase in river pollution, it becomes pertinent that the other water sources be developed and tapped in order to either augment or serve as alternative sources of water supply. Shah Alam and Sandakan have introduced rainwater harvesting in new housing developments [2]. Kelatan, Perlis, Pahang, Selangor, and Terengganu States have also been combining surface water with groundwater for their integrated water resources management [3].

Ex-mining pools have basically been used for irrigation, recreational activities, and retention ponds except for drinking due to the view that they are highly polluted. Study has shown that ex-tin mining lakes located in Bestari Jaya area, Selangor, are highly polluted with heavy metals [4]. More lakes have to be analyzed in other parts of Malaysia in order to assess their suitability for public water supply. The objectives of this study were to examine the water quality of ex-mining lakes in Perak; and to compare and contrast them to the water quality of the rivers used for Perak drinking water supply.

2. Materials and Methods

2.1. Sample locations

Twelve water samples, ten from ex-mining lakes and two from rivers, were collected for analyses on water quality. Details of the sampling sites are shown in Table1. The sample collection was done on the 29th of July and 2nd of August, 2012. River Perak and River Kinta are currently used by the Metropolitan Utilities Corporation Sdn Bhd (MUC) as intake points for water supply in the State of Perak.

Table 1. Sampling information with sampling sites descriptions

No	Date /Time	Site/Source	Site Description	Coordinates	Elevation (*m*)
S1	29/7/12 11:30am	Lalua Tronoh Mines/Ex – Mining Pool	Inactive mining site that is partly surrounded by residential buildings and forests. The vicinity of the other part is used for cattle grazing.	4°25'47.4"N 100°58'53"E	30.79
S2	29/7/12 12:01pm	Coop. Sec. of State, Tronoh/Ex–Mining Pool	It has the same future as in S1.	4°25'05.9"N 101°00'59"E	27.13
S3	29/7/12 6:31pm	Bemban 1, Batu Gajah/ Ex–Mining Pool	A lake with muddy shore & surrounded by forests. It is located near the roadside and a construction site.	4°28'05.9"N 101°00'59"E	42.07
S4	30/7/12 4:16pm	Bemban 2, Batu Gajah/ Ex–Mining Pool	A mangrove area. Sewage is discharge into it through a drainage busy road. Some restaurants and stores in the vicinity.	4°28'08.4"N 101°01'31"E	40.85
S5	31/7/12 4:31pm	Train Station, Batu Gajah/Ex–Mining Pool	It is located adjacent to a busy road and train station. It is almost fully surrounded by forests and used for fishing by the villagers.	4°27'44.7"N 101°03'04"E	20.12
S6	30/7/12 6:17pm	Taman Bemban, Batu Gajah/Ex–Mining Pool	It lies by the side of road, agricultural area and residential buildings. Wreckages of building materials deposited close to shoreline of it.	4°27'57.0"N 101°00'56"E	42.38
S7	31/7/12 7:05pm	Kampung Bemban G.G.1, Batu Gajah/ Ex – Mining Pool	It lies by the side of forests, agricultural area and residential buildings. It is basically used as irrigation water.	4°27'52.0"N 101°00'54"E	42.99
S8	31/7/12 6:10pm	Kampung Bemban G.G.2, Batu Gajah/ Ex – Mining Pool	It has the same future as in S7.	4°27'48.2"N 101°00'53"E	41.46
S9	02/8/12 7:05pm	Taman Tasik Putra, Opp. ESSO Mobil/ Ex – Mining Pool	It is located opposite a filling station and surrounded by forests.	4°23'56.9"N 100°59'09"E	18.90
R10	31/7/12 5:34pm	Kepayan Bay, Bota/Sungai Perak	A river with fishing site. Turtle breeding center, market, restaurants and abattoir are also located within the vicinity.	4°21'18.1N 100°52'25"E	18.29
R11	02/8/12 6:00pm	Batu Gajah/ Sungai Kinta	The river cuts across residential area. The bank faces erosion. Malls, restaurants, and busy rods are within its vicinity.	4°28'25.9"N 101°03'08"E	25.00
S12	02/8/12 6:30pm	Universiti Teknologi Petronas, Tronoh/ Ex – Mining Pool	This is an open mining lake that is surrounded by few trees, and close to a filling station and hostels.	4°23'04.0"N 100°58'43"E	18.60

2.2. Methods

Temperature, pH, turbidity, and Electrical Conductivity (EC) were measured in the field at each sampling point. The Total Dissolved Solids (TDS) was obtained by multiplying the EC values by a conversion factor of 0.67. Temperature and pH were measured with pH meter (HACH – Sension 4, USA), Turbidity with turbidimeter 2100P (HACH, USA), and EC with EP meter (MYRON, USA). The samples were measured in triplicates and the average value calculated. All analyses were of analytical grades and complied with the standard methods [5]. SO_4^{2-} and Cl^- in the samples were measured using Ion Chromatography (Metrohm) and other cations by Atomic Absorption Spectrophotometer (AAS, AA-6800, SHIMADZU, Japan). Chemical Oxygen Demand (COD) was determined by digestion using DRB 200 and colorimetric method using Spectrophotometer DR 2800 (HACH, USA).

3. Results and Discussion

Physico-chemical parameters from the 12 sampling locations are summarized in Table 2.

Table 2. Physico-Chemical Parameters in the Water Samples in the Study Area

No	Temp. °C	pH	Turbidity NTU	EC µs/cm	TDS mg/l	COD mg/l	SO_4^{2-} mg/l	Cl^- mg/l	Na^+ mg/l	As mg/l	Fe mg/l	Pb mg/l
S1	30.5	9.0	85	140	93.8	14	0.205	1.014	1.215	ND	ND	0.047
S2	29.8	8.8	39.3	200	134	22	0.051	1.103	1.471	ND	0.14	0.047
S3	34.1	8.8	34	275	184	31	15.31	1.747	0.947	ND	ND	0.047
S4	33.7	7.4	41.3	160	107	10	3.030	3.339	3.071	ND	ND	0.075
S5	29.1	7.5	28.5	120	80	9	0.336	1.360	0.947	ND	ND	0.070
S6	29.8	7.7	19	344	230	51	0.273	1.550	1.710	0.004	ND	0.056
S7	28.1	7.8	51.4	355	238	20	0.692	2.517	2.738	0.002	ND	0.042
S8	30.6	8.8	69.4	400	268	38	0.176	3.148	2.869	0.005	ND	ND
S9	31.7	7.5	5.6	115	77	11	1.029	2.895	0.669	0.002	ND	ND
R10	29.9	6.2	74.2	55	36.8	40	0.977	0.545	2.230	0.003	ND	0.019
R11	30.8	7.1	11.8	340	228	35	0.956	2.253	3.668	0.004	ND	0.037
S12	30.6	7.4	32.9	100	67	4	2.408	0.483	1.123	0.004	ND	0.019
WHO	-	6.5-8	5	1000	500	10	250	250	200	0.001	0.3	0.01

ND = Not Detected

3.1. Physical Parameters

3.1.1. Temperature

Temperatures of the samples fell within the range (27.8 – 35.3°C). Similar range was reported by Yap et al, 2009 [6].

3.1.2. pH

The pH ranged from 4.96 to 9.81, similar to the one reported by Yap et al, 2009 [6]. R10 was most acidic water. This is in line with the findings by Chau and Jiang, 2002 [7], which argued that natural rivers are always slightly acidic because they normally originate from rain water and incorporate with acid released from the leaves of the forests within the area. S1 had the highest alkaline in its water. This could be due to high algal concentrations in the pool.

3.1.3. Electrical Conductivity (EC)

The results showed that the EC of all the water samples were low and met up with the WHO standard [8].

3.1.4. Total Dissolved Solutes (TDS)

The highest range of TDS was found in S8. This could be due to the runoff of agricultural activities in the area and leaching of soil contaminant. However, TDS in all the samples were still within the WHO limit.

3.1.5. Turbidity

All the water samples fell above the WHO standard. Highest turbidity was recorded in S1 and could be caused by the high presence of fine clay, silt particles, and plankton observed in the lake.

3.2. Chemical Parameters

Sulfate, Chloride, Sodium, Arsenic, and Iron in the water samples were found to be within the prescribed limit. The concentrations of Pb ranged from 0.00 to 0.075mg/l. This result was close to the ones obtained by Yap et al, 2009 [6] and Yap et al, 2011 [9], wider than that reported by Shuhaimi et al, 2012, [10], and far wider than the report obtained by Muhammad et al, 2010 [11]. The concentrations of Pb in all the samples except S8 and S9 are slightly above the prescribed limit.

3.3. Organic Pollutant

S6 had the highest COD value. This could be as a result of the indiscriminate discharge of agricultural wastes into the water. Reports have shown that agricultural, industrial, and domestic wastewaters are considerable sources of high COD [12].

4. Conclusions

Analyses showed that the rivers and the mining lakes are highly turbid, have high concentrations of Pb, and have common water quality problem. This implies that water from the ex-mining lakes can also be used for water supply after treatment since the rivers are already used by the MUC for treatment. However, future work should carry out more tests on other water quality parameters and at different seasons.

Acknowledgements
The authors are thankful to the management and authorities of the Universiti Teknologi PETRONAS, especially the Department of Civil Engineering for the facilities provided during this study.

References

[1] Tuncok, I. K., Mays, L. W., and Briscoe, J. Water Resources Management in Developing Countries and the role of World Bank 1999 *29th Annual Water Resources Planning and Management Conference, ASCE*

[2] Mohd-Shawahid, H. O., Suhaimi, A. R., Rasyikah, M. K., Jamaluddin, S. A., Huang, Y. F., Farah, M. S. Policies and Incentives for Rainwater Harvesting in Malaysia 2007 *Rainwater Utilization Colloquium*

[3] Azuhan, M. Utilization of Groundwater Resources in Peninsular Malaysia 2012 *J. of Groundwater and Geoscience*

[4] Muhammad, A. A., Mohd, J. M., Ismail, Y. Study of Water Quality and Heavy Metals in Soils and Water of Ex – Mining Area Bestari Jaya, Peninsular Malaysia 2010 *Int. Journal of Basic and Applied Sci.*, **10** 7 – 27

[5] APHA – AWWA Standard Methods for Examination of Water and Wastewater 2005 *American Public Health Association/American Water Works Association.*

[6] Yap, C. K., Fairuz, M. S., Yeow, K. L., Hatta, M. Y., Ismail, A., Ismail, A. R., and Tan, S. G. Dissolved Heavy Metals and Water Quality in the Surface Waters of Rivers and Drainages of West Peninsular Malaysia 2009 *Asian Journal of Water, Environment and Pollution* **6** 51 – 59

[7] Chau, K. W. and Jiang, Y. W. Three – Dimensional Pollutant Transport Model for the Pearl River Estuary 2002 *Journal of Water Research* **36** 2029 – 2039.

[8] World Health Organization (WHO) Guidelines for Drinking-Water Quality: 4th Edition, WHO Press, Geneva, Switzerland. 2011

[9] Yap, C. K., Chee, M. W., Shamrina, S., Edward, F. B., Chiew, W., and Tan, S. G. Assessment of Surface Water Quality in the Malaysian Coastal Waters by Using Multivariate Analyses 2011 *Journal of Sains Malaysiana* **40** 1053-1064

[10] Shuhaimi-Othman, M., Nadzifah, Y. and Azmah, M. Metal Concentrations in Sungai Sedii Kecil, Johor, Peninsular Malaysia 2012 *Journal of Tropical Marine Ecosystem* **1** 15-23.

[11] Muhammad, A. A., Mohd., J. M., and Ismail, B. Y. Study of Water Quality and Heavy Metals in Soil and Water of Ex-Mining Area of Jaya, Peninsular Malaysia 2010 *Int. Journal of Basic & Applied Sciences* **10**

[12] Mthembu, M. S., Djarova, T. G. and Basson, A. K. Effect of Agricultural and Industrial Developments on the Quality of Water at UMhlathuze River (Northern Coast of Kwa-Zulu Natal RSA) 2011 *African Journal of Microbiology Research* **5** 5780-5786

Generalized equivalent circuit model for ultra wideband antenna structure with double steps for energy scavenging

Oon Kheng Heong[1], Goh Chin Hock[2], Chandan Kumar Chakrabarty[3] and Goh Tian Hock[4]

[1,2,3,4]UNITEN Putrajaya Campus, Jalan IKRAM-UNITEN, 43000 Kajang, Selangor.

Email: [1]oonkh87@yahoo.com, [2]chinhock@uniten.edu.my, [3]chandan@uniten.edu.my, [4]hock321@hotmail.com

Abstract. There are various types of UWB antennas can be used to scavenge energy from the air and one of them is the printed disc monopole antenna. One of the new challenges imposed on ultra wideband is the design of a generalized antenna circuit model. It is developed in order to extract the inductance and capacitance values of the UWB antennas. In this research work, the developed circuit model can be used to represent the rectangular printed disc monopole antenna with double steps. The antenna structure is simulated with CST Microwave Studio, while the circuit model is simulated with AWR Microwave Office. In order to ensure the simulation result from the circuit model is accurate, the circuit model is also simulated using Mathlab program. The developed circuit model is found to be able to depict the actual UWB antenna. Energy harvesting from environmental wirelessly is an emerging method, which forms a promising alternative to existing energy scavenging system. The developed UWB can be used to scavenge wideband energy from electromagnetic wave present in the environment.

1. Introduction

Electromagnetic wave sources always present in the environment such as wireless local area network, global positioning system, global system for mobile communications and 4th generation mobile communication. According to federal communication commission definition, UWB refers to a wireless technology that employs a bandwidth larger than 500 MHz or 20% of the center frequency. Ultra Wideband (UWB) antenna is able to scavenge [1] the available energy into the operating bandwidth.

Until now, there are many equivalent circuit models proposed by antenna researchers from the aspect of input impedance or admittance matching. For examples, Wang [2] introduced a degenerated Foster canonical form for electric and magnetic antenna model; Wang and Li [3] circuit refinement method consists of a narrow band model augmented with a macro model; Ansarizadeh [4] circuit topology for rectangular microstrip patch antenna by using non-linear curve-fitting optimization technique to determine exact the parameters of the equivalent circuit model; Ma [5] lumped equivalent circuit model for antenna by using parallel resonant circuit and Zhou [6] conceptual circuit model by connecting input impedance with a parallel and a series LC resonant circuit.

The equivalent circuit of a rectangular narrowband microstrip patch antenna is been related to the physical dimensions of the antenna. Cavity models were being used to model the geometry of a rectangular microstrip patch antenna. Most of the microstrip patch antennas are usually modeled as a simple parallel resonant RLC circuit. There are a lot of work has been done to calculate the RLC values in the circuit model. For UWB antenna, its radiating element can be seen as several parallel RLC circuit in series due to its matching bandwidth are the result of several adjacent resonances.

2. Research Methodology

The steps taken to accomplish this research work for energy scavenging are as shown in Figure 1.

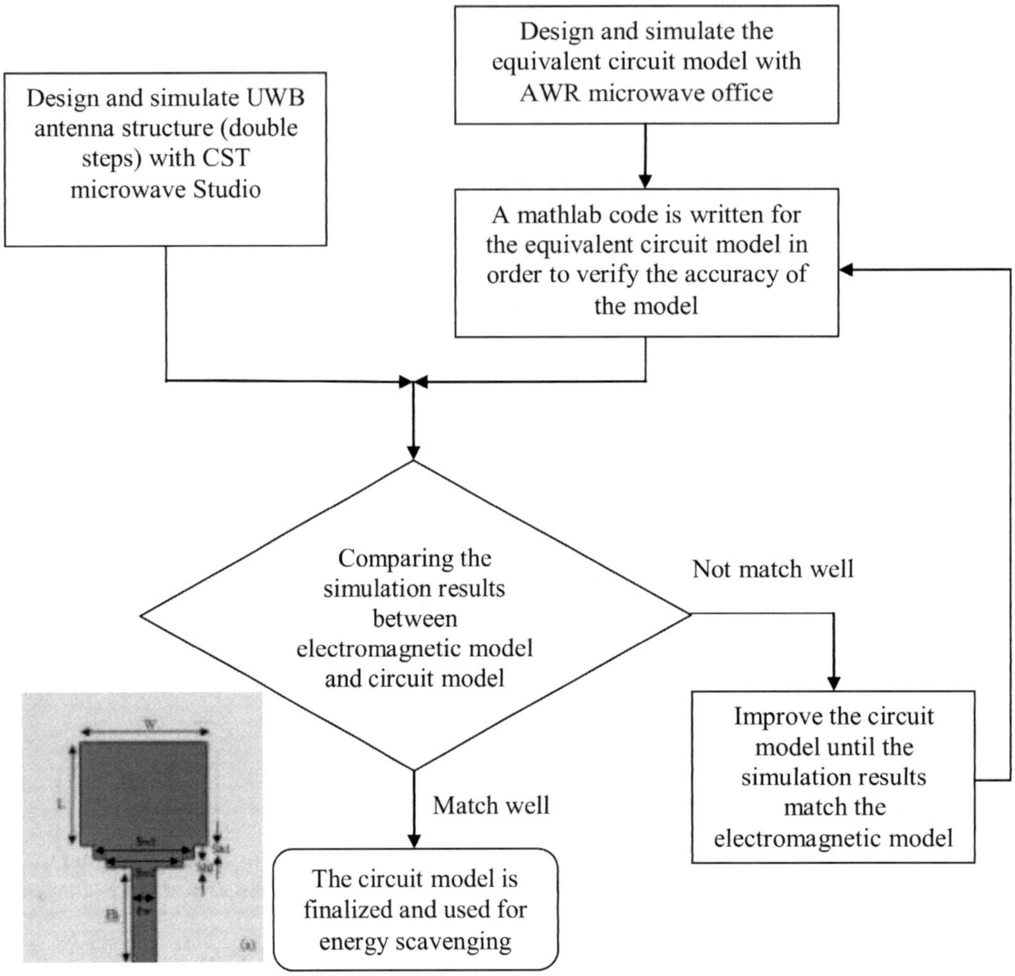

Figure 1. Flowchart of research workflow for UWB antenna with double steps

Figure 1 the design of antenna structure and circuit model simultaneously. To design the ultra wideband antenna for energy scavenging, it is essential to determine the material used for the antenna structure. In this case, the material used is FR4 board due to its ease of fabrication. The printed circuit board has the substrate thickness of 1.6mm and dielectric constant of 3.8. While the patch shape for the antenna is chosen to be rectangular in order to ease the designing process. In designing the UWB antenna with the bandwidth of 3.1 to 10.6GHz, several parameters such as patch shape, patch dimension, feed line width, steps, slots and ground plane need to be taken into account. In this research

IOP Conf. Series: Earth and Environmental Science **16** (2013) 012073 doi:10.1088/1755-1315/16/1/012073

work, steps and slots are introduced to design the antenna to obtain good return loss result. Moreover, the width of the feed line is also fine tuned using CST microwave studio for a better impedance matching in order to increase the efficiency of energy scavenging.

After several trials with different prototype circuit model, patch cavity model is chosen. It is used as the basic design of the equivalent circuit model for ultra wideband rectangular printed disc monopole antenna. It consists of a series resistor and inductor which represent the surface resistance (R_S) and surface inductance (L_S) with combination of parallel inductor and capacitor (LC) circuit or also known as resonant circuit to oscillate at its resonant frequency. Figure 2 shows the general ultra wideband equivalent circuit model for rectangular printed disc monopole antenna with double steps. A tuning capacitor (Ct) is also introduced into the equivalent circuit model to set the resonance frequency for a better impedance matching with the antenna. In order to determine the accuracy of the equivalent circuit model, a Mathlab program code is written for the circuit model as shown in Figure 3. By simplifying the circuit model, the input impedance Z_{in} and the reflection coefficient can be calculated using the impedance at the ports, and then the return loss is plotted. The simulated return loss of the Mathlab and AWR microwave office is compared in order to verify the accuracy.

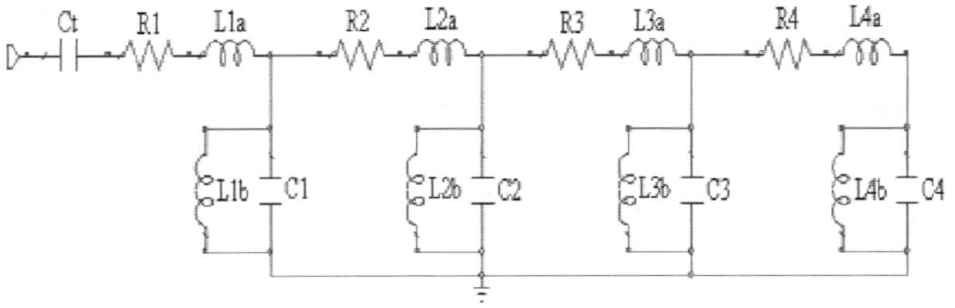

Figure 2. Equivalent circuit model for ultra wideband rectangular printed disc monopole antenna with double steps

3. Results and Discussion

The coupling effects between inductors for same patch cavity model is improved by introducing inductor coupling coefficient (k) between L1a and L1b, L2a and L2b, L3a and L3b, L4a and L4b as shown in Figure 3.

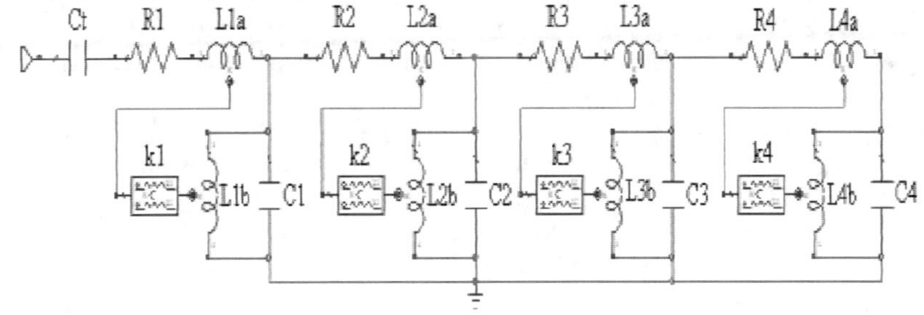

Figure 3. Improved Equivalent circuit model for ultra wideband rectangular printed disc monopole antenna with double steps

Figure 4 illustrates the comparison of return loss result between the circuit model with inductor coupling coefficient (k) and electromagnetic model with CST. It shows that the equivalent circuit model with inductor coupling coefficient (k) match very well with the electromagnetic model. The operating frequency and depth of resonant frequency from both simulation results are quite close to each other. There is no shifting in frequency between circuit model and antenna structure simulation result. Hence, it shows that there are coupling effect exist between inductor which need to take account during circuit simulation. By tuning the values of inductor coupling coefficient (k) and other LC component values of the model is sufficient to solve the frequency shifting which happens with basic model as shown in Figure 2. From the Figure 2, the developed UWB antenna has a excellence performance in scavenging energy into bandwidth of 3.1 GHz to 10.6 GHz.

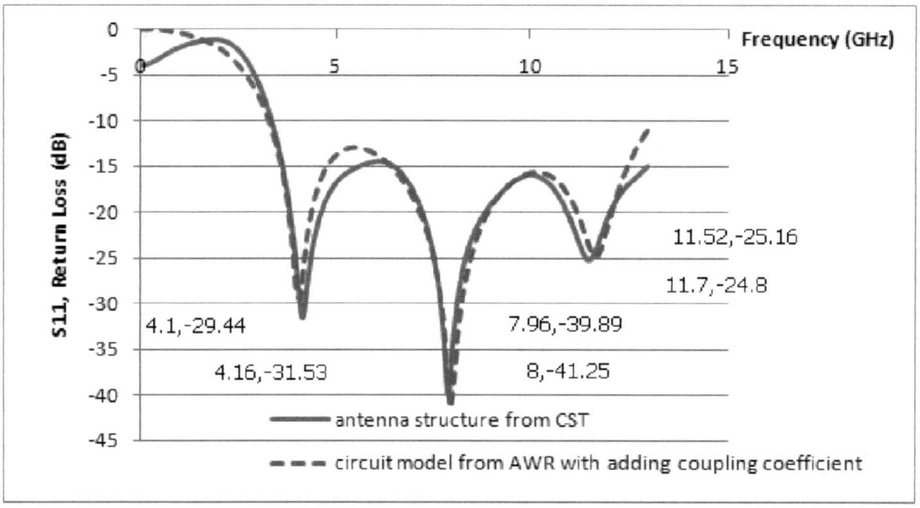

Figure 4. Comparison of return loss result between circuit model and electromagnetic model

References

[1] A. Georgiadis, G. Andia, A. Collado Rectenna Design and Optimization Using Reciprocity Theory and Harmonic Balance Analysis for Electromagnetic (EM) Energy Harvesting 2010 *Antennas and Wireless Propagation Letters, IEEE Journal* **9** 444-446

[2] S. Wang, A. Niknejad, and R. Brodersen Circuit Modeling Methodology For UWB Omnidirectional Small Antennas 2006 *IEEE Journal on Selected Area in Communications* **24** 871-877

[3] Y. Wang, J. Z. Li, and L. X. Ran An Equivalent Circuit Modeling Method for Ultra-Wideband Antenna 2008 *Progress In Electromagnetic Research, PIER* **82** 433-445

[4] M. Ansarizadeh, A. Ghorbani, and R. A. Abd-Alhameed An Approach To Equivalent Circuit Modeling of Rectangular Microstrip Antenna 2008 *Progress In Electromagnetic Research B* **8** 77-86

[5] Ma, T.G. and Wu, S.J. Ultrawideband Band-Notched Folded Strip Monopole Antenna 2007 *IEEE Trans. Antennas Proprag.* **55** 2473-2479

[6] H. J. Zhou, B. H. Sun, Q. Z. Liu and J. Y. Deng Implementation And Investigation Of U-Shaped Aperture UWB Antenna With Dual Band-Notched Characteristics 2008 *Electronics Letters* **44** 1387-1388

IOP Conf. Series: Earth and Environmental Science **16** (2013) 012074 doi:10.1088/1755-1315/16/1/012074

Laminar flame speed of methane- air mixtures at atmospheric conditions

[1]Alaeldeen Altag Yousif, [2]Shaharin A. Sulaiman

Department of Mechanical Engineering, Universiti Teknologi PETRONAS, Bandar Seri Iskandar, 31750 Tronoh, Perak, Malaysia

Email: alatelecom76@yahoo.com.

Abstract. An experimental study on spherically expanding flames propagation of methane- air mixtures was conducted at constant pressure to measure unstretched laminar flame speeds, laminar burning velocity and flame stretch. The mixtures were ignited at equivalence ratios of 0.7, 0.9 and 1.0, under ambient pressure and temperature. It was found that the unstretched laminar burning velocity increased with the equivalence ratio. The flame propagation speed showed different trends at different equivalence ratio for tested mixtures.

1. Introduction

Many combustor design tools require data on various fundamental gas combustion properties in order to design an efficient fuel flexible combustor [1,2]. Laminar burning velocities data are fundamentally important for understanding fuel behavior in combustor device and play an essential role in determining several important aspects of a combustible mixture for both practical and theoretical applications. On a practical level, it affects the fuel burning, performance and emissions in internal combustion engines [3,4]. Furthermore it is an important target for the validation of chemical kinetic mechanisms for alternative and traditional fuels [4,5].

Several studies have been focused on measurements of flame speeds for methane enrichment by ozone, hydrogen and diluted gases mixtures. Z.H et al. [6] investigated the ignition enhancement of CH4-air mixtures by ozone additive using the heat flux method. The results showed that around 8% increased of burning velocity in fuel rich mixtures and 3.3% burning velocity increased for stoichiometric mixtures. Furthermore, a few research work [7,8,9] investigated intensively the effects of hydrogen addition and nitrogen dilution on the laminar flame characteristics of premixed methane-air flames. The laminar burning velocity was found to increase linearly with hydrogen mass fractions for all dilution ratios while the burned gas Markstein length decreased with the increase in hydrogen amount in the mixture except for high hydrogen mole fractions (> 0.6). A flamelet model was carried out by Sivaji et al. [10] to study the effect of exhaust gas recycling for oxy-methane flames. The study indicated that the presence of CO_2 in the oxidant has profound effect on the reaction mechanism through two reactions, one involving the H radical and another involving the H_2O radical. Zeng Chen [11] studied numerically on the extraction of laminar flame speed and Markstein lengths of methane – air mixture. The accuracy of the models was found to strongly depend on the Lewis number.

The purpose of the present study was to investigate experimentally the flame propagation speed characteristics of methane- air mixtures, and necessary experimental data for validation and development of physical model of combustor device. The study covered the lean extend to stoichiometric operating limit of methane - air mixtures in atmospheric conditions, to offer an advantage operation, such as decreasing the temperature and therefore decrease of No_x emissions.

Laminar burning velocity and Markstein lengths can be derived from schlieren photographs as described in [12]. The stretched flame speed relative to the burned gases, S_b is derived from the flame radius versus time by:

$$S_b = \frac{dr}{dt} \quad (1)$$

where r is the radius of the flame in schlieren photograph and t is the elapsing time from spark ignition. The total flame stretch rate α is defined as:

$$\alpha = \frac{d(LnA)}{dt} = \frac{1}{A}\frac{dA}{dt} = \frac{2}{r} S_b \quad (2)$$

where A is the flame surface area. In respect to the early stage of flame expansion, a linear relationship between the flame speeds and flame stretch rate is given by [12]:

$$S_b = S_u - L_b \propto \quad (3)$$

where S_u is the un-stretched flame propagation speed obtained when flame stretch rate $\alpha = 0$. The quantity L_b is defined as the effect of stretch on flame speed and is called the burned gas Markstein length.

2. Experimental setup

A 360-mm inner diameter cylindrical combustion chamber was designed and fabricated to perform experiments with difference gaseous and liquid fuels which can withstand high pressure and temperature generated from explosion of initial condition. Two orthogonal quartz windows were fitted on both flat side of the vessel with diameter 150 mm for the inspection of the flame propagation. Type-K thermocouple was fixed inside the chamber to measure the initial temperature of the mixture. A pressure transducer was fixed outside the vessel to measure the pressure of incoming mixture. Two electrodes were located at the center with 2 mm gap and the ignition was initiated with a 12 V automotive ignition coil system. Two valves were mounted on the chamber body for the inlet of mixture and air as well as for the exhaust. Fuel and dry air were supplied into the chamber through the inlet valve with the corresponding partial pressures. Experiments were conducted at initial pressure of 0.1 MPa and initial temperature of 303 K. Purities of methane and air in the study were 99.5% and 99.995%, respectively. To ensure the homogeneity and motionless of the mixtures the ignition was started after fifteen minutes of the supply of the mixture and air.

The experimental setup is shown in Figure 1. The flames image growth were taken with schlieren photography which contains high speed camera, 50 W halogen lamp, two Plano convex lenses with 1000 mm focal point and two pinholes. The images were captured using Phantom 9.2 high speed camera with 10000 fps. The resulting images were processed using Adobe Photoshop CS 5.2 to measure the flame propagation to get best images in black and white colour for a set of flame frontal area pictures. All pixels lighter than the threshold value were converted to white, and all pixels darker than the threshold were converted to black. The electrodes were removed manually and flame radius was calculated as those of a circle area equal to that of the imaged flame.

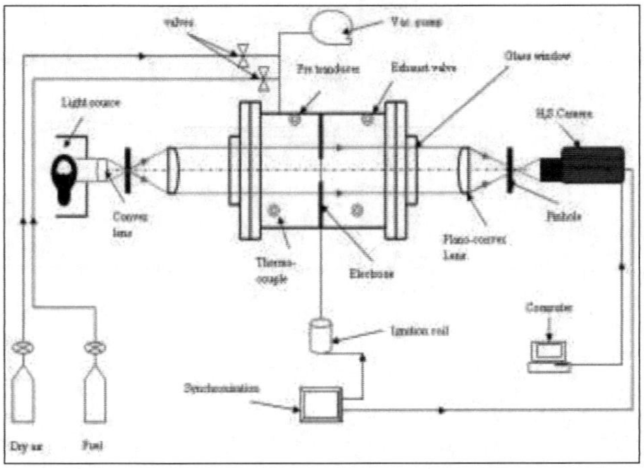

Figure 1: Schematic of the experiment set-up.

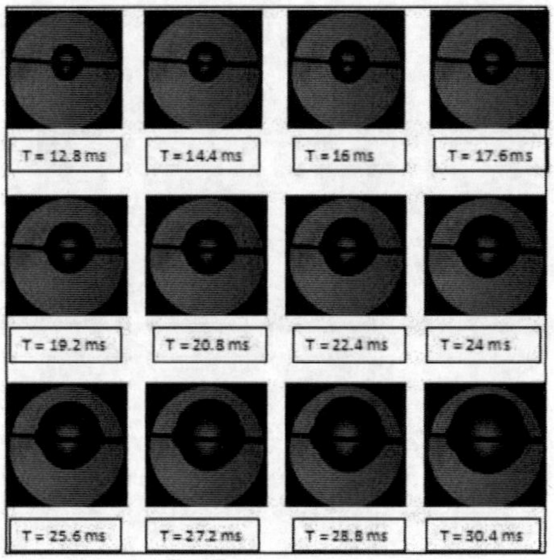

Figure 2: Images captured by using H.S camera.

3. Results and discussion

One fuel was investigated in this work: Methane- air mixtures. Measurements of stretched and laminar flame speed for methane- air flames were carried out for various equivalence ratios ($\Phi = 0.7, 0.9, 1.0$) at atmospheric pressure and temperature. Around 22 frames of images were taken to cover the flame propagation duration from ignition and chamber wall effect.

Figure 2 shows the images of methane–air flames at stoichiometric equivalence ratio obtained by schlieren photograph. Spherically expanding flames propagated smoothly from the electrode gap in the centre of the combustion chamber when the flame radius was developed to a certain value ($r >$ 5mm) and the effect of the electrodes becomes inconsiderable. Figure 3 shows the variation of flame speed as function of stretch rate for methane/air mixture. For all investigated mixtures, the stretch rate decreased as flame front velocity increased. Figure 4 shows the variation of flame radius with the time from start of ignition for methane-air mixtures at equivalence ratios of 0.7, 0.9 and 1.0. As shown in the figure, time and flame radius exhibited a linear relationship, and the radius of the flame increased as time increased. The stretched propagation speeds for methane-air flames for three different mixtures are illustrated in Figure 5. As shown in the figure, the flame propagation speed increased as equivalence ratio increased, and the stretched flame propagation speed increased gradually with the time for all represented mixtures. Figure 6 shows the stretched flame propagation velocities versus flame radius. The flame propagation speed shows different trends at different equivalence ratios for methane–air mixtures, and the stretched flame propagation speed increased slightly with the increase of flame radius (Figure 6).

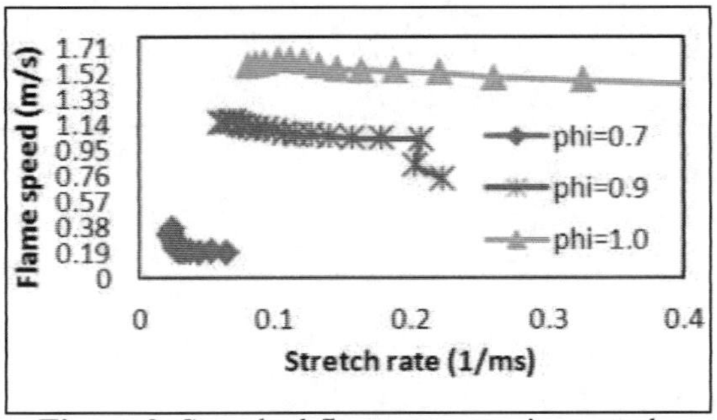

Figure 3. Stretched flame propagation speed versus stretch rate, for methane- air mixture.

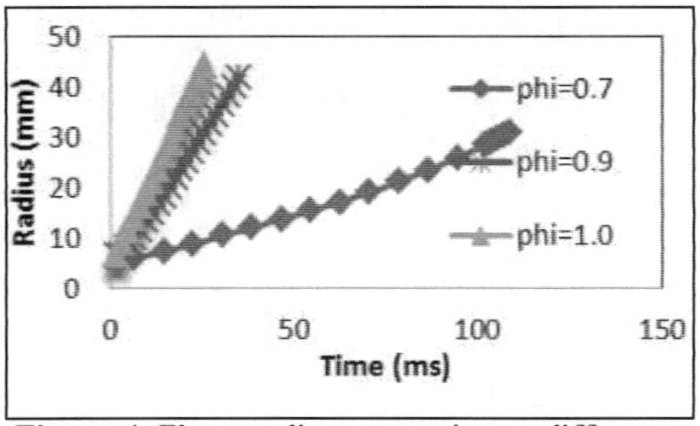

Figure 4. Flame radius versus time at different equivalence ratios for methane- air mixtures.

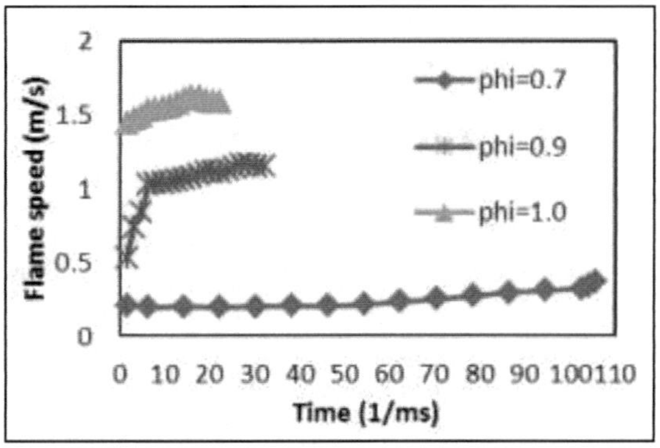

Figure 5. Flame radius versus time at different equivalence ratios for methane- air mixtures.

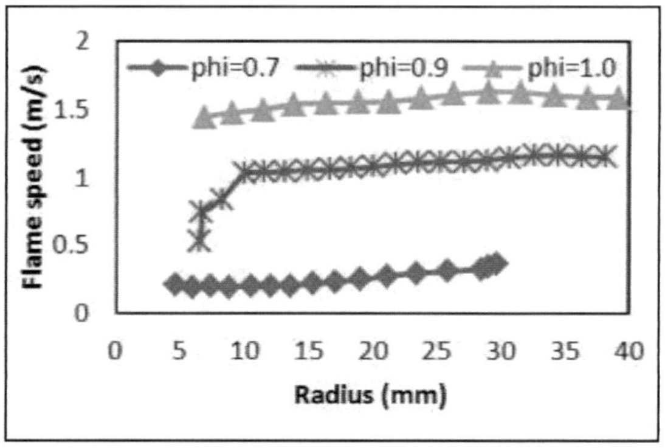

Figure 6. Flame propagation speed versus flame radius, for methane- air mixture.

4. Conclusions

The experimental data of the present work investigated the effect of stretch on laminar flame speeds of methane-air mixtures in different equivalence ratios. Spherically expanding flame technique was used to measure the flame speeds. Thus, from the current study the following conclusions are derived:
1- The stretched burning velocities increased with decreased stretch rate.
2- The unstretched flame speeds increased with the equivalence ratios.
3- The slope of the radius–time line increased with equivalence ratios, and this indicates that the flame propagation speed increased as equivalence ratios increased.
4- The flame propagation speed showed different trends at different equivalence ratio for methane–air mixtures.

References

[1] Huang,J.A. 2003 *J. Fuel. Technol.* **82** 835-842
[2] Metghalichi, M. and Keck, J.C 1980 *J. Combustion and Flame.* **38** 143
[3] Cheng Tung Chong, Simone Hochgreb 2011 *J. Comb.Inst.* **33** 979-986
[4] Law, C.K,Sung,C.J.Wang,H.and Lu.T.F 2003 *J. AIAA.* **41** 1629-1645
[5] Emilien Varea, Vincent Modica, Alexis Vandel, Bruno Renou 2012 *J.Combustion and Flame.* **159** 577-590
[6] Z.H.Wang, L. Yang, B.Li,Z.S. Li, Z.W. Sun, M. Alden, K.F. Cen, A.A. Konnov 2012 *J. Comb and Flame.* **159** 120-129
[7] Tahtouh, F. Halter, E. Samson, C. Mounaim-Rousselllea 2009 *J. Hydrogen Energy.* **34** 8329-8338
[8] R.T.E. Hermanns, A.A.Konnov, R.J.M. Bastiaans 2010 *J. Fuel.* **89** 114-121
[9] Zhenlog Zhao, Zheng Chen 2012 *J. Hydrogen Energy.* **37** 691-697
[10] Sivaji Seepana, Sreenivas Jayanti 2012 *J. Energy and Fuel.* **93** 52-58
[11] Zheng Chen 2011 *J. Combustion and Flame.* **158** 291-300
[12] Bradley D,Gaskell. P, Gu X 1998 *J. Combustion and Flame.* **104** 98-176

Influence of sinusoidal flow on the thermal and hydraulic performance of microchannel heat sink

N I Om[1], P. Gunnasegaran[1] and S. Rajasegaran[1]

[1]Centre for Advanced Computational Engineering (CACE), Universiti Tenaga Nasional, Putrajaya Campus, Jalan IKRAM-UNITEN, 43000, Kajang, Selangor, Malaysia
E-mail: Irmawati@uniten.edu.my

Abstract. In this paper, the effect of sinusoidal flow on the thermal and hydraulic performance of microchannel heat sink (MCHS) is numerically investigated. This investigation covers Reynolds number in the range of $100 \leq Re \leq 1000$ and pure water is used as a working fluid. The three-dimensional steady, laminar flow and heat transfer governing equations are solved using finite volume method (FVM). The water flow field and heat transfer performance inside the sinusoidal microchannels is simulated and the results are compared with the straight microchannels. The effect of using sinusoidal microchannels on temperature distribution, Nusselt number, friction factor and thermal resistance is presented in this paper. It is found that with same rectangular cross-section, sinusoidal microchannels have a better heat transfer performance compared to the straight microchannels.

1. Introduction

Lately number of research based on microchannel have grown broad and boosted the technology to grow rapidly. The introduction of MCHS has even increased with attention to study this newfound discovery more in detail and broad in order to increase the efficiency of it. At the very beginning the MCHS cooling concept was proposed by Tuckerman and Pease [1] in 1981, who pointed out that decrease in liquid cooling channel dimensions to the micron scale, will lead to increase in heat transfer rates.

Al-Bakhit and Fakheri [2] noted that in microchannel heat exchangers, short lengths and comparatively thick walls through which heat is conducted avoid the existence of fully thermal developed flow over a large portion of the heat exchanger. They concluded that there is a significant change in the overall heat transfer coefficient in the developing region. Gamrat et al. [3] performed three-dimensional numerical analysis of convective heat transfer in microchannels. By considering the thermal entrance effect, the numerical analysis did not find any significant effect on the heat transfer in microchannels with the hydraulic diameter down to 100 lm. Peng and Peterson [4] studied experimentally the pressure drop and convective heat transfer for water flow in rectangular straight microchannels. They found that the cross-sectional aspect ratio has a great influence on the flow friction and convective heat transfer in both laminar and turbulent flows. T.L Ngo [5] performed numerically and experimentally investigation on the curvy and zigzag microchannel heat exchanger (MCHE). The results showed that the Nusselt number of zigzag microchannels is 24–34% higher than that of the curvy microchannels, but the friction factor is four to five times larger depending on the Reynolds number. Numerical analysis on the fluid flow and heat transfer characteristics in rectangular cross-section MCHS with distinct inlet and outlet arrangements was carried out by Chein and Chen [6]. They mentioned that better performance in velocity and temperature can be found in the heat sinks having coolant supply and collection vertically via inlet/outlet ports opened on the heat sink cover plate. Numerical analysis on the laminar forced convection and thermal radiation in a participating medium inside a helical pipe was carried out by Zheng [7]. The effects of thermal radiation on the convective heat transfer were investigated. He found and deduced that the thermal radiation could enhance the total heat transfer rate. By using finite element method Zhang [8] executed numerical analysis on the hydrodynamic and thermal characteristics of microchannel networks. A comparison of the cooling performance among step, parallel, and spiral microchannels was also conducted based on the same heat flux and flow rate. They found that step microchannel alleviate the lowest fluid pressure drop and the most uniform temperature distribution over the substrate.

It should be noted that all the previous studied of MCHS are reported in the open literature used conventional microchannel which generally engaged with straight microchannel. Furthermore, to the best knowledge of the authors, the case of sinusoidal MCHS has not been given great attention by the researchers in the past and this has motivated the present study. The present study numerically investigates the three-dimension heat transfer and fluid flow characteristics in sinusoidal MCHS with rectangular cross-section using water as a working fluid. The Reynolds number is ranged from 100 to 1000 and heat flux applied at the bottom plate of MCHS is 1000000 W/m^2.

1. MCHS configuration and mathematical formulation

In this paper, sinusoidal MCHS with a constant rectangular cross-section is considered. The physical configuration of the sinusoidal MCHS is schematically shown in Figure 1 and schematic of the each unit of microchannel i.e. a) conventional channels; b) sinusoidal channels are shown in Figure 2. Heat supplied to the aluminum MCHS substrate through a bottom plate, was removed by flowing water through a number of 24 micro channels. In this study, the effect of sinusoidal flow of the MCHS on the heat transfer and fluid flow characteristics is examined.

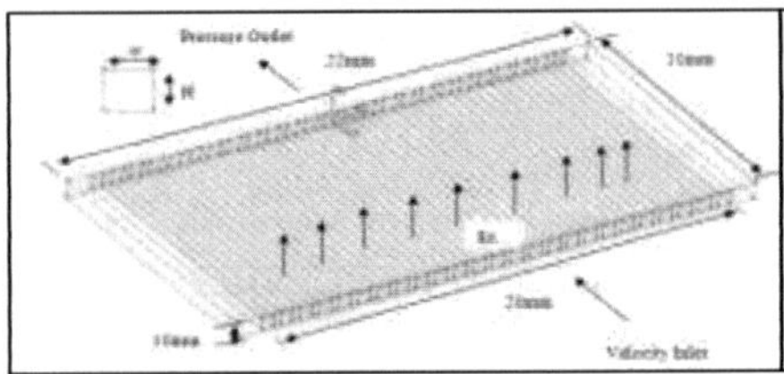

Figure 1. Schematic diagram of computational domain of MCHS

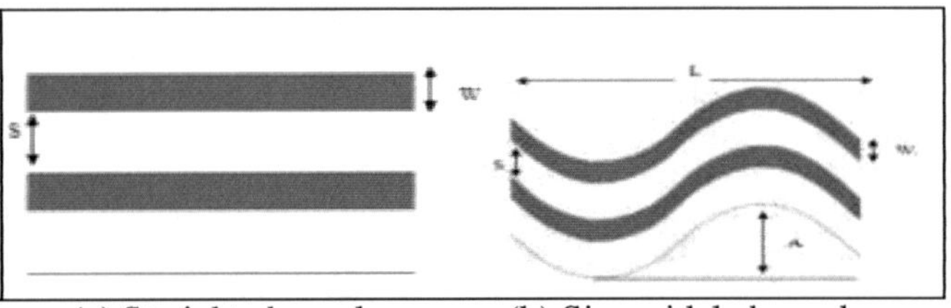

(a) Straight channel (b) Sinusoidal channel

Figure 2. Schematic of the each unit of microchannel

In order to solve the Navier–Stokes and energy equations to investigate the effect of sinusoidal flow on the MCHS performance, the following assumptions were made: (i) both fluid flow and heat transfers are in steady state; (ii) the flow is isochoric which means the volume of the water remains constant throughout the process; (iii) the water flow is laminar; (iv) the water that flows in and out of the MCHS does not change phase during the process; (v) properties of both fluid and heat sink material are temperature-independent;(vi) fluid flow and heat transfer are three-dimensional; (vii) all the surfaces of heat sink exposed to the surroundings are assumed to be insulated except the bottom plate of heat sink where constant heat flux boundary condition simulating the heat generation from external sources is specified.

The governing equations and its boundary conditions in Cartesian coordinates for three-dimensional laminar incompressible flow for the current problem are:

Continuity Equation:

$$\frac{\partial U}{\partial X}+\frac{\partial V}{\partial Y}+\frac{\partial W}{\partial Z}=0 \qquad \text{Eq. 1}$$

X- Momentum equation: $$U\frac{\partial U}{\partial X}+V\frac{\partial U}{\partial Y}+W\frac{\partial U}{\partial Z}=-\frac{dP}{dX}+\frac{1}{Re}\left(\frac{\partial^2 U}{\partial X^2}+\frac{\partial^2 U}{\partial Y^2}+\frac{\partial^2 U}{\partial Z^2}\right)$$ Eq. 2

Y- Momentum equation: $$U\frac{\partial V}{\partial X}+V\frac{\partial V}{\partial Y}+W\frac{\partial V}{\partial Z}=-\frac{dP}{dX}+\frac{1}{Re}\left(\frac{\partial^2 V}{\partial X^2}+\frac{\partial^2 V}{\partial Y^2}+\frac{\partial^2 V}{\partial Z^2}\right)$$ Eq. 3

Z- Momentum equation: $$U\frac{\partial W}{\partial X}+V\frac{\partial W}{\partial Y}+W\frac{\partial W}{\partial Z}=-\frac{dP}{dZ}+\frac{1}{Re}\left(\frac{\partial^2 W}{\partial X^2}+\frac{\partial^2 W}{\partial Y^2}+\frac{\partial^2 W}{\partial Z^2}\right)$$ Eq. 4

Energy Equation: $$U\frac{\partial \theta}{\partial X}+V\frac{\partial \theta}{\partial Y}+W\frac{\partial \theta}{\partial Z}=-\frac{dP}{dZ}+\frac{1}{RePr}\left(\frac{\partial^2 \theta}{\partial X^2}+\frac{\partial^2 \theta}{\partial Y^2}+\frac{\partial^2 \theta}{\partial Z^2}\right)$$ Eq. 5

At the microchannels sections, the inlet water temperature is taken as 293K and the inlet water velocity is calculated based on the required Reynolds number. The Reynolds number considered is ranged from 100 to 1000. The transverse velocities at the inlet are assumed to be zero. On the aluminum substrate, the velocities are zero, and it is assumed to be an adiabatic surface. The heat flux that applied at the bottom plate is 1000000 W/m^2. The properties of water and solid used in the computation are ρ = 998.2 kg/m^3, c_p=4182 J/kg.K, μ=0.001003 kg/m.s, κ=0.6W/m.K, and κ_s=202.4 W/m.K. The numerical computation is carried out by solving the governing conservation equations (Eqs. 1-5) along with the boundary conditions. From numerical computation, the temperature distribution, Nusselt number, friction factor and thermal resistance were determined. This leads to identify the thermal and hydraulic performance of the sinusoidal flow in MCHS.

2. Results and Discussion

The variation of average wall temperature for sinusoidal and straight MCHSs is presented in Figure 3. For both of MCHSs, it is found that the high-temperature region occurs at the edge of the heat sinks since there is no heat dissipation by fluid convection. The low temperature region occurs in the region where microchannels are placed, especially near the center of the heat sink due to the high heat transfer coefficient. It can clearly be seen from Figure 3 that the temperature of straight conventional MCHS is higher compared to the sinusoidal MCHS due to the poor fluid mixing. From the results obtained in Figure 3, lower thermal resistance is expected in sinusoidal MCHS compared to straight MCHS.

The Figure 4 presented the Nusselt numbers with the number of channels for straight and sinusoidal MCHSs at the Reynolds number of 1000. It can be seen from Figure 4 that the middle channel (channel 12) has the highest Nusselt number value. The Nusselt number for other channels is seen to decrease depending on their distances from the wall. The Nusselt number for both types of MCHSs is almost symmetrical with respect to the centerline of the heat sink. In sinusoidal MCHS, it is found that the Dean vortices can quickly develop along the flow direction and disturb the boundary layer. Thus, better heat transfer performance is found in sinusoidal MCHS compared to straight MCHS as shown in Figure 4.

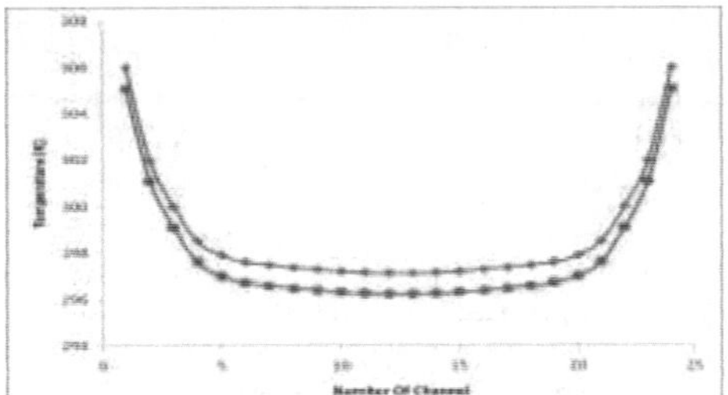

Figure 3: Temperature distribution

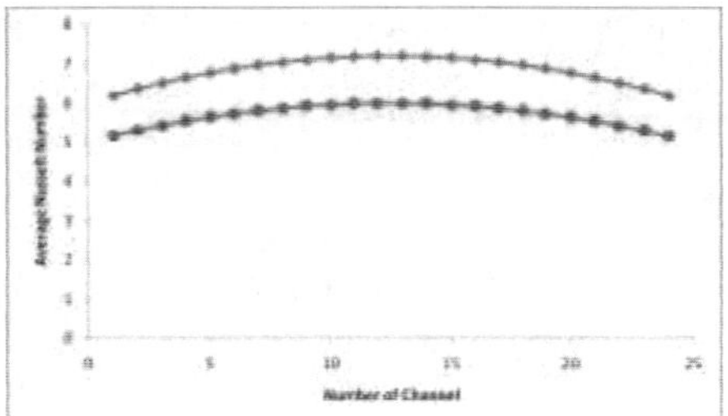

Figure 4: Nusselt number

The variation of the friction factor with Reynolds number for both sinusoidal and straight MCHSs is shown in Figure 5. The results revealed that the friction factor decreases with the increase of Reynolds number for both types of MCHSs. It can be inferred that the changing of Dean vortices patterns along the flow direction through the sinusoidal MCHS appears to give an increase in the friction factor compared to straight MCHS for all values of Reynolds number. The thermal resistance for sinusoidal and straight MCHSs as function of pumping power is shown in Figure 6. It is observed that sinusoidal MCHS reduces the thermal resistance due to its lower wall temperature distribution compared with that of the straight MCHS. Therefore, the sinusoidal MCHS could greatly enhance the cooling performance of MCHS compared with the traditional straight MCHS.

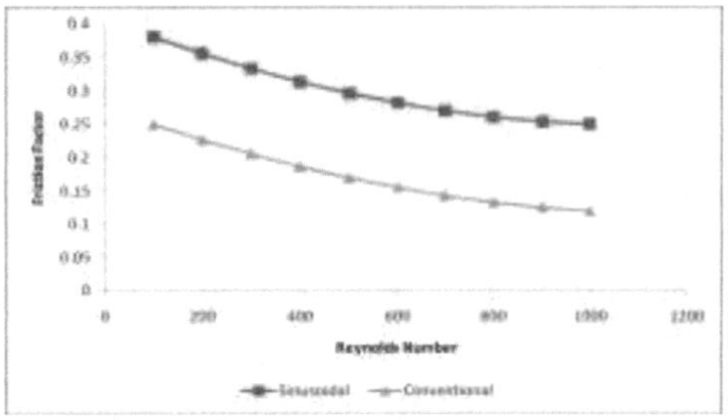

Figure 5: Friction factor

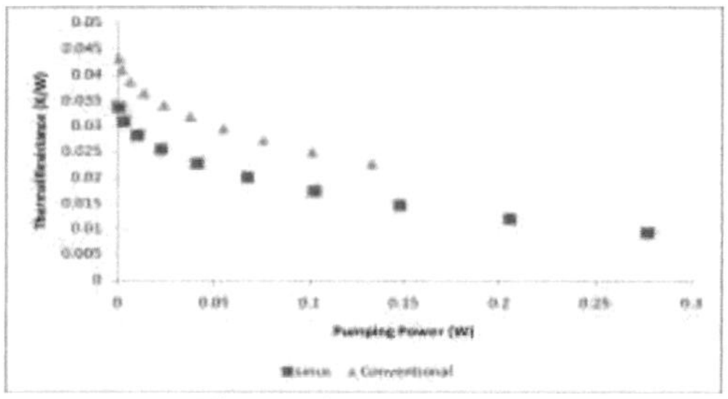

Figure 6: Thermal resistance

3. Conclusion

Numerical simulations on the heat transfer and fluid flow characteristics in sinusoidal MCHS are presented in this paper. The effect of sinusoidal flow in MCHS on the thermal and flow fields was intensively investigated and compared with the straight MCHS. It was found that the heat transfer performance in sinusoidal MCHS is much better compared to straight MCHS and can be proposed as next invention cooling equipment for removing high heat flux instead of using conventional straight microchannel.

Acknowledgement

The authors would like to sincerely thank the Ministry of Higher Education (MOHE) of Malaysia for the provision of a Grant with code No. 12012011ERGS to support this work.

References

[1] D.B. Tuckerman, R.F. Pease, High Performance Heat Sinking for VLSI 1981 *IEEE Electron. Devices Lett.* **2** 126-129.

[2] Albakhit, H. & Fakheri, A., A Hybrid Approach For Full Numerical Simulation of Heat Exchangers 2005 *ASME Conference Proceedings, Heat Transfer Summer Conference (HT)*

[3] Gamrat, C.X. Lin, M.A. Ebadian, Combined Laminar Forced Convection and Thermal Radiation in Helical Pipe 2006 *Int. J. Heat Mass Transfer* **43** 67-78

[4] Peng and Peterson, Reduced Navier–Stokes simulation of Incompressible Microchannel Flows 2008 *Numer. Heat Transf. A* **53** 71-87.

[5] T.L. Ngo, Y. Kato, K. Nikitin, T. Ishizuka, Heat Transfer And Pressure Drop Correlations of Microchannel Heat Exchangers With S-Shaped And Zigzag Fins for Carbon Dioxide Cycles 2007 *Exp. Therm. Fluid Sci.* **32** 560–570

[6] R. Chein and J. Chen, Numerical Study Of The Inlet/Outlet Arrangement Effect on Microchannel Heat Sink Performance 2009 *International Journal of Thermal Sciences* **48** 1627-1638

[7] Zheng, Forced-Convection, Liquid-Cooled Microchannel Heat Sinks, Master Thesis 1987 *Massachusetts Institute of Technology, Cambridge, MA*

[8] J. Zhang, Y.F. Zhang, M. Miao, Y.F. Jin, S.L. Bai, J.Q. Chen, Simulation of Fluid Flow and Heat Transfer in Microchannel Cooling for LTCC Electronic Packages 2009 *IEEE Electronic Packaging Technology and High Density Packaging, ICEPT-HDP* **9** 327–330

IOP Conf. Series: Earth and Environmental Science **16** (2013) 012076 doi:10.1088/1755-1315/16/1/012076

Output improvement of Sg. Piah run-off river hydro-electric station with a new computed river flow-based control system

Razali Jidin[1], Bahari Othman[2]

[1] Centre for Automation and Embedded Computing System (CAECS), College of Engineering, Universiti Tenaga Nasional (UNITEN), Jalan IKRAM-UNITEN, 43000 Kajang, Malaysia

[2] Generation Research Department, Tenaga Nasional Research Centre (TNBR), 43000 Kajang, Selangor, Malaysia.

Email: [1]razali@uniten.edu.my, [2]bahari@tnbr.com.my

Abstract. The lower Sg. Piah hydro-electric station is a river run-off hydro scheme with generators capable of generating 55MW of electricity. It is located 30km away from Sg. Siput, a small town in the state of Perak, Malaysia. The station has two turbines (Pelton) to harness energy from water that flow through a 7km tunnel from a small intake dam. The trait of a run-off river hydro station is small-reservoir that cannot store water for a long duration; therefore potential energy carried by the spillage will be wasted if the dam level is not appropriately regulated. To improve the station annual energy output, a new controller based on the computed river flow has been installed. The controller regulates the dam level with an algorithm based on the river flow derived indirectly from the intake-dam water level and other plant parameters. The controller has been able to maintain the dam at optimum water level and regulate the turbines to maximize the total generation output.

1. Introduction

Tenaga Nasional Berhad (TNB) being the biggest power utility company in Malaysia owns all the electrical transmission and distributions networks, and a number of power stations in Peninsular Malaysia. TNB delivers a total peak demand of about 15,000MW electricity at the end of year 2012 for Peninsular Malaysia that has a population of about 25 million. Almost all of the hydro-electric stations in Peninsular Malaysia are owned by TNB and they are located in the northern part of the country. One of the small ones that located in the state of Perak is the cascaded Sg. Piah run-off river hydro-electric station that has maximum generating capacity of 70MW. The lower Sg. Piah hydro-electric station has two Pelton turbines (almost same runner as shown in Figure 1) to harness energy from water flowing through a 7km tunnel after an intake dam. Each turbine is capable of producing maximum 27.5MW when sufficient river flow is available. Its output is connected to Sg. Siput's 132KV double circuits to feed to the national grid. As the station has a small reservoir such as shown in Figure 2, the feature of run-off river hydro scheme, the operation strategy is to maximize its output according to the available river flow by minimizing the dam spill to the allowable limits.

As the price of fossil fuel increases over the years and often spikes when events such tensions between countries or shortages of gas occur, hydro-electric generating stations can assist to lower overall fuel costs by maximizing their outputs. Therefore to maximize its output, both units of the

Lower Sg. Piah station were equipped with an Online Load Controller or simply known as Online Controller (OC) to harness the energy according to the dam water level. The practice is to regulate the turbines according to water level at the dam [1 – 5]. However the OC had developed a "control hunting" problem at early age of the power station. Since then the OCs for both units were never put in service that the practices were to turn off the OC. Consequently the station was not able to produce energy optimally most of the time during its operation. Most of the time much water were being spill over the dam instead being used to generate additional energy.

Figure1: Turbine Runner at Upper Station almost similar to the Lower Sg. Piah's

Figure 2: Lower Sg. Piah's Intake Dam

Based on the annual rainfall data, computed river flow, tests performed on the station and observations of operation practices at the stations during 2010, it can be concluded that about USD 3 million additional revenues annually can be attained if the station is operated according to the available river flow. To improve the station annual energy output, a new controller called automatic online controller (AOC) has been installed. The controller regulates the dam level with an algorithm based on the river flow instead of the depending on solely on the dam water level. The river flow is derived indirectly from the intake-dam water level and other plant parameters.

2. AOC System Architecture

2.1. AOC System Design

The new AOC has been implemented according to the following specifications:

a. The AOC major components are Siemens S7-315 Programmable Logic Controller (PLC) and Touch Screen Human-Machine Interface (HMI). The PLC and HMI are given Figure 3 and 4 respectively. All control algorithms including flow computation and the logic sequences are implemented within the PLC. AOC acquires plant parameters both from the Supervisory Control and Data Acquisition (SCADA) remote terminal unites (RTU) and control cabinets.
b. Interface between operators and control system either to initiate commands or enter the coefficients is realized on the HMI. The commands such as AOC ON and AOC OFF buttons (to turn on or off the AOC) will be realized on the HMI instead of using the OC ON & OC OFF buttons on the control desk. The touch screen buttons on the HMI will alter the appropriate memory bits within the PLC. Also the generation load demand can be optionally raised or lowered from HMI.
c. Also, the HMI touch screen allows operators to enter coefficients needed for targeted load reference calculations. The floating point coefficients or constants such as K0, K1 and K2 can be

entered by personnel via HMI page similar to the one given in Figure 5. The constants are then forwarded to PLC for turbine discharge flow computations.

Figure 4: Program Logic Controller (PLC)

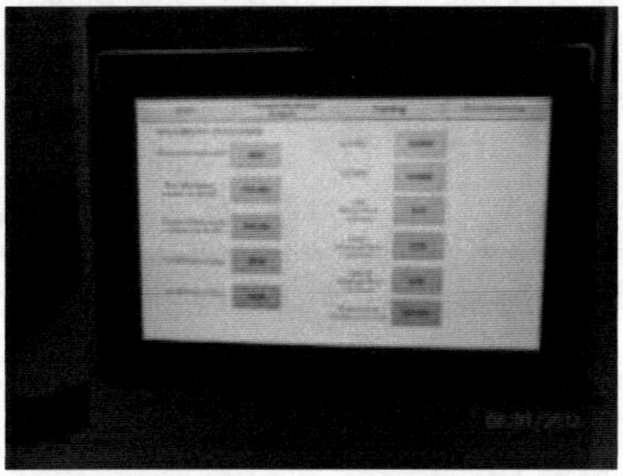

Figure 5: HMI: One of HMI graphical page

2.2. Remote Operation and Drive Level

In addition to the control buttons on HMI, operator can control the station remotely from Bersia group regional control center (BGCC) via the SCADA system. The interface cabinets for SCADA at Lower Sg. Piah station are RTU 1 and RTU 2 which contained terminals points for interfacing to the AOC. At the drive level, the electro-hydraulic governor will regulate the turbine nozzles and deflectors in response to speed setters and load reference setters. At this station, each turbine unit is equipped with two pair of nozzles (Pair 1 & 3 and Pair 2 & 4). The second pair of nozzle is opened when the load is about half of maximum load.

3. AOC Dam Level Control Algorithm

3.1. Computed River Flow-based Dam Level Control Algorithm

The main function of the AOC is to regulate turbine automatically, mainly based on the computed river flow. It can regulate either both or one units only. As for the feedback, the controller's load setter is compared to the turbine load reference rather than MW, to overcome the effect of turbine nozzle hunting zones.

The main computations within the AOC include river flow calculation *Qrf* as given by equation Equation1, optimal/target flow *Qopt* (equation Equation2), regulation periods, and protection schemes to ensures safe operation of the Upper Sg. Piah station.

Average river flow (*Qrf*) for a given time interval (t_f) can be calculated using Equation (1) as follows:

$$\int_0^{t_f} Q_{rf}\,dt = \int_0^{t_f} Q_{tur} + Q_s dt + (WL_c - WL_p)A_p \qquad \text{Equation (1)}$$

Where

Q_s = Spillage (m^3/s)

Q_{tur} = Turbine discharge (m^3/s) for both turbines

$WL1_p$ = Previous dam water level (mSLE meter sea level elev.)

$WL1_c$ = Current dam water level (mSLE)

A_p = Average water surface area for intake pond (m^2)

t_f = Final time for the designed regulation interval (s)

Optimal total turbine discharge (Q_{opt}) is calculated as follows

$$Q_{opt} = \frac{1}{t_f}\int_0^{t_f} Q_{rf}\,dt + (WL1_c - WL1_{tr})A_p \qquad \text{Equation (2)}$$

Where

$WL1_{tr}$ = Targeted dam water level (mSLE)

3.2. PLC Ladder Logic Program

All the flow computations including Q*turb*, Q*s*, Q*rf* and Q*opt*, logic sequences, and protection logics are realized in the form of a ladder logic program which is about two hundred pages long. Six pages of HMI graphics including a trending, AOC commands, monitor and data entry pages have been created.

4. Conclusion

During the two weeks performance tests, the controller has been able to regulate both the dam level and the turbine output successfully. Each turbine was tested to operate up to 25MW when the river flow permits such operating conditions. In near the future we need to conduct further optimization to search for optimum variable target dam water levels versus turbine outputs to optimally generate energy from the varying river flow.

Acknowledgments

The authors would like to thank TNB generation department and Sg. Perak hydro-electric station management for funding this research. Also appreciations go to Sg. Piah and Bersia control centre's personnel for their support and cooperation.

References

[1] Sarasúa, J.I., Fraile-Ardanuy, J.J., Pérez-Díaz, J.I., Wilhelmi, J.R. and Sánchez, J.A. Control of a run of river small hydro power plant 2007, *International Conference on Power Engineering, Energy and Electric Drives, Setubal, Portugal*

[2] Legros, J.M., Mondon, C. Maurin, S. and Libaux, A. EDF Hydropower plants operation: new challenges and technical solutions 2009 *Waterpower XVI, Spokane, WA, USA,*

[3] Yadav Omkar, Kishor Nand, Fraile-Ardanuy Jesus, Mohanty Sourna R., Perez Juan I., Sarasua Jose I. Pond head level control in a run-of-river hydro power plant using fuzzy controller 2011, *IEEE 16th International Conference Intelligent System Application to Power Systems (ISAP), 1-5*

[4] K. Natarajan, Robust PID Controller Design for Hydro-turbines 2005, *IEEE Transactions on Energy Conversion, **20** 61-667*

[5] Karl Heinz Fasol , Historical Perspectives: A Short History of Hydropower Control 2002, *IEEE Control Systems Magazine*

IOP Conf. Series: Earth and Environmental Science **16** (2013) 012077 doi:10.1088/1755-1315/16/1/012077

Used cooking oil as a green chemical admixture in concrete

B. Salmia[1], Zakaria Che Muda, Md. Ashraful Alam, L.M. Sidek[2], B. Hidayah[3]
[1] *Senior Lecture* [2] *Associate Professor* [3] *Lecturer*
E-mail: Salmia@uniten.edu.my

Abstract. According to National Statistics Approximately 1.35 billion gallons of used oil are generated yearly. With the increasing of the concrete usage, a more cost effective and economic new type of admixtures may give positive impacts on the Malaysian construction building as well as worldwide concrete usage. To objective of this is study is to investigate the effect of used cooking oil in terms of slump test, compressive strength test and rebound hammer. By adding the used cooking oil to the concrete, it increases the slump value from 4% to 72%. And the compressive strength have an increment from 1% to 16.8%.The used cooking oil obtains the optimum contribution to the concrete mix proportion of containing used cooking oil of 1.50% from the cement content. The result of used cooking oil from experimental program of slump value and compressive strength proved that used cooking oil have positive effects on replacement of commercially available superplasticizer.

1. Introduction

Over the years, it has been increasing of industrial wastes and by-products either solid based and/or liquid based chemical in all over the world. Even though the environmental agencies over the country have their laws and regulations regarding the safe handling or disposal of them, but there was a big chunk of such waste are illegally disposed off, which may cause severe environmental disorder. Pollution towards environment can caused unstable ecosystem, harm humans and caused disorder in living organisms [1]. In some rural areas, we can still find people cooking and bathing using the water at a river which also cause pollution. Oil is one of the contributors towards water pollution which causes nearly a quarter of all pollution incidents [2]. Oil is one of the most commonly reported types of water pollution and causes nearly a quarter of all pollution incidents. A rapid growth of food service sector in Malaysia has led to a sudden increase of used cooking oil as wastes. Some of this used cooking oil has been properly disposed but most of them are illegally disposed. Nowadays improper disposal of used cooking oil has been a serious issue. According to the Malaysia HRI Food Service Sector Report in 2009 [3],Trade sources comment that the food service market has been growing at a rapid average rate of around 7 percent per annum over the past five years. They indicated that this market is likely to grow at between 7 percent and 10 percent per annum over the next three to five years. Used cooking oil will increase the loading in water treatment plant process. Is required higher amount of energy to treat the wastewater. One of the most popular solutions nowadays is by recycling the used cooking oil and converts it into biodiesel [4]. In Malaysia, there are companies which have a valid recognized license to collect the used cooking oil from community and processed the oil into biodiesel [5]. Previous research was reported that used oil such as used engine oil and soy bean oil were having a potential to act as concrete admixture to improve concrete performance. According to Beddu (2012) [6], a certain dosage of used engine oil can improve the fresh and hardened properties of various types of concrete. In the other hand, their performance of used engine oil in concrete

IOP Conf. Series: Earth and Environmental Science **16** (2013) 012077 doi:10.1088/1755-1315/16/1/012077

containing fly ash (FA) [7] and Microwave incinerated Rice Husk Ash (MIRHA) did not impart any negative effects and comparable with commercially available superplasticizer [8].Bilal et. al [9] have studied the effects of used engine oil on fresh and hardened properties and also structural performance of concrete and it gave the positive effects comparable with the available commercially air entraining agent. Used engine oil also suggested by Aravind and Animesh [10] to be use in highway construction but requires well organized oil collection system. The further research is noteworthy to produce the good quality and performance of concrete. The effects of soybean as a curing agent for concrete pavement have been done [11].

2. Methodology

Seven mixes were made with water/binder ratios (w/b) of 0.55. Water and admixture was measured in percentage by weight proportion of cement used. Ordinary Portland Cement (OPC) Type 1 was used in this research, according to BS EN 197-1 2000. OPC Type 1 was preferred because the observation on concrete properties can be done in normal hydration process hence the advantages of silica fume usage in concrete can be optimized. Aggregate that used to prepare concrete are confirming to BS 882: 1992. In this experimental program, used cooking oil was used as chemical admixture to improve concrete performance. The specimens were compacted by external vibration, varying the vibration time according to its consistency. After casting the specimens in the moulds, they were covered with polythene sheet to prevent evaporation and left for 24 hours. Eventually, all specimens were removed from the moulds and were transferred into the water bath at room temperature for curing until the desired age of testing at 3, 7, 28 and 56. The tested properties of fresh concrete included slump that carried out conforming BS 1881: Part 102: 1983. The tested properties of hardened concrete were the compressive strength, and flexural test were measured at 3, 7, 28 and 56 days for compressive strength test while 28 and 56 days for rebound hammer test also 28 days for flexural test.

3. Results and Discussions

3.1. Concrete Workability

As shown as Figure 1, the slump value for the control test was 25mm, and by addition of different mixed proportion of UCO, the slump value was increased from 25mm to 43mm as for mixed proportion of 2.00% UCO. From control test to 0.25% UCO, there is 0% of increment for the slump test. There are increment of 4% of slump value from 0.25% UCO to 0.50% Of UCO. From 0.50% to 1.50%, the increment of slump value was similar until it comes to 2.0% UCO. From the increasing of 25mm to 43mm, the slump value was increase by percentage of 72%. Until 1.50% UCO, the slump value only increase averagely by 16%, but from 1.50% UCO to 2.00% UCO, the slump value increase by 48% which was a high increment compare to previous mixed proportion of UCO.

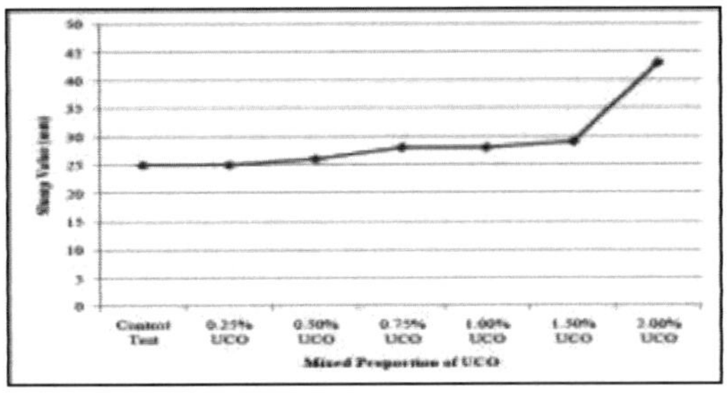

Figure 1: Slump Value of Concrete containing UCO

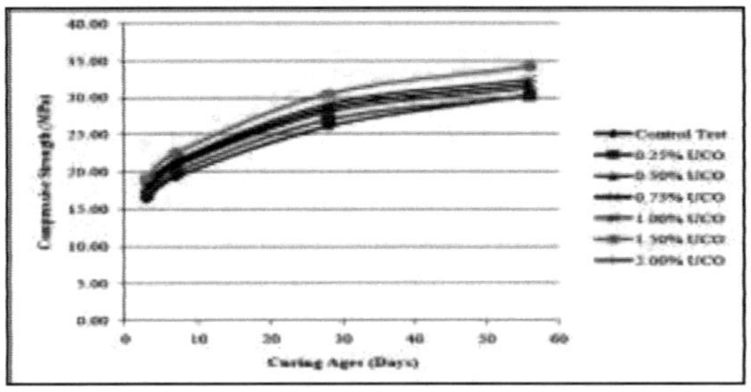

Figure 2: Compressive strength for concrete containing UCO

3.2. Concrete Compressive Strength

For the early curing ages of 3 days, concrete strength of control test was 16.38MPa. From control test to 0.25%, there is an increment of 3.42% for the compressive strength. While from 0.25% to 0.50%, the increment compare to control test is 7.26%. For the following 0.75% and 1.00%, the increments of compressive strength are respectively being 9.22% and 11.34%. The 1.50% of UCO achieves the highest increment which is 16.80% and the final 2.00% of UCO was the lowest increment of 3.30%. The graph of result goes up compare to the control test and shows the positive increment. For the following 7 days, the hardened concrete continues the development of concrete strength, while the strength for control test was 19.33MPa. The concrete strength for 1.50% UCO was also higher than control test with increment of 16.80% and the other mixed proportion obtained similar increment percentage from curing ages of 3 days. All of the mixed proportion achieved higher concrete strength than the control test. The graph pattern of increment for curing ages of 7 days was similar to the result of 3 days. Figure 2 show the different of compression test result from curing ages of 3 days to 56 days. For the compressive strength test of 28 days curing ages, the final strength of concrete strength was 26.09MPa. The highest concrete strength for different mixed proportion of UCO was 1.50% UCO which is 30.48MPa while the concrete strength for control test is 26.09MPa. The same result with curing ages of 3 days and 7 days, the range of increment for compressive strength from 3.30% to 16.80%. While for the compression test of 56 days of curing ages, the increment of the strength was similar with previous curing ages and 1.5% UCO stills the highest strength with 34.19MPa than 30.30MPa of control test.

3.3. Flexural Strength

From Figure 3, it can be clearly seen that the flexural strength increase from 1.8MPa of 0% used cooking oil to 2.01MPa of 1.5% used cooking oil. The flexural strength is increased by 11.7%. Based on the graph, it shows that further addition of used cooking oil had decreased the flexural strength from 2.01MPa of 1.5% used cooking oil to 1.91MPa of 2.0% used cooking oil. The flexural strength value had decreased by 5%. However, the flexural strength of 2.0% used cooking oil is still higher than control test (0% used cooking oil). The flexural strength of 0% used cooking oil to 2.0% used cooking oil is increased by 6%.

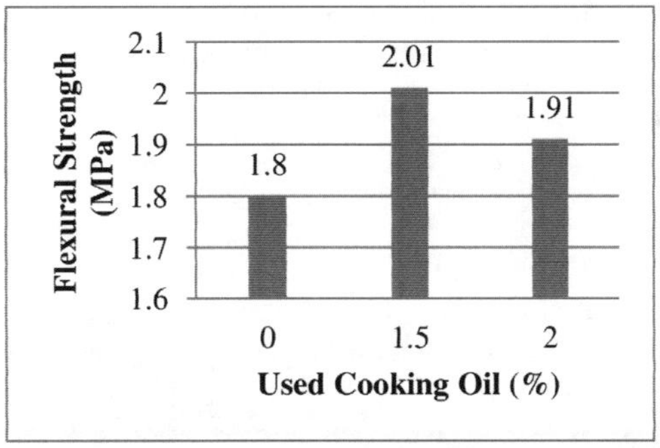

Figure 3: Graph of Flexural Strength VS Percentage of UCO

Based on the results, it is positively shows that the addition of used cooking oil in OPC concrete can actually increase the flexural strength of concrete. But this increasing will come to a limit where further additions of UCO can actually split the flexural strength. Thus, the strength started to decrease. Overall, it shows that used cooking oil can actually behave like super plasticizer. The UCO added in OPC concrete has increased the flexural strength which at the same time has improved the stiffness of the concrete.

4. Conclusion

UCO show a positive result on the concrete performance. When the proportion and dosage of UCO increased from 0.25% to 2.00%, the slump value had increment from 1mm to 18mm and as 72% of increment compare to the control test. When it comes to the compressive strength, the result also shows the used cooking oil with increment of 16.8% on the strength compare to control test. So it is show that the effects of UCO increase the workability and improve the compressive strength up to 2.00% compare to the control test. For the experimental program, total of 7 samples was prepared according to different proportion of UCO from control test, 0.25%, 0.50%, 0.75%, 1.00%, 1.50%, and 2.00%. Based on the performance of UCO, it achieves the maximum contribution at 1.50% when it obtains maximum increment of the compressive strength of 16.8% compare to the control test with 0.00% of UCO.

Acknowledgement
The authors would like to extend their acknowledgement to the Universiti Tenaga Nasional for providing the facilities for accomplishing the present research. Acknowledgement also due to Technical Team civil engineering Laboratory, Universiti Tenaga Nasional for their great contribution and assistance to complete this work.

References

[1] S. Chinchalkar, "An Upper Bound for the Number of Reachable Position 1996 *ICCA Journal* **19** 181-183
[2] Tendring District Council. 2011. Waste Cooking Oil
[3] Flexnews Website. 2009. Malaysia HRI Food Service Sector Food Service Malaysia 2009
[4] Wild Asia. 9 February 2011. Biofuel: Deriving From Recycling Used Cooking Oil.
[5] Man Jang Bio Sdn.Bhd. Biodiesel
[6] B. Salmia. The effects of used engine oil on various types of concrete 2012 PHD Thesis, Universiti Teknologi PETRONAS, Malaysia
[7] C.S. Chin. Performance of Used Engine Oil in Fresh and Hardened States of Normal and Blended Cement Concretes 2007 MSc Dissertation, Universiti Teknologi PETRONAS, Malaysia
[8] Nuruddin, F., Shafiq, N. & Beddu, Salmia. Effects of Used Engine Oil on MIRHA Concrete 2010 *International Conference on Sustainable Building and Infrastructure*
[9] Bilal S. Hamad, Ahmad A. Rteil, Muttassem El-Fadel. Effect of Used Engine Oil on Properties of Fresh and Hardened Concrete 2003 *Elsevier Science Ltd, Construction and Building Materials* **17** 311-318

IOP Conf. Series: Earth and Environmental Science **16** (2013) 012078 doi:10.1088/1755-1315/16/1/012078

Performance of new generation pole light

K C Foo, S Karunanithi, G Thio

Centre of Communication Services Convergence Technologies,
College of Engineering, Universiti Tenaga Nasional, Km 7,
Jalan Ikram-Uniten, 43000 Kajang, Selangor, Malaysia.

Email: sarveswaren@uniten.edu.my

Abstract. This paper describes the design and implementation of a standalone photovoltaic power supply which caters for garden lighting scheme. New Generation Pole Light (NGPL) consists of three parts which are light dependent resistor (LDR) and pyroelectric infrared (PIR) sensors, microcontroller and light emitting diode (LED) and finally, solar charging system. During the night, LED is switched on with two operating modes which are ultra-bright lighting for a predetermine period (when human presence is detected) and dim lighting. Meanwhile, LED is switched off at day time and solar charging system will recover the capacity of discharged battery. NGPL provides portable, sustainable, environmental friendly and requires minimal maintenance for outdoor lighting scheme for both urban and rural areas.

1. Introduction

The alternating current (AC) power supply from TNB has provided great convenience to the customers. However, there are limitations as AC supply isn't available at all locations. Outdoor lighting (for example garden area) using AC power supply is not feasible due to expensive cost of wiring installation. Nowadays, there are a lot of commercial off-the shelf solar powered garden light available. Disadvantages of these products is light is switched on and off at predetermine time rather than triggered by surrounding light intensity. Based on observation on street lights along Jalan Istana, the existing lighting scheme is less efficient with variation of sunrise and sunset timing throughout the year. Another disadvantage is the light is still switched on although there is no presence of human which addresses the need of sustainability and energy saving. Therefore, New Generation Pole Light (NGPL) was developed in order to address the need of intelligent outdoor lighting scheme. Light intensity detection and control is feasible with the usage of microcontroller, pyroelectric infrared sensor and light sensor [1] [2].

2. System Overview

NGPL consists of three parts which are light dependent resistor (LDR) and pyroelectric infrared (PIR) sensors, microcontroller and light emitting diode (LED) and finally, solar charging system as shown in Figure 1.

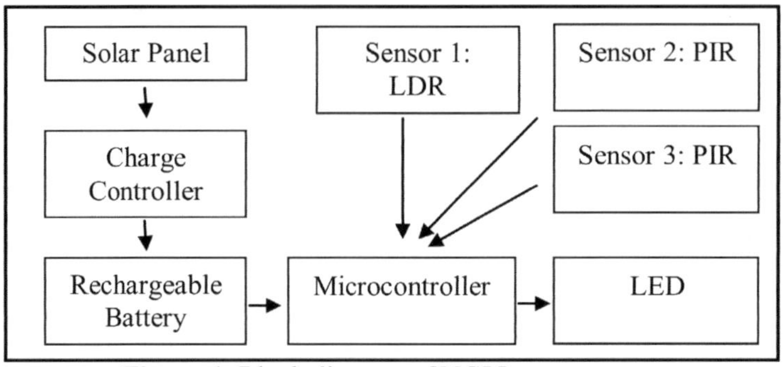

Figure 1. Block diagram of NGPL system.

2.1. Sensors

Two PIR sensors were used to detect reflected infrared signals from the surrounding (human or animal). The PIR sensor used has a detection range of 7 m and deflection angle of 120°. The digital output pin of PIR sensor remains low (0), when there are no motions detected. When a warm body (human or animal) passes within the PIR detection range, output pin becomes high (1). Meanwhile, LDR was identified as a low cost solution for detecting light intensity (brightness). Resistance of LDR changes depending on amount of light shining on its surface. As light intensity decreases, the resistance of LDR will increase. LDR resistance measured at 7pm, 7.10pm and 7.20pm respectively are 60.5 kΩ, 0.343 MΩ and 1.44 MΩ as shown in Figure 2.

Figure 2. Measurement of LDR resistance at various times.

2.2. Microcontroller and Light Emitting Diode

Microcontroller, PIC16F690 receives signals from LDR circuitry and PIR sensors. During the day, the output voltage across the LDR is less than the predefined voltage level; which in return turns off the system (LED). Once the day is dark, voltage across the LDR is greater than the predefined voltage level where the LED is turned on. However, the voltage supplied to the LED is being limited in order to produce dim lighting. When PIR sensors detect presence of human, the microcontroller turns on the LED brighter for a predetermine duration. LED lighting was used instead of light bulb due to its low power consumption; thus, increasing the efficiency of the system without compromising the light intensity. 12 V, 3 W LED was used which emits warm white lighting without any UV or IR radiation. This is essential to avoid interference with PIR sensing.

2.3. Solar Charging System

LDR, PIR sensors, microcontroller and LED are powered using a 12 V, 4 Ah sealed-lead acid battery (manufactured by Innovation Battery Technology). Usage of sealed-lead acid batteries is common in standalone power supply due to its maintenance free nature. However, the battery will need constant charging depending on how frequent the system is being used. This raises the need of using renewable energy for battery charging purpose. Solar power was identified to be used for battery charging purpose as Malaysia is located along the equator line. Hence, SPM010-M monocrystalline photovoltaic module with rating of 10 W, 18 V was used (manufactured by Solar Power Mart). A charge controller circuit was built to regulate solar panel voltage to the battery voltage.

2.4. System Integration

The solar panel was mounted on top of the pole in order to obtain maximum amount of sunlight as shown in Figure 3. LDR was placed in the mid-section of the pole which enable optimal light intensity sensing. Both PIR sensors were respectively placed on left and right side of the pole which maximizes detection of infrared signals. Finally, the rechargeable battery and circuitry were fixed into a lab scaled waterproof container.

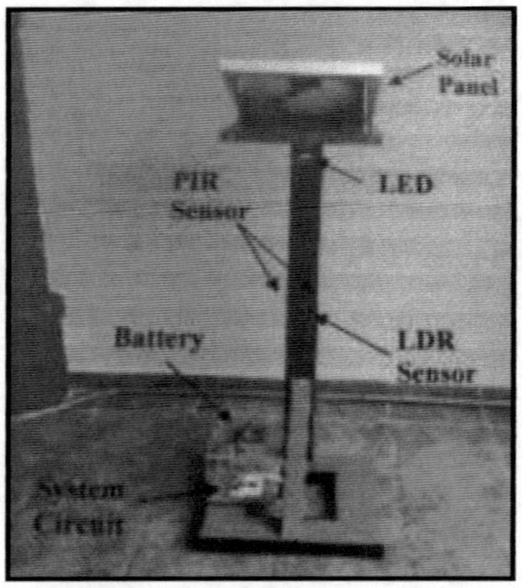

Figure 3. NGPL prototype.

3. Preliminary Results

NGPL was tested in a lab scale environment where output of PIR circuitry was observed using an oscilloscope. PIR sensor outputs digital pulse when human presence is detected as shown in Figure 4; else, there is no change in output level. Current LED lighting used has a beam angle of 45° which limits the system application to narrow lighting area. By replacing the current lighting to surface mount ultra-bright LED, the system is able to be installed for wide area lighting scheme.

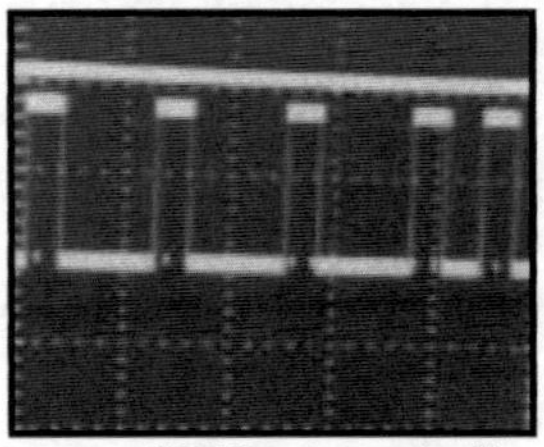

Figure 4. Output of PIR sensor indicating motion.

4. Conclusion

NGPL provides portable, sustainable and environmental friendly lighting scheme for both urban and rural areas. Using solar renewable energy is also cost efficient in the long run as it requires only annual maintenance to desulphate the battery terminals.

References

[1] Bai Y W and Ku Y T 2008 Automatic room light intensity detection and control using a microprocessor and light sensors *IEEE Trans. On Consumer Electronics* **54** issue 3 1173-76

[2] M Moghavvemi and C S Lu 2004 Pyroelectric infrared sensor for intruder detection *IEEE Region 10 Conf. (TENCON 2004)* vol 4 p 656-659

IOP Conf. Series: Earth and Environmental Science **16** (2013) 012079 doi:10.1088/1755-1315/16/1/012079

Investigation of voltage swell mitigation using STATCOM

N A Abdul Razak[1], S Jaafar[1], I S Hussain[1]

[1] Department of Electrical Power Engineering, College of Engineering, Universiti Tenaga Nasional, 43000 Kajang, Selangor.

Email: sabarina@uniten.edu.my, idasuzana@uniten.edu.my

Abstract. STATCOM is one of the best applications of a self commutated FACTS device to control power quality problems in the distribution system. This project proposed a STATCOM model with voltage control mechanism. DQ transformation was implemented in the controller system to achieve better estimation. Then, the model was used to investigate and analyse voltage swell problem in distribution system. The simulation results show that voltage swell could contaminate distribution network with unwanted harmonic frequencies. Negative sequence frequencies give harmful effects to the network. System connected with proposed STATCOM model illustrates that it could mitigate this problems efficiently.

1. Introduction

Static Synchronous Compensator (STATCOM) is an influential controller device for electric transmission and distribution network. STATCOM is employed to provide voltage flicker compensation, voltage stability and power factor control thus increases power quality [1]. Applications of STATCOM also include fast supply of voltage to support the power system. Weak system gives instability to the system since disruption will caused unbalanced voltage. This results in severe damaging impact to both load and power system equipments.

This project aims to develop a STATCOM model with voltage control system using proportional integral, PI, controller using MATLAB Simulink software as described in reference [2]. Then the proposed model will be tested by studying impact of voltage swell in network in terms of total harmonic distortion, THD and harmonic contents.

2. Methodology

The overall test system setup is shown in Figure 1. The settings for the components are given in [2]. The difference between the STATCOM model in [2] and the STATCOM model in this paper was the power electronic device used in the voltage source converter, VSC. This paper used IGBT instead of ideal switch to represent practical implementation of STATCOM. The VSC control system has 4 inputs and 1 output with 2 control subsystems. The controllers are based on PI and DQ transformations.

The Re+jwL represents the transformer that isolates the network from the STATCOM. There are two loads in this test system, labelled as Line 1 (Re+jwL1) and Line 2 (Re+jwL2). The voltage swell was generated using a three phase programmable voltage source. The three phase programmable voltage source is placed beside Line 1. Two simulations were conducted. The first simulation will represent an ideal system with no fault, named as the Reference System. The other simulation, named as the Case Study System, was simulated with voltage swell injected in the test system.

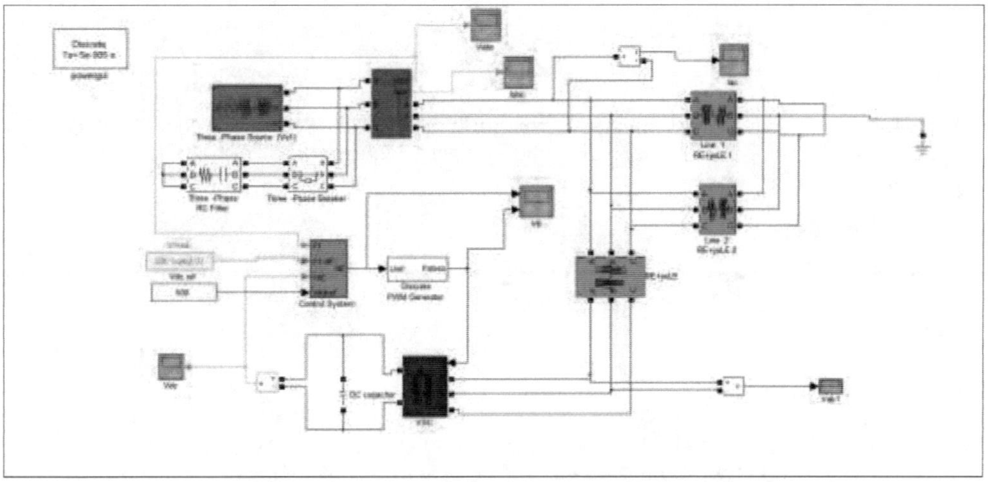

Figure 1. Test system with STATCOM model

2.1. *PI controller*

In this project, PI is used to tune and modify output of VSC in achieving the expected outcome. The control scheme operates in voltage control mode [3] as shown in Figure 2. The first PI will determine the amplitude modulation (m) and the second PI will determine the angle modulation (α). Combination of outputs from magnitude and phase modulation control systems will produce, *Vcontrol*,

$$V_{control}(VE) = modulation * \cos(\omega t+\alpha+\theta) \tag{1}$$

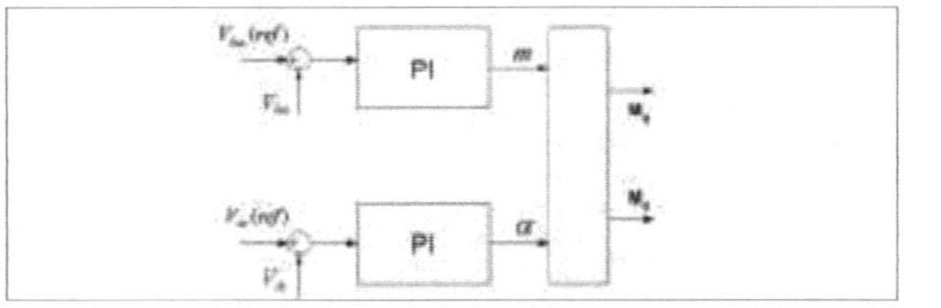

Figure 2. STATCOM control configuration [3]

2.2. *DQ transformations*

DQ transformation is mostly employed for three phase converters' analysis and control design. The DQ controller provides superior transient response and low output voltage THD under nonlinear load operation [4]. In the amplitude modulation subsystem, the V_{bus} is converted into DQ components as follows,

$$V_d=\frac{2}{3}\left(V_a \sin(\omega t)+V_b \sin(\omega t-2\pi/3)+V_c \sin(\omega t+2\pi/3)\right) \tag{2a}$$

$$V_q=\frac{2}{3}\left(V_a \cos(\omega t)+V_b \cos(\omega t-2\pi/3)+V_c \cos(\omega t+2\pi/3)\right) \tag{2b}$$

$$V_0=\frac{1}{3}(V_a+V_b+V_c) \tag{2c}$$

2.3. *Analysis methods*

Harmonic analysis was conducted to evaluate the effect of voltage swell on the network. The analysis was done by analyzing the bus voltage (V1) and the bus current (I_{ac}). The two analyses conducted were computation of THD and the harmonic contents of the signals. THD is given by

$$THD_x = \frac{\sqrt{\sum_{n=2}^{\infty} X_{(n)}^2}}{X_1} \quad (3)$$

Harmonics frequencies can be group according to symmetrical system sequence which is positive sequence harmonic, negative sequence harmonic and zero sequence harmonic. The harmonic contents of a signal are determined as follows:

$$h = np \pm 1 \quad (4)$$

where:

h = order of harmonics, n = an integer 1, 2, 3..., p = number of current pulses per cycle

3. Results and Discussions

Table 1 lists the parameter settings for the simulation. Figure 3-4 show the THD and harmonic contents for V1 and I_{ac} respectively. For the Reference System, the THD_{V1} is at 0.31% and THD_{Iac} is at 0.43%. The THD for the Case Study System are very high, exceeding the 5% THD limit set by the IEEE [5], with THD_{V1} at 54.9% and THD_{Iac} at 13.39%.

Table 1. Simulation Settings

Parameter	Unit
Max Step Size	0.02
Frequency	60 Hz
Time offset Period	0.4 s
Vrmsphase-phase	220 V
Dicrete Time Power Gui	50e-006 s

The Case Study System produced higher magnitude of harmonic frequencies. The fundamental frequency contributed the highest frequency component as expected. This is followed by the sidebands for the switching frequency of 900 Hz, the 13th and the 17th harmonic frequencies. The 13th harmonic frequency is a positive sequence harmonic which can assist the torque production in the system. However, the 17th harmonic frequency is a negative sequence harmonic that can cause inverse rotation of the three phase system resulting in torque pulsations. The results clearly show that the problem needs to be addressed to avoid power quality problem to the system.

4. Conclusion and Future Work

In this paper a STATCOM model with voltage control mechanism developed using Matlab Simulink has been presented. Simulations were carried out to verify the functionality of the proposed model with no fault and with fault presence in the system. The simulation results were analyzed in terms of THD and harmonic contents. The STATCOM model was shown to be able to mitigate the voltage swell problem. In future, similar study could be done with other power switching devices and with different filter configurations. The control system can be designed to have real (P) and reactive power (Q) as inputs in the control system.

IOP Conf. Series: Earth and Environmental Science **16** (2013) 012079 doi:10.1088/1755-1315/16/1/012079

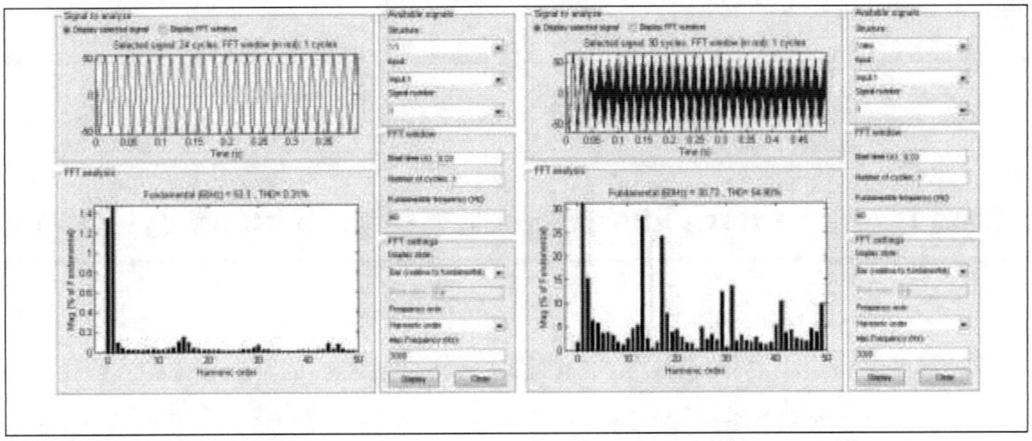

(**a**) Reference System (**b**) Case Study System

Figure 3. THD and Harmonic Content of V1

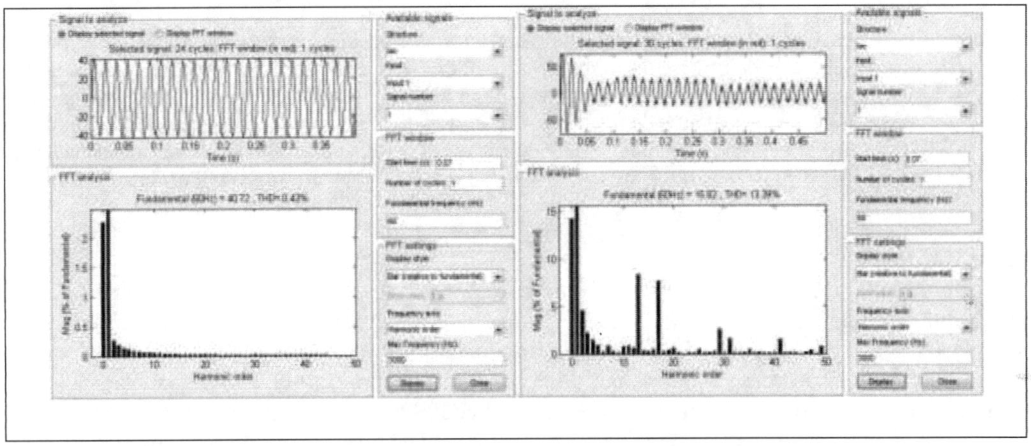

(**a**) Reference System (**b**) Case Study System

Figure 4. THD and Harmonic Content of Iac

5. References

[1] Ben-Sheng Chen and Yuan-Yih Hsu An Analytical Approach to Harmonic Analysis and Controller Design of a STATCOM 2007 *IEEE Transactions of Power Delivery* **22** 423-432

[2] Juan S R and A Medina 2012 Analysis of Power Electronic Controllers in Electric Power Systems Using Simulink *Technology and Engineering Applications of Simulink* (InTech)

[3] Bhaskar M and Ghadir R A Study of Interaction between Dynamic Load and STATCOM 2006 *38th southeastern Symposium on System Theory Tennessee Technological* (University Cookeville) 392-396

[4] Gonzalez M, Cardenas V and Pazos F DQ Transformation Development For Single-Phase Systems To Compensate Harmonic Distortion And Reactive Power 2004 *(IEEE International)* 177–182

[5] IEEE Recommended Practice for Monitoring Electric Power Quality 1995 *IEEE Standard* 1159

IOP Conf. Series: Earth and Environmental Science **16** (2013) 012080 doi:10.1088/1755-1315/16/1/012080

Simulation of solar chimney power plant with an external heat source

Azeemuddin Islamuddin, Hussain H Al-Kayiem and Syed I Gilani

Department of Mechanical Engineering, Universiti Teknologi Petronas, Malaysia

Email: hussain_kayiem@petronas.com.my

Abstract. Solar chimney power plant is a sustainable source of power production. The key parameter to increase the system power output is to increase its size but the plant cannot operate during night hours. This study deals with simulation work to validate results of pilot plant at Manzanares and include the effects of waste heat from a gas turbine power plant in the system. The effects show continuous night operation, a 38.8 percent increase in power at 1000 W/m^2 global solar irradiation at daytime and 1.14 percent increase in overall efficiency.

1. Introduction

Solar chimney power plant runs truly on renewable energy source that is the sun. Many parts of the world that have flat land properties and receive massive amount of solar radiation, can be utilized for such system. It can work under diffuse solar radiation and is durable and require less maintenance. The system has three main components a) collector b) chimney and c) turbine. The collector absorbs heat energy from solar radiations, the air inside the collector moves due to buoyancy effects and finally reaches the center of the chimney with a certain velocity. The turbine located there rotates by extracting the kinetic energy of the air. Finally this kinetic energy is converted to electric power. One of the drawback of this system is that it cannot operate during night hours. Although previous researches have shown by using means to make it work at night. In this study, a simulation is carried out that not only shows night operation of the plant but also power output enhancement and efficiency. The simulation includes the effects of exhaust gases from a nearby gas turbine power plant on the system that originally operates only on solar energy.

A detailed numerical analysis was carried out by Roozbeh [1], showing with increasing solar radiation values, the power increases. Simulation with energy storage layer for different solar radiations was carried out by Tingzhen [2], which resulted that heat loss from the storage medium increases with increase in solar radiations. According to Bernardes [3], the power output can be increased by increasing chimney height, collector area and transmittance. Another study by Jiakuan [4], show that power output of power generating system increases with global solar radiation intensity, collector area and chimney height. It also concluded that the larger the chimney height, the greater would be the driving force in air. The use of water storage system in collector was presented by Kreetz [5], his study indicated a continuous day and night operation of the solar chimney power plant.

The main purpose of this research is to model the system so that it will work at night and also enhance the power output. The results of Manzanares pilot plant were validated by simulating a mathematical model. Later, the effects of exhaust gases were incorporated with the mathematical model. This simulation uses ANSYS Fluent to calculate the heat transfer effects of exhaust gases on the system.

2. Proposed model

A model was needed to implement the exhaust gases effect into the classical solar chimney power plant. The proposed model consists of hollow rectangular channels beneath the collector cover. The orientation of these channels depend upon the equally divided circular collector (360 degrees) by the number of channels (in this study four, each at 90°). The exhaust gases are passed into the channels by means of pipes and exits into the chimney base. Inside the channels the gases released through pipe holes. The walls of the channel becomes hot and a heat transfer mechanism takes place between the walls and the air flowing beneath the collector cover. Consequently, the air will gain more kinetic energy and thus increasing the power output through the turbine. The exhaust gases after useful energy transferred to the collector air, are allowed to escape through upper pipes located at the extreme end of the channels top and then into chimney as shown in Fig.1.

Table 1. Channel and exhaust gas specifications.

	Parameter	Value
Channel	Material	Carbon steel
	Length	117 m
	Thickness	0.015 m
	Inner width	0.17 m
	Height at collector inlet	2 m
	Height at collector outlet	6 m
Exhaust gas	Fuel	Natural gas
	Mass flow	24.5 kg/s (each channel)
	Temperature	643.15 K

The assumptions that were taken into account to utilize the exhaust gases in the system are as follows,

a. The exhaust gases from the turbine exit are just at the inlet of the channels.
b. The temperature of these gases at the inlet of channels is the exit temperature of the gas turbine.
c. The mass flow rate of these gases is equally distributed into the channels i.e. mass flow rate from the exit of the turbine divided by number of channels.

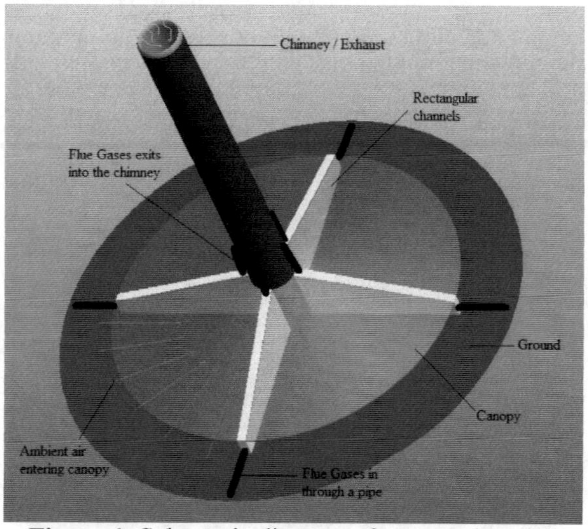

Figure 1. Schematic diagram of proposed model.

3. Governing equations

The mass flow rate of air entering the collector is evaluated as,

$$\dot{m} = \rho_{in} V_{in} A_{in} \tag{1}$$

Where, ρ_{in} is air density, V_{in} inlet velocity and A_{in} is the inlet area.
The actual useful heat gain by the collector is given by [6],

$$Q_u = A_{coll}[S - U_L(T_g - T_{amb})] \tag{2}$$

Where, S is the heat absorbed by the collector, A_{coll} is the exposed collector area to solar radiations, U_L is the overall heat transfer coefficient, T_g and T_{amb} is the ground and ambient temperature respectively. The heat transfer coefficient for ground to collector air is given by [3],

$$h_{c,g-air} = \frac{0.2106 + 0.0026 V_{in}\{\frac{T_m \rho}{\mu g(T_g - T_{fm})}\}^{1/3}}{\{\frac{\mu T_m}{(T_g - T_{fm}) g \rho^2 C_p k^2}\}^{1/3}} \tag{3}$$

Where,
$h_{c,g-air}$ = convective heat transfer coefficient from ground to collector air.
T_{fm} = mean temperature of air.
T_m = mean temperature of T_g and T_{fm}.
C_p is air specific heat, μ is air dynamic viscosity, k is air thermal conductivity and g is gravitational acceleration.
Similarly, for cover to collector air,

$$h_{c,c-air} = \frac{0.2106 + 0.0026 V_{in}\{\frac{T_m \rho}{\mu g(T_{fm} - T_c)}\}^{1/3}}{\{\frac{\mu T_m}{(T_{fm} - T_c) g \rho^2 C_p k^2}\}^{1/3}} \tag{4}$$

Where,
$h_{c,c-air}$ = convective heat transfer coefficient from canopy to collector air.
T_m = mean temperature of T_c and T_{fm} , T_c is the cover temperature.
The maximum power output from the turbine is calculated as [7],

$$P_{out} = \frac{2}{3} \frac{\{\rho_{ch,in} A_{ch} V_{ch,in} H_{ch} g(T_{c,out} - T_{amb})\}}{T_{amb}} \tag{5}$$

Where, $\rho_{ch,in}$ is the air density at chimney inlet, $V_{ch,in}$ is the air velocity at the chimney inlet and A_{ch} is the chimney inlet area. The velocity at chimney inlet is given by [8]

$$V_{ch,in} = \sqrt{2gH_{ch}\frac{(T_{c,out} - T_{amb})}{T_{amb}}} \tag{6}$$

H_{ch} is chimney height and $T_{c,out}$ is collector outlet temperature. The grid power can be found if the turbine efficiency (which is normally ranging between 50 to 90 percent) and efficiency η_f (that include the frictional losses) is taken into account.

4. Validation and results

The simulation was run in two parts, 1) Solar mode 2) Hybrid mode (referred here as the proposed model). Manzanares pilot plant was used as reference plant in the simulation. Solar mode resulted power output of 47.8 KW and 12.8 m/s updraft velocity at 1000 W/m^2 global solar irradiation. This is the validation of mathematical model with Manzanares pilot plant results. In hybrid mode, the heat transfer effects in the system were calculated by using ANSYS Fluent software. The geometry from the proposed model was used consisting of collector section with two channels only. The final results show a 38.8 % increase in power at 1000 W/m^2. The average day efficiency of the hybrid system resulted as 1.25 % compared with 0.11 % of solar mode. The power comparison (Fig. 2) shows that during night hours solar mode delivers negligible power which is due to some heat energy released by the collector that is had absorbed during the daytime. A significant power value can be seen when the plant is running on hybrid mode. Besides, during the daytime it also enhances the power. Power output largely depends upon the temperature and pressure differences between the ambient and collector outlet respectively as represented by eq. (5). The trend of collector pressure difference follows same as that of power (Fig. 3).

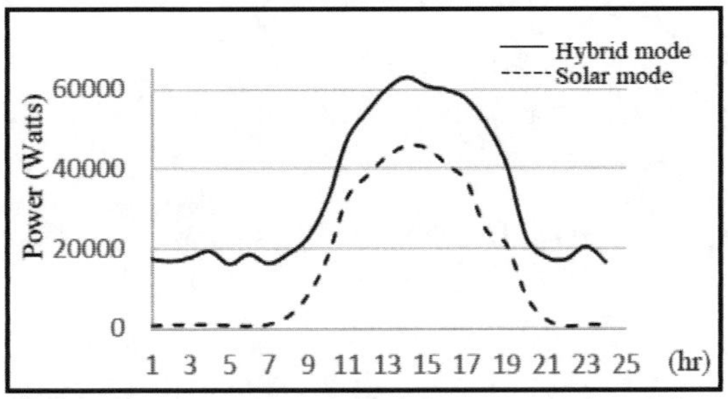

Figure 2. Power comparison between solar and hybrid mode

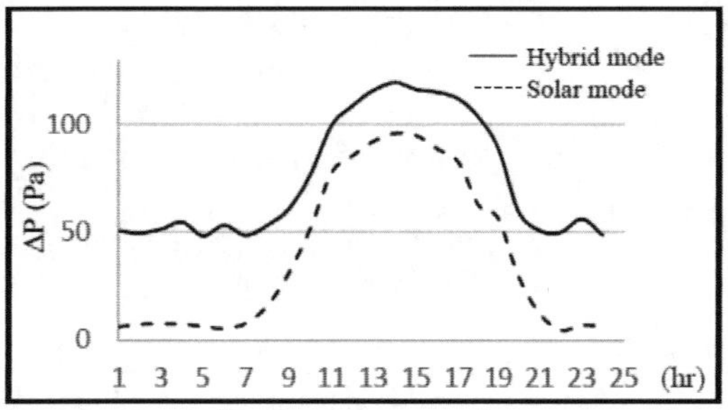

Figure 3. Collector pressure drop comparison between solar and hybrid mode

5. Conclusion

An enhancement technique to improve the performance of the solar chimney power plants has been investigated by mathematical simulation. The simulated model was validated by comparing the results with the operational results of Manzanares pilot prototype plant. The proposed model which based on charging heat to collector air has shown considerable enhancement in the performance. The system under the new proposed thermal hybrid technique becomes able to deliver electricity all around the day.

References

[1] Roozbeh, S, 2011, "Modelling and numerical simulation of solar chimney power plants", Solar Energy, vol. 85, pp. 829-838.

[2] Tingzhen, M, 2008, "Numerical analysis of flow and heat transfer characteristics in solar chimney power plants with energy storage layer", Energy Conversion and Management, vol. 49, pp. 178-185.

[3] Bernardes, 2009, "Analysis of some available heat transfer coefficients applicable to solar chimney power plant collectors", Solar Energy, vol. 83, pp. 264-275.

[4] Jiakuan, Y, 2007, "Simulation of a pilot solar chimney thermal power generating equipment", Renewable Energy, vol. 32, pp. 1637-1644.

[5] Kreetz, H, 1997, "Theoretische Untersuchungen und Auslegung eines temporären Wasserspeichers für das Aufwindkraftwerk", diploma thesis, Technical University Berlin.

[6] Duffie J. A, 1991, "Solar Engineering of Thermal Processes", 2ND Ed, John Wiley & Sons.

[7] Dai, Y. J, 2003, "Case study of solar chimney power plants in northwestern regions of China", Renewable Energy, Vol. 28, pp. 1295-1304.

[8] Haaf, W and Schlaich, J, 1983, "Solar chimneys, part I: Principle and construction of the pilot plant in Manzanares".

IOP Conf. Series: Earth and Environmental Science **16** (2013) 012081 doi:10.1088/1755-1315/16/1/012081

Design and development an insect-inspired humanoid gripper that is structurally sound, yet very flexible

S. Hajjaj[1], and N. Pun[2]

[1]College of Engineering, University Tenaga Nasional,
Jalan IKRAM-UNITEN, 43000 Kajang, Selangor. Tel: 603-8921 2020 (ext: 6402)
E-mail: ssalama@uniten.edu.my, 03-892020 (ext: 6402)

[2] Faculty of Engineering and IT, INTI-International University,
Persiaran Perdana BBN, Putra Nilai, 71800 Nilai, N. Sembilan, Malaysia.

Abstract. One of the biggest challenges in mechanical robotics design is the balance between structural integrity and flexibility. An industrial robotic gripper could be technically advanced, however it contains only 1 Degree of Freedom (DOF). If one is to add more DOFs the design would become complex. On the other hand, the human wrist and fingers contain 23 DOFs, and is very lightweight and highly flexible. Robotics are becoming more and more part of our social life, they are more and more being incorporated in social, medical, and personal application. Therefore, for such robots to be effective, they need to mimic human performance, both in performance as well as in mechanical design. In this work, a Humanoid Gripper is designed and built to mimic a simplified version of a human wrist and fingers. This is attempted by mimicking insect and human designs of grippes. The main challenge was to insure that the gripper is structurally sound, but at the same time flexible and lightweight. A combination of light weight material and a unique design of finger actuators were applied. The gripper is controlled by a PARALLAX servo controller 28823 (PSCI), which mounted on the assembly itself. At the end, a 6 DOF humanoid gripper made of lightweight material, similar in size to the human arm, and is able to carry a weight of 1 Kg has been designed and built.

1. Introduction

When it comes to the mechanical design of robotic grippers, designers are faced with the following dilemma; make the gripper very flexible, while keeping it light weight and structurally sound. The problem is that these two requirements are just the opposite of each other.

To make the robot more flexible, one needs to add more DOFs, which means more actuators, connectors, cables, housing, etc. which will the gripper heavier and more complicated. That is why most industrial grippers have limited number of DOFs [1]

However, when one observes the human hand, one can see not only is it highly flexible (23 DOFs for the wrists and fingers of one hand), but it is also light weight, especially when it is compared to the weight of industrial grippers. The same can be observed in insects' legs and other extremities. Each leg is very light and yet very complex. [1, 2]

Therefore, this work attempts to build a gripper that is closer to the human form, as much as possible by trying to mimic human and insect form.

IOP Conf. Series: Earth and Environmental Science **16** (2013) 012081 doi:10.1088/1755-1315/16/1/012081

2. Literature Review

Attempting to achieve flexibility and structural integrity is not new. There were differences in methods and technologies, each with its advantages. (So and so) combined the use of ultrasonic motors and elastic elements. In this design, the gripper size the same as the human hand, it has almost the same number of DOFs of the human hand [2,3].

The motion is achieved the ultrasonic motor and the elastic elements in it. The ultrasonic motor is powered by ultrasonic vibration of the stator against the rotor, which allows this motor to achieve a large rotation angle. The advantage of the ultrasonic motor is its compact size and high torque that allows it be mounted on the linkage and be used to drive the robot's fingers. This allows the gripper to be reduced in size and dimensions [4].

Another technique builds the gripper out of elastic material and is powered by compressed air. This was developed by Virginia Tech's Robotics and Mechanisms Laboratory of the College of Engineering. It uses microcontroller commands to control the movement on the fingers. This robot hand is mainly operated by using the compressed air.

The material of the robot hand is mainly elastic ligaments. The fingers will hold tight together when the air inside the fingers is being compressed. This robot hand has 1 degree of freedom on each finger. The pressure of the air can be adjusted to manipulate different object, lower pressure is required when the robot hand is holding a soft fragile object while high pressure is used to hold an object tightly [5].

R. D. Howe, et. al discovered that a cockroach leg is very flexible and acts like a spring., which allows the cockroach to walk or run on uneven surface. The robot fingers are made by using plastic springs that could deflect naturally and so are flexible to grasp a wide range of objects. Besides, the material is light in weight and cheap. The main components that operate inside the hand are cables and pulleys which run by motors. [6]

3. Design and Methodology

The design of this gripper is divided into three main categories, 1) design of the fingers, 2) design of the palm, 3) the design of the actuators, and, 4) programming the motion of the gripper.

3.1. Design of the fingers.

The fingers were made of a rubber tubing that has been cut off in strategic places, then a nylon line was fixed to the tip of the finger, and the other end to be connected to the servo motors.

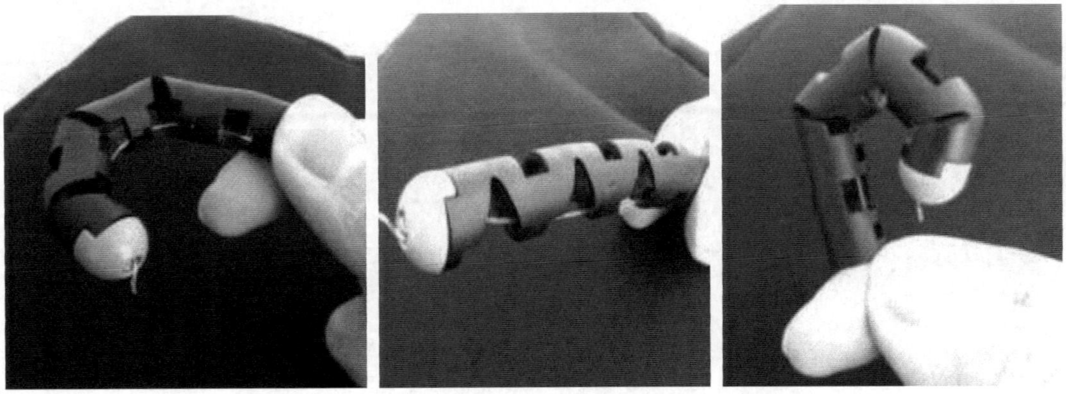

Figure 1. The Design of the fingers used in this project

As it can be seen, the design is very elastic, light weight, yet structurally sounds. The range of its motion can be easily modified by redesigning the locations of the cuts in the rubber tube. Three basic cuts were made to create the three knuckles.

3.2. Design of the Palm, and mounting of the Actuators.

The rubber tubing was then mounted on wooden rods, which in turn were mounted on the palm. Next the whole assembly is mounted on the arm, where the servo motors, controller, and power source were mounted.

This Arrangement greatly mimics the design of the human arm, where the muscles powering our fingers are actually "stored" in the forearms. When you flex your fingers you will notice your muscles at the forearms contract and expand in correspondence to your fingers.

Figure 2. The Complete design of the humanoid gripper

3.3. Controlling the motion of the gripper.

To control the motion of the gripper, the PARALLAX 16 channel servo controller is used. This servo controller uses USB port to communicate with the controlling intelligence. It allows 16 channels of servo controlling by sending commands from a microcontroller via PC or USB connection.

Figure 3. Mounting of the controller unit in the forearm of the gripper.

Since this work focuses on the mechanical aspects of the design, little effort was directed towards programming of motion. Therefore, the method of teaching was utilized. The PARALLAX 16 channel servo controller comes with its own software package, named PSCI.

This software allows the user to control the motion of each motor of the assembly, i.e the motion of each finger. A complex motion can then be created by creating the motion of each finger separately, saving it, and then playing it together at once.

4. Results and Findings

The original objective of this work is to build a light weight, structurally sound robotic gripper that is inspired by human and insect designs.

To demonstrate the flexibility and agility of this gripper, it was programmed to grip several objects and perform several hand gestures, such as "Fist, Rock, Pointing" and others.

4.1. Gripping several objects
As it can be seen from figures, the gripper was able to grab a tissue package as well as soda can. As a bonus, rubber has a high coefficient of friction, which helps in enhancing the gripper's strength.

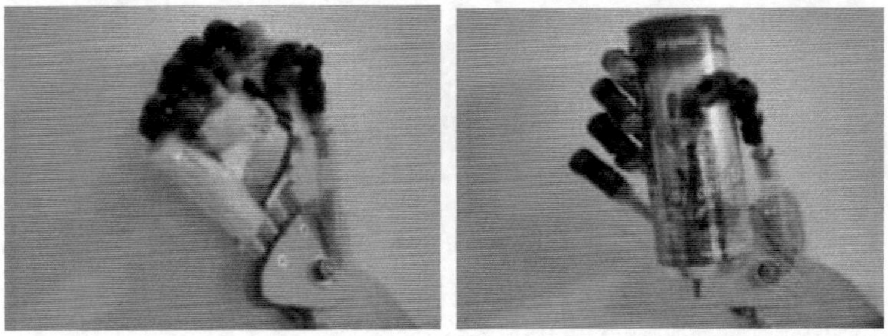

Figure 4. The Humanoid Gripper in action

4.2. Hand Gestures.
As it can be seen from figures, the gripper was programmed to make several gestures, in each of these gestures; the controller would move all the fingers concurrently.

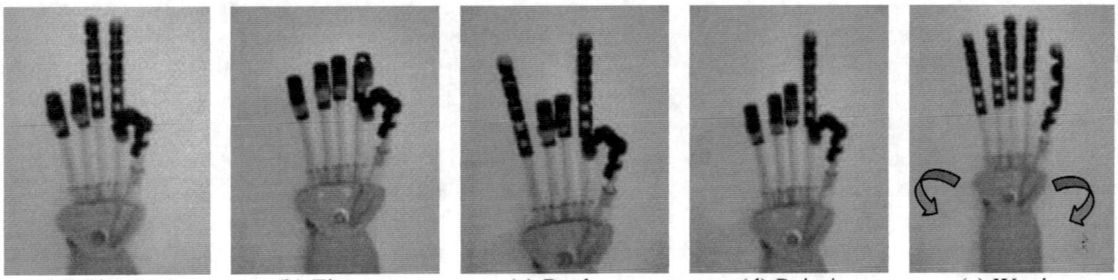

(a) Piece (b) Fist (c) Rock (d) Pointing (e) Waving

Figure 5. The Different gestures achieved by the gripper.

As it can be seen in this section, the objectives of the work has been met, this gripper is light weight, relatively flexible, and is able to grip up to 1kg of weight.

References

[1] Kolluru R, Valavanis K and Tsourveloudis N 2000 *Proc. Int. Conf. on Robotics & Automation* **2** 1988
[2] Russell G, Brown and Randy C 1999 *IEEE Trans. on Robotics and Automation* **15** 174
[3] Raghav Vand Senger S 2012 *Proc. Nat. Conf. on Trends and Advances in Mechanical Engineering* 97 – 400
[4] Zaki A, Mahgoub O, El-Shafei A and Soliman A 2010, Design and Implementation of Efficient Intellegent Robotic Gripper data *WSEAS Transactions on Systems* **9** 1130 - 42
[5] Alqasemi R, Mahler S and Dubey R 2007 *Proc. 10th IEEE Int. Conf. on Rehabilitation Robotic* **1** 432 – 7
[6] OHOL S and KAJALE R 2009 *Int. Conf. on Machine Learning and Computing* **3** 213 – 21

IOP Conf. Series: Earth and Environmental Science **16** (2013) 012082 doi:10.1088/1755-1315/16/1/012082

Development of a bench scale biomass torrefier

M.N. Mohd Ja'afar *, **A. Abd Rahman, A. H. Shamsuddin**

Centre for Renewable Energy, Universiti Tenaga Nasional, Putrajaya Campus, Jalan IKRAM-UNITEN, 43000 Kajang, Selangor, Malaysia
nasharuddin@uniten.edu.my

Abstract. Cofiring biomass with coal has become very popular with power utilities to reduce fossil fuel carbon dioxide (CO_2) emission. It is relatively easy to implement on most common pulverised coal plants. However, raw biomass is difficult to utilise and requires upgrading to a higher quality fuel to substitute coal. Upgrading by torrefaction can improve the properties of biomass close to low rank coals suitable for cofiring. In this study, a bench scale torrefier was developed to produce torrefied biomass samples for further studies of its properties and combustion behaviour. The torrefier was developed from a domestic 1600W electric oven. Biomass pellets was then torrefied at 250 °C for 1 hour using this torrefier. Proximate analysis and gross calorific value (GCV) of the torrefied biomass were carried out. The results showed that GCV of the torrefied biomass had increased when compared to raw. The moisture content and volatile matter had decreased, and ash content and fixed carbon had increased as expected.

1. Introduction

Malaysia's primary energy supply has been steadily growing at an average of 2.6 % from 2006 to 2010 [1]. Furthermore, a significance increase of over twice as much is seen in the use of coal and coke as the source of primary energy from 7,299 ktoe in 2006 to 14,777 ktoe in 2010. This dependence on coal and coke had resulted in a significant increase in carbon dioxide (CO_2) emission from 23.028 million tonnes in 2006 to 39.781 million tonnes in 2010 [2]. Cofiring biomass with coal is a relatively easy, safe, more economical and more efficient utilization of biomass as well as to reduce the amount of fossil fuel CO_2 emission [3, 4]. Combustion of biomass considered as zero net CO_2 emitted because the amount of CO_2 released during combustion equals with the amount of CO_2 the biomass absorbed during its growth [5].

Malaysia is abundance in biomass especially from the oil palm industry that can be used for cofiring at its coal power plants [6]. However, raw biomass is a difficult fuel to utilize mainly due to its heterogeneous nature and uneven composition, quality decay by biodegradation, lower calorific value, high ash content and the thermal instability resulted from high oxygen/carbon (O/C) ratio when compared to coal [7]. Torrefaction is a biomass pre-treatment that could be used to minimize these drawbacks and in the process upgrades the biomass to a higher quality fuel close to low rank coals making it suitable for cofiring. Coal power plant operators can utilize torrefied biomass as a direct thermal substitution, with only minor modification to its existing plant and practices [8].

This paper presents work done in developing a bench scale biomass torrefier to produce samples of torrefied biomass to be used for further fuel characterization studies such as gross calorific value (GCV), proximate and ultimate analyses. Torrefaction process occurs in an inert environment at a temperature between 200 °C to 300 °C [9, 10]. Previous work on torrefaction on Malaysian biomass had been done using thermogravimetric analyzer (TGA) or tube furnace apparatus as the torrefier. Such apparatus can only torrefy small amount of biomass, a maximum of 5g, at a time which produce insufficient sample size for further characterization studies. Torrefying several times to produce enough samples not only is time and energy consuming, it can also lead to a heterogeneous sample when recombined. Hence the work presented in this paper was to develop a bench scale torrefier that could handle a larger biomass sample size.

2. Methodology

2.1. Bench Scale Torrefier
A domestic 1600W electric oven was selected as a base for the torrefier due to its built in heating elements and a basic temperature controller housed in a safe, working enclosure. The 20-litre oven is also large enough to accommodate a sizable sample of biomass. Another advantage of the domestic oven is that it has a rotating mechanism built in which was later modified as an agitator for the biomass sample. The oven has a maximum temperature rating of 250 °C but it was found to be consuming too much time to reach this temperature. Insulation using aluminium lined ceramic fibre improved this, managing to achieve over the 300 °C upper temperature limit. It was also found that the basic bimetal temperature switch could not maintain a steady temperature required for torrefaction. A temperature microcontroller with fuzzy logic control is used to replace the bimetal switch. The fuzzy logic reduces the heating time and control the supply current to heating element to reduce the temperature fluctuation as shown in Figure 1. Biomass sample is placed in a reactor made of a thin-walled aluminium container sealed with aluminium tape. Figure 2 shows the schematic of the bench scale torrefier highlighting the modifications done to the oven.

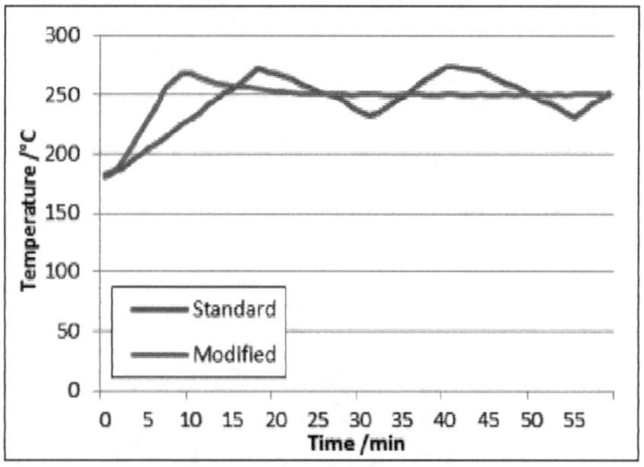

Figure 1. Temperature response comparison between standard oven and modified oven.

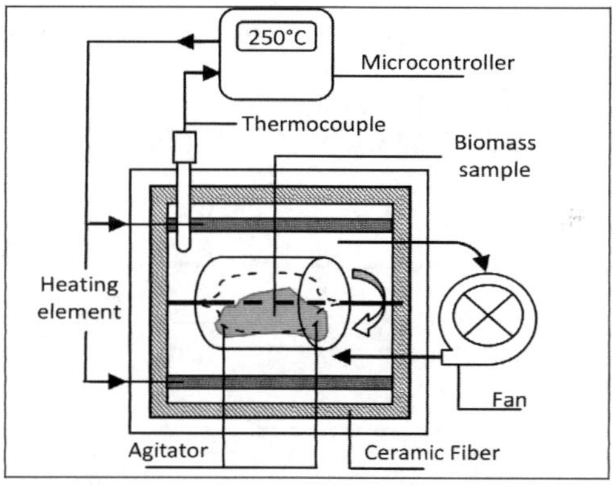

Figure 2. Modifications to 1600 W electric oven.

2.2. Torrefaction Experiment and Analyses
Pelletized empty fruit bunch (EFB) was used for the torrefaction experiment. The EFB pellets were sourced from palm plantations and commercially pelletized at Tanjung Karang, Selangor. The pellets were of 8mm in diameter with a moisture content range of between 15 % – 18 %. A sample of raw pellets was sent for GCV and proximate analyses at a certified laboratory.

A sample of approximately 200g of EFB was placed in the reactor for a single run of the torrefaction experiment. The aluminium tape seal was visually checked to make sure the tape was fixed securely. The torrefier was pre heated at 20 $°Cmin^{-1}$ until 250 °C was reached. This temperature is held for 5 minutes to achieve uniform condition inside the torrefier. The reactor was rolled a few times slowly to make sure the sample inside the reactor is distributed evenly. The reactor was placed inside the torrefier using the agitator rotating latch. Close the door, turn on the agitator and set the timer to one hour. As the time reached one hour, the agitator switch is turned off, and the reactor was taken out. The reactor was then left cool naturally to temperature less than 50 °C. The torrefied sample was removed from the reactor the weight was recorded. The sample was then kept in an airtight glass bottle and sent to the certified lab for GCV and proximate analyses. The ratio of calorific value, CV_{ratio}, mass yield, Y_{mass} and energy yield, Y_{energy}, were then defined by (1), (2) and (3) respectively based on equations suggested by Bridgeman et al. [11].

$$CV_{ratio} = \text{GCV after torrefaction / GCV before torrefaction} \quad (1)$$

$$Y_{mass} = (\text{Mass after torrefaction / Mass before torrefaction}) \times 100 \quad (2)$$

$$Y_{energy} = CV_{ratio} \times Y_{mass} \quad (3)$$

3. Results and Discussion

The GCV for raw EFB was 17.5 MJ/kg and increased to 18.82 MJ/kg after torrefaction which gives a CV_{ratio} of 1.08. This value is found to be slightly lower than other previous work. Comparison of proximate analyses on dry basis before and after torrefaction showed a slight decrease in volatile matter, slight increase in fixed carbon content and insignificant change in ash content. Table 1 summarises the proximate analyses results.

Observation of the torrefied samples showed that it has changed to a very dark brown, more easily breakable and homogeneous in terms of appearance and brittleness as shown in Figure 3. Figure 4 shows the mass yield and energy yield after torrefaction of EFB. Both values showed a decrease with mass yield slightly being lower than energy yield as expected with torrefied biomass. The result was also reflected in the small increase in CV_{ratio} value.

These changes from both observation and analyses confirmed that torrefaction occurred, although in a mild form. This could be due to the densely packed particles of the pelletized EFB slowing heat transfer throughout itself. A higher rate of torrefaction would be achievable by increasing the reactor's temperature or prolonging the residence time similar to previous work on biomass pellets [12]. This can be easily set using the installed microcontroller. Additionally, no ash was found in the torrefied sample indicating that combustion did not occur in the reactor.

Table1. Proximate analysis results of EFB.

	Raw /%	**Torrefied** /%
Volatile matter	77.74	76.36
Fixed carbon	15.66	17.02
Ash	6.60	6.62

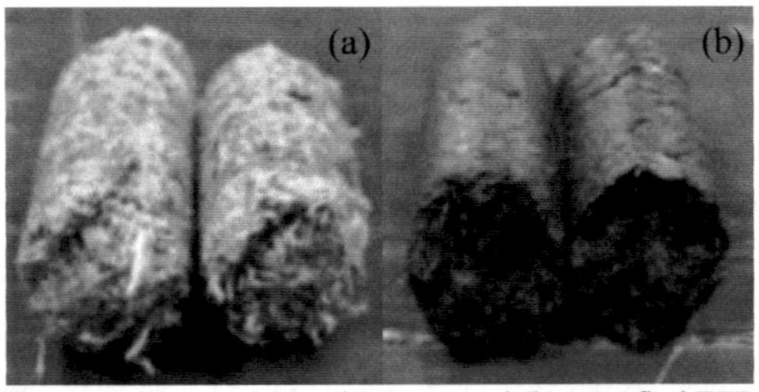

Figure 3. Comparison of (a) raw and (b) torrefied EFB.

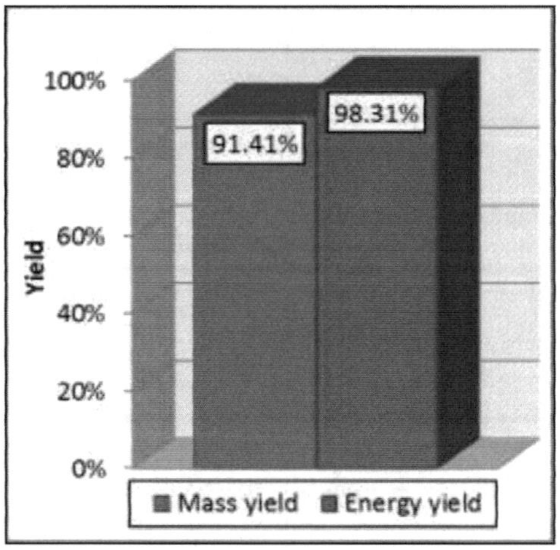

Figure 4. Mass and energy yields of torrefied EFB.

4. Conclusion

A bench scale torrefier was successfully developed from a domestic oven for use to study biomass upgrading by torrefaction. It has been proved that it can be used to produce relatively large, homogenous sample of torrefied fuel. Mild torrefaction of pelletized EFB was achieved, confirmed by both physical observation and analytical results. The result showed that torrefying at 250 °C for one hour has increase the GCV by a factor of 1.08 and slight decrease in volatile matter and increase in fixed carbon content. Mass yield was lower than energy yield, similar with other previously reported work on torrefying biomass. The bench scale torrefier was also able to torrefied the sample evenly.

References

[1] Energy Comission, "National Energy Balance 2010," Energy Commission Malaysia, Putrajaya, 2012.

[2] U. S. Energy Information Administration, "International Energy Statistics," [Online]. Available: http://www.eia.gov/cfapps/ipdbproject/IEDIndex3.cfm?tid=90&pid=44&aid=8. [Accessed 1 3 2013].

[3] M. Sami, K. Annamalai and M. Wooldridge Co-Firing of Coal and Biomass Fuel Blends 2001 *Progress in Energy and Combustion Science* **27** 171-214

[4] L. Baxter Bomass-Coal Co-Combustion: Opportunity for Affordable Renewable Energy 2005 *Fuel* **84** 1295-1302.

[5] L. I. Darvell, J. M. Jones, B. Gudka, X. C. Baxter, A. Saddawi, A. Williams and A. Malmgren, Combustion Properties of Some Power Station Biomass Fuels 2010 *Fuel* **89** 2881-2890

[6] A. B. Nasrin, A. N. Ma, Y. M. Choo, S. Mohamad, M. H. Rohaya, A. Azali and Z. Zainal Oil Palm Biomass as Potential Substitution Raw Materials for Commercial Biomass Briquettes Production 2008 *American Journal of Applied Sciences* **5** 179-183,

[7] Y. F. Huang, W. R. Chen, P. T. Chiueh, W. H. Kuan and S. L. Lo Microwave Torrefaction of Rice Straw and Pennisetum 2012 *Bioresource Technology* **123** 1-7

[8] J. Li, A. Brzdekiewicz, W. Yang and W. Blasiak. Co-Firing Based on Biomass Torrefaction in a Pulverized Coal Boiler with Aim of 100% Fuel Switching 2012 *Applied Energy* **99** 344-354

[9] Y. Uemura, W. N. Omar, T. Tsutsui and S. Yusup Torrefaction of Oil Palm Wastes 2011 *Fuel* **90** 2585-2591

[10] A. A. Jaafar and M. M. Ahmad Torrefaction of Malaysian Palm Kernel Shell 2011 *World Academy of Science, Engineering and Technology* **60** 554-557

[11] T. G. Bridgeman, J. M. Jones, I. Shield and P. T. Williams Torrefaction of Reed Canary Grass, Wheat Straw and Willow to Enhance Solid Fuel Qualities and Combustion Properties 2008 F*uel* **87** 844-856

[12] S. H. Larsson, M. Rudolfsson, M. Nordwaeger, I. Olofsson and R. Samuelsson Effect of Moisture Content, Torrefaction Temperature and Die Temperature in Pilot Scale Pelletizing of Torrefied Norway Spruce 2013 *Applied Energy* **102** 827-832

IOP Conf. Series: Earth and Environmental Science **16** (2013) 012083 doi:10.1088/1755-1315/16/1/012083

Spectrum survey for reliable communications of cognitive radio based smart grid network

Wan Farah Aqilah[1], Shanjeevan Jayavalan[1], Norazizah Mohd Aripin[1], Hafizal Mohamad[2], Aiman Ismail[1]

[1]Department of Electronics & Communication Engineering, Universiti Tenaga Nasional, 43000 Kajang, Selangor, Malaysia.

[2]Wireless Communication Cluster, MIMOS Bhd, Technology Park Malaysia, 57000 Kuala Lumpur, Malaysia.

E-mail: norazizahm@uniten.edu.my, hafizal.mohamad@mimos.my

Abstract. The smart grid (SG) system is expected to involve huge amount of data with different levels of priorities to different applications or users. The traditional grid which tend to deploy propriety networks with limited coverage and bandwidth, is not sufficient to support large scale SG network. Cognitive radio (CR) is a promising communication platform for SG network by utilizing potentially all available spectrum resources, subject to interference constraint. In order to develop a reliable communication framework for CR based SG network, thorough investigations on the current radio spectrum are required. This paper presents the spectrum utilization in Malaysia, specifically in the UHF/VHF bands, cellular (GSM 900, GSM 1800 and 3G), WiMAX, ISM and LTE band. The goal is to determine the potential spectrum that can be exploit by the CR users in the SG network. Measurements was conducted for 24 hours to quantify the average spectrum usage and the amount of available bandwidth. The findings in this paper are important to provide insight of actual spectrum utilization prior to developing a reliable communication platform for CR based SG network.

1. Introduction

In Malaysia, the electricity demand is expected to increase up to 3.5% annually for the next 10 years [1]. However, the local natural gas resources are depleting and there is serious concern on CO_2 emission. As national agenda, Malaysia aims to revolutionalized the existing grid to become a smart grid. Smart grid is a term referring to modernization of generation, transmission and distribution of a power grid system with the integration of advanced ICT infrastructure. The smart grid is envisioned to comprise major features such as advanced metering infrastructure (AMI), demand-side management (DSM), fault-tolerance and self healing system [2]. The concept of SG can be visualized as shown in figure 1.

The key determining factor in establishing these features is timely access to information via a reliable communications infrastructure. According to [3], among the five identified key technology areas in smart grid is the implementation of integrated communications. Currently, utilities company operate extensive private internal communications to support the safe, efficient and reliable delivery of essential services to the public at large [5]. However, the current communication systems are not suited to support the large scale of SG network, which need to support many interconnected components and

IOP Conf. Series: Earth and Environmental Science **16** (2013) 012083 doi:10.1088/1755-1315/16/1/012083

huge amount of data to be transmitted, managed and analyzed. Most systems either lack of bandwidth, limited coverage area, latency or security.Therefore, a revolutionary communication system is urgently demanded, such as cognitive radio.

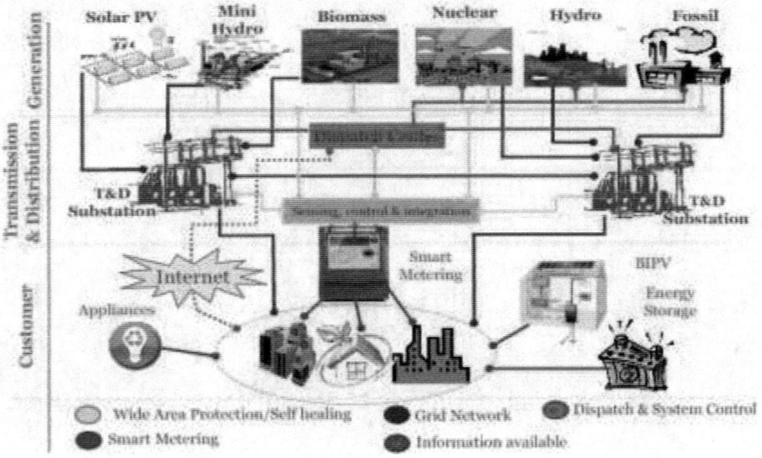

Figure 1. SG Concept

Cognitive radio enables dynamic spectrum acces, which allows the SG to exploits the underutilized frequencies in opportunistic manner. By leveraging CR technology, the radio agility allows the SG devices to sense all potentially unused (idle) spectrum opportunities in the surroundings and utilized them subject to interference constraints. The motivations behind why CR network is a promising communication platform for SG are summarized as follows [3, 5]:

(a) The existing smart grid devices for Home Area Network (HAN) is usually operate in the ISM band (ie: Zigbee, Bluetooth and WiFi). Thus, there exist significant coexistence interference among the devices. Therefore, the spectrum in HAN is becoming overcrowded and contaminated that will endanger reliable SG communications. In view of this, it is beneficial to adopt CR based radio technology in HAN, which will intelligently coordinate the power and schedule the transmission.
(b) The volume of SG data is expected to increase tremendously in the near future. CR network is a workable approach to increase communication capacity by utilizing the idle spectrum efficiently.
(c) SG covers a large geographical area, and therefore different spectrum regulations can exist. CR can enables heterogeneous network through hardware reconfigurability and context awareness.

In view of this, deep and thorough investigations on the current frequency spectrum are required for a successful implementation of CR for reliable SG communications. By doing so, it enables us to understand the dynamic characteristics of spectrum in order to meet the QoS of SG.

The rest of this paper is organized as follows. Section 2 presents the measurement setup, which was conducted in UNITEN and MIMOS. Section 3 discusses the measurement results to determine the potential spectrum for CR based SG network. Finally, conclusions and future recommendations are drawn in Section 4.

2. Measurement Setup

An outdoor spectrum measurement has been carried out at MIMOS located in Technology Park Malaysia, Kuala Lumpur and UNITEN Putrajaya Campus, Selangor (as shown in figure 2). The AOR DA3200 discone antenna which covers the frequency range of 25 MHz to 3 GHz or a tri-band dipole (XPO2V-880-2175/1355) antenna is connected to the ADVANTEST U3741 spectrum analyzer through a low loss coaxial cable. The frequency range of the spectrum analyzer is 9 kHz to 3 GHz. The spectrum analyzer is then connected to a computer with LABVIEW installed, which is used to configure the required parameters for the measurement and to automatically capture the data. The spectrum survey was conducted for a period of 8 weekdays covering frequency range of two TV broadcasting bands (VHF and UHF TV), three cellular bands (GSM 900, GSM 1800 and 3G), two broadband wireless access bands (WiMAX and LTE) and ISM 2.4 GHz band. The discone antenna is used to measure the TV broadcasting band, while the tri-band dipole is used for the other measured bands.

(a) (b) (c)

Figure 2. Measurement setup at UNITEN: (a) Discone antenna (b) Tri-band dipole antenna (c) ADVANTEST U3741 spectrum analyzer and computer installed with LABVIEW.

Decision threshold is a very important parameter to determine whether the spectrum is idle or busy. If the received signal power is above a certain threshold value, the frequency channel is considered as occupied. However, if the received signal is lower than the threshold value, the frequency channel is classified as idle. Selecting very high decision threshold value will cause the frequency channel to be classified as unoccupied even though the licensed signal is present. Selecting very low received power threshold value will result in the frequency channel to be classified as occupied when noise is mistakenly accounted for licensed signal. In this study, the decision threshold value is determined by adding certain dB above the average noise level as recommended by ITU [6]. Table 2 tabulates the frequency bands and their decision threshold selected above the noise level for all eight bands.

Table 2. Received power decision threshold selected for each band.

Service	Frequency Range (MHz)	dB Above Noise Level	Received Power Decision Threshold (dBm)
VHF TV	174 - 230	10	- 85.7
UHF TV	470 - 798	10	- 85.8
GSM 900	880 - 960	3	- 92.8
GSM 1800	1710 - 1880	3	- 90.7
3G (IMT-2000)	1885 - 2200	3	- 89. 2
WiMAX	2300 - 2390	10	- 81.4
ISM 2.4 GHz	2400 - 2500	10	- 81.4
LTE	2500 - 2700	8	-81.9

3. Results & Analysis

This section discusses measurement results and analysis for all 8 bands mentioned earlier. The average duty cycle for both locations is presented in table 3. Generally, the spectrum utilization at MIMOS is slightly higher compared to at UNITEN. It is noted that the spectrum utilization is low for TV bands and medium for cellular bands at both locations. However, the spectrum utilization for broadband wireless access bands and ISM 2.4 GHz are very low at both locations. This is due to the measurement equipment which may not be sensitive enough to capture the broadband wireless access signals. The

LTE service is just recently launched and this explains why the utilization is very low at present. Based on the findings, UHF/VHF TV broadcasting band is recommended as a very potential candidate for the CR based SG communication network in Malaysia, due to low spectrum utilization and high unused bandwidth. Furthermore, the characteristic of TV signal which covers wide are range, longer propagation waves and penetration of buildings is suitable for smart grid application such as connecting the grid to the rural area or backhaul communication platform (ie: WAN) which requires larger bandwidth. Due to medium spectrum utilization in the cellular band, it can be opportunistically used for neighborhood area network (NAN) of SG network. The NAN is basically collecting data from several HANs to be transmitted to the WAN.

Table 3. Summary of average duty cycle

Service	Frequency Range (MHz)	Bandwidth (MHz)	Average Duty Cycle (%)	
			UNITEN	**MIMOS**
VHF TV	174 – 230	56	4.03	9.50
UHF TV	470 – 798	328	3.02	4.35
GSM 900	880 – 960	80	35.31 (UL = 13.31, DL = 63.47)	36.41 (UL = 23.23, DL = 53.41)
GSM 1800	1710 – 1880	170	9.59 (UL = 0, DL = 18.30)	32.38 (UL = 25.42, DL = 41.16)
IMT-2000 (3G)	1885 – 2200	315	26.08 (FDD UL = 11.52, FDD DL = 74.54, TDD =26.29)	30.02 (FDD UL = 22.07, FDD DL = 51.54, TDD = 39.96)
WiMAX	2300 – 2390	10	0.13	0.56
ISM 2.4 GHz	2400 – 2500	100	0.096	0.3
LTE	2500 – 2700	200	0.12	0

4. Recommendations and Conclusions

This paper presented an investigation of idle spectrum that can be exploited for CR based SG network. The findings shows that the spectrum utilization is low for TV band and thus recommended for CR based SG network supporting the WAN. On the hand, the cellular band is a potential spectrum supporting the NAN of CR based SG network. This can provide useful information for the energy utilities and the spectrum regulators to consider opportunistic access of the underutilized spectrum for smart grid network.

References

[1] Mohd Yusof Rakob, Planning for Smart Grid in TNB System, *IEEE Conference on Power and Energy*, 2010
[2] A. Ghassemi, S. Bavarian and L. Lampe, Cognitive Radio for Smart Grid Communications 2010 *1st IEEE International Conference on Smart Grid Communications* 297-302
[3] Vehbi Cagri Gungor and Dilan Sahin, Cognitive Radio Networks for Smart Grid Applications 2012 *IEEE Vehicular Technology Magazine* 41-46
[4] V.C Gungor, B. Lu and G. P. Hancke, Opportunisties and Chalennges of Wireless Sensor Networks in Smart Grids 2010 *IEEE Transactions in Industrial Electronics* **57** 3557-3564
[5] Brett Kilbourne and Klaus Bender, Spectrum for Smart Grid: Policy Recommendations Enabling Current and Future Applications *2010 IEEE International Conference on Smart Grid Communications* **1** 578-582
[6] *Radiocommunication Bureau, HANDBOOK Spectrum Monitoring*, International Telecommunication Union (ITU), 2011.

IOP Conf. Series: Earth and Environmental Science **16** (2013) 012084 doi:10.1088/1755-1315/16/1/012084

Study on two stage activated carbon/HFC-134a based adsorption chiller

K Habib[1*], M Amin B A Majid[1] and Shaharin Anwar B Sulaiman[1]
[1]Mechanical Engineering Department, Universiti Teknologi PETRONAS
Bandar Seri Iskandar, 31750 Tronoh, Perak Darul Ridzuan
Malaysia

[*]E-mail: khairul.habib @petronas.com.my

Abstract. In this paper, a theoretical analysis on the performance of a thermally driven two-stage four-bed adsorption chiller utilizing low-grade waste heat of temperatures between 50°C and 70°C in combination with a heat sink (cooling water) of 30°C for air-conditioning applications has been described. Activated carbon (AC) of type Maxsorb III/HFC-134a pair has been examined as an adsorbent/refrigerant pair. FORTRAN simulation program is developed to analyze the influence of operating conditions (hot and cooling water temperatures and adsorption/desorption cycle times) on the cycle performance in terms of cooling capacity and COP. The main advantage of this two-stage chiller is that it can be operational with smaller regenerating temperature lifts than other heat-driven single-stage chillers. Simulation results shows that the two-stage chiller can be operated effectively with heat sources of 50°C and 70°C in combination with a coolant at 30°C.

1. Introduction

Due to the severity of the ozone layer destruction in the atmosphere, a central challenge in cooling science today is the development of freon-free refrigeration and air-conditioning systems, which occurs partly due to the production and utilization of CFCs and HCFCs in refrigeration. Heat driven adsorption cooling systems seem to be an alternative option as these systems do not utilize electricity as the driving heat sources and mainly non-toxic and environmentally friendly refrigerants. Several heat-pumping and refrigeration applications have been studied using various adsorbent and refrigerant pairs. Some representative examples are silica gel-water [1], zeolite-water [2] activated carbon-ammonia [3], activated carbon-methanol [4] etc. Most of the adsorption cooling cycles require medium and/or high temperature heat sources to act as the driving sources. However, many researchers reported that silica-gel/water and activated-carbon (AC)/methanol adsorption cooling systems can be driven by heat of relatively low, near-ambient temperatures, so that the waste heat below 100°C can be recovered and used [5-6].

This study deals with the utilization of unexploited, near ambient temperature between 50°C and 70°C as the driving heat source with a cooling source of 30°C, and presents the theoretical analysis of a two-stage advance adsorption cooling system where activated carbon (Maxsorb III) and HFC-134a are used as adsorbent-refrigerant pair. Temperature profiles of all the system components (adsorber,

[1] Khairul Habib.

desorber, condenser and evaporator) have been investigated. The influences of heat transfer fluid temperature and flow rate as well as the cycle time on the system performance in terms of cooling load and COP are determined theoretically.

2. Working principle of two-stage adsorption cooling cycle

Figure 1 shows the conceptual Dühring diagram (*P-T-x*) of a conventional and an advanced two-stage adsorption chiller. For practical utilization of renewable energy heat sources such as solar energy, an advanced (two stage) adsorption cooling system is presented here. The operation of this cycle is based on reducing the temperature left of adsorbent ($\Delta T_{reg} = T_{cond} - T_{eva}$) into two smaller temperature lifts. The refrigerant pressure is therefore rises into two progressive pressure steps from the evaporator pressure to the condenser pressure. To achieve this condition an additional two sorption elements are added to the conventional system as shown in Figure 2. The two-stage four bed adsorption cooling system then comprises of six heat exchangers, namely, a condenser, an evaporator and two pairs of sorption elements (SEs). The details of the working principle of two-stage adsorption cycle have been described elsewhere [5].

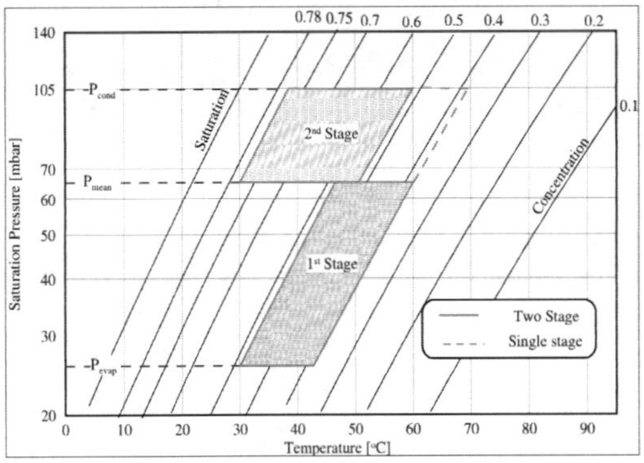

Figure 1. Conceptual Dühring diagram of two-stage adsorption chiller

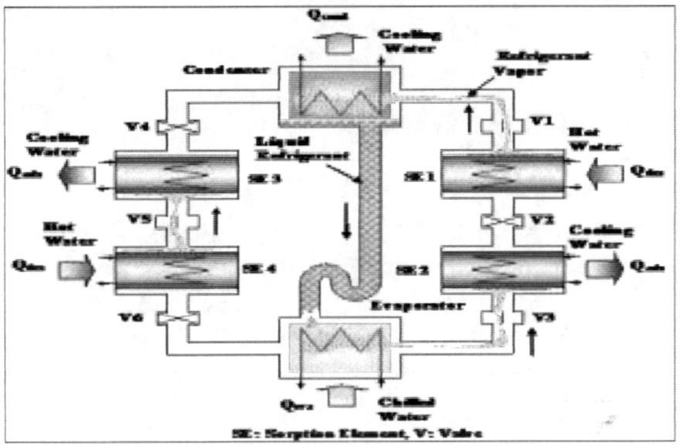

Figure 2. Schematic diagram of two-stage adsorption chiller.

3. Mathematical modelling

3.1. Adsorption isotherms

Dubinin-Astakhov (D-A) model, which is expressed by Eq. (1), is used to estimate the equilibrium uptake of the AC/HFC-134a

$$W = W_0 \exp\left[-\left\{\frac{RT}{E}\ln\left(\frac{p_s}{p}\right)\right\}^n\right] \tag{1}$$

Saha et al. [7] measured the adsorption characteristics of AC-R134a pair and obtained the numerical values of W_0, E and n as 1.66 x 10^{-3} m^3/kg, 82.9 x 10^3 J/kg and 1.3, respectively.

3.2. Adsorption kinetics

In the current adsorption chiller model, the rate of adsorption or desorption is governed by the Fickian diffusion model [8].

3.3. Isosteric heat of adsorption

The isosteric heat of adsorption for AC-R134a has been expressed elsewhere. [9]

3.4. Adsorption or desorption energy balances

The adsorption bed comprises activated carbon, the heat exchanger fins and tubes, and the energy balance equation is given by [10],

$$\left[m_{ac}C_{p,ac} + m_{ac}C_p^{ref}x^{ref} + m_{Al}C_{p,Al} + m_{Cu}C_{p,Cu}\right]\frac{dT_{bed}^{ref}}{dt} = \delta m_{ac}Q_{st}^{ref}\frac{dx_{bed}^{ref}}{dt} + \dot{m}_wC_{p,w}\left(T_{w,in}^{ref} - T_{w,out}^{ref}\right) - \lambda m_{ac}C_{p_v}\left(T_{ads} - T_{eva}\right)\frac{dx}{dt} \quad (2)$$

The outlet temperature of the source is sufficiently accurate to be modelled by the log mean temperature difference (LMTD) method and it is given by:

$$T_{w,out,bed}^{R134a} = T_{bed}^{R134a} + \left(T_{w,in,bed}^{R134a} - T_{bed}^{R134a}\right)\exp\left[-\frac{(UA)_{bed}^{R134a}}{(\dot{m}C_p)_w}\right] \quad (3)$$

Similarly, the energy balance equations of the evaporator, condenser and cooling capacity and COP can be expressed [10].

4. Results and discussion

4.1. Chiller transient response

Figure 3 shows the chiller temporal histories for the reactor beds (adsorber/desorber heat exchangers), evaporator and condenser for the two-stage cycle where hot water temperature has been chosen as 60 °C along with a coolant at 30 °C. The values used in the present simulation are shown in Table 1 [11]. The chilled water inlet temperature is taken as 14 °C. It can be observed from Figure 3 that the adsorption cycle in two-stage operation is able to reach from transient to nearly steady state within 1800 s.

4.2. Heat source temperature

Figure 4 indicates the effects of regeneration temperature on cooling capacity for both single stage and two-stage cycles with a fixed cooling water inlet temperature at 30°C. It is visible from Figure 4 that cooling capacity increases linearly from 0.6 to around 8 kW as the hot water inlet temperature is increased from 45 to 80°C. As observed in Figure 4, the two stage cycles gives comparable cooling capacities when the hot water inlet temperature is between 60 and 70°C. However, when the hot water inlet temperature is less than 55°C, the two-stage cycle could still be operated, a region that could not be reached by the single stage.

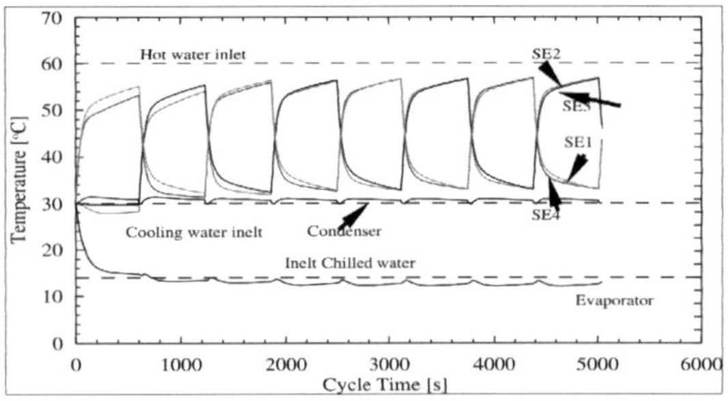

Figure 3. Temperature profiles of different components of AC-HFC 134a cycle at rated conditions

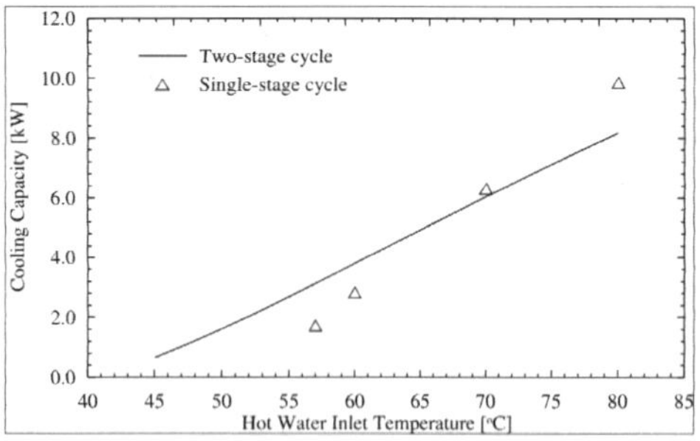

Figure 4. Effect of cooling capacity on heat source temperature.

Figure 5 shows the change of COP of single-stage and two-stage system with the hot water inlet temperature. Being operated by the lower range of inlet hot water temperature, it is expected that the COP of the two stage cycle is lower. The superiority of two-stage cycle is again demonstrated when the inlet hot water temperature is lower than 55°C.

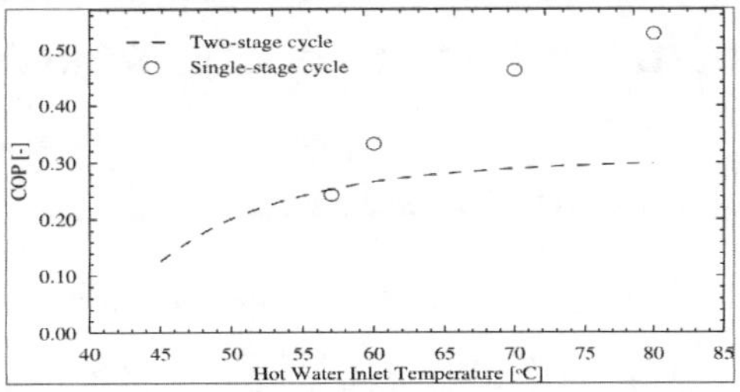

Figure 5. Effects of COP on heat source temperature.

5. Conclusions

The main advantage of the AC/HFC-134a based two-stage adsorption chiller is that it can operate even when the driving heat source temperature is low (typically between 55 to 70°C). The simulation results show that the single-stage chiller is superior when the heat source temperature is relatively higher (above 75°C). However, when the regeneration temperature is at or below 55°C, the two-stage chiller could still be operated, a region that could not be reached by single-stage chiller.

Acknowledgements

Authors would like to thank Ministry of Science, Technology and Innovation (MOSTI) and UTP for the financial support to conduct this work.

References

[1] Chua H T, Ng K C, Malek A, Kashiwagi T, Akisawa A and Saha B B 1999 *Int. J. Refrig.* **22** 194

[2] Wang D C, Xia X X and Wu J Y 2006 *Energ. Conv. Manage.* **47** 590

[3] Miles D J and Shelton S V 1996 *Appl. Therm. Eng.* **16** 389

[4] Anyanwu E E and Ezekwe C I 2003 *Energ. Conv. Manage.* **44** 2879

[5] Saha B B, Akisawa A and Kashiwagi T 2001 *Renew. Energ.* **23** 93

[6] Ferreira Leite A P and Daguenet M 2010 *Energ. Conv. Manage.* **41** 1625

[7] Saha B B, Habib K, El-Sharkawy I I and Koyama S 2009 *Int. J. Refrig.* **32** 1563

[8] Habib K, Saha B B, Rahman K A, Chakraborty A, Koyama S and Ng K C 2010 *Int. J. Refrig.* **33** 706

[9] El-Sharkawy I I, Saha B B, Koyama S and Srinivasan K 2007 *Int. J. Heat and Mass Trans.* **26** 902

[10] Habib K, Saha B B and Chakraborty A, El-sharkawy I I and Koyama S 2008 *Eng. Science Reports Kyushu Univ.* **30** 383

IOP Conf. Series: Earth and Environmental Science **16** (2013) 012085 doi:10.1088/1755-1315/16/1/012085

System design optimization for stand-alone photovoltaic systems sizing by using superstructure model

M A M Azau[1], S Jaafar[1], K Samsudin[2]

[1] Department of Electrical Power Engineering, College of Engineering, Universiti Tenaga Nasional, 43000 Kajang, Selangor.
[2] Department of Computer and Communication Systems Engineering, Faculty of Engineering, Universiti Putra Malaysia, 43400 UPM Serdang, Selangor.

Email: AzrinM@uniten.edu.my, sabarina@uniten.edu.my, kmbs@upm.edu.my

Abstract Although the photovoltaic (PV) systems have been increasingly installed as an alternative and renewable green power generation, the initial set up cost, maintenance cost and equipment mismatch are some of the key issues that slows down the installation in small household. This paper presents the design optimization of stand-alone photovoltaic systems using superstructure model where all possible types of technology of the equipment are captured and life cycle cost analysis is formulated as a mixed integer programming (MIP). A model for investment planning of power generation and long-term decision model are developed in order to help the system engineer to build a cost effective system.

1. Introduction

There are considerable efforts and studies have been made to find methods to optimize the design of the systems so that they are cost effective. Life cycle cost analyses are commonly adopted to size the investment costs of the system, to estimate the operation and maintenance (O&M) costs and to find the costs to replace the equipment over the system's life span [1-3]. The mentioned costs for each equipment; PV modules, inverters, batteries, are usually quantified in kWh which do not depict the individual cost. Therefore there is no explicit relationship between the investment costs and the number of units purchased or the physical dimensions of the unit. This relationship is important in real designs, in which physical dimensions are part of the constraints.

This work aims to produce a cost effective system with an optimized design subjects to the energy demand and control system perspective so that maximum power generated can be harvested. At this stage, this work employs a static superstructure design optimization with life cycle cost analysis (LCCA) approach that captures not only individual investment costs of the available equipment but also the physical dimensions of the equipment such as weights, lengths and widths. These physical dimensions are not addressed in many previous works. The problem is formulated as a mixed integer linear problem (MILP) and developed in General Algebraic Modelling Software (GAMS) in order to find the most optimized combination of the equipment subject to energy demand determined by system designers [4-5]. The results from the superstructure design such as the technology of the PV module, converter and inverter topologies, battery technology and electrical connection are key information to describe the system behaviour and useful for the control system development.

In this paper, the methodology section discusses the superstructure design framework detailing the problem formulation and optimization using GAMS. The next section summarizes the results and discussion, in which some selected results are presented and explained. The concluding remarks and research work that can be extended from this work are briefly discussed in conclusion and future work section.

IOP Conf. Series: Earth and Environmental Science **16** (2013) 012085 doi:10.1088/1755-1315/16/1/012085

2. Methodology

This research is developed in stages. The first stage is to develop the profile data base in which all the physical dimensions cost and total efficiency of the equipment are gathered and tabulated. The static superstructure model is then design enlisting all sets of the equipment. Next, the system specification is obtained and formulated in mixed integer programming in which a single objective function and constraints, both equality and inequality are expressed mathematically. The mathematical expression of the objective function is then linearized and GAMS is used to realize the mixed integer problem programming.

2.1. Superstructure Design

The primary objective of the model is to obtain an optimal expenditure ~~a~~ system. Previously, a designer has to invest in order to build up a PV system that meets the energy demand. In this model only one type of PV module, one type of battery technology and one type of inverter are allowed to be selected for the construction of a system. Figure 1 shows the interactions between the components. All the dimensions and costs of the equipment are tabulated and transformed to the GAMS programming language.

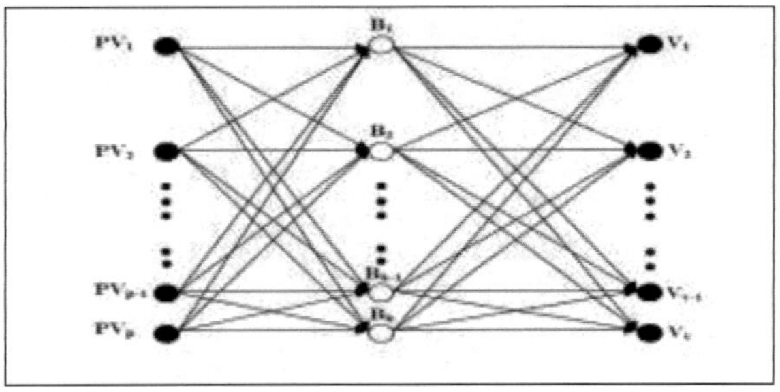

Figure 1. The superstructure framework

2.2. Mixed Integer Programming (MILP)

The general form of the Mixed-Integer-Linear programming is

$$min_{x,y} f(x,y) \tag{1a}$$

subject to

$$h(x,y)=0 \tag{1b}$$

$$(x,y)\leq 0 \tag{1c}$$

$$x \in R^n ,\ y \in \{0,1\}^n \tag{1d}$$

Where $f(x,y)$ is the objective function, $h(x,y)$ is the equality constraints, $g(x,y)$ is the inequality constraints x is the continuous variables and y is the binary variables.

The objective function formulated is to minimise the life cycle cost (LCC) of an installed stand-alone PV system. The simple LCC consists of initial costs of purchasing the equipment, equipment replacement (ERP) and operation and maintenance (OM). However in this report the ERP focuses on the battery replacement cost (BRP) since the battery life span is relatively shorter than any other components, which requires frequent replacement and the cost to assemble a battery bank is the second expensive after the PV module. The objective function and its breakdowns are summarized in equations (2)-(4) [6-8],

$$min^{LCC} = InitialCost_{equip} + \sum_{equip,t}^{T} \frac{costOM_{equip,t} \times (1+e)^t}{(1+i)^t} + \frac{costREP_{equip,t} \times (1+e)^t}{(1+i)^t} \tag{2}$$

$$LCC = Equipment\ Cost + Operation\ and\ Maintenance + Replacement\ Cost \tag{3}$$

$$InitialCost_{equip} = costPV_p + costBat_b + costInv_v + BOS \tag{4}$$

The inequality equations (5) and (6) are describing the energy generation constraints that the PV array must produce and the maximum weight that the PV array can have in order to be mounted on the rooftop,

$$NumberOfPV \times \eta \times P_{max} \geq Energy\ Demand \quad (5)$$

$$NumberOfPV \times ModuleWeight \leq W \quad (6)$$

The selection of the battery and inverter technology are described by inequality equations (7) and (8)

$$NumberOfBattery \times (AhCapacity \times RatedVoltage) \geq Energy\ Demand \times (\frac{DoA}{DoD}) \quad (7)$$

$$NumberOfInverter \times RatedContPower \geq LoadACPower \quad (8)$$

The inequality equations (9) and (10) are important so that the system will not having high voltage loss which will affect the energy supplied to the end user.

$$V_{drop} \leq 0.03 \times V_{sys} \quad (9)$$

$$V_{drop} = \sum I_{sc} \times (\frac{\rho \times l}{A}) \quad (10)$$

3. Results and Discussions

The financial planning normally emphasizes on the initial costs or the investment costs. Many individuals prefer to have low start up costs and they are ready to compromise on long term maintenance and replacement costs. The model developed selects the best components according to constraints set in the problem minimizing the initial purchase of each component. This research only considers 20 types of PV modules, 20 types of battery, 5 types of inverter and 5 sizes of wire. Table 1 shows the total start up cost of a PV system that has the capacity to supply 12kWh energy. It is obvious that the cost to purchase the batteries is the most expensive, exceeding the total cost of the PV modules. It is expected because the system is developed in such that it can continuously supply the energy to the load for five no solar irradiance days. Since the system only allows 50% of the battery capacity to be discharged in order not to reduce the battery life span, a larger storage capacity is required, which means more units are needed to be purchased.

Table 1. Total start-up cost

Component	Component Selected	Total number of component to be purchased	Total Cost (USD)
PV module	PV13	14	23940.00
Battery	B16	78	28938.00
Inverter	V2	3	7395.00
Wire	W4	30	334.50
		Total	**61676.5**

Interestingly, the operating life time of a battery is comparatively shorter than any other component in the PV system developed. In table 1, battery B16 was selected based on its initial cost. However, B16 is a flooded lead acid battery that has a maximum life cycle of 1000 at 50% DoD. From table 2 it can be seen that battery B19 is 20 thousands cheaper than B16. B19 comes from a sealed gel battery class with a higher life cycle at 50% DoD compared to typical flooded lead acid batteries. In addition to that, sealed gel batteries do not use water as charging-discharging medium, which significantly reduces the cost to maintain unlike flooded lead acid batteries whose water levels need to be checked regularly.

Table 2. Selection comparison of the battery based on initial cost and life-cycle criteria

Type of battery	BRP (times)	Initial Purchase Cost (USD)	Purchase + Replacement Cost (USD)
B16	13	28938	350398
B19	4	70900	338748

The system designer may prefer to mount the PV array on a rooftop to save the ground area. However the rooftop has some limitations one of which is how much of weight of the PV modules it can withstand. Let assume a rooftop of a building can only withstand 250kg of auxiliary weight with ample space to place the modules. If a constraint is introduced in the model formulation in such that the total weight of the PV array is not exceeding 250kg, then the model selects PV1 whose total weight is 215.6kg. Table 3 shows the comparison of the weight and total cost between PV1 and initially selected PV module, PV13.

Table 3. Selection comparison of the photovoltaic module based on initial cost and weight

PV module	Total Weight (kg)	Total Cost
PV1	215.600	24318.000
PV13	259.000	23940.00

4. Conclusion and Future Work

In this paper the superstructure design optimization with life cycle cost analysis for standalone photovoltaic systems has been presented. The static mathematical model of the system has been formulated which includes the PV modules, inverters, batteries and wires. Mixed integer linear programming, which has been developed using GAMS, is employed to find the optimum total costs for installing the PV system subject to the energy demand required by the user and other physical dimensions constraints. The developed static model considers a constant load profile assuming the peak sun hour and temperature do not vary throughout the operation.

References

[1] Azzopardi B, Mutale J and Kirschen D 2008 *Cost boundaries for future PV solar cell modules. Sustainable Energy Technologies* 589-594

[2] Barringer H P 2003 A Life Cycle Cost Summary *International Conference of Maintenance 3* Societies (Perth –Australia)

[3] Bloch H P and Geitner F K 1995 Simplified Life-Cycle Cost Computations Applied in the Hydrocarbon Processing Industries *Fourth International Conference on Process Plant Reliability* (Houston, TX: Gulf Publishing Company)

[4] Floudas C A 1995 *Nonlinear and mixed-integer optimization: Fundamentals and applications* (US: Oxford University Press)

[5] Liu P, Gerogiorgis D I and Pistikopoulos E N 2007 Modelling and optimization of polygeneration energy systems *Catalysis Today* **127**(1–4) 347–359

[6] Ekren O, Ekren BY and Ozerdem B 2009 *Applied Energy* Vol.86 Issues 7-8 1043-1054

[7] Mellita A, Benghanemb M and Kalogirouc S A 2007 Modeling and simulation of a stand-alone photovoltaic system using an adaptive artificial neural network: Proposition for a new sizing procedure *Renewable Energy* **32** 285–313

[8] Indulkar C S 2008 Monte Carlo Planning Technique for Renewable Energy Sources *Power System Technology and IEEE Power India Conference* 1-7

IOP Conf. Series: Earth and Environmental Science **16** (2013) 012086 doi:10.1088/1755-1315/16/1/012086

Devolatilization studies of oil palm biomass for torrefaction process optimization

D. Daud*, A. Abd Rahman, A. H. Shamsuddin

Centre for Renewable Energy, Universiti Tenaga Nasional, Putrajaya Campus, Jalan IKRAM-UNITEN, 43000 Kajang, Selangor, Malaysia
Email: darsheena@uniten.edu.my

Abstract. Torrefaction of palm biomass, namely Empty Fruit Bunch (EFB) and Palm Kernel Shell (PKS), was conducted using thermogravimetric analyser (TGA). The experiment was conducted in variation of temperatures of 200 °C, 260 °C and 300 °C at a constant residence time of 30 minutes. During the torrefaction process, the sample went through identifiable drying and devolatilization stages from the TGA mass loss. The percentage of volatile gases released was then derived for each condition referring to proximate analysis results for both biomass. It was observed an average of 96.64% and 87.53 % of the total moisture is released for EFB and PKS respectively. In all cases the volatiles released was observed to increase as the torrefaction temperature was increased with significant variation between EFB and PKS. At 300°C EFB lost almost half of its volatiles matter while PKS lost slightly over one third. Results obtained can be used to optimise condition of torrefaction according to different types of oil palm biomass.

1. Introduction

Utilization of biomass nowadays is considered as a near term energy solution in reducing dependence on fossil fuel. Biomass is a source of renewable energy and can be considered as a part of the natural carbon cycle as carbon dioxide (CO_2) released from biomass combustion has been absorbed from the atmosphere during its lifetime. Globally, biomass energy such as agricultural waste and crop residue has been replacing fossil fuel and increasingly becoming their prime energy source [1].

In 2010, 93.7% of Malaysia's electricity generation of 125,045 GWh was sourced from fossil fuel, mainly natural gas and coal, while biomass stood at only 1.01% [2]. This fact does not reflect the nation's abundance of biomass that can be sourced from its forestry and agricultural sector. According to the National Biomass Strategy, the palm oil industry alone produced almost 80 million dry tonnes of solid biomass in 2010 and this is predicted to grow to 85 – 110 million dry tonnes by 2020 [3]. Increasing biomass utilization in Malaysia's energy mix will directly reduce its overly dependence on fossil fuel and hence reduce its greenhouse gas (GHG) emission intensity.

Oil palm biomass is a difficult fuel to use mainly due to its low bulk density and high moisture content resulting in low calorific value, and heterogeneous nature and uneven composition [4]. The structure of the Malaysian oil palm biomass is mainly tough and fibrous in nature making it difficult to grind for further utilization. Hence, it is essential that oil palm biomass undergoes pretreatment in order to utilize it. A few techniques for upgrading biomass as a better solid bio-fuel have been introduced and one of the techniques is the torrefaction process.

Torrefaction is one form of thermal treatment and can be regarded as a process before gasification and combustion stages [5]. In this process, biomass is heated in an inert atmosphere within the temperature range of 200 °C and 300 °C [6]. The advantages of torrefaction process is that it removes moisture content, decrease light organic volatile component and increase energy density in mass basis as well as increasing the grindability of biomass through degradation of hemicellulose [7]. Torrefaction of biomass also homogenizes its physical nature which is a desirable characteristic in a fuel for power generation. Biomass upgrading via torrefaction would be able to enhance the structure of solid biomass fuel and at the same time minimizes the drawbacks of biomass utilization.

Different biomass will react differently during torrefaction due to the varying compositional moisture content and volatile matter [8]. The presented work studies the effect during devolatilization of two different oil palm biomass at varying degrees of torrefaction conditions.

2. Methodology

Two types of oil palm biomass, namely empty fruit bunch (EFB) and palm kernel shell (PKS), were selected as a raw material for this experiment. The samples were sourced from an oil palm mill in Bidor, Perak, Malaysia. Each biomass sample underwent the fuel preparation steps of weighing, drying, shredding, blending and sieving before it is analyzed. Initially, proximate analysis was carried out according to ASTMD5142-04 to establish the baseline moisture content and volatile matter for both types of biomass. The proximate analysis results are as shown in Table 1.

Torrefaction experiments were carried out using a Perkin Elmer Pyris 1 thermogravimetric analyzer (TGA) as shown in Figure 1. The inert condition is provided by purging a 20mLmin^{-1} flow of Nitrogen gas throughout the experiment. For each experiment, a sample of approximately 10 mg of biomass is placed in a ceramic crucible in a hanging position at the centre of the furnace as shown in Figure 2. The sample is heated from ambient temperature at rate of 10 °Cmin^{-1} to 100 °C where it was held isothermally for 30 minutes to drive off moisture. The sample was then heated again at rate of 5 °Cmin^{-1} until the desired torrefaction temperature is reached and held isothermally for a further 30 minutes. Three different torrefaction temperatures were used; 200 °C, 260 °C, and 300 °C. The TGA recorded the sample mass throughout the experiment and graphs of temperature and percentage mass loss against time were generated using its Pyris software. Mass loss due to drying and devolatilization was then derived using the TGA plots compared to the proximate analysis results as the baseline.

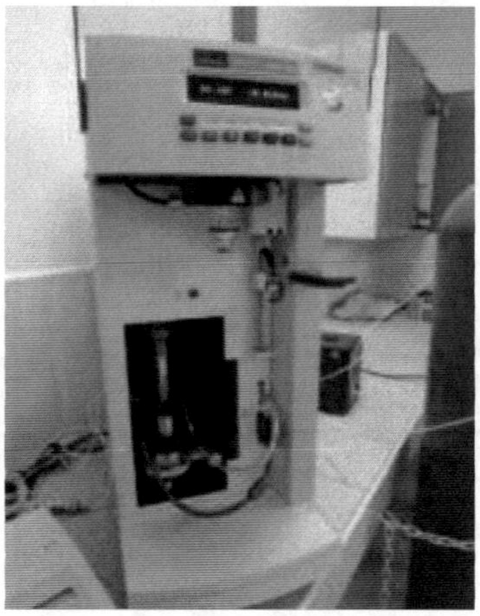

Figure 1. Perkin Elmer Pyris 1 TGA

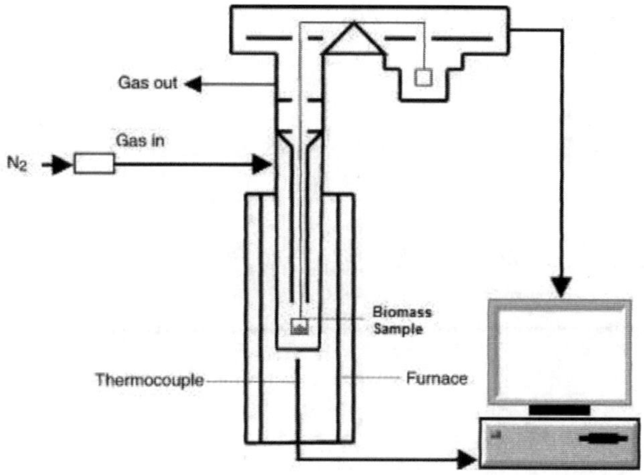

Figure 2. Torrefaction experiment setup

3. Results and Discussion

It was observed that TGA analyses for all samples shown two sifnificant regions of mass loss which corresponded to moisture loss during the first holding temperature of 100 °C and followed by subsequent volatile release during the second holding temperatures for torrefaction.

During the first stage, in the EFB samples, all three cases shown an average of 96.64% of all moisture was lost when compared to the initial moisture from the proximate analysis results. It can be assumed that the samples were fully dried during this stage as mass loss stop occurring within the 30 minutes holding period. The very small difference observed could be due to the samples had actually dried very slightly while in storage. The PKS samples showed a lower percentage of moisture loss at an average of 87.53%. It was determined that mass loss was still ocurring very slightly at the end of the 30 minutes holding time. Therefor the results showed that not all moisture was released during this first stage. Thus a higher holding time for drying is recommended for PKS and is planned for the next phase of this study.

Varying degrees of devolatization was observed during the torrefaction stage where in general, more volatiles was released during the 30 minutes holding time at the higher temperatures. Percentage volatile yield when compared to initial proximate analysis results is shown in Figure 3. Both sample loses only 1.86 % and 1.66 % for EFB and PKS of its volatile matter at a torrefaction temperature of 200 °C. The result confirmed that this is the lower limit of torrefaction for oil palm biomass where little or no devolatilization would occur below this temperature.

Torrefying at of 260 °C for 30 minutes resulted in the amount of volatile matter being released at 16.70 % and 14.67% for EFB and PKS respectively. This small difference could have been due to the release of similar light volatile components shared by both type of biomass. At a torrefaction temperature of 300 °C PKS loses 36.23% of its volatile matter while EFB was significantly higher at 49.66%. The significant difference showed that EFB releases volatile matter easier compared to PKS and can be attributed to the difference in physical nature of the biomass. PKS fibres are more densely packed compared to EFB making it has less surface area exposed to heat for devolatilization to occur. Torrefying EFB at or close to this upper temperature limit would result in a low energy yield as too much energy containing volatile matter would be lost. Even if the resultant torrefied biomass have a high calorific value, this is undesirable for torrefaction as the total energy available of a particular biomass is lost with the released volatile matter. Further study on the effect on resultant calorific value is proposed for the next phase of this study. This results also confirmed the easiness of volatile matter release in EFB compared to other types of biomass found in the author's previous research [8].

Table 1. Proximate analysis of oil palm biomass

	EFB	PKS
Inherent Moisture	8.70 %	5.61 %
Volatile Matter	74.30 %	68.35 %
Fixed Carbon	15.37 %	18.79 %
Ash Content	1.57 %	7.22 %

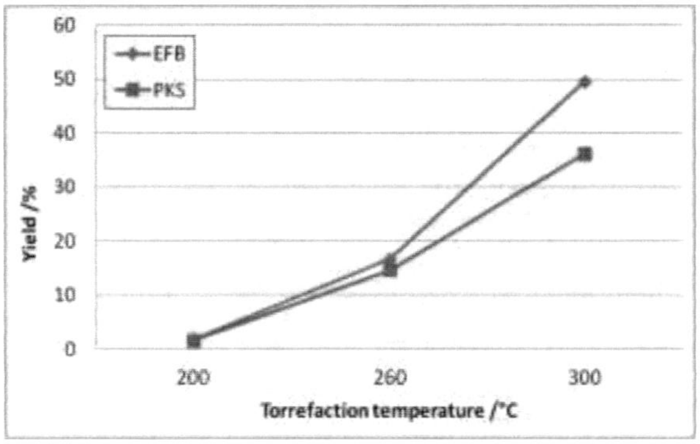

Figure 3. Volatile yield of oil palm biomass torrefied at various temperatures

4. Conclusion

Torrefaction of EFB and PKS has been successfully conducted using a TGA at various torrefaction temperatures in order to identify the mass loss and the volatile yield. An average of 96.64% and 87.53% of the total moisture is released for EFB and PKS respectively. Volatile yield increased as the torrefaction temperature is increased with significant differences of volatile matter released between the two types of biomass. The study showed that EFB releases volatile matter easier, thus dictates a lower torrefaction temperature compared to PKS to produce quality biofuel. It was also confirmed that little torrefaction occurred at 200 °C. Further study on the effect on calorific value would further optimize torrefaction process condition for both types of biomass.

Acknowledgement
Authors would like to acknowledge for the financial supports given: Long Term Research Grant (01201102 LRGS) from Ministry of Higher Education (MOHE) and Universiti Sains Malaysia (USM).

References

[1] O. Edenhofer, R. Pichs-Madruga, Y. Sokona, K. Seyboth, P. Matschoss, S. Kadner, T. Zwickel, P. Eickemeier and S. S. C. v. S. Gerrit Hansen, "Renewable Energy Sources and Climate Change Mitigation: Special Report of the Intergovernmental Panel on Climate Change," Cambridge University Press, Cambridge, 2011.

[2] Energy Commision Malaysia, "Electricity Supply Industry in Malaysia: Performance and Statistical Information 2010," Energy Commision Malaysia, Putrajaya, 2012.

[3] Agensi Innovasi Malaysia, "National Biomass Strategy 2020: New Wealth Creation for Malaysia's Palm Oil Industry," Agensi Innovasi Malaysia, Kuala Lumpur, 2011.

[4] L. S. Kheang, V. Subramaniam and M. Ngatiman, "Oil Palm Biomass Energy Resource Data," Malaysia Palm Oil Board, Bangi, 2012.

[5] P. Bergman, A. Boersma, J. Kiel, M. Prins, K. Ptasinski and F. Janssen, "Torrefaction for Entrained-Flow Gasification of Biomass," Energy Research Centre of the Netherlands (ECN), Petten, 2005.

[6] J. J. Chew and V. Doshi Recent Advances in Biomass Pretreatment – Torrefaction Fundamentals and Technology 2011 *Renewable and Sustainable Energy Review* **15** 4212-4222

[7] M. Wang, Y. Huang, P. Chiueh, W. Kuan and S. Lo Microwave-Induced Torrefaction of Rice Husk and Sugarcane Residues 2012 *Energy* **37** 177-184

[8] D. Daud, A. A. Rahman, A. H. Shamsuddin, N. Abidin, I. I. Zulkepli and M. N. Ja'afar, "Malaysian Biomass Upgrading by Torrefaction 2012 *International Conference on Environmental Research and Technology (ICERT 2012)*

[9] J. S. Tumuluru, S. Sokhansanj, J. R. Hess, C. T. Wright and R. D. Boardman A Review on Biomass Torrefaction Process and Product Properties for Energy Appilcation *Industrial Biotechnology,* 2010.

[10] E. Biagini, F. Lippi, L. Petarca and L. Tognotti Devolatilization Rate of Biomasses and Coal – Biomass Blends: An Experimental Investigation 2002 *Fuel* **81** 1041-1050

IOP Conf. Series: Earth and Environmental Science **16** (2013) 012087 doi:10.1088/1755-1315/16/1/012087

Environmental zoning for service life prediction of building components in Malaysia

Fathoni U, Rohayu C O and Zakaria C M

Centre for Storm Water and Geohazard Management, Universiti Tenaga Nasional

Email: Fathoni@uniten.edu.my

Abstract. Interactions between building materials and environment are very complex. More often than not, deterioration of building component is due to environmental loads with the result of reducing its durability. The environmental loads are less overlooked or underestimated in early design phase. Objective of this study is to distinguish the difference of environmental load for different area in Malaysia. The environmental data consist of climate and pollution data which extracted from the Malaysia Meteorology Department's data bank. From this study, six different environmental zones in Malaysia are distinguished. There are different characteristic of environmental parameters found where some of them dominant to others. This information can assist the engineer to justify the use of different technology, material or construction method in construction industry.

1. Introduction

The surrounding environment condition influenced the deterioration process of building materials. Corrosion is one of the common deterioration resulted by this complex interaction of materials and the nature. The environmental load is described as the deterioration agent from atmosphere (i.e. the emission of gaseous pollution, acid deposition and the acidity of rain water) and its local climate (i.e. the relative humidity, temperature and amount of precipitation). Those agents will accelerate the deterioration process. Degradation can be subjected by behaviour or dreadful conditions but the terminology of deterioration is more on worsening or weakening of particular object. In this study the work deterioration will be used as an expression of a process to decline the performance or the aging process that lower the quality and performance of particular building components.

The amount of pollutant agents in the atmosphere and its significant effect to accelerate the deterioration process need to be studied. Together with the climatic factors which differentiate the characteristic of degradation processes of building materials in different zone can be determined. It will help in understanding the effect of local environmental factors to the deterioration of the building components.

2. Methodology

Historical data have used in this study. It consisted of air pollution data and climatic data. The data was recorded from 17 main measurement stations located in peninsular Malaysia. The data from each measurement stations were recorded by Malaysia Meteorology Department (MMD). For the purpose of this study, data that extracted from the MMD data bank were the monthly weather report which record the received rainfall, temperature, humidity and air pollution data. The pollution data are in the form of wet fall out (WFO). In the tropics, the total amount of acid that return to the earth through precipitation accounts for only half and the other half is deposited in dry form compare to the northern hemisphere, the total amount of wet deposition for chloride is higher than sulphate and nitrate [1, 2]. For this study deposition of salt ion of chloride (Cl^-), sulphate (SO_4^{2-}) and nitrate (NO_3^-) were used. The selections of these salt ions is based on its prominent and different sources from where it was emitted. There are locations without environmental data. Its dispersion generated in the form of GIS maps to determine each deposition level from the area which have no measurement station. These maps are generated using Inverse Distance Weighing (IDW) method. The IDW method can efficiently apply and it has reliable computational process [3].

3. Analysis and discussion

The data that used to generate the GIS map was based on measured by JMM for ten years measurement (1996 - 2005). The Klang Valley and south of Johor were experiencing lowest level of pH. During north-east monsoon season the east coast area has slightly higher pH level. The increasing level of pH was leaded by highly concentration of sulphate that dissolved in rainwater to perform acidic precipitation. The level of pH in south of Johor was decreasing during north-east monsoon. The high amount of received rainfall washed-out the concentration of pollutant in the atmosphere.

Generally the northern and east peninsular received rainfall with pH between 4.8 and 5.2 while the other parts of the peninsular received rainfall with pH between 4.4 and 4.8. The areas that experience high levels of acidity were located in and around Klang Valley and southern part of Johor. These areas are rapid growth centers and industrialized with high population density. These areas were examined received rainfall with pH between 3.8 and 4.9. Figure 1 shows the monthly level of pH and the acid deposition of SO_4^{2-} and NO_3^- in industrial area. Level of pH for industrial area is influenced by gaseous pollutant of SO_2 and NO_2 in the atmosphere. It was examined by the fluctuation concentration of SO_4^{2-} and NO_3^-. Level of pH was become lowest follow by the increase of concentration of SO_2 and NO_x. Level of pH and temperature were the variable of oxidation reaction time for SO_2. Xu have stated in his study on deposition of SO_4^{2-} in East Asia that when the level of pH in the cloud during condensation less than 4 the dissolved H_2O_2 was predominant in oxidizing the gaseous form of SO_2 to SO_4^{2-}. Ozone (O_3) was the predominant oxidant agent when the level of pH greater than 5 and cause the reaction of oxidation 10 times faster [4].

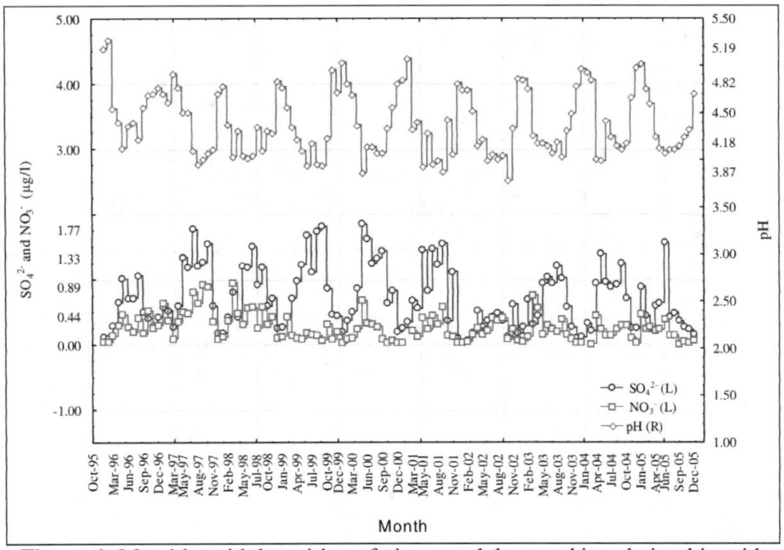

Figure 1. Monthly acid deposition of nitrate, sulphate and its relationship with level of pH in the rain water

Chloride (Cl^-) has been used widely as a tracer in the hydrologic cycle since it was often considered not influenced from any chemical weathering, mineral formation or biological process. The primary source of chloride in the atmosphere was the sea salt ion which was generated by wind stress on the ocean surface that burst at the surface and emit sea water aerosols into the atmosphere. The additional source probably included HCl emission from coal fired generating stations and HCl from sea salt dechlorination. High concentration of Cl^- was found on coastal area.

The east coast of Peninsular Malaysia was experienced with highly concentration of Cl^- during north-east monsoon season. The spread coverage of Cl^- in the precipitation was leaded by the wind blow of the two monsoon seasons. Figure 2 shows the average concentration of Cl^- for ten years measurement. From the figure, it was examined that east coast of Peninsular Malaysia experienced with higher concentration of Cl^-. Using the IDW interpolation in the GIS maps, it has shown that the monsoons seasons extended the dispersion of Cl^- to the land area. During the south-west monsoon season, wind blow from the Indian Ocean and the west coast area experienced with higher Cl^- concentration.

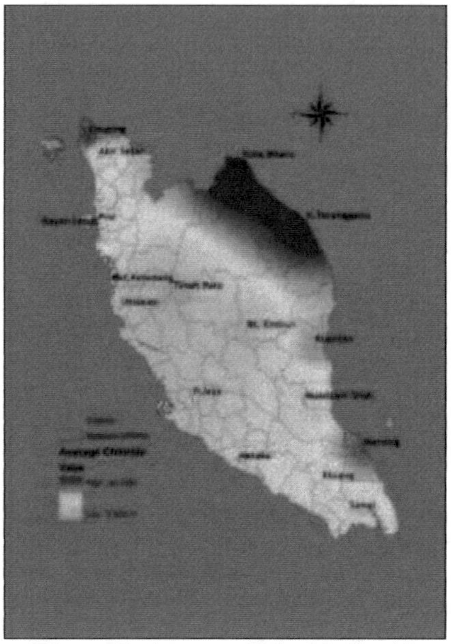

Figure 2. Average Chloride for 10 Years in Peninsular Malaysia

Based on the level of concentration of different level of environmental and pollutant data generated in GIS map and analysed, the zoning of the environmental load was determined. Figure 3 show the different level of concentration and range of maximum and minimum values for each salt ion deposition and level of pH from received rainfall. It is shown that sulphate which was dissolved in rainfall and deposited as wet fall out seems predominant in industrial area and it increased the acidity of rain water to the pH range under 5. Nitrate as result mostly from combustion of vehicles is experienced fairly same range for all zones. Chloride is a dominant pollutant agent for the coastal area and island especially during monsoon seasons.

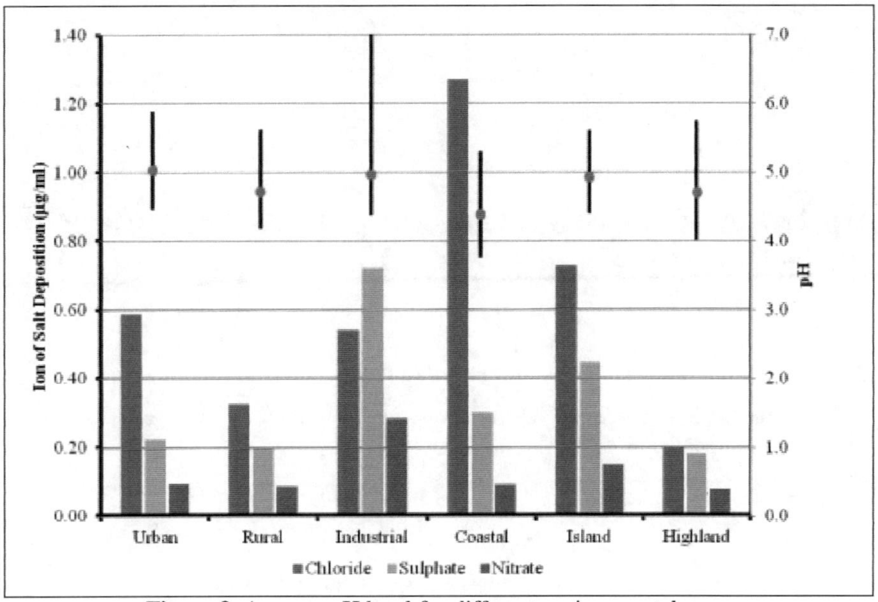

Figure 3. Average pH level for different environmental zone.

4. Conclusion

The ranges of the concentration of pollutant are slightly different for different environmental zoning. Some of pollutant depositions are found dominant for particular zone. The clear determination of characteristic of area will give different view and understanding on how a building should be designed and built. Thus, it can be good information for designer and engineer to justify their design and to propose new technology to produce better quality building.

References

[1] Ch. Anatolaki, and R. Tsitouridou Atmospheric deposition of nitrogen, sulfur and chloride in Thessaloniki, Greece, 2010 *Journal of Atmospheric Research,* **85** 413–428

[2] Chotimongkol, L. et. al. Deterioration By Atmospheric Pollutants Of Historical Places In Thailand 2003 *Proceeding of the EUROMAT 2003*

[3] Cole, I.S. et. Al. Mechanism of Atmospheric Corrosion in Tropical Environments 2000 M*arine Corrosion in Tropical Environments, ASTM STP 1399*, American Society for Testing and Materials, West Conshohocken, PA.

[4] Xu Y. *The Investigation of the Mechanism of Sulphur Deposition in Asia*, 1998 UMI Microform 9834537: University of Iowa, USA, Ph.D. Thesis.

IOP Conf. Series: Earth and Environmental Science **16** (2013) 012088 doi:10.1088/1755-1315/16/1/012088

Development of corrosion risk map for Peninsular Malaysia using climatic and air pollution data

Fathoni U, Zakaria C M and Rohayu C O

Centre for Forensic Engineering, College of Engineering, Universiti Tenaga Nasional, Selangor Malaysia

Email: fathoni@uniten.edu.my

Abstract. Malaysia has catapulted to an era of major transition. This rapid transition has also cause impact to the environment. The human activities contribute to pollutions. Buildings and it component's performances are affected directly or indirectly by air pollutions and climate factors. It has triggering and accelerating degradation processes. When deterioration start, service life of the buildings and its components will decrease. This paper presents initial development of corrosion risk map for Peninsular Malaysia using Geographical Information System (GIS). The air pollution and climate data obtained from Malaysia Meteorology Department (MMD). The air pollution data was the salt ion deposition of nitrate, chloride and sulphate in a form of wet fall out (WFO). The corrosion risk map generated using geographical information system (GIS) using inverse distance weighing (IDW) and weighted overlay method. It found that the corrosion risk map can be generated with further site verification and it can be used by engineers for further prediction of service life of building components in achieving sustainable construction design.

1. Introduction

The surrounding environment condition has influenced the deterioration process of building materials. Corrosion is one of the common deterioration resulted by this complex interaction of metallic materials and the nature. The environmental load is described as the deterioration agent from and its local. Those agents will accelerate the deterioration process. Degradation can be subjected by behavior or dreadful conditions but the terminology of deterioration is more on worsening or weakening of particular object. In this study the work deterioration will be used as an expression of a process to decline the performance or the aging process that lower the quality and performance of particular building components.

Deposition of pollutants on the building surface generally depends on atmospheric concentrations of the pollutants and the local climate. Once the pollutants are on the surface, interactions will vary depending on the amount of exposure, the reactivity and the amount of moisture present. The transformation reactions may take place both in gas phase and in aerosol phase. For most of the materials, SO_2 is the main corrosive agent in the air [1]. Research has discovered that when NO_2 is presented with SO_2, increased corrosion rate occur. The NO_2 oxidizes the SO_2 to sulphate thereby promoting further SO_2 absorption. As a result, SO_2 is considered as a major contributor to deterioration of metallic materials [1, 2].

Moisture conditions are strongly correlated with relative humidity and temperature in absorption process deep in to the exposed building components. The moisture or wetness of surface did not depend only on relative humidity but also on other parameters such as salt deposition, sunlight radiation, wind, and absorption of ambient heat [3]. With the relative humidity greater than 50% for all regions in Malaysia and average temperature of 27°C, average yearly time of wetness (TOW) in Peninsular Malaysia is equal to 0.783 fractional hours. It is classified by ISO 9223 the level of corrosion as class 5 where the layer on the surface of the building components can reside slightly longer and deterioration process will become more effective [4].

The most understandable influence of temperature is on the rate of the chemical reaction resulting in deterioration. If their surface temperature falls below the dew point, the surface becomes moist and in the presence of corrosive pollutants whose concentrations are increasing under the stable influence of the temperature inversion, conducive to certain types of deterioration to materials.

A research study on concrete deterioration in the Malaysian environmental conditions has been embarked by Malaysia Public Work Department (PWD) [5]. Ten locations in the Peninsular were chosen consisting major towns near the coastal area. A building was selected at each site for the study to be carried out with a ranging between 18 years and 50 years of age. The Corrosive Risk Rating (CRR) was a measurement of concrete corrosion risk represented by a range of 1 to 5 numerical scales. The CRR value was dependent towards the sulphate content, chloride content at a depth of 20 mm and the carbonation depth on a concrete building. From the study Climatic Corrosive Index (CCI) was formulated based on the monthly mean temperature and the monthly mean Relative Humidity (RH). A formula was proposed corresponds to the Scheffers's index format.

2. Methodology

In attempting to achieve the objective of this study, historical data have used. It consisted of air pollution data and climatic data. The data was recorded from 17 main measurement stations located in Peninsular Malaysia. The data from each measurement stations were recorded by Malaysia Meteorology Department (MMD). For the purpose of this study, data that extracted from the MMD data bank were the monthly weather report which record the received rainfall, temperature, humidity and air pollution data for 10 years from 1996 to 2005. The pollution data are in the form of wet fall out (WFO). Deposition of salt ion of chloride (Cl^-), sulphate (SO_4^{2-}) and nitrate (NO_3^-) were used. GIS maps have generated to determine data from the area which have no measurement stations. These maps are generated using inverse distance weighing (IDW) method. IDW method can efficiently apply and it has reliable computational process [6].

The process of development Corrosion Risk Map for Peninsular Malaysia was using ArcGIS 9.2. It was started with importing the data, generating the interpolation map using IDW method, reclassifying the IDW map into 5 levels scale and merge the generated map using weighted overlay method. Basically this study was producing initial map of corrosion risk map. The process of merging the generated IDW maps to produce the corrosion risk map was based on an assumption that the corrosion risk (CR) at a particular location is a function of environmental loads factors. It can be expressed in equation (1) where ion Chloride (C), ion Nitrate (N), ion Sulphate (S), received rainfall (Rain) and the time of wetness (TOW) are contributing to the level of corrosion risk.

$$CR = f(N, C, S, Rain, TOW) \quad (1)$$

3. Analysis and Discussion

The following Figures from Figure 1 to Figure 5 show the Chloride (C), Nitrate (N), Sulphate (S), received rainfall (Rain) and time of wetness (TOW) which has generated from respective average IDW map. The map also have classified to 5 different level from 1 to 5 where the class 1 represent lowest value and class 5 represent highest value of data.

It is very obvious for the area where it is more developed and have more industrial activities, these area is emitting high concentration of pollutant to the atmosphere. Those three areas are Klang Valley, southern part of Johor and Penang. The time of wetness for Peninsular Malaysia based on ISO 9223 is classified as class 5 but for the purpose of this study, it reclassified locally to a range of scale from 1 to

5. The purpose of reclassify the time of wetness IDW map is to distinguish locally which area having highest and lowest time of wetness as shown Figure 5.

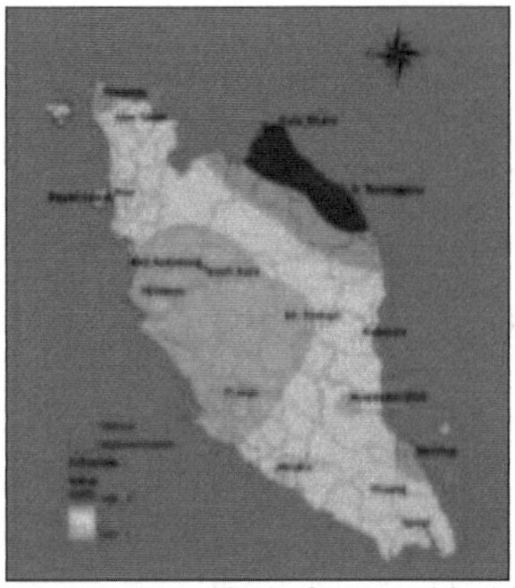

Figure 1. Classified of IDW Chloride Map

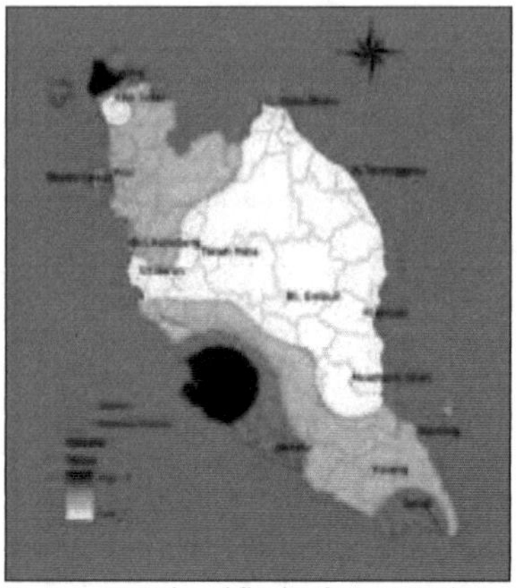

Figure 2. Classified of IDW Nitrate Map

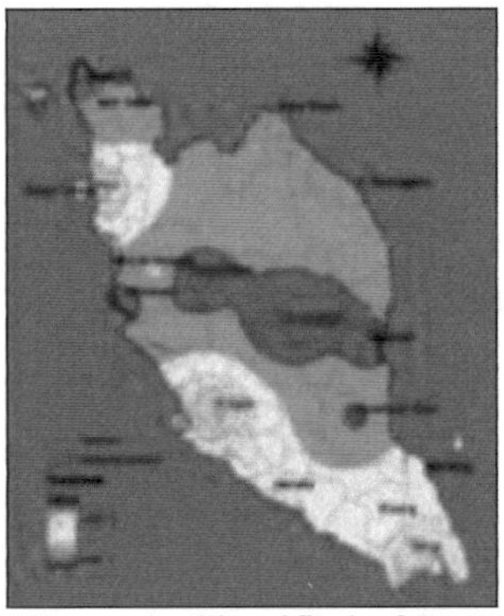

Figure 3. Classified of IDW Sulphate Map

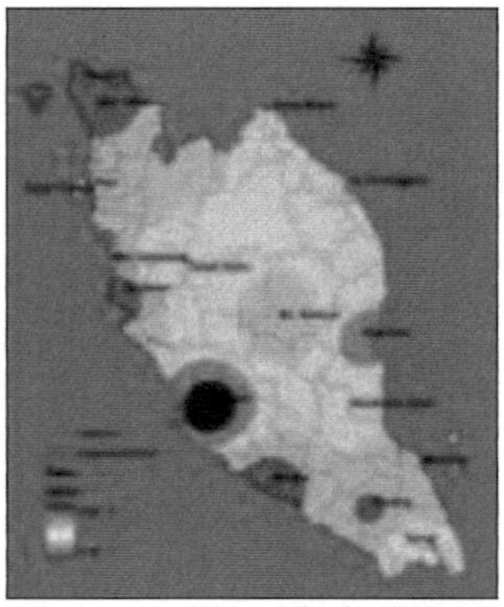

Figure 4. Classified of IDW Rainfall Map

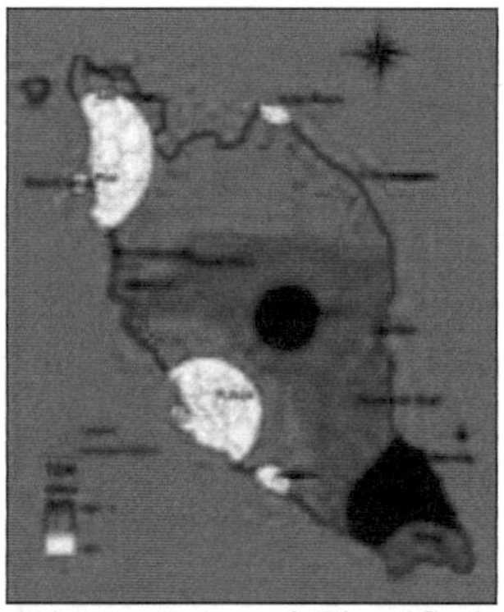

Figure 5. Local Classified IDW time of wetness map for Peninsular Malaysia

Figure 6 shows the corrosion risk map for Peninsular Malaysia. By merging the five layers of the environmental loads (i.e. N, C, S, Rain and TOW), a corrosion risk map generated. There are 5 different classes of corrosion risk that was represented in a scale from 1 to 5. Class 1 indicates the lowest corrosion risk and class 5 indicates the highest corrosion risk. From the generated corrosion risk map, Petaling Jaya has the highest rate of corrosion risk compared to other places. This is obvious, Petaling Jaya has the highest level of Nitrate and Sulphate. After Petaling Jaya, Kuantan and Kuala Terengganu are the area with level of corrosion risk 4. Although the concentration of Nitrate and Sulphate are low, the concentration of Chloride is very high compared to other areas. Kluang, Sitiawan and Alor Setar are the areas which have the lowest risk of corrosion. These areas are not much polluted by Chloride, Nitrate and Sulphate. The time of wetness at these areas are also low, therefore they fall under class 1, which is low corrosion risk. From this study, corrosion risk for Melaka is found slightly contradict with the ground condition. It was found from the previous study on building assessment, many building and infrastructure with metallic materials severely subjected to corrosion. Verification work is required to validate and improve the generated corrosion risk map [7].

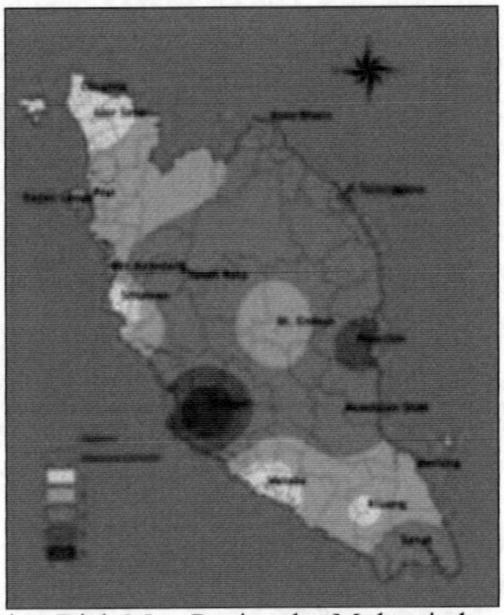

Figure 6. Corrosion Risk Map Peninsular Malaysia based on 10 Year of Data

4. Conclusion

Corrosion map have developed from this study. The result is based on the historical data from MMD for 10 years measured climatic and air pollution data from 17 measurement stations in Peninsular Malaysia. It shows that the generated corrosion map can be used for future study on corrosion cost to assist designer, engineer and maintenance officer also owner of building asset and infrastructure for estimating the maintenance cost. It also can be used in deciding alternative method of coating for corrosion prevention and using alternative technology and materials in construction. Site measurements are required to verify the result. So thus, its result can be reliable.

References

[1] Haagenrud S E 1997 *Environmental Characterisation including Equipment for Monitoring*, CIB W80/RILEM 140-PSL SubGroup 2 Report.
[2] Boubel R W, Douglas L F, Turner D B, Stern A C Fundamentals of Air Pollution 1994 3rd Ed., San Diego, CA, 126-135: Academic Press.
[3] Chotimongkol L Deterioration By Atmospheric Pollutants of Historical Places In Thailand 2003 *Proceeding of the EUROMAT 2003,* 1-5 September 2003, Lausanne, Switzerland.
[4] Tapsir S H and Fathon U Strategy for Sustainability in Affordable Housing – A Challenge to Malaysian Construction Industry 2005 *Proceeding of the World Sustainable Building Conference, Tokyo, Japan.*
[5] Maziah M, Kamaluddin A R, Lim C C and Ng S K, Corrosivity Map For Peninsular Malaysia *Jabatan Kerja Raya Malaysia.*
[6] Cole, I S et. al. Mechanism of Atmospheric Corrosion in Tropical Environments 2000 *Marine Corrosion in Tropical Environments, ASTM STP 1399*, American Society for Testing and Materials, West Conshohocken, PA.
[7] Fathoni U, Tapsir S H and Rizal A M Affordable Housing Research Project - Life Cycle Costing Approach for Residential Housing in Malaysia 2006 *Final Report, Ministry of Housing and Local Government Malaysia, Malaysia.*

4th International Conference on Energy and Environment 2013 (ICEE 2013) IOP Publishing
IOP Conf. Series: Earth and Environmental Science **16** (2013) 012089 doi:10.1088/1755-1315/16/1/012089

Value engineering awareness study for sustainable construction in Malaysia

Fathoni U, Zakaria C M and Rohayu C O

Centre for Forensic Engineering, College of Engineering, Universiti Tenaga Nasional, Selangor Malaysia

Email: fathoni@uniten.edu.my

Abstract. Construction process has often been described as a highly complex process because of the number of disciplines involved during the conceptual, design and construction stage. With the emergence of latest technology and concern for environment, increasing attention in construction industry is given on sustainability. Balance in quality and sustainability has become a major challenge to the construction industry. This paper presents a study that has conducted to determine the acceptance and application of Value Engineering (VE) and Life Cycle Cost Analysis (LCCA) in Malaysia construction industry. A set of questionnaire have distributed to different practitioners in construction industry and the result has reflect the fact that the application of VE and LCCA are still very low.

1. Introduction

Total population of Malaysia, according to the 2000 Census, was 23.27 million compared to 18.38 million in 1991. Malaysia's population have reached 27.6 million by 2010 with population growth 2.7% (2000 – 2010). See to back in the late 1970s and early 1980s, Malaysia, was developing the industry moved from import-oriented industry, begin to export oriented development, industries and factories. The construction sector is visualized to record marginal growth of 0.5% in 2004, supported largely by residential construction activity. Approximately 111,000 units of public affordable housing built during the 9th Malaysia Plan. During 10th Malaysia Plan 78,000 units of affordable public housing will be constructed by Federal Government across the nation. The housing sector would continue to play an important role in bringing development and growth for the country's construction industry. Construction industry in Malaysia contributes to the growth of the nation with the market industry of RM 91.58 billion and 22.66 billion for residential development based on the Construction Industry Board quarterly report 2012 [1].

Value Engineering (VE) was first applied mainly in the process and manufacturing industry. VE was introduced to the construction industry in 1968 as the United State Armed Service Procurement. Elias also mentioned that VE is one of the leading techniques in the study of projects unnecessary costs, setting the project requirements and functions. It is defined by the Society of American Value Engineering as "the systematic application of recognized techniques which identify the function of a product or service, establish a monetary value for that function, and provide the necessary function reliability at the lowest overall cost" [2]. While integration of VE in the construction projects elsewhere, the concepts and application of value engineering do not seem to be well embraced by the Southeast Asian (SEA) construction industry. In the studies conducted on 2004, most of the respondents had previously misconceived VE as just another usual cost cutting exercise. Apparently, the lack of knowledge and awareness about VE is a major cause for its limited application in Southeast Asian construction industry [3].

2. Methodology

This study was conducted to not only have an overview of the VE application in the Malaysian construction industry but also to determine the application of Life Cycle Cost Analysis (LCCA) for VE. Data for this study was collected through questionnaire survey form. The questionnaire was focused on the respondents' knowledge and understanding as well as their experience in the application of Value Engineering (VE) and the concept of Life Cycle Cost Analysis (LCCA) in the construction industry. All the data have stored in the online database from Google Documents as the survey distribution method was through email.

3. Analysis and Discussion

There were 61 set of questionnaires sent to different respondents with different background in the construction industry. However, only 29 respondents made response. Therefore only 29 set of questionnaires were collected. This means that the questionnaire collected only constitute 47.54% of the total amount of the questionnaires sent. The main target of the respondents are from the consultancy firms, such as the Civil and Structural (C&S) engineering consultants as well as quantity surveying consultants, and finally the contractor. This is done to have an overview as well as to make comparisons between the different construction disciplines. However, only 48.0% and 36.4% of the questionnaire response are from the contractor and quantity surveying consultants respectively. The largest numbers of questionnaire response are from the consultants (52%), which make up 45% of the respondents.

3.1. Degree of Understanding about Value Engineering

Questions about the degree of understanding about Value Engineering (VE) were asked. The results and analysis of the questions are further elaborated in the Figure 1. The first question was to assess whether the respondents have knowledge in the concepts of VE, however not to ask the question if they have heard of VE. Figure 1 shows 59% of the respondents know the concepts of VE, while 24% of the respondents replied "No", and 17% of the respondents replied "Not Sure". The results suggest that a majority of the respondents is knowledgeable in VE. However, it is noted that the return rate of the survey form is only at 47.54%, since most of the targeted respondents who had never heard of these terms might not return their questionnaires. Figure 2 shows the questionnaire response to the second question. It shows that a majority, which is 69% of the respondents, replied "No" while only 28% of the respondents had formal VE training. 3% of the respondents replied "Do not Know" for the second question. The response shows that while a majority of the respondents have an understanding of VE (59%), only 28% of the respondents have any formal VE training.

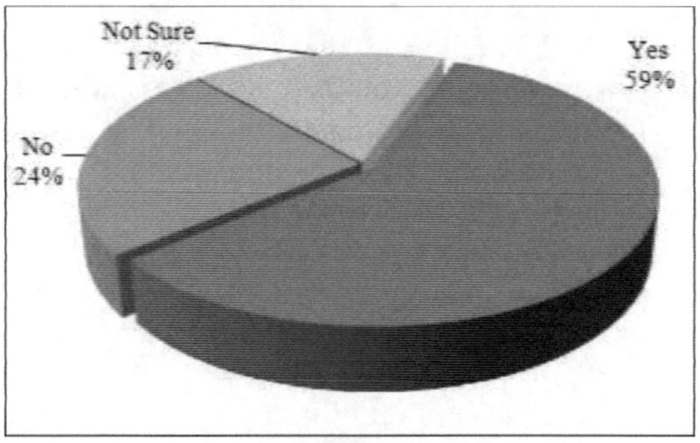

Figure 1. Questionnaire response to question, "Do you know the concept of Value Engineering?"

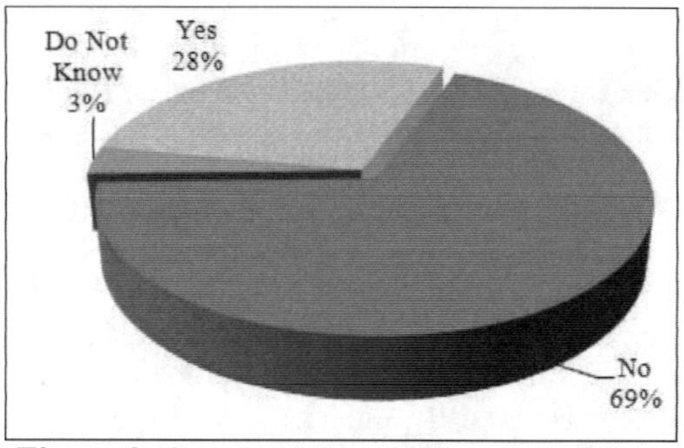

Figure 2. Questionnaire response to question, "Did you have formal Value Engineering training?"

IOP Conf. Series: Earth and Environmental Science **16** (2013) 012089 doi:10.1088/1755-1315/16/1/012089

3.2. Frequency of Applying Value Engineering in Construction

In Figure 3, 34% of the respondents either replied "Do not know" or "Never" for the question on application of VE in the construction industry, while only 38% of the respondents frequently apply the concepts of value engineering in their work. There are 14% of the respondents said that they very frequently apply VE concepts in their work. 28% of the respondents replied "Rarely" in Question 3. The result shows that the application of VE in the construction industry is low, as only 38% of the respondents replied either "Quite Frequently" or "Very Frequently". The mean index value for this question is 3.03, which is "average".

3.3. Degree of Understanding about Life Cycle Cost Analysis

One of the techniques that incorporated in Value Engineering is Life Cycle Costing (LCC) or Life Cycle Cost Analysis (LCCA). This technique is used to estimate or evaluate the sums over a given study period, which would include the operational, maintenance and disposal cost. From LCC, VE is applied as to have the most value over a given time period. Figure 4 shows the result of the questionnaire to have an overview to the application as well as the awareness of one of the VE techniques, LCC from the respondents. 33% of the respondents have not heard of Life Cycle Costing, while 22% of the respondents heard of the term, but were not sure what it is. The results suggest that a majority of the respondents do not know what LCC is, as only 33% of the respondents are knowledgeable of LCC and 11% of the respondents replied "Quite Knowledgeable".

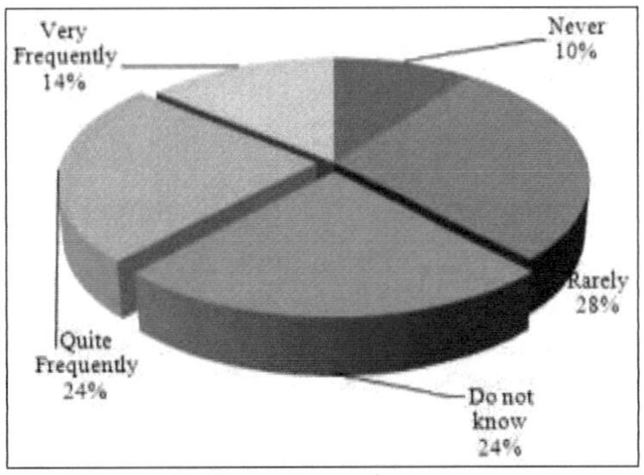

Figure 3. Questionnaire response to question, "How often is the application of VE in your work?"

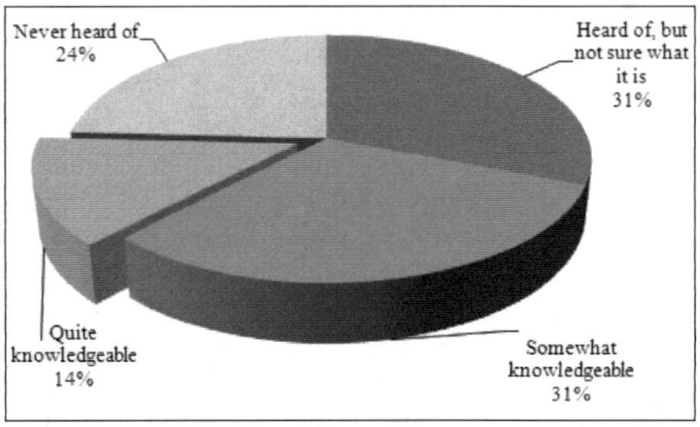

Figure 4. Questionnaire response to question, ": What is your prior knowledge on Life Cycle Costing (LCC)?"

3.4. Factors that Hinders the Implementation of Value Engineering in the Malaysian Construction Industry

The results as in Figure 5 suggest that the lack of application of VE and/or LCC is mainly due to either lack of VE and LCC knowledge (66.67%) and the lack of tools or training to apply VE and LCC techniques (55.56%). This shows that training is critical in the awareness as well as application of value engineering concepts in the construction industry. In fact, the lack of knowledge and training for VE and LCC is an expected answer from the respondents, as half of the respondents do not have an understanding of value engineering. Therefore, training as well as courses is required as to focus on the implementation of Value Engineering. Furthermore, 38.89% of the respondents agree that the lack of communication among project stakeholders is the reason why VE and LCC is not applied in projects, while 27.78% of the respondents replied that the lack of time to implement VE in the project as the reason. Lastly, only 22.22% of the respondents agree that the conflict of interest among project stakeholders is the reason why VE and LCC are not applied in the construction projects.

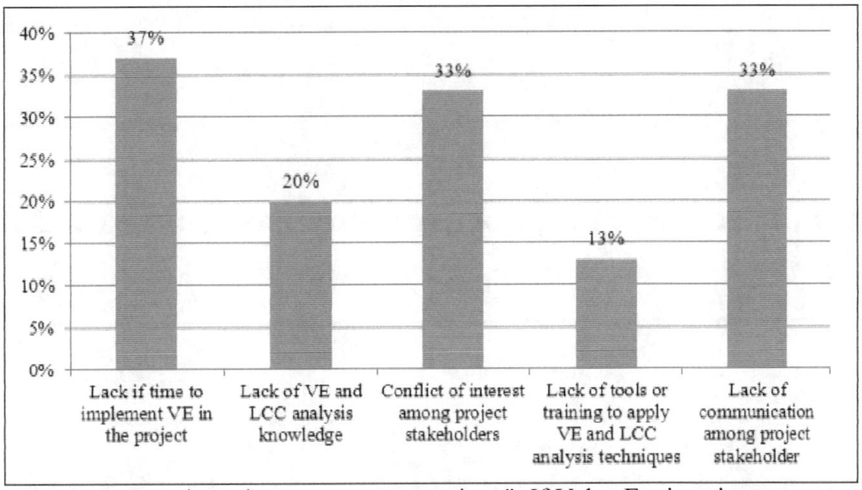

Figure 5. Questionnaire response to question, ": If Value Engineering concepts and/or Life Cycle Cost Analysis is not applied in projects, it is mainly due to..."

4. Conclusion

The questionnaire survey have distributed to Civil and Structural Engineering (C&S) consultancy firms as well as Quantity Survey (QS) consultancy, and contractors. The respondents' knowledge and understanding of Value Engineering (VE) is average, as 50% of the respondents understand what VE is. However, the questionnaire results suggested that the implementation of VE is low, considering the awareness of VE in the construction industry, as the average index shows an average frequency of VE implementation. Other findings include the reason why VE is applied in projects, which is upon clients' request as the main reason. The questionnaire also highlighted the factors that hinder the application of VE as well as life cycle cost analysis as, the lack of VE and LCC knowledge as well as the lack of tools to or training to apply VE and LCC. The results indicate that that not only the awareness and understanding of VE and LCC is required, training to apply VE is vital to increase the use of VE and LCC techniques in the Malaysian construction industry.

References

[1] Construction Industry Development Board (2012), *Construction Quarterly Statistical Buletin 2012.*

[2] Elias, S. (1998). Value Engineering, a Powerful Productivity Tool. Elsevier Science Ltd.

[3] Cheah, C. and Seng, K.T. Appraisal of Value Engineering in Construction in Southeast Asia 2005 *International Journal of Project Management* **23** 151-158

4th International Conference on Energy and Environment 2013 (ICEE 2013) IOP Publishing
IOP Conf. Series: Earth and Environmental Science **16** (2013) 012090 doi:10.1088/1755-1315/16/1/012090

Chemical procedures to detect carcinogenic compound in domestic wastewater

Abd Manan T. S. and Malakahmad A.

Civil Engineering Department, Universiti Teknologi Petronas, Bandar Sri Iskandar, 31750 Tronoh, Perak, Malaysia

E-mail: tehsabariah@gmail.com

Abstract. This review presents chemical methods to detect carcinogenic compound in wastewater. Atomic absorption spectroscopy (AAS), high performance liquid chromatography (HPLC) and gas chromatography mass spectroscopy (GCMS) and their alternative attached equipments were discussed. The application of each method is elaborated using related studies in the field.

1. Introduction

Cancer is a leading cause of death worldwide. The death rates due to cancer are very common in the developing countries. It is estimated that the mortality rates could further increase to more than 13.1 million people per year by the year 2030 [1]. Carcinogen is a compound that has the potential to cause cancer to humans under certain concentrations or periods of exposure [2, 3]. The water-borne contaminants such as heavy metals ion, polycyclic aromatic hydrocarbons (PAHs), N-nitroso compounds, mycotoxins and Heterocyclic Amines (HCAs) are carcinogenic upon consumption [4-9]. Domestic wastewater is wastewater of residential areas origin. It is mainly from the substances of human metabolism and household activities. Over 90% of the carcinogenic loading comes from the domestic areas. These areas can achieve loading values greater than industrial zones with standard loading over discharging rate of more than 10^9 L/d. The strong correlation between cancer and the presence of pollutants in wastewater have been proclaimed in previous studies [10-13]. Therefore, standards supervision and accurate measurement of carcinogenic compounds in domestic wastewater is highly in demand. This review gives descriptions on the detection and measurement methods of carcinogens in domestic wastewaters via chemical approaches. Therefore, the selection on the most appropriate method of detection and measurement can be made. The evaluations are based on advantages and disadvantages, accuracy, speed and time factor of methods and experimental procedures.

2. Detection Methods

Extraction, purification and identification of organic contaminants analyses demands the use of high technology equipments such as atomic absorption spectrophotometer (AAS), high performance liquid chromatography (HPLC) and gas chromatography mass spectrometer (GCMS).

2.1 Atomic Absorption Spectroscopy (AAS)

Atomic Absorption Spectroscopy (AAS) is an analytical procedure to determine the types of metals or metalloids and their concentrations in environmental samples. It employs the absorption of optical radiation of detected metal or metalloid at its gaseous phase. The heavy metals and metalloids contamination can be found in geothermal sources, volcanic and industrial effluents from mining, petrochemical industry, landfills resulting in pollution to the groundwater, soil and aquatic ecosystems. Analytical methods such as flame atomic absorption spectrometry (FAAS), graphite furnace atomic absorption spectrometry (GFAAS), electrothermal atomic absorption spectrometry (EAAS) and inductively coupled plasma optical emission spectrometry (ICP OES) and inductively coupled plasma mass spectrometry (ICP MS) have been applied to direct determination of heavy metal. The GFAAS is the most sensitive technique.

Analysis of arsenic was carried out in West Bengal in India and Bangladesh. Atomic absorption spectrophotometry with vapour generation assembly (AAS VGA) technique was used for the detection method on arsenic. Total arsenic analysis [As(III) + As(V)] requires reduction of As(V) to As(III). As(III) is reduced to AsH_3 vapours and finally to free As atoms, which are responsible for absorption signal in AAS. The vapour generation assembly attached to AAS has acid channel filled with 10 M HCl and the reduction channel with sodium borohydride. The tst conditions for the reduction of the sample in the instrument directly and introduction of KI in the reduction channel along with the sodium borohydride. It is observed that the acid concentration of the sample plays an important role. Further reduction in acid concentration and analysis time is achieved for the arsenic analysis using modified method. Analysis in both of the methods is sensitive at ppb level [14].

The levels of some heavy metals in groundwater, mixed water and wastewater in Alahsa farms, Saudi Arabia were examined. The total contents of Mn, Fe, Cu, Zn, Cd and Pb were determined using GFAAS. The level of heavy metals in groundwater and mixed water were found suitable for irrigation purpose based on the recommendation by the Food and Agriculture Organization. However, the spatial distribution study of Mn and Fe in groundwater in the area under this study was found to increase from southeast to northwest [15].

2.2 High Performance Liquid Chromatography (HPLC)

High performance liquid chromatography (HPLC) instruments consist of a reservoir mobile phase, a pump, an injector, a separation column and a detector. Compounds are separated by injecting a plug of the sample mixture onto the column. The different components in the mixture pass through the column at different rates due to differences in their partitioning behaviour between the mobile liquid phase and the stationary phase. HPLC is one of the most powerful tools in analytical chemistry. It has the ability to separate, identify and quantify the compounds in any sample that can be dissolved in a liquid. Compounds in trace concentrations as low as parts per trillion (ppt) may easily be identified.

HPLC with UV detector is a method used for detection of HCAs in wastewater. HPLC with fluorescence detection is used to detect PAHs and also used for the analysis of pyrolytic mutagen type heterocyclic amines [16]. HCAs have UV characteristics spectra and high entinction coefficients. Therefore, these compounds can be measured with UV detector. UV detector is the most preferable and HCAs can be simultaneously detected at 260-275 nm. HCAs such as Trp-P-1 and Trp-P-2 can be detected using this UV detector. The column chromatography like tSK gel ODS-120A column that used in HPLC provides the best peak symmetry and separation efficiency result [16-18].

Some researchers combined solid phase extraction with HPLC and UV detector and combined HPLC with fluoresce with Gas Chromatography (GC) and AAS. The HPLC with UV detector is a prominent method for analytes detection in research areas due to its frequent availability in the market and its capability of multi samples analysis concurrently with a high accuracy [17].

2.3 Gas Chromatography Mass Spectrometry (GCMS)

Gas chromatography mass spectrometry (GCMS) is a method that combines the features of gas liquid chromatography and mass spectrometry to identify different substances within a test sample. GC can separate volatile and semi-volatile compounds with great resolution but it cannot identify the components. MS can provide detailed structural information on most compounds but it cannot readily separate them. GC and MS are highly compatible techniques which solve the above mentioned barriers.

A new methodology for the separation, identification and quantification of eight *N*-nitrosamines was developed using raw sewage and final-discharge wastewater. The method was based on solid phase extraction (SPE) and a purification technique followed by GC analysis equipped with a triple-quadrupole MS gave an effective separation of the targeted compounds. The limits of detection of this method for *N*-nitrosamines ranged from 0.1 to 1.0 ngL^{-1} and the limits of quantification ranged from 0.4 to 3.3 ngL^{-1} [19].

A multiclass screening method for organic contaminants in surafe water, groundwater and effluent wastewater was developed by Portolet et al. [20]. The screening is based in the use of GCMS/ToF, and the sample procedure involved solid phase extraction with C_{18} cartridges. Around 150 organic contaminants from different chemical families were investigated, including PAHs, octyl/nonyl phenols, PCBs, PBDEs and a notable number of pesticides and several relevant metabolites. Surface water, ground water and effluent wastewater were found contained all target analkytes at three concentration levels (0.02, 0.1 and 1 μgL^{-1}). Influent wastewater and raw leachate from a municipal solid waste treatment plant were spiked at two levels (0.1 and 1 μgL^{-1}). The identification of compounds in influent wastewater was more problematic at concentration of 0.1 μgL^{-1}. Overall, many contaminants could be properly identified at the lowest level 0.02 μgL^{-1} in cleaner matrices. The procedure was applied to the screening of water samples of different origin and matrix composition and allowed the detection of several target contaminants [20].

3. Conclusion

Carcinogenic compounds are a group of chemicals that can cause chromosomal aberrations, resulting in cancer. The presence of carcinogens has been reported in domestic wastewater. Sewage treatment plants inefficient performance cause release of these hazardous compounds to surrounding watercourses. Chemical methods are precise and fast approach to detect the carcinogens presents in the water and wastewater. Carcinogens can be identified accurately using an appropriate method chosen from the available alternatives. Although chemical methods such as AAS, HPLC and GCMS are recommended as detection tools, the possession, operations and maintenances cost of these methods is the main drawback for their wide application.

References

[1] World Health Organization (WHO) 2012 *Fact Sheet: Cancer.* Website: http://www.who.int/mediacentre/factsheets/fs297/en/ Date: 19th June 2012

[2] Weinberg R A 2007 The biology of cancer Garland Science, Taylor & Francis Group, LLC, an informa business 270 Madison Avenue, New York, NY 10016 USA and 2 Park Square, Milton Park, Abingdon, OX14 4RN, UK. ISBN 0-8153-4078-8

[3] Kitchin K T 1999 Carcinogenicity: testing, predicting and interpreting chemical effects *Marcel Dekker, Inc. 270 Madison Avenue, New York, NY 10016.* ISBN 0-8247-9893-7

[4] Wescoat J L and White G F 2003 Water for life: Water management and environmental policy The Press Syndicate of the University of Cambridge, the Pitt Building, Trumpington Street, Cambridge, United Kingdom ISBN 0-521-36211-3

[5] White P A, and Rasmussen J B The genotoxic hazards of domestic wastes in surface waters *Mutat. Res.* **410** 223–236

[6] Saxena P B 2007 Chemistry of polycyclic and heterocyclic compounds *Discovery Publishing House, 4831/24, Ansari road, Prahlad Street, Darya Ganj, New Delhi-110002 (India)* ISBN 81-8356-281-7

[7] Luch A 2004 The carcinogenic effects of Polycyclic Aromatic Hydrocarbon *Imperial College Press, 57, Shelton Street, Covent Garden, London WC2H 9HE* ISBN 1-86094-417-5

[8] Sugimura T, Wakabayashi K, Nakagama H, and Nagao M. Heterocyclic Amines:

Mutagens/carcinogens produces during cooking of meat and fish. 2004 *Cancer Sci.* **95** 290-299

[9] Clayson D B 2000 Toxicological Carcinogenesis *Lewis Publishers, Corporate Blvd., Boca Raton, Florida 33431* ISBN 1-56570-508-8

[10] Sadiq R., Rajani B, and Kleiner Y 2004 Risk analysis for water quality deterioration in distribution, evaluation and control of water loss in urban water networks, Valencia, Spain, June 21-25, 2004, *NRCC* **47067** 1-26

[11] Thakur J S, Rao B T, Rajwanshi A, Parwana H K, and Kumar R 2008 Epidemiological study of high cancer among rural agricultural community of Punjab in Northern India International *Int. J. Environ. Res. Publ. Health.* **5** 399-407

[12] Koivusalo Men, Jaakkola J J K, Varliainen T, Hakulinen T, Karialainen S, Pukkala E, and Tuomisto J 1994 Drinking water mutagenicity and gastrointestinal and urinary tract cancers: An ecological study in Finland *Am. J. Public Health* **84** 1223-1228

[13] Kraybill H F 1978 Carcinogenesis induced by trace contaminants in potable water *A Symposium on Carcinogens held by Blue Cross and Blue Shield of Greater New York at the New York Academy of Medicine May 25, 1977* **54** 413-427

[14] Behari J R, and Prakash R 2006 Determination of total arsenic content in water by atomic absorption spectroscopy (AAS) using vapour generation assembly (VGA) *Chemosphere* **63** 17–21

[15] Assubaie F N 2011 Assessment of the levels of some heavy metals in water in Alahsa Oasis farms, Saudi Arabia, with analysis by atomic absorption spectrophotometry Arabian Journal of Chemistry King Saud University. doi:10.1016/j.arabjc.2011.08.018

[16] Kataoka H 1997 Methods for the determination of mutagenic heterocyclic amines and their applications in environmental analysis *J. Chromatogr. A* **774** 121–142

[17] Cassella R J, Sant'Anab O D, and Santelli R E 2002 Determination of arsenic in petroleum refinery streams by electrothermal atomic absorption spectrometry after multivariate optimization based on Doehlert design *Spectrochim. Acta Part B At Spectrosc.* **57** 1967–1978

[18] Neue U D 1997 HPLC Columns Theory, technology and practice *John Wiley & Sons, Inc., 605 Third Avenue, new York, NY 10158-0012* ISBN 0-471-19037-3

[19] Snyder L R, Kirkland J J and Dolan J W 2010 Introduction to Modern Liquid Chromatography *John Wiley & Sons, inc., Hoboken, New Jersey* ISBN 978-0-470-16754-0

[20] Tania Portoles, Elena Pitarch, Francisco J L, and Felix Hernandez 2011 Development and validation of a rapid and wide-scope qualitative screening method for detection and identification of organic pollutants in natural water and wastewater by gas chromatography time-of-flight mass spectrometry *J. Chromatogr. A* **1218** 303–315

Effects of reclaimed asphalt pavement on indirect tensile strength test of foamed asphalt mix tested in dry condition

Herda Yati Katman[1], Mohd Rasdan Ibrahim[2], Mohd. Yazip Matori[3], Shuhairy Norhisham[1], Norlela Ismail[1]

[1]Centre for Sustainable Technology and Environment, Civil Engineering Department, College of Engineering, Universiti Tenaga Nasional, 43000 Selangor, Malaysia

[2]Centre for Transportation Engineering, Department of Civil Engineering, Faculty of Engineering, University of Malaya, 50603 Kuala Lumpur, Malaysia

[3]Kumpulan IKRAM Sdn. Bhd, 43000 Kajang, Selangor, Malaysia

Email: herda@uniten.edu.my

Abstract: Indirect tensile strength (ITS) test was conducted to analyse strength of the foamed asphalt mixes incorporating reclaimed asphalt pavement. Samples were tested for ITS after cured in the oven at 40°C for 72 hours. This testing condition known as dry condition or unconditioned. Laboratory results show that reclaimed asphalt pavement (RAP) contents insignificantly affect the ITS results. ITS results significantly affected by foamed bitumen contents.

1. Introduction

Pavements are designed to last 10 to 40 years depending on type of highway. Once pavement reach the terminal serviceability index means that pavement no longer can be use, rehabilitation is needed. In some cases the road surfacing needs to be milled creating a milling waste material known as Reclaimed Asphalt Pavement (RAP). The RAP material which contains aggregate and asphalt binder are recyclable and has been used widely as granular base course material in asphalt and Portland cement concrete pavement construction as well as incorporated back into asphalt concrete mixtures for new pavement construction. Incorporating RAP to the asphalt concrete mixtures for highway construction is one of significant ways to ensure sustainable development since aggregate materials are non-renewable natural resources, therefore using RAP helps to reduce demand for extraction of new aggregate and helps to reduce deposits at the landfills.

This paper focuses on the investigation of RAP contents and foamed bitumen contents on tensile strength of dry conditioned samples. With accordance to Malaysia specification REAM-SP 1/2005 [1], the minimum strength for indirect tensile test for samples tested in dry condition is 0.2 MPa.

2. Methodology

Preparation of samples was followed closely to Marshall Procedure in accordance with ASTM D6926 [2] and tested with accordance to D6931 – 07 (2007) [3]. Samples were compacted with the Marshall Compactor with 75 blows for both surfaces and sample was leave in the mould for 24 hours, allowing sufficient strength to develop before extracting. Samples were cured in the oven at 40°C for 72 hours and allowed to cool to ambient temperature. Samples were then tested for Indirect Tensile Test (ITS). The tensile strength of samples was determined by Equation (1).

$$S_t = \left(\frac{2xP_{ult}}{\pi x d x t} \right) \quad (1)$$

Where:

S_t = tensile strength of specimens
P_{ult} = applied load to fail specimens
t = thickness of the specimens
d = diameter of the specimens

3. Materials and mix design

The proportion of RAP, crusher run (CR) and foamed bitumen used in this study are shown in Table 1. The RAP used in this study was obtained from the rehabilitation project of Jalan Pandamaran Klang in Selangor, Malaysia. In this study, samples prepared with 50% RAP were compared with the control sample prepared with 100% CR. Different percentages of foamed bitumen namely 1%, 2%, 3% and 4% were incorporated in the mixes. A 2% Ordinary Portland cement was used as mineral filler. Aggregate gradation with accordance to REAM-SP 1/2005 was used in sample preparation.

Table 1. Mix Designation according to different proportion of RAP, crusher run and foamed bitumen

Mix Designation	Mix Proportion	Foamed Bitumen Content, (%)
Mix A	0% RAP + 100% CR	1,2,3,4
Mix B	50% RAP + 50% CR	1,2,3,4

4. Results and analysis

4.1. Indirect Tensile Strength Test

ITS results were averaged based on results of three specimens. Figure 1 shows result of the ITS test for Mix A and Mix B prepared with different percentages of foamed bitumen. Generally the results show that the indirect tensile strength increases with an increase in foamed bitumen content to an optimum value, however a further increase in foamed bitumen content results in a decreasing tensile strength. For both mixes, optimum ITS was achieved by the mixes prepared with 2% foamed bitumen.

It can be found from Figure1 that ITS value for Mix A is greater than Mix B at any foamed bitumen content. This finding showed by comparing Mix A (0% RAP) and Mix B (50% RAP + 50% CR) at each foamed bitumen content. It seems that ITS roughly decreases when RAP content increases. However, effect of RAP content to ITS is not significant as showed by two-way analysis of variance in Table 2. The result shows that the probability value obtained is more than the predetermined alpha value (p = .323 > .05), thus RAP contents provide insignificant difference on the ITS. Besides of RAP content, interaction factors (RAP * FOAMED BITUMEN) seems to have no effects at all on the ITS which exhibit a probability value greater than the predetermined alpha value (p = .977 > .05). This implies that the application of RAP and interaction between factors (RAP*FOAMED BITUMEN) are not significant to effects the ITS value of the mixes. In addition, all mixes achieved the minimum strength requirement of 0.2 MPa. Statistical analysis shows only foamed bitumen give a significant effect to ITS as the probability is less than predetermined alpha value (p = .00 < .05).

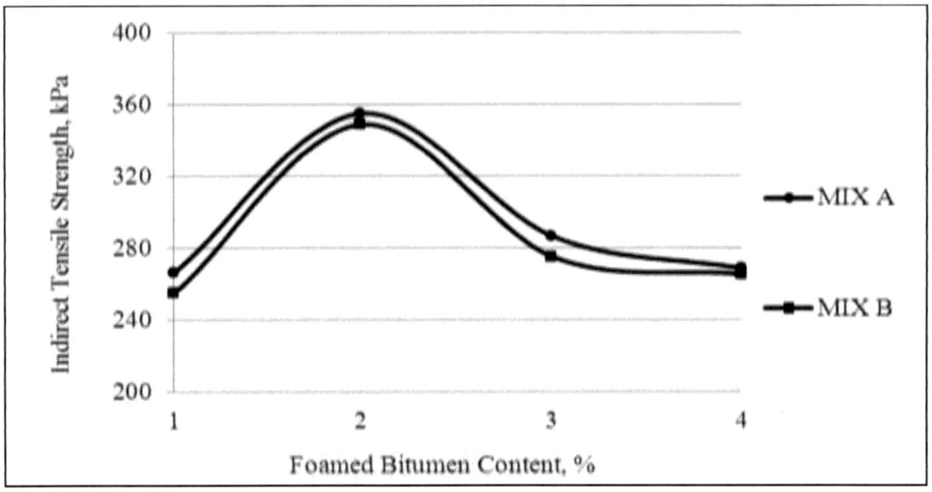

Figure 1. Indirect tensile strength for unconditioned samples

Table 2. Single and interaction effects on indirect tensile strength for unconditioned samples

Source	Type III Sum of Squares	df	Mean Square	F	Sig.
Corrected Model	32253.960[a]	7	4607.709	12.274	.000
Intercept	2019618.184	1	2019618.184	5379.814	.000
RAP	391.234	1	391.234	1.042	.323
FOAMED BITUMEN	31787.655	3	10595.885	28.225	.000
RAP * FOAMED BITUMEN	75.071	3	25.024	.067	.977
Error	6006.507	16	375.407		
Total	2057878.650	24			
Corrected Total	38260.466	23			

a. R Squared = .843 (Adjusted R Squared = .774)

5. Conclusions

Incorporating RAP to foamed asphalt mix and tested with indirect tensile strength test under dry condition shows that RAP contents did not offer significant effect on tensile strength of the mixes. This shown by statistical analysis two-way ANOVA with probability value obtained is more than the predetermined alpha value (p = .323 > .05).This reveals that incorporating RAP to the mixes offers similar strength as samples prepared with virgin aggregate. Since RAP contents do not affect strength of the mixes, therefore this is very advantageous in the context of the environment and sustainability to develop sustainable pavement based on green technology and concept. In this study, foamed bitumen contents significantly influence the tensile strength of the mixes as shown by statistical analysis $p = .00 < .05$. Therefore optimum amount of foamed bitumen in sample preparation is essential to produce superior mixes.

Acknowledgement

The authors are grateful for the cooperation of Kumpulan Ikram Sdn. Bhd in conducting this study and sharing the information.

References

[1] ASTM D6926: 10, "Standard Practice for Preparation of Bituminous Specimens Using Marshall Apparatus". American Society for Testing and Materials, 2007.
[2] D6931 – 07, "Standard test method for indirect tensile (IDT) strength of bituminous mixtures". American Society for Testing and Materials, 2007.
[3] REAM-SP 1 "Specification for cold in-place recycling". Road Engineering Association of Malaysia, 2005.

Voltage variation due to solar photovoltaic in distribution network

H. I. Azad[1], V. K. Ramachandaramurthy[1], Hesamaldin Maleki[1]

[1]Power Quality Research Group, Department of Electrical Power Engineering, Universiti Tenaga Nasional, 43000, Kajang, Selangor, Malaysia

E-mail: h.maleki@ieee.org

Abstract. Grid integration of solar photovoltaic (PV) plant offers reduction in greenhouse emissions and independence from fossil fuels for power generation. The integration of such forms of power generation also brings with it a variety of policy and technical issues. One of the technical issues is the variation in grid voltages in the presence of solar photovoltaic (PV) plant, resulting in degradation of power quality. In this paper, the application of a dq current controller to limit the voltage variation at the point of common coupling (PCC) due to a 2 MW solar photovoltaic (PV) plant will be discussed. The controller's goal is to ensure that the voltage variation meets the momentary voltage change limits specified in TNB's Technical Guidebook for the connection of distributed generation. The proposed dq current controller is shown to be able to limit the voltage variation.

1. Introduction

Direct generation of electricity from fossil fuels has various negative impacts on the environment as the CO_2 emissions released contribute towards the greenhouse effect, deteriorate air quality, and cause other harms to various ecosystems. With increased awareness and electricity demands, renewable sources of power generations (RE) are attracting the attention of policy makers, utilities and businesses around the world. As of 2010, installed capacity for solar PV is estimated to be around 40 GW [1], accounting for only 0.78% of the overall global installed capacity of 4,742 GW.

Though solar PV only represents a small percentage, it's rapid growth is demonstrated by over 17.5 GW of solar PV installations in 2010, as compared to 7.5 GW of installations in 2009 [2]. The rapid growth can be accredited to policy measures such as Feed in Tariffs (FIT), tax incentives and other incentives encouraging innovation, resulting in reduced costs for PV panels and other relevant equipment. The integration of RE in the electricity network however presents various challenges ranging from power quality (PQ), power flow control, voltage and frequency stability, new requirements for simulation tools, and the revision of Grid Codes and Standards. The integration of solar PV is addressed in the grid codes from developed countries, such as the German utility, BDEW, which lists the requirements for solar PV integration and operation in the medium voltage (MV) network [3], while PV integration is not included in codes / guidelines from developing countries, such as the Malaysian TNB guidelines [4]. Solar PV plants are connected to the grid via inverters utilizing pulse width modulation (PWM). The switching nature can result in degradation of power quality in the interconnection network, potentially resulting in grid code violations. As such, control systems need to be designed to take into account the network parameters, grid code regulations, and the state of the PV plant.

For application of the above mentioned, the Putrajaya's medium voltage (MV) network is modeled in PSCAD. A 2 MW PV plant is introduced and the effects on the voltage change, i.e. steady state and momentary, are observed. The simulations are performed with the controller integrated, and the results compared.

2. PV System Modelling and Controller Design

An overview of a grid connected PV plant is shown in Figure 1.

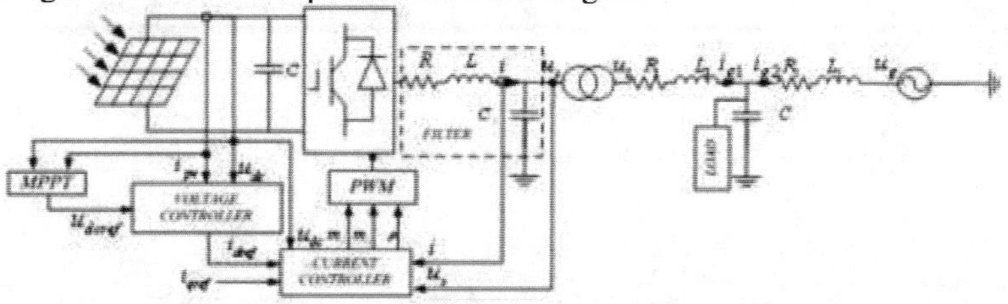

Figure 1. Overview of Grid connection PV plant [6]

The DC output voltage of a PV panel must be regulated at V_{MPP} in order to extract maximum power from the array. The MPPT block "finds" the V_{MPP} by using Maximum Power Point Tracking (MPPT) algorithms. The V_{MPP} is then used as a reference voltage, (u_{dref}) in Figure 1, for the voltage controller, which controls the switching of the IGBTs in the buck / boost converters to regulate the capacitor voltage, also known as the DC link voltage. For independent control of real and reactive power, dq0 transformation of the PV plant and grid parameters is used.

2.1. dq0 representation of Inverter

The DC link voltage is governed by the principle of power balance [7]:

$$\frac{C}{2}\left(\frac{dV_{DC}}{dt}\right)^2 = P_{PV} - P_{DC}. \quad (1)$$

where P_{PV} is the power generated by the PV array, and P_{DC} is the power drawn by the DC side of the inverter. Assuming no losses in the inverter, P_{DC} can be equated to the power output of the inverter:

$$P_{DC} = P_{inv} = \left(\frac{3}{2}\right)Re\{V_t I^*\} + \left(\frac{3}{2}\right)Re\left\{L\frac{dI}{dt}I^*\right\} = P_D + P_L \quad (2)$$

Substituting (2) into (1):

$$\frac{C}{2}\left(\frac{dV_{DC}}{dt}\right)^2 = P_{PV} - \left(\frac{3}{2}\right)Re\{V_t I^*\} + \left(\frac{3}{2}\right)Re\left\{L\frac{dI}{dt}I^*\right\} \quad (3)$$

Equation (3) establishes the relationship between the real power output and the DC link voltage. Equation (2) can be transformed to the dq0 reference frame as:

$$P_{inv} = \left(\frac{3}{2}\right)\{V_{td}I_d + V_{tq}I_q\} \quad (4)$$

The inverter AC side current equations are transformed to the dq0 reference frame as:

$$LI'_d = L\omega I_q + m_d\frac{V_{DC}}{2} - V_{td} \quad (5)$$

$$LI'_q = -L\omega I_d + m_q\frac{V_{DC}}{2} - V_{tq} \quad (6)$$

From equations (5) and (6), it can be seen that control of the 'd' and 'q' axis currents can be achieved via control of the modulation index 'm'.

IOP Conf. Series: Earth and Environmental Science **16** (2013) 012092 doi:10.1088/1755-1315/16/1/012092

2.1.1. DC link control and reference current generator

The relationship between the DC voltage and real power output with the integration of the PLL is shown in Equation 7.

$$\frac{C}{2}\left(\frac{dV_{DC}}{dt}\right)^2 = P_{PV} - \left(\frac{3}{2}\right)\{V_{td}\, I_d\} \tag{7}$$

The DC link controller is shown in Figure 2.

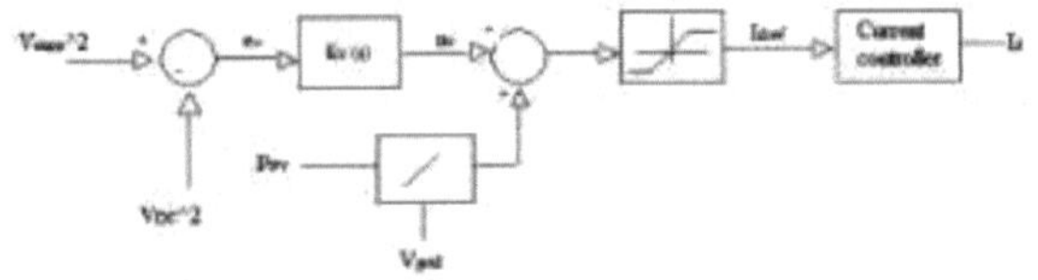

Figure 2. DC link control and reference current generator

2.1.2. Current controller

The current controllers for the 'd' and 'q' axis are shown in Figure 3 and Figure 4.

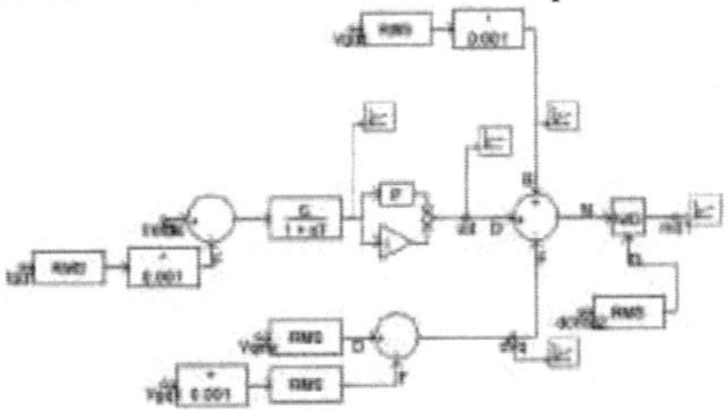

Figure 3. 'd' axis current controller

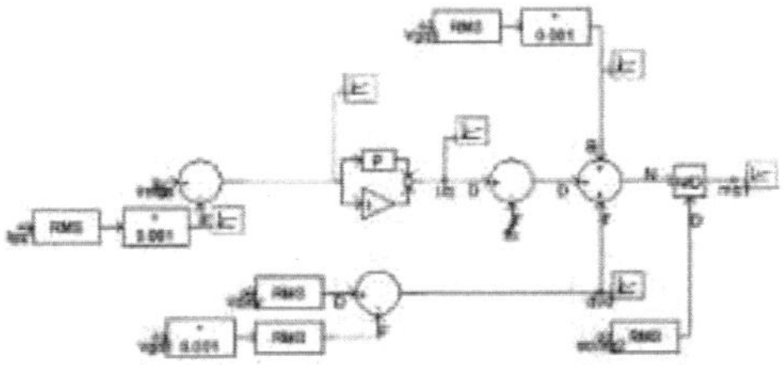

Figure 4. 'q' axis current controller

The performance of the controllers is presented in the sections to follow.

3. Simulation in PSCAD

Simulations were performed in PSCAD for a 2 MW PV plant connected to the Putrajaya MV network and the steady state voltage rise were assessed, for the plant without/with the controller.

3.1. Steady State Voltage Change

For steady state voltage rise, the solar radiation was held constant at 800 W/m^2, as shown in Figure 5.

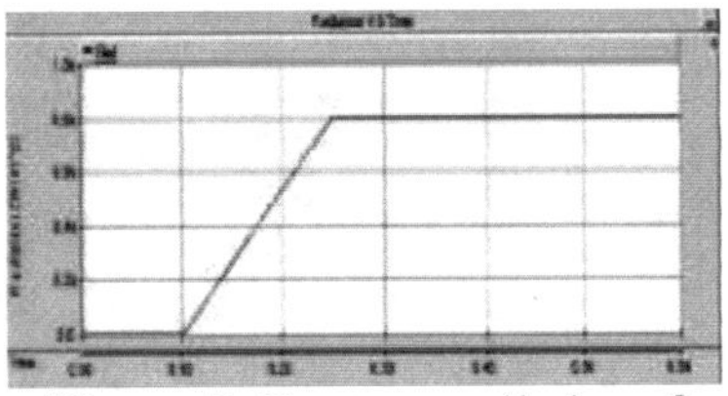

Figure 5. Constant radiation of 800 W/m2

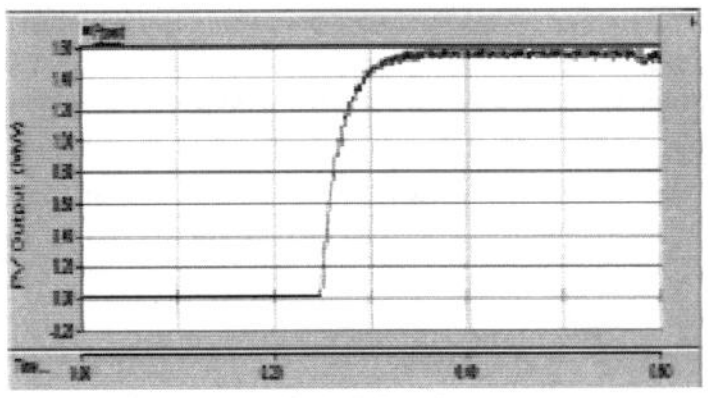

Figure 6. PV plant output power

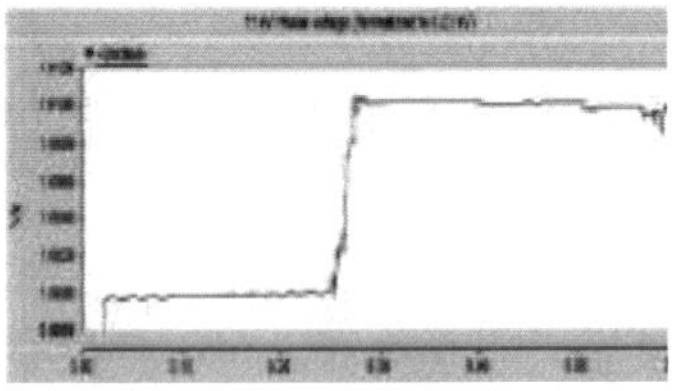

Figure 7. PCC voltage

The plant output, 1.6 MW, shown in Figure 6 is expected at 800 W/m^2. In Figure 7, it can be shown that upon connection of the PV plant, the grid steady voltage rises slightly more than 1% of the nominal voltage. While this is not a violation of the grid code [4], it may limit PV penetration in the network. The PV plant was simulated with the controller integrated and the results are as shown in Figure 8 to Figure 11.

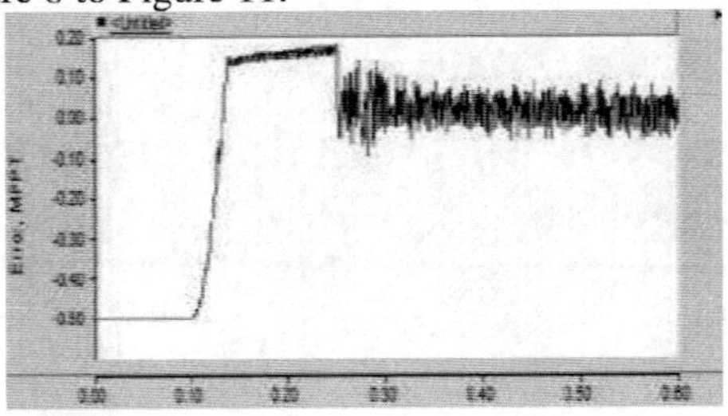

Figure 8. Error between DC link voltage and V_{MPP}

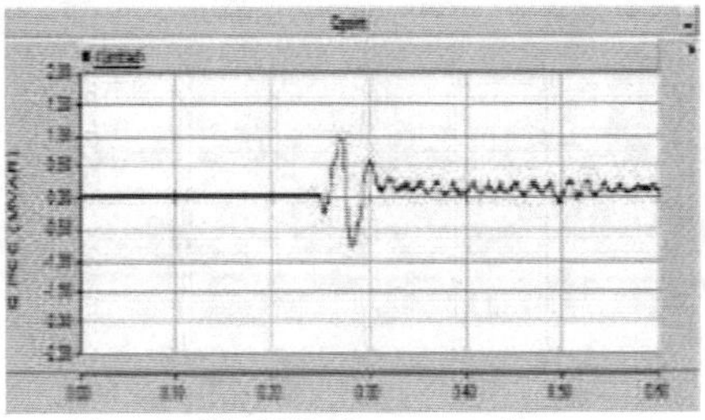

Figure 10. PV plant reactive power output

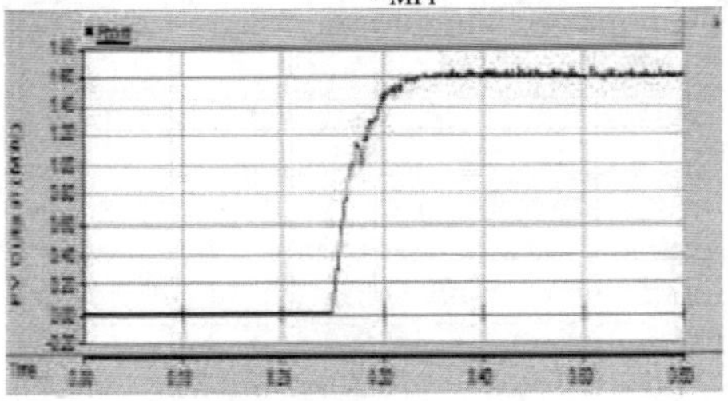

Figure 9. PV plant output power with controller integrated

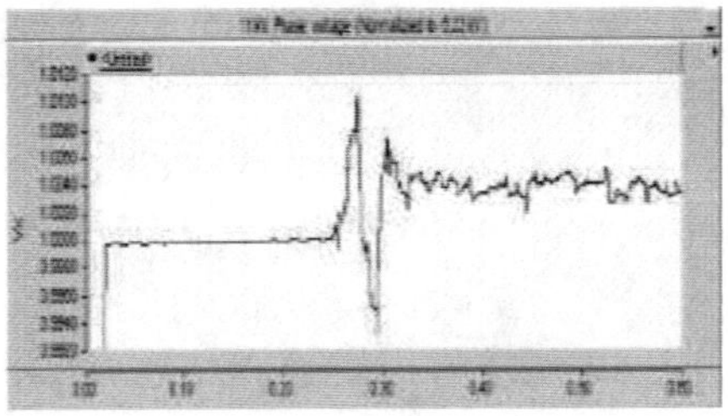

Figure 11. PCC Voltage

From Figure 8, the DC link controller can be seen to track the V_{MPP} output by the MPPT controller, effectively. The real power output, Figure 9 remains unchanged as compared to the case without the controller, Figure 6. The current controller is designed to regulate the power factor at near unity, to prevent the flow of excessive reactive power which results in the increase of PCC voltages. As seen from Figure 10, the reactive power output is almost zero. At the PCC, the steady state voltage change is reduced to 0.4%, as compared to 1% without the controller. It can be concluded that with the controller integrated, the PV penetration can be increased without violating the steady state voltage guidelines by the utility.

4. Conclusion

Solar PV penetration contributes to deteriorated power quality, amongst other negative impacts. Grid codes / guidelines must be modified accordingly to allow for ease of integration while ensuring system reliability and quality of supply. A 2 MW solar PV plant connected to the Putrajaya MV network was simulated and the impact on the steady state voltage and momentary voltage change was observed. Operation of the plant did not result in any violations of TNB's technical guidelines but posed limits on increased PV penetration. Integration of the controller resulted in reduced variations in the steady state and momentary voltage changes.

References

[1] Renewable Energy Policy Network for the 21st Century, 2011, REN21.
[2] Mehta, S. (2011, April), State of the Solar Union: Present and Future, IEEE Power and Energy Magazine, pp. 93-96.
[3] Generating Plants Connected to the Medium-Voltage Network, June 2008, BDEW, Berlin.
[4] Technical Guidebook for the Connection of Generation to the Distribution Network, First Edition, March 2005, TNB Research Sdn. Bhd.

IOP Conf. Series: Earth and Environmental Science **16** (2013) 012093 doi:10.1088/1755-1315/16/1/012093

Transmission tower classification based on landslide risk map generated by Geographical Information System (GIS) at Cameron Highlands

Hazwani NK, Rohayu CO, Fathoni U, Inz Baharuddin
Centre for Forensic Engineering, Universiti Tenaga Nasional, Jalan IKRAM-UNITEN, 43000 Kajang, Selangor, Malaysia.
E-mail: hazwani@uniten.edu.my

Abstract. Transmission tower is usually locates at remote area which is covered by hilly topography. Landslide is mainly occurring at hilly area and causing failure to the tower structure. This phenomenon subsequently will affect the national electricity supply. A landslide risk hazard map is generated using Geographical Information System (GIS). Risk classification is introduced to initiate the monitoring process along Jor-Bintang transmission line, Cameron Highland, Pahang. The classification has been divided into three categories, which are low, medium and high. This method can be applied in slope monitoring activities since all towers have been classified based on their risk level. Therefore, maintenance schedule can be planned smoothly and efficiently.

1. Introduction

Landslide is one of the world's catastrophes that affect enormous economic loss and one of the common interference nowadays. Malaysia is located at latitude 2° to 7° north and 100° to 119° east. Total area of Peninsular Malaysia is about 131,589 km^2 with 60% of the area covered by hilly topography. Every year, many landslide cases reported, either at man-made or natural slopes.

Application of new technology in prediction of slope assessment in Malaysia began in the early 1990's. As an effort to diminish landslide hazard and their consequences, government agencies and private organizations collaborated to deal with the issues. In Malaysia, existing slope assessment scale is normally conducted by dividing into two categories; large-scale and medium-to- small-scale. Lembaga Lebuhraya Malaysia (LLM) applying large-scale assessments in prioritizing slope maintenance along roads and highways. While medium-to-small-scale assessments in controlling development in hilly areas are widely use by Jabatan Kerja Raya (JKR), Jabatan Perancang Bandar dan Desa (JPPD) and Jabatan Mineral & Geossains (JMG) [1].

Slope monitoring guidelines prepared by JKR recommended standard of good practice for the maintenance of man-made slopes, disturbed terrain features and natural terrain hazard mitigation measures. The guidelines are widely used in slope engineering practice especially on new development at landslide prone area [2].

2. Landslide Risk Map Analysis using GIS

GIS is a computer system for managing spatial data. GIS is utilized to yield useful knowledge, application of colored maps and images, statistical graphics, tables, and various on-screen responses for interactive appearance. The software is made up from several interrelated and linked components with different functions. This software provides a comprehensive tool for assessing temporal environment data. A combination of GIS and Remote Sensing is useful for environmental auditing [3].

Landslide is the most common natural hazard these days that impacts the environment and human activities. Most of the occurrences are due to human factors such as extensive earth cutting, land clearing, agricultural activities, burning and uphill developments [4]. Identification of Landslide Susceptibility Zonation (LSZ) is important for safe strategic planning for future development. LSZ can be defined as the division of land surface into near-homogenous zones and rank according to the degrees of actual or potential hazard due to landslides. Methods of producing LSZ can be divided into two categories; qualitative approach and quantitative approach. Qualitative approach is integrated in generating LSZ map since a lot of subjectivity is introduced in preparing various thematic data layers. While quantitative approaches are introduced to minimize subjectivity in the weight assignment process. Common landslide causative factors such as lithology, lineament, slope, aspect, landuse and drainage are important for weights and ratings [5].

Type of approach used for this research is statistical approach (multivariate statistical analysis) because it is suitable for GIS tools and analysis. In this method, each individual thematic data layers are compared to the existing landslide distribution layers. The weight value of each category of causative factors is assigned based on landslide density. This involves overlay of landside distribution layer on each thematic data layers and calculation of respective landslide density values.

3. GIS Modelling

GIS Modelling needs various types of raw data before it comes to analysis stage. Most of the data are collected from previous researchers, government agencies and private sectors. Data sources are summarized in Table 1.

Table 1. Source of Data Collected

Description	Sources
Contour maps of Peninsular Malaysia (2004)	Department of Survey and Mapping Malaysia (JUPEM)
Soil maps	Department of Agriculture (DOA)
Land Use Map	Department of Agriculture (DOA)
Geology Maps (Peninsular Malaysia)	Minerals and Geo-Science Department (JMG)
Weightage value for Landuse Type	Literature Review
Weightage value for Lithology	Literature Review
Weightage value for Slope Category	Literature Review
Weightage value for Aspect	Literature Review
Weightage value for Elevation	Literature Review

For this research, statistical approach is used in producing landslide hazard map. This method is based on observation of relationship between each factor. Weightage values are differentiated based on types of soil, geological properties, landuse type and slope, as listed in the Table 2 [6].

Table 2 Weightage Values

LANDUSE TYPE				ASPECT		
Tin Mine	5	Wet Paddy	3	Flat	-1^{O}	1
Water Tank	5	Rubber	3	North	$0 - 22.5^{O}$	1
Pond	5	Oil Palm	3	Northeast	$22.5 - 67.5^{O}$	1
Primary Forest	5	Vegetables	3	East	$67.5 - 112.5^{O}$	2
Rifle Rang	5	Cash Crops	2	Southeast	$112.5 - 157.5^{O}$	2
Rural Building	5	Rock	2	South	$157.5 - 202.5^{O}$	3
Cemetery	5	Cocoa	2	Southwest	$202.5 - 247.5^{O}$	3
Education	5	Outcrop	2	West	$247.5 - 292.5^{O}$	4
Industrial	5	Annual Crops	2	Northwest	$292.5 - 337.5^{O}$	4

IOP Conf. Series: Earth and Environmental Science **16** (2013) 012093 doi:10.1088/1755-1315/16/1/012093

Jetty	5	Banana	2	North	337.5 – 360°	1
Lake	5	Bush	2			
Major River	5	Coconut	2			
Non Tidal	5	Sand	1			
Others Building	5	Clear Land	1			
Railway Station	5	Grass	1			
SLOPE CATEGORY				LITHOLOGY		
0 – 15°		1		Carboniferous		5
15 – 25°		2		Permian		4
25 – 35°		3		Triassic		3
35 – 45°		4		Tertiary		3
45 – 90°		5		Igneous Activity		2
				Silurian-Ordovician		2
				Devonian		2

4. Landslide Hazard Map and Risk Classification

Landslide Hazard Map is generated using Statistical Approach by considering weightage value for every data layer. Raw datas such as contour map, lithology map and landuse are analyzed using Model Builder application in GIS. All data must be in raster format before it can be layered to each other. The calculation for both study areas is done based on weightage value of every layer as shown in Figure 2.

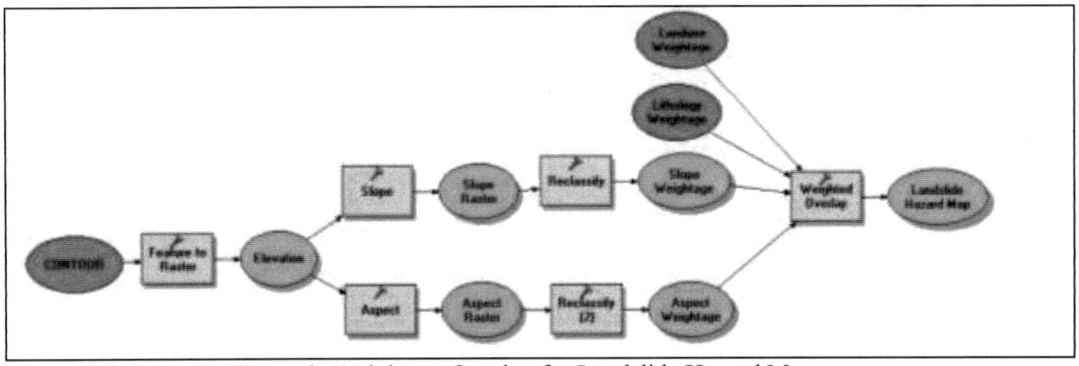

Figure 2. Weightage Overlay for Landslide Hazard Map

Classification of risk has been divided into three stages, which are low hazard (1 to 2), medium hazard (3 to 4) and high hazard (5 to 6). Figure 3 shows the detail layers in producing landslide hazard map for Jor-Bintang transmission line in Cameron Highland. After producing the erosion hazard map, all transmission towers have been arranged into their hazard class by GIS zoom scale 1:20,000. The classification is summarized in Table 3.

IOP Conf. Series: Earth and Environmental Science **16** (2013) 012093 doi:10.1088/1755-1315/16/1/012093

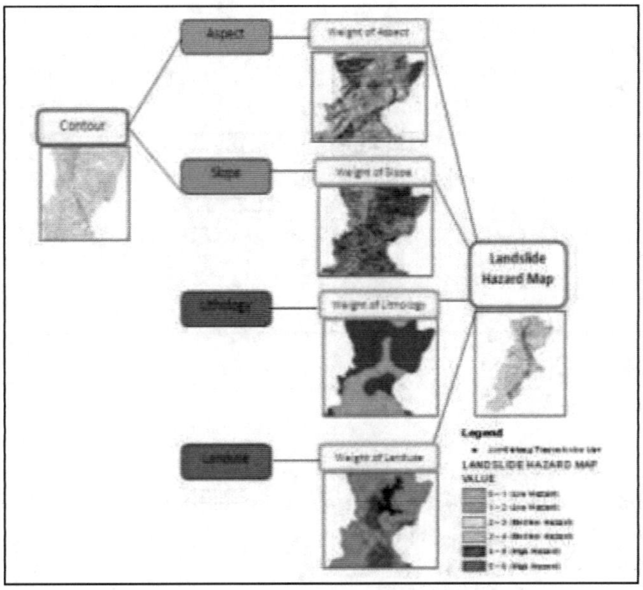

Figure 3. Landslide Risk Map

Table 3. Weightage for Lithology

Risk Level	List of Tower
Low Hazard	T2, T15, T18, T20, T22, T23, T24, T28
Medium Hazard	T5, T6, T12, T13, T3, T9, T16, T17, T19, T21, T25, T26, T29, T30, T31, T32, T33, T34, T36, T37, T38, T39, T40, T45
High Hazard	T4, T7, T8, T10, T11, T14, T27, T35, T41, T42-T44

5. Conclusion

As a conclusion, by referring to the risk level, frequency in monitoring stability of transmission tower can be planned smoothly and efficiently, especially during rainy season.

References

[1] Slope Engineering Branch, S. E. (2006). *Guidelines on Slope Maintenance in Malaysia*. Jabatan Kerja Raya (JKR).

[2] Suhaimi Jamaludin, A. N. (2006). Landslide Hazard and Risk Assessment L: The Malaysian Experience. IAEG 2006 Engineering *Geology for Tomorrow's Cities*, (p.55). Nottingham, United Kingdom.

[3] Kalpana, O. (2006). *Soil Erosion Risk Modelling and Current Erosion Damage Assessment Using Remote Sensing and GIS Techniques*. Dehradun: Indian Institute of Remote Sensing.

[4] F.Ahmad, S. L. Landslide Hazard Zonnation For Paya Terubung-Relau Slopes Terrain in Penang Island, 2006

[5] Kanungo, D. M. Landslide Susceptibility Zonation (LSZ) Mapping - A *Review Journal of South Asia Disaster Studies* 81-105.

[6] Rohayu, C. O. (2006). *Maklumat Geosains Terintegrasi untuk Proses Perancangan Guna Tanah Fizikal di Cyberjaya*. Universiti Kebangsaan Malaysia.

IOP Conf. Series: Earth and Environmental Science **16** (2013) 012094 doi:10.1088/1755-1315/16/1/012094

Conceptual design for transmission line inspection robot

M.F. Abdul Jalal*, K.S. Mohamed Sahari, A. Anuar, A.D. Mohd Arshad, M.S. Idris

Centre for Advanced Mechatronics and Robotics (CAMaRo),
UniversitiTenagaNasional, Putrajaya Campus, Jalan IKRAM-UNITEN, 43000,
Kajang, Selangor, Malaysia
E-mail: mfairuz@uniten.edu.my

Abstract. Power transmission line is used for power distribution purposes due to their cost effective measure compared to underlying cable. However, prolonged exposure to natural weather may cause fatigue stress to the lines as well as induce material failure. Therefore, periodical line inspection is considered uttermost important as a preventive measure to avoid power outage. However, transmission line inspection has always been a high risk and expensive work. Hazardous works that may harm operator as well as routine that requires precise handling can be performed by robots. Various types of robots have been designed and developed for line inspection but only perform well on a straight and continuous line. As these robots encounter an obstacle during the inspection, then the real problem in terms of robot stability and smooth operation arises. In this paper, conceptual design and evaluation for transmission line inspection robot is presented. The inspection robot mobile robot must be able to bypass or avoid obstacles as it travels along the power transmission line.

1. Introduction

Power transmission lines are used worldwide to transmit electricity for power distribution purposes to the customer. The underground transmission system is environmentally preferable but has a significantly higher cost. On the other hand, the overhead lines are cheaper but are easily exposed to the changes in the natural weather. Prolonged exposure of the transmission line to the natural weather may cause stresses to the lines as much as material corrosion. Repetition of expansion and contraction of the transmission line due to temperature changes induce fatigue stress to the line mainly due to the aging of the material. It is vital for the transmission line to be inspected to prevent any fracture or deformation that may propagate along the lines and to locate hot spots to prevent further power lost during the transmission.

There are many methods to inspect the deterioration of the transmission line including conventional inspection using crane and inspection crate, aerial inspection using helicopter or unmanned aerial vehicle, ground inspection by foot or terrace vehicle patrol, as well as inspection robot. Various concepts of line inspection robot are developed in order to replace the presence of human due to high risk during the operation. Although none of the design has been mass produced, the main focus of the design process however is always the same, which is bypassing obstacles on the transmission line. Comparison between diversified concurrent transmission line inspection robots from various institution [1]–[5] is shown in Table 1.

Table 1. Concurrent Technology of Transmission Line Inspection Robot

Robot	Description
Tokyo Electric Power Mobile Robot [1]	Inventor: Tokyo electric power co. inc. (Japan) Dimension: 11kg and 3 meter Mechanism: Roller(transverse), Arc shaped guide rail (to bypass) Slop limit: 30° maximum | Limitation: Only for single fibre-optic overhead ground wire transmission line
LineScout [2]	Inventor: Hydro-Quebec Company (Canada) Dimension: 112 kg and 1.37 meter Mechanism: Roller(transverse), robotic arm walker (bypass) Slope limit: 30° maximum | Limitation: Heavy, not suitable for high tension conductor
Shandong University Mobile Robot [3]	Inventor: Shandong University (china) Dimension: 45kg and 1.2 meter Mechanism: Roller(transverse), robotic arm(bypass) Slope limit: 60° maximum | Limitation: Arm movement not flexible due to short middle arm mechanism.
Expliner [4]	Inventor: Hibot (Japan) Dimension: 80kg and 1.5 meter Mechanism: Roller(transverse), robotic arm (bypass) Slope limit: 30° maximum | Limitation: Stability problem and only can be operating in double line usage
Wuhan University Mobile Robot [5]	Inventor: Wuhan University (China) Dimension: 30kg and 0.76 meter Mechanism: roller(transverse), acrobatic movement (bypass) Slope limit: 30° maximum | Limitation: Unable to bypass large obstacles.

2. Conceptual design

It is essential to understand and study the nature of the obstacles on the transmission line in terms of size and geometrical configuration. The concept generation process is closely linked to the identification of engineering specifications since the concept generated is a combination of various aspects that were derived as a solution for the engineering specifications. The robot must be able to grasp the line properly, avoid or bypass obstacles such as vibration damper, line spacer and aerial marker as well as overcome the jumper wire in order to transfer to another line span. The concepts of the transmission line inspection robot in Figure 1 are generated by means of morphological diagram as shown in Table 2.

Table 2. Morphological diagram for concept generation

Specification	Option 1	Option 2	Option 3	Option 4	Option 5	Option 6	Option 7	Option 8
Main body frame	Rigid *×	Segmented ⌂ Δ	Flexible	Extendable +				
Drive system	Electro Motor *×+⌂ Δ	Bar Linkage	Hydraulic	Pneumatic				
Steering mechanism	Actuator at end joint *×+⌂	Movable arm	Manipulate weight Δ	Movable body				
Cable gripping mechanism	Clipping Roller ×+⌂ Δ	Grooved Wheel	Robotic finger ×	Two large wheels *	Belting	Robot weight		
Movable part mechanism	Gear set *×+⌂ Δ	Slider	Thread	Cable	Chain and sprocket	Belting	Cam and follower	Bar linkage
Bypassing tower via line jumper	Articulated arm ×	Adjustable arm +⌂	Manipulate centre of body mass ⌂ Δ	Swinging	Crawling	Body waving Δ	Wheel /Roller *	Acrobatic movement
Bypassing inline obstacle	Articulated arm ×	Adjustable Arm +⌂	Manipulate centre of body mass ⌂ Δ	Swinging	Crawling	Body waving Δ	Wheel /Roller *	Acrobatic movement
Power Supply	On board (battery)	In line Energy Harvester	Off board	On board (solar)				

Indicator: * Concept 1 × Concept 2 + Concept 3 ⌂ Concept 4 Δ Concept 5

2.1. Concept 1
This concept consists of rigid bodies with flexible joint attached with big wheel. Each of the wheels is controlled separately. Both wheels will provide gripping force exerted to the transmission line. The robot is able to bypass the obstacle just by rolling over on the obstacle.

2.2. Concept 2
For this concept, the robot has one main body attached with three articulated arms. Two motor driven wheels are attached to the gripper for locomotion. The robotic arm will provide more flexibility to the robot for the inspection exercise.

2.3. Concept3
This concept consists of a main body, extendable platform, 2 grippers for support and a gripper for propulsion. The two grippers are attached to the extendable platform and gripper with clipping roller attached to the main body. When an obstacle is detected, the extendable platform will slide out of the obstacle. Two arms temporarily grip the line to give support to the robot while the wheel frame is driven forward to bypass the obstacle.

2.4. Concept 4
This concept consists of 3 arms attached to 3 segmented bodies which enable the robot to move to another transmission line especially for bypassing tower. When an obstacle is detected, the front gripper will disengage from the line and bypass the obstacle. As the front gripper grabs back the line, the middle gripper will be detached from line and the motor driven wheel at the front and back gripper will drive the body forward. The same process is then repeated for the back gripper. The segmented body can rotate horizontally and act as self-balancing system by manipulating the entire body mass of the robot.

2.5. Concept 5
This concept is inspired by leech movement. It consists of only 2 arms attached to segmented body and able to self-stabilize via centre of body mass manipulation. When an obstacle is detected, the front gripper will detach from the line and bypass the obstacle. At the same time, the arm will self-balance itself and grasp back the transmission line. The back gripper will then drive the whole body forward.

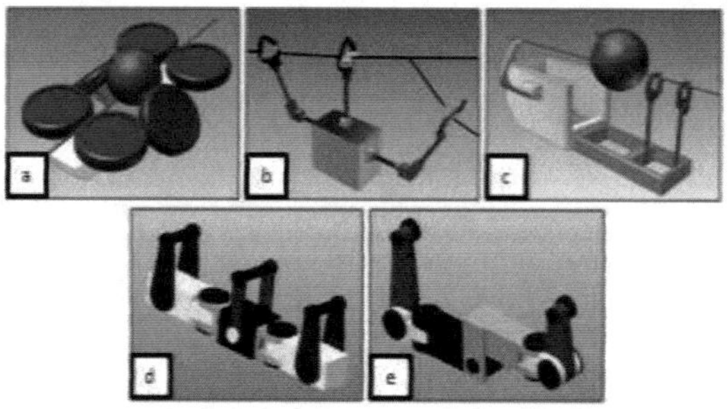

Figure 1: a) Concept 1, b) Concept 2, c) Concept 3, d) Concept 4 and e) Concept 5

3. Discussion
In order to choose the best concept for the transmission line inspection robot, the entire robot performance is evaluated. Each concept is compared to each other relatively as presented in Table 3. Concept 2 appears to be the best solution followed by concept 5 with ability to travel along the line, bypass obstacles and tower albeit high complexity in controls system for precise gripper coordination. On the other hand, concept 1 and 3 are suitable for continuous line and able to bypass in line obstacles. However, due to inflexibility of the main body frame, concept 1 and 3 are unable to bypass towers. Their control system is also considered simple in comparison with other concepts.

Table 3. Performance table

Selection criteria	Concept 1	Concept 2	Concept 3	Concept 4	Concept 5
Bypass tower	×××	√√√	×××	××	√√
Bypass obstacle	√√√	√√√	√√√	√√	√√√
Slope climbing	√√√	√√	√	√	√√
Stability	√√√	√	√	√√√	√
Close-up inspection	√√√	√	√	√	√
Flexibility	×	√√√	×	√	√√√
Slope climbing	√√	√√	√	√	√√
Control complexity	√√√	×××	√√√	×	×××

Indicator: × Average; ×× Poor; ××× Very Poor; √ Good; √√ Very Good; √√√ Excellent

4. Conclusion

The application of robot for transmission line is considered essential due to hazardous nature of inspection routine itself, low operation cost and the need of cautious handling measure during the inspection session. Even though various types of robots have been developed for line inspection, they only perform well on a straight and continuous line. Due to limitation of the robot flexibility and control complexity, the robot will encounter challenges in avoiding obstacle as well as bypass towers via line jumper. The overall robot design should be lightweight so that the transmission line suffers no additional load. The best concept to bypass in-line obstacle and tower is concept 2, followed by concept 5. However, concept 1 is the best concept for continuous straight line and bypass in-line obstacle.

Acknowledgement

The authors would like to sincerely thank the Ministry of Higher Education (MOHE) of Malaysia for the provision of a Grant with code No.20120125FRGS to support this work.

References

[1] J. Sawada, K. Kusumato, T. Munaka, Y. Maikawa and Y. Ishikawa, A Mobile Robot for Inspection of Power Transmission Lines 1991 *Power Engineering Review, IEEE* **11**

[2] S. Montambault and N. Pouliot, LineScout Technology: Development of an Inspection Robot Capable *Transmission & Distribution Construction, Operation and Live-Line Maintenance, 2006. ESMO 2006. IEEE 11th International Conference*

[3] Zhou, Fengyu, Xiao, Hairong and Wu, Aiguo, Control Strategy and Implementation of an Inspection Robot for 110kV Power Transmission Lines. Dalian, China, 2006. *6th World Congress on Intelligent Control.*

[4] P. Debenest, M. Guarnieri, K. Takita, E. F. Fukushima, H. Shigeo, K. Tamura, A. Kimura, H. Kubokawa, N. Iwama and F. S., Expliner – Robot for Inspection of Transmission Lines 2008 *IEEE International Conference on Robotics and Automation, Pasadena*

[5] Wu, Gongping, Xiao, Hua and Xiao, Xiaohui, Transmission Line Inspection Robot and Deicing Robot: Key Technologies, Prototypes and Applications, Montreal, Canada, 2010. *1st International Conference on Applied Robotics for the Power Industry.*

IOP Conf. Series: Earth and Environmental Science **16** (2013) 012095 doi:10.1088/1755-1315/16/1/012095

Adsorption of arsenate from aqueous solution by rice husk-based adsorbent

Taimur Khan[1] **and Malay Chaudhuri**[2]

[1,2] Department of Civil Engineering, Universiti Teknologi PETRONAS, 31750 Tronoh, Perak Darul Ridzuan, Malaysia

Email: Taimurkhan7@gmail.com

Abstract. Rice husk-based adsorbent (RHBA) was prepared by burning rice husk in a muffle furnace at 400°C for 4 h and adsorption of arsenate by the RHBA from aqueous solution was examined. Batch adsorption test showed that extent of arsenate adsorption depended on contact time and pH. Equilibrium adsorption was attained in 60 min, with maximum adsorption occurring at pH 7. Equilibrium adsorption data were well described by the Freundlich isotherm model. Freundlich constants K_f and $1/n$ were 3.62 and 2, respectively. The RHBA is effective in the adsorption of arsenate from water and is a potentially suitable filter medium for removing arsenate from groundwater at wells or in households.

1. Introduction

Arsenic is a highly toxic element and historically has been used as a poison [1]. Due to its toxicity and mobility, water pollution due to arsenic contamination has gained attention of the world and has been recorded by the World Health Organization (WHO) as a top priority issue. Arsenic in the aquatic environment is primarily present in the form of two inorganic species (arsenate and arsenite) and arsenate [(As (V))] is the predominant form of arsenic in oxidizing conditions [2]. The presence of arsenic in the natural water is due to the process of leaching from the arsenic containing source rocks and sediments [3,4]. More recently, arsenic has been used as an insecticide, fungicide, rodenticide and wood preservative. The common applications of arsenic are in the manufacture of pesticides, desiccants, glass, alloys, electronic components, pigments and pharmaceuticals, which create additional arsenic problems [5]. The presence of arsenic in drinking water even at low concentrations poses a threat to human health, as arsenic is a well-known toxic element and a documented human carcinogen. Long-term exposure to arsenic via drinking-water causes cancer of the skin, lungs, urinary bladder and kidney as well as other skin changes such as pigmentation and thickening (hyperkeratosis) [6]. Millions of people worldwide are exposed to naturally occurring arsenic-contaminated groundwater, which they use as their only source of water [7]. A number of large aquifers in various parts of the world have been identified with problems from arsenic occurring at high concentrations. The most noteworthy occurrences are in parts of West Bengal (India) and Bangladesh, Taiwan, northern China, Hungary, Mexico, Chile, Argentina and many parts of the USA [8] and Vietnam [9]. According to WHO and the United States Environmental Protection Agency (US EPA), the maximum allowable concentration of arsenic in drinking water is 10 μg/L. However, in many parts of the world arsenic concentration about 100 times more than the permissible limit is found. According to a WHO survey conducted in 2006, the number of people toxified by arsenic in India and Bangladesh alone are about 700 million, which might be the largest poisoning in the world history [10 - 12]. Several

IOP Conf. Series: Earth and Environmental Science **16** (2013) 012095 doi:10.1088/1755-1315/16/1/012095

techniques such as solvent extraction, chemical precipitation as synthetic coagulants, ferrihydrite precipitation, iron co-precipitation, ion exchange and reverse osmosis can be used to remove arsenic from water. However, the disadvantages associated with these techniques are incomplete metal removal, high cost of reagent and energy requirements [13]. There is an urgent need for low-cost adsorption medium for removing arsenic from drinking water and to reduce arsenic concentration in food grains.

In the present study, rice husk-based adsorbent (RHBA) was prepared from rice husk and adsorption of arsenate [(As (V))] by the RHBA from aqueous solution was examined.

2. Materials and methods

2.1. Preparation of RHBA

Rice husk was cleaned several times with distilled water in order to remove the dust and dried in an oven at 105°C for 24 h. The cleaned and dried rice husk was then subjected to burning in a muffle furnace at a temperature of 400°C for 4 h. The resulting rice husk-based adsorbent was ground to a size of 200-500 µm and used in batch adsorption test.

2.2. Batch adsorption test

Batch adsorption test was carried out by shaking 100 mL of arsenate solution of desired concentration with RHBA in conical flasks at room temperature (22°C) using an orbital shaker at 150 rpm. After a predetermined contact time, a flask was removed from the orbital shaker and the supernatant was filtered through 0.45 µm membrane filter and analysed for residual arsenate concentration. The effects of contact time, initial arsenate concentration, pH and RHBA dose on adsorption were determined. The pH of the solution was adjusted by 0.1 N NaOH or 0.1 N HCl. Adsorption isotherm was determined by batch equilibrium test using optimum contact time and pH for adsorption of arsenate by RHBA.

3. Results and discussion

3.1. Characterisation of RHBA

The values of BET surface area, micropore area, micropore volume and average pore diameter are 134.5 m^2/g, 2.75 m^2/g, 0.000392 mL/g, and 78.11 Å, respectively. Scanning electron micrograph of RHBA (Figure 1) shows presence of macro- and micro-pores.

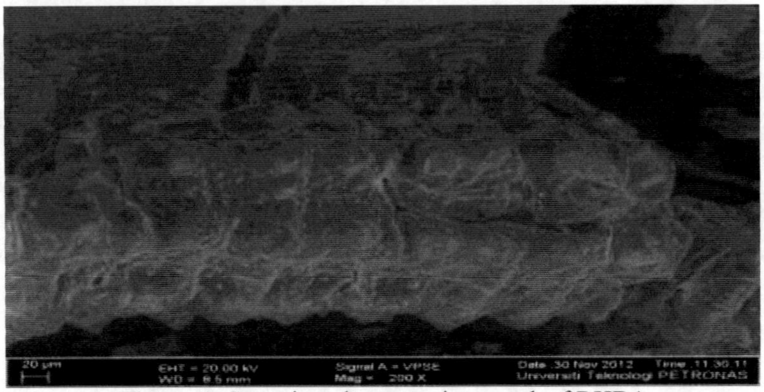

Figure 1. Scanning electron micrograph of RHBA

3.2. Effect of pH

The effect of pH on arsenate adsorption by RHBA was studied in the initial pH range between 5-9 with contact time of 24 h and the results are shown in Figure 2. Maximum adsorption of

89% from a 0.1 mg/L arsenate solution by 1 g/L RHBA dose was achieved at pH 7. All subsequent adsorption tests were conducted at pH 7.

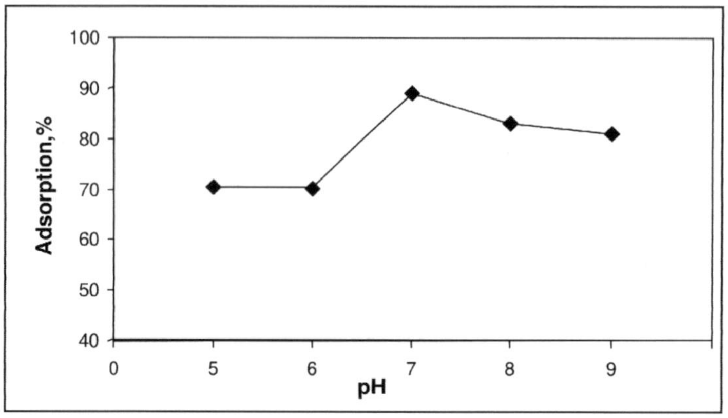

Figure 2. Effect of pH on adsorption of arsenate by RHBA

3.3. *Effect of contact time and initial concentration*

Effect of contact time and initial arsenate concentration on adsorption by RHBA is shown in Figure 3. The contact time was varied in the range 5-120 min at arsenate concentration 0.1, 0.15, 0.2 and 0.25 mg/L, RHBA dose of 1 g/L and pH 7. Equilibrium was attained in 60 min. A contact time of 60 min has also been reported by Bhattacharya *et al.* [14] for the adsorption of arsenate by uncalcined and calcined aluminium hydroxide powder. A contact time of 60 min was used in all subsequent adsorption tests.

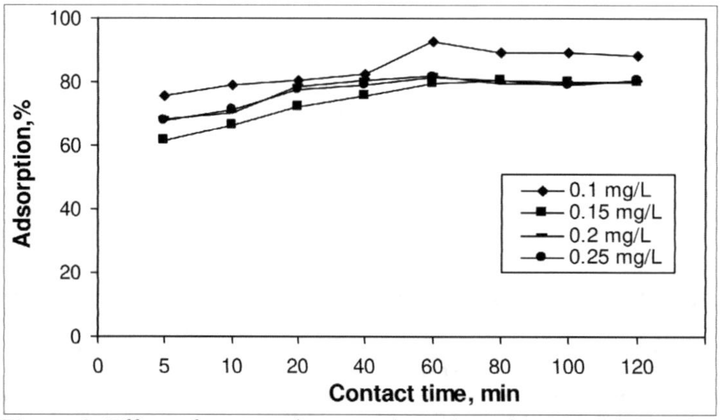

Figure 3. Effect of contact time on adsorption of arsenate by RHBA

3.4. *Adsorption isotherm*

Equilibrium arsenate adsorption data were fitted to the linear form ($log\ q_e = log\ K_f + 1/n\ log\ C_e$) of the Freundlich adsorption isotherm (Figure 4). The values of Freundlich constants K_f and $1/n$ for arsenate RHBA were found to be 3.67 and 2, respectively.

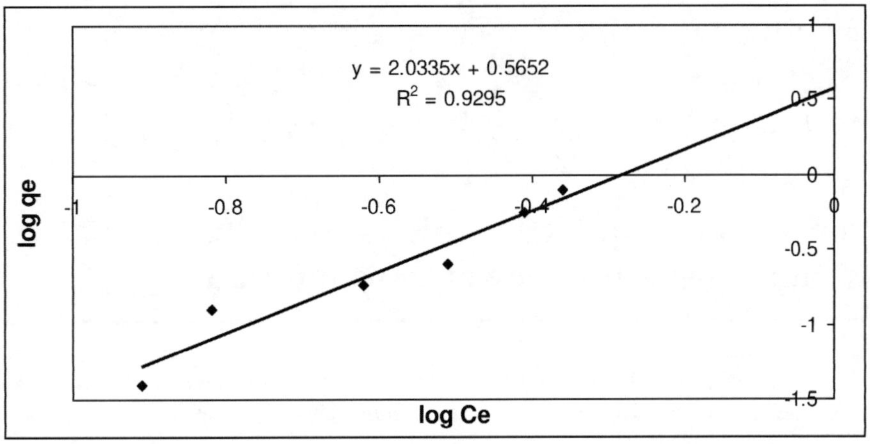

Figure 4. Freundlich isotherm for arsenate adsorption

4. Conclusions

Rice husk-based adsorbent was found effective in the adsorption of arsenate (As(V)) from aqueous solution and maximum adsorption occurred at pH7. Equilibrium was attained in 60 min and equilibrium adsorption data were well described by the Freundlich isotherm model. Freundlich constants K_f and *1/n* were 3.67 and 2, respectively. Being a low-cost agricultural by product, the rice husk-based adsorbent can be implemented as an effective adsorbent for arsenate removal from groundwater at wells or in households.

Acknowledgment

The authors are thankful to the management and authorities of the Universiti Teknologi PETRONAS for providing facilities for this research. Research support through grant FRGS/2/2010/TK/UTP/01/6 is gratefully acknowledged.

References

[1] Pandey P K, Choubey S, Verma Y, Pandey M and Chandrashekhar K 2009 *Bioresource Technol.* **100** 634-637
[2] Cullen W R and Reimer K J 1989 *Chem. Rev.* **89** 713–764
[3] Robertson F N 1989 *Environ. Geochem. Health,* **11** 171–176
[4] Hering J G and Elimelech M 1995 International perspective on arsenic in groundwater: problems and treatment strategies. Proc. AWWA Annual Conference, Anaheim, Ca.
[5] Waugh J L T 1982 *Encyclopedia of Science and Technology* McGraw-Hill, New York
[6] WHO 2011 *Arsenic in Drinking Water,* WHO Fact Sheet No. 210. World Health organization, Geneva, Switzerland
[7] Nordstorm D K 2002 *Science* **296** 2143-2145
[8] Kinniburgh D G and Smedley P L 2001 *Arsenic Contamination of Groundwater in Bangladesh* British Geological Survey Technical Report WC/00/19. British Geological Survey, Keyworth, United Kingdom
[9] Berg M, Tran H C, Nguyen T C, Pham HV, Schertenleib R and Giger W 2011 *Environ. Sci. Technol.* **35** 2621-2626
[10] Niu C H, Volesky B and Cleiman D 2007 *Water Res.* **41** 2473–2478
[11] De M 2005 *Curr. Sci.* **88** 683–684
[12] Hossain M F 2006 *Agric. Ecosyst. Environ.* **113** 1–16
[13] Ranjan D, Talat M and Hasan S H 2009 *J. Hazard. Mater.* **166** 1050-1059
[14] Bhattacharya I N, Gochhayat P K, Paul S and Mitra P K 2005 *EJMP&EP* **5** 48-62

IOP Conf. Series: Earth and Environmental Science **16** (2013) 012096 doi:10.1088/1755-1315/16/1/012096

Detection of VLF and LF emissions of fluorescent light for efficient management of power consumption

Sharifah Suhaila bt. Syed Othman, Goh Chin Hock, Farah Hani bt. Nordin, Nur Badariah bt. Ahmad Mustafa, Nagaletchumi Balasubramaniam and Z.A.M. Sharrif

Centre of Signal Processing and Control Systems (SPaCS), College of Engineering, Universiti Tenaga Nasional, Campus Putrajaya, Jalan IKRAM-UNITEN, 43000 Kajang, Selangor, Malaysia

Email: sheermagle@yahoo.co.uk, chinhock@uniten.edu.my, farah@uniten.edu.my

Abstract. In this research work, a detection probe of Very Low Frequency and Low Frequency (LF) emissions of fluorescent light is developed by using low cost loop antenna. The developed loop antenna is able to operate at VLF and LF bandwidth. The developed antenna is tested and measured with signal generator and oscilloscope in order to verify the usefulness of antenna. The developed antenna is subsequently used to detect the signal emitted by the fluorescent light. The antenna probe is located at different distance in order to obtain the peak voltage of received signal. Besides that, the fluorescent light is switch on and off respectively in order to verify the source of signal. From the oscilloscope, the received signal is operating at approximately 28 KHz. Hence, the developed antenna probe can be used for efficient management of power consumption as 28 KHz signal is detected if the light is on.

1. Introduction

The fluorescent light can be found in office, household, factory and commercial building. It has higher efficiency than the tungsten filament bulbs. The fluorescent tube contains an electrode with a gas fill of argon or krypton mixture and a small amount of mercury. A starter is used to discharge the light and controlled by ballast. It controls the current flowing and operates at particular frequency. When the operation is normal, electromagnetic waves emitted from fluorescent lamp occur because of arcing at the electrodes. The operation of a fluorescent lamp at higher frequency leads to a reduction in detectable visual flicker as well as increased ease of dimming in order to increase the efficiency. Hence, electronic switching ballast is used for higher frequency operation in order to meet the requirements. There are lots of fluorescent lights are switched on everyday and energy consumption is high if they are not managed well.

The field of antenna propagation and engineering [1] offers a lot of potential which could be exploited for efficient management of power consumption. In this research work, a detection probe of fluorescent light is developed by using low cost loop antenna. Nearly all VLF and LF antennas are electrically short, shorter than one quarter of the radiated wavelength. Their low radiation resistance makes them inefficient and a very low resistance grounds and conductors is needed to avoid dissipating transmitter power. These electrically short antennas need loading coils of high inductance to bring them into resonance. The small loop antenna is also known as a magnetic loop antenna since it behaves electrically as a coil with a small radiation resistance. Since the small loop antenna as

IOP Conf. Series: Earth and Environmental Science **16** (2013) 012096 doi:10.1088/1755-1315/16/1/012096

shown in Figure 2 is essentially a coil, its electrical impedance is inductive, with an inductive reactance much greater than its radiation resistance. A ground connection is an alternative to reduce the response to electric fields of the loop antenna. The grounded loop antenna is operating well on the high frequency bands. An ungrounded loop antenna will substantially attenuate high frequency signal by its equivalent circuit of resistance, distributed shunt capacitance and series inductance [2].

2. Research Methodology

The general steps taken to accomplish this research work are as shown in Figure 1.

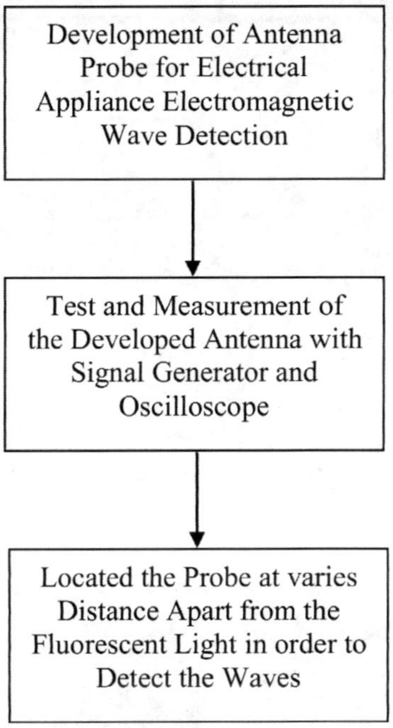

Figure 1. Flowchart of research workflow for electromagnetic wave detection of Fluorescent Light

To design the ultra wideband antenna, it is essential to determine the operating frequency for the antenna structure. In this case, the lowest frequency of low frequency band is chosen. In order to calculate the required wire length the simple wavelength frequency is used as starting point.

$$\lambda = \frac{c}{10f} \tag{1}$$

where f = frequency
c = speed of light
λ = wavelength

3. Test and Measurement

In order to test the developed antenna, function generator and oscilloscope are used as shown in Figure 2. The developed ungrounded loop antenna is tested from 30 KHz to 300 KHz in order to verify its usability in low frequency band.

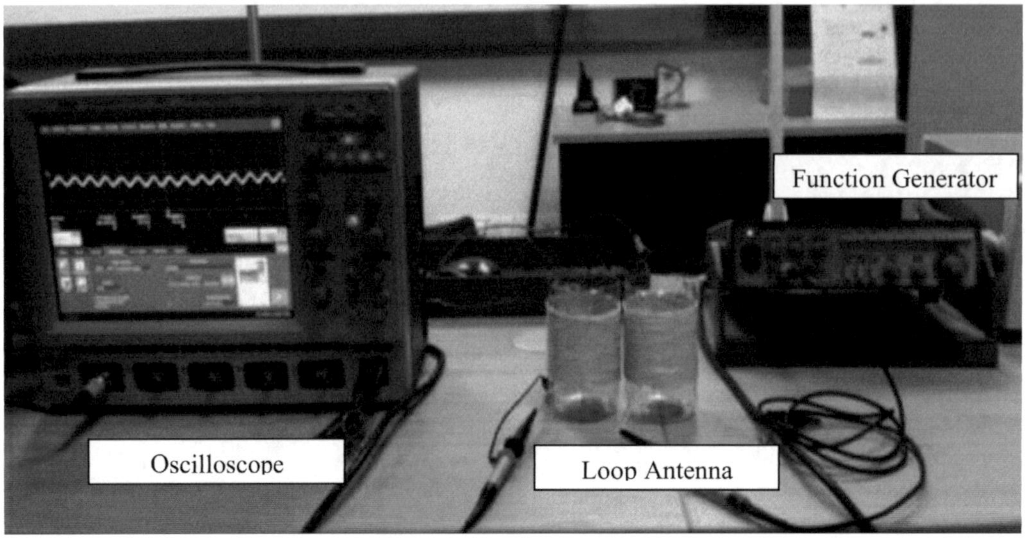

Figure 2. Testing the loop antenna with function generator and oscilloscope

Figure 3 shows the testing and measurement of the fluorescent light by varying it at fixed interval distance of 0 cm, 5 cm, 10 cm, 15 cm, 20 cm and 25 cm vertically (as shown in arrow). Besides that, a table lamp also tested with the developed antenna. The distance of the antenna is varied at fixed intervals of 0 cm, 5 cm, 10 cm, 15 cm, 20 cm and 25 cm horizontally (as shown in arrow) as shown in Figure 3. As shown in Table 1, all the measured results are recorded and documented with spreadsheet.

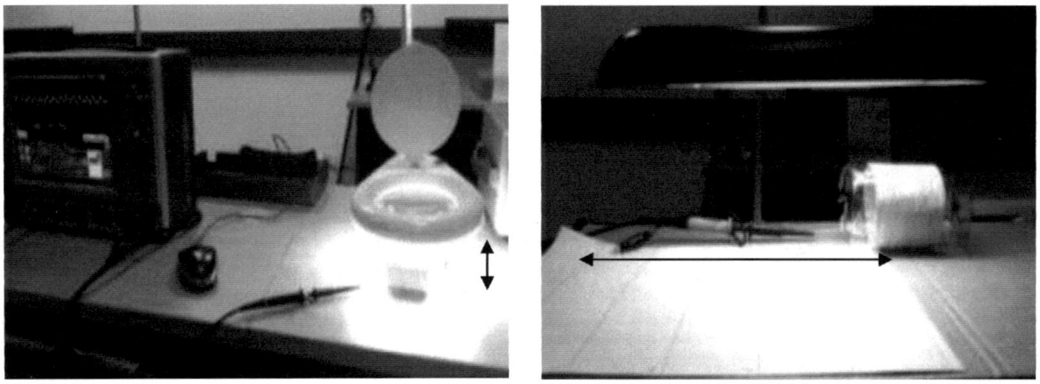

Figure 3. Testing the task lamp with the developed ungrounded loop antenna

Table 1. Results of the task lamp and table lamp with the developed ungrounded loop antenna

		Task Lamp		**Table Lamp**	
Distance	Condition	Frequency (kHz)	Peak to peak voltage, Vpp (mV)	Frequency (kHz)	Peak to peak voltage, Vpp (mV)
0	off	-	-	-	-
5	off	-	-	-	-
10	off	-	-	-	-
15	off	-	-	-	-
20	off	-	-	-	-
25	off	-	-	-	-
0	on	28.70	766	52.62	725
5	on	28.94	756	52.76	719
10	on	28.36	372	52.57	688
15	on	28.35	214	52.86	631
20	on	28.45	123	52.51	619
25	on	28.83	94	52.07	306

4. Conclusion
An electromagnetic detection probe for fluorescent light is developed by using low cost ungrounded loop antenna is successfully developed. From the obtained measured results, the antenna can be used to characterize and detect [3-5] the electrical appliances which are switched on. Hence, the developed antenna can be further improved and integrated with embedded software for management of power consumption efficiently.

Acknowledgement
The authors would like to thank the Ministry of Higher Education Malaysia for the financial support given under the Exploratory Research Grant Scheme (ERGS) in conducting this research.

References

[1] C. A. Balanis, Antenna Theory: A Review 1992, *Proceedings of the IEEE* **80** 7-23
[2] D.M Pozar, Microwave Engineering 1998 2nd Edition, John Wiley & Sons, New York
[3] P. Leelaprute, M. Nakamura, T. Tsuchiya, K. Matsumoto, and T. Kikuno. Describing and verifying integrated services of home network systems 2005 *The 10th Asia-Pacific Software Engineering Conference (APSEC2005)* 549-558
[4] Y. Iwafune, Y. Yagita, K. Ogimoto, Estimation of Appliance Electricity Consumption by Monitoring Currents on Residential Distribution Boards 2010 *Proceedings of 2010 International Conference on Power System Technology (POWERCON)*
[5] R.G. Olsen and C.E. Lyon, Modeling of Extremely Low frequency Magnetic field Sources Using Multipole Techniques 1996 *IEEE Transactions on Power Delivery,* **11** 1563 - 1570

IOP Conf. Series: Earth and Environmental Science **16** (2013) 012097 doi:10.1088/1755-1315/16/1/012097

Electrical field of electrical appliances versus distance: A preliminary analysis

Nur Badariah Ahmad Mustafa, Farah Hani Nordin, Fakaruddin Ali Ahmad Ismail, Ammar Ahmed Alkahtani, Nagaletchumi Balasubramaniam, Goh Chin Hock and Z.A.M. Shariff

Centre of Signal Processing and Control Systems (SPaCS), College of Engineering, Universiti Tenaga Nasional, Campus Putrajaya, Jalan IKRAM-UNITEN, 43000 Kajang, Selangor, Malaysia

Email: NBadariah@uniten.edu.my, Farah@uniten.edu.my

Abstract. Every household electrical appliance that is plugged in emits electric field even if it is not operating. The source where the appliance is plugged into and the components of household electrical appliance contribute to electric field emission. The electric field may cause unknown disturbance to the environment or also affect the human health and the effect might depends on the strength of the electric field emitted by the appliance. This paper will investigate the strength of the electric field emitted by four different electrical appliances using spectrum analyser. The strength will be captured at three different distances; (i) 1m (ii) 2m and (iii) 3m and analysis of the strength of the electrical field is done based on the three different distances. The measurement results show that the strength of the electric field is strongest when it is captured at 1m and the weakest at 3m from the electrical appliance. The results proved that the farther an object is located from the electrical appliance; the less effect the magnetic field has.

1. Introduction

AC power produces electric field and a magnetic field. The difference between electric field and magnetic field is that an electric field is produced by voltage while magnetic field results from the flow of current through wires or electrical devices. These two fields together are referred to as electric and magnetic fields, or EMFs. Electric field is present everywhere in the environment but is invisible to the human eye. This is because all charged objects create an electric field that travels into the space that surrounds it. Electric field increase in strength as voltage is raised and it can be detected even if the equipment is switched off but not unplugged from the main power socket [1].

Studies have shown that magnetic field has a higher potential to affect the human's health [2-5] while electric field is said to be harmless because it is easily shielded and can be avoided by earthing and screening [4]. This is only true if the electric appliances are made to comply with the national guidelines but if the appliance is not complying with the national standard as reported in [1] where the fields emitted by the junction box on the workstation and the desk-light exceeded those suggested guidance levels for computers, then further research on electric field is found necessary.

The strength of the electric field is dependent upon how charged the object creating the field is and upon the distance of separation from the charged objects. Since the distance between electrical device and object is one of the factors that contribute to the strength of electric field emission, it is the main objective of this paper to investigate the strength of electric field from four different household electrical appliances such as stand fan, liquid-crystal display (LCD) television, desk lamp and microwave oven.

2. Data Collection

This paper involves data collection of electric field emitted from electrical appliances using spectrum analyzer, Spectran 1010E, which is capable to detect the strength of the electric field within 0 to 30,000 kHz. The appliance is placed at one fixed point while the spectrum analyzer is placed 1m from the appliance where the strength of the electric field is recorded. Once the strength is recorded, the spectrum analyzer is moved 2m before the strength of the electric field is recorded. Then, the spectrum analyzer is placed 3m from the appliance and again the strength of the electric field emitted by the appliance is recorded. Figure 1 illustrates the data collections methodology.

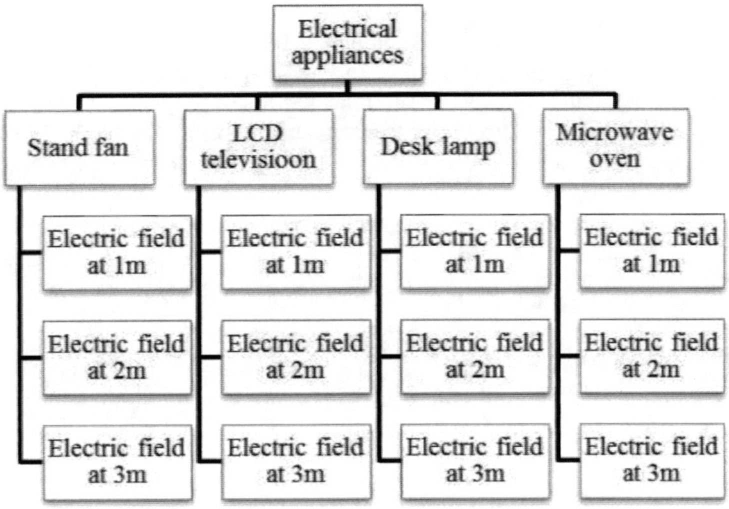

Figure 1. The flow of data collection method.

Even though the data is not collected in a controlled environment, precautions are taken to minimize the possibility of interference of electric field from the environment where all other electrical devices in the laboratory such as air conditioner and fan are turned off. This is to avoid the electrical field detected by the spectrum analyzer is mixed with other unwanted radiation from other appliances. Moreover, the Spectran 1010E device is guarded against any human or biological physical contact when in use. Instead, the commands are made through the LCS Software operated from a laptop which is placed as far as possible from the spectrum analyzer.

Since the electrical appliance delivers non-identical electric field when in off or on mode, the data is collected in three phases: (i) OFF state in 2 minutes (ii) ON state is 5 minutes and (iii) OFF state is 3 minutes. This is to make sure that the data collected is consistent for all appliances.

3. Results and Discussion

The electric field emission is measured for the frequency range between 0kHz to 1200kHz.The electric field emitted by four appliances are measured by varying the distance (1 meter, 2 meter and 3 meter) between appliance and spectrum analyzer. Figure 2 shows that the strength of the electric field emits from the LCD television decreases as the spectrum analyzer is placed farther.

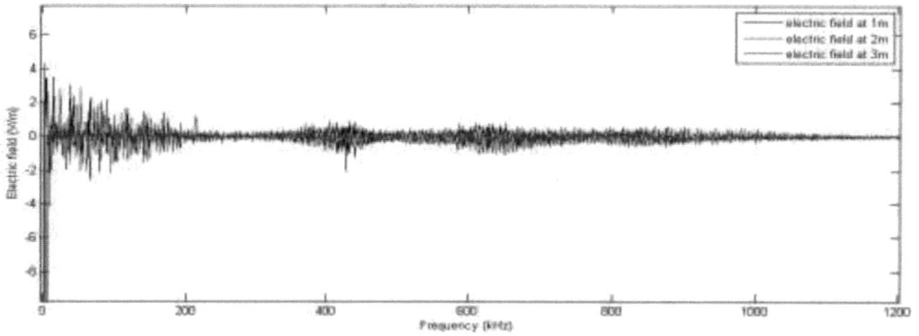

Figure 2. Electric field at 1m, 2m at 3m emitted from a LCD television

Figure 3 shows the strength of the electric field emitted by a stand fan. It shows that the strength of the electric field varies approximately between -4 to 2 V/m when the spectrum analyzer is placed 1m from the stand fan. The strength of the electric field is significantly lower when the spectrum analyzer is placed 2m from the stand fan. At 3m the strength is approximately between 0.4423 to -0.489 V/m. From the observations, LCD television emits weaker electric field compared to the stand fan for the range between 300 to 1200 kHz.

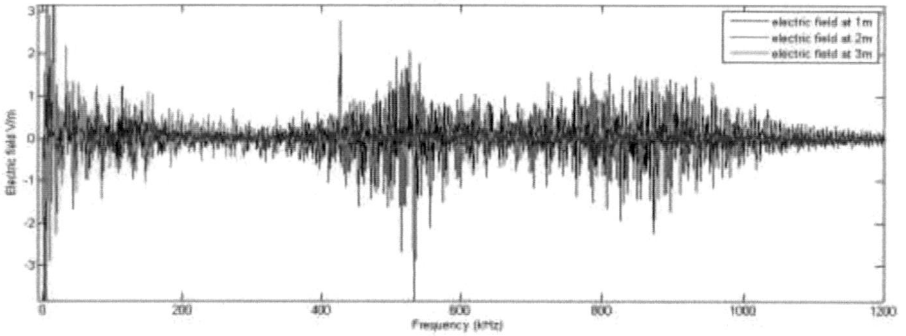

Figure 3. Electric field at 1m, 2m at 3m emitted from a stand fan

The electric filed emitted by the desk lamp is weak as shown in Figure 4, which is approximately between 0.6744 to -0.9288 V/m in frequency range of 500 to 1100 kHz. As the spectrum analyzer is placed farther from the desk lamp, the strength of the electric field still show reduction but not significant. However, at 0 kHz it can be seen that the highest strength of the electric field is 14.53 V/m at 1m, followed by 7.897 V/m at 2m and 2.826 V/m at 3m.

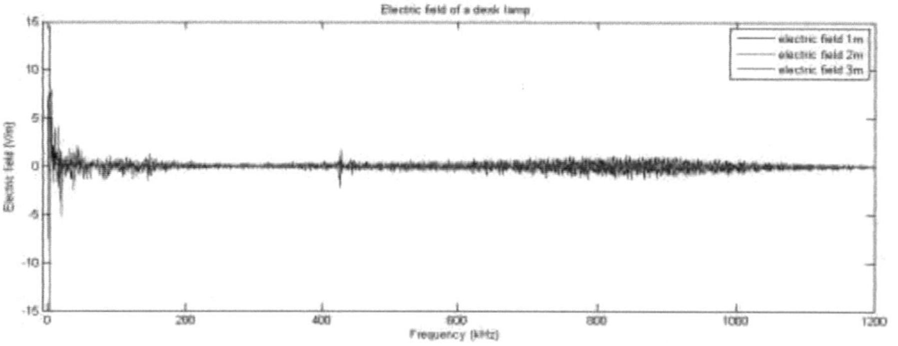

Figure 4. Electric field at 1m, 2m at 3m emitted from a desk lamp

The electric field emitted by the microwave oven is shown in Figure 5 where it does not show a significant reduction as the spectrum analyzer is put farther from the microwave oven at frequency more than 600 kHz. However, the reduction is seen approximately between frequencies 0 to 200 kHz.

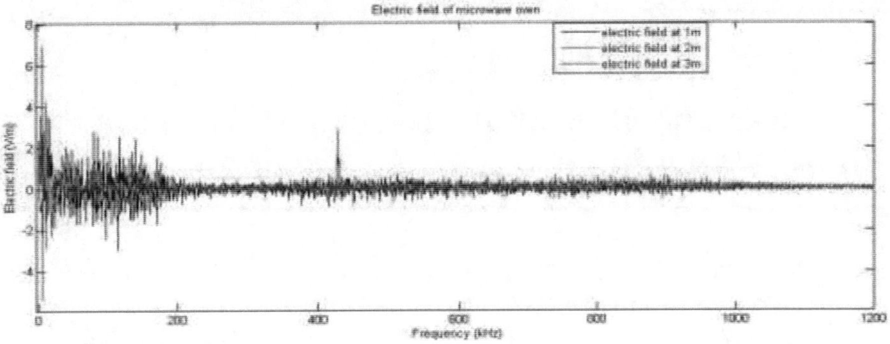

Figure 5. Electric field at 1m, 2m at 3m emitted from a microwave oven

4. Conclusion

From the results, LCD television and desk lamp emitted stable electric field where the strength of the electric field does not vary much between frequencies. However, the reduction of the electric field can still be seen between 0 to 200 kHz for LCD television and desk lamp. The electric fields emitted by the stand fan and microwave oven are more dynamic due to its motor component inside the appliance. For the stand fan, the reduction in the strength of electric field is clearly shown as the spectrum analyzer is put farther. On the other hand, the reduction of the electric field emitted by the microwave oven is clearly seen between 0 to 200 kHz. The same electric field pattern (reduction of electric field) can be seen for all four tested appliances between 800 to 1200 kHz. In conclusion, from the four appliances tested, it shows that the strength of the electric field decreases as the field travels away from the appliance.

5. Acknowledgement

The authors would like to thank the Ministry of Higher Education (MOHE) Malaysia for the financial support given under the Exploratory Research Grant Scheme (ERGS) in conducting this research.

6. References

[1] K.S. Jamiesona et al 2007 The effects of electric fields on charged molecules and particles in individual microenvironments *Atmospheric Environment* **41** 5224–5235

[2] Ammar Ahmed Alkahtani et al 2012 Analysis on RF Radiation of Electrical Appliances *IEEE Int. Conf. on Control System, Computing and Engineering* 549-553

[3] A.S. Farag a et al 1998 Electromagnetic fields in the home *Electric Power Systems Research* **45** 73–89

[4] Ali Zamanian and Cy Hardiman 2005 Electromagnetic Radiation and Human Health: A Review of Sources and Effects *High Frequency Electronics (Summit Technical Media)* 16-26

[5] National Radiation Laboratory Ministry of Health New Zealand 2011 Electric And Magnetic Fields And Your Health Information On Electric And Magnetic Fields Associated With Transmission Lines,Distribution Lines and Electrical Equipment

IOP Conf. Series: Earth and Environmental Science **16** (2013) 012098 doi:10.1088/1755-1315/16/1/012098

Effects of reclaimed asphalt pavement on indirect tensile strength test of conditioned foamed asphalt mix

Herda Yati Katman[1], Mohd Rasdan Ibrahim[2], Mohd. Yazip Matori[3], Shuhairy Norhisham[1], Norlela Ismail[1]

[1]Centre for Sustainable Technology and Environment, Civil Engineering Department, College of Engineering, Universiti Tenaga Nasional, 43000 Selangor, Malaysia

[2]Centre for Transportation Engineering, Department of Civil Engineering, Faculty of Engineering, University of Malaya, 50603 Kuala Lumpur, Malaysia

[3]Kumpulan IKRAM Sdn. Bhd, 43000 Kajang, Selangor, Malaysia

Email: Herda@uniten.edu.my

Abstract. This paper presents the results of Indirect Tensile Strength (ITS) Test for samples prepared with reclaimed asphalt pavement (RAP). Samples were conditioned in water at 25°C for 24 hours prior to testing. Results show that recycled aggregate from reclaimed asphalt pavement performs as well as virgin aggregate.

1. Introduction

Many works related to sustainable environment have been done by researchers and engineers including turning waste materials from civil engineering construction and maintenance works to something valuable. Due to greater demands for natural resources conservation and sustainability, there are many attempts to recycle aggregates obtained from milled pavement to be use in new pavement construction and rehabilitation.

G.-P. He and W.-G. Wong [1] revealed that bitumen grade and RAP content affect the indirect tensile strength of the sampel. ITS decreases with an increase of RAP content tested under dry and soaked conditions. Results also showed that bitumen grade and ageing of RAP material have significant impact on moisture susceptibility in permanent deformation. Other researchers found that incorporating RAP improves some engineering properties like tensile strength, rutting performance, fatigue and stiffness [2, 3]. However there is also reported that hot mix asphalt incorporating RAP provides inferior fatigue and thermal performance when compared with virgin mixes [4].

This paper focuses on the investigation of reclaimed asphalt pavement contents and foamed bitumen contents on tensile strength of conditioned samples. Samples were conditioned by immersing in water at 25°C for 24 hours. Minimum strength requirement for Indirect Tensile Test on conditioned samples with accordance to Malaysia specification REAM-SP 1/2005 is 0.15 MPa [5].

2. Methodology

Samples compacted 75 blows per side were prepared with accordance to ASTM D6926 [6] and tested with accordance to D6931 – 07 [7]. Gradation limits as shown in Figure 1 was used in samples preparation. Samples were prepared with different percentages of RAP and different percentages of foamed bitumen as shown in Table 1. Three replicates samples were prepared for each mixture. Samples were cured at 40°C for 72 hours. After samples cool to room temperature, samples were then immersed in water at 25°C for 24 hours prior to testing. The ITS of bituminous mixtures is conducted by loading a cylindrical specimen across its vertical diametral plane at 50mm/min deformation rate and 25°C test temperature. The peak load at failure is recorded and used to calculate the indirect tensile strength of the specimen. Statistical analysis two-way ANOVA was utilized to analyze the effect of RAP and foamed bitumen to ITS value. The tensile strength of samples was determined by Equation (1).

$$S_t = \left(\frac{2xP_{ult}}{\pi x d x t} \right) \quad (1)$$

Where:

S_t = tensile strength of specimens
P_{ult} = applied load to fail specimens
t = thickness of the specimens
d = diameter of the specimens

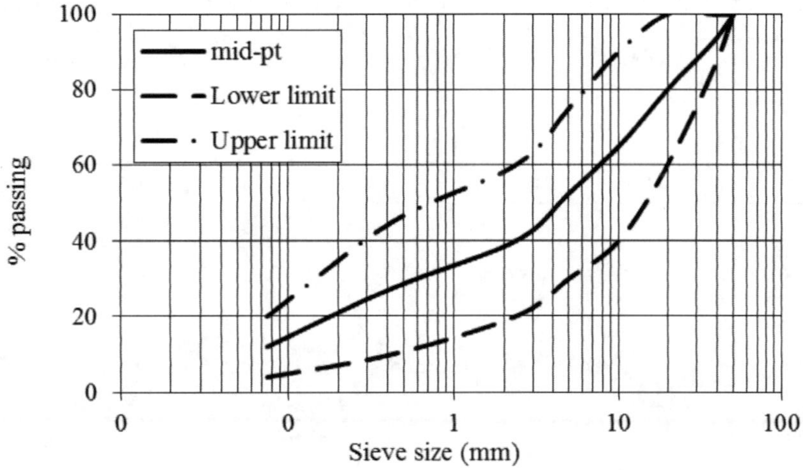

Figure 1: Aggregate gradation (REAM, 2005)

Table 1: Mix Designation according to different proportion of RAP, crusher run and foamed bitumen

Mix Designation	Mix Proportion	Foamed Bitumen Content, (%)
Mix A	0% RAP + 100% CR	1,2,3,4
Mix B	50% RAP + 50% CR	1,2,3,4

3. Results and analysis

The ITS result for conditioned samples is depicted in Figure 2. Results show that all mixes exceed the minimum strength requirement, 0.15 MPa. Similar profile was observed for both mixes. It is shown that the ITS increases with increase of the foamed bitumen content to a maximum value, further increase in foamed bitumen results in a decrease in the ITS value. Mixes prepared with 2% foamed bitumen showed the highest ITS value. This finding shows the importance of bitumen content on the ITS value. Adding excessive bitumen would affect the aggregate interlocking hence affects the performance of the samples.

Generally, ITS value for Mix A is similar with Mix B at any bitumen contents as shown in Figure 2. This indicates that ITS value not affected by incorporating RAP to the mixes. Even by conditioning the samples in water at 25°C for 24 hours, the performance of Mix B is in par with Mix A. This shows that absorption capacity of samples prepared with RAP (Mix B) is similar with samples prepared with virgin aggregates (Mix A).

Effects of foamed bitumen content and RAP contents to ITS value were further analysed using statistical two-way analysis of variance (ANOVA). Table 2 presents the results of ANOVA which shows the effects of factors (RAP and foamed bitumen content) and interaction between factors on the ITS of conditioned samples. The results show that the probability value obtained by foamed bitumen is less than the predetermined alpha value ($p = .000 < .05$) whereas probability value obtained by RAP ($p = .275 > .05$) and interaction factors ($p =.899 > .05$) are greater than the predetermined alpha value, thus only foamed bitumen content provide a significant difference on the ITS. This statistical analysis shows that incorporating reclaimed asphalt pavement insignificantly affects the ITS value. This means that mixes prepared with 50% RAP (Mix B) provide similar strength with mixes prepared with 100% virgin aggregates (Mix A).

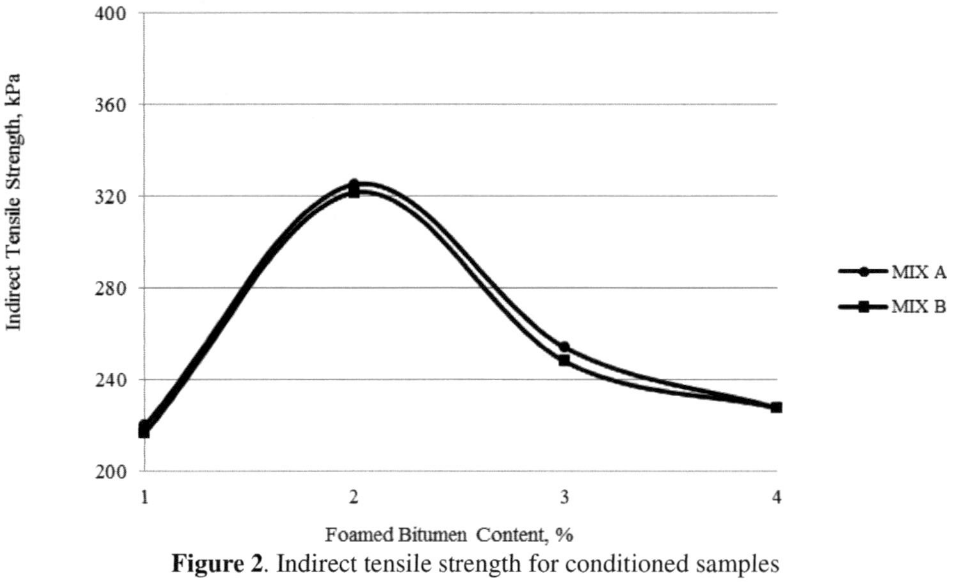

Figure 2. Indirect tensile strength for conditioned samples

Table 2: Single and interaction effects on indirect tensile strength for conditioned samples

Source	Type III Sum of Squares	df	Mean Square	F	Sig.
Corrected Model	40921.213[a]	7	5845.888	119.315	.000
Intercept	1562436.540	1	1562436.540	31889.443	.000
RAP	62.727	1	62.727	1.280	.275
FOAMED BITUMEN	40830.083	3	13610.028	277.782	.000
RAP * FOAMED BITUMEN	28.403	3	9.468	.193	.899
Error	783.927	16	48.995		
Total	1604141.680	24			
Corrected Total	41705.140	23			

a. R Squared = .981 (Adjusted R Squared = .973)

4. Conclusions

Samples prepared with different percentages of RAP contents and foamed bitumen contents conditioned in water at 25°C for 24 hours were tested for indirect tensile strength test. Results show that RAP contents provide insignificant effects whilst foamed bitumen contents have significant effects on the strength of the mixes. Based on the results, it seems that water absorption for both mixes gives similar effects to ITS. Therefore it can be conclude that recycled aggregate from reclaimed asphalt pavement performs as well as virgin aggregate.

Acknowledgement

The authors are grateful for the cooperation of Kumpulan Ikram Sdn. Bhd in conducting this study and sharing the information.

References

[1] G.-P. He and W.-G. Wong, Effects of moisture on strength and permanent deformation of foamed asphalt mix incorporating RAP materials 2008 *Construction and Building Materials,* **22** 30-40

[2] Huang, B., Zhang, Z., and Kinger, W. Fatigue crack characteristics of HMA mixtures containing RAP 2004 *5th International RILEM Conference on Cracking in Pavements*, Limoges, France

[3] Mohammad, L.N., Negulescu, I.I., Wu, Z., Daranga, C., Daly, W.H. and Abadie, C, Investigation of the use of recycled polymer modified asphalt binder in asphalt concrete pavements 2003 *Journal of the Association of Asphalt Paving Technologists* **72** 551-594

[4] McDaniel,R., and Shah, A. Use of reclaimed asphalt pavement (rap) under superpave specifications, *Journal of the Association*, 2003.

[5] REAM-SP 1 Specification for cold in-place recycling 2005. *Road Engineering Association of Malaysia*

[6] ASTM D6926: 10 Standard Practice for Preparation of Bituminous Specimens Using Marshall Apparatus.

[7] D6931 – 07, Standard test method for indirect tensile (IDT) strength of bituminous mixtures. American Society for Testing and Materials, 2007.

IOP Conf. Series: Earth and Environmental Science **16** (2013) 012099 doi:10.1088/1755-1315/16/1/012099

Efficiency improvement of the maximum power point tracking for PV systems using support vector machine technique

Ameer A. Kareim[1, 2], Muhamad Bin Mansor[1]

[1]Department of Electrical Power Engineering, College of Engineering, University Tenaga Nasional, 43000 Kajang, Selangor D.E., Malaysia.
[2]Department of Engineering Affairs, Presidency of Kufa University, Iraq
E-mail: safawi.ameer@yahoo.com, muhamadm@uniten.edu.my

Abstract. The aim of this paper is to improve efficiency of maximum power point tracking (MPPT) for PV systems. The Support Vector Machine (SVM) was proposed to achieve the MPPT controller. The theoretical, the perturbation and observation (P&O), and incremental conductance (IC) algorithms were used to compare with proposed SVM algorithm. MATLAB models for PV module, theoretical, SVM, P&O, and IC algorithms are implemented. The improved MPPT uses the SVM method to predict the optimum voltage of the PV system in order to extract the maximum power point (MPP). The SVM technique used two inputs which are solar radiation and ambient temperature of the modeled PV module. The results show that the proposed SVM technique has less Root Mean Square Error (RMSE) and higher efficiency than P&O and IC methods.

1. Introduction

As a fact, the operation voltage (V_{pv}) and operation current (I_{pv}) of the PV module are very critical in order to obtain the maximum efficiency of the PV modules. In other words, each photovoltaic cell has an optimum operation point, called maximum power point (MPP).

Many methods proposed in the literature to track the MPP for a PV system. In [1] a modified P&O based MPPT is presented to track the maximum power point of PV systems. The proposed algorithm is bifurcated; maximum power computation and direct power control of the power drawn from the PV module. The optimum power is calculated online, with the enhanced perturb and observe algorithm. The optimum power computed, is compared with instant actual PV power. In [2] an incremental conductance (IC) algorithm has been proposed to solve the problem of the P&O algorithm, under swiftly changing environmental conditions

The main goal of this paper is to improve the performance of the traditional controller used in MPPT such as P&O and IC algorithms by introducing Support Vector Machine (SVM) algorithm in order to increase the tracking response and then increase the tracking efficiency of the PV system.

2. Proposed PV System

A support vector machine (SVM) based controller is used to predict the instantaneous optimum voltage V (optimum) of the PV system in order to ensure the maximum power operation.

2.1 PV model

The PSAR CSUN 220-60P PV module is proposed for a MATLAB simulation model [3]. The data of this research (solar radiation and ambient temperature) was provided by Centre for Electric Vehicles, UNITEN.

2.2 Proposed Maximum Power Point Tracker (MPPT)

A boost converter is a type of switching-mode power supply, which contains a minimum of two semiconductor switches [4]. The output voltage is intended to be the PV output voltage (V_{pv}), whereas, the optimum voltage ($V_{optimum}$) of the PV module is:

$$\frac{V_{pv}}{V_{pv\ (optimum)}} = \frac{1}{1-D} \quad (1)$$

A new MPPT controller system is proposed in this research using SVMs as a regression tool. The SVM regression is simulated by using LIBSVM MATLAB toolbox [5].

3. Evaluation of the PV System

3.1 Evaluation of the PV module

Figure 1 and Figure 2 show the I-V and P-V characteristic of the proposed PV module.

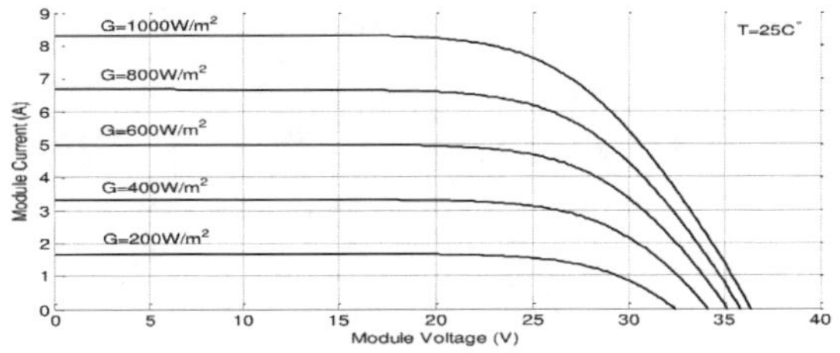

Figure 1. I-V Characteristic curves of the simulated PV module

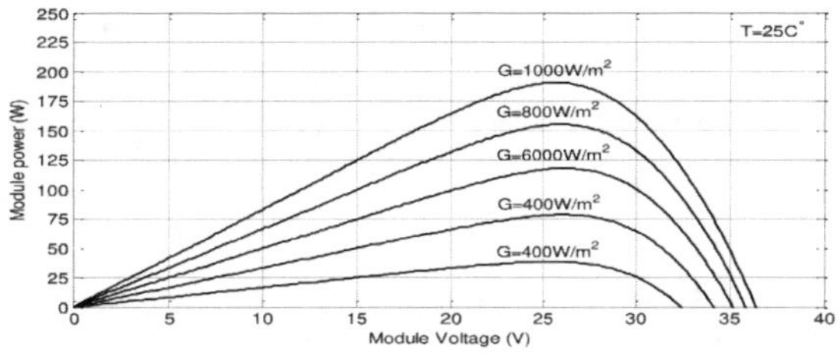

Figure 2. I-V Characteristic curves of the simulated PV module

The results of Figures 2 and 3 are produced comparison as shown in Table 1 below.

Table 1. Comparison of electrical performance between PV simulation results & datasheet

Specifications	**Datasheet**	**Simulation**
Maximum Power (P_{max})	220 W	190.34 W
Voltage at P_{max} (V_{mpp})	29 V	25.73 V
Current at P_{max} (I_{mpp})	7.59 A	7.398 A
Open Circuit Voltage (V_{oc})	36.4 V	36.38 V
Short Circuit Current (I_{sc})	8.32 A	8.32 A
Maximum Power (P_{max})	220 W	190.34 W

3.2 Evaluation of the traditional and proposed MPPT algorithms

To evaluate the traditional and proposed MPPT algorithms, a statistical error and efficiency are used: This error is Root Mean Square Error (RMSE).The RMSE is given by:

$$\eta = \frac{Total\ Energy\ (Simulated)}{TotalTheoretical\ Energy} \times 100\ \% \quad (2)$$

Figures 3, 4, 5, 6, and 7, show the simulated voltage and power of traditional and SVM algorithms.

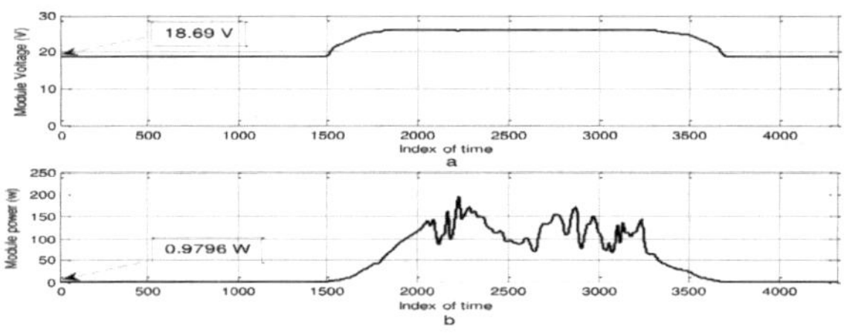

Figure 3. Theoretical voltage and power respectively of MPP

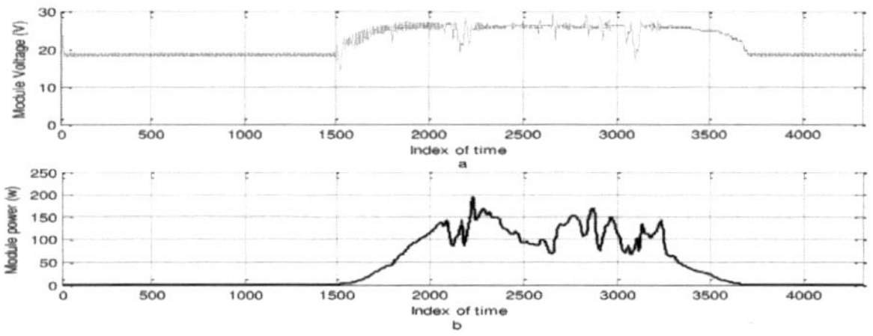

Figure 4. Practical voltage and power from P&O

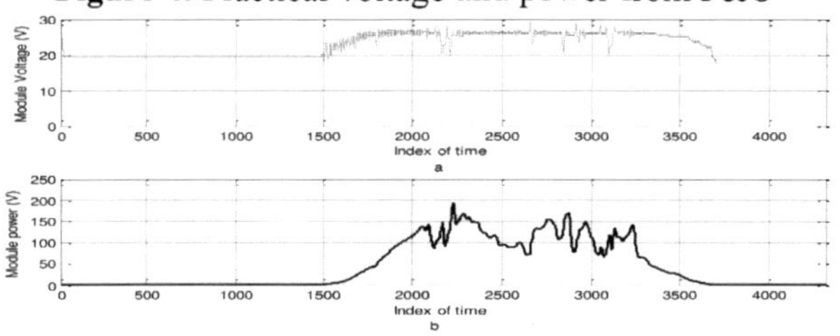

Figure 5. Practical voltage and power obtained from IC

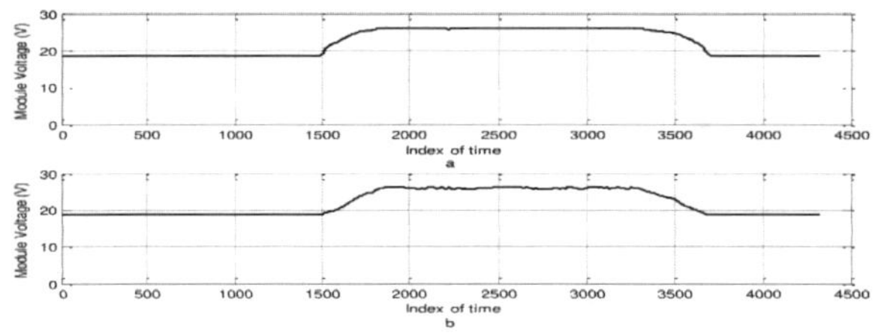

Figure 6. (a) Theoretical voltage of MPP and (b) Predicted voltage by SVM

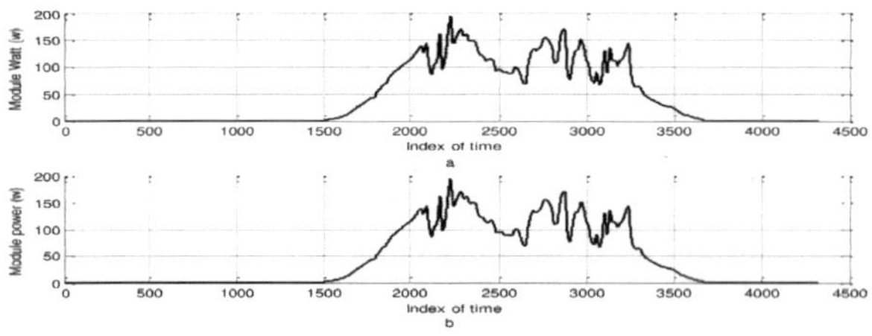

Figure 7. (a) Theoretical power of MPP and (b) Predicted power by SVM

The RMSE values of voltages and power values for traditional and proposed algorithms are tabulated as in Table 2.

Table 2. Comparison of RMSE values of traditional and proposed algorithms

Type of Error \ Type of Algorithm	P & O Algorithm	IC Algorithm	SVM Algorithm
RMSE of voltage	1.105256	1.083662	**0.64293**
RMSE of power	2.603093	1.853943	**0.14425**

The MPPT tracking efficiency for any maximum power point tracker is defined as the ratio of the total simulated energy to the total theoretical energy. It is measured by [Total Energy (Simulation)] ÷ [Total Theoretical Energy] × 100%.

$$\eta\,(P\&O) = \frac{51.8123}{52.4202} \times 100\,\% = 98.8404\% \quad (4)$$

$$\eta\,(IC) \;\; = \frac{51.9561}{52.4202} \times 100\,\% = 99.1146\% \quad (5)$$

$$\eta\,(SVM) = \frac{52.3888}{52.4202} \times 100\,\% \;=\; 99.9401\% \quad (6)$$

4. Conclusions

The PV module and the traditional algorithms by SVM using MATLAB models had been developed. The enhanced MPPT algorithm had extracted the maximum power point from the PV module, established from solar irradiation and surrounding temperature. The outcomes of the simulation had revealed that, the response of the proposed SVM algorithm was better than the traditional algorithms. Furthermore, the typical tracking effectiveness of the developed algorithm was greater than the traditional algorithms. Future work recommended using other MPPT algorithms in order to get high efficiency from PV system.

References

[1] Mohamed Azab 2008 A New Maximum Power Point Tracking for Photovoltaic Systems *Journal of World Academy of Science Engineering and Technology (WASET)* **44** 571-574

[2] K. H. Hussein, I. Muta, T. Hoshino and M. Osakada 1995 Maximum Photovoltaic Power Tracking: An Algorithm for Rapidly Changing Atmospheric Conditions *IEEE Proceedings, Generation, Transmission and Distribution* **142** 59-64

[3] CSUN Energy For Future, CSUN. *Polycrystalline Solar Module.* [online] Available from: http://www.chinasunergy.com. [Accessed 22 November 2012]

[4] Tamer T. N. Khatib, A. Mohamed, N. Amin and K. Sopian 2010 An Efficient Maximum Power Point Tracking Controller for Photovoltaic System Using New Boost Converter Design and Improved Control Algorithm *WSEAS Transactions on, power systems* **5** 53-63

[5] Chih-Chung Chang and Chih-Jen Lin. *LIBSVM: a library for support vector machines.* *"ACM Transactions on Intelligent Systems and Technology.* [online] Available from: http://www.csie.ntu.edu.tw/~cjlin/libsvm. [Accessed 26 December 2012]

IOP Conf. Series: Earth and Environmental Science 16 (2013) 012100 doi:10.1088/1755-1315/16/1/012100

A compact ultra wideband antenna with WiMax band rejection for energy scavenging

Y E Jalil[1], B Kasi[2] and C K Chakrabarty[1]

[1]Department of Electronics and Communication Engineering, Universiti Tenaga Nasional, Kajang, Selangor 43000, Malaysia

[2]Department of Electrical and Electronics Engineering, Universiti Infrastruktur Kuala Lumpur, Kajang, Selangor 43000, Malaysia

E-mail: yantierana@uniten.edu.my

Abstract. Radio Frequency (RF) energy harvesting has been rapidly advancing as a promising alternative to existing energy scavenging system. A well designed broadband antenna such as ultra-wideband (UWB) antenna can be used as one of the major components in an RF energy scavenging system. This paper presents a compact UWB antenna showing good impedance matching over a bandwidth of 2.8 to 11 GHz, suiTable for broadband RF energy scavenging. Nevertheless, the antenna usage in wireless communication has a limitation due to the problem of interference between UWB system and other narrowband systems. Thus, the proposed antenna is successfully designed with a single band-notched at the targeted WiMAX operating band of 3.3 to 3.6 GHz.

1. Introduction

Radio Frequency (RF) energy harvesting aims at converting ambient RF energy extracted from propagating radio waves into storable electrical energy to power electronics [1]. A typical RF energy scavenging system consists of an antenna with impedance matching circuit, rectifier, voltage booster and a charging circuit. Therefore, broadband antennas such as UWB antennas play an important role in RF energy harvesting systems, since they are able to collect energy from various sources operating in their matching band. However, their usage in UWB wireless communication system has a limitation due to the consequences of having to share the spectrum with a number of other established narrowband applications such as WiMAX (3.3 – 3.6 GHz) and WLAN IEEE 802.11a (5.15 to 5.85 GHz). The interference from a strong narrowband signal, could affect the overall systems. To overcome this unwanted problem, it is desirable to design UWB antennas integrated with band rejection characteristics in the affected frequency bands. In recent times, numerous researchers have proposed diverse antenna geometries, design methods and structures in order to achieve band-notched features [2-4].

In this paper, an octagon-shaped microstrip antenna with ultra-wideband impedance matching from 2.8 to 11 GHz is proposed. A frequency band-notched from 3.3 to 3.6 GHz can be achieved by simply employing a slanted inverted U-shaped slot in the radiating patch. The outline of this paper is as follows. The geometry of the proposed UWB antenna is presented in section 2 while the characteristics of the antenna are presented in section 3. The conclusions are summarized in Section 4.

IOP Conf. Series: Earth and Environmental Science **16** (2013) 012100 doi:10.1088/1755-1315/16/1/012100

2. Antenna Geometry

The proposed UWB antenna configuration and its geometrical parameters are shown in Figure 1. This proposed antenna design consists of an octagon-shaped radiator fed by a microstrip line printed on a partial grounded substrate. The microstrip line feed is designed to match a 50 Ω characteristic impedance. With the aim to increase the impedance bandwidth, both the radiating plane and the ground plane have tapered shapes at both ends. Apart from that, a rectangular slit is added on top of the ground plane to further enhance the impedance bandwidth of the antenna. The proposed antenna is designed on a standard Taconic TLC-30 substrate with a dielectric substance of 3, a loss tangent of 0.003 and with thickness of 1.575 mm. The final optimized parameter values are tabulated in Table 1.

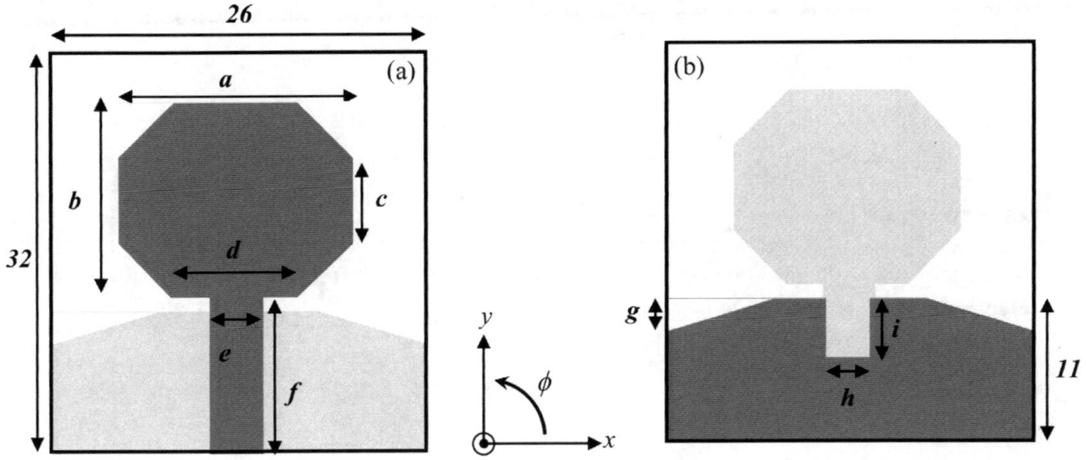

Figure 1. Geometry of the proposed antenna (units in mm). (a) Front view (b) Rear View

Table 1. Antenna Parameters

Parameter	Value (mm)	Parameter	Value (mm)
a	20.0	*f*	12.0
b	18.0	*g*	2.0
c	8.0	*h*	3.65
d	12.0	*i*	3.40
e	3.9		

The geometry and dimensions of the UWB antenna with band-notched design (will be referred to as antenna 2) is illustrated in Figure 2. A slanted inverted-U shaped slot is etched onto the radiating patch in order to create a frequency band-notch at the targeted WiMAX band. The previously optimized design parameters of the UWB antenna need no additional retuning work when the band-notched design is applied. In general, the design concept of the notch function is to adjust the total length of the U-shaped slot to be approximately half-wavelength at the desired unwanted frequency [5]. The final design parameters of the slot are tabulated in Table 1 as well.

Figure 2. Geometry of the proposed antenna with band notched design (antenna 2)

3. Results and Discussion

The simulations of the proposed antenna performance are performed using CST Microwave Studio. Figure 3(a) shows the calculated return loss curves of the proposed antenna with and without the notched band. It can be seen that the calculated return loss curve is less than -10 dB from 2.8 GHz to beyond 11 GHz for antenna 1, giving a clear indication that the impedance bandwidth is more than 8.2 GHz covering the entire UWB frequency range. Furthermore, the required frequency notch at WiMAX band from 3.3 to 3.6 GHz is successfully achieved as displayed in the return loss curve for antenna 2. A notch band with a peak of –3.3 dB is clearly demonstrated by the return loss curve for antenna 2.

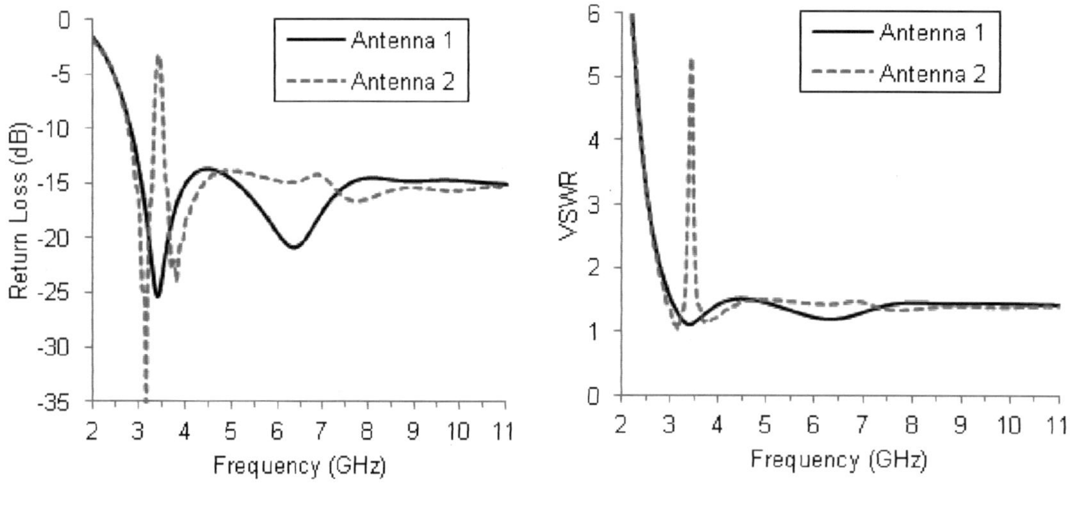

Figure 3. Simulated antennas performance (a) Return loss (b) VSWR

The comparison of simulated VSWR for the proposed antenna designs with and without slot is illustrated in Figure 3(b). From the Figure, it is evident that the addition of the slot onto the radiating patch is indeed initiates the desired filtering property. Compared to antenna 1 design, the targeted frequency band has been successfully blocks out by the antenna 2 with VSWR > 2 and still maintains good impedance matching at other frequencies in the UWB frequency band

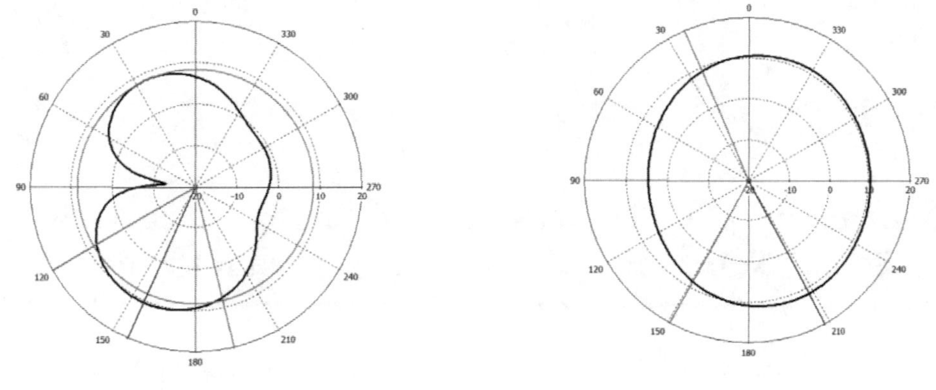

(a) (b)

Figure 4. Simulated radiation patterns for proposed antenna 2 in (a) E-plane (b) H-plane

Figure 4 depicted the simulated far-field radiation pattern of antenna 2 in the E-plane and the H-plane at 3.4 GHz. It can be observed that, the patterns obtained are typical of those for monopole antennas. In the H-plane the pattern is quite omni-directional while the bidirectional pattern can be seen in the E-plane. It can be stated that the proposed band notched structure do not have any significant influence on the radiation patterns of the UWB antenna.

4. Conclusions

This paper proposed a compact UWB microstrip antenna design. The antenna has a total size equal to 26 x 32 mm^2 and has the capability of achieving an input impedance bandwidth from 2.8 to 11 GHz. With these features, the antenna is a promising candidate for compact and miniaturized RF energy scavenging system. Additionally, a slanted U-shaped slot is introduced on the radiating patch to create a notched band from 3.3 to 3.6 GHz to minimize the potential interferences from WiMAX system. The modified antenna is suiTable for numerous UWB applications.

References

[1] Ding Y, Arslan T and Hamilton A 2012 Broadband Antenna for RF Energy Scavenging System *Antennas and Propagation Conference (LAPC)(Loughborough, UK, 12 - 13 Nov 2012*) 1-4

[2] William, J. and Nakkeeran, R. 2009 CPW-Fed UWB Slot Antenna With Band Notched Design *Microwave Conference, 2009. APMC 2009. Asia Pacific (Singapore, 7-10 Dec. 2009)* 1833-1836

[3] Y. Y. Liu and W.X. 2011 Dual Band-Notched Antenna With The Parasitic Strip for UWB *Progress in Electromagnetics Research Letters (2011)* (vol 25) 21–30

[4] [4] Wei Hu, Jianhua Zhang and Fengge Hu 2010 Design of An Ultra-Wideband Antenna With Dual Band-Notched Characteristic *9th International Symposium on Antennas Propagation and EM Theory (ISAPE) (Guangzho, China, 29 Nov – 2 Dec 2010)* 57-59

[5] Qing-Xin Chu and Ying-Ying Yang 2008 A Compact Ultrawideband Antenna With 3.4/5.5 GHz Dual Band-Notched Characteristics *Antennas and Propagation, IEEE Transactions* **56** 3637-3644

IOP Conf. Series: Earth and Environmental Science **16** (2013) 012101 doi:10.1088/1755-1315/16/1/012101

A hybrid genetic algorithm and linear regression for prediction of NOx emission in power generation plant

Muhammad Afif Bunyamin[1], Keem Siah Yap[1], Nur Liyana Afiqah Abdul Aziz[1], Sheih Kiong Tiong[1], Shen Yuong Wong[1], Md Fauzan Kamal[2]

[1] Department of Electronics and Communication Engineering, College of Engineering, Universiti Tenaga Nasional

[2] Enviroment Unit, TNB Research Sdn. Bhd.

Email: yapkeemuniten.edu.my

Abstract. This paper presents a new approach of gas emission estimation in power generation plant using a hybrid Genetic Algorithm (GA) and Linear Regression (LR) (denoted as GA-LR). The LR is one of the approaches that model the relationship between an output dependant variable, y, with one or more explanatory variables or inputs which denoted as x. It is able to estimate unknown model parameters from inputs data. On the other hand, GA is used to search for the optimal solution until specific criteria is met causing termination. These results include providing good solutions as compared to one optimal solution for complex problems. Thus, GA is widely used as feature selection. By combining the LR and GA (GA-LR), this new technique is able to select the most important input features as well as giving more accurate prediction by minimizing the prediction errors. This new technique is able to produce more consistent of gas emission estimation, which may help in reducing population to the environment. In this paper, the study's interest is focused on nitrous oxides (NOx) prediction. The results of the experiment are encouraging.

1. Introduction

The emissions of the acidifying substances are increasing from time to time [1]. A lot of efforts have been taken to measure and to reduce these acidifying gases. Electricity productions sectors produce tons of these harmful gases due to the gas and fossil fuel combustion for the electricity generations. Carbon oxides, sulphur oxides, and nitrous oxides are the main source of the acidifying emissions produced [2]. Previously at power plant, a conventional method is used. The gases emissions are measured by using sensors. However, the costs and the inefficiency occur during the maintenance duration bring the needs of artificial intelligence solution [3].

When other methods may have faced difficulties in solving the problems as number of input features (parameter of power generation plant) maybe huge, Genetic Algorithm (GA), with its independency characteristic to the specific knowledge of the problem, is used to ease the solutions [4] [5]. The main objective of GA is to get the least number of selected features with the minimum percentage error. Once GA selected the feature, the Linear Regression (LR) will be used as prediction mode. This paper is organized as follows: Section 2 describes both algorithms GA and LR. The results of both LR and GA-LR are presents and discuss in Section 3. Section 4 presents the conclusion and suggestion of further works.

2. The hybrid GA-LR model

In the LR model, the dependant variable y_i is assumed linear to the regression vector $\boldsymbol{X}_i$ with the assumption of n data points $(y_i, \boldsymbol{X}_i)$, for. One popular LR equation is in the form of $y_i = \boldsymbol{X}_i\boldsymbol{\beta}_i$, where $\boldsymbol{X}_i$ is the input variables which in independent while the dependent variable y_i also known as response variable which the best set for the data points is produced [6]. The $\boldsymbol{\beta}_i$ stands for regression coefficient represent the inference of the function 7]. LR is one of the methods that estimate data by using linear predictor functions and unknown model parameters [8]. The solutions of LR can be found by moore-penrose pseudo inverse matrix method [9].

Genetic Algorithms (GA) have been intensively studied during the past few decades and many applications have benefited from the use of GA [10]. GA is adaptive and robust computational methods that can be used to solve search and optimization problems. The dynamics of GA based on the mechanics of natural selection and genetic processes of living organisms. Any GA starts with a population of randomly generated solutions (chromosomes) and advances toward better solutions by applying genetic operators, modeled on genetic processes occurring in nature. From one generation to another, populations evolve according to the principles of natural selection and the survival of the fittest individuals. The fittest strings are chosen to reproduce, crossover, and occasionally mutate, eventually evolve into a population of solutions that are highly adapted to the desired environment. By imitation of the natural process, GA is capable of developing solutions to the real problems. GA is being applied to a variety of problems and becoming an important tool in machine learning and function optimization. A simple GA that has given good results in a variety of engineering problems is composed of three operators, i.e., selection, crossover, mutation and fitness function. A brief description of the three operators is detailed as follows.

Selection -The selection process copies individual strings (called parent chromosomes) into a new tentative population (known as mating pool) for genetic operations. According to Darwin's evolution theory the best ones should survive and create new offspring. There are many methods how to select the best chromosomes, for example roulette wheel selection, Boltzman selection, tournament selection, rank selection, steady state selection and some others.

Crossover - The crossover operator oversees the mating process of two chromosomes. Two parent chromosomes are selected from the mating pool randomly and the crossover rate, which is a real number between zero and one, determines the probability of producing a pair new chromosome from the parents, which are called the offspring. Offspring inherits complementing genetic material from its parents. Some of the commonly used techniques are one-point crossover, two-point crossover and multipoint crossover.

Mutation - The main aim of mutation is to introduce genetic diversity into the population. Sometimes, it helps to regain the information lost in earlier generations. Like natural genetic systems, mutation in GA is made occasionally. A random position of a random string is selected and replaced by another character from the alphabet. In the case of binary representation it negates the bit value and is known as bit mutation. Normally, mutation rate is kept fixed in order to sustain diversity, (which may be lost due to crossover and very low mutation rate) into the population. Mutation is not always worth performing. High mutation rate can lead to genetic search to a random one. It may change the value of an important bit and thereby affect the fast convergence to a good solution. Moreover, it may slow down the process of convergence at the final stage of GA.

Fitness function - In order to evaluate how good the different individuals in the population are, a fitness function needs to be defined. In achieving the objectives to minimize the error of the predicted output with the targeted output, and to encourage the inclusion of *Don't Care* (DC) features that minimizing the number of input feature, GA fitness function employed is defined as $fitness = pf_1 + qf_2$, where f_1 is the mean square error of prediction, f_2 is the percentage of features that has been assigned as DC (i.e., compression rate of input feature), p and q are user select parameters.

3. Experiments and Results

The NOx dataset consists of 3405 data samples. Each of the data consists of 155 input features that take from the parameter of the plant. Meanwhile, there is one single target output which represents the concentration of the NOx. The dataset are divided into three subsets; 50% for training, 25% for prediction and 25% for test. In this experiment, the GA settings are

- Population Type = 50
- Population Size = 150
- Generation Limit = 500

First, the dataset is tested using LR method, the mean square error for validation and test sets are 0.063 and 0.144 respectively. These results are used as benchmarks for comparison with results of GA-LR.

In second stages, the GA-LR is used to learn the training set with different values of user select parameter p and q. The respective mean square errors of validation set as well as number of feature used are recorded in Table 1. The smallest error is 0.028 (when p = 1.00 and q = 0.00) with 80 features. The features are not much reduced compared to the accepTable error, 0.058 (when p = 0.80 and q = 0.15) with only 11 features. After the parameters are determined (p = 0.80 and q = 0.15), GA-LR algorithm predicts that the error of the testing data is 0.133.

In third stage, further experiments have been conducted the test the sensitivity of the performance of GA-LR with different setting (numbers of population and generation) of GA. Table 2 indicates that GA-LR is insensitive to the change of GA's parameters setting. The difference of the features selected and the errors are rather small.

Table 1. Mean square errors and number of features selected (in parenthesis) of validation set based on p and q

q \ p	**0.5**	**0.6**	**0.7**	**0.8**	**0.9**	**1.0**
0.00	0.045 (18)	0.053 (14)	0.051 (14)	0.041 (20)	0.038 (23)	**0.028 (80)**
0.05	0.053 (18)	0.057 (15)	0.052 (23)	0.049 (16)	0.049 (16)	0.034 (89)
0.10	0.060 (17)	0.054 (17)	0.055 (21)	0.052 (17)	0.048 (18)	0.032 (83)
0.15	0.059 (16)	0.059 (15)	0.058 (14)	**0.058 (11)**	0.050 (25)	0.043 (84)
0.20	0.060 (12)	0.058 (14)	0.060 (12)	0.059 (17)	0.048 (27)	0.053 (88)
0.25	0.058 (13)	0.055 (14)	0.055 (15)	0.057 (19)	0.050 (22)	0.037 (83)
0.30	0.059 (16)	0.058 (17)	0.055 (18)	0.055 (17)	0.050 (27)	0.042 (90)

Table 2. Results of GA-LR for different GA's parameters

Generation	**Population**	**Number of Selected Features**	**Mean Square Error**	
			Train	**Testing**
150	50	11	0.058	0.113
	100	8	0.060	0.111
	150	7	0.062	0.135
300	50	8	0.060	0.141
	150	8	0.061	0.132
	250	9	0.056	0.108
600	250	7	0.062	0.123
	450	7	0.062	0.123
	600	7	0.062	0.123

4. Summary

In this research, a new algorithm (i.e., GA-LR) has been proposed which perform prediction to the NOx emission in power generation plant. The GA-LR is conceived to be a better NOx emission predictor with the least features selection while maintaining the lowest possible error. The proposed GA-LR allows reliable prediction of NOx emission which can help the monitoring of the gas pollution. Early action can be taken into measure to reduce the gas pollution from the power generation plant. Further works can be focused on other type of gases and improving the error by different methods.

References

[1] Azid I. A.,Ripin Z. M., Aris M.S., Ahmad A. L., Seetharamu K. N. and Yusoff R. M. 2000 Predicting combined-cycle natural gas power plant emission by using artificial neural networks *in Proc. Of IEEE TENCON (Kuala Lumpur, Malaysia, Sep. 2000)* 512-517

[2] Tronci S., Baratti R. and Servida A. 2002 Monitoring pollutant emissions in a 4.8MW power plant through neural networks *Neurocomputing (Mar. 2002)* **43** 3-15

[3] Zheng L., Zhou H.,Wang C. and Cen K. 2008 Combining Support Vector Regression and Ant Colony Optimization to Reduce NOx Emissions in Coal-Fired Utility Boilers *ACS Publictions: Energy Fuels* **22** 1034-1040

[4] Yang J. and Honavar V. 1998 Feature Subset Selection Using A Genetic Algorithm *IEEE Intelligent Systems and Their Application* **13** 44-49

[5] Jain A. and Zonker D. 1997 Feature Selection: Evaluation, Application, and Small Sample Performance *IEEE trans. on Pattern Analysis and Machine Intelligence* **19** 153-158

[6] Neter J., Kurtner M., Nachtsheim C. J. and Wasserman W. *Applied Linear Regression Model* (McGraw Hill/Irwin Series Operations and Decision Sciences)

[7] Afifi A. A. and Elashoff R. M. 2012 Missing Observation in Multivariate Statistic II, Point Estimation in Simple Linear Regression *Journal of American Statistical Association* **62** 10-29

[8] Fang Y., Hans-George M. and Jane-Ling W. 2005 Functional Linear Regression Analysis for Longitudal Data *Project Euclids: Annals of. Statistics* (vol 33) 2873-2903

[9] Huang G. B., Wang D. H., and Lan Y. 2011 Extreme Learning Machines: A Survey, International *Journal of Machine Leaning and Cybernetics* **2** 107-122

[10] Wei H., Tong Z., and Jian S. 2004 Use of a genetic algorithm to optimize multistage erbium-doped fiber-amplifier systems with complex structures *US National Library of Medicine National Institutes of Health* **12** 531–544

4th International Conference on Energy and Environment 2013 (ICEE 2013) IOP Publishing
IOP Conf. Series: Earth and Environmental Science **16** (2013) 012102 doi:10.1088/1755-1315/16/1/012102

A hybrid fuzzy logic and extreme learning machine for improving efficiency of circulating water systems in power generation plant

Nur Liyana Afiqah Abdul Aziz, Keem Siah Yap, Muhammad Afif Bunyamin

Department of Electronics and Communication Engineering, College of Engineering, Universiti Tenaga Nasional

Email: yapkeem@uniten.edu.my

Abstract. This paper presents a new approach of the fault detection for improving efficiency of circulating water system (CWS) in a power generation plant using a hybrid Fuzzy Logic System (FLS) and Extreme Learning Machine (ELM) neural network. The FLS is a mathematical tool for calculating the uncertainties where precision and significance are applied in the real world. It is based on natural language which has the ability of "computing the word". The ELM is an extremely fast learning algorithm for neural network that can completed the training cycle in a very short time. By combining the FLS and ELM, new hybrid model, i.e., FLS-ELM is developed. The applicability of this proposed hybrid model is validated in fault detection in CWS which may help to improve overall efficiency of power generation plant, hence, consuming less natural recourses and producing less pollutions.

1. Introduction

In power generation plant, there are several factors that influenced the efficiency of the power generated; type of fuels used, plant age and type, capacity utilization as well as heat sink system [1]. The CWS or heat sink system is one of the main factors in optimizing the efficiency of the power generated [2]. Heat is a form of energy lost by the system. By cooling the turbines, there will be less heat loss resulting in increases of the efficiency rate [3]. Power can be generated efficiently if the turbine works efficiently when turbine backpressure is maintained at low level [4]. This state can only be fulfilled with an efficient heat transfer process by the water circulating system. Sufficient cooling water is channeled to the turbine condensers continuously. Then, the condensing steam is discharged from the turbine through the same system. Heat transfer will become inefficient when there are blockages in the condenser due to gibberish like sand, small shells and seaweeds that able to escape the water filtering process [5]. The condenser needs to be cleaned in order to maintain the efficiency of the heat transfer to the cooling water from the exhaust steam.

This paper is organized as follows: Section 2 describes FLS and ELM as well as the hybrid FLS-ELM. The results of ELM and FLS-ELM are presents discuss in Section 3. Lastly, section 4 presents the summary and suggestion of further works.

2. The hybrid FIS-ELM model

The ELM is proven to have higher generalization performance due to its capability of having smallest training error and smallest norm of weights [6]. Almost all piecewise-continuous functions can be used

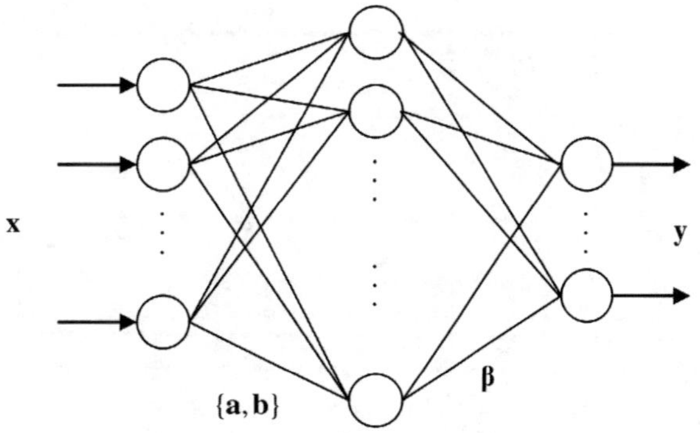

Figure 1. Architecture of an ELM

as activation functions in ELM including differential and non-differential functions as well as discontinuous functions. This feature allows ELM to be suiTable for any nonlinear activation functions in addition to the ability to use fully complex functions for its activation functions. In the existing literature, it has better performance when compared to existed traditional algorithm over regression and general classification problems [7]. In general, ELM is equal to a three layer Feedforward neural network as shown in Figure 1. Consider a set of N training samples (with a input vector and respectively target output vector), $(\mathbf{x}_j, \mathbf{t}_j) \in \mathbf{R}^n \times \mathbf{R}^m$, is used to training a ELM that with L number of hidden nodes. In a perfect case, the output of this ELM respectively to $\mathbf{x}_j$ should be

$$f(\mathbf{x}_j) = \sum_{i=1}^{L} \beta_i G(\mathbf{a}_i, b_i, \mathbf{x}_j) = \mathbf{t}_j \text{ for } j = 1, \ldots, N \quad (1)$$

where $\mathbf{a}_i$ and b_i are the input weights and bias (learning parameters) of the hidden nodes, β_i is the output weights, and $G(\mathbf{a}_i, b_i, \mathbf{x}_j)$ is the output of the ith hidden neuron respectively to the input vector $\mathbf{x}_j$. Equation (2) shows the definition of the $G(\mathbf{a}_i, b_i, \mathbf{x}_j)$ hidden neuron.

$$G(\mathbf{a}_i, b_i, \mathbf{x}_j) = \exp\{-b_i \left\| \mathbf{x}_j - \mathbf{a}_i \right\|^2\}, \; b_i \in R^+ \quad (2)$$

The algorithms of ELM are summarized as follows.

Step 1: Randomly assign the input weights $\mathbf{a}_i$ and b_i for $j = 1, \ldots L$.

Step 2: Calculate the initial hidden layer output matrix, $\mathbf{H}$, as follows.

$$\mathbf{H} = \begin{bmatrix} G(\mathbf{a}_1, b_1, \mathbf{x}_1) & \cdots & G(\mathbf{a}_L, b_L, \mathbf{x}_L) \\ \vdots & \cdots & \vdots \\ G(\mathbf{a}_1, b_1, \mathbf{x}_{N_0}) & \cdots & G(\mathbf{a}_L, b_L, \mathbf{x}_{N_0}) \end{bmatrix}_{N \times L} \quad (3)$$

Step 3: Estimate the initial output weights, $\boldsymbol{\beta}$ by following equations,

$$\boldsymbol{\beta} = (\mathbf{H}^T \mathbf{H})^{-1} \mathbf{H}\mathbf{T}, \quad (4)$$

where $\mathbf{T} = [\mathbf{t}_1, \ldots, \mathbf{t}_N]$ is the respective targeted output vectors.

For the proposed FLS-ELM model has the similar algorithms as ELM butthe equation (3) is redefined as

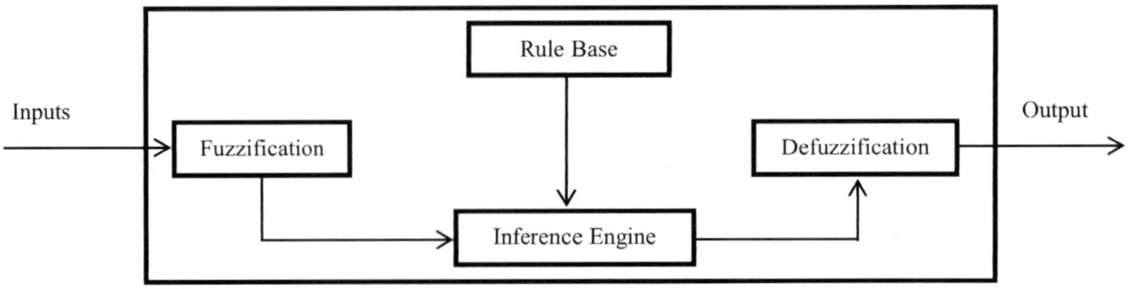

Figure 2. Fuzzy Inference System (FIS)

where function $F(.)$ is represents the operation of a Fuzzy Inference System (FIS) (refer to Figure 2). A FIS can be summarized 4 sub modules, i.e., fuzzification, rule based, inference engineer and defuzzification. Note that for the fuzzification module of FIS, the membership for an input attribute, x, is defined as

$$u(x) = \exp\left[-\frac{(x-c)^2}{2\sigma^2}\right], \tag{6}$$

where c and σ are the centre and standard deviation respectively. The full details of a FIS can be found in [5] and [8].

3. Experiments and Results

The collected dataset of CWS consists of 2500 data samples, each of them with 12 input features (parameters of CWS, mainly temperatures and pressures along CWS) and two output classes (normal and fault). They are divided into three subsets, i.e., 1000 data for training, 500 data for validation and 1000 data for test. Experiments are conducted for both ELM and FLS-ELM. For ELM, the best validation accuracy rate was obtained for L = 1000 (number of hidden neuron), i.e., 97.132%. With the ELM trained with the best validation rate, the accuracy rate to the test set is 96.610%.

For FIS-ELM, searching of the best validation accuracy rate is through a grid search method, i.e., number hidden neuron (number rules) and standard deviation (σ) of membership function. Table 1 shows the validation results of FLS-ELM.

Table 1. Validation Accuracy rates of FLS-ELM for difference number of rules and standard deviation

σ = 2^n, n =		1	2	3	4	5
No. of Rules =						
200	97.156	97.168	97.268	97.216	97.220	97.244
210	97.200	97.288	97.228	97.356	97.332	97.248
220	97.248	97.320	97.204	97.376	97.284	97.324
230	97.236	97.212	97.224	97.344	97.336	97.228
240	97.300	97.244	97.352	97.284	97.216	97.364
						97.260

The validation results suggested the best number of rules (number of hidden neuron) is 250 and the standard deviation of membership function is 2^1 = 2. The FLS-ELM with the highest validation accuracy rate is further use to classify test data. The test accuracy rate is 96.988%. In conclusion, FLS-ELM is better than ELM, i.e., has better test accuracy rate (96.988% vs. 96.610%) and smaller number of hidden neuron (250 vs. 1000).

4. Summary

In this research, a new hybrid FLS-ELM model has been developed and applied in the fault detection of power generation plant. The experimental results show the proposed model able to perform well in fault detection which will increase the efficiency of power generation plant. With the higher efficiency, both consumption on natural resources and pollutions due to power generation expected to be reduced. The results of the proposed model is slightly higher than ELM, however it has a new feature, i.e., potentially to be used for discovery the explicit knowledge from the raw data into the form of "If-then" rules as it was developed based on fuzzy logic. Other further works can be focus on the implementing proposed model on other application domains to further evaluate its applicability.

Acknowledgement

The authors would like to thank Dr. S. C. Tan (Multimedia University, Malaysia) for sharing the dataset.

References

[1] See K.F and Coelli T 2012 An Analysis of Factors that Influence the Technical Efficiency of Malaysia Thermal Power Plants *Energy Economics* **34** 677-685

[2] Arocena P and Waddams-Price C 2002 Generating Efficiency: Economic and Environmental Regulation of Public and Private Electricity Generators in Spain *Int. J. Ind. Organ.* **20** 41-69

[3] Barros C.P and Peypoch N. 2008 Technical Efficiency of Thermoelectric Power Plants *Energy Economics* **30** 3118-3127

[4] Russell and R.R. 1990 Continuity of Measures of Technical Efficiency *Journal of Economic Theory* **51** 255–267

[5] Yap K. S., Lim C. P., and Au M. T. 2011 Improved GART Neural Network Model With Application to Power Systems *IEEE Transactions on Neural Networks* **22** 2310-2323

[6] Huang G. B., Wang D. H., and Lan Y. 2011 Extreme Learning Machines: A Survey, International *Journal of Machine Leaning and Cybernetics* **2** 107-122

[7] Yap K. S. 2011 A New Multi Agent System Based On Online Sequential Extreme Learning Machines and Bayesian Formalism *The 8th International Conference on Networking, Sensing and Control (Delft Conference Centre Delft, Netherland, 11-14 April 2010)* 74-79

[8] Jang J. S. R. 1993 ANFIS: Adaptive-Network Based Fuzzy Inference System *IEEE Trans. on Systems, Man and Cybernetics* **23** 665–685

IOP Conf. Series: Earth and Environmental Science **16** (2013) 012103 doi:10.1088/1755-1315/16/1/012103

Electromagnetic spectrum survey of the environment in a locality in Kuala Lumpur, Malaysia

Wafa Ali Abood[*, 1,2], Norashidah Md Din [1], Aiman Ismail [1], Hafizal Mohamad [3]

[1] Centre for Communications Service Convergence Technology, College of Engineering, Universiti Tenaga Nasional, Jalan IKRAM- UNITEN, 43000 Kajang, Selangor, Malaysia

[2] Dura Power Plant, Ministry Of Electricity, Al-Mahdia Quarter, 1901 Alintisar, Baghdad, Iraq

[3] Wireless Communication Cluster, MIMOS Bhd, Technology Park Malaysia, 57000 Kuala Lumpur, Malaysia

E-mail: wafa_ali_2009@yahoo.com, norashidah@uniten.edu.my

Abstract. The electromagnetic spectrum in the environment is becoming a scarce resource with the emergence of a high number of wireless communications services Cognitive radio (CR) is viewed as a possible solution to the spectrum bottleneck which work in a premise that at an any given time and spatial region there are frequency bands that has no signal occupancy. The CR technique utilizes a temporarily unoccupied licensed band by allowing secondary users to exploit opportunistically the underutilized spectrum licensed to primary users without any harmful interference. Before investigating the technical and political implications of CR, it is necessary to know to what extent the licensed bands are temporally unoccupied. In this paper a spectrum occupancy measurements is conducted to study the utilization of RF spectrum in an environment. The measurements are performed on UHF TV, GSM900 and GSM1800 frequency bands in an urban area in Kuala Lumpur, Malaysia. The evaluation made is based on the power detection principle. From the measurements, the spectrum holes are identified. The obtained results show that the spectral usage is 7.37% for UHF TV band, 12.8% for GSM900 and 5.3% for GSM1800 band leading to the conclusion that a significant amount of spectrum is available for deployment of cognitive radio.

1. Introduction

The demand for wireless transmission has placed great pressure on the scarce radio spectrum. However, some engineers argue that this problem is the consequence of an inefficient spectrum access method [1], [2] and [3], rather than the physical scarcity of spectrum. Currently, regulatory agencies issue spectral licenses to authorize exclusive use of a specified frequency band. In many cases, the licensed user only occupies the spectrum for certain durations, and the desired transmission only covers a particular geographical area. As a result, unoccupied licensed spectrum, or 'spectral holes', can be found at different combinations of frequency, time, and geographical area. This suggests that the radio spectrum could be better exploited. Cognitive Radio (CR), introduced by Mitola [4], is a new radio system concept. It is an intelligent radio which is able to sense its environment, and adapt its physical operating parameters accordingly, to satisfy its system requirements. CR is viewed as a technology to exploit the underutilized radio spectrum. In this paper, we present a spectrum study that have been conducted in an outdoor and indoor environments, i.e. 24 hours measurements for the TV band and a 72 hours measurements for both GSM800 and GSM900, at a building in University

Malaya, Kuala Lumpur, Malaysia. The objective of these measurements is to determine the spectrum occupancy for each measured band in an environment as well as identifying the spectrum holes. The organization of this paper is as follows. Section 2 describes the methodology and Section 3 the results and discussion. The conclusion is given in Section 4.

2. Methodology

The measurements were conducted in a building in Univ. of Malaya at two locations (i) indoor, on the 16th floor which faced a large window as in Figure 1; and (ii) outdoor, 3rd floor of the same building, as shown in Figure 2. The equipment used consists of an Ultra Wideband Discone antenna, a spectrum analyzer, lab view and matlab software. Table 1 shows the settings for the spectrum analyzer. The indoor measurements were conducted for a continuous 72 hours for the GSM1800, GSM900 bands and 24 hours for the outdoor TV band measurements. The detected energy level at the receiver is compared to a threshold value ‘ ’ to classify the received data into either as a signal denoted by ‘1’ or as a noise denoted by ‘0’ as follows:

$$P_c(f_i,t_j) = \begin{cases} 1, & P(f_i,t_j) \geq \gamma \\ 0, & P(f_i,t_j) < \gamma \end{cases} \quad (1)$$

Where (f$_i$,) is the energy of elements received data matrix in a specific frequency and time. The threshold value can be estimated by replacing the antenna with a 50 ohm matched load to determine the receiver noise floor plus some margin value. The margin value used here is 3dB.

(a) Indoor (b) Outdoor

Figure 1. Measurement locations: indoor and outdoor

Table 1. Spectrum analyzer settings

Parameter	Value
Start frequency – end frequency(MHz)	TV: 470-798 GSM900 :880-960 GSM1800: 1700-1800
Resolution / Video Bandwidth (RBW /VBW)	10 kHz / 10 kHz
Sweep Time	50 ms
Detection type	Average Detector
Number of samples	1000

3. Results and Discussion

The indoor spectrum occupancy plot consists of three sub-plots. The first sub-plot presents the average power received during a 24-hours period for the TV band and a 72- hours period for GSM900 and GSM1800 cellular services. The middle sub-plot presents the spectrum occupancy versus frequency and time. The black points indicate that the frequency channel at the specific time is occupied. The last sub-plot shows the duty cycle versus frequency. Duty cycle is defined as the fraction of time the signal is above the threshold. For the indoor location, Figure 2 shows the occupancy of the UHF TV band, which is from 470 MHz -798 MHz, i.e. the Analog TV & Terrestrial TV Broadcasting range. We can observe that strong signals appear in some channel as in channel 27, 28, 33, 44 and 46 due to the height of the measurement location. Furthermore, some of the TV transmitters are located near the measurement site. Some stations appear to switch off their broadcasting during night hours. There are some weak signals detected and these can be attributed to the multi path effects and blockages. Average Duty cycle of this band is 7.37% which indicates that only 24.174MHz from the whole bandwidth is occupied.

IOP Conf. Series: Earth and Environmental Science **16** (2013) 012103 doi:10.1088/1755-1315/16/1/012103

Figure 3 shows the occupancy of the GSM900 band; i.e 880 MHz – 915 MHz are uplink channels and 925MHz – 960MHz are downlink channels. The Average Duty Cycle of this band is 12.8 %, which indicates that only 10.24 MHz is occupied from the whole bandwidth. It appears that downlink channels are most occupied because it is for point to multipoint mobile applications and active control channels are used to maintain the coverage of the cellular service. The uplink channels show very low activity because it depends on the number of active mobile users at the measurement site. Furthermore, it is clear that the usage of this band is less during midnight hours. The difference in the occupancy patterns between upper band channels and lower band channels can be explained as follows. The transmitted power for the base station is higher than the transmitted power of the user's mobile phone, so the higher power can be detected easily. Furthermore, the location of the antenna used in the measurements is considerably high from the ground level where the mobile phone users are concentrated. In Figure 4, the Average Duty Cycle is 5.3% for GSM1800 cellular service which is allocated the frequencies 1710 MHz – 1785 MHz for uplink channels and 1805 MHz – 1880 MHz for downlink channels. The difference between the downlink and uplink channels is also observed due to the same factors explained in GSM900. In contrast with GSM900, no uplink signal appeared in GSM1800 .This is due to the fact that transmitter power for GSM900 is larger than transmitted power of GSM1800 and it is difficult to detect the mobile user activity since the location of the measurement site is high from the ground level as well as less direct line of sight between the antenna and the users. Table 2 summarizes the results obtained for the measured band as well as the amount of spectrum holes.

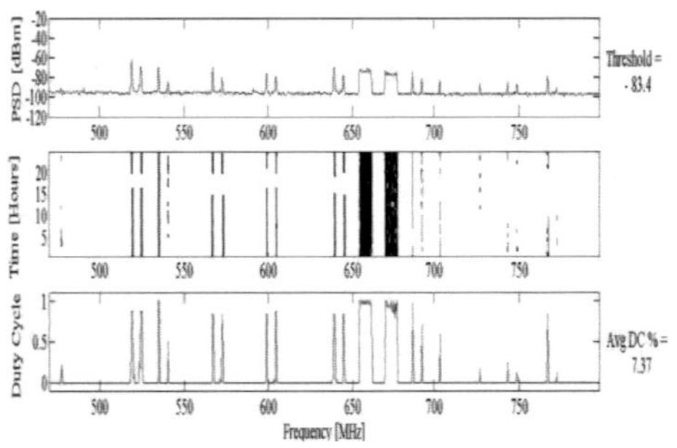

Figure 2. Occupancy of UHF TV Band

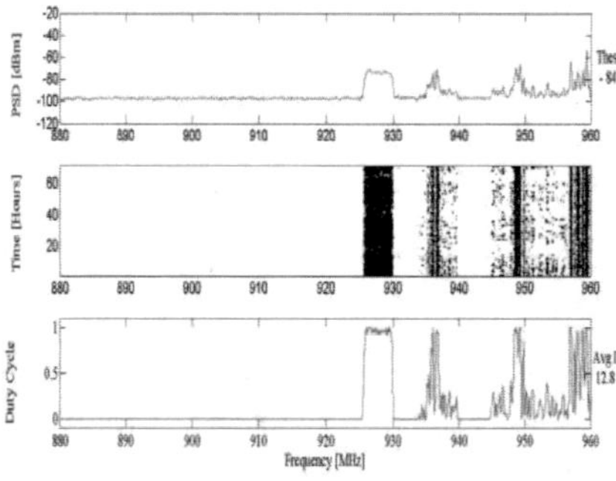

Figure 3. Occupancy of GSM 900

Table 2. Summary of results

Measured Band	BW MHz	Duty Cycle %	Spectrum (MHz)	Spectrum holes (MHz)
UHF TV band	328	7.37	24.17	303.83
GSM 900	80	12.8	10.24	69.76
GSM 1800	170	5.3	9.01	160.99

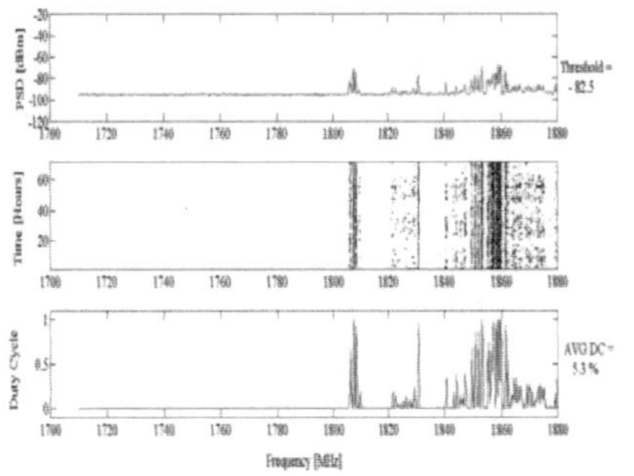

Figure 4. Occupancy of GSM 1800

The comparisons between the outdoor and indoor measurements are shown in Figure 5. The Power Spectral Density received in an outdoor environment is found to be greater than in indoor environment for both GSM 900 and GSM 1800 services due to either the location of the antenna in the outdoor environment in level three which is nearer to base stations or due to signal penetration from outdoor-indoor environments which results in lower received power in the indoor environment. The observed difference between powers received can be noted for the UHF TV band in some channels due to the fact that the indoor measurement is in a high location and faced to some TV transmitters resulting in high power received.

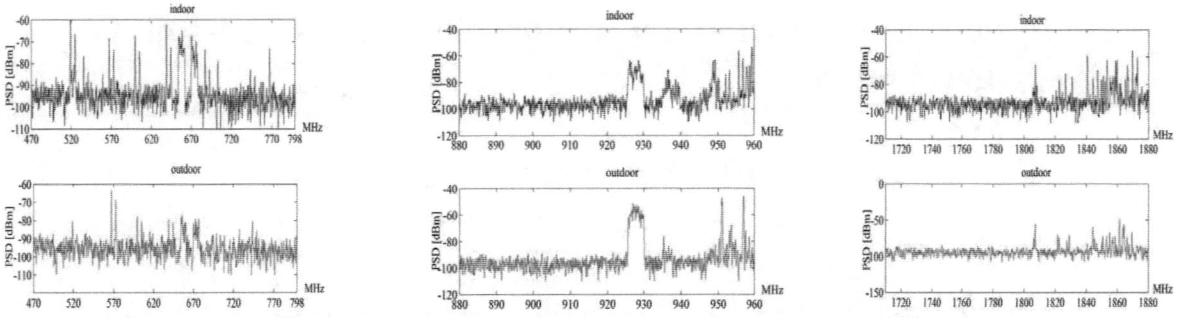

(a) **(b)** **(c)**

Figure 5. Received Power in both indoor and outdoor environments (a) UHF TV Band (b) GSM900 service (c) GSM1800 service

4. Conclusion

In this work, we studied the utilization of the electromagnetic communication spectrum in an environment in urban Kuala Lumpur, Malaysia at measured bands of UHF TV broadcasting, GSM 900 and GSM 1800 cellular services. The obtained results indicate that there are significant amount of unused frequencies that may open opportunities to be used in the future using CR technology. The indoor measurement indicates that the power received is highly dependent on the location of measurement, so care must be taken to avoid the effects of multi-path and hidden terminal problems. Further spectrum studies are recommended for longer period of times and at different locations especially for the bands which show less utilization. The obtained results demonstrate that there is a significant amount of the spectrum that is available for further use.

References

[1] Haykin S and Fellow L 2005 Cognitive Radio: Brain-Empowered Wireless Communications, *IEEE Journal on Selected Areas in Communications* **23** 201-220

[2] R I C, Rowe G B and Sowerby K W 2007 A Quantitative Analysis of Spectral Occupancy Measurements for Cognitive Radio *IEEE 65th 2007 Vehicular Technology Conference (VTC2007-Spring) (Dublin, Ireland, 22-25 April 2007)* 3016-3020

[3] Yin S, Chen D, Zhang Q, Liu M and Li S 2012 Mining Spectrum Usage Data: Large –Scale Spectrum Measurement Study *IEEE Transactions on Mobile Computing* **11** 1033-64

[4] Mitola, J. and Maguire G. Q. 1999 Cognitive Radio: Making Software Radios More Personal *IEEE Personal Communications* **6** 13-18

Improvement of engineering properties of liquefied soil using Bio-VegeGrout

I N Z Baharuddin[1, 2], R C Omar[1, 2] and Y Devarajan[1]

[1]Department of Civil Engineering, College of Engineering, UNITEN, 43000 Kajang, Malaysia

[2]Centre for Forensic Engineering (CeFE), College of Engineering, UNITEN, 43000, Kajang, Malaysia
E-mail: intan@uniten.edu.my

Abstract. Lab tests were carried out in order to analyze and investigate the liquefaction behavior of marine deposit soil which was placed in the region of LCCT, Sepang. The soil was found exposed to liquefaction. Step to improvise the soil was to inject Biovege grout into the soil sample. Soil sample was prepared with three different grout amount, 10%, 25% and 50%. The relation between percentage of Biovege grout injected and soil improvement were observed and recorded. The void ratio and the permeability of the soil sample decreased with increasing grout percentage. The soil becomes stiffer as the amount of grout used increase. The results obtained indicate, higher amount of grout injection reflects better soil improvement in term of cohesion, friction angle, shear stress and void ratio. The Biovege grout increases the resistance of soil against liquefaction

1. Introduction

Soil liquefaction occurs when the soil goes sudden massive loading which eventually forces the soil to flow. This phenomenon will eventually lead to instability of soil and causes fatal damage to human lives as well as properties. A lot of researches were conducted to develop sustainable or green materials as well as the methodology in hope of reducing carbon foot prints. As for soil improvement, research on strengthening soil using microorganisms (i.e. bacteria) has been the focus as it has very low carbon footprint while improving the properties of weak soil with minimal cost. Van Paassen (2009) [4] proved that bacteria can be used as the source to carbonate precipitation, which changes the mechanical properties of soil. Whilst, the winner of 2010 Next Generation Award [3], G.K. Dosier of American University of Sharjah, UAE found that the mixture of common bacteria, calcium chloride and urea could transform sand into sandstone using Microbial Induced Carbonate Precipitation (MICP) technique. This technique uses microbes to produce links or bridges between soil particles a chain of chemical reactions. Meanwhile another study conducted by H. Kamarudin (2012) [1] shows that bacteria to form carbon precipitation could also be derived from vegeTables waste also known as VegeGrout has significant impacts on the properties of remoulded soil samples. Therefore, this study was conducted to evaluate the improvement of engineering properties of liquefied soil using VegeGrout.

2. Soil Modification

2.1 Bio-Mediated Soil Improvement

The Bio-Vege Grout contains microorganism which precipitate the calcite or calcium carbonate ($CaCO_3$) around each and every particle of the soil by going through some chemical reactions. Thus this method is economical and at the same time has low impact on enviroment. Bacteria and fungi are most commonly found in the waste as the waste starts to decompose. This formation of calcite on the grain will increase the stiffness and the cohesiveness among particles. Thin layer of calcite will form on the particle and the gap and voids between the particles which water can fill during an earthquake leading to liquefaction. Indirectly the strength of the soil will increase to withstand the loading.

3. Methodology

3.1 Soil Sampling and Preparation

Soil samples for this study were taken from site nearby Low Cost Carriage Terminal (LCCT), Sepang, Selangor. According to the site investigation report these soils are susceptible to liquefaction phenomenon which has SPT N value less than 10. Both disturbed and undisturbed samples were taken for laboratory testing. All samples were tested to obtain their index and mechanical properties.

3.2 Preparation of Bio-Vegegrout

Four types of vegeTable wastes namely water spinach, cucumber, spinach and long beans were collected from Pasar Borong Selangor to produce five different Bio-VegeGrout. These vegeTable wastes with mass of five kg each were stored in five different air tight containers for four weeks. As the results, five different VegeGrout liquid were produced i.e. spinach, long beans, water spinach, cucumber and mix vegeTables. It is important to ensure the liquid produced from vegeTable wastes is free from any impurities and dirt, therefore it is required to filter the liquid before applying it to the soil.

3.3 Application of Bio-Vegegrout Onto the Soil

In order to study the performance of Bio-VegeGrout on soils, three samples weigh 2000 grams with different amount i.e. 10%, 25% and 50% of Bio-VegeGrout were injected and incubated for four weeks. The percentage of Bio-VegeGrout used was calculated based on the mass of soil used. To ensure the Bio-VegeGrout is spreading equally throughout the soil layers, the soils were split into three layers with 6 cm in depth and Bio-VegeGrout liquid was equally injected through 1.5 cm spacing drinking straw that penetrate the soil layers. Details of the test sample set up are shown in Figure 1.

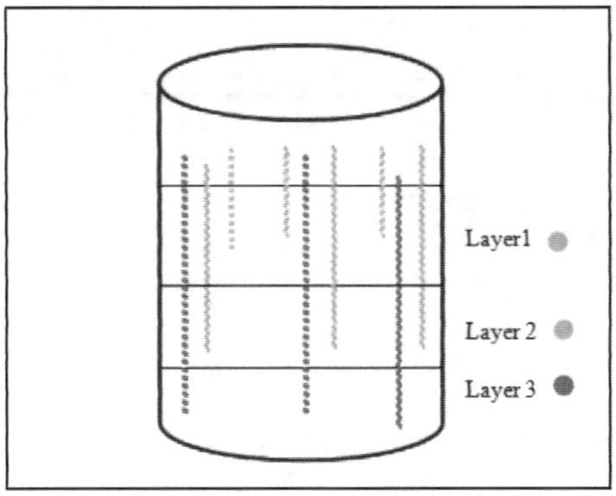

Figure 1. Schematic diagram of test sample

IOP Conf. Series: Earth and Environmental Science **16** (2013) 012104 doi:10.1088/1755-1315/16/1/012104

3.4 Index and Engineering Properties
A series of laboratory testing was conducted in determining the index and engineering properties of the soil. The tests consist of initial moisture content, sieve analysis, shear strength namely Unconsolidated Undrained triaxial testing (UU), consolidation and permeability tests. A series of testing to obtain before and after treatment with Bio-VegeGrout were recorded.

4. Result Analysis and Discussion
From the laboratory testing, index properties of the soil are considered as slightly silty gravelly SAND with 9.5% and 2.2 as its moisture content and specific gravity respectively. The result obtained from the stress strain graphs in Figure 2 proves that the injection of Bio-VegeGrout has increased the shear stress of the soil. The Figure 2 indicates the effect of grout on the soil stress under three different confining pressures. The amount of grout used reflects the concentration of bacteria presence, thus more bacteria can produce a higher amount of calcium carbonate precipitation. With the addition of 50% grout, the soil is able to withstand twice the load of the actual soil can support. Consequently, the addition of Bio-VegeGrout will reduce the exposure of soil to liquefaction.

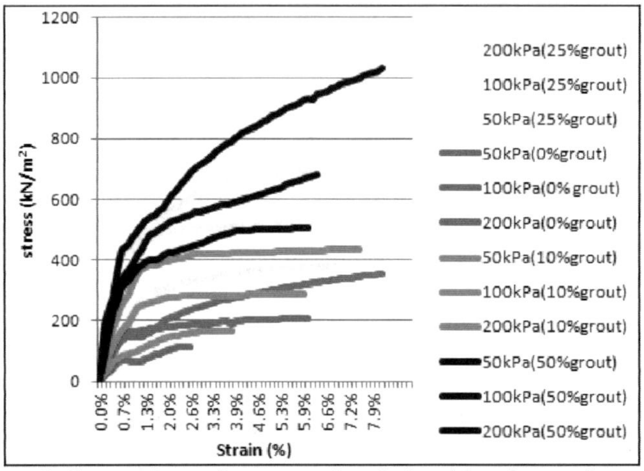

Figure 2. Comparison of stress strain relationship of treated samples

From the laboratory testing and result analysis the shear strength parameters are increased by at least 7% for friction angle and more than 100% for cohesion value after 4 weeks of curing period which eventually increase the shear strength of the soil and its bearing capacity. Other engineering properties such as permeability and consolidation value are increased. The result shows that the permeability value of the soil, *k* reduce as the percentage of Bio-VegeGrout increase. This is due to *the* process of calcite precipitation which reduces the drainage condition of the soil. The grout has a very low viscosity and small size of bacteria cell which allows the grout to seep through low permeability soil and penetrate into the voids by the gravity force only. This situation takes place as the voids in the soil are being filled by the calcium carbonate leaving less space for the water to flow through the soil. In real life condition, biogrouts are used for soil cementation and also for leakage control or prevent seepage of toxic pollutants into permeable soil (Ivan B. Gratchev, 2006) [2]. Consequently results in reduction of void ratio of the soil which can be advantages in term of avoiding any harmful pollutants to seep into the soil and it can be used to avoid landslides which are triggered by high level of ground water Table. Moreover, results for consolidation test indicate that the rate of consolidation becomes slower as the permeability of soil decreases. This relation can be used to reduce the amount of soil settlement for building construction. A summary of properties of soil before and after the injection of the Bio-VegeGrout is shown in Table 1.

Table 1. Summary of engineering properties of soil using Bio-VegeGrout

Parameter	**Amount of Bio-VegeGrout (%)**			
	0%	**10%**	**25%**	**50%**
Cohesion, c (kN/m^2)	15	38	53	61
Friction angle, φ (°)	26	28	31	38
Coefficient of Permeability, *k* (m/s)	$3.36x10^{-6}$	$1.1x10^{-6}$	$5.7x10^{-7}$	$2.99x10^{-7}$
Initial Void Ratio	0.33	0.3	0.26	0.19

5. Conclusion and Recommendation

As a conclusion, the application of Bio-VegeGrout via injection does improve the engineering properties of the soil with the increment of Bio-VegeGrout percentage. However, it is recommended to include the works; i.e. perform Scanning Electron Microscope test (SEM) and X-Ray Diffraction test (XRD) as well as to monitor pH and temperature during curing period for future work.

Acknowledgement

The author would like to thank Malaysian Ministry of Higher Education (MOHE) of Malaysia for Granting the Project under code ERGS/1/2011/TK/UNITEN/02/40.

References

[1] Huzaifah 2012 *Microbiologically Induced Carbonate Precipitation for Ground Improment* (Degree's Thesis, Universiti Tenaga Nasional)

[2] Ivan B. Gratchev and Kyoji Sassa 2006 How Reliable is the Plasticity Index for Estimating the Liquefaction Potential of Clayey Sands? *Journal of Geotechnical and Geoenvironmental Engineering* **132** 124-127

[3] LaBarre S. *The Better Brick: 2010 Next Generation Winner.* [online]. Available from: http://www.metropolismag.com/story/20100512/the-better-brick-next-generation-winner [Accessed 01 January 2013]

[4] Van Paassen and L.A. 2009 *Biogrout, Ground Improvement By Microbially Induced Carbonate Precipitation* (PhD thesis, Delft University of Technology)

Transmission Tower Environment Monitoring Using UAV

Redia Mohd Redzuwan[1], Norashidah Md Din,Mohd Zafri Baharuddin, Intan Shafinaz Mustafa, Rohayu Che' Omar

College of Engineering, Universiti Tenaga Nasional, Jalan IKRAM-UNITEN, 43000 Kajang, Selangor, Malaysia.

[1]redia@uniten.edu.my

Abstract. Power utility engineers used to conduct ground survey to collect topographic data. Therefore, they can get detailed and accurate information, but these techniques take a lot of labors and expenses, and spending times for the surveying.An attractive solution to the ground survey is using images taken using Unmanned Aerial Vehicle (UAV). Images captured from UAV can be collected quickly and efficiently over the same area covered in the land survey, in a fraction of the time. The purpose of this research is to mosaic the large numbers of spectral images together into a region wide panoramic image which allows experts to analyze the data for transmission tower monitoring purposes.

1. Introduction

Power utility engineers take into account numerous factors when physically designing a new transmission line. There are several design options available to minimize costs and impacts on the communities and environment. Construction costs, repair time and environmental impact, among other factors, generally lead to placing transmission lines place above ground (known as "overhead") rather than underground. There are many types of structures that can be used to support overhead power lines. These structures differ in cost, span length (the distance between structures) and in height and width. In the end, the structure design must fit within the available corridor, minimize environmental impacts and be cost effective for customers and utility.

Usually, field workers used to conduct land survey to collects topographic data for design purposes. Land surveying is conducted on the ground, so generally it is very close to observation targets. Therefore, they can get detailed and accurate information, but these techniques take a lot of labors and expenses, and spending times for the surveying. Also, it is not always safe in the case of the stricken or rural area.

Hence, to save cost and time, Unmanned Aerial Vehicle (UAV) is a promising approach in the process of collecting topographic data for area of interest. UAV is a flight is either controlled autonomously by computers in the vehicle, or under the remote control of a navigator, or pilot on the ground or in another vehicle. When an UAV is flying in certain height to perform a task of remote sensing, the serial images are shot by a digital camera mounted on UAV.

The purpose of this research is to mosaic the large numbers of spectral images together into a region wide panoramic image which allows experts to analyze the data.

2. Related Work

Nowadays, data from UAV technology can be used in mapping applications but it needs some modification from certain aspect in order to fulfill requirement of mapping. An UAV based monitoring system is developed to obtain the wide-area coverage and the high levels of detail and accuracy of ground surveying [1]. The accuracy and resolution of the image are equal to images that taken from the ground though the survey are conducted from the sky [2][3].

In [4], Jiayuan Lin, Heping Tao, Yangchun Wang and Zhou Huang claim that the UAV imagery has very high spatial resolution which can be adjusted by changing flight altitude. Not only that, they also proposed UAV as a great potential platform for mountain hazards survey with the advantages of region-scale coverage and moderate timeliness.

U. Niemather, S.Rothmund, M.R. James, J. Traveletti and M. Joswig proposed a low cost remote sensing approach based on UAV and digital compact camera to enables high-resolution acquisitions of landslides [5].

3. Field Work

In this study, the minimal flight time and associated travel route of a UAV is determined by using Google Earth and AirPhotoNavi software. AirPhotoNavi software is able to calculate image overlap calculation and flight path determination while Google Earth application could easily be able to determine the region of interest and displays the feasible flight path for UAV.

3.1. *Overlap calculation.* The optimum flight path condition will be executed by keying in the camera information and shooting condition. For camera information requirement, consider a NDC (Near Infrared Digital Camera) with a focal length 28 mm, vertical sensor size, Sensor Size (V) 14.9mm and horizontal sensor size, Sensor Size (H) of 22.3mm are being used. The most suitable condition for UAV to fly with a flight speed of 166km/h and altitude of 1005.8m is highlighted with a yellow in colour as shown in Figure 1.

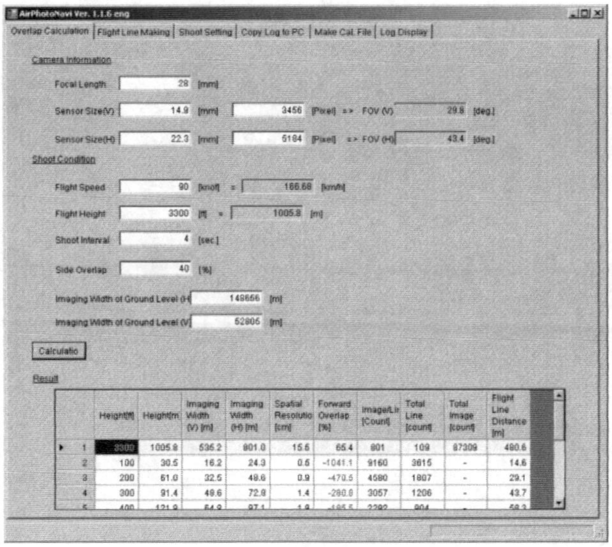

Figure 1. Overlap Calculation Tab

3.2. *Flight Path* Using AirPhotoNavi software, in Flight Line Making tab (Figure 2); the coordinates of the Baseline Start Point and Baseline End Point will be shown automatically once import button has been clicked and a feasible 3D (three dimensional) flight path will be executed in Google Earth (Figure 3) simultaneously.

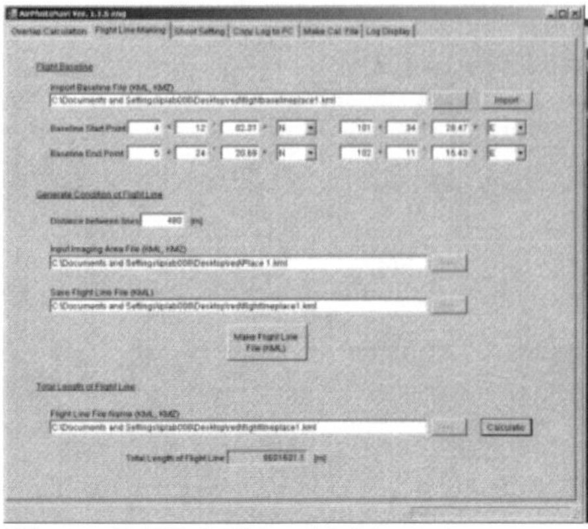

Figure 2. Flight Line Making tab

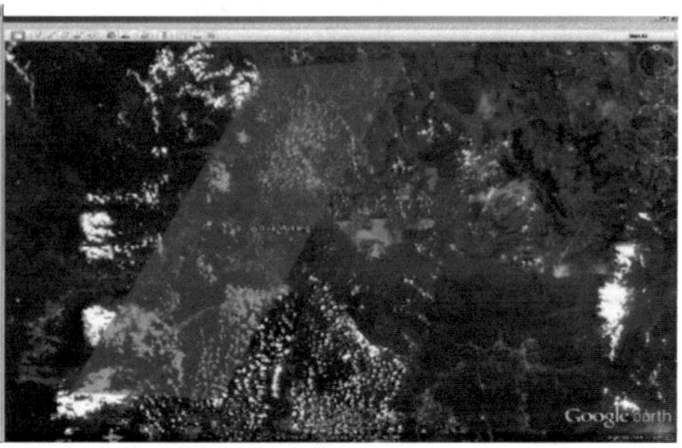

Figure 3. Feasible Flight Path

4. Result and Analysis

All the sequences images taken from UAV have been mosaicked using EnsoMosaic software to create 2-D planimetric mapping covering the test area (Fig. 3). Every image taken has 60% overlapped region with a ground resolution 15.5cm and 1000 x 2752 pixels. The mosaicked image is free from any distortion and represents the whole simulation model.

Figure 4. The mosaicked image

From Table 1, it can be concluded that the conventional method has longer duration of project planning, field work and image processing compared to the UAV method. Since less time is required in UAV method, it makes this approach to be more cost effective.

Table 1. Time Estimation for transmission tower monitoring

	Land Survey	**UAV**
Project Planning	1 month	1 month
Field Work	1 month	1 day
Image Processing	1 month	2 weeks
Labour	4-5 person	2 person

5. Conclusion and Future Work

This paper has proved that the UAV low altitude remote sensing system could provide the same result as ground survey with a low-cost, less-security risk and simple-operation. Based on the result obtained, it can be concluded that this approach can be used effectively for transmission tower environment monitoring since the images captured from UAV can be collected quickly and efficiently over the same area covered in the land survey, in a fraction of the time. Further work would focus on calculate and reduce cumulate errors and residuals errors of spectral images.

References

[1] A reference Masahiko Nagai, Apichon Witayangkurn, Kiyoshi Honda, and Ryosuke Shibasaki, "UAV-Based Sensor Web Monitoring System," International Journal of Navigation and Observation, vol. 2012, Article ID 858792, 7 pages, 2012.

[2] Niethammer, U., James, M.R., Rothmund, S., Travelletti, J., Joswig, M. "UAV-based remote sensing of the Super-Sauze landslide: Evaluation and results" (2012) Engineering Geology, 128, pp. 2-11.

[3] Liu, C., Li, W., Lei, W., Liu, L., Wu, H. "Architecture planning and geo-disasters assessment mapping of landslide by using airborne LiDAR data and UAV images" (2011) Proceedings of SPIE - The International Society for Optical Engineering, 8286, art. no. 82861Q,

[4] Jiayuan Lin; Heping Tao; Yangchun Wang; Zhou Huang; , "Practical application of unmanned aerial vehicles for mountain hazards survey," Geoinformatics, 2010 18th International Conference on , vol., no., pp.1-5, 18-20 June 2010 More references

[5] U. Niethammer, M.R. James, S. Rothmund, J. Travelletti, M. Joswig, UAV-based remote sensing of the Super-Sauze landslide: Evaluation and results, Engineering Geology, Volume 128, 9 March 2012, Pages 2-11, ISSN 0013-7952, 10.1016/j.enggeo.2011.03.012.

IOP Conf. Series: Earth and Environmental Science **16** (2013) 012106 doi:10.1088/1755-1315/16/1/012106

CFD study of temperature distribution in full scale boiler adopting in-furnace coal blending

S S A Fadhil[1], H Hasini[1] and N H Shuaib[2]

[1]Centre for Advanced Computational Engineering (CACE), UnivesitiTenagaNasional, Jalan IKRAM-UNITEN, 43000, Kajang, Selangor, Malaysia
[2]TNB Research Sdn. Bhd., Lorong Ayer Hitam, 43000, Kajang, Selangor, Malaysia
E-mail: sitisarah@uniten.edu.my

Abstract. This paper describes the investigation of temperature characteristics of an in-furnace combustion using different coals in a 700 MW full scale boiler. Single mixture fraction approach is adopted for combustion model of both primary and secondary coals. The primary coal was based on the properties of Adaro which has been used as the design coal for the boiler under investigation. The secondary blend coal was selected based on sub-bituminous coal with higher calorific value. Both coals are simultaneously injected into the furnace at alternate coal burner elevations. The general prediction of the temperature contours at primary combustion zone shows identical pattern compared with conventional single coal combustion in similar furnace. Reasonable agreement was achieved by the prediction of the average temperature at furnace exit. The temperature distribution is at different furnace elevation is non-uniform with higher temperature predicted at circumferential "ring-like" region at lower burner levels for both cases. The maximum flame temperature is higher at the elevation where coal of higher calorific value is injected. The temperature magnitude is within the accepTable limit and the variations does not differ much compared to the conventional single coal combustion.

1. Introduction

Coal blending method has been widely applied to many coal-powered power stations around the world. Common method used is by mixing two or more coals of different properties in a bunker or pulveriser. In this way, the properties of mixed coals could be pre-determined prior to the combustion. However, additional coal mixing procedure needs to be established and uncertain coal properties delivered to power station could create difficulties. Alternatively, the in-furnace blending method could be used where different coals are injected into the furnace at different burners depending on the type of furnace. The method is most suiTable for tangential fired boiler which will provide good heat distribution at all sides of furnace walls.In the open literature, extensive studies have been done to analysecoal combustion characteristics in furnace [1-2]. In recent years, many alternatives methods were developed to utilise limited amount of resources available without compromising the boiler efficiency. The quest for new method of fuel blending has been increasingly important with the increase of fuel price. Studies on in-furnace blending in lower rated furnace capacity have been attempted [3-4]. The prediction of burnout rate and the amount of unburned carbon resulting from furnace blend combustion was the primary concern of these studies. It has been demonstrated numerically and experimentally that the method could reduce the unburned carbon in furnace. The adoption of in-furnace blending to large capacity boiler pose challenging task in determining the resulting effect to various boiler parameters. Rapid development of computing power has made CFD a promising tool in virtually emulate the combustion process in complex furnace geometry. Detail flow visualization and temperature distribution are made possible with CFD thus allowing online virtual access to hazardous regions in furnace which was experimentally impossible. This paper presents the preliminary calculation of combustion properties and temperature distributioninside a 700 MW full scale boiler furnace when in-furnace blending of different coals is adopted. Comparison of temperature distribution at different furnace elevation is made between single and dual coal combustions.

2. Boiler model and simulation setup

The boiler model was developed based on the full scale tangential fired furnace. Simplifications were made on a few boiler components where detail dimensions were absent. In addition, simplifications on complex geometry were necessary to avoid highly skewed mesh which could cause divergence in the solution. The simplifications were made at the regions where flowproperties are not critical.The schematic diagram of the furnace modelis illustrated in Figure 1. The furnace adopts tangential firing system at which imaginary fireball is formed at the furnace centre. Heat is distributed to the surrounding wall when combustion flame creating the fireball dissipates radially outwards towards the walls. The furnace can be generally divided into four distinctive regions namely bottom ash hopper, windbox, superheater/reheaterand rear pass regions. Figure 2 illustrates the mesh scheme employed by the furnace. Coal and air are injected into the furnace through the windbox, which consists of several injection elevations inclusive of the over-fire air ports. The flow and combustion characteristics are critical in this region and therefore, higher mesh concentration is applied in order to sufficiently resolve the flow field.

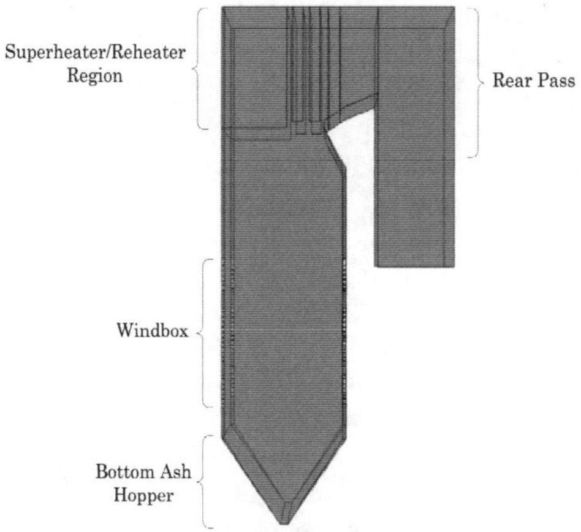

Figure 1. Schematic diagram of the furnace

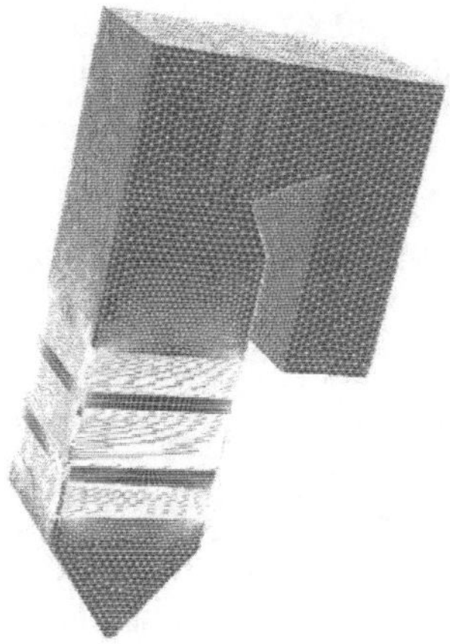

Figure 2. Mesh scheme of the furnace.

In this work, the standard k-εmodel is used as the closure for turbulence Reynolds equation. SIMPLE method is employed with Lagrangian particle trajectory to track the coal particle.Species and chemical reactions are modelled using the mixture-fraction/probability density function (PDF)approach, where the turbulence-chemistry interaction is modelled using a double delta PDF. The boundary condition is set according the actual operating conditions of the boiler. Table 1 summarizes the boundary conditions used in this analysis.Coal inlet conditions are set as injection at which the particle is assumed to be combusting and its diameter distribution is assumed to follow the Rosin-Rammler distribution. To create the fireball effect, direction vector of both coal and air inlets areresolved in x-, y- and z- components according the design condition. The secondary air (SA) inlets are set as velocity inlet with the velocity of 15 m/s and temperature of 550 K. The pulverized coal (PC) inlets are set as velocity inlet with the velocity of 50 m/s. The calculations were executed until approximately 10,000 iterations to achieve convergence.

Table 1: Boundary conditions of Adaro coal and blend coal

Parameters	Primary coal	Secondary coal
Density (kg/m3)	1,300	1,400
Cp (J/kg.K)	1,000	1,680
Thermal Conductivity (W/m.K)	0.0454	0.0454
Vaporization Temperature (K)	400	400
Volatile Component Fraction (%)	28	12.9
Burnout stoichiometric ratio	2.67	2.67
Combustible fraction (%)	64	79.1

3. Results and Discussion

Simulations on single and dual coal combustion have been performed. Coal properties based on Adaro was assumed for primary coal while for the blend coal case simulations, a typical secondary coal with higher calorific value was used. Bothprimary and secondary coals are injectedat alternate furnace elevations. From the analysis, it is found that the temperature is slightly higher in the in-furnace blending case. This could be attributed to the higher calorific value used for the secondary coal in the blend case. The distribution of temperature contour at a few furnace vertical iso-planes is illustrated in Figure 3.Regions of higher temperature are obvious at circular "ring-shaped" area enclosing fireball centre. This could be clearly seen at the lower burner levels. This trend is applied to both cases with overall higher temperature magnitude on the blend case. The trend is also dominant at lower burner elevations PC1-PC3. As the combustion gas flows upwards towards the highest burner level, the temperature distribution becomes more uniform with no distinct fireball shape predicted. It can also be observed that the temperature distribution for in-furnace blending seems uniform throughout the burner elevations despite being injected with two types of coal. The temperature distribution for blending case is almost identical with the single coal combustion with slightly higher temperature. Thus, from the thermal point of view, it has been demonstrated that the in-furnace blend method is a promising way to mix different coals during combustion. Nevertheless, further investigation needs to be carried out to look at detail unburned carbon content, pollutant emission and many other factors that might result from the in-furnace blend of different coals.

4. Conclusion

The simulation of full scalein-furnace coal blending simulation has been carried out with the aim to analyse the distribution of temperature inside the furnace. The simulation of single coal and blend coal combustion were carried out based on the two different coal properties. Results from the simulation showslightly higher temperature distribution for in-furnace blend case due to higher calorific value for the secondary coal. The temperature distribution along furnace height at different burner elevation is almost identical for both cases with higher magnitude for the blend case. Regions of higher temperature are obvious at circular "ring-shaped" area and this pattern is significant at lower burner elevations.At higher burner elevations, more uniform temperature distribution is predicted with accepTable temperature magnitude. Despite having slightly higher temperature in the blend case, the range of maximum temperature is within an accepTable limit for coal combustion.

References

[1] Hasini H, Yusoff M Z, Shuaib N H and Boosroh M H 2009 Analysis of Flow and Temperature Distribution in a Full Scale Utility Boiler Using CFD *Proceeding of ICEE 2009 3rd International Conference on Energy and Environment (Malacca, 7-8 December 2009)* 208-214

[2] Fadhil S S A and Hasini H 2011 The Effect of Loading Condition on the Formation of Residual Swirl in Tangentially-fired Boiler *The 4th International Meeting of Advances in Thermofluids (IMAT2012)(Malacca, 3-4 October 2011)* **1440** 1049-1058

[3] Lee B H, Kim S G, Song J H, Chang Y J and Jeon C H 2011 Influence of Coal Blending Methods on Unburned Carbon and NO Emission In A Drop-Tube Furnace *Energy Fuels* **25**

IOP Conf. Series: Earth and Environmental Science **16** (2013) 012106 doi:10.1088/1755-1315/16/1/012106

5055-62
[4] Shen Y S, Guo B Y, Yu A B and Zulli P 2009 A Three-Dimensional Numerical Study of The Combustion of Coal Blends in Blast Furnace *Energy Fuel* **88** 255-63

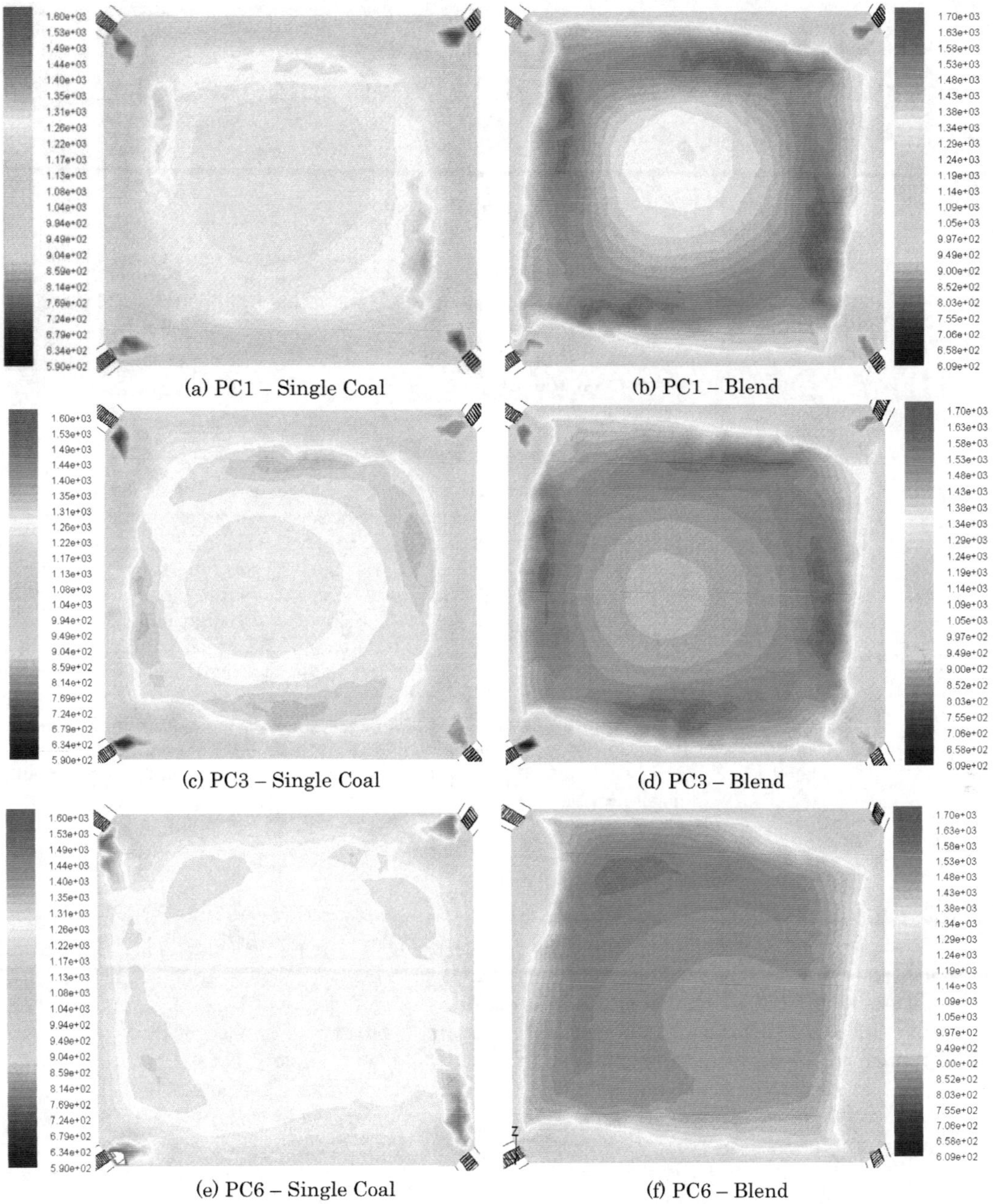

Figure 3. Comparison of temperature contours between single and blend coal injections at different burner elevation (PC1,PC3,PC6).

IOP Conf. Series: Earth and Environmental Science **16** (2013) 012107 doi:10.1088/1755-1315/16/1/012107

Numerical simulation of fluid flow and heat transfer in enhanced copper tube

M M Rahman[1], **T Zhen, A K Kadir**

Centre for Advanced Computational Engineering, College of Engineering, Universiti Tenaga Nasional, Jalan IKRAM-UNITEN, 43000 Kajang, Selangor, Malaysia

E-mail: mujibur@uniten.edu.my

Abstract. Inner grooved tube is enhanced with grooves by increasing the inner surface area. Due to its high efficiency of heat transfer, it is used widely in power generation, air conditioning and many other applications. Heat exchanger is one of the example that uses inner grooved tube to enhance rate heat transfer. Precision in production of inner grooved copper tube is very important because it affects the tube's performance due to various tube parameters. Therefore, it is necessary to carry out analysis in optimizing tube performance prior to production in order to avoid unnecessary loss. The analysis can be carried out either through experimentation or numerical simulation. However, experimental study is too costly and takes longer time in gathering necessary information. Therefore, numerical simulation is conducted instead of experimental research. Firstly, the model of inner grooved tube was generated using SOLIDWORKS. Then it was imported into GAMBIT for healing, followed by meshing, boundary types and zones settings. Next, simulation was done in FLUENT where all the boundary conditions are set. The simulation results were observed and compared with published experimental results. It showed that heat transfer enhancement in range of 649.66% to 917.22% of inner grooved tube compared to plain tube.

1. Introduction

Inner groove tube has become more popular because its heat transfer efficiency can be increased by 20-30% and energy can be saved up to 15% as compared to the traditional plain tube. This inner grove tube is usually used in air conditioning which requires high efficiency of heat transfer and low energy consumption [1]. Tube parameters that involved are geometry of inner grooved shape, type of refrigerant, fin height, tube wall thickness, apex angle and helix angle [2]. Several studies have been done by previous researchers on the effect of the parameters of inner grooved tube to heat transfer rate in order for them to optimize the heat transfer performance. Kadir et al [3] studied relationship between different types of grooved tubes and surface heat transfer using fully developed turbulent air flow. The grooved tubes used in this experiment were circular, trapezoidal and rectangular. Parameters such as length of tube, diameter of tube and Reynolds's number for the fluid flow were fixed in this experiment so that they wouldn't affect the results of heat transfer enhancement. From

[1] Corresponding author: Assoc. Prof. Dr. Md. Mujibur Rahman.

the experimental results using smooth tube as the comparison, heat transfer enhancement showed 63% for circular groove, 58% for trapezoidal groove and 47% for rectangular groove.

Akio et al [4] conducted an experiment to determine the effects of fin height and helix angle on condensation inside a herringbone micro-finned tube by using R410 as refrigerant. They stated that herringbone micro-fined tube is able to increase the heat transfer coefficient 2 to 4 times compared to the conventional helical micro-fined tube. The highest heat transfer can be achieved by 0.18 mm; any higher fin would get the same heat transfer enhancement at fin height of 0.18 mm but it would result in increasing of pressure drop. Chamra et al [5] conducted research of R22 condensation of heat transfer and pressure drop for single grooved tube and cross-grooved tube with outer diameter of 15.88 mm and mass flow rate of 45181 kg/h. The parameters set for single-grooved are internal fins of 74-80, 0.35 mm fin height, and 30° fin included angle whereas for cross-grooved tube, it has two sets of grooves where first set is the same as the single grooved with the same helix angle and second set was the opposite angular direction as the first set. Data was collected for the condensation at 24.0°C in a 2.44 m long test section. Throughout the research, they found out that R22 condensation heat transfer performance in cross-grooved tube is higher as compared to single-grooved tube. They also reported that cross-grooved tube has the highest heat transfer performance at helix angle of 27°. Other than that, Uchida et al [6] also did an experiment to study the evaporative and condensation heat transfer performance on R407C on smooth tube, single-grooved tubes and cross-grooved tubes. The results showed that cross-grooved tube heat transfer coefficients are three times and 20-40% higher than the other two.

Investigation in heat transfer and pressure drop characteristics in a spirally grooved tube with and without twisted tape insert using water as the medium was done by Bharadwaj at al [7]. By comparing with the normal smooth tube, they found that the spirally grooved tube without twisted tape is able to achieve maximum heat transfer enhancement up to 400% in the laminar range and 140% in the turbulent range by supplying constant pumping to the tube. Colombo et al [8] did an investigation on flow boiling and convective condensation for oil-free R134 with three tubes which are two micro-fined tubes and a plain tube. This experiment operated horizontally and the tubes were cooled or heated using water stream. The micro-fined tubes' specifications were set to have fins with an apex angle of 40° and fin number of 54 and 82, respectively. For evaporative conditions, it was set to have 5°C nominal temperature and 0.2941 kg/m^{2}s mass flux whereas for condensation condition, the nominal temperature of 35°C and mass flux of 0.2273 kg/m^{2}s were set. During the experiment, average heat transfer coefficient were monitored and measured. In comparison between two micro-fined tubes and plain tube, it showed that micro-fined tubes has higher heat transfer coefficient for both evaporative and condensation processes. However, differences showed between these two micro-fined tubes when in condensation process. A micro-fined tube with smaller number of fins showed a better performance than that in micro-fined with larger number of fins.

It can clearly be observed from the overview above that all the research works reported are conducted experimentally to evaluate the performance of enhanced tubes. Development of experimental facilities is very tedious and expensive at the same time running the experimentation is also time consuming hence there is a need to evaluate the performance of enhanced tube through numerical simulation. Therefore, the objective of this paper is to report the outcomes of a numerical simulation study in evaluating the heat transfer characteristics of inner grooved copper tube.

2. Methodology

2.1. Model generation

SOLIDWORKS was used in modelling inner grooved tube. The specifications of the inner grooved tube are taken according to previous researcher [2] in order to validate the model. Modification has been done to the length of the tube whereas it has been reduced from 3.67 m to 0.5 m and number of grooves from 50 grooves to 40 grooves. These modifications are due to limitation of computer facility and time. Tube specifications are as shown in Table 1 and the model of inner grooved tube is shown in Figure 1.

Table 1. Specifications of enhanced tube.

Tube parameter	Value
Outer diameter, d_o	7.00 mm
Inner diameter, d_i	6.50 mm
Groove height, h	0.21 mm
Bottom wall thickness, bwt	0.25 mm
Total wall thickness	0.46 mm
Apex angle, γ	40.00°
Helix angle, α	18.00°

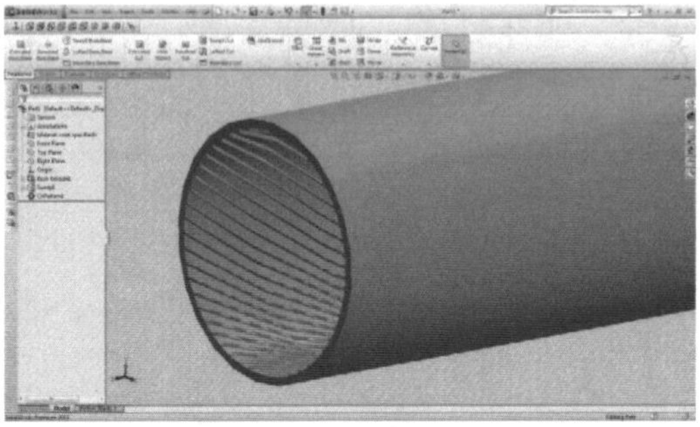

Figure 1. Model of inner grooved

2.2. Computational Fluid Dynamics (CFD) Simulation

The working fluids for this simulation were water and refrigerant R22 where the heat exchanger was counter flow (water and refrigerant flow in opposite direction) as shown in Figure 2.

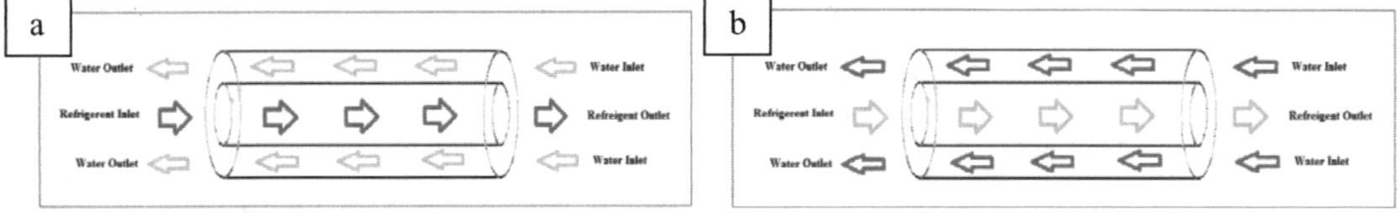

Figure 2. Schematic diagram of counter flow heat exchanger (a) condensation (b) evaporation

3. Results and Discussion

The pressure contours and pressure drop in enhanced tube are shown in Figure 3. It is evident that the pressure drop in refrigerant region is much higher compared to the water region.

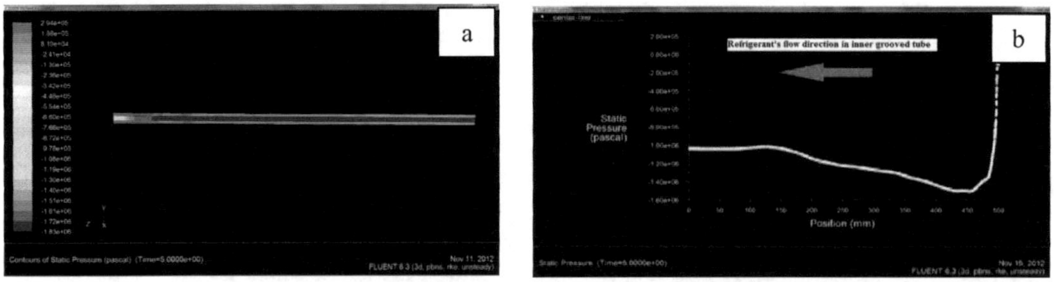

Figure 3. Pressure contours and pressure drop in enhanced tube (a) pressure contour (b) pressure drop

It can be observed (Figure 3b) that there is a drastic drop in pressure (starting from right side) for the first 50 mm (450 mm-500 mm) which is due to the existence of grooves inside the tube. This phenomenon occurred because of friction between the working fluid and inner tube surface. As the friction increases, the resistance of working fluid to move along the tube increases. The pressure after the drastic drop is very low. Hence, it is assumed to be no pressure change afterward.

A total of four cases were simulated and the results are shown in Table 2. The results are also compared with experimental results reported in [2]. The differences are in the range of 9.43% to 11.87%. Theoretically, the total heat transfer coefficient should be higher in simulation because in simulation it is assumed that the tube is well insulated and no heat loss to the surrounding.

Table 2. Overall heat transfer coefficient of enhanced tube

Cases	Overall heat transfer coefficient, U_0 (W/m^2.K)		Difference (%)
	Experiment [2]	Simulation	
1	2511	2212.87	11.87
2	2696	2574.25	4.51
3	2211	2425.60	9.70
4	3543	3877.11	9.43

4. Conclusions

Flow of refrigerant (R22) through enhanced copper tube for two different conditions, i.e., condensation and evaporation was simulated by CFD and the results were compared with published experimental results. The overall heat transfer coefficient of enhanced tube is found to be higher compared to plane tube. A moderately good agreement was observed between simulation and experimental results.

Acknowledgement

Financial support from Universiti Tenaga Nasional for enabling the author to present this paper in ICEE 2013 is greatly appreciated.

References

[1] Kasai K, Fujino H, Ebisu T and Torikoshi K 1999 Proc. *36th National Heat Transfer Symposium of Japan* **1** [in Japanese] 141-142

[2] Rahman M M, Ling Y M, Soon, G W and Kuan G. A 2012 *International Journal of Energy and Environment* **3** 49-60

[3] Kadir B, Murat C, Hasan G and Tuba B 2008 *Applied Thermal Engineering* **29** 753-761

[4] Akio M, Yusuke O, Satoshi O and Yoshihiko M 2002 *International Journal of Refrigeration* **26** 417-424

[5] Chamra R W and Randlett M 1996 *International Journal of Heat Mass Transfer* **39** 1839-1846.

[6] Uchida M, Itoh M, Shikazono, N and Kudoh, M 1996 *Proceedings of 1996 International Refrigerant Conference (Purdue)* pp 133-138.

[7] Bharadwaj, P, Khondge A D and Date A W 2008 *International Journal of Heat and Mass Transfer* **52** 1938-1944.

[8] Colombo L P M, Lucchini A and Muzzio A 2012 *International Journal of Refrigeration* **35** 2150-2165.

IOP Conf. Series: Earth and Environmental Science 16 (2013) 012108 doi:10.1088/1755-1315/16/1/012108

Computational Fluid Dynamics (CFD) investigation onto passenger car disk brake design

Kannan M. Munisamy, Shangkari K.Kanasan Moorthy

College Of Engineering Universiti Tenaga Nasional, Jalan IKRAM-UNITEN, 43009, Kajang, Selangor
E-mail: kannan@uniten.edu.my

Abstract. The aim of this study is to investigate the flow and heat transfer in ventilated disc brakes using Computational Fluid Dynamics (CFD). NACA Series blade is designed for ventilated disc brake and the cooling characteristic is compared to the baseline design. The ventilated disc brakes are simulated using commercial CFD software FLUENT^{TM} using simulation configuration that was obtained from experiment data. The NACA Series blade design shows improvements in Nusselt number compared to baseline design.

1. Introduction

The brake is the most essential part of a vehicle regardless of it physical properties because brake system are considered as a safety control of a vehicle during emergency or any other normal conditions usage [1]. Efficient braking system enables a vehicle to stop smoothly during driving on condition that the vehicle provide essential braking torque to the wheel and at the same time dissipate heat generated by friction between the brake rotor and the pad [2].

As a result of demand on increasing vehicle speed, weight and acceleration, the necessity for brake disc cooling is also increasing [3]. Efficiency of the disc brake need to be achieved in order to full fill customers demand. Firstly, the cause of this problem should be identified. During braking, vehicles kinetic and potential energy are converted into thermal energy by the friction between brake pads and rotor faces. When the brake disc temperature rises, the braking performance reduce which may lead to deformation, judder, squeal, increased wear and complete damage of brake. Larger amount of heat generated will be absorbed by brake parts in a very short period of time during braking. As a result, the brake disc must store and dissipate energy quickly during braking process [4].

Therefore, ventilated disc brake is becoming more essential for better cooling efficiency as it consume a larger area to dissipate heat [5]. Ventilated disc brake help to improves the convection cooling by separating the braking surfaces using air passages. The rotor acts as a centrifugal fan to draw cool air from the inboard side, passing through the rotor passages and exhausting at the outer diameter. Other than that, ventilated disc brake help in storing energy and dissipating it quickly and efficiently during braking since braking process occur very fast [6].

2. CFD Simulation of Flow and Heat Transfer.

The baseline design is a commercial-vehicle ventilated cast iron disc and this design has 36 equally spaced blades. The design of this blade is done in two dimensional and the commercial CFD software

FLUENT^{TM} is used for simulation. The simulation configuration is tabulated in Table 1 below as boundary condition to CFD model.

The flow is periodical model consist of 55188 computational mesh for baseline design and 66790 computational mesh for NACA Series design. A mesh dependency test was carried out to make sure that the selected mesh will generate a mesh independent result. The RNG k-ε turbulent model with standard wall treatment is employed. The disc is simulated with six different velocity and rotational speeds. The brake wall is set to be isothermal at 300°C to simulate the braking heat.

Table 1. The simulation boundary condition

Velocity in (m/s)	Rotational Speed (rad/s)	Pout
2.46	67.34	0 gage
2.89	84.18	0 gage
3.28	101.01	0 gage
4.00	117.85	0 gage
4.41	134.68	0 gage
5.54	168	0 gage

Figure 1 compares the Nusselt number for both baseline and NACA Series design blade design. Both Nusselt number increases as the rotational speed increases and NACA Series blade design disc brake has higher Nusselt number than baseline design disc brake. The NACA Series blade design has cooling improvement of Nusselt number approximately by 14.4% compared to baseline design disc brake. Figure 2 illustrates about the temperature contour of both designs in the highest rotational speed. NACA Series design shows a better temperature distribution near the blade than the baseline design. This proofs the finding of the Nusselt number above.

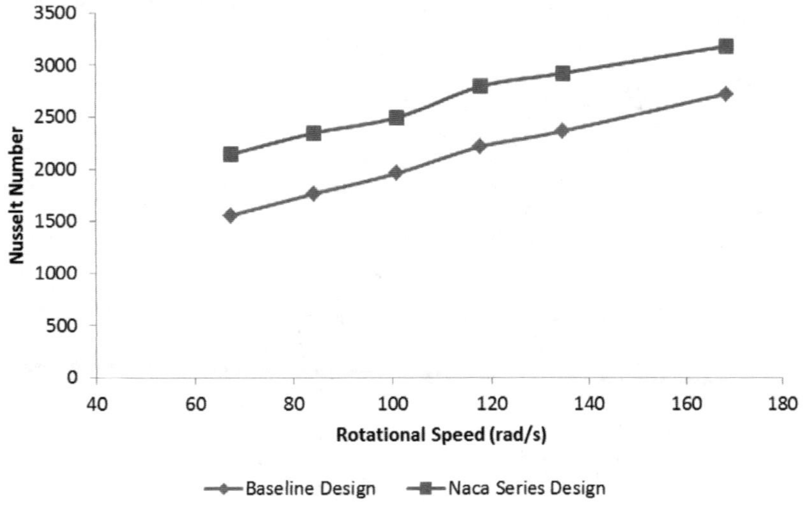

Figure 1. Nusselt Number against Rotational Speed

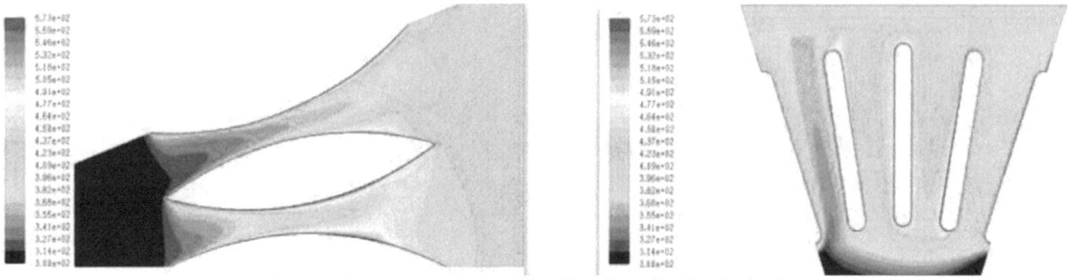

Figure 2. Temperature distribution for both designs

3. Conclusion

The investigation concludes that the NACA Series design is superior in cooling compared to the baseline design ventilated disc brake. From CFD simulations, the Nusselt number coefficient increases as the roational speed increases for both designs.

References

[1] Kannan M. 2008 Aerodynamics of car disc brake *The monthly Bulletin of The Institution of Engineers,Malaysia* (JURUTERA)

[2] A.D. McPhee and D.A. Johnson 2007 Experimental heat transfer and flow analysis of a vented brake rotor *International Journal of Thermal Sciences* **47** 458-467

[3] Kannan M. Munisamy, Norshah Hafeez Shuaib, Mohd. Zamri Yusoff and Savithry K.Thangaraju 2011 Flow Streamlining for Cooling Improvement In Ventilated Brake Disc Using CFD As Design Tool *Journal - The Institution of Engineers, Malaysia* **72** 27-35

[4] M.Z Akop, R.Kien, M.R.Mansor, M.A Mohd Rosli 2009 *Thermal Stress Analysis of Heavy Truck Brake Disc Rotor* (Degree's Thesis, Universiti Teknikal Melaka, Malaysia)

[5] Dr Marko Tirović 2007 Development of Ventilated Brake Discs *JUMV International Automotive Conference with Exhibition (Beograd Serbia, 23-25 April 2007)*

[6] D. Parish and D. G. MacManus 2005 Aerodynamic Investigations of Ventilated Brake Disks *Proceeding of the Institution of Mechanical Engineers, Part D, Journal of Automobile Engineering* **219** 471-486

IOP Conf. Series: Earth and Environmental Science **16** (2013) 012109 doi:10.1088/1755-1315/16/1/012109

Disk brake design for cooling improvement using Computational Fluid Dynamics (CFD)

Kannan M. Munisamy, Ramel Shafik

College Of Engineering Universiti Tenaga Nasional, Jalan IKRAM-UNITEN, 43009, Kajang, Selangor
E-mail: kannan@uniten.edu.my

Abstract. The car disk brake design is improved with two different blade designs compared to the baseline blade design. The two designs were simulated in Computational fluid dynamics (CFD) to obtain heat transfer properties such as Nusselt number and Heat transfer coefficient. The heat transfer property is compared against the baseline design. The improved shape has the highest heat transfer performance. The curved design is inferior to baseline design in heat transfer performance.

1. Introduction

Disk brake is the component in the wheel to slow up vehicles when moving. It is essential function to slow rotation of the wheel by the friction caused by pushing brake pads against a brake disc with a set of callipers [1]. The downside of it, in fast cars like in racing cars, they push the car to the limit whereas the cars always speed up at high speed. This leads to a lot of braking in use during a race. As we know, friction created by the disk brake to slowing down the cars causing increasing in temperature on the disk brake [2]. As the temperature increases the functionality of the brake faded as high temperature causing the disk brake to lose it friction [3]. Less friction ends up making the racing cars to slow down more difficult. To have a good racing line before cornering the car needed to slow. So less grip to brakes making the drive brakes earlier before corner hence optimum time cannot be optimize end up having a slow lap time.

It is interesting that by doing a small improvement to the cooling system [4] on the disk brake can make the car more easier to slow down and also can increase the optimization of the racing car to get a better lap time.

Until now the problem is not completely solve but many things has been implemented to get the optimum cooling which can decrease the problem [5]. Some of the implement that had been taken was by replacing the material of the disk itself with material of name such as ceramics, mixing metal materials to succeed in creating new material and etc. Brake will fade due to high temperature because it will expand and losses the strength and surface friction. So by replacing it with material that has a higher faded point it will increase the time it loses it function. But to do this task, it cost a lot of money because it is expensive. So to avoid this problem, the aim is to report the effect on cooling of blade with streamlining [6] the conventional unidirectional racing disk brake passage flow. The improvement in the cooling will directly affects the car performance especially for racing cars in critical corners and achieving top speed. Only with specific aerodynamic manipulation can be achieved without any additional of mass of extra gadget at the brake system.

2. CFD Simulation of Flow and Heat Transfer

All of the disk designs are cast iron disc and has 32 equal spaced blades. The designs are modelled in 2D and the commercial CFD software FLUENT^{TM}.

Turbulence equation that was selected for this study is the RNG k-ε with standard wall treatment is employed. The disc is simulated with six different velocity and rotational speeds as given in Table 1. The brake disk is set to be in isothermal temperature of 600°C. The simulation is run till convergence is achieved.

Table 1. The simulation boundary condition

Velocity in (m/s)	Rotational Speed (rad/s)	Pout
2.46	67.34	0 gage
2.89	84.18	0 gage
3.28	101.01	0 gage
4.00	117.85	0 gage
4.41	134.68	0 gage
5.54	168	0 gage

The Figure 1 below illustrates the comparison of the three designs for Nusselt Number against rotational speed.

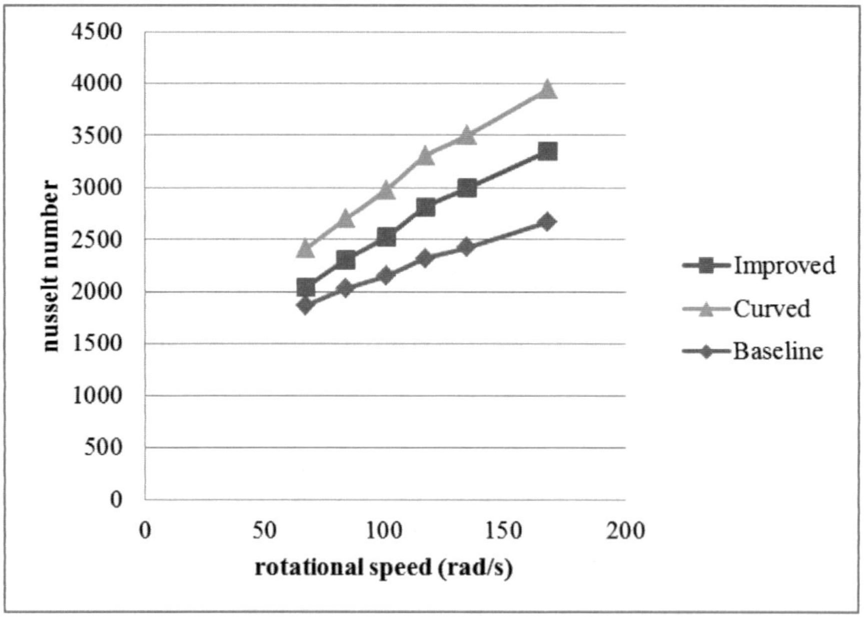

Figure 1. Nusselt Number against Rotational Speed

IOP Conf. Series: Earth and Environmental Science **16** (2013) 012109 doi:10.1088/1755-1315/16/1/012109

From Figure 1, it can be deduced that by increasing the rotational speed the Nusselt Number is increasing almost linearly. Curved design is the most superior in heat transfer performance. It is about 40% better than baseline design. The improved design is 20% better than the baseline design. Figure 2 below shows the temperature distribution in Kelvin. The curved design shows lowest temperature distribution in the flow passage compared to all the other designs.

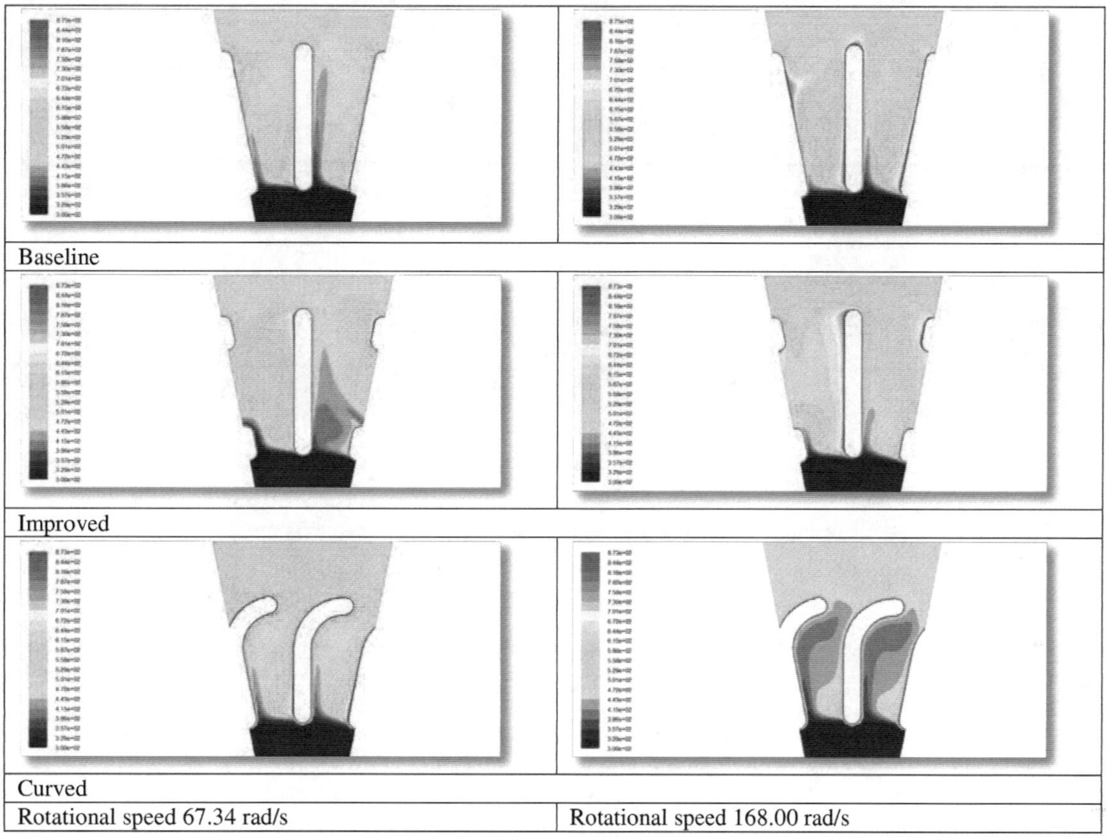

Figure 2. Temperature distribution for baseline, improved baseline, and curve model

3. Conclusion

The Nusselt number of the brake disk is increasing with increasing rotational speed. The blade design has significant effect on the heat transfer property. The curved design above with specific profile has the best heat transfer property compared the other two blade designs.

References

[1] Kannan M. 2008 Aerodynamics of car disc brake *The monthly Bulletin of The Institution of Engineers,Malaysia* (JURUTERA)

[2] Incropera, Dewitt, Bergmann and Lavine 2007 *Fundamental of Heat and Mass Transfer, Heat Convection* (New York: Wiley) 6-8

[3] Limpert, R. 1975 The Thermal v Performance of Automotive Disc Brakes - (750873) *Automobile Engineering Meeting* (Detroit: Michigan)

[4] Limpert, R. 1975 Cooling analysis of Disc Brake Rotors - (751014) *Truck Meeting* (Philadelphia: Pa)

[5] Kannan M. Munisamy, Norshah Hafeez Shuaib, Mohd. Zamri Yusoff , Savithry K. Thangaraju 2011 Flow streamlining for cooling improvement in ventilated brake disc using CFD as design tool *Journal - The Institution of Engineers, Malaysia* **72** 27-35
[6] Hudson, M.D. and R.L. Ruhl. 1997 Ventilated Brake Rotor Air Flow Investigation - (971033) *International Congress and Exposition* (Detroit: Michigan)
[7] Chaniotis, A.K,D. Poulikakos and Y.Ventikos 2003 *J Heat Transfer* **125** 575

IOP Conf. Series: Earth and Environmental Science **16** (2013) 012110 doi:10.1088/1755-1315/16/1/012110

The influence of chamfering and corner radiusing on the discharge coefficient of rotating axial orifices

A Idris[1], **K Pullen**[2]

[1]Department of Mechanical Engineering, College of Engineering, Universiti Tenaga Nasional,43000, Kajang, MALAYSIA

[2]Energy Systems & Engines Group, City University London, Northhampton Square, EC1V 0HB, London, UNITED KINGDOM

E-mail: azree @uniten.edu.my

Abstract. The effects of chamfering and corner radiusing on the discharge coefficient of rotating axial orifices are presented in this paper. Both experimental and CFD results show that chamfering and corner radiusing improve the discharge coefficient of rotating orifices. For non-inclined rotating orifices, the discharge coefficient reduces with increasing speed, but chamfered and radiused orifices manage to have higher discharge coefficient (C_d) than the straight edge orifices. Comparing between chamfering and corner radiusing, the radiused corner orifice has the highest C_d at every rotational speed. This is because the inlet radius helps guiding the flow into the orifice and avoiding flow separation at the inlet.

1. Introduction

The characteristic of the flow through a set of rotating orifices for particular conditions is quantified in terms of a discharge coefficient which is the ratio of the actual flow divided by the flow for the ideal case.

$$C_d = \frac{\dot{m}_{actual}}{\dot{m}_{ideal}} \quad (1)$$

The actual flow can be measured experimentally whereas the ideal flow is calculated according to equations derived by analysis according to stated assumptions. If the discharge coefficient is known for a particular geometry, shaft speed and flow boundary conditions such as fluid properties and pressure, the mass flow through the orifices can be calculated.

The earliest study on rotating orifices was by Meyfarth and Shine [1]. A disc with several holes drilled in it was mounted in front of a glass window. It was then placed in the exhaust stream of a J-75 turbojet engine and it was found that the glass temperature was significantly reduced. Wittig *et al.* [2] used the discharge coefficient derived from energy conservation, entropy and ideal gas relations. They conducted tests similar to Meyfarth and Shine [1], but with longer orifices. Their results showed that a longer orifice gave a higher discharge coefficient than a shorter orifice. Maeng *et al.* [3] had a similar rotating axial orifice setup to Wittig *et al.*, and varied the rotational speed from 0 to 10000 rpm. They used the discharge coefficient derived from Steady Flow Energy Equation. The discharge coeefficient is written as:

$$C_{d,Maeng} = \frac{\dot{m}_{actual} \cdot \dfrac{\sqrt{T_1}}{p_1 A}}{\sqrt{\dfrac{2\gamma}{(\gamma-1)R}\left[\left(\dfrac{1}{\Pi}\right)^{\frac{2}{\gamma}}\left(1+\dfrac{T_{rot}}{T_1}\right)-\left(\dfrac{1}{\Pi}\right)^{\frac{\gamma+1}{\gamma}}\right]}} \tag{2}$$

Idris (2004) [4] performed experimental and computational investigation of the discharge characteristics and behavior of rotating axial and radial orifices and highlighted that for rotating orifices, the most important parameter influencing the discharge coefficient is the angle of incidence. The highest discharge coefficient is attained when the incidence angle is bordering on zero.

2. Experimental setup

The objective of the experimental study was to gather discharge coefficient data for rotating orifices with combinations of flow and geometric parameters. Figure 1 shows the front view of the experimental rig. Chamfer is defined by the chamfer angle, β_w, and the ratio of *w/d*, while corner radius is represented by the ratio of *r/d*, Figure 2.

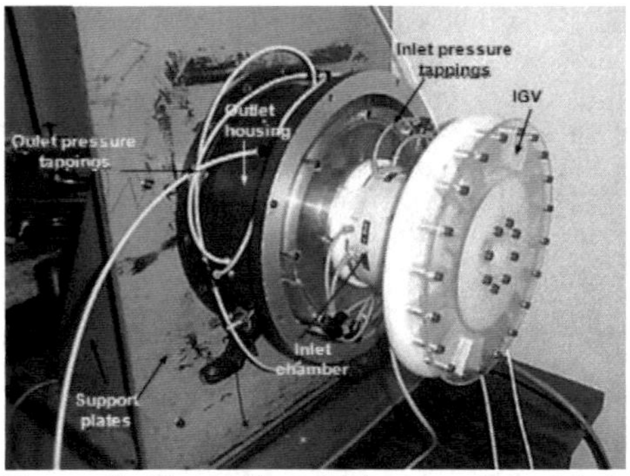

Figure 1. Front view of the experimental rig

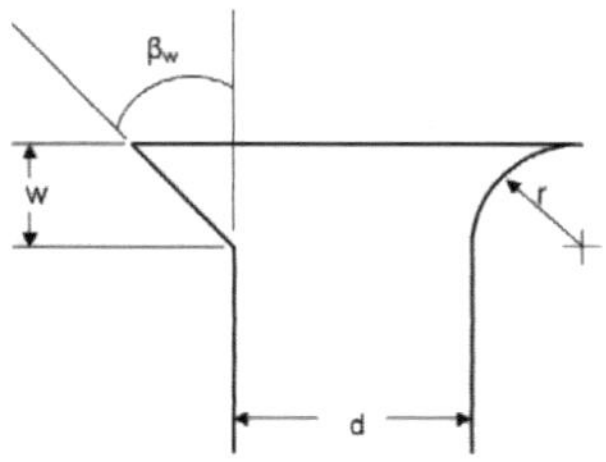

Figure 2. Definition for chamfer (*w/d*) and corner radius (*r/d*)

3. CFD Modelling

A proprietary software called STAR-CD is used in the present study. The process of fine tuning and validating CFD codes and results are performed by comparing with experimental data. The algorithm used in this study to resolve the pressure-velocity linkage in the transport equation is SIMPLE while the turbulence model used is the RNG form of the k - ε for its improved resolution of separation and swirl. Figure 3 shows the variable parameters used in building the axial orifice model. To form a chamfer, a patch is first created as shown in Figure 4a. This patch is then extruded around the local coordinate to form the chamfer, Figure 4b. The chamfer is then attached to the orifice and the inlet and outlet regions to form the complete model, Figure 4c. A similar approach is used to form the corner radius, Figure 4d.

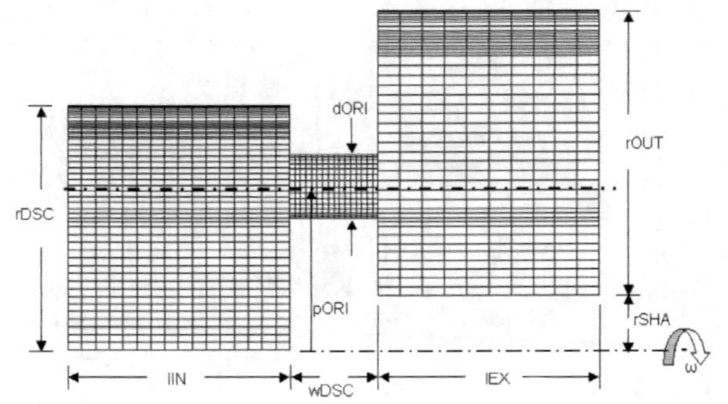

Figure 3. Variable parameters used in the axial orifice model

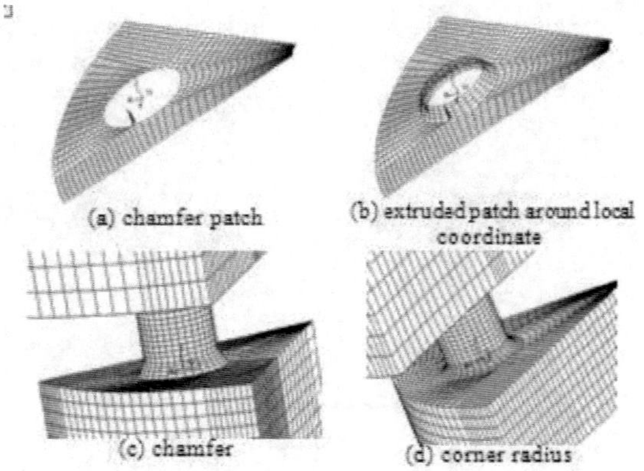

Figure 4. CFD models for chamfering and corner radiusing

4. Results and discussion

Figure 5 shows the experimental results of discharge coefficient versus incidence angle for the present study and others. Hay and Spencer [5], and an ASME nozzle have data for stationary orifices while the present study shows results for both rotating and non-rotating cases. A straight orifice with an L/d = 1.4 for the present study gives a C_d of 0.81 at 0 rpm and pressure ratio of 1.06, while Hay and Spencer [5] with L/d = 1.0 has C_d = 0.95 for a chamfered orifice at 30°, and C_d = 0.88 for a chamfered orifice at 45° and C_d = 0.94 for a radiused corner orifice. The pressure ratio used in their study was 1.2. For chamfered orifices, the discharge coefficient depends on the angle of the chamfer, with a 30° angle giving higher C_d than the 45° angle chamfer.

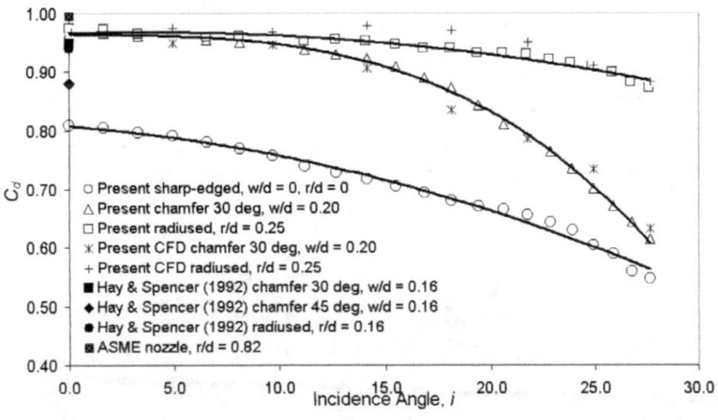

Figure 5. C_d versus Incidence Angle for chamfering and radiusing by various studies

Numerical simulations were carried out in STAR-CD to find the chamfer angle that can give the best C_d for the range of speed from 0 rpm to 21000 rpm. The CFD results are compared with the experimental data at 30° chamfer. The highest C_d can be achieved for chamfered angles from 20° to 45° for 0 rpm. When the speed is at 6000 rpm the chamfered angles from 25° to 40° can still maintain a high C_d around 0.94. For speed of 12000 rpm a 40° chamfer gives the highest C_d at 0.86. The CFD results indicate that a 40° chamfer gives the best C_d for the speed up to 12000 rpm. For radiused orifices, the discharge coefficient can be maintained at a high value for $r/d \geq 2.0$. The CFD results confirm that radiusing the inlet allows the orifice to withstand the effect of rotation on the discharge coefficient. Figure 6 shows the CFD velocity vectors for sharp-edged, chamfered and corner radiused at 0 rpm, 12000 rpm, and 21000 rpm. The radiused corner orifice has the highest C_d at every rotational speed because the inlet radius helps to guide the flow into the orifice and avoid separation.

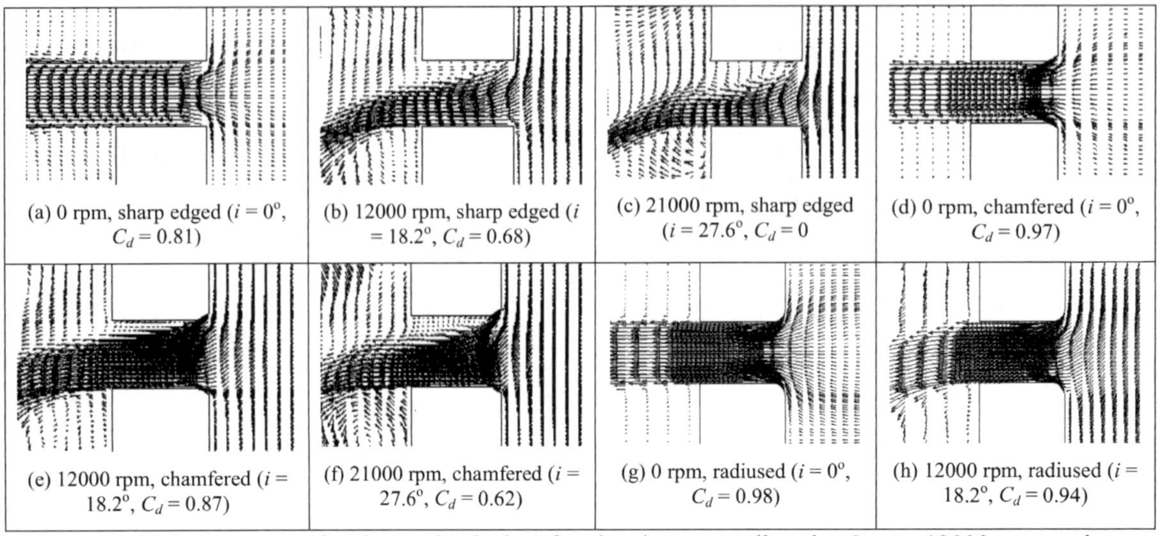

(a) 0 rpm, sharp edged ($i = 0^o$, $C_d = 0.81$)
(b) 12000 rpm, sharp edged ($i = 18.2^o$, $C_d = 0.68$)
(c) 21000 rpm, sharp edged ($i = 27.6^o$, $C_d = 0$
(d) 0 rpm, chamfered ($i = 0^o$, $C_d = 0.97$)
(e) 12000 rpm, chamfered ($i = 18.2^o$, $C_d = 0.87$)
(f) 21000 rpm, chamfered ($i = 27.6^o$, $C_d = 0.62$)
(g) 0 rpm, radiused ($i = 0^o$, $C_d = 0.98$)
(h) 12000 rpm, radiused ($i = 18.2^o$, $C_d = 0.94$)

Figure 6. Velocity vectors for sharp-edged, chamfered and corner radiused at 0 rpm, 12000 rpm, and 21000 rpm

5. Conclusions

Both experimental and CFD results show that chamfered and radiused orifices give higher C_d than the straight edge orifices. Radiused orifices, however can sustain the discharge coefficient much higher over a wider range of the incidence angle compared to sharp-edged and chamfered orifices. For chamfered orifices, the experimental results show that a 30° angle giving higher C_d than the 45° angle chamfer. Simulations between 30° and 45° are carried out using CFD, and the results show that a 40° chamfer gives the best C_d for the speed up to 12000 rpm. Even at 21000 rpm, it can still give a C_d which is higher compared to much of the others. The CFD results for radiused orifices show that the discharge coefficient can be maintained at a high value for $r/d \geq 2.0$. Radiusing the inlet allows the orifice to withstand the effect of rotation on the discharge coefficient.

References

[1] Meyfarth, P. F. and Shine, A. J. 1965 Experimental Study of Flow Through Moving Orifices *ASME Journal of Basic Engineering* **87** 1082-1083

[2] Wittig, S., Jakoby, R. and Weissert, I. 1996 Experimental and Numerical Study of Orifice Discharge Coefficients in High-Speed Rotating Disks *ASME Journal of Turbomachinery* **118** 400-407

[3] Maeng, D. J., Lee, J. S., Jakoby, R., Kim, S., and Wittig, S. 1999 Characteristics of Discharge Coefficient in a Rotating Disk System *Transactions - ASME Journal of Engineering for Gas Turbines and Power* **121** 663-669

[4] Idris, A., Pullen, K. and Barnes, D. 2004 An Investigation Into the Flow within Inclined Rotating Orifices and the Influence of Incidence Angle on the Discharge Coefficient *Proc. Instn Mech. Engrs, Part A: J. Power and Energy* **218** 55-69

[5] Hay, N. and Spencer, A. 1992 Discharge Coefficients of Cooling Holes With Radiused and Chamfered Inlets *Journal of Turbomachinery* **114** 701-706

The effects of area contraction on the performance of UNITEN's shock tube: Numerical study

A M Mohsen, M Z Yusoff[1], A Al-Falahi

[1]Centre for Advanced Computational Engineering, College of Engineering Universiti Tenaga Nasional, UNITEN, Jalan IKRAM-UNITEN, 43009 Kajang, Selangor, Darul Ehsan, Malaysia.

E-mail: zamri@uniten.edu.my

Abstract. Numerical study into the effects of area contraction on shock tube performance has been reported in this paper. The shock tube is an important component of high speed fluid flow test facility was designed and built at the Universiti Tenaga Nasional (UNITEN). In the above mentioned facility, a small area contraction, in form of a bush, was placed adjacent to the diaphragm section to facilitate the diaphragm rupturing process when the pressure ratio across the diaphragm increases to a certain value. To investigate the effects of the small area contraction on facility performance, numerical simulations were conducted at different operating conditions (diaphragm pressure ratios P_4/P_1 of 10, 15, and 20). A two-dimensional time-accurate Navier-Stokes CFD solver was used to simulate the transient flow in the facility with and without area contraction. The numerical results show that the facility performance is influenced by area contraction in the diaphragm section. For instance, when operating the facility with area contraction using diaphragm pressure ratio (P_4/P_1) of 10, the shock wave strength and shock wave speed decrease by 18% and 8% respectively.

1. Introduction

A short duration high speed flow test facility is an apparatus used to generate gas flow conditions of high enthalpy and high Mach number for a very short duration (in the order of milliseconds). Such facility includes shock tube, shock tunnel, plasma jet, arc heated tunnel, free piston tunnel, and gun tunnel [1]. Shock tube is one of the most versatile and economical high speed flow test facilities. It is used to generate homogenous, high enthalpy and high pressure gas flow conditions for a very short duration. It allows scientists to perform experimental and numerical studies on shock waves, their interaction, and other Mach number phenomena [2]. Shock tube is an excellent tool for investigating the nature of shock wave propagation in addition to high temperature phenomena in gases [3].

In short duration test facilities, it is quite difficult to fully examine all aspects of flow. This is due to extreme flow conditions and severe time constraint involved. Therefore, numerical studies have become very important tool to provide a quantitative description on the flow field conditions in such facilities [4, 5]. For example, the transient flows in shock tunnel facility have been investigated by Chue et al. [6] using a numerical analysis. Moreover, a numerical study using a finite volume based code was performed by Jacob et al. [7]. In their study, they simulated the shock reflection process in an axi-symmetric shock tube and a high Mach number nozzle using a perfect gas with no boundary layers applied on the shock tube walls. Although, many works were published on the development of shock tubes, it is essential to mention that far too little attention has been paid to the effects of a small area change in the diaphragm section on the transient flow process in a shock tube. The present study involved numerical simulations on the facility with and without area contraction.

IOP Conf. Series: Earth and Environmental Science **16** (2013) 012111 doi:10.1088/1755-1315/16/1/012111

2. Two-Dimensional CFD Solver

A two dimensional time accurate Navier-Stokes solver was developed and programmed based on the dimensions and configuration of the test facility which has been recently built at the Universiti Tenaga Nasional (UNITEN) [8]. The solver uses second order accurate cell-vertex finite volume spatial discretization and forth orders accurate Runge-Kutta temporal integration. The current program was developed and validated against exact solution (the Sod's tube problem) and experimental measurements in an earlier work by Al-Falahi et al. [9]. For more details on the solver, refer to References [9, 10].

3. Results and Discussion

The physics of the flow in the shock tube with and without area contraction have been analyzed and presented. In order to study the transient pressure in the facility, two stations have been selected to record and store the data of the interesting parameters as shown in Figure 1.

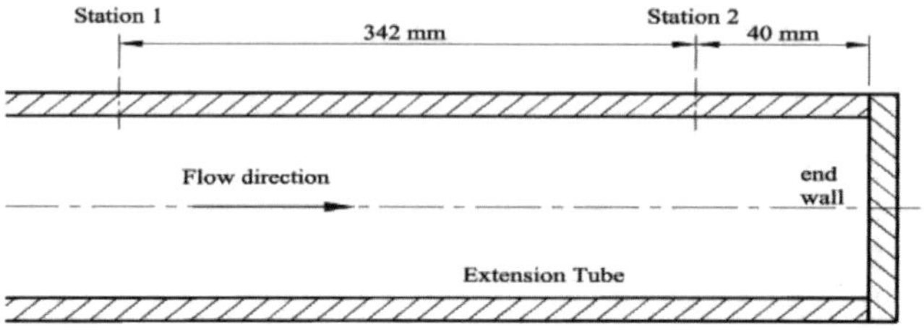

Figure 1. The two stations on the extension tube

3.1. Pressure Transient

The pressure history at station 1 and 2 obtained with area contraction for diaphragm pressure ratio (P_4/P_1) of 10 is plotted in Figure 2(a). From this Figure, it is possible to follow the physics of the flow inside the shock tube. When one traces the pressure history at station 1, the first jump represents the shock wave, while the second jump represents the reflected shock wave. In the first jump, the pressure inside the shock tube increases rapidly from 100 KPa to about 225 KPa. The shock wave compresses and heats the test gas as it is traveling towards the end of the shock tube, and subsequently reflects off at the closed end of the driven section to further compresses the test gas and increases its pressure to around 450 KPa. Following that, the reflected shock wave interacts with the contact surface and this process results in further increase in the pressure until it reaches its maximum value of 520 KPa.

The pressure history at the two stations obtained after removing the bush from the facility is depicted in Figure 2(b). This Figure shows similar trend as for area contraction. The shock wave increases the pressure at station 1 from 100 KPa to around 275 KPa and then reflects and further increases the pressure to 660 KPa. Finally, it interacts with the contact surface to achieve a peak pressure of 800 KPa. It is interesting to note that the shock wave strength, reflected shock strength, and peak pressure are higher than those obtained with area contraction represented earlier.

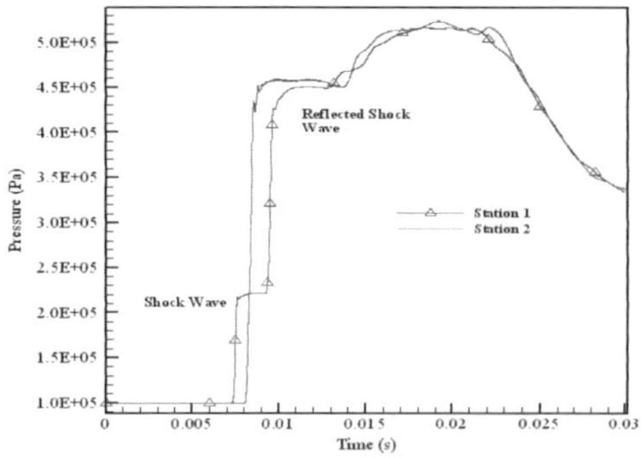

(a) with area contraction

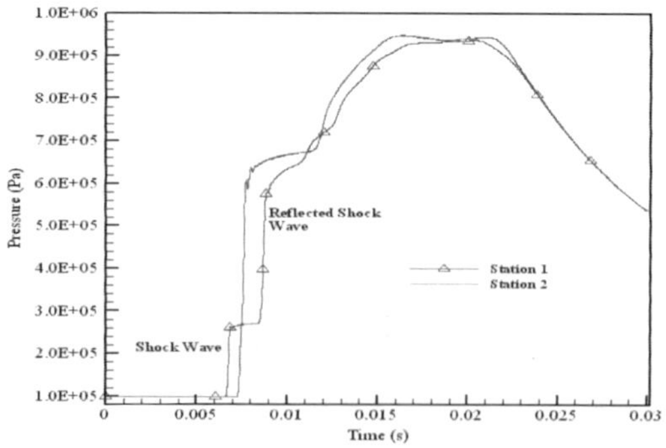

(b) without area contraction

Figure 2. Pressure history with and without area contraction (*$P_4/P_1=10$, Air-Air*)

3.2. Density Transient

Time-distance diagram is a useful tool used to provide an overall view of the flow process inside the shock tube. Figure 3(a) depicts the x-t diagram for density profile in the facility for shot with area contraction. From this Figure, it can be noted that after diaphragm rupture, a shock wave propagates along the driven section, followed by the contact surface. At the same time, a rarefaction waves are generated and propagate in the compression chamber. Following that, both waves reflect off at the closed ends of the shock tube. The shock interacts with the contact surface. Due to the existence of area contraction in the diaphragm section, the rarefaction wave undergoes another reflection within the driver section of the facility as shown in the same Figure. This process is undesirable since it is useful to examine the flow conditions at the region between the reflected rarefaction wave and the contact surface, which is more commonly known as Region 3.

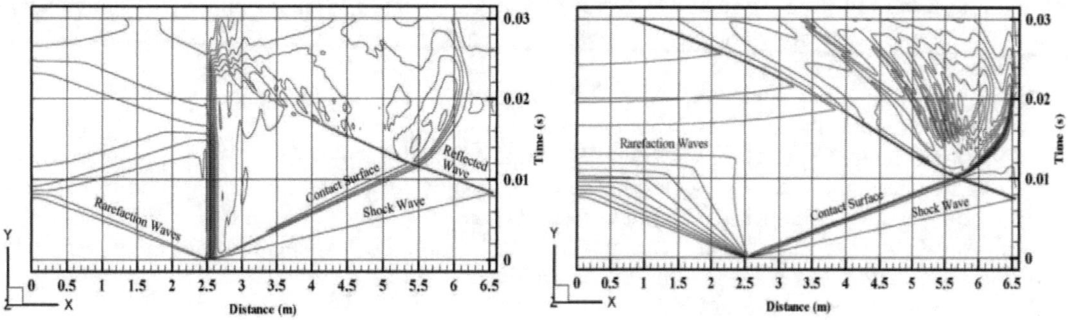

(a) with area contraction (b) without area contraction

Figure 3. *x-t* diagram for density profile

The x-t diagram for density profile obtained with no area contraction in the diaphragm section is also plotted in Figure 3(b). This shows that removing the bush from the diaphragm section has allowed both reflected waves passing into the other sections of the facility and interacting with each other.

3.3. Shock Wave Mach Number

In order to summarize the effects of area contraction in the diaphragm section on shock Mach number, the obtained values of shock wave Mach number at different operating conditions are gathered in Figure 4. It is shown that higher shock Mach numbers are obtained for shots with no area contraction as the same operating conditions are applied. For example; shock Mach number of 1.6 is obtained for shot with area contraction at diaphragm pressure ratio (P_4/P_1) of 20. While at the same diaphragm pressure ratio, the obtained shock Mach number has increased to about 1.72 when the facility was operated with no contraction. This Figure also shows that the shock wave Mach number has increased by increasing the pressure ratio across the diaphragm.

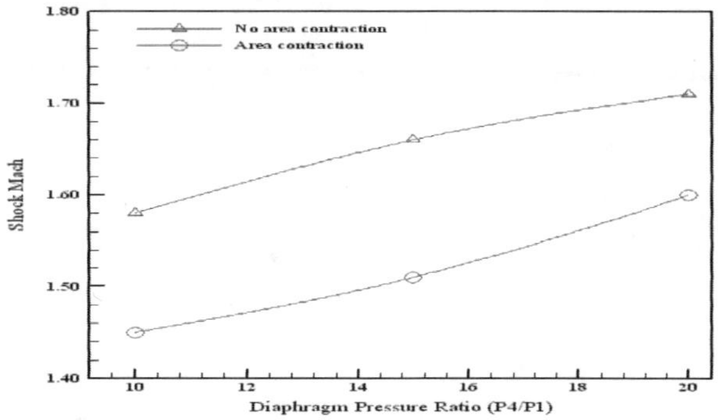

Figure 4. Shock Mach number at different operating conditions

4. Conclusion

The paper reported the flow process inside a short duration high speed flow test facility built at the college of engineering- The Universiti Tenaga Nasional in Malaysia. The effects of area contraction on the facility performance have been investigated numerically. The present code showed good capability of providing the x-t diagram successfully. From this diagram the useful duration or test time can be determined. Results also show that two-dimensional modeling of the high speed flow test facility is an effective way to obtain facility performance data. However, it has been shown that the instantaneous pressure history and shock Mach number are significantly affected by the presence of area contraction in the facility.

References

[1] Lu F K and Wilson D R 1994 Survey of Short Duration, Hypersonic and Hypervelocity Facilities *18th AIAA Aerospace Ground Testing Conference AIAA (Univercity of Texas, Alington)* (vol 94-2491) pp 2-12

[2] Jiang Z and Takayama K 1998 Reflection and Focusing of Toroidal Shock Waves from Coaxial Annular Shock Tubes *J.Computers and Fluids* **27** 553-562

[3] Nagai H, Yang H S, Ueta Y, Murakami M, Takano N and Yanaka K 2001 Application of Super Fluid Shock Tube Facility To Experiments of Highly Transient Low-Temperature Thermo-Fluid Dynamic Phenomena *J.Cryogenics* **41** 421-428

[4] Wilson G J, Sharma S P and Gillespie W D 1993 Time Dependent Simulations of Reflected Shock/Boundary Layer Interaction *Report Number: AIAA 93-0480*

[5] Goozee R J, Jacobs P A and Buttsworth D R 2006 Simulation of a complete reflected shock tunnel showing a vortex mechanism for flow contamination. J.*International Journal of Shock wave* **15(3-4)** 165-176

[6] Chue R S M. and Eitelberg G 1998 Studies of the transient flows in high enthalpy shock tunnels *J.Experiments in fluid* **25** 474-486

[7] Jacob P A 1991 Simulation of transient flow in a shock tunnel and a high mach number nozzle *NASA Contractor Report 187606 ICASE Report* (No. 91-60)

[8] Al-Falahi A, Yusoff M Z and Yusaf T 2008 Development of a Short Duration Hypersonic Test Facility at Universiti Tenaga Nasional *J.Institute of Engineers, Malaysia* **69(1)** 32-38

[9] Al-Falahi A, Yusoff M Z and Yusaf T 2008 Numerical Simulation of Inviscid Transient Flows in Shock Tube and Its Validations *International Journal of Engineering and Natural Sciences* **4** 52-62

[10] Al-Falahi A 2008 *Design, Construction and Performance Evaluation of Short Duration High Speed Flow Test Facility* (Ph.D. thesis, Univesity Tenaga Nasional-Malaysia)

Trend and future of diesel engine: Development of high efficiency and low emission low temperature combustion diesel engine.

R J Ho[1], M Z Yusoff [2], K. Palanisamy[2]

[1]JKM, Politeknik Kuching Sarawak, 93050 Kuching, Sarawak, Malaysia

[2]College of Engineering, Universiti Tenaga Nasional (UNITEN), Malaysia

E-mail: rj_5847@hotmail.com

Abstract. Stringent emission policy has put automotive research & development on developing high efficiency and low pollutant power train. Conventional direct injection diesel engine with diffused flame has reached its limitation and has driven R&D to explore other field of combustion. Low temperature combustion (LTC) and homogeneous charge combustion ignition has been proven to be effective methods in decreasing combustion pollutant emission. Nitrogen Oxide (NO_x) and Particulate Matter (PM) formation from combustion can be greatly suppressed. A review on each of method is covered to identify the condition and processes that result in these reductions. The critical parameters that allow such combustion to take place will be highlighted and serves as emphasis to the direction of developing future diesel engine system. This paper is written to explore potential of present numerical and experimental methods in optimizing diesel engine design through adoption of the new combustion technology.

1. Introduction

One of the main sources of pollutant emission are contributed by diesel combustion from either power generation or transports. Compressed ignition engine such diesel engine is one of the most efficient power plant for transportation due to its relatively high compression ratios and lack of throttling losses [1]. As result, wide applications of such diesel engine for transportation are one of the most noTable sources of pollution emission. It is reported that excess of 30% of total passenger automobile in European countries are powered by diesel engine [2]. Research on effects of diesel combustion indicates emissions of nitrogen oxide emission (NO_x) and particulate matter (PM) that has adverse effect on human health [3]. Overall, it is estimated that 6%-32% of NO_x is contributed by passenger automobile. Exhaust NO_x emission is about 300 times stronger greenhouse effect if compared to carbon dioxide(CO_2) [3]. For the past decades, extensive research has been done for NO_x and PM reduction. Increasing engine thermal efficiencies while suppressing pollutant emission are very challenging as it has contradicting objectives. Thermal efficiency, NO_x and PM are strong functions of temperature and equivalence ratio. Engine efficiency and NO_x emission increases with combustion temperature while reduction of flame temperature reduces soot oxidation increasing soot emission [4].

To suppress NO_x, a relatively new combustion technology known as low temperature combustion (LTC) has been introduced. Application for burners and furnaces has been shown to have the potential of reducing NO_x and soot formation [6]. This mode of combustion has been researched and developed under different names in different countries such as High temperature air combustion (HTAC) in Japan, Flameless Oxidation (FLOX) in Germany, Low Nitric Oxide Injection (LNI) in United State of America and Mild Combustion in Italy [7]. This mode of combustion in diesel engine is further researched under HCCI combustion system [8]. However, there are limited successes due to complex fluid and chemical interaction under engine dynamic state condition.

2. Background Studies

Compressed ignitions are the most fuel-efficient engines ever developed for transportation industries mainly attributed by its relatively high compression ratios and reduced throttling losses. However, diesel engine comes with drawback of high NO_x and PM has made it a challenge to the industry to comply to stringent emission legislation. Numerous studies have been done on this area where efficiencies have increased and pollutant emissions have decreased substantially [1]. However, improvement to conventional diesel engine has come to a limitation as describe in Figure1. A reduction of pollutant emission is limited with conventional diesel combustion. Piston compression stroke rises combustion air above diesel fuel autoignition temperature. Fuel injection takes place during combustion stroke with fuel ignited when fuel vaporization achieves ignition equivalence ratio producing lifted, partially premixed turbulent jet surrounded by a diffusion flame [4]. Stoichiometric mixture is oxidized and combusted with high temperature diffused flame resulting in NO_x formation and reduces the available oxygen for remaining fuel entrained in the flame resulting in soot formation. To meet increasing demand of stricter emission and efficiencies, researches are directed towards low temperature combustion.

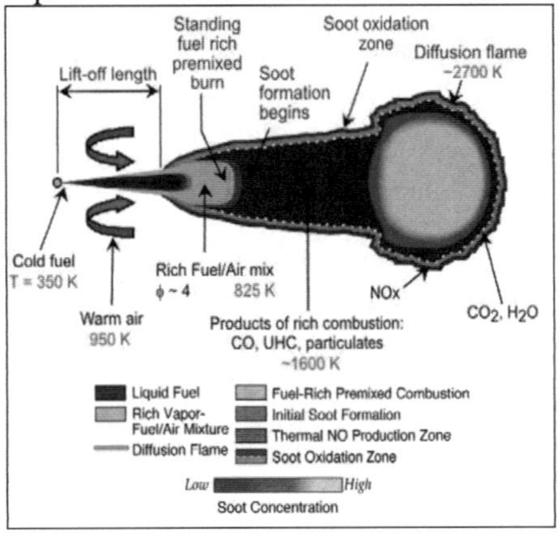

Figure1. Conventional direct injection diesel spray combustion schematics[1].

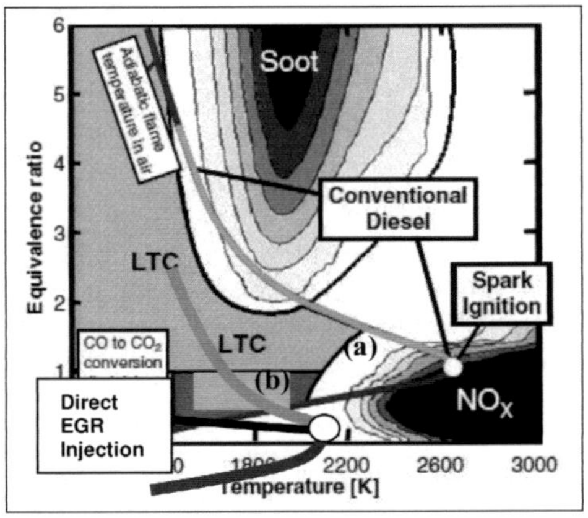

Figure2. NO_x and PM formation region corresponding to equivalence ratio at temperature[1].

A systematic approach from numerical simulation indicates that highest efficiency were attained from increase of compression ratio, decrease of equivalence ratio and increase of EGR[9].

2.1. Homogeneous Charge Compressing Ignition (HCCI)

HCCI is based on pollutant emission reduction concept by suppressing peak flame temperature by having self- ignition of premixed homogeneous charge in combustion chamber for homogeneous combustion. Maximum temperature of the charge is kept low by means of low equivalence ratio with or without exhaust gas recirculation (EGR) dilution [8]. This mode of combustion offers significant amount of pollutant emission reduction. HCCI are also classified under LTC but distinction is HCCI suppresses combustion temperature with lean combustion with low equivalence ratio. Applications of HCCI on diesel engine are limited to the ability of diesel engine in-cylinder flow process to replicate mixture homogeneity throughout the combustion volume prior to multiple point autoignition. Before

HCCI combustion can be fully applied to automotive diesel engine, main challenge has to be solved. They are poor low-loads combustion efficiency, limited operation range, poor understanding of limiting factors and poor combustion control under transient condition [1]. This limitation is primarily due to complex chemical kinetic controlled combustion coupled with complex in-cylinder fluid interaction and condition that determines the autoignition timing and combustion processes [10]. Optical imaging on HCCI system indicates that soot formation is dependent on oxygen concentration of fuel and OH* radical is responsible for NO_x formation [11].

An experimental and numerical investigation was conducted on correlation on the effect reactant temperature and mixture stratification to HCCI combustion characteristics [10]. The research indicates HCCI combustion consisted of two stages. First stage of low temperature heat release combustion and later followed by high temperature heat release combustion. Different port injection strategies infer that autoignition advances with mixture in-homogeneity. This result in large temperature stratification causes by low mixing duration from short ignition delay. Chemilumisnescence intensity increases with shorter ignition delay indicating OH* formation concentrated at different localised stratified region. Turbulence generated from higher engine speed produced higher homogeneous temperature distribution. Mixture autoignition is also dependant on thermal stratification. Lower wall temperature causes large thermal stratification at wall region. Strong heat transfer to lower wall temperature to have lower reaction rate. Comparison with different wall temperature indicates that chemical kinetic reaction has been "frozen" due to low temperature condition that is similar to blow-off during engine rapid expansion stroke producing high exhaust unburned hydrocarbon emission. Large thermal stratification has lower heat release rate which is beneficial at high load application while low thermal stratification avoids quenching that benefits low load condition.

2.2. Low Temperature Combustion

Discovery of such combustion mode dated back to as early as 1989 with discovery of combustion with carbon monoxide below 1ppm with negligible NO_x formation which was initially thought to be malfunction of sensors [5]. Review on low temperature combustion research under different names shares similarities of homogeneous fuel oxidation without development of flame produces extremely low NO_x and PM emission. This can be achieved by combusting the fuel at special condition by mixing fuel with inert gas, mainly EGR, prior to contact with oxidizer for combustion providing that EGR with reactant temperature are sufficiently high[6]. Such flameless LTC involves strong mixing of exhaust gas recirculation to keep excess air ratio low enough for flameless combustion regime with temperature higher than mixture autoignition point. This can be validate by the methods used to produce flameless combustion conducted with fuel injected into strong EGR zone at high temperature preheated air [7, 6]. The basis for such combustion pollutant emission is to avoid locally rich regions by enhancing pre-combustion mixing and dilution to yield locally low equivalence ratio with EGR to reduce peak combustion temperatures. Similar finding was discovered for diesel spray combustion. As mixture oxygen concentration reduced and combusting the elevated temperature, neither flame luminosity nor OH chemiluminescence is evident. A low temperature reaction took place without any formation of soot. Although such reaction did indicate a slower ignition process, but it still shows certain agreement with apparent heat release rate (AHRR) curves.

3. Proposed Combustion Technology

Limited successes for diesel engine have been reported since as it has strong dependencies on mixture concentration and chemical kinetic [10]. A novel combustion control is proposed based on the hypothesis of direct exhaust gas recirculation injection at high pressure after fuel injection is able to suppress flame development, promoting fuel evaporation and mixing as a mean of combustion control for high efficiency and low pollution LTC diesel engine. Numerical method will be applied together with existing mathematical model to simulate EGR injection parameters to produce desired LTC with OpenFOAM. Parametric optimization will be done using OpenFoam simulation platform which later compared with experiments as prove of concept and model validation. EGR injection after fuel injection is intended to improve fuel evaporation and mixing promoting LTC to reduce both NO_x and soot formation at the same time. Figure2 is an equivalence ratio versus combustion temperature graph together with nitrogen oxide and soot formation region respectively. Operation of conventional diesel engine falls along the line noted by (a) with ignition taking place at stoichiometric equivalence ratio of 1 forming high temperature flame combustion with thermal NOx and finally ends with extremely rich mixture region entrained in the flame with depleted oxygen producing soot. The proposed method of direct EGR injection is intended to shift the operation of combustion with diffused flame from line (a) to low temperature flameless combustioon at line (b). This could possibly be achieved with accurate direct EGR injection timing, amount and duration after fuel injection. Fuel is first allowed to undergo HCCI combustion with excess air and later followed by direct EGR injection to suppress flame development when the mixing reaches stoichiometric ratio. Fuel injection and combustion can continue with high EGR LTC reducing both NOx and soot formation regardless of load condition with high turbulence mixing of LTC chemical kinetic reaction and reduce unburned hydrocarbon by reducing wall wetting at the same time.

4. Conclusion

There has been great interest in LTC and HCCI technology for high efficiency and low pollution emission diesel engine application. However, the application of both combustion modes in diesel engine has limited success due to complex combustion control and narrow engine operation range. The novelty of this research comes from the proposed hypotheses of direct EGR injection as a mean of combustion control to expand the range of engine operation under low temperature flameless combustion mode. This novel combustion method has the potential of alleviating challenges faced by both LTC and HCCI technology making these high efficiency and low pollution emission combustion mode to be successfully implemented to conventional diesel engine with minimal modification.

References

[1] Dec J E 2009 Advanced Compression-Ignition Engines Understanding the In-Cylinder Processes *Proc. Combustion Institute* 2727-2742
[2] Nelson P F, Tibbett A R and Day S J 2008 Effects of Vehicle Type and Fuel Quality on Real World Toxic Emissions From Diesel Vehicles *Atmospheric Environment* **42** 5291-5303
[3] Karavalakis G, Poulopoulos S and Zervas E 2012 Impact of Diesel Fuels on the Emissions of Non-Regulated Pollutants *Fuel* 85-91
[4] Tree D R and Svensson K I 2007 Soot Processes in Compression Ignition Engines *Progress in Energy and Combustion Science* **33** 272-309
[5] Wunning J 2005 Flameless Oxidation *6th HiTACG Symposium* (Essen – Germany)
[6] Wunning J and Wunning J 1997 Flameless Oxidation to Reduce Thermal NO-Formation *Prog. Energy & Combustion* 81-94
[7] Weber R, Smart J P and Kamp W V 2005 On the (MILD) Combustion of Gaseous, Liquid, and Solid Fuels in High Temperature Preheated Air *Proc.Combustion Institute*
[8] Cavaliere A and Joannon M 2004 Mild Combustion *Progress in Energy and Combustion Science* **30** 329-366
[9] Caton J A 2012 The Thermodynamic Characteristics of High Efficiency, Internal Combustion Engines *Energy Conversion and Management* 84-93
[10] Liu H, Zheng Z, Yao M, Zhang P, Zheng Z, He B and Qi Y 2012 Influence of Temperature and Mixture Stratification on HCCI Combustion Using Chemiluminescence Images and CFD Analysis *App. Thermal Eng.* **33-34** 135-143
[11] Mancaruso E and Vaglieco B M 2010 Optical Investigation of Combustion Inside the Engine in HCCI Mode and Using Alternative Diesel Fuel *Exp.Thermal and Fluid Sci.,* **34** 346-351

4th International Conference on Energy and Environment 2013 (ICEE 2013) IOP Publishing
IOP Conf. Series: Earth and Environmental Science **16** (2013) 012113 doi:10.1088/1755-1315/16/1/012113

Dynamic indoor thermal comfort model identification based on neural computing PMV index

K S Mohamed Sahari, M F Abdul Jalal, R Z Homod, Y K Eng

Centre for Advanced Mechatronics and Robotics (CAMaRo), Universiti Tenaga Nasional, Putrajaya Campus, Jalan IKRAM-UNITEN, 43000, Kajang, Selangor, Malaysia

E-mail: khairuls@uniten.edu.my

Abstract. This paper focuses on modelling and simulation of building dynamic thermal comfort control for non-linear HVAC system. Thermal comfort in general refers to temperature and also humidity. However in reality, temperature or humidity is just one of the factors affecting the thermal comfort but not the main measures. Besides, as HVAC control system has the characteristic of time delay, large inertia, and highly nonlinear behaviour, it is difficult to determine the thermal comfort sensation accurately if we use traditional Fanger's PMV index. Hence, Artificial Neural Network (ANN) has been introduced due to its ability to approximate any nonlinear mapping. Using ANN to train, we can get the input-output mapping of HVAC control system or in other word; we can propose a practical approach to identify thermal comfort of a building. Simulations were carried out to validate and verify the proposed method. Results show that the proposed ANN method can track down the desired thermal sensation for a specified condition space.

1. Introduction

Thermal comfort can be defined as the state of mind which expresses satisfaction with the thermal environment, and therefore it depends on the individual's physiology and psychology [1]. As each building presents its own constructional characteristics and occupant's habit, most common HVAC systems are incapable of providing thermal comfort by taking into account these variations, further increasing energy consumptions. To solve the problems, selection of suitable control system such as neural computing Predictive Mean Vote (PMV) indices is important to consider the indoor ambiences and occupants' habits to achieve optimal thermal comfort, hence minimize the percentage of dissatisfaction and subsequently reduce energy consumption.

HVAC control system has the characteristic of time delay, large inertia, and highly nonlinear behaviour, making it is difficult to determine the thermal comfort sensation accurately if we use traditional Fanger's PMV index [2]. Hence, in this paper, Artificial Neural Network (ANN) has been introduced due to its ability to approximate any nonlinear mapping. Using ANN to train, we can get the input-output mapping of HVAC control system or in other word; we can propose a practical approach to identify thermal comfort of a building.

2. Modelling Approach

PMV indices are adopted in this paper. In the calculation, six variables were taken into consideration: two personal dependant variables which are clothing insulation and occupant's activity level; four environmental dependent variables which are air temperature, air humidity, mean radiant temperature, and air velocity. Based on these, the following equations were derived for dynamic thermal comfort sensation model:

$$M_H cp_H \frac{dT_{sk,t}}{dt} = M - W \mp R \mp C \mp K - E - RES \tag{1}$$

$$\begin{aligned} M_H cp_H \frac{dT_{sk,t}}{dt} &= (M-W) - \left\{3.9 f_{cl}(t_{cl} - \bar{t}_r)\right\} - \left\{f_{cl} \times (12.1\sqrt{V_{ar}}) \times (t_{cl} - t_a)\right\} - \\ &- \left\{(3.05\times10^{-3})(256T_{sk,t} - 3373 - p_a) + 0.42\left[(M-W) - 58.15\right]\right\} \\ &- \left\{0.0014M(34 - t_a) + (1.72\times10^{-5})(M)(5867 - p_a)\right\} \end{aligned} \tag{2}$$

$$t_{cl} = T_{sk,t} - y \tag{3}$$

$$y = 3.96\times10^{-8} f_{cl} l_{cl}\left[(t_{cl} + 273)^4 - (\bar{t}_r + 273)^4\right] + 2.38 f_{cl} l_{cl} (t_{cl} - t_a)^{1.25} \tag{4}$$

Substitute (3) into (2) yields:

$$\begin{aligned} M_H cp_H \frac{dT_{sk,t}}{dt} &= (M-W) - \left\{3.9 f_{cl}(T_{sk,t} - y - \bar{t}_r)\right\} - 12.1 f_{cl}\sqrt{V_{ar}}(T_{sk,t} - y - t_a) - \\ &- \left\{(3.05\times10^{-3})(256T_{sk,t} - 3373 - p_a) + 0.42\left[(M-W) - 58.15\right]\right\} \\ &- \left\{0.0014M(34 - t_a) + (1.72\times10^{-5})(M)(5867 - p_a)\right\} \end{aligned} \tag{5}$$

Taking Laplace Transform with zero initial condition:

$$\begin{aligned} M_H cp_H ST_{sk}(s) &= (M-W) - \left\{3.9 f_{cl}(T_{sk}(s) - y - \bar{t}_r)\right\} - 12.1 f_{cl}\sqrt{V_{ar}}(T_{sk}(s) - y - t_a) - \\ &- \left\{(3.05\times10^{-3})(256T_{sk}(s) - 3373 - p_a) + 0.42\left[(M-W) - 58.15\right]\right\} \\ &- \left\{0.0014M(34 - t_a) + (1.72\times10^{-5})(M)(5867 - p_a)\right\} \end{aligned} \tag{6}$$

Collect the T_{sk} to the left hand side of (6) and rearrange:

$$M_H cp_H ST_{sk}(s) + 3.9 f_{cl} T_{sk}(s) + 12.1 f_{cl}\sqrt{V_{ar}} T_{sk}(s) + 0.781 T_{sk}(s) = L \tag{7}$$

$$\left[3.9 f_{cl} + 12.1 f_{cl}\sqrt{V_{ar}} + 0.781\right] \times T_{sk}(s)\left[\left(\frac{M_H cp_H}{3.9 f_{cl} + 12.1 f_{cl}\sqrt{V_{ar}} + 0.781}\right)S + 1\right] = L \tag{8}$$

$$XT_{sk}(s)\left[\left(\frac{M_H cp_H}{X}\right)S + 1\right] = L \tag{9}$$

$$PMV(s) = \left(\frac{L}{\tau S + 1}\right)\left(\frac{1}{X}\right) \tag{10}$$

where

$$\tau = \frac{M_H cp_H}{X} \tag{11}$$

$$X = 3.9 f_{cl} + 12.1 f_{cl} \sqrt{V_{ar}} + 0.781 \tag{12}$$

$$L = (M - W) + (y + \bar{t}_r) + 12.1 f_{cl} \sqrt{V_{ar}} (y + t_a) - \left\{ \frac{y}{0.155 I_{cl}} \right\} + 3.05 \times 10^{-3} (3373 + p_a) + 0.42[58.15 - (M - W)] + 0.0014 M (t_a - 34) + (1.7 \times 10^{-5}) M (p_a - 5867) \tag{13}$$

where S is heat rate storage in a body, M is metabolism, W is External work, R is heat exchange by radiation, C is heat exchange by convection, K is heat exchange by conduction, E is heat loss by evaporation, RES is heat exchange by respiration, M_H is mass of human body in (kg), cp_H is specific heat of human body in (J/kg.°C), $T_{sk,t}$ is skin's temperature at time t in (°C), t_{cl} is surface temperature of clothing in (°C), f_{cl} is ratio of the surface area of the clothed body to surface area of nude body and p_a is water vapor pressure in (kPa).

It is noted that the value of t_{cl} in equation (3) is to be calculated iteratively in finding the roots of nonlinear equation. A first guessed value of t_{cl} will be introduced to the second member in equation (3). If the initial guess value of t_{cl} is far from root, it will consume a long time to converge to the root. Therefore, the initial guessed value in this research was taken by averaging the air temperature t_a and skin temperature T_{sk}.

To make sure the equation used for each term in the transient heat balance equation correct or not, sample input will be supplied and heat transfer for each term has to be calculated out and the result has to be validated using the excel spread sheet of ISO 7730 PMV calculator [3]. Based on comparison between manual calculations and values obtained using ISO 7730 PMV calculator, it can be shown that heat transfer for each term in equation (1) is quite similar for both approaches. Hence, the equations are validated and can be used for model derivation.

3. Application of Artificial Neural Network (ANN)

From equations (1) to (13), we can observe that the PMV value calculation is very complicated if using manual calculation. In addition, the determination of t_{cl} value using equation (6) may take long computing time due to iterative native of the equation and hence it is not practical to determine PMV value in real time application. In order to solve this problem, a nonlinear identification model, Artificial Neural Network (ANN) is introduced to approximate nonlinear relationship between input and output value.

Since PMV value is affected most by the six thermal variables; indoor air temperature, mean radiant temperature, indoor air humidity, human metabolism, air velocity, and thermal clothing insulation, therefore the input layer dimension chosen is 6-dimensional. For output layer, there is only one PMV value hence the dimension for output layer chosen is one-dimensional. 3 hidden layers are used to reduce the complexity of calculation. To ensure the output is not fall into the flat area at the beginning of study, the initial weight values are randomly generated, and the values are relatively small so that each neuron can learns from the biggest change in the local activation function. Learning rate of 0.10 is selected for this research.

4. Simulation Results

All simulations are conducted using our previous building model in Matlab environment [4]. Figure 1 (a) shows the graph of dynamic PMV thermal comfort value against 43000 samples data. As can be seen from the graph, the black colour curve is the model or dynamic PMV output while the red colour curve is the plant or neural network model output. The result from this graph showed that neural network can learn well or mimic dynamic PMV thermal comfort since both the curves are quite similar. Based on the specified minimum desired error of 0.0001, the network training achieved the desired goal error at epochs

200. The derived dynamic PMV equation was then validated by comparing the trained dynamic neural computing PMV index with TS-Fuzzy Logic model as shown in Figure 1 (b). The result showed that human thermal comfort evaluation model established in this paper agreed well with existing research results, and may directly be used in future research work.

Figure 1 (c) shows the result from practical approach of PMV thermal comfort identification using Artificial Neural Network, the training and validation data. The dashed lines in each graph indicate the perfect result. The correlation coefficient R value is an indication of the relationship between outputs and targets. The R value of training and validation are 0.99969 and 0.99971 respectively. This indicates that there is an exact linear relationship between outputs and targets. It can be said that for this simulation, the overall performance indicates a good fit.

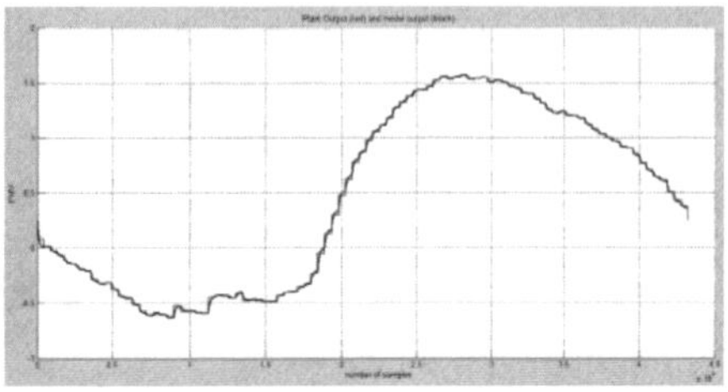

(a) Dynamic PMV value

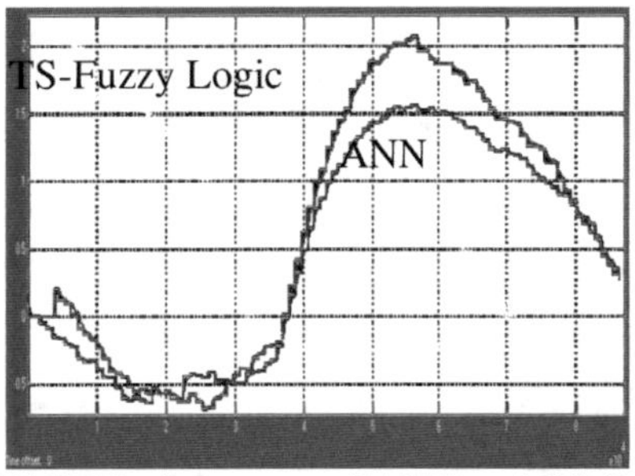

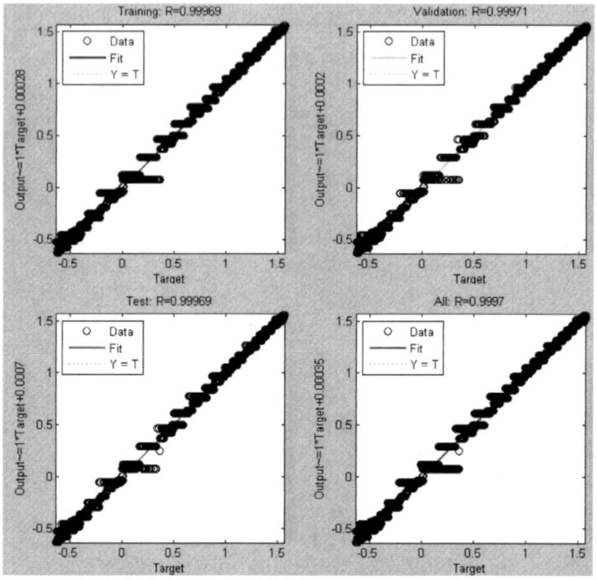

(b) PMV Performance of ANN and TS-Fuzzy Logic Model (c) Regression Analysis

Figure 1. Simulation results for validation of proposed method

5. Conclusion

The results proved that human thermal comfort evaluation model established in this paper agreed well with existing research results, and may directly be used in future research work. Results also show that the proposed ANN method can track down the desired PMV thermal comfort for a specified condition space.

6. Acknowledgement

The authors would like to sincerely thank the Ministry of Higher Education (MOHE) of Malaysia for the provision of a FRGS grant to support this work.

References

[1] ISO7730 2005 Ergonomics of the thermal environment-- Analytical determination and interpretation of thermal comfort using calculation of the PMV and PPD indices and local thermal comfort criteria

[2] Fanger, P O 1972 *Thermal comfort analysis and applications in environment engineering* (New York: Mcgraw-Hill)

[3] ISO7730 PMV calculator from http://www.eat.lth.se/fileadmin/eat/Termisk_miljoe/PMV-PPD.html

[4] Homod, R Z, Sahari, K S M., Almurib, H A F., Nagi, F H. 2011 *Energy and Buildings* **43** 2046-2049.

IOP Conf. Series: Earth and Environmental Science **16** (2013) 012114 doi:10.1088/1755-1315/16/1/012114

Friction measurement system using load cell for Tribotronic system on Pin-On-Disc (POD) tribometer

M Shaparuddin Bahrudin [1], Siti Fazlili Abdullah[1] and M Reyasudin Basir Khan[2]

[1]Center for Micro and Nano Engineering (CeMNE), College of Engineering, Universiti Tenaga Nasional, Jalan IKRAM – UNITEN, 43000 Kajang, Selangor, Malaysia.

[2]Center for Automation and Embedded Computing System (CAECS), College of Engineering, Universiti Tenaga Nasional, Jalan IKRAM – UNITEN, 43000 Kajang, Selangor, Malaysia.

E-mail: shaparuddin@uniten.edu.my, siti@uniten.edu.my, reyasudin@uniten.edu.my

Abstract. This paper discusses part of Tribotronic system on Pin-On-Disc (POD) Tribometer focusing on friction measurement using load cell. The analysis conducted using MATLAB software to measure Coefficient of Friction (CoF) of 2 sample types in order to prove the validity of the measurement. The 2 samples are lubricant (Shell Helix Fully synthetic oil) and lubricant (Shell Helix Fully synthetic oil) with additive. It is revealed that the measurement of the lubricant with additive shows better CoF (0.006) than lubricant without additive, CoF of 0.02.

1. Introduction

Tribology plays a key role in implementation in reduction of emission in modern-day machine design. Tribology based machine design caused major contribution in development of compact and low weight machinery [1-4]. Tribotronics, applies to the integration of tribology and electronics. Tribotronics or active tribology based on adaptive performance is thought of as being critical in the implementation of smart machine concepts.

The conditions of a tribological system are monitored by sensors that provide information on i.e. temperature, pressure, friction, vibration, oil properties such as total acid number or additive depletion and other parameters of interest. The signals from these sensors are processes and transmitted to the control unit. In the computational or decision making part, real-time software based on tribological algorithms calculates the required action which is then implemented by actuators. Such a system is thus autonomous and self-adjusting. This allows for on-line tuning of the tribological system for the best performance. The key principle of tribology is the practice of additional so-called loss outputs such as friction, wear and vibration [5]. This paper is focus on developing simple method of measuring friction using load cell and analysis on the friction pattern using MATLAB software.

1.1 Coefficient of Friction

The ratio between the this frictional force (F) and normal load (W) is known as Coefficient of Friction (CoF), denoted with symbol 'μ' [6]. This relation led to the formulation of First Law of Friction which is expressed by (1):

$$F = \mu W \tag{1}$$

IOP Conf. Series: Earth and Environmental Science **16** (2013) 012114 doi:10.1088/1755-1315/16/1/012114

1.2 Load Cell

This project utilizes strain gauge type single point load cell that consist of four strain gauges. It has capacity to up to 100 kg of load and output voltage up to 10 V with of sensitivity of 2.0 mV/V [7].

2. Methodology

2.1 Electrical system

Electrical components involve for the experiment including the load cell itself is as shown in Figure 1. The cell is fixed to an aluminium arm that is elastically swingable. The cable attach on the load cell is connected to the signal conditioner. The signal conditioner will normalize the voltage output from the load cell between 0 to 2.5 V. Normalization is important in order to comply data logger input specification that can only capture signal between specified voltage [8].

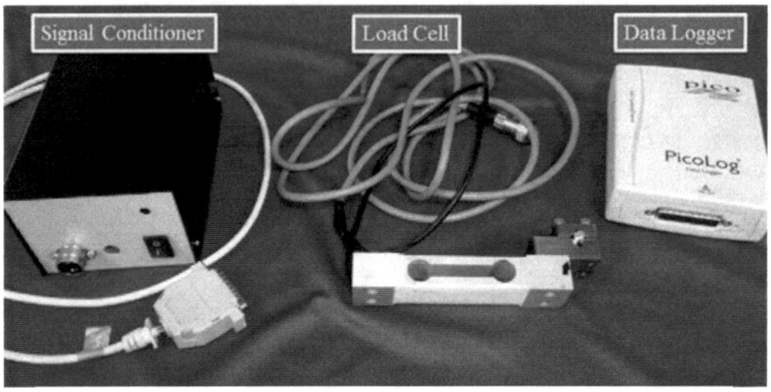

Figure 1. Electrical component involved

2.2 Mechanical arrangement

Arrangement of the load cell on top of the POD is shown in Figure 2. Basically a stationary contact pin hold by an elastic type arm that can be extended to almost half diameter of a disc underneath it. The arm is also connected by high tension string wire attached with a weight holder. The disc is rotated electrically and controlled by motor drive. The disc rotation is measured in RPM. The contact pin is hold into a holder that is attached to an arm that can be slide the arm is fixed with a load cell to detect the sliding force. When the disc is rotating and with a load applied on the pin, the contact between the pin bottoms with the disc surface will cause a sliding force that is detected by the load cell and read by data logger.

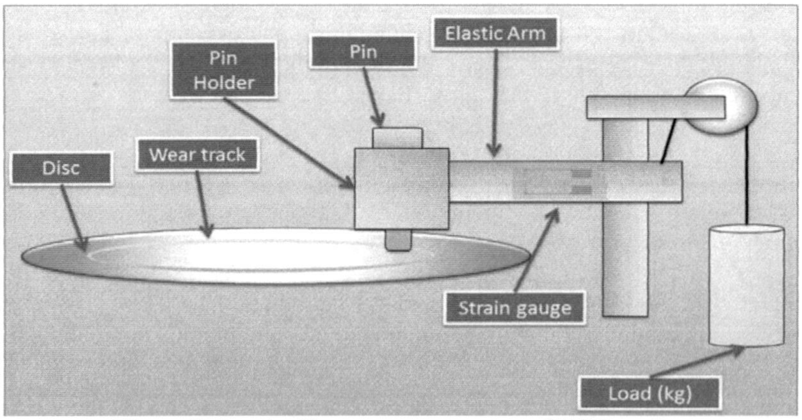

Figure 2. Load cell arrangement on top of POD system

IOP Conf. Series: Earth and Environmental Science **16** (2013) 012114 doi:10.1088/1755-1315/16/1/012114

2.3 Control system and software

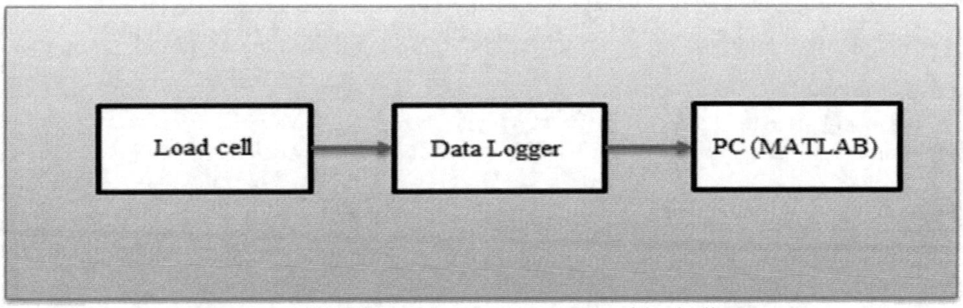

Figure 3. Friction measurement system block diagram

Initially the load cell is calibrated for 20 N/V then the output signal from the load cell in term of voltage variation is captured by the data logger and transmitted to personal computer (PC). This output signal is translated into friction force using equation (1) with W (N) is pre input as 20 kg. The experiment runs on 2 types of samples.

3. Results and discussion

Figure 4 shows the tribological behaviour of Coeficient of Friction (CoF) with respect to the Time (s) in lubricant and lubricant with additive. As a comparison we choose the value of CoF from the range of 200 s to 1200 s. The average value of CoF for the lubricant and without lubricant are 0.02 and 0.006 respectively. That means the CoF can be reduced significantly, i.e. about 70 %.

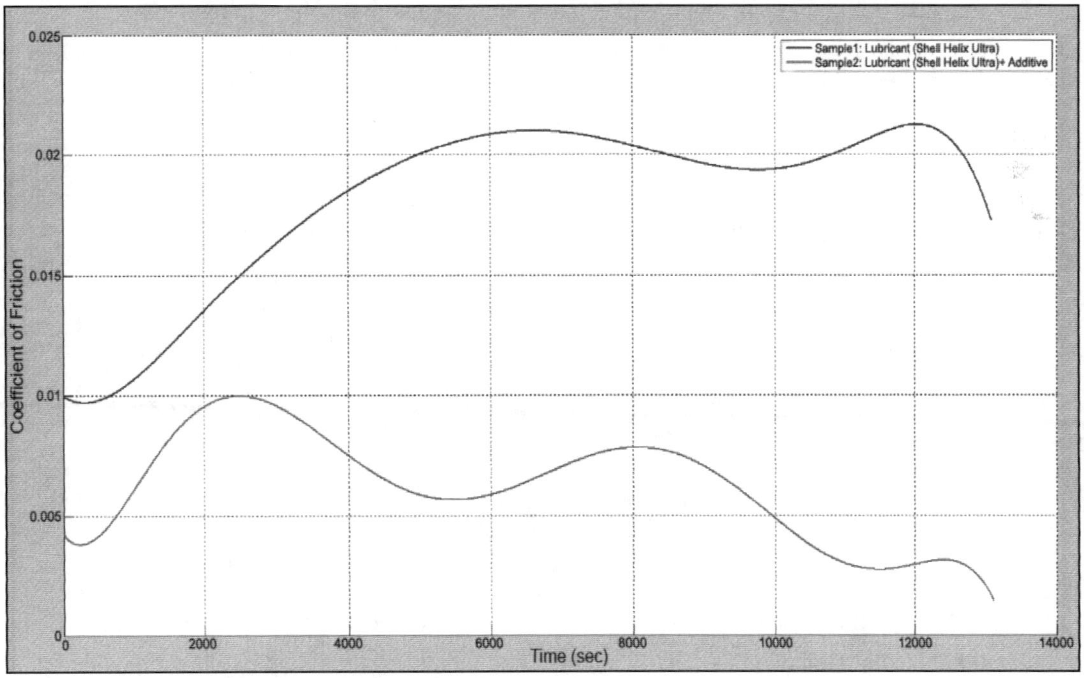

Figure 4. Coefficient of Friction (CoF) versus Time (sec) for the lubricant and lubricant with additive

4. Conclusion

The load cell arrangement and analysis with MATLAB has succeeded measuring CoF for both samples. Results show that with the addition of additive, the CoF reduced remarkably.

Acknowledgments

The author would like to thank Ministry of Higher Education (MOHE) for funding this research (MOHE grant no: 20110105 FRGS) and UNITEN seed fund (grant no:JS10050449)

References

[1] Rana,Prashant 2010 *Real Time Lubricant Ageing Analysis-First Step Toward a Tribotronic System* (Master's Thesis, Lulea University of Technology,Lulea)
[2] Glavatskih,S and Erik,H 2012 Tribotronic - Towards Active Tribology *Tribology International* **41** 934-939
[3] Nair,RP ,Griffin, D and Randall, XN 2009 The Use of The Pin-On-Disc Tribology Test Method To Study Three Unique Industrial Application *17th International Conference on Wear of Materials (Las Vegas,Nevada, 19-22 April 2009)* **267** 823-827
[4] I,Tzanakis 2012 Future Perspective on Sustainable Tribology *Renewable and Sustainable Energy Review* **16** 4126-4140
[5] Jost, HP 1966 *Jost Report,HMSO* (Department of Education and Science)
[6] Ian, HM 1992 *Tribology : Friction and Wear of Engineering Materials* (Arnold: London)
[7] Kuanyi, Xn. *Single Point Load Cells Specification* [online]. Available from:
[8] http://www.kuanyi.com/html/products_detail.asp?mainID=138&subID=6&detailID=39 [Accessed 7 February 2013]
[9] Pico Technology 2009 *PicoLog 1000 Series : User's Guide* (James House: Cambridgeshire)

The study on modelled and measured weather data for building energy simulation programs for Malaysia.

B Faizal[1,*], N M D Ulang[1]

[1] School of Housing, Building and Planning, Universiti Sains Malaysia, 11800 Penang, Malaysia.

baharumfaizal@gmail.com

Abstract. This paper will study the usability of future weather data generated from climate model for Malaysia. Detailed future weather data is required for the building energy assessment as input parameters. The future weather data required is normally achieved from climate prediction models. The purpose of this study is to examine the gaps between weather data generated by climate model and the data measured by weather station in Bayan Lepas, Penang. Furthermore, this studies also to establish the modelled weather data for the use for future building energy simulation program. In order to achieve this purpose, simulated weather data sets HadCM3 were supplied by the Hadley Centre in the United Kingdom. The measured weather data was supplied by Malaysian Meteorological Department for Bayan Lepas, Penang. The period of analysed time was 18 years from 1990 to 2007 where the available data overlaps between HadCM3 and measured data. Several major weather variables were used in these studies such as Dry-bulb temperature, Solar radiation and Wind speed. The outcome from this studies shows a good match between HadCM3 data and measured data indicates that HadCM3 model is suitable for the purpose of future building energy simulation for Malaysia.

1. Introduction

Future building simulation requires simulated weather data from climate models. Currently, there are numbers of climate models used globally. Climate models are used for future predictions on climate changes and the study of climate dynamics. The development of climate models and their connection with climate change prediction was identified in the early 20th century as discussed by Edwards in his works 'History of climate modeling' [1] and Weart; 'The development of general circulation models of climate' [2].

Basically, there are two main categories of global climate models. Atmosphere General Circulation Models (AGCMs) and Ocean General Circulation Models (OGCMs) account the changes within the atmospheric and ocean changes respectively. These two types of models can be coupled together as Atmospheric-Ocean General Circulation Models (AOGCMs) giving a comprehensive model for future climate prediction [3].

AGCMs combine the atmosphere together with land surface models and impose the sea's surface temperature. AGCMs normally consist of a 'dynamical core' that integrates the large-scale fluid motion usually to simulate surface pressure; temperature and water vapor in layers and so on. OGCM has basically the same concept as AGCMs. The difference is that OGCMs simulate the ocean together with the impose of fluxes from the atmosphere and can include the sea ice model [1].

The coupling of AOGCMs has meant progression towards a comprehensive model since 1980s [2]. There are two types of OAGCMs. Earth System Models (ESMs) OAGCMs models climate components such as land surface, cryosphere, hydrology (lakes, evaporation, rivers and rainfall), and vegetation. Meanwhile, Integrated Assessment Modes (IAMs) OAGCMs comprise the impact of climate change, economic responses, and policy scenarios in a single modeling structure [1].

Climate models require complicated computing processes and are usually developed at meteorological and research centers. There are a number of research centers developing climate models globally such as the Max Planck Institute in Germany, Japan's Earth Simulator Centre, and the Goddard Institute for Space Studies in the United States [1]. In the United Kingdom, the Hadley Centre at the UK Met Office developed the Hadley Model which is used widely by various works such IPCC climate impact assessment. For this research, Hadley Coupled Model Version Three (HadCM3) data was employed. HadCM3 was the key model used in the development of IPCC Third Assessment Report (TAR) [4].

2. Weather data for building energy simulation

2.1. Special Report Emissions Scenarios (SRES)

The IPCCs Special Report Emissions Scenarios (2000) contained a set of new projections of future Green House Gases emissions. Each projection was a "storyline", describing the manner of world population, political, economies and lifestyle changes over the next few decades [4]. The storylines are summarized into four scenario families with the severest scenario being A1. The scenario describes a very rapid economic growth with increasing globalization, an increase in general wealth, with convergence between regions and reduced differences in regional per capita income. The scenario also describes rapid technological change. In addition, the A1 scenario is divided into three variants: the emphasis of fossil fuels (A1F1), non-fossil fuels (A1T) or a balance across all sources (A1B).

The A2 scenario has slower economic growth than A1's scenario but has a faster population growth due to less convergence of fertility rates. The scenario also portrays that economic and technological changes are regionally oriented.

The B1 scenario portrays the same rapid economic growth with the A1 scenario and the more integrated world, and is more environmentally sustainable. Clean and efficient tech-nologies are introduced. Meanwhile, B2 scenarios have a lower population rate than A2 but it is still higher than A1 and B1. Future developments are locally oriented in terms of environmentally, economically and socially.

2.2. The Hadley Centre Model (HadCM3)

HadCM3 is a coupled AOGCM developed at the UK Hadley Centre in 1999. HadCM3 is a sophisticated climate model, a compilation of two models, the atmospheric model (HadAM3) and ocean model (HadOM3). It also contains the sea-ice model. The atmospheric component divides the global surface into 96 x 73 grid cell, where a grid cell has the resolution of 2.5 degrees of latitude and 3.75 degrees of longitude and also has 19 levels. The grid cell has a surface resolution of about 417 kilometres x 278 kilometres at the Equator, reducing to 295 kilometres x 278 kilometres at 45 degrees of latitude [5]. The oceanic model has 20 levels with a horizontal resolution of 1.25 x 1.25 degrees [6].

The weather data from HadCM3 is available from years 1860 to 2099, while every year is split conveniently into 360 days with30 days in each month. In addition, HadCM3 is a first integrated model without flux adjustments. The flux adjustment was required for previous models in order to avoid the simulation drifting into unrealistic states of climate [6].

Meanwhile, HadCM3 simulates 28 weather components in a daily-average format. The variables accessible are such dry-bulb temperature (DB), maximum (Tmax) and minimum (Tmin) temperature, relative humidity (rH), Wind Speed, Down Short-Wave Flux (DSWF) and so on [7].

For this study, the data provided by the National Climatic Data Centre (NCDC) [8] were in ASCII data format. Data extraction for Malaysia's grids was performed for these raw data sets.

2.2.1. HadCM3 grid on Malaysia.

Malaysia's peninsular can be found between latitudes 1.5°N and 7°N and longitudes 99.5°E and 104°E. Moreover, there are two HadCM3 grid points situated in or near Malaysia's peninsula, 101.25°E longitude and 5.0°N latitude, and 101.25°E longitude and 2.5°N latitude, with resolution of approximately 300 kilometres.

Figure 1 shows the HadCM3 grids that covering most of Malaysia's peninsular. The first point, 101.25°E longitude and 5.0°N latitude is centred in Perak and covered most area of Northern region. It covers more land areas than the second grid point, 101.25°E longitude and 2.5°N latitude, which is centred on Malacca Strait (sea areas).

2.3. Result and Discussion

These analyses were undertaken to examine the gap or dissimilarities between HadCM3 and measured data supplied by the Malaysian Meteorological Department (MMD) [9]. The data used in these analyses for HadCM3 focused on grid 101.25°E, 5.0°N which is inland, whereas the measured data are for Bayan Lepas, Penang which is on coast as shown in Figure 1. There may therefore be some difference in climate between the two sites. The period of analysed time was 18 years from 1990 to 2007 where the available data overlaps between HadCM3 and measured data.

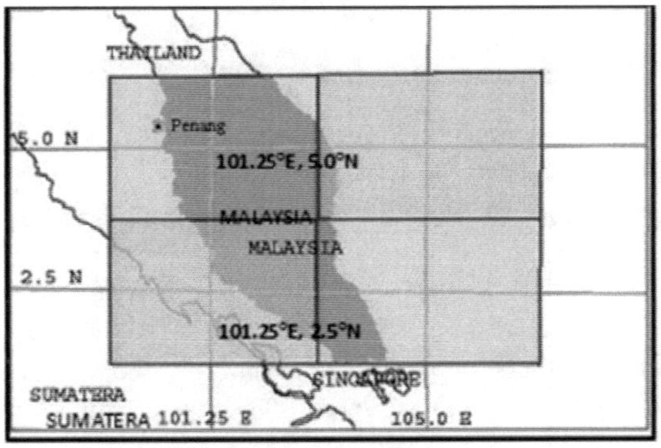

Figure 1. HadCM3 Grids in Peninsular Malaysia. [10]

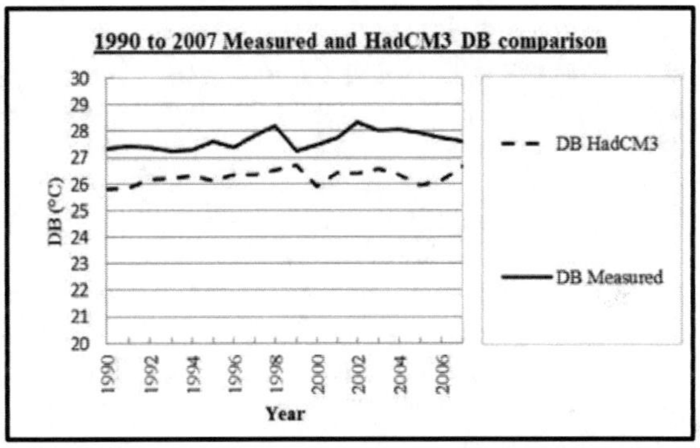

Figure 2. Comparison of measured and HadCM3 Dry-Bulb temperature from 1990 to 2007.

Several major weather variables were used in these studies such as Dry-bulb temperature, Solar radiation and Wind speed. The annual means of Dry-bulb temperature analysis shows a consistent difference like that shown in figure 2. The average difference shows that the measured Dry-bulb temperature was 1.5°C warmer than HadCM3 data. The differences are expected as HadCM3 weather data summarised across the whole grid area while measured data supplied was station or point data.

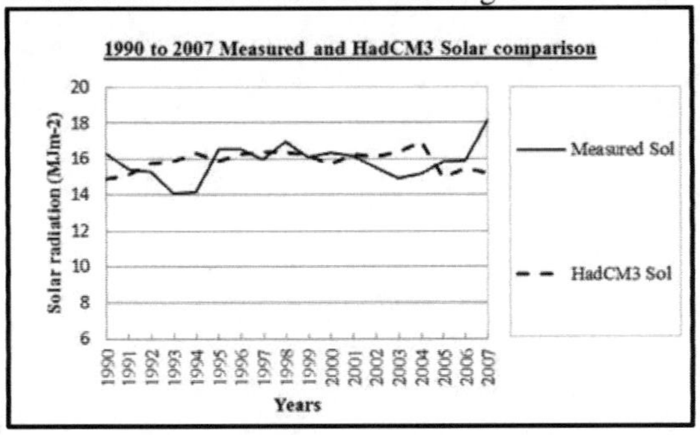

Figure 3. Comparison of annual averages for measured and HadCM3 SOL from 1990 to 2007.

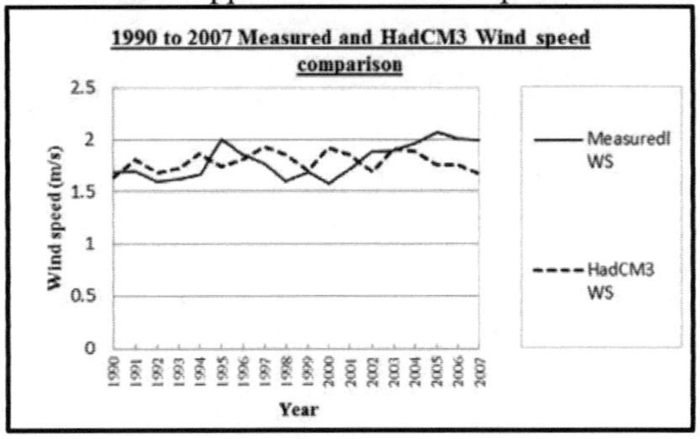

Figure 2. Comparison of annual averages for measured and HadCM3 WS from 1990 to 2007.

IOP Conf. Series: Earth and Environmental Science **16** (2013) 012115 doi:10.1088/1755-1315/16/1/012115

On the other hand, the comparisons for Solar radiation (figure 3) and wind speed (figure 4) shows a close relation between measured and HadCM3 weather data. From Figure 3 it can be seen that the HadCM3 solar radiation shows close to measured data.

3. Conclusion

This paper discussed the background of weather data required for future building simulation. The weather data sets were obtained from simulated global climate models then compared with real weather data sets obtained from the Malaysian Met Office. The matching results show the similarity or close relation between modeled and real weather data. This result also established the usability of HadCM3 modeled weather data generated from Hadley Model for building energy simulation program.

References

[1] Edwards P N 2011 History of climate modeling *Wiley Interdiscipli-nary Reviews: Climate Change* **2** 128

[2] Weart S 2010 The development of general circulation models of climate *Studies in History and Philosophy of Modern Physics* **41** 208

[3] Solomon S Qin D Manning M Chen Z Marquis M Averyt K B Tignor M and Miller H L 2007 *IPCC The Physical Science Basis: Contribution of Working Group I to the Fourth Assessment Report of the Intergovernmental Panel on Climate Change* (Cambridge, UK) (Cambridge University Press) pp 20

[4] Nakicenovic N and Swart Rob 2000 *IPCC Emission Scenarios: A Special Report of Working Group II of the Intergovernmental Panel on Climate Change* (Cambridge University Press)

[5] Hulme M Mitchell J Ingram W Lowe J Johs T New M Viner D 1999 Climate change scenarios for global impacts studies *Global Environment Change* **9** S3

[6] Met Office 2011 Hadley Centre at UK Met Office http://www.metoffice.gov.uk/climate-change/resources/hadley

[7] Gordon C Cooper C Senior C A Banks H Gregory J M Johns T C Mitchell J F B and Wood R A 2000 The simulation of SST, sea ice extents and ocean heat transports in a version of the Hadley Centre coupled model without flux adjustments *Climate Dynamics* **16** 147-168

[8] More references National Climatic Data Centre 2011 US Department of Commerce Http://www.ncdc.noaa.gov/oa/ncdc.html.

[9] Malaysian Meteorological Department 2011 Official Portal of Malaysian Meterological Department (MMD): Ministry of Science, Technology and Innovation (MOSTI) http://www.met.gov.my.

[10] More references Met Office (2011) Hadley Centre at UK Met Office http://www.metoffice.gov.uk/climate-change/resources/hadley.

IOP Conf. Series: Earth and Environmental Science **16** (2013) 012116 doi:10.1088/1755-1315/16/1/012116

Biological carbon fixation: A study of *Isochrysis sp.* growth under actual coal-fired power plant's flue gas

Liyana Yahya[1], Muhammad Nazry Chik[1] and Mohd Asyraf Mohd. Azmir Pang[1]

[1] *Researchers, Emission and Waste Management Technology, Generation and Environment Unit, TNB Research Sdn. Bhd.*
Email: liyana.yahya@tnbr.com.my (Corresponding author)

Abstract: Preliminary study on the growth of marine microalgae *Isochrysis sp.* was carried out using actual flue gas from a coal-fired power station. The species was cultured using a 2x10-L customized bubble column photobioreactor skid under specified culture conditions. With an initial culture density of 0.459 Abs (optical density at 560 nm wavelength), the species was found able to survive - observed by increases in optical densities, number of cells and weights - in the presence of actual coal-fired flue gas containing on average 4.08 % O_2, 200.21 mg/m^3 SO_2, 212.29 mg/m$_3$ NO_x, 4.73 % CO_2 and 50.72 mg/m^3 CO. Results thus add value to the potential and capability of microalgae, especially for *Isochrysis sp.,* to be the biological carbon fixer in neutralizing carbon emissions from power plants.

1. Introduction

Many literatures have identified phototropic microalgae as an alternative method among available carbon capture technologies [1],[2]. The use of its natural, more efficient photosynthesis process, and the potential of valuable downstream products it can generate, make it eligible to be a more sustainable method [3].

Several experiments and demonstration projects have been documented using microalgae to fix CO_2 from actual flue gas. Microalgae species used in these works include *Chlorella sp.* [4], [5], *Dunaliella tertiolecta* [6] and *Botryococcus braunii, Scendesmus sp.*, [7].

As a first step in selecting suiTable species for biological carbon fixation, a marine microalgae species *Isochrysis sp.* has been screened and found dominant in sea waters in the vicinity of a coastal coal-fired power plant in Manjung, Perak [8]. Comparison of its carbon fixation ability, under simulated flue gas in laboratory conditions, has proved that it has a better fixation rate over *Tetraselmis sp.* and *Nannochloropsis Oculata* [9], [10]. This work thus offers a preliminary extended study to verify the species's suitability to be used as a biological carbon fixer in a coal-fired power station, while enriching the database on species growth under actual flue gas conditions.

2. Materials And Method

2.1 Site Location

The experiment was carried out at the Sultan Azlan Shah Power Station. This is a 3 x 700 MW coal-fired power plant located in the district of Manjung, Perak, along the Straits of Malacca, west coast of Malaysia. Approximate geographical coordinates are 4° 9'30.52" North and 100° 38' 27.63" East.

2.2 Photobioreactor

A customized skid-mounted photobioreactor (PBR) was used in the experiment. The PBR is of bubbling column type, having double cylindrical columns of 10-L each as the reactor vessel. The material of the reactor column is of polycarbonate having dimension 3 mm thick, 200 mm diameter and 800 mm high. The PBR skid was placed inside the flue gas stack structure, as depicted in Figure 1 below.

Figure 1. The setup PBR inside the stack structure house

2.3 Source of Flue Gas

Portions of actual flue gas were tapped after the station's Continuous Emission and Monitoring System's (CEMS) analyzers. Small bore polypropylene tubes was used to route this spent gas into the PBR, where the flow was regulated with a needle valve to be at around 0.15 ± 0.03 l/min.

2.4 Microalgae Species

Isochrysis sp. was used in the study. This species's taxonomy, in accordance to *Phylum / Class / Order / Family / Genus / Species* sequence is *Haptophyta / Prymnesiophyceae / Isochrysidales / Isochrysidaceae / Isochrysis.*

Its physical appearance can be characterized by having slightly elongated motile cells, a stigma and two smooth flagella. Mainly benthic and typically in spherically round shape. It measures about 3 x 6 μm. The species has been known to be used as aquaculture diet and has been proposed as a source of decosahexanoic acid (DHA) for nutraceuticals. To a lesser extent too, it has been tried for lab-scale biodiesel production [11].

2.5 Culturing Conditions

The species was injected with 10 mL nitrate as nutrients. The culture temperature was left unregulated to ambient temperature, which ranged from 26.10 °C to 30.96 °C. Since it was rather dark inside the stack structure house, artificial illuminance, in the form of five units of T5-type tubular fluorescent bulbs were used with 12-hour timer-controlled activated. Average peak illuminance recorded was about 15 μmol/m^{-2}.s^{-1}. pH of the culture was maintained to be between 6 – 7 with the help of an air pump.

2.6 Procedures

The culture was introduced into the reactor column at an initial optical density of 0.459 Abs, measured at 560 nm wavelength. The culture was left for six consecutive days where 30 mL samples were withdrawn daily for optical density, weight and cell-count measurements. Culture's pH was not regulated however on the last two days in the pure interest to observe the pH change and its effect on species growth.

2.7 Tools and Instrumentation

Culture's pH was measured using EUTECH INSTRUMENT sensor (pH Range : -2.00-16.00, Relative accuracy : ±0.01). Temperature was measured using type-K thermocouple. Dissolved CO_2 was read using METTLER TOLEDO's InPro 5000(i) potentiometric sensor (accuracy ±0.1). Illuminance was recorded using a Skye's high output Photosynthetically Active Radiation (PAR) sensor with linearity error of less than 0.2 %. All pH and temperature were automatically logged every 10-minute intervals using DEWE-43 data acquisition system. Culture's optical density was measured using HACH's DR 2800 spectrophotometer. Cell count was made possible upon use of Neuber-improved haemocytometer slide. Samples were filtered using magnetic vacuum filter, then dried-up in room temperature for 24 hours before weights were recorded using analytical balance. Flue gas compositions were extracted from station's Plant's Information System.

3. Results and Discussions

Figure 2 and Figure 3 below demonstrate that the species was able to survive in the culture supplied with the actual flue gas. Except for the last two days, the optical density increased from the initial 0.459 Abs to 0.638 Abs (Figure 2). Similarly, the number of cells has registered an increase by 18 % from the initial count of 290,000 cells (Figure 3). The dry weights of the culture were also observed to be on the rise, where it increased from about 0.003 mg/l to the maximum of 0.007 mg/l.

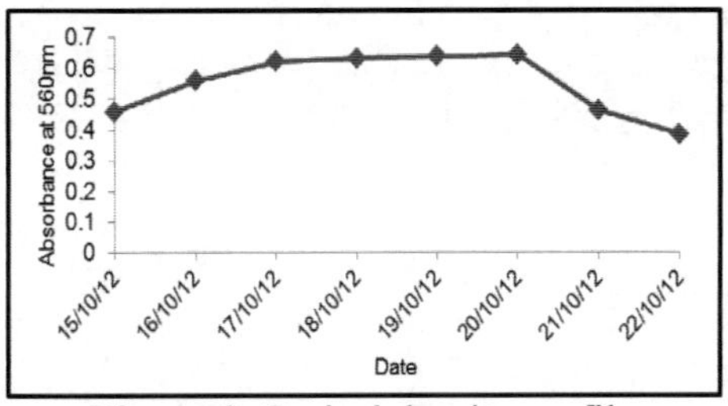

Figure 2. Optical density profile

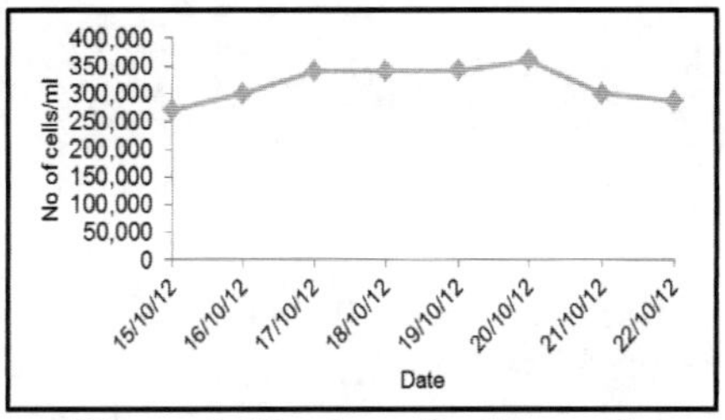

Figure 3. Number of cells profile

On average, the flue gas compositions were 4.08 % O_2, 200.21 mg/m^3 SO_2, 212.29 mg/m^3 NO_x, 4.73 % CO_2 and 50.72 mg/m$_3$ CO throughout the period. There were only minimal amount of particulates found accumulated at the bottom of the reactor vessel only on the first day of experiment.

As expected, the growth retards upon unregulated pH. Figure 5 illustrates the increase of accumulated dissolved CO_2 upon unregulated pH. The high amount of dissolved CO_2 – up to the maximum of 15.17 % - in last two days happened as the flue gas was supplied continuously without controlling the pH, deliberately done as part of experiment. Figure 4 below depicts that the pH was dropped to and lingered around pH 6 for about two consecutive days, which further dampened the growth. The cells' density and count declined by about 40 % and 43 %, respectively, as the consequence.

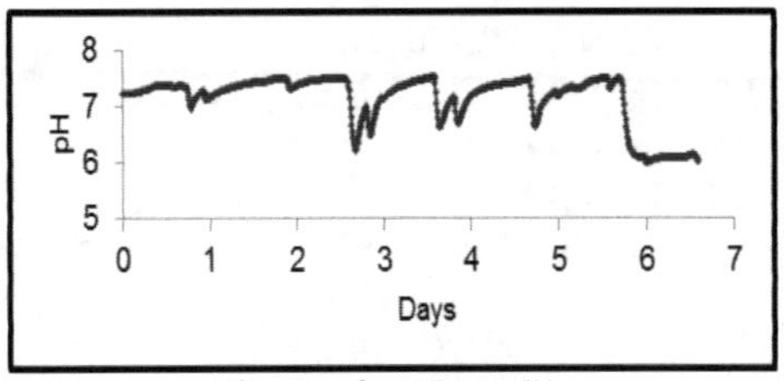

Figure 4. pH profile

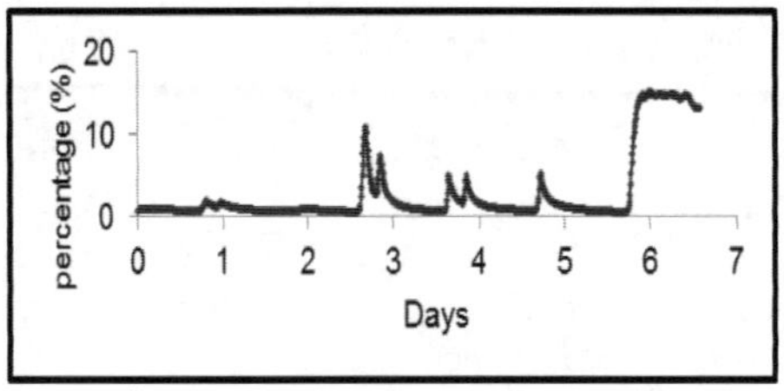

Figure 5. Dissolved CO_2 profile

4. Conclusions and Further Works

In this brief study, a marine microalgae species *Isochrysis sp.* was put into a field trial, cultured with CO_2 source from an actual flue gas from a coal-fired power station. It can be seen that with regulated aeration, the pH was managed to be maintained at slightly above pH 6 and 7, giving a right culture condition for the species to survive and reproduce, within the flue gas compositions of 4.08 % O_2, 200.21 mg/m^3 SO_2, 212.29 mg/m^3 NO_x, 4.73 % CO_2 and 50.72 mg/m$_3$ CO. This conducive environment is expected to diminish once the culture's pH dropped to 6 and lower. Despite of a short term conduct, valuable lessons has been acquired especially in the importance of the need and scheme of pH control meant for a bigger scale facility. In this experiment, aeration has proved to be one of effective and cheap means of pH control; however, its application might release the CO_2 bubbles that are supposedly to be dissolved and consumed by microalgae, other than introducing detrimental shear stress to the cells. Thus, enclosure design for a bubble column photobioreactor and regimes of flow velocity shall be further studied. Frequency of aeration can also be optimized to be inversely proportional to a denser culture, as to minimize electricity consumption of air pump/blower.

Acknowledgement
The authors extend greatest gratitude to the Management of Sultan Azlan Shah Power Station for granting access and co-operation for the study be conducted at the power station. Also to its Chemical and Environment Services for allowing their laboratory facility is used. Special appreciation goes also to Muhamad Termizi bin Abu Bakar, Muhammad Zul'izuddin bin Muda and Mohd Faisal bin Mohamad Darus of TNBR for their kind technical supports and logistic matters. Finally, many thanks to the personnel of Alam Sekitar Malaysia Sdn. Bhd. (ASMA) for their co-operations during the study.

References
[1] M. Packer 2009 Algal Capture of Carbon Dioxide; Biomass Generation As A Tool For Greenhouse Gas Mitigation With Reference To New Zealand Energy Strategy and Policy *Energy Policy* **37** 3428–3437
[2] Shih-Hsin Ho, et al. 2011 Perspectives On Microalgal CO2-Emission Mitigation Systems – A Review *Biotechnology Advances* **29** 189-198
[3] Bei Wang, Yanqun Li, Nan Wu and Christopher Q. Lan. 2008 CO2 Bio-Mitigation Using Microalgae Applied Microbiology Biotechnology **79** 707–718
[4] K. Maeda, et al. 1995 CO_2 Fixation From The Flue Gas On Coal-Fired Thermal Power Plant by Microalgae *Energy Conversion and Management: Proceedings of the Second International Conference on Carbon Dioxide Removal (Kyoto, Japan, June-September 1995)* **36** 717-720
[5] Ji. Doucha, Frantisek Straka and Karel Livansky 2005 Utilization of Flue Gas for Cultivation of Microalgae (Chlorella sp.) in An Outdoor Open Thin-Layer Photobioreactor *Journal of Applied Phycology* **17** 403–412
[6] J.C.M. Pires et al. 2012 Carbon Dioxide Capture From Flue Gases Using Microalgae: Engineering Aspects and Biorefinery Concept *Renewable and Sustainable Energy Reviews* **16** 3043–3053
[7] E. Suali and R. Sarbatly 2012 Conversion of Microalgae To Biofuel *Renewable and Sustainable Energy Reviews* **16** 4316–4342
[8] Muhammad Nazry Chik, Liyana Yahya, Mohd Hariffin Boosroh and Lokman Shamsuddin 2012 Scanning of Microalgae Species For Biological CO_2 Fixation At A Malaysian Coastal Coal-Fired Power Station *OIDA International Journal of Sustainable Development* **04** 19-28
[9] Muhammad Nazry Chik and Liyana Yahya 2012 Comparison of Carbon Fixation Abilities of Selected Marine Microalgae Species *Association of Muslim Scientist E-Journal* **2** 1-11
[10] Liyana Yahya, Muhammad Nazry Chik and Mohd Hariffin Boosroh 2012 Carbon Fixation Ability Between Marine Isochrysis Sp. and Nannochloropsis Oculata *Proceedings for the 12th Malaysian Society of Applied Biology*
[11] Aquafuels 2011 *Algae and Aquatic Biomass for A Sustainabnle Production of 2nd Generation Biofuels: Taxonomy, Biology and Biotechnology*. Aquafuels FP7 241301-2 [online] Available from: www.aquafuels.be. [Accessed 2011].

IOP Conf. Series: Earth and Environmental Science **16** (2013) 012117 doi:10.1088/1755-1315/16/1/012117

An Optimized 2.4GHz RF Power Amplifier Performance for WLAN System

Mohammed H Ali[1], C K Chakrabarty[1], Ahmed N Abdalla[2], Goh C Hock[1]
[1] Dep.of Electr. and Comm. Eng., Uni. Tenaga Nasional, Selangor43009, Malaysia
[2]Faculty of Elec. and Electr. Eng., Uni. Malaysia Pahang, Pekan 26600, Malaysia.

E-mail: mha9969@yahoo.com

Abstract. Recently, the design of RF power amplifiers (PAs) for modern wireless systems are faced with a difficult tradeoff for example, cellphone; battery lifetime is largely determined by the power efficiency of the PA and high spectral efficiency which have ability to transmit data at the highest possible rate for a given channel bandwidth. This paper presents the design a multi stage class AB power Amplifier with high power added efficiency (PAE) and acceptable linearity for the WLAN applications. The open-circuited third harmonic control circuit enhances the efficiency of the PA without deteriorating the linearity of class-AB mode of the PA. The voltage and current waveforms are simulated to evaluate the appropriate operation for the modes. The effectiveness of the proposed controller has been verified by comparing proposed method with another methods using simulation study under a variety of conditions. The proposed circuit operation for a WLAN signals delivers a power-added efficiency (PAE) of 37.6% is measured at 31.6-dBm output power while dissipating 34.61 mA from a 1.8V supply. Finally, the proposed PA is show a good and acceptable result for the WLAN system.

1. Introduction

Nowadays, RF Power amplifier efficient are becoming more important as technology advances integration of wireless communication systems requires stringent capabilities in terms of linearity and efficiency. It is difficult to employ these RF linearization techniques as a complete transmitter for a base station due to the complicated circuits and/or insufficient linearization which need to amplify signals efficiently to increase the battery life that typically deliver their highest efficiency at maximum output power in linearity becomes significant [1].

The most common power amplification techniques with inherently more power efficient have been researched to satisfy the requirements [2], [3]. Such techniques are included load modulation technique, switching mode amplifiers [4], envelope elimination and restoration [5], the multi-modes and multiband of operation and envelope tracking [6]. The Doherty Power Amplifier is a promising candidate with advantages of high PAE, low cost and simple construction [7]. Resistive feedback [8] involves good wide band matching and flat gain but suffer from poor noise figure and large power dissipation, it has normally features of wideband but consumes large current because of several distributed amplifiers which is not proper for low power applications. This technique has previously

[1] Department of Electr. and Comm. Eng., University TenagaNasional, Selangor43009, Malaysia.

been implemented in integrated multi-band multi-standard receivers [9] and LNAs [10] where a matched input is required, but also in multi-band multi-standard transmitters [11] and compact power amplifier drivers [12].

This paper focus on control the shape and overlap of the drain voltage and/or current waveforms for class-AB power amplifier using appropriate gate biasing and harmonic terminations. First, exploit the harmonic contents so the loss in the active device is minimized. Second, bias the amplifier toward class AB/A (AB or A) to improve the linearity. Finally, the simulation result show that proposed PA are significantly improvement the linearity, efficiency and output power can be achieved with proper harmonic terminations and proper biasing voltage.

2. Proposed Circuit Design

For design and simulation purposes a faithful mathematical model which closely describes the actual behaviour of this inherently nonlinear power amplifier is required. Figure 1 shows the schematic circuit diagram of a typical class AB RF power amplifier with linearization network. Some of the key specifications in the design of this PA are its frequency of operation, gain, output power level, and bias voltages are adaptively modified of the input/output impedances. In addition, all the impedances of any RF block are terminated with 50 Ω, especially to aid in the testing of these blocks. Finally, it is easy to see that with the load modulation can be carried out in two stages with different control voltage profiles.

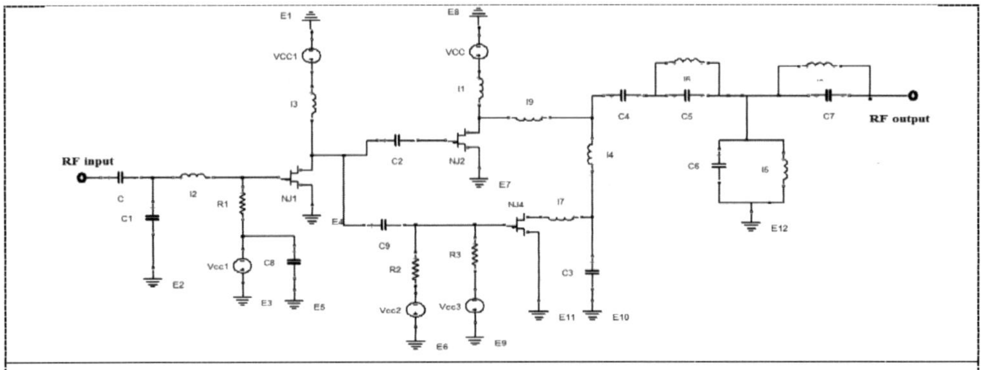

Figure 1. Optimized PA schematic

The maximum output power is determined by the maximum voltage swing at the output since the output impedance is fixed through the 50Ωs. This results in a limited current in the output stage since the single-ended output impedance is set to 50Ω and the supply voltage is 1.8V. A higher output power can be achieved by reducing the output impedance. This will increase the output current, but the influence of wire resistance will then be more significant.

3. Simulation Results

To enable to test the proposed class AB amplifier, closer examination of the voltages and currents of WLAN amplifier is required. Figure 2 shows the voltage and current waveforms at the drain terminal for the class AB output stage of 2-stages WLAN PA.

From figure 2, the propose circuit design achieved a linear power gain of 11.3dB, with an output power of 19.8dBm at 2.4GHz, which provides few dB of margin over the design requirements, and 1-dB output compression at 15.3dBm which is defined as the 9dBm in an input level axis, which means a 0dB back-off level in this work. At that point, the PAE is measured to be 73.4%. Also, the PAE is measured to be 68.3% at 6dB back off, 54.2% at 12dB back off, and 29.9% at 18dB back off, which represents a PAE improvement of 1.58 times, 2.245 times, and 2.986 times respectively, compared to a single-stage class AB design.

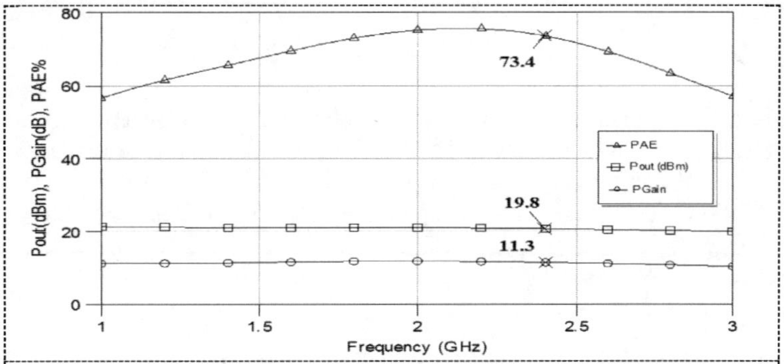

Figure 2. Output power, power gain and PAE

From figure 3, once the auxiliary amplifier is turned ON, the current from the auxiliary amplifier through the inductor partially cancels that coming from the main amplifier, causing the inductor loss to the output power ratio to decrease. Since the loss at the first efficiency peak is significantly higher, it causes the first efficiency peak to be somewhat much lower than the second peak. This concept is the same in between the auxiliary #1 and the auxiliary #2, which causes the second efficiency peak to be somewhat much lower than the third peak and so on.

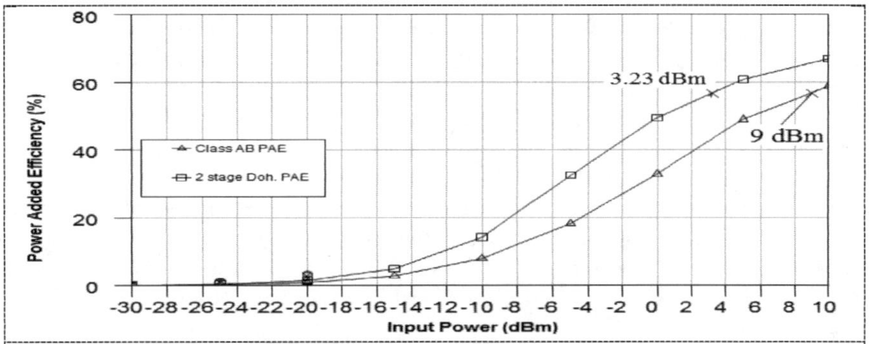

Figure 3. PAE versus input power

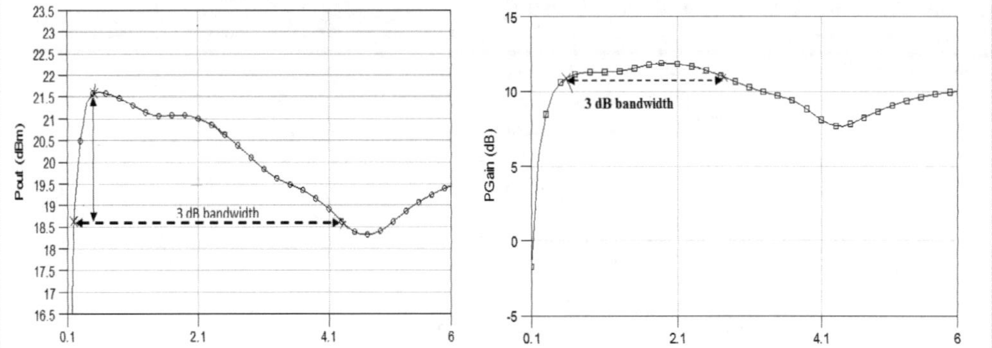

Figure 4. a) Output power versus frequency **b)** Power gain versus frequency

From figure 4a, output stage design the 3-dB bandwidth is measured to be 4.232GHz with about ±2GHz from the center frequency 2.4GHz of WLAN standard as shown. Also there is a flatter gain over a wide frequency range with only ±0.5dB gain variation for 2.13GHz bandwidth as shown in figure 4b.

4. Conclusion

In this paper, Class AB is studied the ideal theoretical solution for the efficiency-linearity tradeoff and have proposed a new concept of a class AB PA 2.4 GHz using multi-stage technique. However, the performance of this amplifier relies heavily on the realization of the harmonic traps and the input biasing voltage. For the third-harmonic resonator to generate a third-harmonic component of voltage, it requires third-harmonic current which does not exist. The simulation result showed that the proposed significantly improved the efficiency over a broad average output power compared to the conventional balanced amplifier.

5. References

[1] Kumar G, Sitaraman S, Sridharan V, Sankaran N, Fuhan L, Kumbhat N, Nair V, Kamgaing T, Juskey F, Sundaram V, and Tummala R 2012. Modeling and design of an ultra-miniaturized WLAN sub-system with chip-last embedded PA and digital dies. *IEEE 62nd Electronic Components and Technology Conference*, 1015–1022.

[2] Yargholi M, and Tarighat A P 2012. Resistive feedback LNA with dual band notch filter for suppressing WLAN signals in UWB receivers. *National Conference on Communications*, 1–4.

[3] Bakkaloglu B and Fontaine P A 2005 Multi-mode, multi-band RF transceiver circuits for mobile terminals in deep-submicron CMOS processes. *IEEE Radio Freq. integrated Circuits Symp.*, 483–486.

[4] Nam J, Shin JH, and Kim B. 2005. A handset power amplifier with high efficiency at low level using load modulation technique. *IEEE Trans. Microw. Theory Tech.,* 53, 2639-2644.

[5] Raab, FH, Sigmon BE, Myers RG, and Jackson RM. 1998. L-band transmitter using Kahn EER technique. *IEEE Trans. Microw. Theory Tech.,* 46, 2220-2225.

[6] Tsai Pi H, Metzger AG, Zampardi PJ, Iwamoto M, and Asbeck PM. 2005. Design of high efficiency current-mode class D amplifier for wireless handsets. *IEEE Trans. Microw. Theory Tech.,* 53,144-151.

[7] Kimball DF, Jeong J, Hsia C, Draxler P, Lanfranco S, Nagy W, Linthicum K, Larson L E, and Asbeck P M 2006. High-efficiency envelope tracking W-CDMA base-station amplifier using GaN HFETs. *IEEE Trans. Microw. Theory Tech.,* 54, 3848-3856.

[8] Joo S, Choi T, and Jung B 2009. A 2.4-GHz Resistive Feedback LNA in 0.13um CMOS. IEEE J. Solid-State Circuits, 44, 3019–3029.

[9] Joo T, Koo B, and Hong S 2013. A WLAN RF CMOS PA With Large-Signal MGTR Method. *IEEE Transactions on Microwave Theory and Techniques*, PP, 1–8.

[10] Hwann-Kaeo C, Kuei-Cheng L, Wei-Hsien C, and Ying-Zong J 2012. A 1-V 5-GHz Self-Bias Folded-Switch Mixer in 90-nm CMOS for WLAN Receiver. IEEE Transactions on Circuits and Systems I: Regular Papers, 59, 1215–1227.

[11] Point R 2003. An RF CMOS Transmitter Integrating a Power Amplifier and a Transmit/Receive Switch for 802.11b Wireless Local Area Network Applications. *IEEE Radio Frequency Integrated Circuits Symposium*, 431-434.

[12] SungWon C, and Dawson J L 2011. Digital predistortion using quadrature ΔΣ modulation with fast adaptation for WLAN power amplifiers. *IEEE MTT-S International Microwave Symposium Digest*, 1–4.

4th International Conference on Energy and Environment 2013 (ICEE 2013) IOP Publishing
IOP Conf. Series: Earth and Environmental Science **16** (2013) 012118 doi:10.1088/1755-1315/16/1/012118

Harvesting biogas from wastewater sludge and food waste

K. H. Chua, W.L. Cheah , C.F. Tan, [1]Y.P. Leong

Institute of Energy Policy and Research, Universiti Tenaga Nasional
Jalan IKRAM- UNITEN, 43000 Kajang, Selangor, Malaysia.
[1] TNB Research Sdn, Bhd
E-mail: ckh@uniten.edu.my

Abstract. Wastewater sludge and food waste are good source of biogas. Anaerobic treatment of slude and food waste able to produce biogas which is a potential renewable energy source. This study looks into the potential biogas generation and the effects of temperature on biogas generation. A lab scale reactor was used to simulate the biogas generation. The results show that wastewater sludge able to produced upto 44.82 ml biogas/kg of sludge. When mixed with food waste at a ratio of 30:70 (food waste), the biogas generated were 219.07 ml/kg of waste. Anaerobic of food waste alone produced biogas amount to 59.75 ml/kg of food waste. Anaerobic treatment also reduces the volume of waste. The effect of temperature shows that higher temperature produces more biogas than lower temperature.

1. Introduction

Wastewater treatment and management is an essential part of preserving good public health for a country. Strict regulation had been enforced for proper treatment of wastewater and disposal of treated effluent. Since 1994, the management of wastewater in most of the states Peninsular Malaysia is under a private company named Indah Water Konsortium IWK Sdn Bhd. In recent sustainable report by IWK, it has operates and maintained 14342 km of public sewer networks, 5605 public sewerage treatment plants and 778 network pump stations [1]. The generation of wastewater sludge ranges from 60 to 90 g dried solids per population equivalent or PE in European Union [2]. Concentrated wastewater sludge contains high organic contents and has potential as biogas source if treated anaerobically. Previously, the treated sludge are used as fertiliser for crops or simply landfilled. Anaerobic digestion however can convert part of the organic sludge to methane, carbon dioxide and traces of other gases. The composition of the biogas and rate of generation is a function of the type and concentration of organic waste, reaction environment such as pH, temperature and moisture contents. The composition of biogas from wastewater sludge ranges from 50-75% CH_4 and 25 - 50% CO_2 plus traces of H_2S [13].

Generally, the anaerobic digestion follows: hydrolysis, acidogenesis, acetogenesis and methanogensis. The hydrolysis stage degrades both insoluble organic material and high molecular compounds such as fat and grease into soluble organic substances. The acidogenic bacteria further breakdown the souluble organic compound into volatile fatty acid. During the acetogenic stage, the volatile fatty acid was converted by acetogens into acetic acid and other gases such as CO_2 and H_2. The methanogenis bacteria digest the acetate into methane and CO_2. Within an anaerobic environment, parameters such as pH, temperature and retention time influence the biogas generation. Methanogenic bacteria prefer optimum pH between 6.5 and 7.2 [3]. Temperature affects substrate digestion, influences microorganism growth rates and metabolism and microbes population dynamic. High temperature enhances the biological and chemical reaction rates. The inhibition factor includes the presence of ammonia, sulphide, sodium, potassium and heavy metals. Co-digestion of sludge with animal manure,food waste, fruit and vegeTable waste and glycerol is another new approach to waste digestion. In a study by Chua et al, it was found that mixing food waste with chicken manure produces more biogas than anaerobic digestion of food waste alone [4]. Fountoulakis et al found that addition of glycerol in wastewater sludge at 1% v/v increases the CH_4 production [5]. Recent report by IMeche had indicated that about 30 to 50% or 1.2 to 2 billion tonnes of the food produced globally never reach the consumers. 4.4 million Tonnes of avoidable food waste being dumped in UK [6]. In Malaysia, it was reported that Malaysian disposed 930 tonnes of food waste a day [7]. In a study by Kim et al shows that it is possible to treat the food waste anaerobically and generate 254 to 282 litre CH_4 per kg COD degraded [8]. Higher temperature further improves the biogas generation from treating food waste anaerobically [9]. Co-digestion of cattle manure with food waste with sewage sludge in 70:20:10 ratios at 36°C yielded maximum biogas of 603 l CH_4/kg Volatile solid feed [10].

This study looks into the co-digestion of sewage sludge with food waste taken from a site that operated hospitality and tourism business which also operate its own wastewater treatment plants. Co-digestion of sewage sludge and food waste presents a environmental friendly option to solve the waste disposal problem and also an opportunity to harvest the biogas for direct heating or other usage.

2. Methodology and Results

Samples were collected from site and delivered to an experimental laboratory to be analysed. Some of the samples chemical characteristics were given in Table 1. A series of laboratory scale reactor consists of circular PVC containers with volume of about 20 litres were prepared. Each reactor had a gas outlet and tube for gas collection. The chemical analysis followed *Standard Method for the examination of water and wastewater by APHA.* COD was measured using Hach Spectrophotometer 2100. BOD measurement was measured by YSI BOD meter. All the samples were subjected dilution as the concentration of BOD_5 and COD are extremely high.

Table 1. Sample Characteristic

Parameters	Sewage sludge	Food waste
Biochemical oxygen demand, BOD_5, mg/l	29,524	162,086
Chemical oxygen demand, COD mg/l	135,711	682,847
Ratio of BOD:COD	1: 4.60	1: 4.21
Moisture content,%	68.9	94.3
Biochemical oxygen demand, BOD_5, mg/l	29,524	162,086
Chemical oxygen demand, COD mg/l	135,711	682,847
Ratio of BOD:COD	1: 4.60	1: 4.21

The proportion of co-digestion of food waste and sludge (w/w) and initial mass are given in Table 2. All the reactors were placed in a shaded area except *Sample 3* which was placed under direct sunlight. The average temperature for shaded area and direct sunlight were 28.5°C and 30.5°C respectively. The reactors were monitored over 65 days.

Table 2. Proportions of food waste and sewage sludge

	Food waste	Sewage sludge	Total initial mass, kg
Sample 1	100	0	7.13
Sample 2	70	30	8.16
Sample 3 (O)	70	30	7.97
Sample 4	30	70	9.77
Sample 5	0	100	9.75

The effects of anaerobic digestion on BOD_5 and COD removal after 65 days are given in Table 3 below. It shows that the percentage removal of BOD_5 ranges from 93.24 to 97.63%. In term of COD, the percentage removal ranges from 35.4 to 59.42%. It shows that the percentage removal for food waste in terms of BOD_5 and COD from anaerobic digestion is higher than sewage sludge.

Table 3. BOD_5 and COD Concentration and percentage removal

	Initial BOD_5 con. mg/l	BOD_5 con. after 65 days, mg/l	% removal	Initial COD con, mg/l	COD con. after 65 days, mg/l	% removal
Sample 1	162086	5510	96.60	682847	277129	59.42
Sample 2	122298	5890	95.18	518706	335067	35.40
Sample 3 (O)	122560	2906	97.63	519340	310000	40.31
Sample 4	69287	4650	93.29	299851	147500	50.81
Sample 5	29524	1997	93.24	135711	62500	53.94

This is expected as the sewage sludge is the residue from the biological treatment of wastewater which contents 0.8 to 1.2% dried solids of which 59% to 88% volatile solids and 32 to 41 % protein [11]. Food waste however has higher biodegradable organic matter that allows microbes to digest. Comparison between *Sample 2* and *Sample 3* indicates that BOD_5 and COD removal correlate with ambient temperature. Higher removal for higher temperature indicates higher biological activities. The biogas generation is presented in the Figures below and Table 4. The accumulated biogas per mass of waste ranges from 13.41 ml/kg of waste to 219.07 ml/kg of waste. These are empirical values based on experimental set-up. On site, the generation of biogas ranges from 0.75 to 1.2 m^3/kg volatile solid [11]. The laboratory scale result shows that sewage sludge able to generate more biogas per mass of total solids than food waste. Comparison between *Sample 2* and *Sample 4* indicated similar trend. *Sample 5* has much higher moisture content than *Sample 1* hence in terms of dried solids; it yields much higher gas generation.

Table 4. Biogas Generation

	Initial Moisture content %	Volume of biogas accumulated, ml	Gas generation, ml /kg mass of waste	Gas generation ml /kg mass of total solids
Sample 1	68.87	426	59.75	191.92
Sample 2	75.48	138	16.91	68.97
Sample 3 (O)	76.12	1746	219.07	917.38
Sample 4	85.76	131	13.41	94.17
Sample 5	92.8	437	44.82	622.5

The highest biogas generation is at *Sample 3* where the reactor was under direct sunlight as compare with *Sample 2*. It is able to increase biogas generation by 13-folds. The proportion of sewage sludge to food waste was 30:70. Higher temperature enhances the microbial growth dynamic and allows more waste being converted to biogas.

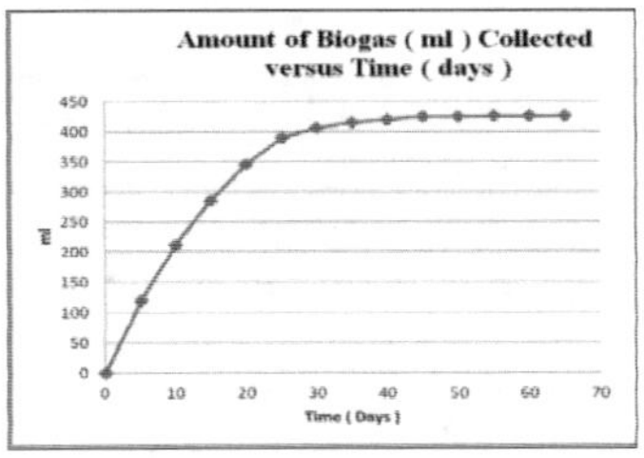

Figure 1. Biogas Generation for *Sample 1*

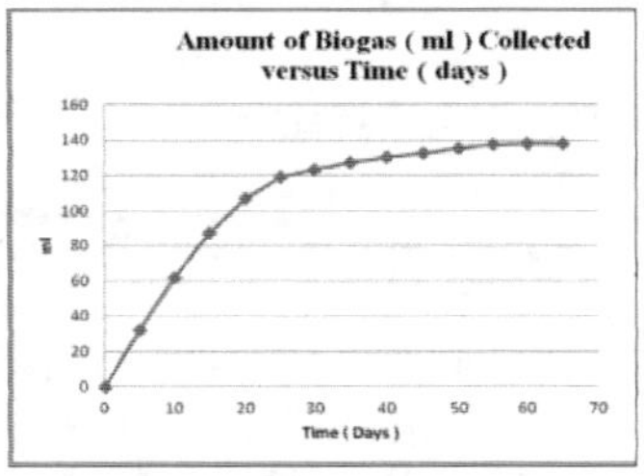

Figure 2. Biogas generation for *Sample 2*

The graphs indicate that the rate of biogas formation reaches plateau after almost 30 days. The generation rate per day shows that biogas generation from sewage sludge is faster than food waste. It takes almost 10 days to reach the 50% maximum potential biogas generation. The generation rate can be modelled using Monod equation as quoted by Appels et al [12].

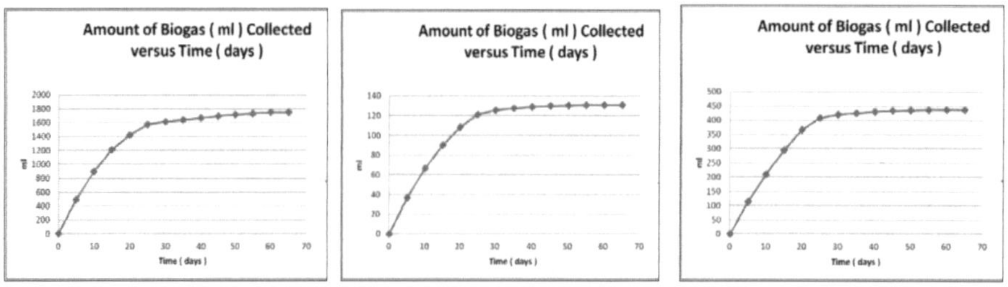

Figure 3. Biogas Generation for *Sample 3, Sample 4* and *Sample 5*

3. Conclusion

This empirical study shows that temperature influence greatly the biogas production. Co digestion food waste with sewage sludge allows the microbial population in sludge to digest the food waste and enhance waste degeneration. If similar reactor is to be located at cooler temperature environment, a "greenhouse" concept approach is plausible where the reactor is exposed to direct sunlight in a confined place. It allows heat retention from direct sunlight without exposing to cooler air.

Acknowledgment

The authors would like to thank Genting World Resort especially Mr. Chia Chin Kuan, Vice President of Engineering Department for granting us permission to collect samples. The authors also like to thank Energy Commission for partial funding of the project.

References

[1] IWK Sustainability Report 2010 *Achieveing Excellence Through Innovation* (IWK press)
[2] European Commision 1986 *Council Directive 86/278/EEC On The Protection of The Environment and Particular Soil When Sewage Sludge is Used In Agriculture* (European Council)
[3] Turovskiy, IS and Mathai PK 2006 *Wastewater Sludge Processing* (New York: Wiley)
[4] K.H. Chua et al 2008 A Case Study of On The Anaerobic Digestion of Food Waste and Gas Formation *International Conference on Construction and Build Technology (Kuala Lumpur, Malaysia, 16-20 June 2008)* **29** 311-316
[5] M.S. Fountoulakis, I. Petousi and T. Manios 2010 Co-Digestion of Sewage Sludge With Glycerol To Boost Biogas Production *Waste Management* **30** 1849-1853
[6] Waste and Resources Action Programme *WRAP Welcomes IMechE Food Waste Report* [online] Available from: http://www.wrap.org.uk/content/wrap-welcomes-imeche-food-waste-report [Accessed 10 January 2013]
[7] Soo Ewe Jin Food *Waste A Reflection of Our Lifestyle* [online] Available from: thestar.com.my/columnists/story.asp?file=/2011/6/13/columnists/mondaystarters/8878522 [Accessed 13 June 2011]
[8] J.K. Kim et al 2007 Volumetric Scale-Up of A Three Stage Fermentation System for Food Waste Treatment *Bioresource Technology* **99** 4394-4399
[9] J.K. Kim et al 2006 Effects of Temperature and Hydraulic Retention Time on Anaerobic Digestion of Food Waste *Journal of Bioscience and Bioengineering* **102** 328-332
[10] E. Maranon 2012 Co-Digestion of Cattle Manure With Food Waste and Sludge To Increase Biogas Production *Waste Management* **32** 1821-1825
[11] Metcalf and Eddy 2004 *Treatment, Resuse, and Disposal of Vsolids and Biosolids: Wastewater Engineering: Treatment and Reuse* (McGraw Hill)
[12] Appel et al 2008 Principles and Potentials of The Anaerobic Digestion of Waste-Activated Sludge *Progress in Energy and Combustion Science* **34** 755-781
[13] Shefaliverma 2002 *Anaerobic Digestion of Biodegradable Organics in Municipal Solid Wastes* (M.S. Thesis, Columbia University)

IOP Conf. Series: Earth and Environmental Science **16** (2013) 012119 doi:10.1088/1755-1315/16/1/012119

Self-Efficacy and green entrepreneurship

K L Tan, S Suhaida, Y P Leong
Institute of Energy and Policy Research
Universiti Tenaga Nasional,

Email: tan@uniten.edu.my

Abstract. The objective of this study is to investigate empirically the extent to which self-efficacy contributes to the development of green entrepreneurial intention. The measurement constructs of self-efficacy were classified into market opportunities, innovative environment, initiating relationships, defining purpose, coping with challenges, and developing human resources. The study comprises 252 usable convenient samples through structured questionnaires. The coefficient of determination R^2 shows that the variance of intention to entrepreneurship is explained by the variance of the independent variables. It was also found that the model is fit for prediction.

1. Introduction

Entrepreneurship has been defined in various ways. One of the most cited definition is "the creation of new enterprise" [1]. Green entrepreneurship refers to enterprise that involves processes, activities or products that promote eco-friendly environment. There has been a growing awareness these days on the importance of these green enterprises in an economic development process. This growing awareness has led many public administrations from all political ideologies and all administrative levels to establish agencies to stimulate and to favor the creation of green enterprises.

Various antecedents to entrepreneurship such as personal traits and characteristics have been investigated but they have not been found to be reliable predictors of entrepreneurial behavior [2]. There are some other factors which might possibly and significantly influence the intention to entrepreneurship. One of these factors identify by [3] is self-efficacy. This study attempts to look at how self-efficacy affects the intentionality of green entrepreneurship.

2. Literature Review

Entrepreneurial intention is a state of mind that influences the cognitive thinking of the entrepreneur towards the forming of an entrepreneurial activity. It is a process that focuses on a complex relationship among entrepreneurial ideas and theirs resulting outcome. [4] posit that entrepreneurial intentions predict entrepreneurial behavior and attitudes successfully influence intentions.

Self-efficacy is defined as a person's belief in his or her capability to carry out a task [5] and has a significant influence on the process of development of entrepreneurial intention. [6] found a significant weighted average correlation between self-efficacy and new ventures. [3] found that self-efficacy helps in bolstering positive thoughts and controlling negative thoughts that are relevant to business start-ups.

[7] conducted a study of entrepreneurial self-efficacy (ESE) by developing constructs such as marketing, innovation, management, risk-taking, and financial control skills. One of the most significant findings was that innovation and risk-taking appeared to be important cognitive capabilities in the ESE study. [8] found that individuals with high self-efficacy are more willing to make an effort to overcome obstacles in business ventures. Similarly, [9] found that self-efficacy is a significant contributing factor to start one's business. [10] found that abilities such as recognizing opportunities and driving the business venture are critical in the entrepreneurial process. These entrepreneurial skills have been used to create an expanded measure of ESE developed by [11]. He contended that self-efficacy to undertake new venture

initiatives will positively influence entrepreneurial intentions. He further developed and classified a self-efficacy measurement constructs into *market opportunities, innovative environment, initiating relationships, defining purpose, coping with challenges, and developing human resources.*

To empirically analyze the impact of entrepreneurial self-efficacy constructs on entrepreneurial intentions among university students, the following hypotheses were formulated for the study.

H_1: There is a significant relationship between the awareness of market opportunity and green entrepreneurial intention.

H_2: There is a significant relationship between the awareness of innovative environment and green entrepreneurial intention.

H_3: There is a significant relationship between the awareness of initiating relationship and green entrepreneurial intention.

H_4: There is a significant relationship between the awareness of defining purpose and green entrepreneurial intention.

H_5: There is a significant relationship between the awareness of challenges and green entrepreneurial intention.

H_6: There is a significant relationship between the awareness of developing human resources and green entrepreneurial intention.

3. Methodology

3.1 Research Design

The samples were collected at various strategic locations in Universiti Tenaga Nasional, Putrajaya Campus. Structured questionnaires which consist of demographic characteristics and six measurement constructs were then administered to the final year and graduate student respondents who were conveniently selected. A total of 252 usable sample subjects were obtained.

3.2 The Model

$$Y = \beta_0 + \beta 1\ X_1 + \beta 2\ X_2 + \beta 3\ X_3 + \beta 3\ X_3 + \beta 4\ X_4 + \beta 5\ X_5 + \beta 6\ X_6 + \varepsilon_t \quad (1)$$

where Y is the Entrepreneurial Intention as dependent variable and the independent variables are $X_1 =$ Opportunity, $X_2 =$ Environment, $X_3 =$ Relationship, $X_4 =$ Purpose, $X_5 =$ Challenges, $X_6 =$ Human Resources. β_0 is the constant term of the model. β_i, i=1 to 6 as coefficients of the respective variables for the model and ε_t is the error term.

4. Output Analysis

4.1 Demographic Profiling

From the respondents' profile, it shows that male respondents contributed a larger percentage which is 72.4 percent as compared with female respondents (27.8%). In terms of ethnicity group, Malays constituted 52.4 percent, followed by 15.9 of Indian respondents. Chinese respondents stood at 9.9 percent. There are 71.8 percent respondents who were aged between 21 to 25 years old. Respondents aged between 16-20 constituted 4.42 percent. The respondents with ages above 26 was 13.9 percent. Also from the table, most (74.6 percent) respondents were from the undergraduate program, followed by 25.4 percent of graduate respondents.

4.2 Reliability Test

Table 1. Reliability of scales

Dimension	No. of Items	Cronbach Alpha
Intention	6	0.943
Opportunity	7	0.934
Environment	4	0.913

Relationship	3	0.918
Purpose	3	0.899
Challenge	3	0.851
Human Resource	3	0.902

Table 1 shows the reliability coefficients of the main variables were higher than 0.7 and they all indicated that the reliability among items was consistent [12]. Such a result is consistent with the study conducted by [11] which found internal consistency with reliability coefficients for the measurement constructs were higher than the cut-off point of 0.7. It therefore serves as a reliable foundation for further testing and subsequent analysis.

4.3 Regression Result

Table 2. Standardized (simultaneous) regression between entrepreneurial intention and the measurement variables

Dependent Var: Intention Independent variables	β	t	Sig.	VIF
Opportunities	0.325***	3.470	0.001	2.917
Environment	0.303**	1.895	0.039	3.458
Relationships	0.098	0.489	0.625	2.700
Purpose	0.141*	0.615	0.053	3.340
Challenges	0.445**	2.343	0.020	2.895
Human Resources	0.201	0.952	0.342	3.184

Notes: The regression coefficients shown in the table are standardized regression coefficients (beta coefficients), ***, ** and * indicates that the estimated coefficient is significantly different from zero at 1 percent, 5 and 10 percent respectively. R= 0.591, R^2 = 0.35, F-Value = 26.145 ($P<0.01$)

The general results for the linear multiple regression analysis of entrepreneurial intention arising from market opportunities, innovative environment, initiating relationships, defining purpose, coping with challenges, and developing human resources are reported in Table 2. Multicollinearity test of the six independent variables has been done. Using a cut-off value of VIF less than 5, no multicollinearity among the variables is found. The coefficient of determination R^2 shows that 35 percent of the variance of entrepreneurial intention is explained by the variance of the independent variables. And the F-value is statistically significant at the 1 percent level implying that the regression model is reliable for prediction.

The estimated coefficient of correlation (R = 0.591) shows a relatively strong linear correlation between the independent and dependent variables. The regression result shows that 'market opportunities' has a positive effect on entrepreneurial intention as the estimated coefficient is 0.325 with the confidence level of 99 percent. The 'initiating environment' also has positive association with the entrepreneurial intention as the estimated coefficient is 0.303 and is significant at 5 percent level. Variable 'challenge' has also shown significant with 95 percent confidence and it is positively related to entrepreneurial intention with the estimated coefficient 0.445. Another variable that has association with the entrepreneurial intention is defining purpose. It is positively and statistically significant with entrepreneurial intention at 90 percent confidence level as the estimated coefficient is 0.141.

5. Conclusion

The general linear regression result indicates that the dimensions 'market opportunity', 'initiating environment' and 'challenge' were by far the strongest and positive predictors of entrepreneurial intentionality. The variable 'defining purpose' has also shown statistically significant. Based on this regression result, H_1, H_2, H_4, H_5, are well supported. These findings are consistent with that of De [11]. Variables such as 'initiating relationship' and 'developing human resources' are not significant could possibly be due to respondents recognizing such variables as more likely to be relevant only after the business start-up. The fact that the intentionality examined here is depended only on self-efficacy may simplify the phenomenon excessively. There is likely that other factors that exist apart from self-efficacy could be deemed as significant. A second limitation in this study is that it only involves one particular sample from UNITEN campus and the generalizability of the findings is limited. Future research needs to include a wider and more representative spectrum of respondents including existing green entrepreneurs for greater representation.

Acknowledgement
This research is funded by Suruhanjaya Tenaga. The authors wish to thank Pn Nurullaidah Bte Md. Ali and Pn. Azura Bte Muhammad for their coordinating effort during the questionnaire survey.

References

[1] Low M B, MacMillan I C 1988 Entrepreneurship: past research and future challenges *Journal of management* **14(2)** 139-161.
[2] Ajzen I 1991 The theory of planned behavior *Organizational Behavior and Human Decision Processes* **50** 179-211.
[3] Drnovšek M J, Wincent and Melissa S C 2010 Entrepreneurial self-efficacy and business start-up: developing a multi-dimensional definition *International Journal of Entrepreneurial Behaviour & Research* **16(4)** 329 – 348.
[4] Kim M S, and Hunter J E 1993 Relationships Among attitudes, Behavioral Intentions, and behavior. A meta-analysis of Past Reserach Part 2 *Communication Reserach* **20(3)** 331- 364.
[5] Bird B 1988 Implementing entrepreneurial ideas: The case of intentions *Academy of Management Review* **13** 442-453
[6] Stajkovic A D, & Luthans F 1998 Self-efficacy and work-related performance A meta-analysis *Psychological Bulletin* **124** 240-261.
[7] Chen C C, Greene P G, and Crick A 1998 Does entrepreneurial self-efficacy distinguish entrepreneurs from managers? *Journal of Business Venturing* **13** 295-316.
[8] Krueger N F, Reily M D, and Carsrud A L 2000 Competing models of entrepreneurial intentions *Journal of Business Venturing* **15(5/6)** 411-432.
[9] Nuepert K E, Krueger N F, Chua B L 2004 Entrepreneurial self-efficacy and its relationship to business plan competitions: an international examination In *International Entrepreneurship: The Globalization of SME's Orientation, Environment and Strategy* (Hamid Etemad, Ed), Edward Elgar Publishing Ltd 126-141.
[10] Chandler G N, and Jansen E 1992 The Founder's Self-assessed Competence and Venture Performance *Journal of Business venturing* **7** 223-236.
[11] De Noble A F, Jung D, and Ehrlich S B 1999 Entrepreneurial self-efficacy: The development of a measure and its relationship to entrepreneurial action *Frontiers Entrepreneurship Research* Boston: Babson College.

IOP Conf. Series: Earth and Environmental Science **16** (2013) 012120 doi:10.1088/1755-1315/16/1/012120

Enhancing energy security in Malayia: the challenges towards sustainable environment

E J M Sahid[1], C Ch Siang[2] and L Y Peng[3]
[1, 3] Institute of Energy Policy and Research (IEPRe), Universiti Tenaga Nasional
[2] The Institute of Energy Economics, Japan (IEEJ)

[1]E-Mail: endang@uniten.edu.my

Abstract. Energy is known as one of the essential ingredients for economic development and security of energy supply is crucial in ensuring continuous economic development of a country. Malaysia's proven domestic oil reserves are estimated to last for another 25 years, while that of gas for another 39 years as of 2011. Despite the depleting indigenous energy resources, the primary energy demand has continued to grow robustly, at an annual rate of 6.3 percent per year from 1990 to 2010, while the primary energy import has grown 7.2% per year and the primary energy export has grown at a slower rate of 1.9% per year. This worrying trend is further compounded by the faster rate of primary oil import averaging 10.5% per year while the primary energy export has shrink at a rate of 1.4% per year. This paper has identified two main concerns namely overdependence on fossil fuel and increasing energy import dependency in creating a precarious position towards energy self-sufficiency. The study will analyse the energy security of the country and explore possible options and challenges in enhancing the energy supply security toward sustainable environment.

1. Introduction

Energy is known as one of the essential ingredients for economic development and security of energy supply is crucial in ensuring continuous economic development of any particular country. Malaysia's energy supply security has been thrust into the attention of energy industry players and policy makers in the recent years due to unrelenting energy demand and diminishing indigenous energy sources.

This paper will attempt to discuss on the energy security concerns of the country also explore the possible options and challenges in enhancing the energy supply security towards sustainable environment.

2. Policy Evolution

Energy policy for Malaysia could be grouped into two different period of formulation; the first period was from 1970 to 2000 and the second period was from 2001 to 2011. The period from 1970 to 2000 provides the earliest policies direction of the energy sector. Among the energy policies formulated during this period are National Petroleum Policy (1975), the National Energy Policy (1979), the National Depletion Policy (1980) and the Four-Fuel Diversification Policy (1981). The National Energy Policy outlined the three-pronged objectives namely; supply objective, utilization objective and environment objective. The Four-Fuel Diversification Policy was then introduced to complement the National Depletion Policy to ensure the security and reliability of energy supply.

The period between 2001 and 2011 focuses on the role of renewable energy sources as one of the energy mix and few policies has been introduced to emphasize on the green technology and environment aspects of energy development. Among the policies introduced are the Five-Fuel

Diversification Policy (2001), the National Green Technology Policy (2009), Renewable Energy Policy (2009) and the National Policy on Climate Change (2010). [5]

3. Energy Use Structure

Malaysia is well endowed with conventional energy sources as well renewable energy sources. Malaysia has the 14th largest gas reserve and 27th largest proven crude oil reserve in the world. As of January 2011, the total reserve of oil and gas is reported to be 19.9 billion barrels of oil equivalent, with 78 percent share of gas and 22 percent of oil. As for coal, Malaysia's reserve is only 19.38 trillion tonnes which located in Sarawak [9].

Total final energy demand has tripled from 1990 to 2010, increasing from 14 Mtoe in 1990 to 43 Mtoe in 2010 (IEA 2012). The industry and transport sector accounted for the largest share in the total final energy demand at around 30 percent and 33 percent in 2010 respectively. Oil has accounted for a large portion of final energy demand albeit at a decreasing trends; oil share in the final energy demand was about 66 percent in 1990 reducing slightly to 57 percent in 2010. The high percentage share of oil demand was mainly from the transport sector which consumed around 60 percent of the total oil demand. The high demand of energy in the transport sector is due to high level of motorization mainly resulting from growing population, automobile dependent urban sprawl, subsidized transport fuels price, and promotion of automobile industry in the country.

Despite the introduction of the Five-Fuel Diversification Policy, Malaysia is still very dependent on fossil fuel as the major source energy supply. Oil and gas has contributed a major share of more than 80 percent of the primary energy supply from 1990 through to 2010. However, the share of oil and gas has decreased to about 74 percent in 2010. The share of coal in primary energy demand however, has experienced a slight increase from 6 percent in 1990 to 20 percent in 2010 [6].

Coal and gas dominates the electricity generation mix for Malaysia in the recent years. In 2011, 45 percent of Malaysia electricity is generated by coal and 43 percent of electricity is generated by gas [9]. The remaining is generated by hydro, fuel oil/diesel and biomass.

4. Energy Security and Challenges

There are various definition of energy security from literatures, such as '*continuity of energy supplies relative to demand*' [10], '*ability of an economy to guarantee the availability of energy resource supply in a sustainable and timely manner with the energy price being at a level that will not adversely affect the economic performance of the economy*' [1]. For this study, the authors have used the definition specified by the Asia Pacific Energy Research Centre (APERC). APERC has identified three main elements of energy security, namely: physical aspect that are availability and accessibility of the resources; economic aspects that is the affordability of the resource acquisition and energy infrastructure; and environmental aspect of the resources namely acceptability of the resource supply.

The analysis on the historical trends of energy demand and supply shows that Malaysia's energy supply security concerns are also due to the over-dependence on fossil-fuel which has in turn caused an increasing energy imports in order to fulfil the domestic demand.

4.1. Over-dependency to Fossil Fuel.

The share of fossil fuels in the primary energy supply has remained around 90 percent from 1990 through to 2010. This highly dependency of fossil fuels coupled with the dwindling domestic fossil fuel reserves constraint will definitely force Malaysia to import more fossil fuel at high market price where the fuel price is volatile if thorough fuel mix diversification is not carefully consider.

For power generation sector, concerns are towards energy supply security for the fuels for the power plants. As stated in section above, more than 80% of electricity is produced using coal and natural gas with almost equal share among the two fuels. However, the declining domestic gas production and over stretching of supply system has led to frequent supply interruptions in the power generation industry. As for coal, almost all of the coals used in the power plants are imported. The coal is imported mainly from South Africa, Australia and Indonesia. The amount of imported coal has increased steadily for Peninsular Malaysia, increasing from 11.9 million tonnes in 2009 to 19.2 million tonnes in 2011 [9]. In term of price, coal is subjected to market forces. The prices of coals even though has been stable for decades have increased beyond 2003 and surged to a new level due to the increase of global demand of the fuel source.

4.2. Increasing dependence on energy imports

Increasing demand for fossil fuels, coupled with the depleting domestic fossil fuel reserves, is expected to result in increasing the country's net import. The total final energy demand has grown at a fast rate of 5.8 percent per year, increasing from 14 Mtoe in 1990 to 43 Mtoe in 2010. To fulfill the demand the primary energy import has grown at a fast rate of average 7.2 percent per year while the primary energy export has grown at a slow rate 1.9 percent per year.

Malaysia's is showing vulnerability on its energy security particularly due to increasing dependency on oil import and compounded by the nature of oil market and contract that normally allow for greater price volatility. Oil production has remain almost constant from 1990 to 2010 at 0.2 percent average growth rate per year, on the other hand the oil primary demand has grown at a fast rate of 4.8 percent per year from 1990 to 2010. In order to fulfill the demand the primary oil import has grown at a fast rate of average 10.5 percent per year while the primary energy export has shrink at a rate of 1.4 percent per year [6].

5. Towards Sustainable Future

The strategic options to achieve sustainable energy development could also further enhance energy supply security. The options are could be segregated into four main segments namely:

i. Diversifying sources of energy supply.

By intensifying indigenous gas and hydro resources development, securing more gas from foreign sources, strengthening and expanding supply infrastructures to facilitate regional interconnection and exploring and building capacity for the nuclear options.

ii. Reducing carbon content of energy.

By developing centralized renewable energy, study on the low carbon electricity such as application of carbon capture and storage system at coal-based power plants, and increasing use of alternate fuels and cleaner source of energy for transport sector.

iii. Efficient utilisation of energy for low carbon growth.

By enhancing energy efficiency (EE) in the residential and commercial sector, reducing demand for personalised modes of transport and increasing share of public transport, planned public transport scheme, promoting cogeneration in industrial facilities and tackling energy inefficiency in Small Medium Enterprise (SMEs)

iv. Facilitating low-carbon industries and service development to promote economic growth.

By manufacturing of high energy efficiency products such as LEDs lights, promoting energy efficiency products and developing new energy industries.

6. Conclusion

This study has identified two major issues concerning energy supply security in Malaysia, namely over-dependency to fossil fuel and increasing of energy import dependency. This paper discussed the evolution of energy policies, the recent energy demand trend, energy security and challenges and subsequently relevant strategic options towards achieving sustainable energy development which could also enhance energy supply security for the country.

7. Acknowledgement

The authors are grateful to Energy Commission's Chair in Energy Economics Grant 01090001 IEPR which has supported the study. The authors also appreciative to the Institute of Energy Policy and Research (IEPRe) for sharing the necessary information required for the study. Any of the opinions, findings and conclusions expressed in this paper are those of the authors and do not necessarily reflect the views of the Energy Commission, IEPRe nor IEEJ.

References

[1] Asia Pacific Energy Research Centre, Japan 2007 *A Quest for Energy Security in the 21st Century: Resources and Constraints, 2007* Asia Pacific Energy Research Centre.
[2] Chua S C & Oh T H 2010 Review on Malaysia's national energy developments: Key policies, agencies, programmes and international involvements *Renewable and Sustainable Energy Reviews* **14** 2916-2925
[3] Department of Statistics 2011 *Malaysia Economic Statistics - Time Series, 2011* Putrajaya, Department of Statistics
[4] *Economic Transformation Programme - A Roadmap for Malaysia* Putrajaya, Prime Minister's Department
[5] Endang J M S, Dr Tan C S, Dr. Leong Y P 2012 *Sustainable Energy Development- A case study for Malaysia* 16th International Conference on ISO & TQM 2012. Vaxjo Sweeden, July 2012
[6] International Energy Agency 2012 *Energy Balance for Non-OECD Countries – 2012 Edition* International Energy Agency
[7] Ministry of Energy, Green Technology and Water 2010 *National Renewable Energy Policy & Action Plan 2010* Putrajaya, Kementerian Tenaga, Teknologi Hijau dan Air
[8] Ong H C, Mahlia T M I, Masjuki H H 2011 A review on energy scenario and sustainable energy in Malaysia *Renewable and Sustainable Energy Reviews* **15(1)** 639–47.
[9] UNITEN 2012 Proceedings for National Energy Security Conference 2012
[10] Winzer 2011 *Conceptualizing Energy Security* EPRG Working Paper. University of Cambridge

IOP Conf. Series: Earth and Environmental Science **16** (2013) 012121 doi:10.1088/1755-1315/16/1/012121

Green buildings in Malaysia towards greener environment: challenges for policy makers

M S Suhaida, K L Tan and Y P Leong
Institute of Energy and Policy Research
Universiti Tenaga Nasional,

E-Mail: suhaida@uniten.edu.my

Abstract. The launch of the National Green Technology Policy (NGTP) in 2009 is a manifesto of the government's seriousness in implementing "green" initiatives for the country. Specifically for buildings, the government promotes the application of renewable energy (RE) and energy efficiency (EE) and the application of green building index. With the introduction of Low Carbon Cities Framework, Green Pass, Green Neighbourhood, Green Building Index by various agencies and organisations in Malaysia, it is time to look back and see how all these tools could come together. This paper attempts to identify the challenges in harmonising the green initiatives for policy makers toward greener environment for sustainability.

1. Introduction

Green building initiatives in Malaysia is a concern that has received bigger attentions from government agencies, private organizations and the public at large for almost a decade now. With the launch of National Green Technology Policy (NGTP) in 2009, focus on green buildings has intensified with the promotion of application of renewable energy (RE) and energy efficiency (EE), as well as the green building index.

2. Policies and programmes

The launch of the National Green Technology Policy (NGTP) in 2009 is a manifesto of the government's seriousness in implementing "green" initiatives for the country. These include among others intensification of green technology research and innovation towards commercialization, promotion and public awareness of green technology. Specifically for buildings, the government promotes the application of renewable energy (RE) and energy efficiency (EE) in buildings such as solar photovoltaic (PV), rainwater harvesting, phasing out of incandescent lights, and the application of green building index [1].

The National Policy on Climate Change (2009) through the Strategic Thrust for RE and EE champions the energy efficiency through promotion of green buildings in commercial / institutional, industrial and residential sector. This is achievable through application of low or zero energy concept in the design and construction of new buildings; retrofitting of efficient ventilation and cooling systems as well as lighting systems; energy conservation practice in buildings; retrofitting existing buildings to include EE features and generate RE; and development of a green building index.

The government of Malaysia has taken several pro-active actions in promoting energy efficiency through demonstration buildings including the Energy Commission's Diamond Building that could encourage private sector to also construct and design low energy buildings [2].

In summary, Table 1 shows the various building related policies and initiatives championed by the government together with various stakeholders.

YEAR	POLICIES/ PROGRAMMES	AGENCY
FUTURE	• Green PASS (Green Performance Assessment System In Construction) • *In consulation phasewith stakeholders* • *Developed by Construction Industry Development Board of Malaysia (CIDB)* • *Covers construction phase and operational phase of the building.* – *Construction phase 5 elements- assessment of the construction site, building materials, energy, water and waste.* – *Operational phase 3 elements - building indoor environmental quality (IEQ), energy and water*	CIDB
2012	• PWD Green Rating Scheme (pH JKR Malaysia) • Rating tool for government buildings • 5 criteria - Energy efficiency, Indoor environmental quality, Sustainable site planning & management, Materials & resources, Water efficiency	PWD
2011	• Low Carbon Cities Framework & Assessment System • *Covers four aspect of township – Environment, Infrastucture, Transportation & Building* • *Low Carbon Building: 5 criteria – Energy (EE+RE), indoor environment quality, site construction management, materials, water management* • *PILOT projects – Miri City Council, University of Malaya, Pulau Sahbesar in Kenyir, Port Dickson Municipal Council and Hang Tuah Jaya in Melaka* • Green Township – Putrajaya & Cyberjaya • Green Neighbourhood Guidelines • *Promoted by Ministry of Housing and Local Government (MHLG)* • *for local council enforcement use in evaluating plan submission* • *in line with LCCF with 4 criteria - Smart Location, Neighbourhood Pattern & Design, Green Infrastructure, Green Communities Network* • Building Sector Energy Efficiency Project (BSEEP) • *5-year project - collaboration between United Nations Development Program (UNDP), Global Environmental Facility (GEF) and the government (Public Work Department (PWD) as the implementer)*	MEGTW MEGTW MHLG PWD
2010	• National Energy Efficiency Master Plan Study 2010 • *Replacing Incandescent to Compact Florescent Lamp (CFL)* • *Replacing Inefficient Refrigerators with 5-star Refrigerators,* • *Raising Air Conditioner Temperature to 25°C,* • *Replacing T8 to T5 lamp for Government Offices* • *Energy Auditing for Commercial Buildings* • Economic Transformation Programme • *Energy Performance Management System (EPMS) for government entities*	MEGTW
2009	• National Green Technology Policy • *National Energy Centre (PTM) restructuring to Malaysian Green Technology Corporation (MGTC),* • *Green Technology Financing Scheme(GTFS) (2010 -2015)* – *Approved GT Value for Financing (RM): 1,118,895,495.00* – *Balance of GT Value for Financing (RM): 2,381,104,505.00* • *Green Township in Putrajaya and Cyberjaya,* • *International Greentech and Eco Products Exhibition and conference Malaysia (IGEM).* • National Policy on Climate Change • Green Building Index (GBI) • *Developed by Malaysian Institute of Architects (PAM) and the Association of Consulting Engineers Malaysia (ACEM) supported by Malaysia Green Building Confederation (MGBC). Separates between Residential & Non-residential – the non-residential rating tools are customised by nature of whether they are commercial, industrial or institutional - including the Industrial Rating Tool* • *6 criteria measuring energy efficiency (EE) – indoor environment quality, sustainable site and management, materials and resources, water efficiency, and innovation*	MEGTW MNRE private initiative

YEAR	POLICIES/ PROGRAMMES	AGENCY
2001- 2008	• The Efficient Management of Electrical Energy Regulation 2008 • Uniform Building By-Laws (UBBL) (1984), amended 2007 • Code of Practice on EE and Use of RE for Non-residential Buildings – MS 1525: 2001, revised 2007 – by SIRIM • Guidelines for Conducting Energy Audits in Commercial Buildings (2004) • Design Strategies for Energy Efficiency in New Buildings (Non-Domestic) (2004)	
2000s - 1970s	• National Policy on the Environment (2002) • The Electricity Supply Act 1990 and the Electricity Supply Act (Amended) 2001 • Four Fuel Diversification Policy (1981) • National Depletion Policy (1980) • National Energy Policy (1979)	

3. Green Building Ratings

A UNEP expert, Dr Arab Hoballah has said in one of many climate change related venues in 2009, "…almost no country in the world can hope to achieve carbon dioxide-reduction targets without including the building sector into their plan of action."[1]. Improving energy efficiency in buildings is one of the most cost-effective measures for reducing carbon dioxide (CO2) emission.

To do just that, one must go back to basic, defining what green building is. According to U.S. Environmental Protection Agency, green building or sustainable building (sustainable building) refers to a structure and using process that is environmentally responsible and resource-efficient throughout a building's life-cycle: from design, construction, operation, maintenance, renovation, and demolition [3]. To determine whether a building meets all the criteria of a green building, a rating system is applied. Another question arises, which rating system should one adopt? Are they all the same?

Generally, there are two approaches to measure the "greenness" of a building. First approach is by measuring the level of energy consumption and usage per square metre of the building. A building is then labelled as Low, Zero or Green Energy Buildings. The measurement uses the balance or equation between reducing energy demand for a building with its optimisation of renewable energy resources that could be exported into external grids. [4]. On the other hand, Low and Zero Carbon term uses carbon emission calculation for a building, which largely relates back to energy consumption.

Referring to the previous table, one will notice the issue, "Which one should be used for building in Malaysia?" The regulator is then posed with additional question, "to fully regulate or allow for self-regulation?

4. Conclusion and recommendations

With the aforementioned questions on hand, the policy makers in Malaysia should try harmonise the various rating or measurement systems for the building sector. Adopting a review by Lombard, et. al. (2009) on building energy labelling, rating for green building in Malaysia must have (1) definition of green building index, (2) development of green building calculation tool, (3) setting a threshold value for the performance index, (4) definition of the comparison scenario, (5) definition of the scale for green building rating, (6) identification of potential green building measures and (7) gathering information in the certification process.

Next consideration is to fully regulate the building sector or leave some space for self-regulation. This action could be implemented in stages based on the rate of awareness and acceptance by the stakeholders. Malaysia policy makers should (1) review all existing policies – to better fit to country's needs & position, (2) assess policies implementation , i.e. programmes, incentives etc – must be based on problem-solving, potential, & dedicated resources, (3) focus on proven programmes –to have better yields, and (4) focus on major improvements in high potential and critical areas. Efficiency, effectiveness, and sustainability must be evaluated at the systemic and programmatic level. While one particular program form may seem to be effective, there is potential for this program to have minimal or even negative consequences on a system's design intent [6].

It is hoped that as Malaysia heads towards a more comprehensive implementation of green development in the country, the building sector could champion the cause, by maturing fast enough. However, the green initiative requires strong support from all the stakeholders. The awareness campaigns need to be intensified.

References

[1] Suhaida M S, Chua K H, & Leong Y P 2011 Sustainable Development in the Building Sector *Building Green Framework in Malaysia* 15th International Conference on ISO & TQM (1-ICIT) 1–8 Retrieved from http://www.uniten.edu.my/newhome/uploaded/admin/research/centres/iepre/2012/papers/Sustainable Development in the Building Sector Green Building Framework in Malaysia.pdf

[2] Tan C S, Suhaida M S, & Leong Y P 2011 Sustainability Development through Energy Efficiency Initiatives in Malaysia 15th International Conference on ISO & TQM (15-ICIT). Retrieved from http://www.uniten.edu.my/newhome/uploaded/admin/research/centres/iepre/2012/papers/Sustainability Development through Energy Efficiency Initiatives in Malaysia.pdf

[3] U.S. Environmental Protection Agency, Basic Information *Green Building* Available from http://www.epa.gov/greenbuilding/pubs/about.htm Accessed on 1 March 2013.

[4] Voss K, Musall E, & Lichtmeß M 2011 From Low-Energy to Net Zero-Energy Buildings: Status and Perspectives *Green Building* **6(1)** 46–57.

[5] Luis P L, José O, Rocío G, Ismael R M March 2009 A review of benchmarking, rating and labelling concepts within the framework of building energy certification schemes *Energy and Buildings* **41(3)** 272-278 ISSN 0378-7788.

[6] Andrew P, Arno K, Raymond L April 2012 A roadmap for navigating voluntary and mandated programs for building energy efficiency *Energy Policy* **43** 415-426 ISSN 0301-4215.

IOP Conf. Series: Earth and Environmental Science **16** (2013) 012122 doi:10.1088/1755-1315/16/1/012122

Towards greener environment: Energy efficient pathways for the transportation sector in Malaysia

M S Indati[1], A T Ghate[2], Y P Leong[3]
[1,3]Institute of Energy Policy and Research (IEPRe), Universiti Tenaga Nasional
E-Mail: [1] indati@uniten.edu.my, [3] Ypleong@uniten.edu.my
[2] The Energy and Resources Institute (TERI), India

Email: akshima@teri.res.in

Abstract. Transportation sector is the second most energy consuming sector after industrial sector, accounting for 40% of total energy consumption in Malaysia. The transportation sector is one of the most energy intensive sectors in the country and relies primarily on petroleum products, which in total account for nearly 98% of the total consumption in the sector. Since it is heavily reliant on petroleum based fuels, the sector contributes significantly to the greenhouse gas (GHG) emissions. The need to reduce the greenhouse gas emission is paramount as Malaysia at Conference of the Parties (COP15) pledged to reduce its carbon intensity by 40% by 2020 from 2005 level subject to availability of technology and finance. Transport sector will be among the first sectors that need to be addressed to achieve this goal, as two-thirds of the emissions come from fuel combustion in transport sector. This paper will analyse the factors influencing the transport sector's growth and energy consumption trends and discuss the key issues and challenges for greener environment and sustainable transportation in Malaysia. The paper will also discuss the policy and strategic options aimed towards energy efficient pathways in Malaysia.

1. Introduction

Malaysia has experienced tremendous economic success in the last three decades, which has triggered massive urbanization and led to a rise in the per capita income of its population. Transportation sector one of the most energy intensive sectors in the country and relies primarily on petroleum products, which in total account for nearly 98% of the total energy consumption in the sector. As this sector is heavily reliant on petroleum, it contributes significantly to the greenhouse gas emissions. The need to reduce the greenhouse gas emissions is paramount as Malaysia has set a voluntary target to reduce its carbon intensity by 40% by 2020 from 2005 level. To achieve these levels of CO_2 emission reductions, transport sector will have to be among the priority sectors that need to be addressed, as two-thirds of the greenhouse gas emissions come from fuel combustion in the sector. This paper will analyse the past trends of transport sectors' growth and the key issues and challenges faced for sustainable transportation in Malaysia. The paper will also discuss the policy and strategic options aimed towards energy efficient pathways for transport sector in Malaysia.

2. Energy Consumption and GDP Trends in Transportation Sector

Transport sector in Malaysia relies heavily on petroleum products for meeting its energy needs. Among all the fuels, motor petrol dominates the energy consumption in the sector. Motor petrol accounted for the largest share in fuel consumption (53.60%) in 2008, followed by diesel (32.22%) and aviation fuels (ATF) with a share of 12.88%. Usage of natural gas (1.18%) and electricity (0.09%) in the transport sector was minimal. The heavy reliance of transport sector on petroleum products (98%) is a worrying trend for the future. The lack of diversification of the fuel basket in the sector has triggered some policy action; the country is currently pursuing a five-fuel diversification policy, however its impact on the transport sector has been insignificant.

Within transport sector, road transport is the main consumer of energy. The share of road transport in the total energy consumption of the transport sector in 2008 was the highest at 78.6%, followed by air transport (12.9%), maritime transport (8.1%) and railways (0.4%).

Rapid population growth coupled with economic growth has increased the demand for transport services in Malaysia. Consequently, transport sector's Growth Domestic Product (GDP) has grown rapidly from RM 5,910 million in 1987 to RM 46,185 million in 2009, an increase of about eight times. However, the share of transport sector in the country's GDP has remained almost constant at about 7% during this period. The cumulative annual growth rate (CAGR) of transport sector's GDP was 12.1% from 1987 to 2000. From 2000 onwards, the CAGR of the sector's GDP has declined to 7.1%.

3. Motorization and growth in Vehicle Ownership

The total number of vehicles in Malaysia has increased from about 5 million in 1991 to 19 million in 2009, an average annual growth rate of about 8%. The growth in the number of vehicles in the country has been much faster than the growth in population; while the total population has increased by one and a half times from 1991 to 2009, the total number of vehicles in the country has increased by about four times figure 1.

Vehicle ownership is defined as the number of vehicles per thousand people. The vehicle ownership levels in Malaysia have increased from about 266 to 672 vehicles per thousand population from 1991 to 2009, an increase of about 2.5 times. Car ownership has shown the highest increase of about three times from 98 to 300 cars per thousand population during this period, followed by two-wheelers which showed an increase of about 2.3 times (140 to 316 two-wheelers per 1000 population). Ownership of public transport modes in Malaysia is very small as compared to personal modes; it had only 6 public transport vehicles per 1000 persons in 2009 figure 2.

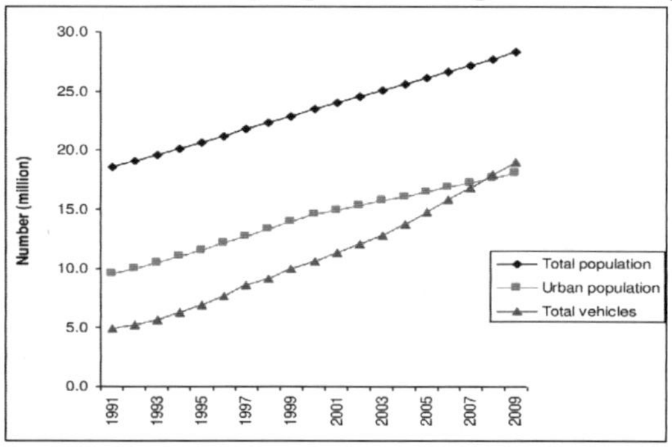

Figure 1. Growth in total population and urban population vs. growth in number of registered vehicles in Malaysia (1991-2009)

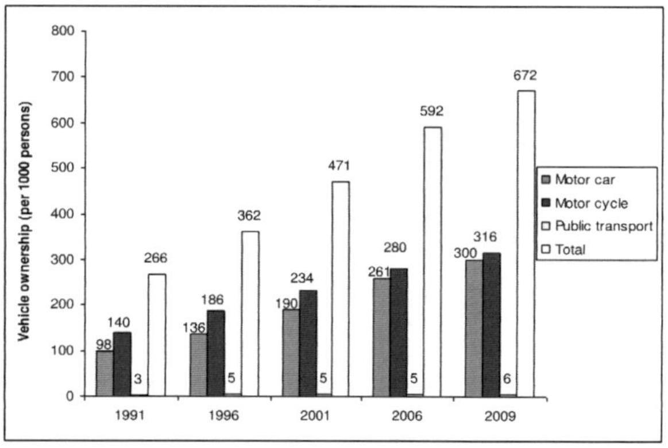

Figure 2. Growth in vehicle ownership levels in Malaysia (1991-2009)

4. Higher Growth Rate of Private Vehicles

Cars have shown the highest growth rate in Malaysia during the period 1991 to 2009 with an average annual growth rate of about 9%, followed by two-wheelers (7%), whereas public transport modes (buses, taxis and hire and drive cars) have registered a much slower growth rate of about 5% during this period Cars and two-wheelers have had the highest share in the total vehicles registered in Malaysia in the last two decades.

5. Key Findings

5.1 Fast pace of motorization in Malaysia

The key drivers of rapid motorization in Malaysia, among others, include, i) increasing per capita GDP, ii) growing population and urbanization, iii) increasing subsidies on transport fuels, and iv) promotion of domestic automobile industry by the government.

5.2 Increasing share of private modes and declining share of public modes

Motorization in Malaysia is characterised by higher growth of private vehicles i.e. motor cars and motor cycles, which have the highest share in transport vehicles registered in the country; together the two modes had a high share of about 92% of the total vehicles in the country in 2009. While private vehicles have a high share and fast growth rate, public transport modes in Malaysia (buses, taxis and hire and drive cars) have a very small share in the total registered vehicles (about 1% only in 2009). The share of public transport in cities has continuously declined from 34% in 1985 to 20% in 1997 and is now closer to 10–12%.

5.3 Inter-modal shares skewed towards road sector

There has been an increase of about four times in on-road passenger transport activities in Malaysia between 1991 and 2009. In comparison to road traffic, rail traffic in Malaysia has not increased significantly. The railways witnessed an average annual decline of about 4% in the number of passengers from 7,614 thousand in 1992 to 4,267 thousand in 2009. The rail passenger traffic (passenger kilometres) has also shown a declining trend till 2003, after which there has been an upward trend, mainly due to efforts of the government to promote rail movement in urban areas.

5.4 Faster growth of road infrastructure as compared to rail infrastructure

In terms of transport infrastructure, there has been an increase of about eight times in road length in Malaysia from 14,446 km in 1980 to 117,604 km in 2007. However, the growth in railway track length in Malaysia has been quite slow as compared to the growth of road network. The track length increased at an average annual growth rate of about 1% from 1989 to 2009. However, there has been an increase of about 1200 km in the rail track length in the last year (2009 to 2010), which shows the increasing recognition given by government to the rail transportation.

5.5 Lack of adequate policies focusing on fuel diversification, promoting energy efficient and clean use of energy in the transport sector

The consumption of both petrol and diesel has been increasing rapidly with growing motorization and increasing dependence on private modes. There is lack of diversification of fuel basket for the transport sector. One of the ways to diversify fuel mix is by promoting biofuels. Moreover, the energy intensity of Malaysia is high as compared to most of the net energy exporter countries. One of the reasons for the inefficient use of energy has been the subsidised price of energy in the country, especially those of petrol and diesel. There are no major government policies focusing on promoting efficient and clean transportation within the country, especially for road transport sector as it accounts for the bulk of energy consumption in the sector.

6. Policy and Strategic Options aimed towards Energy Efficient Pathways

6.1 Arresting the fast pace of motorization (ownership and utilization of vehicles).

Government can consider adopting policies like *vehicle quota system* and can adopt *fiscal measures* like charging of *fee for using congested parts* of the cities, levy higher *parking fees*, *increase vehicle registration charges, and increase fuel prices* to arrest current trends of rapid growth in vehicle ownership and utilization.

6.2 Increasing share of public and non-motorised modes in urban areas.

Intra-urban public transport projects need to be undertaken in order to encourage a shift from personal to mass modes of transport. This can be done by improving public transport systems in terms of their capacity, coverage and quality. In addition to improve public transport and discourage the use of personal motor vehicles, government should target encouraging use of non motorized modes in cities, as they are 'greener' modes of travel.

6.3 Diversifying the fuel basket of transport sector by encouraging use of fuels other than petrol and diesel.
The government needs to adopt a comprehensive strategy to *diversify the fuel basket* for the transport sector and adopt fuels like *natural gas, electricity and biodiesel* on a large scale.

6.4 Encouraging energy efficient road-based movement.
The government can target developing *mandatory fuel efficiency norms, stringent emission standards and strict inspection and maintenance* regime that will encourage energy efficient and less polluting movement of road based modes.

7. Concluding Remark

Adopting energy efficient pathways for the transportation sector in Malaysia will have a critical role in achieving a greener environment. The paper has highlighted various issues from the analysis of past trends of population and economic growth, energy supply and consumption patterns and transport sector growth. A shift to public and rail-based transportation is one of the ultimate options towards energy efficient pathways. But, it is dependant on a number of factors such as the provision of infrastructure, introduction of mass transportation systems, incentives for shifting to mass transport modes, etc. The road towards energy efficient and sustainable pathways for transportation sector demands *new interventions* from the *policy makers, political leadership, support from individuals, advancement in technology* and major *transformation of the energy market* towards promoting efficient and sustainable transportation in Malaysia.

Acknowledgement
Authors and UNITEN gratefully acknowledge the fund provided by the Energy Commission of Malaysia to undertake the study.

References

[1] The Tenth Malaysia Plan 2011-2015 2010 Economic planning Unit, Percetakan Nasional Malaysia Berhad
[2] Govinda R, Timilsina, Ashish S 2009 Transport Sector CO_2 emissions growth in Asia:Underlying Factors and Policy Options *Energy Policy* **37** 4523-4539
[3] Reducing Vehicle Emission in Asia 2003 Asian Development Bank. Available from: http://www.adb.org/documents/guidelines/vehicle_emissions/reducing_vehicle_emissions.pdf Accessed 26 February 2011.
[4] Vehicle Emissions Standards and Inspection and Maintenance 2003 Asian development Bank. Available from: http://www.adb.org/documents/guidelines/vehicle_emissions/reducing_vehicle_emissions.pdf Accessed 26 February 2011
[5] Ministry of Transport, Malaysia http://www.mot.gov.my/stat/index.htm Accessed 12 December 2010.

IOP Conf. Series: Earth and Environmental Science **16** (2013) 012123 doi:10.1088/1755-1315/16/1/012123

Water security and its challenges for Malaysia

M A Malek [1], M A M Nor [2], Leong Y P [3]
[1, 3] The Institute of Energy, Policy and Research (IPRe)
Universiti Tenaga Nasional, Jalan IKRAM-UNITEN, 43000, Selangor, Malaysia
[2] RPM Engineers Sdn. Bhd. Selangor, Malaysia

E-mail: Marlinda@uniten.edu.my

Abstract. Water Security in Malaysia is a national issue. The Malaysian water services industry faces issues which need to be tackled immediately for it to be viable and sustainable. Among them are the decentralized water services sector, ineffective governance structure, unsustainable tariffs, huge investments required to develop the water supply and sewerage infrastructure, inefficient operation by the operators and high non-revenue water (NRW) losses. In Malaysia, the "Sectorial" approach embedded in the present water management system and its transformation towards "Integrated Water Resources Management" (IWRM), is still in a state of inertia. This paper presents the need to transform, from a "Supply" Management mode (a characteristic of a developing country) to a "Water Demand" Management mode (a characteristic of a developed country). Issues on "Water Demand" Management for the Environment which can be a threat to the need for sustainable development for biodiversity are highlighted here. Reliable water accounting systems are found still lacking in this country, especially in the Agriculture and Environmental Sectors, where figures are still highly based on "traditional" assumptions. Water Quality deterioration remains an issue especially for the Water Supply and Environment Sectors. Available surface water resource is depleting in many regions and states in the Peninsular. Apart from the Reduce, Reuse and Recycle (3R) option for surface water, another option would be to begin a concerted effort for groundwater exploitation. However there are still grey areas of knowledge in the groundwater resources in this country for affirmative decisions and development of appropriate policies. It is also found that, there are no concerted plans to prepare the public for the change from "Supply" Management to "Demand" Management. In a developed nation, this change is through stakeholder platforms and supported by appropriate policies, rules and regulations that are based on validated Sciences, Technologies and Innovations (STI). Transforming from "Supply" Management to "Demand" Management is a formidable task. This requires the wisdom and knowledge of all experts in the Water Resources Sector.

1. Introduction

Water Security is an emerging concept which has gained increasing attention over the past five years. In 2009, the World Economic Forum described Water Security as the "gossamer" that links together the web of food, energy, climate, economic growth and human security challenges that the world economy faces over the next two decades" [1] as illustrated in Figure 1 below.

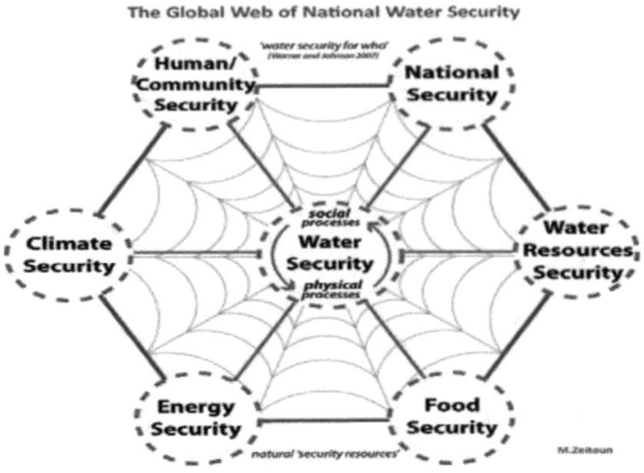

Figure 1. Water Security web .

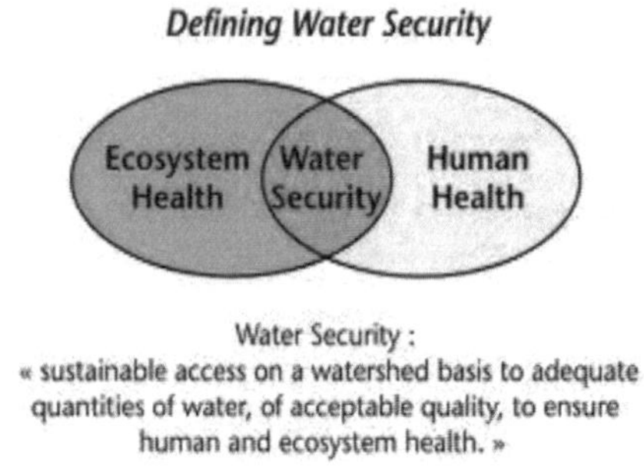

Figure 2. Defination of Water Security

Water Security is define as "sustainable access on a watershed basis to adequate quantities of water, of acceptable quality, to ensure human and ecosystem health". This definition sets baseline requirements for water resources management in a watershed on a continuous basis. There must be an access to adequate quantities of acceptable quality of water for both humans and the environment. Water Security takes a broad look at all demands placed upon a watershed, including quality, quantity (including climate change and allocation), aquatic ecosystem health, human health, risk and adaptive governance.

2. Background

The 5th World Water Forum (WWF) held in Istanbul in 2009 with the theme of "Bridging Divides For Water" ended with mixed conclusions and showed a significant divide in opinions. But the true scope and urgency of the world water crisis became recognized and many engaged to solve water related issues. Water scarcity is among the main problems to be faced by many societies around the world in the XXIst century. Furthermore, water resource management is administratively complicated because it involves legal, environmental, technological, economic and political considerations.

3. The Need for Water Demand Management (WDM)

In Malaysia, WDM for a Developed Nation is in line with the Government Transformation Plan (GTP) launched in January 2010. The notion is to accelerate the pace towards the country's Vision 2020, to be a fully Developed Nation. The key-essentials of this GTP is to improve the service levels, deliveries (People first) and productivity improvements (Performance now). Within this strategy is also the goal for the society to be a scientific and progressive society, contributing to the scientific and technological civilisation in the future. Concurrent with the GTP, is the country's Economic Transformation Plan (ETP) that sets the ambition for the country to be a High Income Nation as a fully developed country. Is the present state of the water resources management in the country able to support the GTP and realise the goals of the ETP? Is the Water Resources Sectors prepared and able to respond to meet the needs of the people and characteristics of a fully developed Nation?

IOP Conf. Series: Earth and Environmental Science 16 (2013) 012123 doi:10.1088/1755-1315/16/1/012123

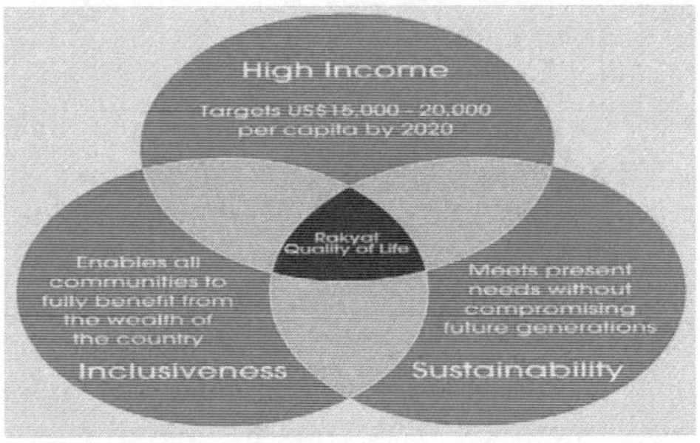

Figure 3. Water Demand Management for a Developed Nation

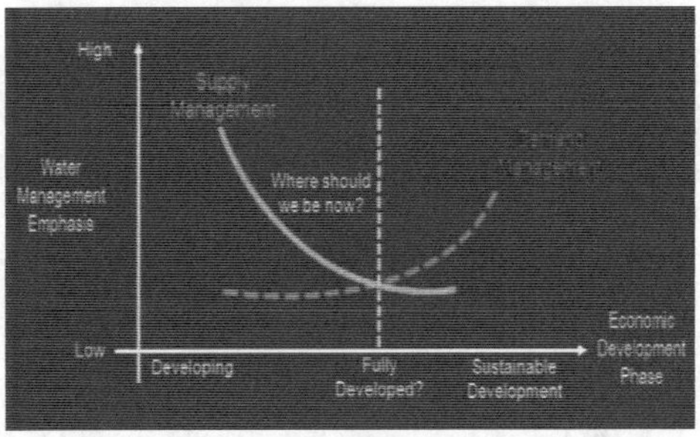

Figure 4. Requirement for Water Demand Management

4. Water Demand Management (WDM)

Is WDM vital for a developed Nation Status? This is currently the most pressing global issue related to the environment, society and technology. Water, being within the ambit of the environment, is important because it is link to the food production. WDM for Agriculture accounting to about 70% of water use and the remainder is for potable water and other uses. To secure food, water supply and distribution, usage or demand of water has to be adequately managed and shared in what is termed as WDM. WDM means reducing the use of water and wastage by better management rather than just providing more supply. Figure 5 illustrates the importance for Malaysia to embark on an effective WDM throughout the country. Table 1 shows statistics on deficit amount of unregulated flows in all states of Malaysia and an increase trend of Total Consumptive Water Demand till 2050.

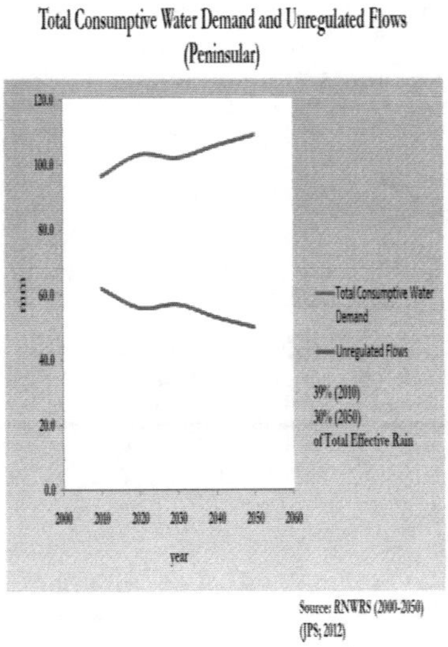

Table 1: Consumption Water Demand and Unregulated Flow Excess/deficit

States	Land Area sq km	Total Consumptive Water Demand (mm)					Effective rain (mm)	Excess/deficit (mm) - Unregulated Flows				
		2010	2020	2030	2040	2050		2010	2020	2030	2040	2050
Perlis	821	372.3	364.2	348.0	345.5	342.3	71	(302)	(294)	(277)	(275)	(272)
Kedah	9,500	307.6	313.2	299.1	302.4	302.8	113	(195)	(201)	(187)	(190)	(190)
Pulau Pinang	1,048	720.4	791.3	797.5	834.4	853.2	120	[illegible]	(671)	(677)	(714)	(733)
Perak	21,035	92.7	91.6	85.5	85.6	86.1	140	47	48	[illegible]	54	53
Selangor	8,396	266.6	296.6	306.0	328.7	348.0	114	(153)	(183)	(192)	(215)	(234)
Negeri Sembilan	6,686	51.0	54.1	53.6	54.7	56.0	74	23	19	20	19	18
Melaka	1,664	193.9	220.1	225.9	246.0	263.4	86	(108)	(135)	(140)	(161)	(178)
Johor	19,210	37.2	45.8	53.8	60.6	67.7	171	134	125	117	110	103
Pahang	36,137	20.1	26.2	24.8	25.2	26.5	165	145	139	140	140	138
Terengganu	13,035	67.8	74.8	74.4	76.7	78.7	254	186	179	179	177	175
Kelantan	15,099	108.1	107.2	105.0	106.0	106.2	176	67	68	70	70	69
Pen Malaysia	132,631	96.5	103.1	102.2	105.9	109.2	159	62	56	57	53	50
Sabah	73,631	12.4	18.4	18.9	19.6	20.0	177	165	159	158	157	157
FT Labuan	91	197.7	264.3	285.0	304.0	318.0	323	125	58	37	19	4
Sarawak	124,450	8.4	17.3	17.0	17.5	18.0	221	212	203	203	203	202
East Malaysia	198,172	10.0	17.9	17.9	18.4	18.9	269	258	251	251	250	250
Total Malaysia	330803	44.7	52.0	51.7	53.5	55.1	225.0	180.3	173.0	173.3	171.5	169.9

Note: Water Deficit States highlighted in yellow colour

Source: NWRS Review 2000-2050

Figure 5. Current situattion in Malaysia.

5. Water Demand Management for Non-Potable and Industrial Water Users

It does not make economic sense to supply Potable Water to users who only need Non-Potable water in the water supply system. Furthermore, the supply is subsidised by most of the State Governments concerned. This water supply system has been in practice since public water supply was introduced over 200 years ago and until the 1980s. However, with changing situation over the past three decades, where in some industrialised areas like the Klang Valley and Penang, the percentage of industrial usage is appreciably high and close to 40% [2, 3]. As agreed by the Heads of States including Malaysia at the recent Rio+20 forum, there has not been much progress in IWRM over the past 20 years. We continue to be bounded by the "Sectorial" approach - the main reason for the persistent issues of water management in this country. WDM requires commitment of sectors to resolve inter-sector needs and allocation. Results of the current WDM practised over a period of more than 30 years have not been encouraging. In Peninsular Malaysia, the lowest and most successful in Non- Revenue Water (NRW) is Penang with 18.2% and the highest is Pahang at 55.3%. Overall, the NRW exercise has not been successful. One possible reason is that the current tariff is too low, uneconomic and needs revision. The low tariff does not encourage new innovations in water reduction technology and runs against the aspirations of Green Technology.

6. Conclusion and Recommendations

Tariffs, pricing, sectorial water management are policy issues that need to be resolved in the water industry of Malaysia. We also need Science, Technology and Innovation (STI) to survive and prosper where knowledge and technology are ready and available here when needed. In Malaysia, are we investing and putting in enough efforts in STI? Are we waiting for a disaster to justify efforts in STI? Are we bounded by the maxim "Necessity is the Mother of Invention" and there is no necessity now? Proposal deliberated and recommendations made in this paper will need to be further studied and pursued in the interest of nation's WDM for sustainable development and wealth creation.

Acknowledgment

The authors wish to acknowledge the assistance and support from the special work of the Water Demand Management Task Force organized by The Academy of Sciences Malaysia (ASM) lead by Dato' Ir. Syed Muhammad Shahabudin, FASc with the support of the administrative staffs and financial support from ASM.

References

[1] World Economic Forum Annual Meeting Report 2009 Davos, Switzerland at http://www.weforum.org/sessions/summary/global-agenda-2009-view-davos
[2] JPS 201 Review of National Water Resources Study (RNWRS) 2000-2050 Report
[3] SMHB Sdn. Bhd. 1999 National Water Resources Study (NWRS) 2000 – 2050 Report

4th International Conference on Energy and Environment 2013 (ICEE 2013) IOP Publishing
IOP Conf. Series: Earth and Environmental Science **16** (2013) 012124 doi:10.1088/1755-1315/16/1/012124

APEC's greener energy outlook

A M Isa, R D Samuelson
Asia Pacific Energy Research Centre, The Institute of Energy Economics, Japan,
Inui Building, Kachidoki 11F, 1-13-1 Kachidoki, Chuo-ku, Tokyo, 104-0054 Japan

E-mail: aishah@aperc.ieej.or.jp

Abstract. The APEC member economies combined accounts for more than 50% of the world's GDP and consume almost 60% of the world's energy. Under the 2011 Honolulu Declaration, APEC Leaders have set an aspirational goal to reduce APEC's aggregate energy intensity by 45 percent by 2035, compared to 2005 levels. This article summarises the results from an APEC-wide study on APEC energy demand and supply outlook from 2010 to 2035. Our business-as-usual projections show that by 2035, APEC energy demand will have increased by 40% of 2010 levels. We also found that historical trends for declining energy intensity will continue and that APEC will likely achieve its aspirational intensity reduction goal. However, our results also suggest that CO_2 emissions will continue to rise and energy security will become less assured. Recognizing these vulnerabilities, APEC has already initiated a broad range of activities to achieve its 'green growth' objectives. While these have been fairly successful in guiding APEC economies towards a path of more sustainable development, these efforts will need to be intensified further to avoid serious environmental degradation.

1. Asia-Pacific Economic Cooperation (APEC)

Asia-Pacific Economic Cooperation, or APEC, is a multi-lateral organization established in 1989 with the vision of achieving stability, security and prosperity for the region by reducing barriers to trade and investment, facilitating the exchange of goods, services, resources, and technical know-how, and strengthening economic and technical cooperation. APEC operates on the basis of non-binding commitments, open dialogue and equal respect for the views of all participants. All decisions made within APEC are reached by consensus and commitments are undertaken on a voluntary basis.

There are currently 21 member economies, all from regions bordering the Pacific Ocean, namely Australia; Brunei Darussalam; Canada; Chile; China; Hong Kong, China; Indonesia; Japan; Korea; Malaysia; Mexico; New Zealand; Papua New Guinea; Peru; Philippines; Russia; Singapore; Chinese Taipei; Thailand; United States; and Viet Nam.

In 2010, APEC member economies collectively had real GDP in purchasing power parity terms of $35.8 trillion (about 53% of world GDP and a population of over 2.7 billion people (about 40% of world population) [1]. In terms of energy, out of the 8 316 Million tonnes of oil equivalent (Mtoe) of final energy consumed worldwide in 2009, about 56% was consumed in the APEC region [2].

All indicators point to further rise in APEC energy demand. This thirst for energy will have significant consequences regionally and globally, not just on the energy markets, but also on geopolitics, finances and the environment. This makes energy cooperation an important APEC agenda item. APEC Leaders are committed to addressing the region's economic and environmental challenges by speeding the transition toward a global low-carbon economy in a way that enhances energy security and creates new sources of economic growth and employment.

2. The APEC energy supply and demand outlook

To foster a common understanding on regional energy issues, the APEC Leaders, under the Osaka Action Agenda, established the Asia-Pacific Energy Research Centre (APERC) tasked with conducting research on the APEC regional outlook, energy market developments and energy policy issues. APERC's flagship project is the *APEC Energy Demand and Supply Outlook* project which is a long-term projection of energy supply and demand trends in APEC member economies and provides a statistically supported review of opportunities and challenges facing the APEC economies, individually and as a region. The fifth cycle of the Outlook project has just been successfully completed and the final report was published online in February 2013. This article presents some of the results and key findings from the *APEC Energy Demand and Supply Outlook-5th Edition* business-as-usual (BAU) scenario.

Historically, the past two decades have seen explosive growth in APEC's GDP. This trend of rapid growth is likely to continue in developing member economies, particularly developing Asian economies like China, Indonesia and Viet Nam. In developed economies, more moderate trends are expected. Combined, APEC as a whole will see about 4% GDP growth annually (See Figure 1). Under our BAU assumptions, population growth will be much slower at 0.4% annually from 2010-2035, almost half of the 0.7% annual growth from 1990-2010. The continuing economic and population growth will drive energy demand in the future, as can be seen in Figure 2.

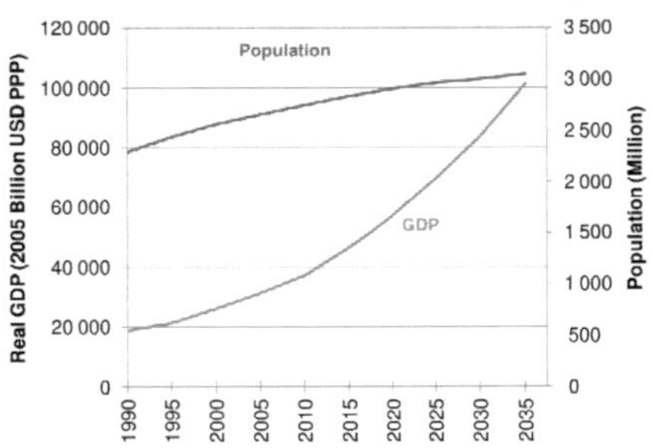

Figure 1. APEC Macroeconomic Trends, 1990-2035 [2][3]

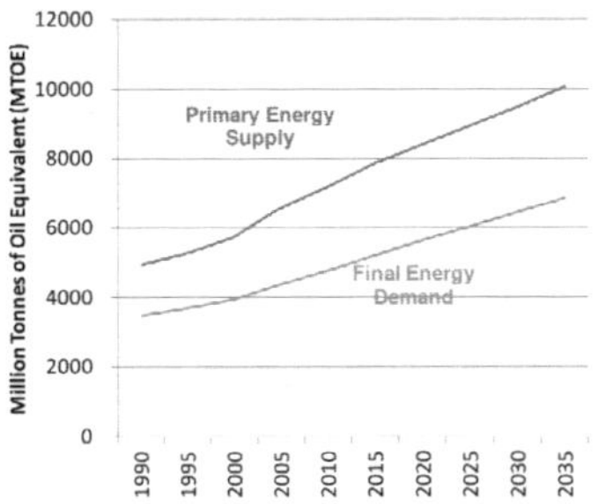

Figure 2. APEC Energy Trends, 1990-2035 [2][3]

In 2009, APEC's final energy demand amounted to 4640 Mtoe. The breakdown for final energy consumption by sector is shown in Figure 3(a) and breakdown by fuel is shown in Figure 3(b). Fossil fuels (coal, gas and oil) dominated the fuel mix for both the final demand and primary supply mix in 2009 (Figure 3), and based on our analysis this trend will continue in the long-term (Figure 4).

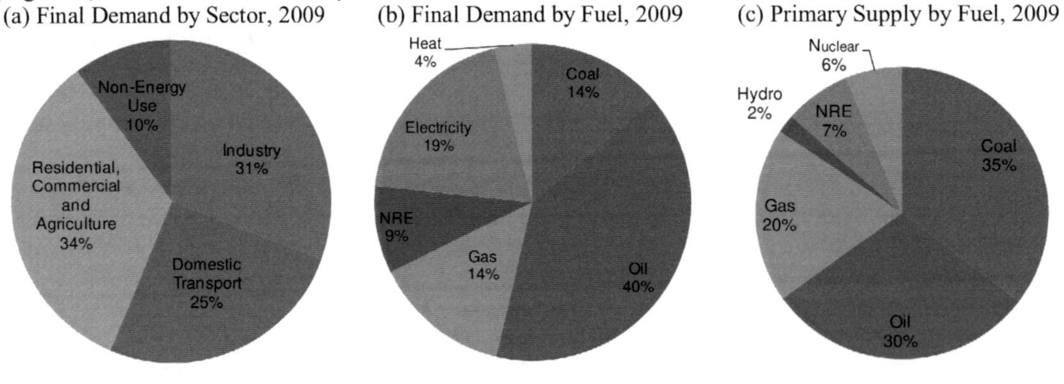

Figure 3. Structure of APEC Energy Demand and Supply, 2009 [2]

The good news is that while fossil fuels will still be the fuel of choice in the future, concern for environmental sustainability and energy security will see economies opting for cleaner fossil fuel applications and implementing more energy efficiency and conservation measures to moderate energy demand growth. As a result, the APEC-wide aspirational goal of 45% reduction in energy intensity by 2035 will probably be met. The bad news is that BAU is still environmentally unsustainable –total CO_2 emissions from fossil fuel combustion, aggregated from 2010 to 2035, is projected to reach 575 billion tonnes of CO_2, equivalent to 32% increase (Figure 5). This is the opposite of what we should be striving for, which is a *reduction* in CO_2 emissions [4]. Therefore there is still much room for improvement if we are to prevent the damaging impacts of climate change.

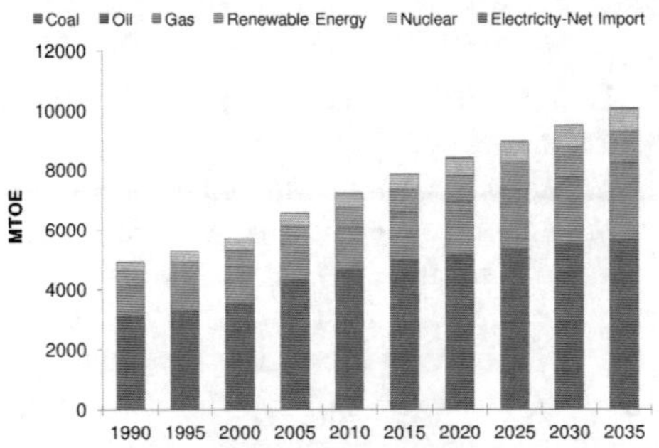

Figure 4:Primary Energy Supply by Fuel, 1990-2035 [2][3]

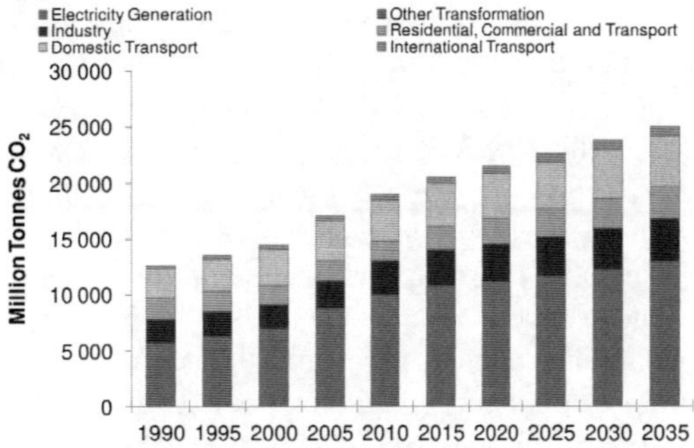

Figure 5:CO_2 Emissions by Sector, 1990-2035 [2][3]

There are basically three methods for reducing CO_2 emissions: using less energy (through energy efficiency and conservation methods); capturing and sequestration of CO_2; and switching to less-carbon-intensive energy sources like nuclear and renewable energy. Our analysis of government policies after the Fukushima Nuclear Accident of March 2011 show that while several APEC economies have reassessed their original policies for nuclear power deployment, over the long-term, nuclear energy will continue to play a modest but important role in the APEC electricity generation mix. Most economies with existing nuclear power plants, like China, are continuing operations but under much more stringent safety rules and review mechanisms. These same upgrades to safety standards, combined with negative public perceptions and increasing construction cost, will necessarily cause some delays and scaling back of new nuclear power plant development in the region.

Renewable energy development, on the other hand, will likely undergo explosive growth in the coming years, driven by supportive government policies and technical improvements which continue to reduce cost and improve system performance. We expect total primary energy supply from renewable energy sources to increase by about 53% from 2010 to 2035, with significant introduction of wind and solar power generation (almost 500% growth in total renewable energy capacity from 2010 to 2035), bio-fuels in the transport sector (about 130% growth in total bio-fuels use from 2010 to 2035) and commercial bio-energy in the residential, commercial and agricultural sector. At the same time, with growing affluence and better access to commercial energy, use of traditional residential biomass will probably decline.

3. APEC 'green growth' strategies

Recognizing the importance of protecting the environment and managing energy resources, APEC Leaders have consistently included sustainable development as part of the APEC agenda since 1993. This issue became a key focus in the 2011 Honolulu Leader Declaration, in which APEC Leaders committed to advancing the shared 'green growth' objectives. APEC defines 'green growth' as environmentally sustainable economic growth and development which help economies to successfully transition to a clean energy future. APEC has introduced several 'green growth' strategies, including:

- *APEC economies will aspire to reduce APEC's aggregate energy intensity by 45 percent by 2035, compared to 2005 levels*. This target, established in the 2011 Honolulu Declaration, is an upward revision from the 2007 Sydney Declaration goal of 25 percent reduction [1]. To assist economies in developing their energy efficiency policies, two mechanisms are being implemented – the Peer Review for Energy Efficiency (PREE) mechanism in which economies invite a team of experts to analyse their policies and provide objective feedback and the Cooperative Energy Efficiency Design for Sustainability (CEEDS) mechanism in which interested economies participate in knowledge-sharing and capacity building workshops on improving energy efficiencies in selected sectors [3]. APEC also has a dedicated Expert Group on Energy Efficiency and Conservation (EGEE&C) that promotes energy conservation and the application of energy efficiency practices and technologies in the APEC region. This includes maintaining the APEC Energy Standards Information System (APEC-ESIS) database that provides up-to-date information about energy efficiency standards and labelling that apply to products in the APEC region [5].

- *Reduce applied tariff rates on environmental goods and services*. It is expected that the lower tariff rates will decrease cost, and thus facilitate deployment across the region.
- *Rationalize and phase out inefficient fossil-fuel subsidies that encourage wasteful consumption, while recognizing the importance of providing those in need with essential energy services.* Realizing that inadequate information about existing subsidies is frequently an impediment to reform; APEC Leaders have agreed to continue building regional capacity for the reform of subsidies and initiated an annual voluntary report mechanism to assess the progress of reforms in each APEC economy [1].
- *Promote technology development and deployment of low-emission energy supply*. This will be facilitated through economic and technical cooperation between member economies and capacity building activities by APEC organizations. Some examples of cooperation are the Biofuels Task Force analyzing the issues related to the introduction of biofuels in APEC economies, the APEC Expert Group on New and Renewable Energy Technologies (EGNRET) that promotes activities and implements projects that facilitates the use of new and renewable energy technologies in the APEC region and the Peer Review on Low Carbon Energy Supply (PRLCE) mechanism that is focused on identifying and overcoming barriers to introducing Low Carbon Energy Supply in a host economy.
- *Promote the creation of low-carbon communities in the region.* Under this initiative, APEC aims to provide a guiding principle on how to define and develop a low-carbon town that successfully integrates the use of smart grids, renewable energy and energy efficient technologies to mitigate rapidly growing energy demand in urban areas and thus reduce fossil fuel consumption. The first low-carbon model town case study was the Yujiapu Central Business District project in China.

4. The way forward

The ‘green growth’ strategies being implemented have prompted APEC economies to pursue a more sustainable economic development pathway. Since 2007, several energy efficiency initiatives have been introduced, particularly for appliances, buildings, industry and transport. At the same time, energy supplies have gradually become more diversified. Our business-as-usual analysis shows that with the current policies in place, APEC economies will likely be able to meet the APEC wide energy intensity reduction goal; however, the projected increase in CO_2 emissions remains a threat to environmental sustainability. Therefore, APEC economies are strongly urged to intensify efforts to meet this challenge in its quest for a clean energy future. APEC as an organization can assist in these efforts through the sharing of best practices, conducting voluntary peer reviews and supporting the education and capacity development for managing climate change challenges in the region.

References

[1] APEC 2013 Asia Pacific Economic Cooperation, at http://www.apec.org
[2] International Energy Agency 2012 World Energy Statistics 2011 retrieved from OECD/IEA CD-ROM Service
[3] APERC 2013 Asia Pacific Energy Research Centre at http://aperc.ieej.or.jp
[4] Intergovernmental Panel on Climate Change 2007 *Climate Change 2007: Synthesis Report* retrieved from www.ipcc.ch/publications_and_data/ar4/syr/en/contents.html
[5] APEC 2013 About APEC-ESIS http://www.apec-esis.org/

Blueprint for the development of low carbon society scenarios for Asian regions- case study of Iskandar Malaysia

C S Ho[1], Y Matsuoka[2], L W Chau[1], B T Teh[1], J J Simson[2] and K Gomi[2]
[1]Faculty of Built Environment, Universiti Teknologi Malaysia, 81310 UTM Johor Bahru, Johor, Malaysia.
[2]Graduate School of Engineering, Kyoto University, Kyotodaigaku Katsura, Nishikyo-ku, Kyoto 615-8549, Japan.

Email: ho@utm.my

Abstract. Malaysian government aims to reduce 40% reduction of carbon emission intensity by the year 2020 using 2005 as the base year. Several mitigation and adaptation strategies in addressing environmental and climate change are formulated at national, regional and local level to mitigate greenhouse gases. This paper aims to examine local and regional resilient policy actions to reduce greenhouse gases using the empirical case of Iskandar Malaysia. The study case is selected because it is one of the fast developing economic corridor regions in Malaysia. In this study, a low carbon society blueprint is initiated to guide the rapid development of this economic corridor towards low carbon green growth. The blueprint provides the sustainable green growth roadmap with major 12 actions for the region. It is done through a bottom-up approach where stakeholder discussions are carried out to allow local communities participation in the plan formulation.

1. Introduction

In the era of sustainability, addressing climate change had been growing as one of the imperative concern for every city government. Cities are the main contributors to greenhouse gases (GHGs) emission and susceptible to global warming. Cities with increasing population growth and continuous economy development consuming vast resources generating enormous waste are producing a large volume of GHGs. Many cities predominantly in developed countries today have established action plan and roadmap to tackle climate change issue. However, the main challenge lies on cities in developing nations; which are vulnerable to climate change. In particular for the cities of developing nations in Asian regions with including Malaysian cities, whereby the urban population increase is high, economy growth is rapid and climate change is new for them. There is still lack of knowledge to tackle on implication of climate change and GHG baseline study methodology at the city level. In line with the Malaysia Government's aspiration for low emission development path and good global citizenship in combating climate change, Iskandar Malaysia (IM) as the first region in Malaysia has initiated climate change blueprint in ensuring the climate resilient development for sustainability. This paper provides an insight on the carbon mitigation roadmap in IM. Hopefully the benefit of this paper is not limited to other Malaysian cities solely as the substantial findings will be disseminated towards other cities in Asian countries.

2. An Overview on Iskandar Malaysia

Iskandar Malaysia is a visionary economic region in Johor that was established in 2006 as one of the catalyst development corridors to spur growth of the Malaysian economy. It has an area of 2,216.3 km^2 and it is the largest single development project ever to be undertaken in Southeast Asia. Strategically located at the southern-most tip of Mainland Asia to tap on a vast market of about 1 billion people within a 6-hour flight radius (Figure 1), IM is set to become an integrated global node that synergises with growth of Singapore and Indonesia. To that end, it has been projected that population in the urban region will more than double from 1.35 million in 2005 to over 3 million by 2025 and GDP will almost quadruple from MYR35.7 billion to MYR141.4 billion over the same period (Khazanah Nasional, 2006). Geographically, IM covers the entire Districts of Johor Bahru and Kulaijaya and several sub-districts of Pontian, comprising five local authorities. Being rapid developing region, it possesses the great opportunity incorporating good governance and low carbon mitigation policies into the urban structure of this developing region. Developed cities are rather saturated and high carbon locked hence, it is found to be difficult and more expensive to retrofit.

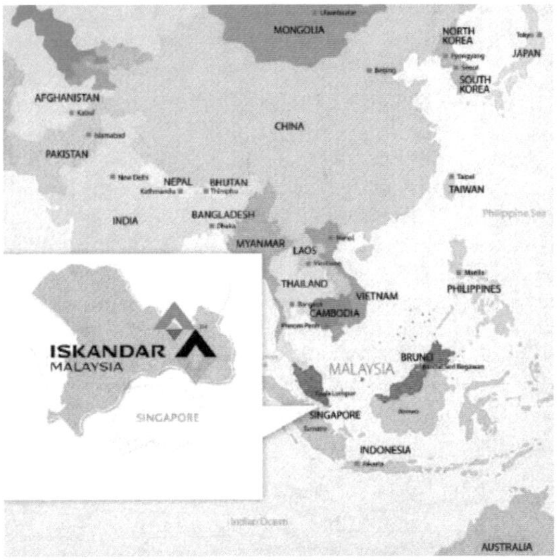

	2005	2025	2025 /2005
Population (1000)	1,353	3,000	2.22
Household (1000)	303	706	2.33
GDP (Bill. RM)	35.7	141.4	3.96
Gross output (Bill. RM)	121.4	438.9	3.61
Primary industry	1.5	2.4	1.59
Secondary industry	86.2	274.0	3.18
Tertiary industry	33.7	162.5	4.82
Passenger transport demand (Mill. passenger-km)	9,565	59,524	6.22
Freight transport demand (Mill. ton-km)	8,269	26,054	3.15

Figure 1. Location and socio-economic scenarios of Iskandar Malaysia (Source: Iskandar Regional Development Authority, 2010)

3. Low Carbon Society Blueprint for Iskandar Malaysia 2025

The Iskandar Malaysia Low Carbon Society Blueprint 2025 (IMLCSBP 2025) is a written document that presents comprehensive climate change mitigation policies (carbon emission intensity reduction actions and sub-actions) and detailed strategies (measures and programs) to guide development of IM towards achieving its vision of ‘a strong, sustainable metropolis of international standing’ by 2025. The integration of two competing goals – ‘strong’ and ‘sustainable’ – in a single development vision poses great challenges to IM’s growth policies and development planning. On one hand, the urban region needs to develop a prosperous, resilient, robust and globally competitive *economy* (the ‘strong’ dimension); on the other (the ‘sustainability’ dimension), it needs to nurture a healthy and knowledgeable *society* that subscribes to low carbon living and at the same time develop a total urban-regional *environment* that enables rapid economic growth but reduces growth’s energy demand and carbon emission intensity. This calls for a holistic and integrated approach, involving policies and strategies on *Green Economy*, *Green Community* and *Green Environment*, to decouple rapid growth from carbon emission in IM. Meeting this challenge has been the primary goal and underlying philosophy of the IMLCSBP 2025. Essentially, the Blueprint comprises two principal components: 1) narrative on growth scenarios, policies, measures and programs to achieve a minimum targeted 50% reduction in carbon emission intensity by 2025 based on the 2005 level; 2) scenario-based modelling and projection of carbon emission reductions achievable.

It was the outcome from unique social inclusive approach between experts, policymakers and communities. The scientific inputs from multidisciplinary local and foreign researchers cum feedbacks from local communities and policymakers are considered in the crafting of blueprint. A series of stakeholder discussions were held by the research team with residents, industries, business communities, local municipalities, public agencies and non-governmental organizations in the process of policy making for blueprint. The public consultations ensure the gap of science and policy making being minimize and the blueprint will be feasible in term of socio-political context and effective implementation.

4. Twelve Actions Transforming Climate Resilient Iskandar Malaysia

The greenhouse gases (GHGs) emission for Iskandar Malaysia in year 2005 is estimated to be 11.4MtCO_2eq and this is projected to triple to 31.3 MtCO_2eq in year 2025 business as usual (Bau) scenario. With the introduction of the 12 countermeasures from the blueprint, increment of GHGs emission has been slowed down, and it is expected to be 18.9 MtCO_2eq for year 2025 countermeasure (CM) scenario. These twelve possible mitigation options that capable to lower carbon emission for the case of Iskandar Malaysia are; (i) Integrated green transportation, (ii) Green industry, (iii) Low carbon urban governance, (iv) Green building and construction, (v) Green energy system and renewable energy, (vi) Low carbon lifestyle, (vii) Community engagement and consensus building, (iix) Walkable, safe and livable city design (ix) Smart urban growth, (x) Green and blue infrastructure and rural resources, (xi) Sustainable waste management and (xii) Clean air environment (Table 1). Action 5- Green energy system and renewable energy, Action 6- Low carbon lifestyle and Action 1- Integrated green transportation are the higher priority strategies to cut carbon emission significantly. They are comprised of 57% of total emission reduction. These three actions should have more attentions by policymakers. Action 3- Low carbon urban governance and Action 7- Community engagement and consensus building does not have any direct contribution yet they are vital in the creation of low carbon society development in Iskandar Malaysia.

Table 1. Carbon reduction contribution of 12 mitigation strategies for low carbon Iskandar Malaysia

Actions	Contribution * ($ktCO_2eq$)		Share	
Green Economy	**6,937**		**54%**	
Action 1 Integrated Green Transportation		1,916		15%
Action 2 Green Industry		1,094		9%
Action 3 Low Carbon Urban Governance **		-		-
Action 4 Green Building and Construction		1,203		9%
Action 5 Green Energy System and Renewable Energy		2,725		21%
Green Community	**2,727**		**21%**	
Action 6 Low Carbon Lifestyle		2,727		21%
Action 7 Community Engagement and Consensus Building**		-		-
Green Environment	**3,094**		**24%**	
Action 8 Walkable, Safe and Livable City Design		263		2%
Action 9 Smart Urban Growth		1,214		10%
Action 10 Green and Blue Infrastructure and Rural Resources		392		3%
Action 11 Sustainable Waste Management		1,224		10%
Action 12 Clean Air Environment**		-		-
Total	**12,758**		**100%**	

*Contribution to GHG emission reduction from 2025BaU to 2025CM ** Action 3, 7 and 12 does not have direct emission reduction, but their effect is included in other Actions. *** Since contribution of Action 10 includes carbon sink by forest conservation and urban tree planting, the total of contribution of the 12 Actions is greater than difference of the GHG emissions between 2025BaU and 2025CM.

5. The Way Forward

During the formulation of the Low Carbon Society Blueprint of Iskandar Malaysia, there is an ever presence of "policy-science gaps" in terms of time-scale (e.g. long-term versus short-term gains), priority (e.g. economic feasibility and budgetary concerns over social and ecological impacts) and practical considerations (e.g. institutional capacity and human capital to translate research into policy) between policymakers and researchers. While not all gaps are able to be patched as well as intended, it is learned that having policymakers onboard the research team helps significantly in identifying these issues early on and in finding middle grounds. Further understanding needs to be gained to identify processes and factors that lead to prevalent uptake of evidence-based policy recommendations within specific contexts.

Effective communication of research evidences is vital; research evidences need to be communicated in languages readable to policymakers. Furthermore, proposed policies need to "appeal" to policymakers through, among others, identification of "quick win" and "low-lying fruits" programs; outlining of direct implementation, resource allocation and benefit/cost implications; and showing of sensitivity to institutional capacity and needs.

Last, it is found that "disciplinary gaps" exist even among academic researchers from different professional (e.g. social science versus pure science and engineering) and academic-cultural backgrounds. While these are virtually inevitable, it is found that working over disciplinary gaps gives rise to perspectives and solutions that otherwise would not be thought of.

Acknowledgement

This work is supported by the Science and Technology Research Partnership for Sustainable Development Program (SATREPS).

References

[1] C40 Cities Climate Leadership Group 2011 *Climate Action in Megacities: C40 Cities Baseline and Opportunities*

[2] Ho C S, Ahmad S, Muhammad Hussein M Z S, Chau L W, Matsuoka Y, Kurata G, Fujiwara T, Shimada K, Gomi K, Yoshimoto K and Janice J S 2009 *Low Carbon City 2025: Sustainable Iskandar Malaysia*

[3] Khazanah Nasional 2006 *Comprehensive Development Plan for South Johor Economic Region 2006-2025* Kuala Lumpur: Khazanah Nasional

[4] Low Carbon Asia Research Center 2012 *Low Carbon Society Blueprint for Iskandar Malaysia 2025 – Summary for Policymakers* Johor Bahru: Low Carbon Asia Research Center

[5] Ministry of Energy, Green Technology and Water Malaysia 2011 *National Green Technology Policy* Malaysia: Ministry of Energy, Green Technology and Water Malaysia

[6] Ministry of Natural Resources and Environment Malaysia 2010 *National Policy on Climate Change* Malaysia: Ministry of Natural Resources and Environment Malaysia

[7] Ministry of Natural Resources and Environment Malaysia 2011 *Second National Communication to the UNFCC* Malaysia: Ministry of Natural Resources and Environment Malaysia.

[8] Teh B T and Ho C S 2011 The Making of Green Malaysia: Building, Neighbourhood, City, Region and Nation *The Second International Conference on Sustainable Future for Human Security* 569-578

[9] The World Bank 2010 *Cities and Climate Change: An Urgent Agenda* Washington: World Bank

[10] UN-Habitat 2012 Developing Local Climate Change Plans *A Guide for Cities in Developing Countries, Cities and Climate Change Initiative Tool Series* Nairobi: UN-Habitat

[11] United Nations Human Settlements Programme 2011 Cities and Climate Change *Global Report on Human Settlements 2011* London: Earthscan.

IOP Conf. Series: Earth and Environmental Science **16** (2013) 012126 doi:10.1088/1755-1315/16/1/012126

Electricity energy outlook in Malaysia

C S Tan[1], K Maragatham[2], Y P Leong[3]
[1, 3] Institute of Energy Policy and Research (IEPRe), Universiti Tenaga Nasional
[2] Malaysia Nuclear Agency, Kajang, Malaysia

[1]E-Mail: chingsin@uniten.edu.my

Abstract: Population and income growth are the key drivers behind the growing demand for energy. Demand for electricity in Malaysia is always growing in tandem with its Gross Domestic Product (GDP) growth. The growth for electricity in Malaysia forecasted by Economic Planning Unit (EPU) has shown an increase of 3.52% in 2012 compared to 3.48% in 2011. This growth has been driven by strong demand growth from commercial and domestic sectors. The share of electricity consumption to total energy consumption has increased from 17.4% in 2007 to 21.7% in 2012. The total electricity production was reported at 122.12TWh in 2012, where gas is still the major fuel source contributing to 52.7% of the total generation fuel mix of electricity followed by Coal, 38.9%, hydro, 7.3%, oil, 1% and others, 0.2%. This paper aims to discuss the energy outlook particularly the electricity production and ways toward greener environment in electricity production in Malaysia

1. Global Energy Status

Energy is the foundation that supports and spurs the socio-economic development of a country. Development is not possible without energy and sustainable development is not possible without sustainable sources of energy. Therefore planning for future energy demand is very crucial. Different countries may have different approaches in planning their future energy demand. Nonetheless, the future energy demand is always influenced by various non-energy policy factors such as population growth, economic growth, energy prices and adoption of technological advancement At the global level, the most fundamental relationship in GDP remains robust – more people with more income means that the production and consumption of energy will rise as shown in Figure 1.

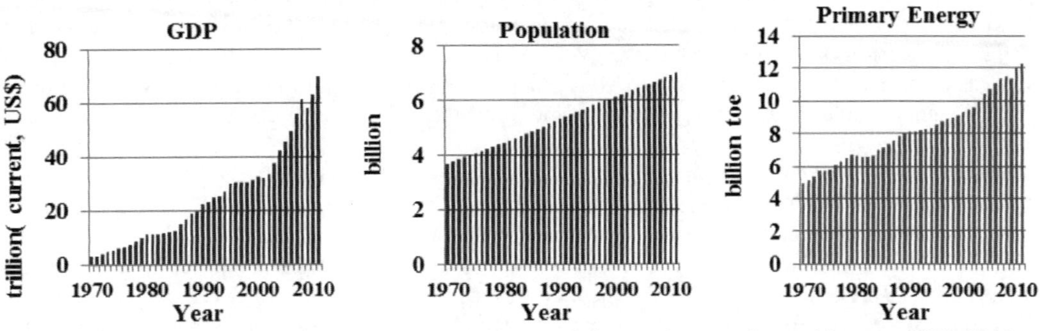

Figure 1. Correlation between economic activity, population, and energy demand between 1970-2010

The primary energy demand is highly dependent upon the price of fuels used as well as the technology utilized in providing the energy to the various sectors. Thus, energy demand is sensitive to the changes in energy price; type of fuels used at different sectors or industry and the pace of technological change which impacts directly or indirectly on energy supply and demand. Nevertheless, government policies and energy prices will have an influence towards the pace of development and deployment of new technologies, taking into account the economic factors and global market conditions.

It is reported that oil is still the major consumption worldwide with 34%, followed by natural gas 22%, coal 26%, nuclear 6%, Biomass/Waste 9%, hydro 2% and other renewable 1% in 2010. The global energy demands will increase to around 35 percent from 2010 to 2040 [2],. Therefore, a diverse, reliable and affordable fuel mix will be needed to provide energy that enables economic growth and societal advancements. Projection shows that oil will still remain the largest single source of energy demand until 2040. However, it is projected that natural gas will displace coal as the second-largest fuel by 2025. Natural gas has been seen as an economical, clean and abundant fuel as compared to other fuel sources. It will help to meet the rising power generation demand in the near future.

Thus, oil, natural gas and coal will continue to play a major role in the long-term energy supply. Together, these three fuels will provide approximately 80 percent of the total global energy supply by 2040. Nuclear will grow significantly, mainly due to rising electricity demand and a desire to reduce CO2 emissions. From 2010 to 2040, the use of nuclear energy is predicted to increase from 5.5 % to 8 %. Another notable shift in the energy fuel mix is the significant growth in renewable energy such as wind, solar and biofuels. These renewable energy technologies will grow rapidly. By 2040, renewable will only make up of 3 to 4 percent of the total world energy supply. Therefore greater advances in technology are needed to increase the commercial viability and associated economics of developing these resources to its full potential.

The projections for fuel mix changes significantly, whereby natural gas and nuclear will become the most economic fuels for power generation. Even though coal is a very competitive economic option for generating electricity, however due to it costs arising from greenhouse gas policies, natural gas becomes increasingly competitive. This is due to the fact that natural gas emits up to 60 percent less CO2 than coal when generating electricity. Renewable energy is already playing a bigger role globally – particularly through wind and solar. However, there is also a potential downside due to the intermittent renewable resources that may have on the cost and reliability of electricity supply.

2. Malaysia Energy Status

There are many challenges and issues with regard to energy- supply and demand in Malaysia namely energy security, fuel supply and pricing, especially gas pricing, renewable energy, energy efficiency and conservation, sensitivities of nuclear option and the restructuring of the electricity supply industry.

Energy security has been always been an important issue highlighted by various agencies in supporting Malaysia's socio-economic growth trajectory towards becoming a high income nation by 2020. However, the underlying concerns in Malaysia with respect to its national energy security is on how to ensure continuous supply of the energy at affordable prices as a result of depleting indigenous oil and gas resources in the country. While Malaysia is expected to revert to being a net oil importer by 2015 [3] such a situation will pose considerable long-term economic, environmental and energy security challenges.

Malaysia energy development is currently guided by the National Energy Policy, which was introduced in 1979 with three objectives namely, supply objective, utilization objective and environment objective. Supply objective is to ensure adequate, secure and cost-effective supply of energy, while the utilization objective is to promote efficient utilization of energy and discourage wasteful and non-productive patterns of energy consumption, and finally the environmental objective is to ensure factors pertaining to environmental protection are not neglected in the production and utilization of energy.

To achieve these three objectives, various related polices were formulated such National Depletion Policy was introduced in 1980 to safeguard the exploitation of natural oil reserves. The Four-Fuel Diversification Strategy policy was introduced in 1981 (i.e., oil, natural gas coal and hydro). These polices accelerated the transition from a dependence on oil as the main energy source to natural gas as a cleaner and cheaper source of energy. This policy was then succeeded by the Five-Fuel Diversification Policy in 2001, with the addition of Renewable Energy (RE) as a fifth fuel source to complement oil, natural gas, coal and hydro.

Although the general objectives of the 1979 National Energy Policy remain unchanged, the Five-Fuel Diversification Policy signaled the intention of the Government of Malaysia for concentrated efforts to usher energy sector development on a greener path, by encouraging the use of renewable sources of energy. As part of its initiative to ensure adequate, reliable and cost-effective supply of energy, the adoption of clean-coal power generation technology was also encouraged. All these policy initiatives were conscious efforts by the Government to diversify fuel sources for power generation to ensure sufficient and reliable power at affordable prices.

Under the Tenth Malaysia Plan (2011-2015), Feed-in Tariff (FiT) is designed and targeted to achieve 985MW and 2080MW by 2015 and 2020 respectively from various types of RE resources. In addition, Malaysia plans for an additional 6 GW of new generation capacity between 2015 and 2020[3] to meet increasing demand.

The total primary energy supply in Malaysia has increased by 2.99% from 74,582ktoe to 76,809ktoe in 2009 and 2010 respectively. While the total final demand has increased by 1.54% from 40,845ktoe to 41,476 ktoe in 2009 and 2010 respectively. The electricity gross production has increased by 2.3% from 105,706GWh to 108,175GWh in 2009 and 2010 respectively. In terms of electricity demand, it has increased by 8.53% from 96,302GWh to 104,519GWh. As for commercial primary energy mix in 2010, natural gas accounted for 46%, followed by crude oil and petroleum for 33%, coal for 19%, and hydropower for 2% while RE is negligible [4,5].

However, the growth for electricity in Malaysia forecasted by Economic Planning Unit (EPU) has shown an increase of 3.52% in 2012 compared to 3.48% in 2011[6]. This growth has been driven by strong demand growth from commercial and domestic sectors. The share of electricity consumption to total energy consumption has increased from 17.4% to 21.7% in 2007 and 2012 respectively. The total electricity production was reported at 122.12TWh in 2012, where gas is still the major fuel source contributing to 52.7% of the total generation fuel mix of electricity followed by Coal, 38.9%, hydro, 7.3%, oil, 1% and others, 0.2%.

Table 1. Electricity Production and its share in 2005, 2010 and 2030

Fuel by Type	Electricity Production (GWh)			Share of Electricity Production (%)		
	2005	2010	2030	2005	2010	2030
Coal	23,134	49,675	154,686	26.5	41.6	49.0
Oil	2,489	2,855	3,107	2.9	2.4	1.0
Natural Gas	55,899	55,700	139,025	64.0	46.6	44.0
Hydro	5,784	11,245	18,166	6.6	9.4	5.7
Total	87,306	119,475	315,984	100.0	100.0	100.0

Based on the projections Malaysia electricity production will increase from 87,306GWh to 314,984GWh in 2005 and 2030 respectively with an annual growth rate of 5.3%. However, the projection does not take into consideration of nuclear and RE is insignificant [7].

3. Ways towards Greener Environment in Electricity Production in Malaysia

In order to build a sustainable greener environment in energy platform for growth, it is important to manage our high dependence on fossil fuels. Therefore, moving towards cleaner power generation such as advanced efficient coal and gas power plant (i.e., Ultra Supercritical boiler for coal power plant and H-Frame turbine for gas power plant technologies) with carbon capture and storage technology and, simultaneously, intensifying the development of Renewable Energy as the fifth fuel resource as well as instituting demand side management, Energy Efficiency and Conservation may be regarded as another substitute energy source for national energy balance in the country. Due to the Fukushima Nuclear Accident in Japan, many countries have started reassessing their nuclear policies including Malaysia due to safety reasons.

4. Conclusion

Globally, the energy projection has shows a continued increase in electricity production for the years to come. Malaysia electricity production is also projected to be increased with a growth rate of 5.3% from 2005 to 2030. Deployment of RE and advances of RE technology will increase with FiT implementation in Malaysia. RE will contribute in the electricity production in the near future. Therefore, in order to ensure diversity of fuel sources, a balanced fuel mix is important.

5. Acknowledgment

The authors are grateful to Energy Commission's Chair in Energy Economics Grant 01090001IEPR which has supported the study. The authors also appreciative to the Institute of Energy Policy and Research (IEPRe) for sharing the necessary information required for the study. Any of the opinions, findings and conclusions expressed in this paper are those of the authors and do not necessarily reflect the views of the Energy Commission and IEPRe.

References

[1] World Bank database
[2] Exxon Mobile The Outlook For Energy *A View To 2040*
[3] Kettha 3 Nov 2010 Green Economy, Clean Technology And Sustainable Development – The Wave Of The Future *The Second World Chinese Economic Forum* Palace Of The Golden Horses, Kuala Lumpur
[4] National Energy Balance 2009 Malaysian Green Technology Corporation Malaysia ISSN No: 0128-6323
[5] National Energy Balance 2010 Energy Commission Malaysia ISSN No: -0128-6323
[6] Economy Planning Unit (EPU) *The Malaysia Economy in Figures 2012.*
[7] IEE Japan March 2009 Energy Supply Security Planning for the ASEAN (ESSPA) *The 2nd ASEAN Energy Demand Outlook* 47-52 Malaysia.

IOP Conf. Series: Earth and Environmental Science **16** (2013) 012127 doi:10.1088/1755-1315/16/1/012127

Sustainable green inner-wall design for flexible floor plan

N M Tawil[1], H A Husaini[1], A I Ani[1], H Basri[2], R M Saleh[1]
[1]Department of Architecture, Faculty of Engineering and Built Environment, Universiti Kebangsaan Malaysia
[2]Department of Civil and Structural Engineering, Faculty of Engineering and Built Environment, Universiti Kebangsaan Malaysia

E-Mail: nmtawil@gmail.com

Abstract. The rises of house price in the market is so drastic that it effects the younger generation nowadays especially young executives and young couples who could not afford to buy their first home. The factors that determine the house price presumably are the interior and exterior structural of the house itself. So to lessen the house price, we have to minimize the usage of wet construction thus the idea of having a sustainable green inner-wall implemented into the house with a flexible floor plan. This concept is user-friendly as it is built on needs and the ownership's affordability. They can design the interior of the house however they want with using minimal cost because it does not involve wet construction.

1. Introduction

The housing policy in Malaysia ensures that the low income groups and the middle income groups have access to adequate, affordable shelter and other related facilities. Various strategies and programmes were plan into action by the Government to serve the people better and so that different group of people can benefit from the housing sector. Such issues that will be included are the expansion facilities, introduction of appropriate incentives and instituting regulatory measure to ensure efficient housing development [1].

The sudden price change in the residential property market in Malaysia has affected the home ownership to some extent [2]. Several aspects that determine the house prices are; buildings' financial cost, legal costs, costs of financing agencies, advertising costs, site costs, financing costs and the most important part is the cost of construction itself [3]. One of the strategies to cut down the house price is by implementing a new housing concept thus this paper is to study how the sustainable green inner-wall design with flexible floor plan house can make it affordable to the middle income group of people that we are targeting.

2. Flexible Floor Plan

Housing that can adapt to the changing needs of the homeowners or users determines the definition of flexible housing and it is definitely a broad definition. The possibility of choosing different house layouts prior to occupation as well as ability to adjust one's housing over time is included in the definition. In addition to that, it comprises the potential to incorporate new technologies over time, to adjust to changing demographics, or even to completely change the use of the building from housing to something else [4].

Flexible housing actually gives the users control to design their own interior according to their life style and configuration. From the economic viewpoint, that type of housing would definitely sustain the house value for a life time, avoid the cost in constructing, renovating and repairing [5]. Loo Chern Wah (2012) made a conclusion that flexible housing is less expensive to renovate rather than conventional built houses [6].

3. Alternate Inner-wall

Three category of inner wall:

Fixed wall

A fixed wall is a wall for the purpose of separating rooms, or dividing a room. It may be constructed with gypsum, bricks or blocks from clay, terra-cotta or concrete, reinforced, or hollow. Timber can also be used. This type of partition consists of a wooden framework either supported on the floor below or by side walls. Metal lath and plaster, properly laid, forms a reinforced fixed wall. Fixed walls constructed from fibre cement sheeting are popular as bases for tiling in kitchens or in wet areas like bathrooms. Plain or reinforced fixed walls may also be constructed from concrete, including pre-cast concrete blocks. Internal wall partitions also known as office partitioning is made using plasterboard (drywall), or varieties of glass. Toughened glass is a common option as it is feasible however there is also low iron glass better known as opti-white glass which increases light and solar heat transmission. Walls are constructed using beads and tracking which are either hung from the ceiling or fixed into the ground. The panels are inserted into the tracking and fixed. [7]

Wood paneling wall liners is intended for Sheet Metal, Blast Resistant and Fiberglass Shelters, and can be furnished in numerous styles and materials. With the addition of an optional 5/8 inch plywood sub-liner, this wall liner becomes suitable for mounting shelving, fixtures, enclosures and equipment. To complete the look and feel, paneling wall liners are trimmed with baseboards and corner molding.

Due to the ease of installation, familiarity, fire resistance, non-toxicity, and sound attenuation, gypsum wall board is universal in construction. Gypsum board Consists of a thin layer of gypsum between two layers of heavy paper and is used or walls and ceilings in most buildings. Gypsum is a mineral found in sedimentary rock formations. It contains chemically combined water that is driven off as steam when subjected to high Temperatures. Gypsum wall board is the most common interior finish used in Canada and the United States. There are several types of gypsum board manufactured for specific purposes. New products are being developed and the industry is always changing. [8]

Figure 1. Modular Arts Interlocking, Rock Tiles (retrieved from www.modtiles.com/servlet/storefront):

a) Furniture wall

A wardrobe or cabinet does not only function as a place to store clothes or other things but it can also function as a divider between different spaces which can also be known as furniture wall. This type of wall could save the cost of building because it does not need to have a concrete wall and furniture separately. Aside from acting as a divider in between rooms, it can also separate space through zones or sections.

Figure 2. Examples of Furniture Wall

b) Multipurpose wall

Today most mobile and flexible wall system - the multipurpose wall - is making it possible to divide one room into several different ones and can separate one room into various small rooms. With movable walls it can divide a large room into a smaller one and turn it back into a large room again. Multipurpose wall can be a rolling, mobile, folding room divider which provide temporary walls in place of traditional non mobile products such as operable walls, accordion partitions, cubicles, pipe and drape, and shoji screens. They are sometimes folded and are on wheels enabling mobility and ease of storage. The Multipurpose wall partition has two full panel end members which provide support, rigidity, privacy and noise reduction. It can divide space quickly where non-mbile permanent room dividers may be unavailable or impracticable. They may also be used as a convenient sight divider to conceal door openings to restrooms, commercial kitchens and other backroom areas. [9].

Figure 3. Examples of Multipurpose Wall

4. Conclusion

The issue of the increase of high price over the years has been a burdened to the low and middle income groups. This is specifically true to those who have yet bought their first home and young executives who just graduated and started to work. The concept of implementing a sustainable green inner-wall with a flexible floor plan helps these people in a way that they can afford to buy their first home and the design evolves according to their needs and family growth. They could not afford to buy a house with a static floor plan because the nature of the interior of the house that uses the conventional method. Using this idea would definitely benefit both parties; the buyers and the developer.

References

[1] Aziz W N, *Housing affordability issues for middle income households: The Malaysian perspective*. 2010. 22nd International Housing Research Conference, URBAN DYNAMICS & HOUSING CHANGE - Crossing into the 2nd Decade of the 3rd Millenium.. Housing And Urban Issues In Developing Countries, ENHR 2010, 4-7 July, ISTANBUL.

IOP Conf. Series: Earth and Environmental Science **16** (2013) 012127 doi:10.1088/1755-1315/16/1/012127

[2] Zainal A H Jan - April 2010 House Price and Affordability in Housing in Malaysia 2010 *Akademika* **78** 37-46.

[3] Ivor H S 1983 Green Building Materials *Building Economics: Appraisal and control of building design cost and efficiency* http://www.calrecycle.ca.gov/greenbuilding/materials/, Last updated March 2012.

[4] Jeremy T & T S 2005 *Flexible Housing: The means To The End* Architectural Research Quarterly.

[5] Friedman Y 1999 *"Functions Follows Form" in Non-Plan, Hughes, Jonathan and Sadler, Simon eds* Oxford: Architectural Press.

[6] Loo C W 2012 *Kajian Penerapan Konsep Perumahan Fleksibel Dalam Menyelesaikan Masalah Penyesuaian Keperluan Pengguna* Thesis Universiti Kebangsaan Malaysia.

[7] Mary L McD October 1989 Repairing Historic Flat Plaster Walls and Ceilings *21 Preservation Brief. Technical Preservation Brief* National Park Service US Department Of The Interior. http://www.nps.gov/history/hps/TPS/briefs/brief21.htm

[8]Sutor Acoustical *Gypsum Wall Board Systems:* (1979) http://www.sutoracoustical.com/Gypsum%20Wall%20Board.htm

[9] Martin K Moveable Walls *Divide your room with movable walls;* http://www.movable-wall.com/ Last updated March 2012.

[10] Tawil N M, Musa A R, Abdullah N A G, Che Ani A I, Basri H & Rahmat R A O K 2012 *An Investigation Of Portable Green Inner Wall For Self Build Remodelled House* .

Faculty of Engineering and Built Environment, Universiti Kebangsaan Malaysia.

[11] Bebco Industries Portable Modular Building Division 2011 Interior & Wall Liners *Wood Paneling Wall Liners* http://www.okbebco.com/portable_modular_building_interior_walls.htm

IOP Conf. Series: Earth and Environmental Science **16** (2013) 012128 doi:10.1088/1755-1315/16/1/012128

Application of Integrated Flood Analysis System (IFAS) for Dungun River Basin

I Hafiz[1], N D M Nor[1], L M Sidek[2], H Basri[2], F K[3], M N Hanapi[4], Livia L[4]

[1]Centre for Stormwater & Geohazard Management, College of Engineering, Universiti Tenaga Nasional, Jalan IKRAM-UNITEN, 43000 Kajang, Selangor, Malaysia.

[2]Centre for Sustainable Technology and Environment (CSTEN), College of Engineering, Universiti Tenaga Nasional, Jalan IKRAM-UNITEN, 43000 Kajang, Selangor, Malaysia.

[3]International Centre for Water Hazard and Risk Management (ICHARM), 1-6 Minamihara, Tsukuba, Ibaraki 305-8516, Japan.

[4]Water Resources Management & Hydrology Division, Department of Irrigation & Drainage KM 7 Jalan Ampang, 68000, Ampang, Kuala Lumpur, Malaysia.

E-mail: muhammadhafizishak@yahoo.com

Abstract. The Northeast monsoon happening during the months of October until January is the major rainy season found in the eastern part of Peninsular Malaysia. The Dungun river basin (1,858 km^2) is exposed to this season thus experiencing characteristically regular flooding due to the prolong rainfall events. The annual rainfall over the river basins are 2,880 mm with great proportion falling in the months of December (19.4%). This study is to apply the Integrated Flood Analysis System (IFAS) model which Dungun river basin has been chosen for this study as the catchments have range of flood and relevant data that can be used to develop the model. The satellite data used in this study is provided by JAXA Global Rainfall Watch. The main feature of this real-time flood analysis model is the satellite-based rainfall data input employed during the model creation phase. The performance of the model for the river basins from satellite and ground-based rainfall data are compared using three error analysis methods.

1. Introduction

Flood has become regular disaster as it happened almost every year in different states. The northeast monsoon, prevailing between November and February, brings heavy rainfall as much as 610mm in 24 hours in extreme cases predominantly to the east coast of Peninsular Malaysia and to Sabah and Sarawak [1]. Rain bearing winds also comes with the southwest monsoon from April to September, though the rainfall during this period are generally less than during northeast monsoon [2]. There are also two transitional period between the monsoon when convectional thunderstorm are common.

Floods usually occurred from November to January as the northeast monsoon brings large volume of runoff to the relatively large catchment areas [3]. Major floods recorded in the study area; especially at lower reach of the catchment are the flood of Dec 2003, Dec 2004, Dec 2005, Feb 2006 and Nov 2009[4]. High rainfall intensity and the properties of the river regime are among the factor of Dungun flood [3]. It will be worsening as the rain occurs during the high tide to the low-lying area e.g. Taman Rakyat, Dungun[5].

2. Study Area

Dungun River is the main river in Dungun district which comprises a catchment area of 1858 km^2 and river length of approximately 110 km, flowing through four mukims from Pasir Raja and flowing to Kuala Jengai, Jerangau and finally to Kuala Dungun (Figure 1). The Dungun River is the second largest river in Terengganu ranging from 50 meters to 300 meters in width [6]. Figure 2 shows the water level at Jambatan Jerangau during the 12th December 2004 flood event.

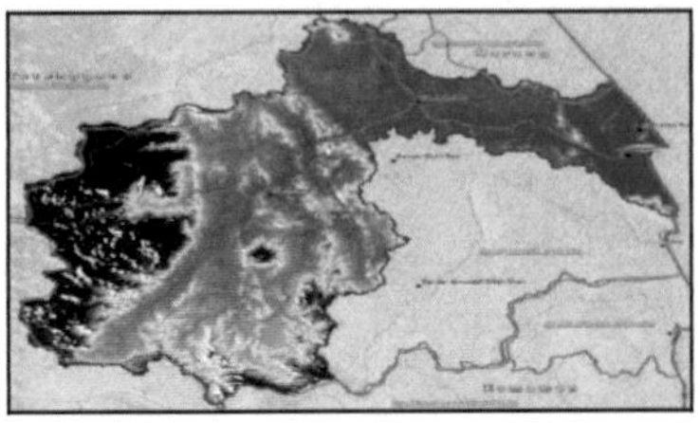

Figure 1. Dungun River Basin

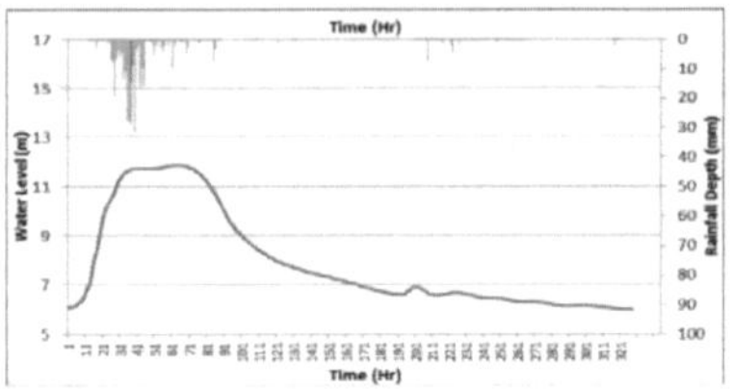

Figure 2. Water Level at Jambatan Jerangau during 12th December 2004 Flood **(4832441)**

3. Methodology

International Centre for Water Hazard and Risk Management (ICHARM) under the auspices of UNESCO has developed a brief flood-runoff analysis system as a toolkit for more effective and efficient flood analysis in developing countries. This system is called Integrated Flood Analysis System (IFAS).

The application of IFAS which includes interface for utilisation of satellite-based rainfall as the input is suitable for the catchment in developing country which the availability of ground rainfall stations is limited. The design concept in developing the flood analysis model are [7] the software can utilize both satellite and ground-based rainfall as an input; implementation of runoff analysis engine; implementation of a model creation and a parameter estimation function; visualisation of flood result and free distribution.

3.1. Model Setting

Digital Elevation Model (DEM) and the landuse data are determined from the Global Data. Global data sets relevant to flood analysis which can be used create run-off analysis model and estimates parameters for the river basin. The elevation within the catchment area (Figure 3) can be determined and the model will then generate the catchment boundary.

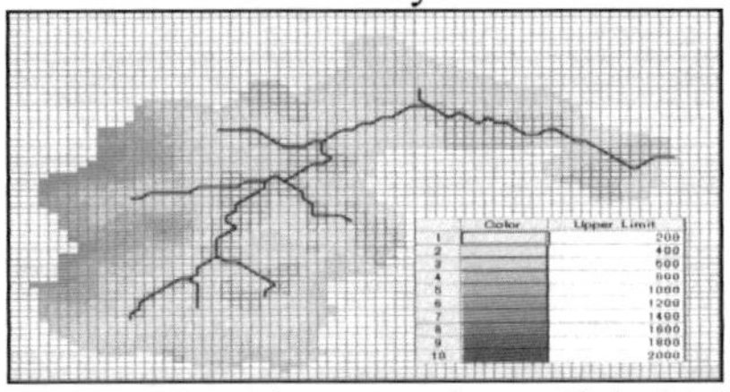

Figure 3. Elevation Map from Global Map

3.2. Rainfall Input

Two types of rainfall data are used as input data to the IFAS model which are ground and satellite-based rainfall data. There are altogether six ground rainfall stations in the catchment area with four stations equipped with telemetry devices as listed in Table 1.

GSMaP (Global Satellite Mapping of Precipitation) is the name of the satellite rainfall data which uses the Microwave radiation device equipped with satellite TRMM/TMI, Aqua/AMSR-E, ADEOS-II/AMSR and DMSP/SSMI to provide hourly rainfall data to the catchment located within 60 degrees north and 60 degrees south longitude. GSMaP rainfall data can be automatically downloaded to IFAS by using the downloading function of rainfall data manager [8]. Table 2 lists the type of GSMaP data and the data availability.

IOP Conf. Series: Earth and Environmental Science **16** (2013) 012128 doi:10.1088/1755-1315/16/1/012128

Table 1. Ground-based Rainfall

Station Name	Station No	Data Transmission
Pasir Raja	4529001	Telemetry
Kuala Jengai	4730002	Telemetry
SM Sultan Omar	4734079	Telemetry
Jambatan Jerangau	4832011	Telemetry
Delong	4833078	Manual
Kuala Abang	4834001	Manual

Table 2. Type of GSMaP

Satellite-Based Rainfall	Data Period	
	From	To
GSMaP MVK+	January 2003	December 2006
GSMaP NRT	December 2006	Present

3.3 Catchment Parameter

Three types of catchment parameter data from Global Map used in calibrating process which are surface, groundwater and river course parameter. The calculated discharge produced by IFAS is compared to the observed discharge data from river station at Jambatan Jerangau. Figure 4 shows the simulated hydrograph before and after calibration compared with the observed data.

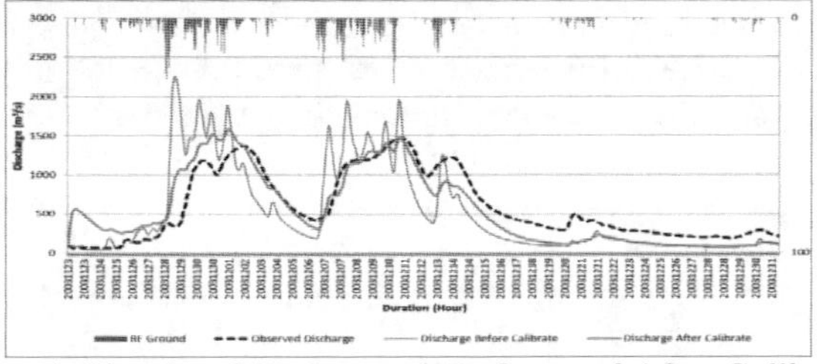

Figure 4. Simulated Hydrograph Before and After Calibration

3.4 Result

The simulated hydrograph from both satellite and ground-based rainfall are compared with the observed hydrograph during flood event in 2003 as shown in Figure 5.

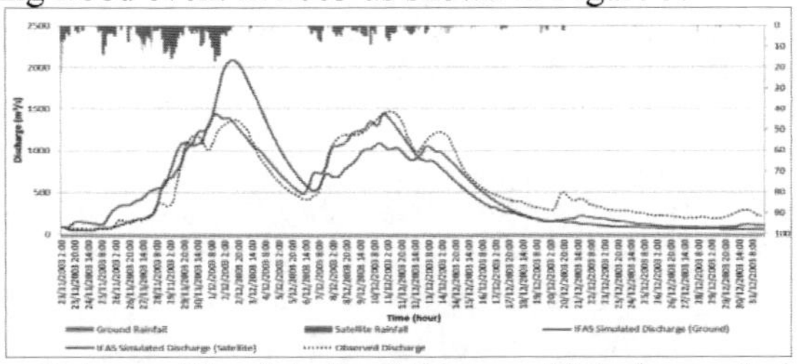

Figure 5. Comparison Between Simulated and Observed Discharge during 2003 Flood Event

3.5 Error Analysis

The calculated discharges are compared with the observed discharge and the calculated errors are shown in Table 3. Three types of error analysis are adopted in comparing the simulated and observed hydrograph.

a) The coefficient of efficiency [9] is define as:

$$E_c = 1 - \frac{\sum_{i=1}^{n} |Qrec - Q\,sim|}{\sum_{i=1}^{n} |Qrec - \bar{Q}rec|} \qquad (1)$$

where $Q_{rec}(t)$ is the recorded discharge at time t, $Q_{sim}(t)$ is the simulated discharge at time t, $\bar{Q}_{rec}$ is the average recorded discharge during the storm event, and n is the number of discharge records during the storm event. The better the fit, the closer CE is to unity.

b) The error of peak discharge [9] is define as:

where $(Q_p)_{sim}$ is the peak discharge of the simulated hydrograph, and $(Q_p)_{rec}$ is the recorded peak discharge.

$$E_p = \left[\frac{(Qp)sim-(Qp)rec}{Qp\,rec}\right]x100\% \quad (2)$$

c) The error of the time to peak discharge [9] is define as:

$$ET_p = (T_p)_{sim} - (T_p)_{rec} \quad (3)$$

where $(T_p)_{sim}$ is the simulated time to peak discharge, and $(T_p)_{rec}$ is the recorded time to peak discharge.

Table 3: Calculated Error Analysis

Event	Type of Rainfall Data	Coefficient of Efficiency	Time to Peak Error (Hour)	Peak Discharge Error (%)
2003	Ground	0.72	9	1.68
	Satellite	0.36	3	41.96

4. Conclusion

The simulated discharge produced by IFAS for all the events are compared to the observed discharge station at Jambatan Jerangau. IFAS software is able to mimic the shape of the observed hydrograph quite satisfactory by using the concept of tank model. However it is noticed that the satellite data provided by GSMaP does not truly represent the ground-based rainfall data at ground level as the satellite data always overestimated result of hydrograph compared to ground-based rainfall. This phenomenon is inherent as the satellite-based rainfall data obtained in the upper atmosphere which represent the overall catchment rainfall while the ground-based rainfall representing the point rainfall where the station located.

Acknowledgement

The authors wish to acknowledge International Centre for Water Hazard and Risk Management (ICHARM) and Department of Irrigation and Drainage of Malaysia (DID) for their support and collaboration in succeeding this paper. Besides, technical guidance and advices from ZHL Engineers Sdn. Bhd are highly appreciated.

References

[1] Chan N W 1995 Final Report *A Contextual Analysis of Flood Hazard Management in Peninsular Malaysia* 7-8

[2] Chia C W 2004 Buletin Ingenieur *Managing Flood Problems in Malaysia* **22** 38-43.ISSN 0128-4347

[3] Barzani G, Jumaat H A, Mohd Ekhwan H T, Sahibin A R, Hafizan H J 2007 Coastal Flood Phenomenon in Terengganu, Malaysia: Special Reference to Dungun *Research Journal of Environmental Sciences* **1(3)** 102- 109. ISSN 1819-3412.

[4] Department of Irrigation and Drainage (DID) 2009 LAPORAN Banjir 2008- 2009. Unit Hidrologi Jabatan Pengairan dan Saliran Negeri Terengganu.

[5] Department of Irrigation and Drainage (DID) 2004 LAPORAN Banjir 2003- 2004. Unit Hidrologi Jabatan Pengairan dan Saliran Negeri Terengganu.

[6] Norhayati M T, Suhaimi S, Noor Azhar M S, Marinah M A, Mohd Shafaril M A, Nik Faris Malik N I A, Wan Nor Hasmozan W S 2008 Behaviour of Water Quality Parameters During EBB Tide in Dungun River Estuary, Terengganu *Journal of Sustainability Science and Management* **3(1)** 1-10.

[7] Sugiura T, Fukami K, Fujiwara N, Hamaguchi N, Nakamura S, Hironaka S, Nakamura K, Wada T, Ishikawa M, Shimizu T, Inomata H, Itou K 2009 Development of Integrated Flood Analysis System (IFAS) And Its Applications *Conference of 7th ISE & 8th HIC, Chile.*

[8] Fukami K, Sugiura T, Magome J, & Kawakami T, 2009 User Manual *Integrated Flood Analysis System (IFAS Version 1.2)* 88-89. ISSN 0386-5878

IOP Conf. Series: Earth and Environmental Science **16** (2013) 012129 doi:10.1088/1755-1315/16/1/012129

Flood forecasting and early warning system for Dungun River Basin

I Hafiz[1], M D Nor, L M Sidek[2], H Basri[2], K Fukami[3], M N Hanapi[4], L Livia[4]
[1]Centre for Stormwater & Geohazard Management, College of Engineering, Universiti Tenaga Nasional, Jalan IKRAM-UNITEN, 43000 Kajang, Selangor, Malaysia.
[2]Centre for Sustainable Technology and Environment (CSTEN), College of Engineering, Universiti Tenaga Nasional, Jalan IKRAM-UNITEN, 43000 Kajang, Selangor, Malaysia.
[3]International Centre for Water Hazard and Risk Management (ICHARM), 1-6 Minamihara, Tsukuba, Ibaraki 305-8516, Japan.
[4]Water Resources Management & Hydrology Division, Department of Irrigation & Drainage KM 7 Jalan Ampang, 68000, Ampang, Kuala Lumpur, Malaysia.

E-mail: muhammadhafizishak@yahoo.com

Abstract. Floods can bring such disasters to the affected dweller due to loss of properties, crops and even deaths. The damages to properties and crops by the severe flooding are occurred due to the increase in the economic value of the properties as well as the extent of the flood. Flood forecasting & warning system is one of the examples of the non-structural measures which can give early warning to the affected people. People who live near the flood-prone areas will be warned so that they can evacuate themselves and their belongings before the arrival of the flood. This can considerably reduce flood loss and damage and above all, the loss of human lives. Integrated Flood Analysis System (IFAS) model is a runoff analysis model converting rainfall into runoff for a given river basin. The simulation can be done using either ground or satellite-based rainfall to produce calculated discharge within the river. The calculated discharge is used to generate the flood inundation map within the catchment area for the selected flood event using Infowork RS.

1. Introduction

Flooding is the most frequent natural hazard that happened in Malaysia which we have suffering great economic losses in form of properties and crops, as well as losses in life. Every year, the government has to spend a lot of money in order to reduce the damage caused by flood. The main reason of the loss is people did not have enough time to make preparation and evacuate when the flood occurs. These loss can be minimize if the flood has been forecasted to determine the time and the duration of the flood before the actual flood happened so that people have enough time to make preparation and evacuate.

The basic cause of river flooding is the incidence of heavy rainfall and the resultant large concentration of runoff, which exceeds river capacity. However, in recent years, rapid development within river catchment has resulted in higher runoff and deteriorated river capacity; this has in turn resulted in an increase in the flood frequency and magnitude [1]. Developments of a flood forecasting & warning system is highly expected as a quick and efficient means to reduce flood disaster and minimize human damage, where river improvements and developments are not necessary sufficient.

2. Flood Forecasting Concept

In order to develop the flood forecasting & warning system, there are two methods used which are by real time flood forecasting and flood forecasting using forecast rainfall data [2]. Real time flood forecasting is a method of forecasting the flood using real time rainfall data. The rainfall data used for real time flood forecasting did not have lead time since it is not forecasted rainfall [3]. The input rainfall data is from either ground rainfall station or real time satellite rainfall. Since the real time flood forecasting use real time rainfall data, the lead time available for flood warning alert is also not very long depending on the lag time of the catchment.

The other type of flood forecasting & warning system uses forecasted rainfall as the input data for the simulation. The forecasted rainfall is numerically calculated based on the historical rainfall event to give lead time for the data before actual rainfall occurs [3]. As the result the flood forecasting & warning system will gives more time to the people before the real flood occurs. However, data from the forecasted rainfall always have problems in term of data accuracy.

3. Study Area

Dungun River (Figure 1) is located in Dungun district which has total catchment area about 1858 km^2 and river length of 110 km [4]. The catchment is influenced strongly by the Northeast monsoon which brings along heavy rainfall in the months of November to March [5]. The long duration of the rain events cause prolonged flooding along some stretches of the river.

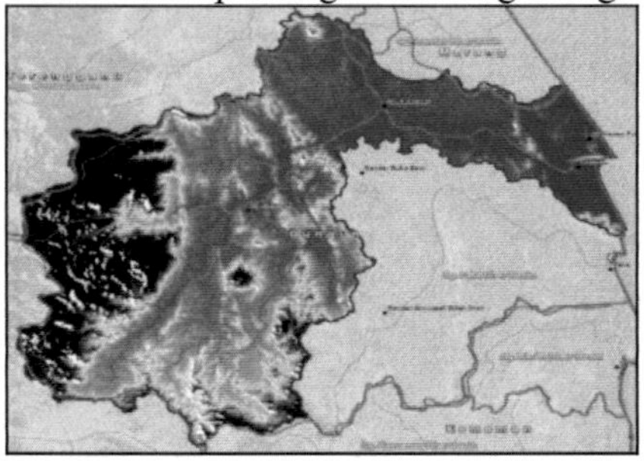

Figure 1. Dungun River Basin

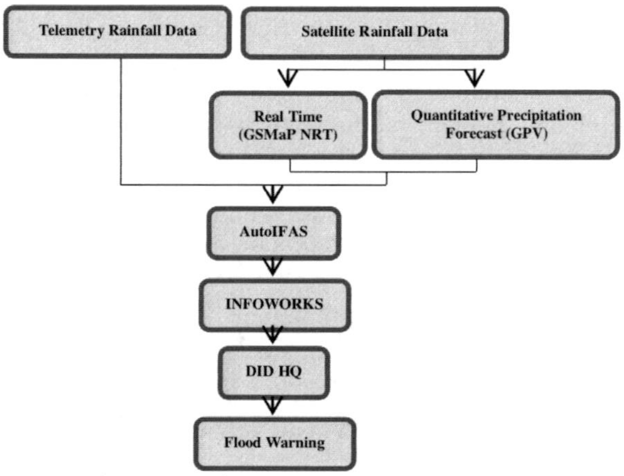

Figure 2. Flood Forecasting Process for Dungun River Basin

4. Methodology

The basic concept of developing the flood forecasting process for Dungun River Basin can be categorized in four stages which are the rainfall data, the runoff analysis model, the flood inundation model and the flood warning. Figure 2 illustrate the stages of flood forecasting process for Dungun River Basin.

4.1. Rainfall Data

The telemetry ground-based rainfall data is used as the input data in the hydrological model. The data which is collected from the Infobanjir portal will give the real-time rainfall input every hour. There are altogether six ground rainfall stations in the catchment area with four stations equipped with telemetry devices as listed in Table 1.

Two types of satellite-based rainfall data used which are the real time and forecasted satellite-based rainfall data. The quality of satellite-based rainfall data is high in term of coverage but low in term of accuracy. Table 2 and 3 show the details of the real time satellite-based rainfall and the forecasted rainfall respectively.

Table 1. Ground-based Rainfall

Station Name	Station No	Data Transmission
Pasir Raja	4529001	Telemetry
Kuala Jengai	4730002	Telemetry
SM Sultan Omar	4734079	Telemetry
Jambatan Jerangau	4832011	Telemetry
Delong	4833078	Manual

Table 2. Real Time Satellite-based Rainfall

Name	Coverage	Spatial Resolution	Time Interval	Delivery Delay
GSMaP	60°N - 60°S	0.1° (10km)	1 hour	4 hour

Table 3. Forecasted Rainfall

Forecast Rainfall Data	Forecast Frequency	Lead Time	Data Interval	Data Resolution	Data Format
JMA GSM	Every 6 hours	192 hours (8 days)	6 hours and 12 hours	0.5°x0.5°x17 levels (originally 20km horizontal resolution)	grib2

4.2. Runoff Analysis Model

IFAS is a runoff analysis model converting rainfall into runoff for a given river basin. It can be classified as conceptual or parametric and physically-based, fully distributed model [2]. The model can produce runoff simulation using both ground and satellite-based rainfall data.

4.3. Flood Warning

AutoIFAS is the model that will download data and run the simulation automatically based on the hydrological analysis in IFAS. The model can simulate the calculated flood discharge (Figure 3) using both telemetry and satellite rainfall data. Figure 4 shows the flood warning issued when the discharge reaches certain threshold level set according to level at Jambatan Jerangau river station as it is the only discharge data that available within Dungun River Catchment.

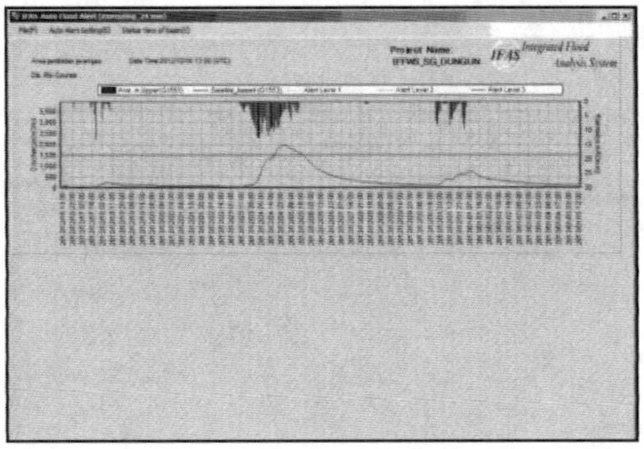

Figure 3. Flood Discharge during 2012 Flood Event

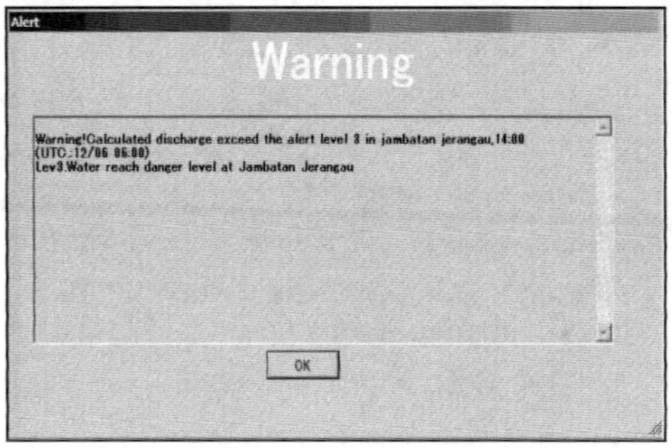

Figure 4. Flood Warning

4.4. Flood Inundation Model

The hydrodynamic analysis of Dungun River based on the results from IFAS will be used to obtain the estimated inundation areas. The discharge of the flood event will be used as the input data for the hydrodynamic model InfoWorks which will used to account for the flood inundation analysis.

The inundation location of flood will be determined along Dungun River from the Jambatan Jerangau to the river mouth. Figure 5 shows the flood inundation map and Figure 6 shows the close up of the flood map at Taman Rakyat, Dungun during Dec 2012 flood event.

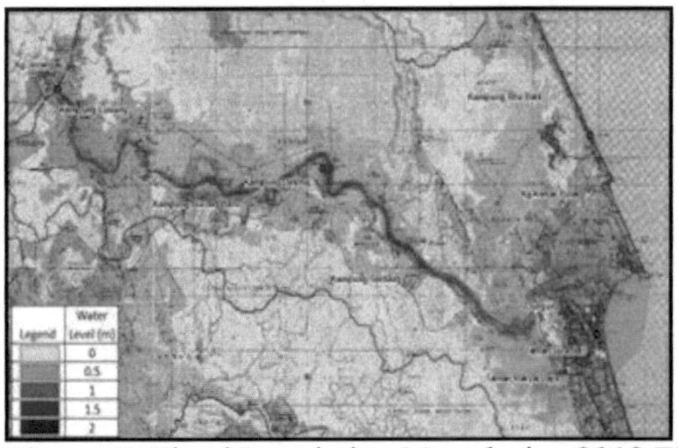

Figure 5. Flood Inundation Map during 2012 Flood

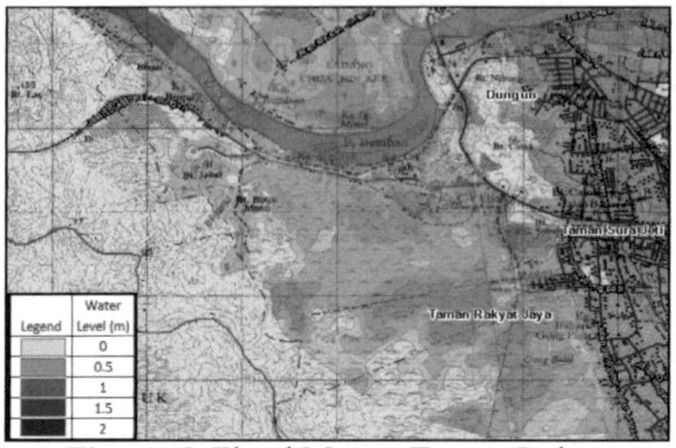

Figure 6. Flood Map at Taman Rakyat during 2012 Flood

4. Conclusion

IFAS software can be used in developing the flood forecasting and warning system with the presence of real time and forecasted rainfall from ground and satellite. The calculated discharge produced from the software used as the input of the hydrodynamic model to generate flood inundation map within Dungun River Basin. Warning will be issued when the calculated discharge reaches the certain value based on the threshold level at Jambatan Jerangau river station. People who lived in the flood prone area will have time to prepare and evacuate, hence the flood damages can be reduced.

Acknowledgement

The authors wish to acknowledge International Centre for Water Hazard and Risk Management (ICHARM) and Department of Irrigation and Drainage of Malaysia (DID) for their support and collaboration in succeeding this paper. Besides, technical guidance and advices from ZHL Engineers Sdn. Bhd are highly appreciated.

References

[1] Sulaiman A H 2007 Flood and Drought Management in Malaysia *Keynote Lecture 2 for National seminar on extreme weather and climate change, 21-27 June 2007, Putrajaya.*

[2] Sugiura T, Fukami K, Fujiwara N, Hamaguchi N, Nakamura S, Hironaka S, Nakamura K, Wada T, Ishikawa M, Shimizu T, Inomata H, Itou K 2009 Development of Integrated Flood Analysis System (IFAS) And Its Applications *Conference of 7th ISE & 8th HIC, Chile.*

[3] Fukami K, Sugiura T, Magome J, & Kawakami T 2009 User Manual *Integrated Flood Analysis System (IFAS Version 1.2)* 88-89. ISSN 0386-5878

[4] Norhayati M T, Suhaimi S, Noor Azhar M S, Marinah M A, Mohd Shafaril M A, Nik Faris Malik N I A, Wan Nor Hasmozan W S 2008 Behaviour of Water Quality Parameters During EBB Tide in Dungun River Estuary, Terengganu *Journal of Sustainability Science and Management* **3(1)** 1-10.

[5] Barzani G, Jumaat H A, Mohd Ekhwan H T, Sahibin A R, Hafizan H J 2007 Coastal Flood Phenomenon in Terengganu, Malaysia: Special Reference to Dungun *Research Journal of Environmental Sciences* **1(3)** 102- 109. ISSN 1819-3412.

IOP Conf. Series: Earth and Environmental Science **16** (2013) 012130 doi:10.1088/1755-1315/16/1/012130

Stage discharge curve for Guillemard Bridge streamflow sation based on rating curve method using historical flood event data.

[1]F C Ros, [2]H Tosaka, [1]L M Sidek, [1]M N Desa, , [1]K Arifin
[1] Dept. of Civil Engineering, Universiti Tenaga Nasional, Jln Kajang-Puchong, 43000 Kajang, Selangor, Malaysia
[2] Dept. of System Innovation, The University of Tokyo, 7-3-1 Hongo, Bunkyo-ku, Tokyo 113-8656, Japan

E-mail: [1]faie@uniten.edu.my

Abstract The purpose of the stage-discharge curves varies from water quality study, flood modelling study, can be used to project climate change scenarios and so on. As the bed of the river often changes due to the annual monsoon seasons that sometimes cause by massive floods, the capacity of the river will changed causing shifting controlled to happen. This study proposes to use the historical flood event data from 1960 to 2009 in calculating the stage-discharge curve of Guillemard Bridge located in Sg. Kelantan. Regression analysis was done to check the quality of the data and examine the correlation between the two variables, Q and H. The mean values of the two variables then were adopted to find the value of difference between zero gauge height and the level of zero flow, "a", K and "n" to fit into rating curve equation and finally plotting the stage-discharge rating curve. Regression analysis of the historical flood data indicate that 91 percent of the original uncertainty has been explained by the analysis with the standard error of 0.085.

1. Introduction

Both the stage and the discharge of a stream vary most of the time and in order to obtain a continuous record of discharge, the stage is recorded and the discharge computed from correlation of stage and discharge. [1] This correlation or calibration is known as the stage discharge relationship. The stage-discharge relationship is very important in surface hydrology, as quality discharge data is highly dependent on a satisfactory head – discharge relationship at the gauging station [2]. For establishing this relationship, the rating curve method is the most widespread technique and has been applied on thousands of gauging stations worldwide for decades. [3]

It is known that in natural river, the stage-discharge relationship does change. Usually it is called shifting controlled. Shifting controlled is usually due to erosion, deposition of sediment at the site or after a big flood. [4]

This paper discusses the result of stage discharge rating curve calculated based on flood historical event for Guillemard Bridge Station of Sg. Kelantan. The data were regressed using the least square method technique and then the stage-discharge rating is plotted based on rating curve method.

2. Theory

The functional relationship between stage and discharge can be established by field measurement of stage and discharge and thereafter can be expressed as a rating curve. Idyllically, a rating curves can be describes as a unique functional relationship between stage and discharge and therefore it is obtained as a smooth and continuous curve with reasonable degree of sensitivity. [4] The stage-discharge relationship may be expressed by

$$Q = K(H - a)^n \quad (1)$$

Where Q is the discharge in cubic metres per second, K is a constant, H is the gauge height, "a" is the gauge height for a zero flow and n is an exponent.[4]

Usually the value of "a", ie the difference between zero gauge height and the level of zero flow is not zero and its magnitude is unknown. In order to solve this, we need to plot Q against H on logarithmic plot. This will show when "a" is not zero will produce a curve. The graph will produce a straight line when the correct "a" is known. Taking the logarithm of both sides of the general equation will give us:

$$\log Q = \log K + n \log(H - a) \quad (2)$$

3. Case Study

Historical flood data from 1960 until 2009 were used to predict the stage-discharge curve for Guillemard Bridge streamflow station in Sg. Kelantan. Based on the historical event shown in Figure 1, it is noted that some major flood event happened in 1965, 1967, 1969, 1972, and 1973. The capacity of the river also decreased over the time especially in between 1967 and 1972 where a huge difference can be seen in the graph. It is believe that the major flood occurred in 1965, 1967 and 1969 causing this shift controls to happen.

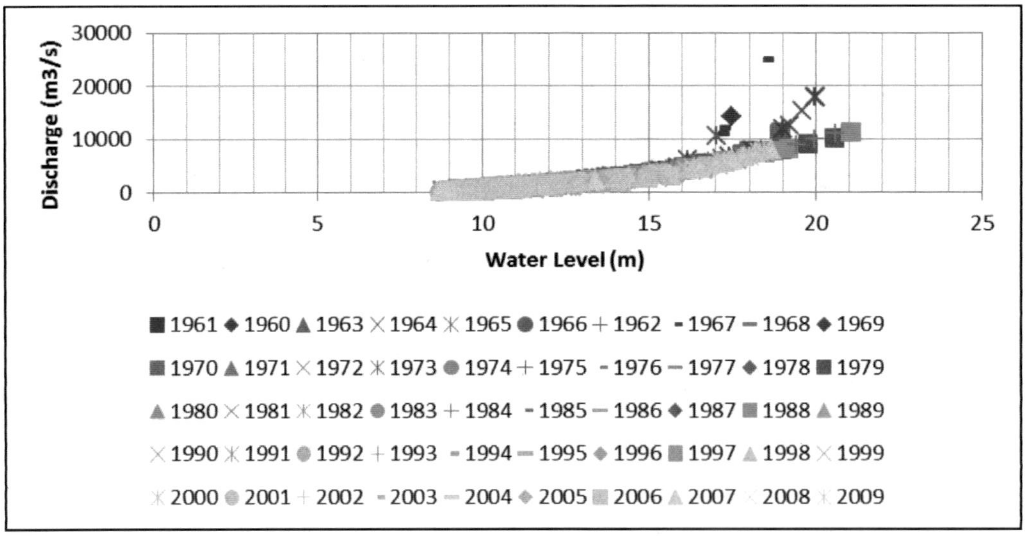

Figure 1. Flood event data of Guillemard Bridge streamflow station for year 1960 until 2009.

3. Analysis and Result

Based from the historical flood event data in Figure 1, power regression type was plotted as shown in Figure 2. The black line representing the mean value of the two variables involved. The logarithm graph then was plotted to find the standard error deviation, coefficient of determination and correlation coefficient of the data as shown in Figure 3.

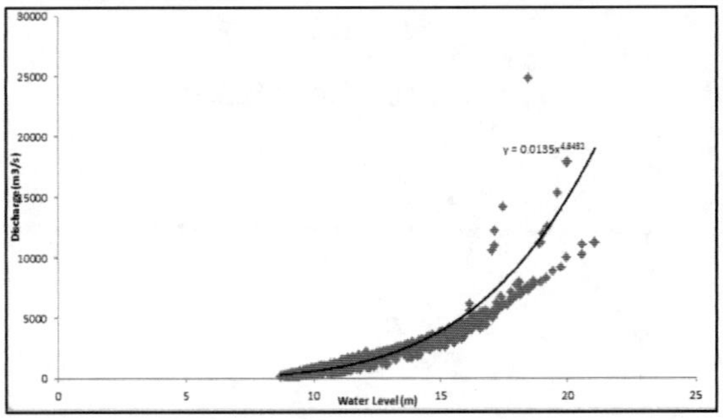

Figure 2. Regression graph of water level vs. discharge for Guillemard Bridge streamflow station.

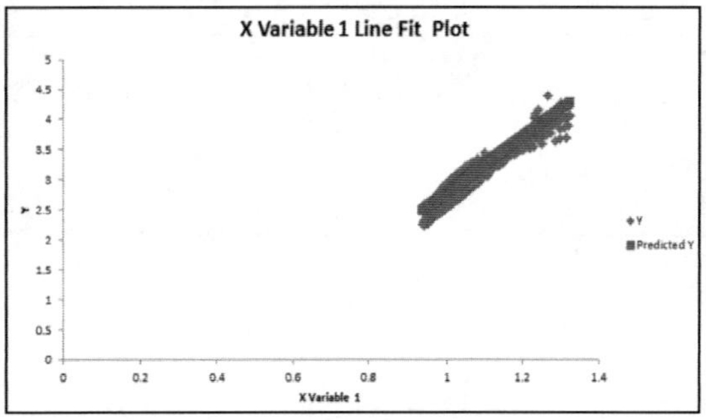

Figure 3. Linear least square regression method to determine the standard error and correlation coefficient.

The mean values of the two variables then were adopted to find the value of difference between zero gauge height and the level of zero flow, "a" and to calculate the rating curve for Guillemard Bridge streamflow station of Sg. Kelantan. Various values of "a" were assumed until the data fit with the mean value of regression analysis. To validate the "a" value, logarithmic graph of Q against (H-a) is plotted until the guessing value gives a straight line as shown in Figure 4. By solving the simultaneous equations, the value of n and K can be found. Figure 5 shows the stage-storage curve derived from the equation (1).

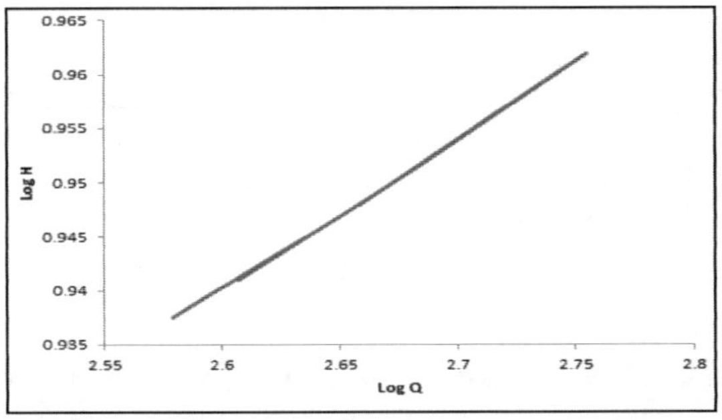

Figure 4. Log Q vs. Log H shows a straight line when a=0.5 is applied to selected random data (1970) for validation process

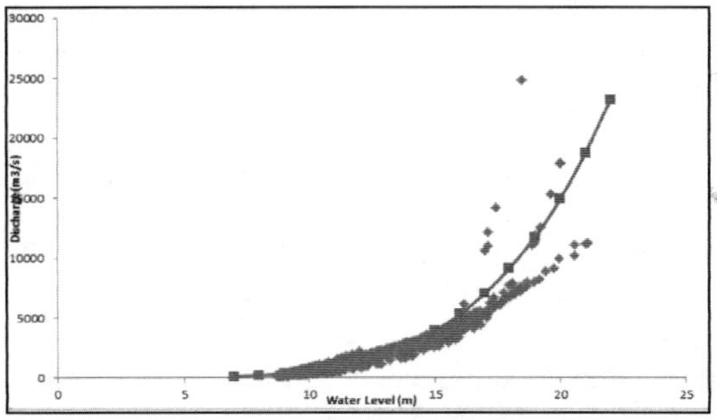

Figure 5. Graph of stage-discharge rating curve (red-line)

4. Discussions

Regression analysis of the data historical flood event data was conducted to find stage-storage rating curve. Based from the statistical calculation, the standard error is 0.085 with coefficient of determination equal to 0.91 and coefficient correlation equals to 0.96 has been obtained. These results indicate that 91 percent of the original uncertainty has been explained by the analysis. This result supports the conclusion that the equation represents an excellent fit.

Once the data for mean values of the two variables has been obtained, stage-discharge rating curve need to be plot in the form of equation (1). After several trial and error and validation process, the value of "a" is 0.5m thus will give the equation of the rating curve equals to:

$$Q = 0.027(H - 0.5)^{4.443} \tag{2}$$

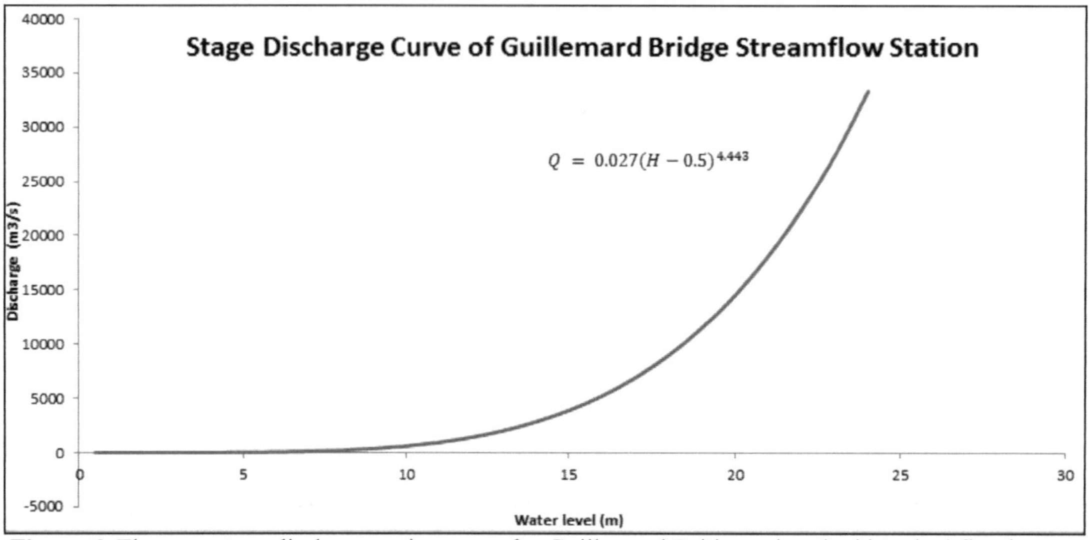

Figure 6. The new stage discharge rating curve for Guillemard Bridge using the historical flood event data.

5. Conclusions

Stage discharge rating curve for Guillemard Bridge streamflow station was calculated based on general rating curve equation using the historical flood event data. Regression analysis of the historical flood data indicate that 91 percent of the original uncertainty has been explained by the analysis with the standard error of 0.085. The "a" value for the difference between zero gauge height and the level of zero flow also has been determined by using trial and error until it fits the regression data. By solving the simultaneous equations the value of n and K were obtained. Hence the equation for stage discharge rating curve $Q = 0.027(H-0.5)^{4.443}$ was obtained using the rating curve method.

Acknowledgment
The author would like to acknowledge the support of Universiti Tenaga Nasional, University of Tokyo and JICA in this work.

References

[1] Aytac G, Ali A, and H Md Azamathulla 2012 A Practical Approach to Formulate Stage-Discharge Relationship in Natural Rivers *Neural Computation & Application Journal, Springer-Verlag London Limited Edition.*
[2] Asgeir P O 2006 Modelling Stage-Discharge Relationship Affected by Hyteresis using Jones Formula and Nonlinear Regression *Hydrological Sciences Journal, IAHS*
[3] Guerrero J L, Ida K W, Sven H, Chong Y X and Lundin L H 2012 Temporal Variability in Stage-Discharge Relationship *Elsevier Journal of Hydrology* 446-447
[4] Hydrological Procedure No. 10, Jabatan Pengairan dan Saliran Malaysia, Kementerian Pertanian Malaysia, 1976.

Integrated site investigation procedure for environment protection toward sustainable development

R C Omar, R Roslan and I N Z Baharuddin
Centre for Forensic Engineering (CeFE), College of Engineering, UNITEN, 43000, Kajang, Malaysia

E-mail: rohayu@uniten.edu.my

Abstract. The spatial configuration of cities and their relationship to the urban environment has recently been the subject of empirical, theoretical and policy research. An awareness of environmental issues can assist policy makers, planners, developers and others to recognize the constraints imposed upon development due the physical environment especially in areas, which are susceptible to erosion, flooding and landslide. This paper highlights the key requirements for considering an assessment to protect our urban environment by incorporating three main factor i.e. policy practice, planning process and engineering investigation. Base on this three main factor the framework of the assessment is carried out. The assessment can be divided into three different categories, namely as investigation for planning, investigation for urban development and specialized investigation and mitigation. The minimum requirements for the planning and urban development investigation are listed. These guidelines suggest the level at which the various types of investigation should be carried out as well as the range of application, the scope and methodology to be used for different investigation. It is hoped that this procedure will provide guidance in the establishment and protection of urban ecosystem toward sustainable development.

1.Introduction

Sensible planning of land use and decisions on planning applications depends on sound information. Much of these concern economic, social and environmental issues. Whilst the physical and mechanical characteristics of ground material are relevant to many land uses, there is often a limited appreciation of this amongst decision makers. In order to satisfy the public's demand for safe land, application, planning and balance urban growth, appropriate geological information are need to be integrated into the planning process. Procedure should be drawn up to ensure that the recently instituted legislative measures are implemented and enforced, accordingly.

2.The legal basic of planning

The planning process in Malaysia is dictated primarily by the Town and Country Planning Act (1976). It is supported complemented by other legislation such as Road, Drainage and Building Act 1974 (A133). There are two levels of planning, land use planning and regulation and control of development. The development planning system was introduced in accordance with the Town and Country Planning Act 1976 (A172) and provides for two tiers of planning, namely Structure Plan (Section 7(3)) and Local Plan (Section 12). The process of regulation and control of development involves consultation and input from several technical departments, prior to the provision of concession of land conversion or planning permission [Act A172 Section 18 and 19(1)]. The development only can be carrying out after it has been granted under Section 22 or extended under

Section 24(3). The respective involved in the consultation departments the local authority before permission is granted include the Town and Country Planning Department (TCPD), the Public Works Department (PWD) and the Drainage and Irrigation Department (DID).

Act A133 is used for Building Plan Approval during the development control process. It includes provisions for the giving of notices, the depositions of plans and information concerning the building development and the inspection of the works. Local Authority is required by statute to administer and enforce the Building Regulation. The basic purpose of this Act 133 is to ensure a high standard of health and safety for persons in or near, buildings and of others who might be affected by building or matters connected with buildings. In Selangor, Malaysia, the Building Plan Approval is also include the approval of earthwork condition Section 4(70(a)) & 4(70(b)) and Earthworks By-Laws 1992 for Selangor Local Authority. As far as ground engineering is concerned, the main regulation governing it is based on Earthworks By-Laws 1992, where erosion and sedimentation covered under Section 2(6)(a); slope stabilization is under Section 2(6)(e)(ix); foundation is under Section 2(8), 2(9), 2(12) and 2(11); and structural stability is under Section 85A in Act A133.

3. Land Use Planning Issues in Malaysia related to Prevention of Geological Hazard

In 1990's, the basic geological information utilized in suitability analysis that constrained to urban development is only relate to the physical features of the area, such as existing landforms with respect to the ruggedness of the terrain, slope analysis and mitigation of cut slope and surface erosion [1-4] The use of geological information first time establish in the preparation of Cyberjaya master plan [5, 6]. There are several of standards for regulation and control of development as example site gradient and land use zone; availability of access road and road reserves, engineering controls for flood prone areas, drainage requirements for low areas and availability of adequate water supply. Unfortunately no proper geological inputs and standards were put forward to the Town and Country Planning Department of Peninsular Malaysia [1]. Earthworks By-Laws 1992 for Selangor State in (Section 4(70(a)) & 4(70(b)) Act A133 described that site investigation and soil investigation need to be considered as part of regulation and control of development. Without specified requirement for the level of site investigation, the best method of construction for sustainable development cannot be implemented [7, 8]. Currently, Section 2(8) and Section 4 Earthworks By-Laws 1992, recommend only detail site investigation on foundation which related to the piling condition, capacity of end bearing loads; type of piling according to engineering characteristic of the soil. Other information such as vulnerability of areas to land degradation during development, the stability of each involved in the development and the level of detail requirements for each earthwork are neglected.

4. Development of Integrated Site Investigation Procedure

The assessment of integrated site investigation procedure can be divided into three different categories, namely as investigation for planning, investigation for urban development and specialized investigation and mitigation (Figure 1).

4.1 Investigation for planning

Investigation of this nature provides the broadest level of detail on the geotechnical conditions of a site. This is the first steps in evaluating a particular area for town planning purpose which consist of regional engineering geological mapping or a feasibility study for development of particular site where regional engineering geological mapping has not ready been done. Investigation for planning will be used for assessment of the area greater than 1,000 ha. Investigation for planning consists of the planning of future development axes of urban areas; assessment of the natural resources of areas earmarked for future development, to reserve and prevent the sterilization of these resources; identification of geological constraints at an early stage in the planned development; and the provision of information for an environmental impact assessment of urban development, which will assist in identifying geological sensitive areas. All the information in this first stage procedure will be used for deciding and formulate the objective and references to cumulative land use development.

IOP Conf. Series: Earth and Environmental Science **16** (2013) 012131 doi:10.1088/1755-1315/16/1/012131

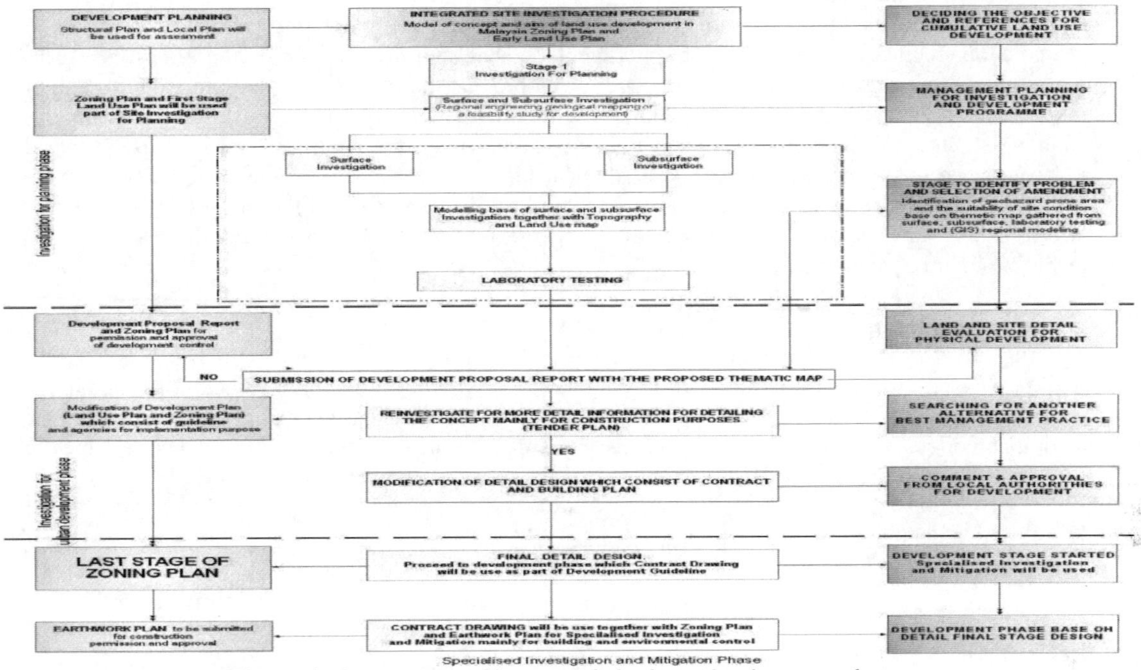

Figure 1. Assessment of integrated site investigation procedure

4.2 Investigation for Urban Development

Typically this type of investigation is required for an area more a less than 10,000 ha in size. Assessment from investigation for planning will be used to evaluate the feasibility of utilising the area for urban development. The purpose of this investigation is to determine and map the detailed engineering geological and geotechnical conditions of the terrain in order to delineate and define areas of geotechnical constraint and also for application of development control and approval from local authorities. Urban development investigation are subdivided into two types, based on the size of the area considered, as the approach and execution of engineering geological and geotechnical investigation of small areas differ significantly from those of larger areas. Urban development investigation sites is carried out to enable classification of the site in terms of the recommended foundation solution and earthworks procedures base on Malaysia Act i.e., Act 172 1976 and Act 133 1974. This type of investigation mainly for zoning of plots or stands for layout plan and earthwork plan for preparation of development control report. Layout plan, earthwork plan and development control report should be accompanied by a map demarcating the ground condition into zones according to Development Control Report Guideline from Malaysia Town and Country Planning Department (TCPD). In this layout plan, earthwork plan and development control report also provides all relevant additional site conditions such as problem soils, possible slope instability base on cut slope, geohazard prone area base on simulation of soil and rock properties, topography, land use and policy practice accordingly to Malaysia planning process.

4.3 Specialized Investigation and Mitigation

These investigations are more specialised and detailed in nature. This investigation is main for site investigation for civil engineering construction and environmental engineering evaluation. This investigation is the last stage of the assessment that will provide a consultant and contractor more detail about the ground base on the zoning and earth work plan.

5. Advantage of the Procedure

Presented with this integrated site investigation procedure on the urban environment protection is able to make the engineering geological input and geotechnical input to the town planning process with an overall view to ensure the planning of future developments that are inherently safe from a geologist point of view. Five aims are listed which are to identify and eliminate potentially disastrous or dangerous geotechnical situations arising; to identify and try to improve inefficient land-use proposals; to provide an engineering geological/geotechnical advisory service to planners and developers; to advise Local Authorities on the geotechnical suitability of land for avoiding geo-hazard problems; to advise Local Authorities on the geological and geotechnical lease or clause conditions required in legal for site investigation assessment.

6. Concluding Remark

Planning may be seen as an exercise in decision making in the use of land. If it has a single major weakness, it is that decisions are based on monitory term. Although one can establish lots of environment information but, the implications to be drawn may are based can one's opinion. Planning recommendations and decisions reflects the values both of planners, engineer and engineering geologist in which they operate.

7. Acknowledgement

The authors would like to express their sincere gratitude to MOSTEE for 08-03-0021-EA001 research funding.

References

[1] Yahya N and Ming L Y 1990. An Approach to Integrate Construction Material Resource as a Component into the Planning Process with special reference to Klang Valley, Region. *Course On Geological Assessment and Evaluation of Construction Material, In Lands Development*. 24-26 July, 1990. Bangi.

[2] JPBD Semenanjung 1997. Physical Planning Guidelines for the MSC Volume 1-8.

[3] Road, Drainage and Building 1974 (ACT A133). Government of Malaysia, Kuala Lumpur.

[4] Pereira J J and Komoo I. 1998. Using geological Information as planning tools for urban centers. The Case of Klang Valley, Malaysia. *Proceeding Eight International Congress International Associaition for Engineering Geology and the Environment*. 21-25 September 1998. Vancouver, Canada.

[5] Omar R C, 2006. Maklumat Geosains Terintegrasi untuk Proses Perancangan Gunatanah Fizikal Pembangunan Lestari: Kajian Kes Cyberjaya. *Tesis Phd. UKM.* Unpublished.

[6] Town and Country Planning Act 1976 (Act 172). Government of Malaysia, Kuala Lumpur.

[7] Noor W M M, 1997. Integrating Geology in Urban Planning-the case of Cyberjaya. Paper presented at the *LESTARI Round Table Dialogue on Planning Response for Geohazard Prevention*, UKM, Bangi, Malaysia. 10 May 1997.

IOP Conf. Series: Earth and Environmental Science **16** (2013) 012132 doi:10.1088/1755-1315/16/1/012132

Real time monitoring of slope condition for transmission tower safety in Kenyir, Malaysia

R. C. Omar, A. Ismail, N. H. N. Khalid, N.M. Din*, H. Hussain, M.Z. Jamaludin, F. Abdullah, A.Z.Arazad[1] & H.Yusop[1]
Centre for Forensic Engineering & Centre for Communications Service Convergence Technologies, College of Engineering, Universiti Tenaga Nasional,
Jalan IKRAM- UNITEN, 43000 Kajang, Selangor, Malaysia
[1] TNB Research Sdn, Bhd

Email: *norashidah@uniten.edu.my

Abstract. The Malaysia national electricity grid traverses throughout the nation over urban and rural areas including mountainous terrain. A major number of the transmission towers have been in existence for over 40 years and some traversed through very remote and high altitude areas like the Titiwangsa range that forms the backbone of the Malay Peninsula. This paper describes the instrumentation and real time monitoring in a transmission tower site in Kenyir, a hilly terrain in the East Coast of Malaysia. The site itself which is between 300-500m above sea level is deep in the rainforest area of Kenyir. The site and surrounding areas has been identified with signs of slope failure. A design concern is the real time slope monitoring sensors reliability and data integrity from the remote area with potential interference to the electronics equipment from the power line. The monitoring system comprised of an automated system for collecting and reporting field monitoring data. The instruments collect readings and transmit real time through GSM to the monitoring office over designated intervals. This initiative is a part of a project on developing an early warning system for monitoring landslide hazards at selected transmission towers. This paper reviews the various instrumentation used and challenges faced and the output received for slope movement warnings.

1. Introduction

Electronic instrumentation for remote monitoring of slope can be an invaluable tool in condition monitoring especially in remote and difficult to access areas [1]. For the electric utility provider such a system is desirable in order to monitor unstable or potentially unstable transmission tower slopes throughout the grid. Water levels, movements and deformation sensors can be deployed and data from them collected to give an indication of the slope condition. All of these sensors can be integrated to send data into a data logging system and sent centrally to the monitoring office. These systems are self-contained unit transmitting data through cellular telecommunications and batteries charged by solar panels. This paper describes the instrumentation and real time monitoring deployment and issues of transmission tower slopes in Kenyir, Malaysia. The location of the towers that are monitored, i.e. Tower A and B are shown in Figure 1.0.

2. The Kenyir Site: Monitoring Methodology and Results

Towers A and B are identified as potentially to be unstable. Some indications of slope failure are seen in the surrounding areas as can be seen in Figures 2 and 3. Tower C has the highest elevation and the only place where there is cellular telecommunication service. No sign of slope failure found here.

Figure 1. Location of slopes **Figure 2.** Tower A **Figure 3.** Tower B

Landslide monitoring involves selecting parameters to observe and monitoring how they change with time. The two most important parameters to observe are groundwater levels and displacement. The slopes at Tower A and B are monitored using piezometers for determination of water levels, inclinometers for monitoring slope movement and data loggers. In between the towers a rain gauge was installed to measure the amount of rainfall over time. Figure 4 shows the location of piezometer and inclinometer placement at Tower B. Whereas Figure 5 shows the communication and data logging modules powered by solar panels.

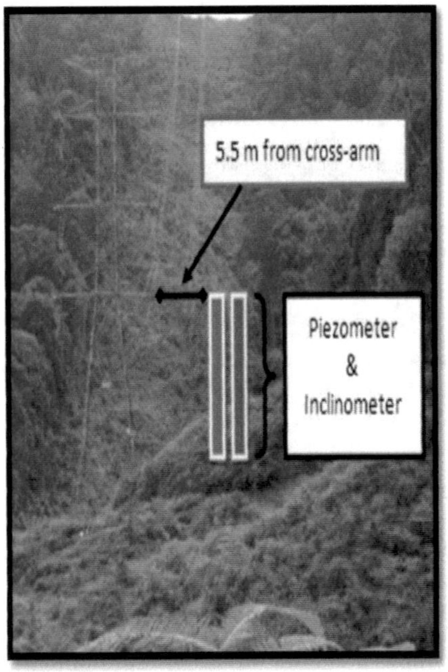

Figure 4. Placement of the piezometer and inclinometer at Tower B

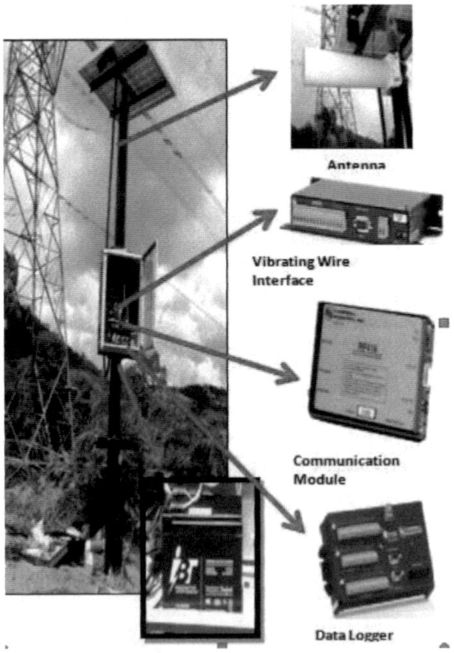

Figure 5. The data logging modules

IOP Conf. Series: Earth and Environmental Science **16** (2013) 012132 doi:10.1088/1755-1315/16/1/012132

The collection of localize data at the respective transmission tower is done through respective data loggers. The data loggers are interfaced using proprietary programming environment and interfaced to the RF communication modules where data are transmitted from the respective transmission tower slope to Tower C where the transceiver located at that tower is connected to the GSM modem for transmission to the monitoring office through the SMS (Short Message Service). The data logging capability at the monitoring office can go to a minimum of collecting data at 1 minute intervals. Normal setting is to collect data at 10 minute intervals. The user interface was developed for easy access to real time data where the data can be retrieved and visualized. Figure 6 depicts the data logging environment on site before transmission over GSM.

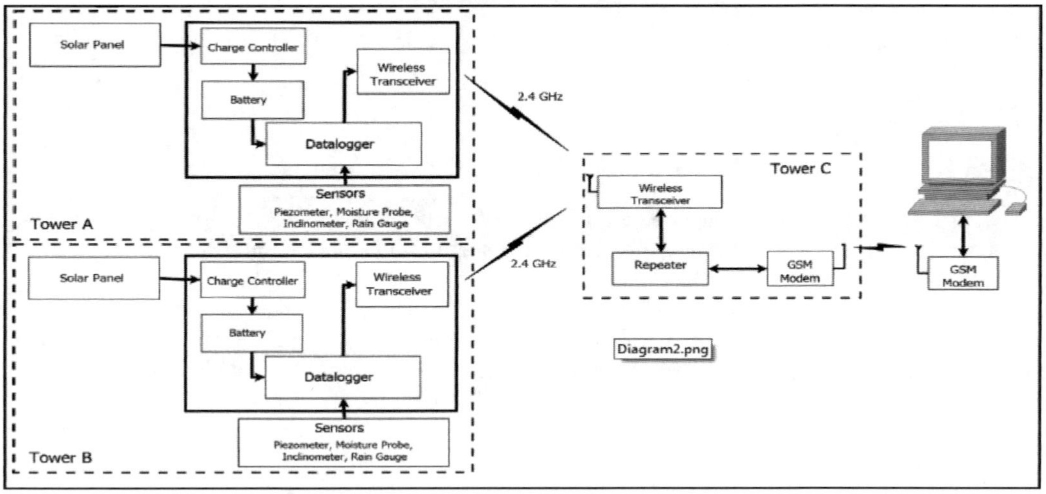

Figure 6. The data logging environment block diagram

Figure 7 presents one of the user interfaces developed for the monitoring environment and Figure 8 and 9 show samples of GUI for sensor values received at the site.

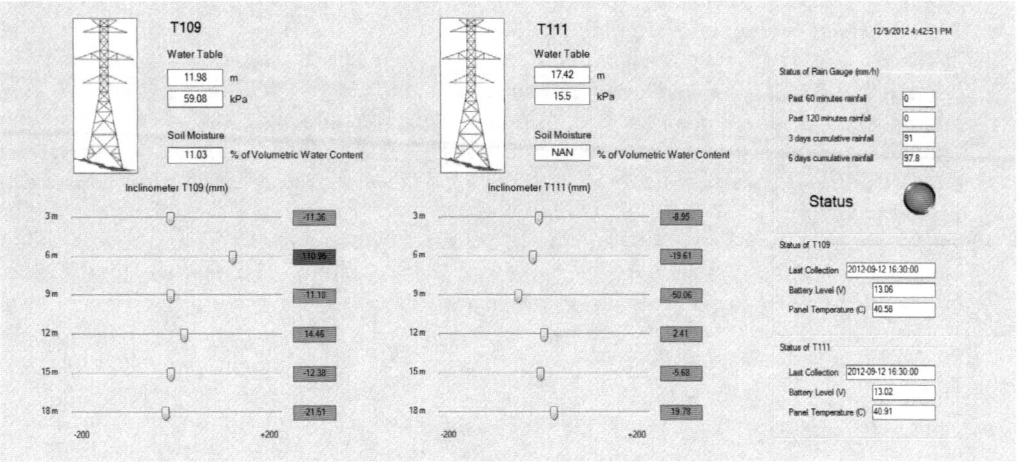

Figure 7. The slope monitoring GUI

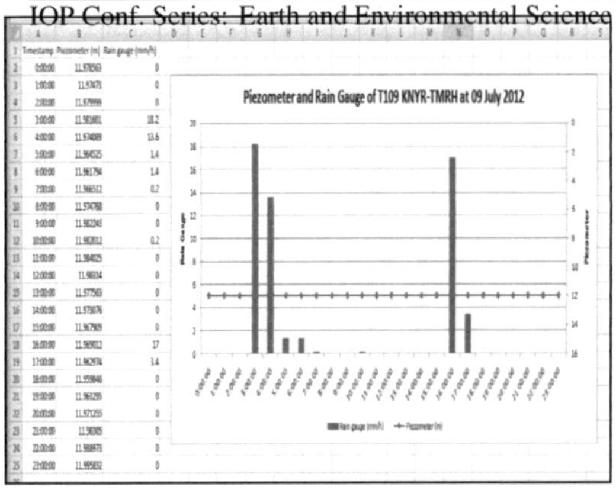

Figure 8. Sample of the piezometer and rain gauge readings

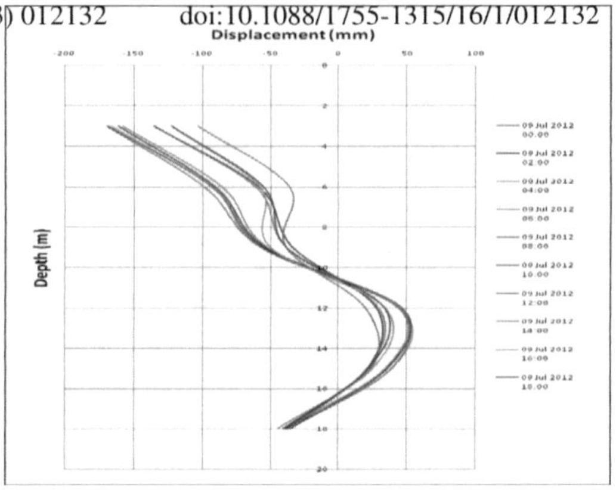

Figure 9. Sample of the inclinometer readings

3. Conclusion

In this work, we identified and developed a real time system for transmission tower slope monitoring. The real challenge is to relate to the data collected for slope failure prediction. Data will be collected over time in developing a base line for the slopes. Targeted and acceptable limit for business continuity for proactive measures and recovery will be identified. The use of digital spatial information [2] and new localization monitoring techniques such as in [3] and [4] will also be explored Ultimately, once this is established, the framework can be extended over a bigger transmission network coverage, albeit the national grid of Malaysia.

Acknowledgment

We would like to thank Tenaga Nasional Berhad for supporting this study.

References

[1] Arnhardt C, Asch K, Azzam R, Bill R, Fernandez-Steeger T M, Homfeld S D, Kallash A, Niemeyer F, Ritter H, Toloczyki M, and Walter K 2007 Sensor based Landslide Early Warning System - SLEWS. Development of a geoservice infrastructure as basis for early warning systems for landslides by integration of real-time sensors *GEOTECHNOLOGIEN Science Report. Early Warning Systems in Earth Management. Kick-Off-Meeting 10 October 2007 Technical University Karlsruhe* 75 - 88.

[2] Van Westen C J 2007 Mapping Landslides: Recent developments in the use of Digital Spatial Information *A.K. Turner and R.L. Schuster (eds.): Landslides and Society, 1st North American Landslide Conference, Vail, Colorado, June 3-8, 2007* 221-238.

[3] Terzis A, Anandarajah A, Moore K, and Wang I J 2006 *Slip Surface Localization in WirelessSensor Networks for Lanslide Prediction*. IPSN'06, April 19–21, 2006, Nashville, Tennessee, USA.

[4] Zheng Y, Yang G, Hu H 2011 Research on deformation characteristics of slope based on slope monitoring *Second International Conference on Mechanic Automation and Control Engineering (MACE)* 15-17 July 2011 3344-3349.

IOP Conf. Series: Earth and Environmental Science **16** (2013) 012133 doi:10.1088/1755-1315/16/1/012133

Transmission tower classification based on landslide risk Map generated by Geographical Information System (GIS) at Cameron Highlands

Hazwani NK[1], **Rohayu CO**[1], **Fathoni U**[1], **INZ Baharuddin**[1], **Azwin ZA**[2]
[1] Centre for Forensic Engineering, Universiti Tenaga Nasional, Jalan IKRAM-UNITEN, 43000 Kajang, Selangor, Malaysia.
[2] TNB Research Sdn, Bhd

E-mail: hazwani@uniten.edu.my

Abstract. Transmission tower is usually locates at remote area which is covered by hilly topography. Landslide is mainly occurring at hilly area and causing failure to the tower structure. This phenomenon subsequently will affect the national electricity supply. A landslide risk hazard map is generated using Geographical Information System (GIS). Risk classification is introduced to initiate the monitoring process along Jor-Bintang transmission line, Cameron Highland, Pahang. The classification has been divided into three categories, which are low, medium and high. This method can be applied in slope monitoring activities since all towers have been classified based on their risk level. Therefore, maintenance schedule can be planned smoothly and efficiently.

1. Introduction

Landslide is one of the world's catastrophes that affect enormous economic loss and one of the common interference nowadays. Malaysia is located at latitude 2° to 7° north and 100° to 119° east. Total area of Peninsular Malaysia is about 131,589 km^2 with 60% of the area covered by hilly topography. Every year, many landslide cases reported, either at man-made or natural slopes.

Application of new technology in prediction of slope assessment in Malaysia began in the early 1990's. As an effort to diminish landslide hazard and their consequences, government agencies and private organizations collaborated to deal with the issues. In Malaysia, existing slope assessment scale is normally conducted by dividing into two categories; large-scale and medium-to- small-scale. Lembaga Lebuhraya Malaysia (LLM) applying large-scale assessments in prioritizing slope maintenance along roads and highways. While medium-to-small-scale assessments in controlling development in hilly areas are widely use by Jabatan Kerja Raya (JKR), Jabatan Perancang Bandar dan Desa (JPPD) and Jabatan Mineral & Geossains (JMG) [1].

Slope monitoring guidelines prepared by JKR recommended standard of good practice for the maintenance of man-made slopes, disturbed terrain features and natural terrain hazard mitigation measures. The guidelines are widely used in slope engineering practice especially on new development at landslide prone area [2].

2. Landslide Risk Map Analysis using GIS

GIS is a computer system for managing spatial data. GIS is utilized to yield useful knowledge, application of colored maps and images, statistical graphics, tables, and various on-screen responses for interactive appearance. The software is made up from several interrelated and linked components with different functions. This software provides a comprehensive tool for assessing temporal environment data. A combination of GIS and Remote Sensing is useful for environmental auditing [3].

Landslide is the most common natural hazard these days that impacts the environment and human activities. Most of the occurrences are due to human factors such as extensive earth cutting, land clearing, agricultural activities, burning and uphill developments [4]. Identification of Landslide Susceptibility Zonation (LSZ) is important for safe strategic planning for future development. LSZ can be defined as the division of land surface into near-homogenous zones and rank according to the degrees of actual or potential hazard due to landslides. Methods of producing LSZ can be divided into two categories; qualitative approach and quantitative approach. Qualitative approach is integrated in generating LSZ map since a lot of subjectivity is introduced in preparing various thematic data layers. While quantitative approaches are introduced to minimize subjectivity in the weight assignment process. Common landslide causative factors such as lithology, lineament, slope, aspect, landuse and drainage are important for weights and ratings [5].

Type of approach used for this research is statistical approach (multivariate statistical analysis) because it is suitable for GIS tools and analysis. In this method, each individual thematic data layers are compared to the existing landslide distribution layers. The weight value of each category of causative factors is assigned based on landslide density. This involves overlay of landside distribution layer on each thematic data layers and calculation of respective landslide density values.

3. GIS Modelling

GIS Modelling needs various types of raw data before it comes to analysis stage. Most of the data are collected from previous researchers, government agencies and private sectors. Data sources are summarized in Table 1.

Table 1. Source of Data Collected

Description	Sources
Contour maps of Peninsular Malaysia (2004)	Department of Survey and Mapping Malaysia (JUPEM)
Soil maps	Department of Agriculture (DOA)
Land Use Map	Department of Agriculture (DOA)
Geology Maps (Peninsular Malaysia)	Minerals and Geo-Science Department (JMG)
Weightage value for Landuse Type	Literature Review
Weightage value for Lithology	Literature Review
Weightage value for Slope Category	Literature Review
Weightage value for Aspect	Literature Review
Weightage value for Elevation	Literature Review

For this research, statistical approach is used in producing landslide hazard map. This method is based on observation of relationship between each factor. Weightage values are differentiated based on types of soil, geological properties, landuse type and slope, as listed in the Table 6 [6].

4. Landslide Hazard Map and Risk Classification

Landslide Hazard Map is generated using Statistical Approach by considering weightage value for every data layer. Raw datas such as contour map, lithology map and landuse are analyzed using Model Builder application in GIS. All data must be in raster format before it can be layered to each other. The calculation for both study areas is done based on weightage value of every layer as shown in Figure 2.

Table 2 Weightage Values

LANDUSE TYPE				ASPECT		
Tin Mine	5	Wet Paddy	3	Flat	-1^{O}	1
Water Tank	5	Rubber	3	North	$0 - 22.5^{O}$	1
Pond	5	Oil Palm	3	Northeast	$22.5 - 67.5^{O}$	1
Primary Forest	5	Vegetables	3	East	$67.5 - 112.5^{O}$	2
Rifle Rang	5	Cash Crops	2	Southeast	$112.5 - 157.5^{O}$	2
Rural Building	5	Rock	2	South	$157.5 - 202.5^{O}$	3
Cemetery	5	Cocoa	2	Southwest	$202.5 - 247.5^{O}$	3
Education	5	Outcrop	2	West	$247.5 - 292.5^{O}$	4
Industrial	5	Annual Crops	2	Northwest	$292.5 - 337.5^{O}$	4
Jetty	5	Banana	2	North	$337.5 - 360^{O}$	1
Lake	5	Bush	2			
Major River	5	Coconut	2			
Non Tidal	5	Sand	1			
Others Building	5	Clear Land	1			
Railway Station	5	Grass	1			
SLOPE CATEGORY				LITHOLOGY		
$0 - 15^{O}$	1			Carboniferous		5
$15 - 25^{O}$	2			Permian		4
$25 - 35^{O}$	3			Triassic		3
$35 - 45^{O}$	4			Tertiary		3
$45 - 90^{O}$	5			Igneous Activity		2
				Silurian-Ordovician		2
				Devonian		2

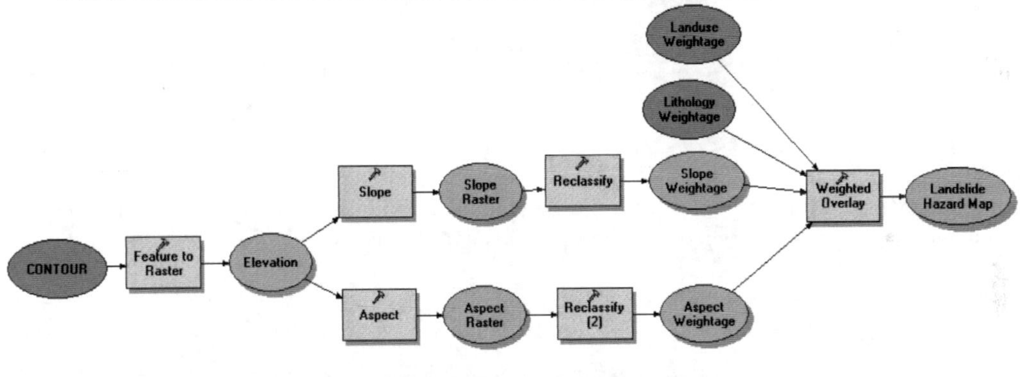

Figure 2. Weightage Overlay for Landslide Hazard Map

Classification of risk has been divided into three stages, which are low hazard (1 to 2), medium hazard (3 to 4) and high hazard (5 to 6). Figure 3 shows the detail layers in producing landslide hazard map for Jor-Bintang transmission line in Cameron Highland. After producing the erosion hazard map, all transmission towers have been arranged into their hazard class by GIS zoom scale 1:20,000. The classification is summarized in Table 3.

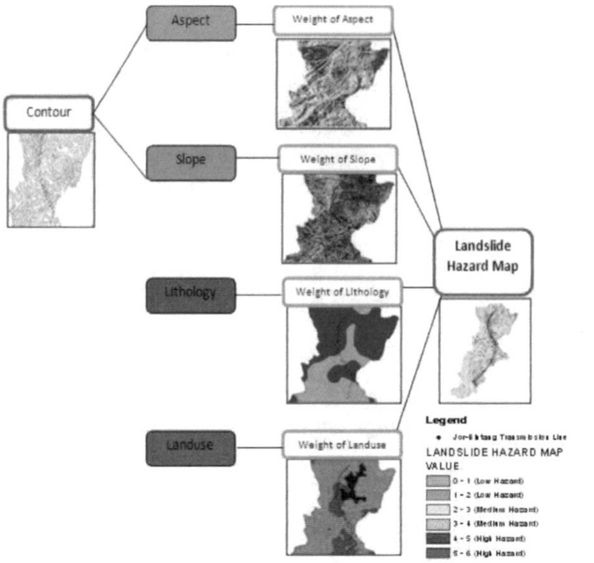

Figure 3. Landslide Risk Map

Table 3 Weightage for Lithology

Risk Level	List of Tower
Low Hazard	T2, T15, T18, T20, T22, T23, T24, T28
Medium Hazard	T5, T6, T12, T13, T3, T9, T16, T17, T19, T21, T25, T26, T29, T30, T31, T32, T33, T34, T36, T37, T38, T39, T40, T45
High Hazard	T4, T7, T8, T10, T11, T14, T27, T35, T41, T42-T44

5. Conclusion

As a conclusion, by referring to the risk level, frequency in monitoring stability of transmission tower can be planned smoothly and efficiently, especially during rainy season.

6. References

[1] Slope Engineering Branch, S. E. 2006 *Guidelines on Slope Maintenance in Malaysia* Jabatan Kerja Raya (JKR).

[2] Suhaimi Jamaludin A N 2006 Landslide Hazard and Risk Assessment L: The Malaysian Experience. IAEG 2006 Engineering *Geology for Tomorrow's Cities* 55 Nottingham, United Kingdom.

[3] Kalpana O 2006 *Soil Erosion Risk Modelling and Current Erosion Damage Assessment Using Remote Sensing and GIS Techniques* Dehradun: Indian Institute of Remote Sensing.

[4] F.Ahmad, S. L. 2006 Landslide Hazard Zonnation For Paya Terubung-Relau Slopes Terrain in Penang Island.

[5] Kanungo D M 2009 Journal of South Asia Disaster Studies *Landslide Susceptibility Zonation (LSZ) Mapping - A Review* 81-105.

[6] Rohayu C O 2006 *Maklumat Geosains Terintegrasi untuk Proses Perancangan Guna Tanah Fizikal di Cyberjaya* Universiti Kebangsaan Malaysia.

IOP Conf. Series: Earth and Environmental Science **16** (2013) 012134 doi:10.1088/1755-1315/16/1/012134

Regional analysis assessment of landslide hazard and zoning map for transmission line route selection using GIS

I N Z Baharuddin[1, 2], R C Omar[1, 2], F Usman[1, 2], M A Mejan[2], M K Abd Halim[2] and M A Zainol[2] and M S Zulkarnain[2]
[1]Department of Civil Engineering, College of Engineering, UNITEN, 43000 Kajang, Malaysia
[2]Centre for Forensic Engineering (CeFE), College of Engineering, UNITEN, 43000, Kajang, Malaysia

E-mail: intan@uniten.edu.my

Abstract. The stability of ground as foundation for infrastructure development is always associated with geology and geomorphology aspects. Failure to carefully analyze these aspects may induce ground instability such subsidence and landslide which eventually can cause catastrophe to the infrastructure i.e. instability of transmission tower. However, in some cases such as the study area this is unavoidable. A GIS system for analysis of route was favoured to perform optimal route predictions based selection by incorporating multiple influence factors into its analysis by incorporating the Landslide Hazard Map (LHM) that was produced on basis of slope map, aspect map, land use map and geological map with the help of ArcGIS using weighted overlay method. Based on LHM it is safe to conclude that the proposed route for Ulu Jelai- Neggiri-Lebir-LILO transmission line has very low risk in term of landslides.

1. Introduction

In the territorial planning stage, analysis and identification of natural geological and geomorphological hazards (geohazards) associated with natural processes has becomes crucial aspect. Detailed assessment on this aspect is also vital and will provide better solution which emphasizes the element of optimization [1]. However, in transmission line route planning, only topography aspect is considered while geology aspect such as its formation and rock type are often overlooked. Therefore, there is a great need to develop a strategy utilizing advanced techniques of geo-informatics for route planning.

The stability of ground as foundation for infrastructure development is always associated with geology and geomorphology aspects. Failure to carefully analyze these aspects may induce ground instability such subsidence and landslide which eventually can cause catastrophe to the infrastructure i.e. instability of transmission tower. It is always been the preference of the engineers to avoid dealing with problematic ground formation such as karstic formation. However, in some cases such as the study area i.e. Ulu Jelai-Neggeri-Lebir-LILO line this is unavoidable. Therefore, analysis of geological risk factors and assessment of potential geohazards is vital in terms of rational management and sustainable of an area [2, 3, 4]. By adopting geographical information system (GIS) in transmission line route planning, geohazard aspects namely landslide hazard and karstic formation maps can be incorporated in the planning stage.

2. Development of Route Selection

The GIS based approaches for transmission route selection utilize relative rankings and weights on considering factors affecting the potential routes. Prior to the actual route design, corridor selection needs to be done. The area to which the proposed route should limit is first identified for collecting and assembling data. The base maps for the corridor selected are collected and the derived maps are generated. Once the derived maps are ready, the suitability analysis is done and a route is found abiding by the governing criteria. Figure 1 shows the schematic representation of the methodology. When a new routing process is initiated, the list of multiple variables has to be decided first and then the criteria by which these variable constraint the routing process. Factors such as land use, type of geology, elevation, infrastructures crossing, and land cost are taken into consideration and are given high priority.

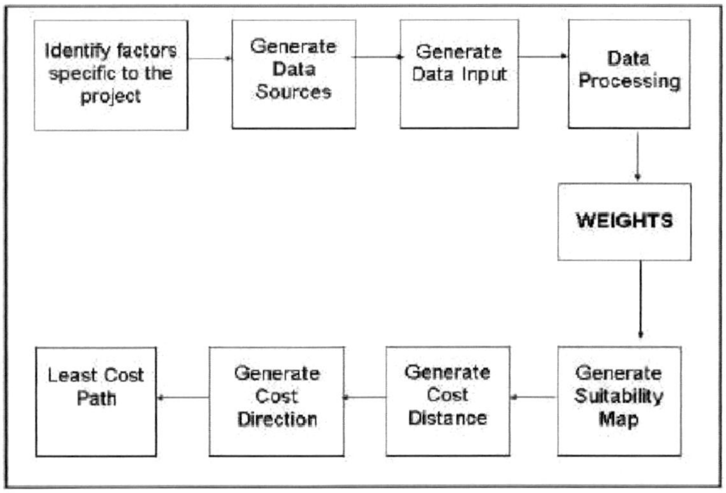

Figure 1 the schematic representation of methodology

A GIS system for analysis of route was favored because it can perform optimal route predictions based selection by incorporating multiple influence factors into its analysis. These influence factors are grouped into common viewpoints or perspectives addressing a common data theme or point of view such as landslide, slope, road and others. In this project, the weightage used to generate route. A higher percentage of influence factors indicate the priority of avoiding area with high percentage of influence. A weighting system has to be devised for weighing each of the map layers. The weighted layers are all summed up to form the suitability layer. This layer forms the basis of the GIS analysis. The Analytical Hierarchy Process has been used to derive weight ages for the thematic layers. The suitability layer is analyzed for the optimal route. For this study, weightage in ascending manner from 1 to 5 are assigned with 5 is considered as the least preferable.

3. Regional analysis of Ulu Jelai- Neggiri-Lebir-LILO

Regional analysis assessment is used to determine landslides hazard zoning by extracting relevant parameters from remote sensing image. Valuable parameters such as land use zone, lithology, lineament (geological structure), groundwater level, weather condition (precipitation), vegetation cover, topography and geomorphology are determined based on satellite image and secondary measured data. Each of the parameters layers will be transferred to ArcGIS to develop the layer for geohazard zoning and risk assessment especially for landslide and erosion feature. This project used the concept proposed by [5, 6] where geohazard zoning and risk maps are used to determine the site specific critical slope.

From analysis of 45 out of 674 towers are located on hilly area and risk of failure this classified into three stages; high, medium and low risk. The risk is measured by the probability and severity of loss to the element at risk which is unit area by calculation and comparison of several parameters such as; slope aspect, groundwater, runoff, erosion rate, and rainfall etc. The risk assessments based on GIS analysis are completed with the help of slope inventory database. Slope inventory database is done based on the priority knowledge that causes the landslide in that area under investigation. The parameters that involved in the slope inventory database were: general slope information; discontinuity data; geotechnical inspection; rainfall inspection; rock slopes: visual assessment; maintenance inspection; existing information for rock slope and inventory and statistical analysis of landslide.

3.1. Landslide Hazard Map (LHM)

Landslide Hazard Map (LHM) as in Figure 2 was prepared on the basis of slope map, aspect map, land use map and geological map with the help of ArcGIS using weighted overlay method. Each cell size used in this study area was 30 m. Slope and aspect map of the study area was developed from Digital Elevation Model (DEM). Landslide hazard map is generated in GIS software where it counts pixels and classify them into five classes; very high risk, high, medium, low and very low risk.

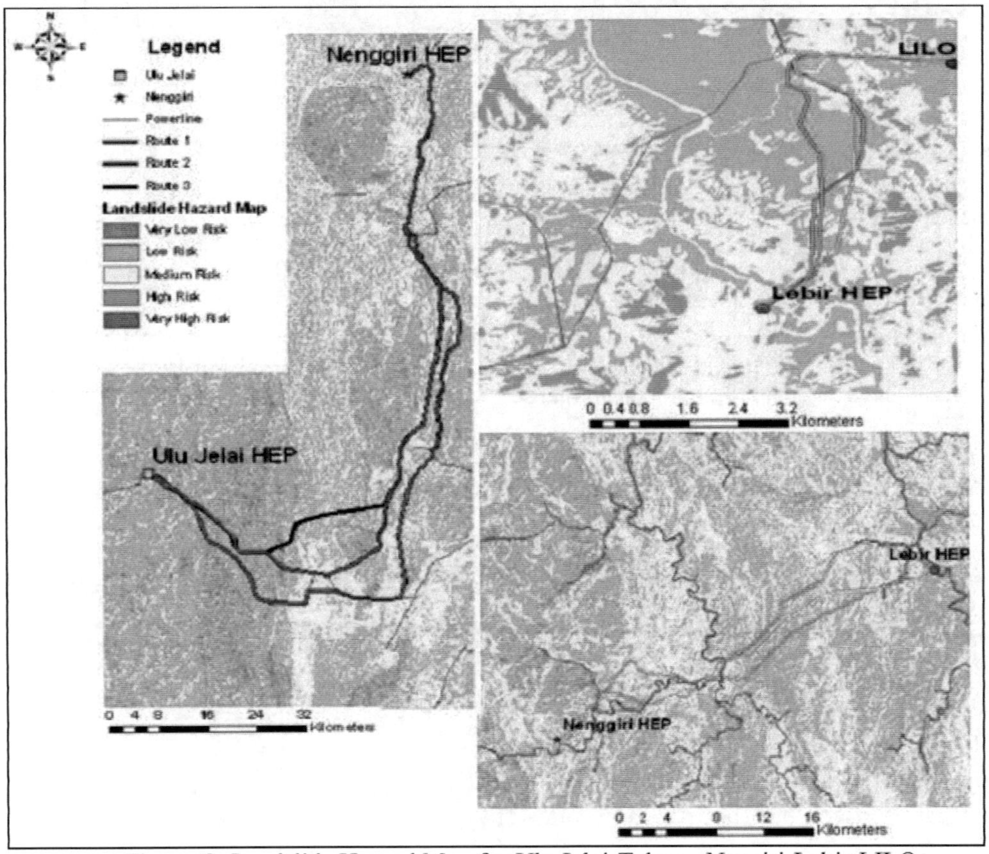

Figure 2. Landslide Hazard Map for Ulu Jelai-Telom - Neggiri-Lebir-LILO

Route options have been generated by GIS in three areas which are LILO – Lebir, Nenggiri – Lebir and Nenggiri – Ulu Jelai with three route options for each area. For regional analysis, each area has a minority area of very high landslide risk which makes it suitable and safe to built transmission line or any other construction. Analysis of landslide risk in terms of percentage is summarized as in Figure 3.

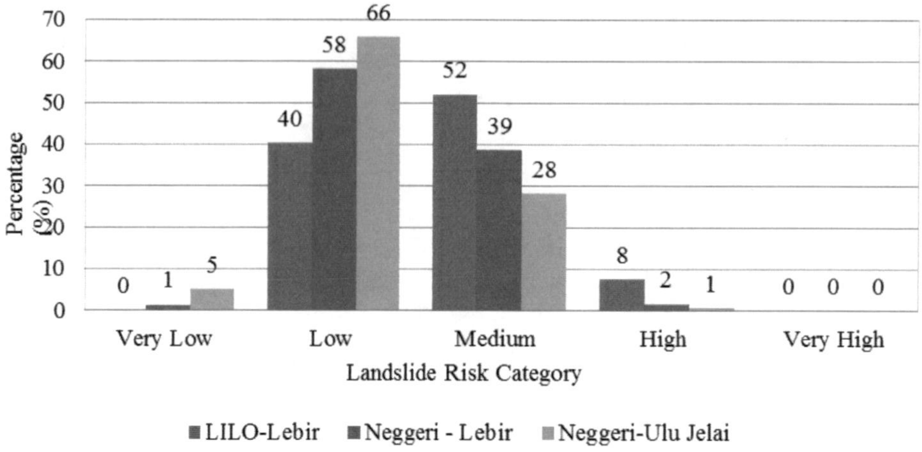

Figure 3. Summary of landslide risk for Ulu Jelai-Neggeri-Lebir-LILO tranmission line

4. Conclusding remarks

Planning may be seen as an exercise in decision making of any development. However, the utilization of latest technology as in GIS would reduce the impact on environment, socio-economy as well as financial. Based on LHM it is safe to conclude that the proposed route for Ulu Jelai- Neggiri-Lebir-LILO transmission line has very low risk in term of landslides.

5. Acknowledgement

The author would like to thank Tenaga Nasional Berhad (TNB) and its subsidiary Tenaga Nasional Berhad Research (TNBR) for granting this project.

References

[1] Campus S, Barbero S, Bovo S, Forlati F 2007 Evaluation and Prevention of Natural Risks *Taylor & Francis, UK* 470.

[2] Crozier M, Glade T 2010 Hazard assessment for risk analysis and risk management *Alcántara-Ayala I., Goudie A. (eds.), Geomorphological Hazards and Disaster Prevention* Cambridge Univ. Press, Cambridge, UK 221-232.

[3] Hassanzadeh N M 2000 Landslide hazard zonation in shalmanrood watershed *M. Sc. thesis, Tehran University.*

[4] Higgitt D 2010 Geomorphological hazards and sustainable development *Alcántara-Ayala I., Goudie A. (eds.), Geomorphological Hazards and Disaster Prevention* Cambridge Univ. Press, Cambridge, UK 257-268.

[5] Nagarajan R, Mukherjee A, Roy A and Khire M V 1998 Temporal remote sensing data and GIS application in landslide hazard zonation of part of western ghat, India *Int. J. Remote sensing* **19(4)** 573-585.

[6] Smith K, Petley D N 2009 Environmental Hazards *Assessing Risk and Reducing Disaster. 5th edit. Routledge, New York, USA* 416.

IOP Conf. Series: Earth and Environmental Science **16** (2013) 012135 doi:10.1088/1755-1315/16/1/012135

Novel configurations of solar distillation system for potable water production

[1]ARiahi,[2] K W Yusof, [3]N Sapari, [4]B S Singh, [5]A M Hashim
[1,2,3,5] Department of Civil Engineering, UniversitiTeknologi PETRONAS, 31750 Tronoh, Perak, MALAYSIA
[4]Fundamental and Applied Sciences Department, UniversitiTeknologi PETRONAS, 31750 Tronoh, Perak, MALAYSIA

E-mail: [1]ali_riahiupm@yahoo.com

Abstract - More and more surface water are polluted with toxic chemicals. Alternatively brackish and saline water are used as feed water to water treatment plants. Expensive desalination process via reverse osmosis or distillation is used in the plants. Thus, this conventional desalination is not suitable for low and medium income countries. A cheaper method is by solar distillation. However the rate of water production by this method is generally considered low. This research attempts to enhance water production of solar distillation by optimizing solar capture, evaporation and condensation processes. Solar radiation data was captured in several days in Perak, Malaysia. Three kinds of experiments were done by fabricating triangular solar distillation systems. First type was conventional solar still, second type was combined with 50 Watt solar photovoltaic panel and 40 Watt Dc heater, while third type was integrated with 12 Volt Solar battery and 40Watt Dc heater. The present investigation showed that the productivity of second and third systems were 150% and 480% of the conventional still type, respectively. The finding of this research can be expected to have wide application in water supply particularly in areas where fresh surface water is limited.
Keywords – Solar distillation; Solar radiation; Evaporation; Condensation

1. Introduction

One of the most plentiful resources on Earth is water which covers 75% of the planet's area. Groundwater, lakes and rivers are different types of freshwater which contains just 3% in Earth water [1]. Obtaining fresh water is the most important challenge in the world. Achieving fresh water quality parameters according to WHO standards [2] is required to remove chemicals and heavy metals from natural water resources by using solar distillation system which is the significant factors for producing drinking water. There are some methods for desalination and providing potable water and the simplest one is using distillation system. During the last years, many researchers studied to find the solutions for developing the traditional solar still performance for producing higher value of potable water. This paper emphasized the development of solar distillation still to enhance the potable water productivity and to investigate the relationship between solar radiation in Malaysia and water productivity by adding solar photovoltaic system.

2. Research Background

Solar distillation still is the most commonly used method of producing drinking water in arid, remote and coastal areas in the world. This method is generally the simplest, most flexible and most economical means of brackish water and groundwater treatment. The problem with solar still however, it produces less fresh water [1].

Currently, to raise the productivity of solar distillation systems many types of solar stills were designed, fabricated and tested. Later development in solar distillation stills operation is moving towards the increase of water temperature for reaching high amount of evaporation inside solar still.

The productivity of pyramid shape solar still was increased by 231% and reached to 4.62 L/m².d using solar thermal by treating 18.4 L saline water expose to peak sun radiation between 700 and 1000 W/m² in Jordan [3], while it reached to 6.6 L/m².d by using 2 flat plate collectors combined with a single sloped solar still with maximum solar radiation intensity of 780 W/m² and taking 7 hours to reach the water to 92.3°C as peak temperature in India [4]. 6.26 L/m².d water productivity obtained by cooling the glass surface cover of double slope solar still using shading screen in Algeria by treating 9.45 L of salty water under 800 W/m² peak sun radiation and 71°C as peak water temperature [5]. However, the amount of productivity increased by 370% in Saudi Arabia by using 2 AC heaters to reach to 11.9 L/m².day treating 60 L of seawater expose to 1000 W/m² as peak solar radiation intensity, but the problem reported was the high cost of electrical consumption [6]. The water production of solar distillation stills was also improved by using forced convention such as fin, sponges, pebbles, black rubber and sand in solar still, utilizing solar collector coupled with solar still [7].

3. Methodology

In this research, a model of triangular double sloped solar still was constructed by using the materials of steel basin has 34cm length and 20cm width, transparent plastic cover and PVC Pipe for fabricating the frame.A 50 Watt solar panel and a 40 Watt Dc heater were combined as the source of solar energy capture for converting it to electrical energy and transferring to the thermal energy, respectively. Three types of experiments were done by combining 50Watt solar photovoltaic panel and 12volt solar battery as an auxiliary system of solar power energy system to the 40Watt Dc heater inside the trough of solar triangular distillation still.

4. Results and discussion

First experiment was organized by exposing the solar still to only sun radiation on 02.12.2012. The values of solar radiation intensity, ambient and water temperatures obtained in the first experiment as well as the relationship between water temperature and sun radiation intensity are shown in Figures 1 and 2.

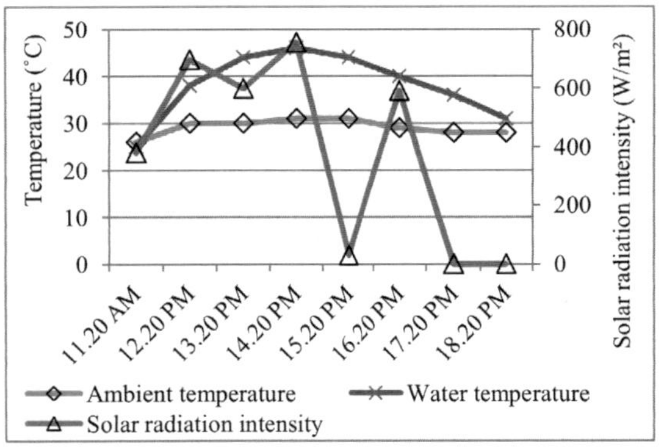

Figure 1. Solar radiation intensity values versus ambient and water temperatures.

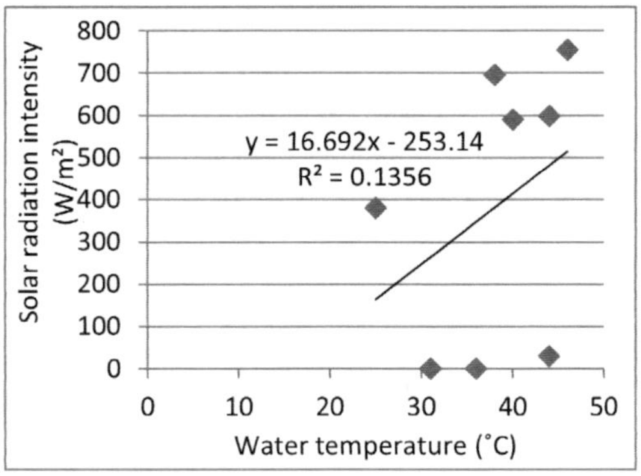

Figure 2. Relationship between water temperature and solar radiation intensity on 02.12.2012.

The value of 100 ml potable water was produced during 7 hours in UTP, Perak, Malaysia by heating 1.5 Liter of brackish water and exposing triangular solar still to only solar radiation on 02.12.2012. The second experiment was done by using 50Watt solar panel and 40 Watt Dc heater in Perak, Malaysia on 03.12.2012 is shown in Figure.3. Solar radiation intensity, ambient and water temperatures measured in second experiment as well as the linear relationship between solar radiation and water temperature are shown in Figures 4 and 5. The value of 150 ml distilled water was produced during 4 hours of heating 2 Liter of brackish water in UTP, Perak, Malaysia on 03.12.2012 by using thermal energy of solar radiation and Dc heater. By using 12Volt solar battery and 40Watt Dc heater in UTP, Perak, Malaysia the last experiment was done on 06.12.2012. The values of hourly water production, solar radiation intensity, ambient and water temperatures and battery voltage consumption obtained in third experiment are shown in Figures 6, 7, 8 and 9. A linear relationship can be observed by obtaining the relation of water production and water temperature variations in Figure.8. By considering the above results it can be shown that by increasing the solar radiation intensity, the ambient and water temperature will be raised, thus the high value of water production can be obtained during the day and by decreasing the intensity of solar energy, the reduction of productivity and surrounding temperatures can be observed.

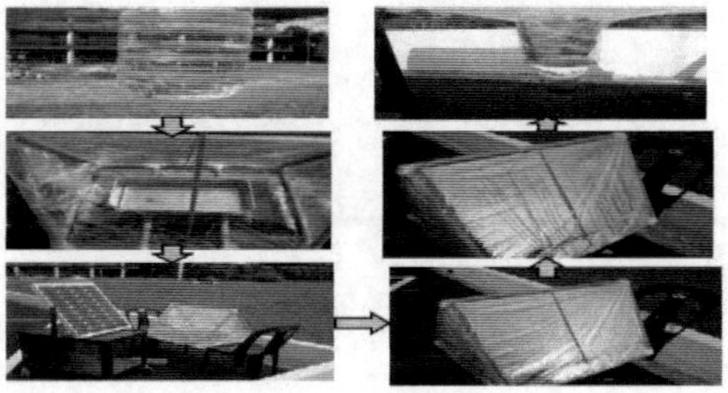

Figure 3. The thermal transfer processes in triangular solar distillation still using solar photovoltaic and Dc heater.

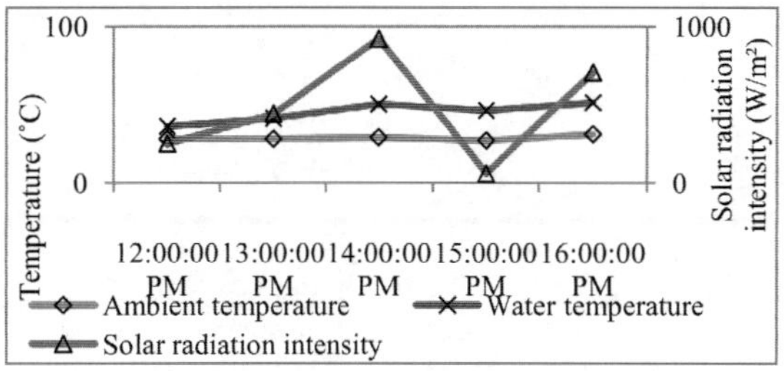

Figure 4. Solar radiation intensity values versus ambient and water temperatures on 03.12.2012.

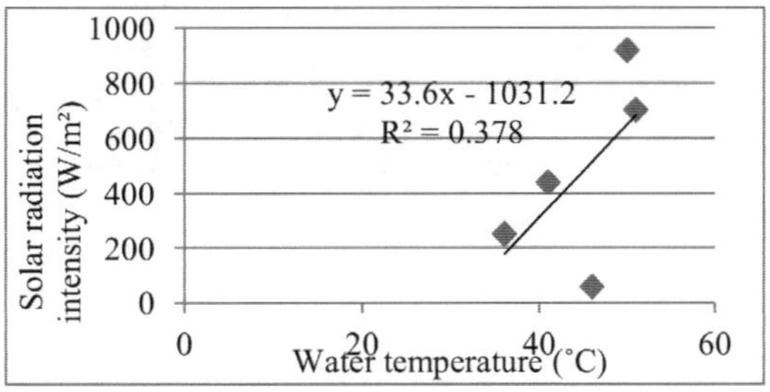

Figure 5. Relationship between water temperature and solar radiation intensity on 03.12.2012.

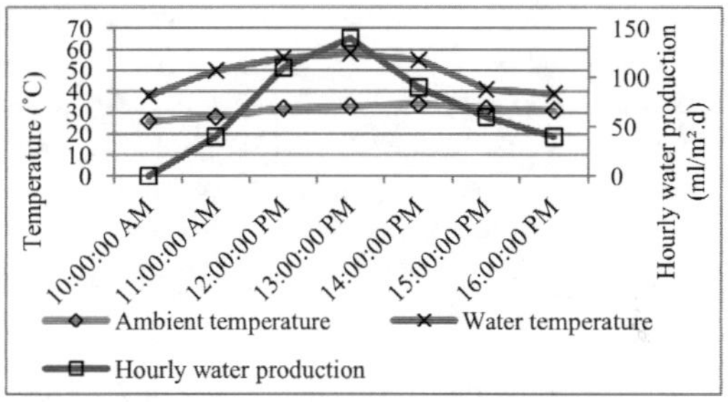

Figure 6. Hourly water production values versus ambient and water temperatures on 06.12.2012.

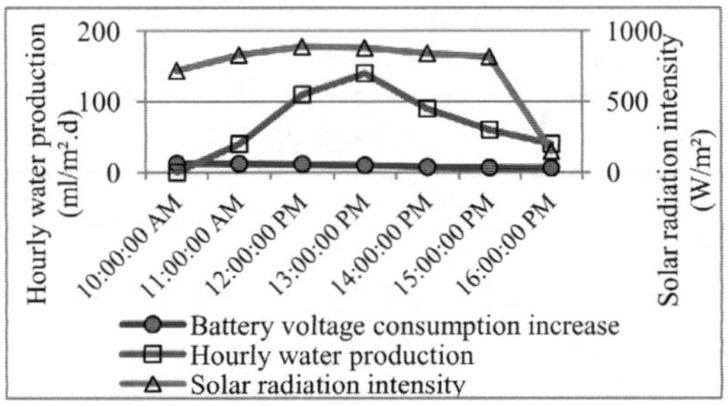

Figure 7. The hourly water production variations versus solar radiation intensity and battery power consumption increase on 06.12.2012.

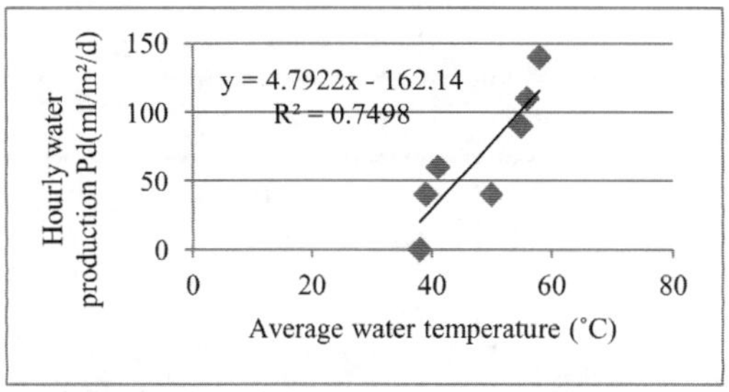

Figure 8. Relationship between average water temperature and hourly water production on 06.12.2012.

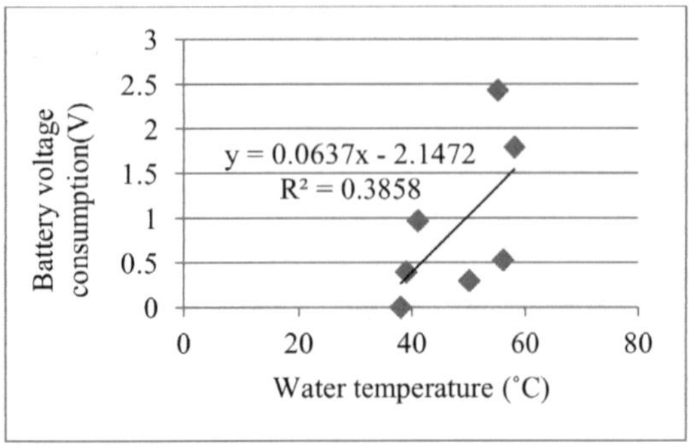

Figure 9. Relationship between water temperature and battery power consumption on 06.12.2012.

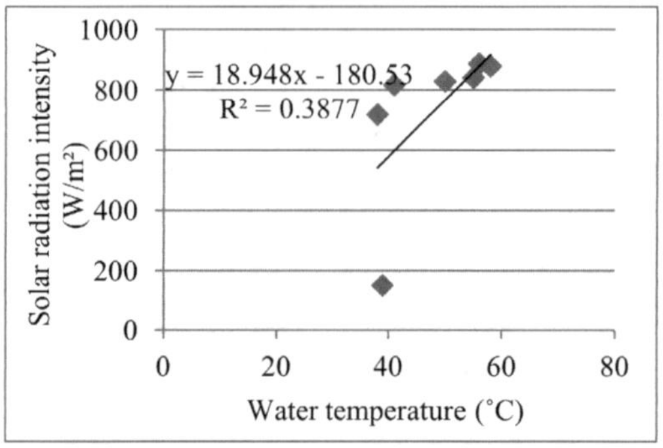

Figure 10. Relationship between water temperature and solar radiation intensity on 06.12.2012.

However, the increase in water temperature and productivity depends on raising the battery power consumption and sun radiation intensity which can be shown in Figures 9 and 10.

5. Conclusion

The present study of triangular solar distillation still showed that the rate of water production is mostly depended on the values of water temperature, ambient temperature and solar radiation intensity as well as solar power supply. These values are most significant factors for developing the evaporation and condensation of thermal transfer process in solar distillation system. The relationship between water temperature inside solar still and sun radiation intensity is more linear by combining the solar power systems than exposing the solar still to only solar radiation. The production rate using second and third systems was 150% and 480% of the conventional type production respectively. The higher absorptivity of sun radiation can be obtained by using glass instead of transparent plastic as cover of solar still.

Acknowledgment
The author would like to acknowledge the support of Universiti Teknologi Petronas in this work.

References

[1] Velmurugan V, Deenadayalan CK, Vinod H, Srithar K 2008 Desalination of effluent using fin type solar still *Energy 33* 1719–1727
[2] World Health Organization 2008 Guidelines for Drinking e Water Quality (Electronic Resource) *Incorporating 1st and 2nd Addenda, Recommendations -third ed* **1** http://www.who.int/water_sanitation_health/dwq/ fulltext.pdf (accessed 15.05.11.) Zaw M, Emett M T 2002 Arsenic removal from water using advanced oxidation processes *Toxicology Letters* 133, 113-118.
[3] Ali A B, Ahmad A. Al-Hallaq, Imad A E S, and Mohammad Z O 2005 A solar still augmented with a flat-plate collector *Desalination 172* 227-234.
[4] Shiv Kumar, Tiwari G N, Gaur M K 2010 Development of empirical relation to evaluate the heat transfer coefficients andfractional energy in basin type hybrid (PV/T) active solar still *Desalination 250* 214–221
[5] Zeroual M, Bouguettaia H, Bechki D, Boughali S, Bouchekima B, and Mahcene H 2011 Experimental Investigation on a Double Slope Solar Still With Partially Cooled Condenser in the Region of Ouargla (Algeria) *Energy Procedia* **6** 736–742.
[6] Ahmed Z Al-Garni August 2012 Productivity Enhancement of Solar Still Using Water Heater and Cooling Fan *Journal of Solar Energy Engineering* **134** / 031006-1
[7] Sampathkumar K, Arjunan T V, Pitchandi P, Senthilkumar P 2010 Active solar distillation—A detailed review *Renew. Sustainable Energy Rev* **14** 1503–1526.

IOP Conf. Series: Earth and Environmental Science **16** (2013) 012136 doi:10.1088/1755-1315/16/1/012136

Nuclear energy in Malaysia – closing the gaps

Malaysian Nuclear Society (MNS)
http://www.nuklearmalaysia.gov.my

E-mail: azizm@uniten.edu.my[1]

Abstract. This article is prepared by the Malaysian Nuclear Society (MNS) to present the views of the Malaysian scientific community on the need for Malaysia to urgently upgrade its technical know-how and expertise to support the nuclear energy industry for future sustainable economic development of the country. It also present scientific views that nuclear energy will bring economic growth as well as technically sound industry, capable of supporting nuclear energy industry needs in the country, and recommend action items for timely technical upgrading of Malaysian expertise related to nuclear energy industry.

1. Background

Nuclear energy, over the last decades, has shown its resiliency in meeting the shortfall of energy demand from fossil fuel. Apart from its natural abundance in nature, with Kazakhstan, Canada and Australia together account for 63% of world uranium production, with Namibia, Russia, Niger, Uzbekistan, and the United States producing in excess of 1000 tonnes per year, the "yellowcake," the dry powder-form material consisting of natural uranium U_3O_8, the price of uranium remains stable in the world market [1]. In Contrast, the prices of petroleum and coal were driven by market speculation and shortage of supply due to political instability in the gulf region.

While Malaysia continue to subside its domestic oil consumption with billions of ringgit per year, the oil resources in Malaysia would not be bottomless (this it is true worldwide, as Arabs countries known for its riches in oil such as United Arab Emirates (UAE) is investing USD20 billion for four 1,400MW civil nuclear power reactors of Generation-3 APR-1400 of type), the conservative use Malaysian oil could be further enhances through the use of economically proven of nuclear-based electricity generation to cover the expected ~30% energy supply-demand gap by 2020.

The use of civil nuclear power reactors for electricity generation needs long term commitment from all parties involved i.e. the regulatory bodies, the government, the utility bodies, the human capital development sectors, the research and development agencies, the nuclear related industrial sectors etc., as a sound nuclear-based industrial sectors could not be build overnight as it requires a technically matured community to sustain and maintained its operation with the safety and operable limit, and further to develop into a technologically driven industry able to generate spin-off industries in Malaysia.

For Malaysia to emulate Japan and other countries practicing nuclear-based electricity generation, it has to leap jump ahead technically in terms of nuclear standardization and practice. The dormant state of its nuclear science and technology for the past 30 years has to be revitalized to produce more experts in nuclear science and technology, and upgrade its current industrial standards further to

[1] To whom any correspondence should be addressed.

nuclear standard. It will also needs to beef up its regulatory bodies to better implement the nuclear standardization in the practicing industry, and thus protecting the public at large from the folly of an incompetent management of the nuclear power industry. The supply of expertise in related nuclear science and technology should be sufficient and well educated enough to cater for the stringent needs the nuclear power industry in Malaysia. The stringent need of nuclear power industry would require the Malaysian education system to upgrade its teaching capability to produced qualified nuclear engineers and scientists to support and sustain the nuclear energy industry in Malaysia.

2. Implementation of Nuclear Energy Programme in Malaysia: Issues and Recommendations

While the introduction on nuclear energy is expected to drive the economic growth of Malaysia to another level, the cascade effect will also unleash high technology industrial development for Malaysia [2]. Time has proven that countries such as France, Japan, China and recently Korea that uses nuclear for its energy generation have emerged stronger economically, despite the turbulence of world oil prices. Not only that, the industrial development in high-tech industry related to the nuclear energy industry, such as automotive, ITC, heavy industry, robotics, semiconductor etc., has also propagated and its products has become more competitive in the world market.

Looking back, Malaysia has been lagging behind 30 years in terms of nuclear science and technology, as compared with Korea when it was almost at the same level with Malaysia in 1960s, when Korea started to commission its first nuclear research reactor. The gap now between Korea and Malaysia, in terms of nuclear science and technology, for the latter to start building its first nuclear power plant for electricity generation, is so wide. Such contrast requires that Malaysia re-examine itself to realign all its resources for effective implementation of nuclear power program in Malaysia. The following chapters will give the results of such re-examination.

2.1 Research and Development (R&D) / Technical Aspects

One of the basic requirements in for establishing and implementing quality assurance programmes for nuclear power plants is the need for research and development (R&D) to enhance nuclear safety by continuously improving the methods employed to achieve quality [3]. The R&D preparedness to support the nuclear power industry requires a matured workforce able to carry out effective R&D programmes towards enhancing the nuclear safety. The personnel needs to be trained so that they are competent and qualified enough to perform their assigned work and understand the safety consequences of their activities.

As Malaysian has been lagging behind, in terms of R&D infrastructure to support the nuclear power industry per se, it needs to catch up very fast what it has loss for the past 30 years. Preparing for the right technical support for the nuclear power industry in Malaysia to prosper in the future needs capable personnel in the right number. If Malaysia were to venture into R&D in nuclear power programmes, it needs to identify some critical issues and gaps facing the R&D community.

One of the critical issues is finding the right manpower to undertake R&D projects related to nuclear energy development programmes. The following actions are suggested to be taken in order to meet the demand of nuclear international codes and practices:

- Training of manpower related to nuclear energy development programmes has to be carried out outside the country. This would give Malaysia disadvantages such as it need a bigger capital investment needed to finance the course in foreign countries, thus limiting the time spent for the course stint, with limited number of trainees that could be send, as compared if the course were to be carried out locally.
- As technology and procedures improved over the times, training programmes has to be carried out continuously to meet the demand of international nuclear standards and practices, more capital investments needed for carrying out such retraining programmes overseas.
- Experts in nuclear science and engineering need to have a hands-on experience before they are certified good enough to become operators of a nuclear power plants, run regular integrity assessments of the plants and carry out operation and maintenance according to international

nuclear standard and practices. Experience has shown that such experts need at least 3 to 5 years of hands-on training on actual site before they are qualified and skill enough to be certified as having skills in compliance with international nuclear standard and practices.

For the above to be addressed effectively in as shortest time as possible, all parties related to research and development (R&D) in nuclear science/physics and engineering i.e. research institutions, institutions of higher learning, standard institutions, governing bodies and policy maker, should work together in synchronization to make sure that Malaysian R&D in nuclear science/physics and engineering would be ready to support the nuclear power industry in Malaysia, in the shortest time possible.

2.2 Human Resource Development

One of the most important aspects in the infrastructure development towards a civilian nuclear power industry in Malaysia is its quality of human resource, able to handle and maintain the nuclear power plant operation within its operation and safety limit conditions, as well as carrying out R&D works to support the nuclear power plant operation and its continuous enhancement. Professional bodies in Malaysia such as the institute of engineers (mechanical, electrical/electronics, civil), architects, advance manufacturing, consultants, chemists, environmentalists, surveyors, planers, chartered institute of building, materials, logistics, bar council, medical, etc., should seriously consider to adopt nuclear standards and practices into their charters. This would ensure that Malaysian professional bodies should have sufficient professional expertise to handle all aspects related to nuclear power industry i.e. architecture and design, construction, project management, installation, component fabrication, transportation, legal, medical etc.

On the national scale, Malaysia needs to have a relook at its education system and revamp the system where necessary, to produce a well rounded and educated nuclear scientists and engineers, able to operate and maintain nuclear power plant operation with its operating and safety limit conditions, as well as carry out R&D works to support the operation of the plant, and to upgrade the nuclear power industry to next level. Higher institution of learning offering good courses in nuclear science and engineering should be given the green lights to foster close partnership with renowned international nuclear science and engineering institutes to produced experts in science and engineering of international outstanding. The setting up centre of excellence in nuclear science and engineering to cater for the needs to produce local expertise of outstanding capability, and to carry out excellent R&D works in these areas should be considered seriously.

2.3 Industrial

While the Malaysian industrial sectors (civil, structural, M&E, architectural, design) have evolve since the 1980 from conventional to oil & gas and aviation industries, its nuclear industry is still at its infantry. The current exposure of local industry to all aspects of nuclear science and engineering is limited, as Malaysia as yet to adopt a strong stand towards the use of nuclear for its electricity generation, although it recently has established the Malaysian Nuclear Power Corporation (MNPC) last January 2011, under the auspicious of the Prime Minister's Department

The industry involvement in Malaysia can be divided into three stages – pre-construction, construction, and operation and maintenance. While the pre-construction would involve site feasibility study (site location, EIA study), design, economics feasibility study, funding, civil, structural, M&E, none of these have ever have components of nuclear plant development before. The construction, operation and maintenance of a nuclear power plants would add a new dimension and experience to the people involved. As nuclear power plant, construction, operation and maintenance is highly technical and specialized, the personnel involved in these key areas needed to be accredited with nuclear standards and practices recognized internationally before they are allowed to venture into construction, operation and maintenance of a nuclear power projects. Malaysia needs to enforce adequate regulation so that these people follow nuclear standards and practices, to protect the civilians from being short-change due to the negligence in nuclear power plant construction.

2.4 Legal

One of the most important aspects in nuclear power plant construction, operation and maintenance is the legal aspects. With the Act 304 Atomic Energy Licensing 1984 currently as the sole act used in enforcing all aspect on the use of atomic energy, radiation use and radioactive sources in Malaysia [4]. This act is not as comprehensive as the Nuclear Power Laws of the USA in regulating the nuclear power industry in the country.

Malaysian jurisprudence, law makers and legal fraternity needs to be updated on current laws and regulation regarding nuclear power plant construction, operation and maintenance. The universities and education institutions should to tie up with the university from foreign countries to improve curriculum on Nuclear Power Laws.

2.5 Public Acceptance and Waste Management

For greater public acceptance, the people involved in the planning of the nuclear power plant construction needs to be transparent and able to educate the public at large, what are the steps to be taken for safe construction and operation of nuclear power plants. The public should be made known that the construction, operation and maintenance of nuclear power plants are in accordance with international nuclear standards and practice that comply with stringent standards of nuclear quality assurance programme. The public also needs to be ensured that adequate regulations and laws are being enforced to protect them from harmful effects of nuclear power plant failure. The fact that there are now a total of 442 of nuclear power plants in operation worldwide shows that nuclear power plant

Another issue is the management of nuclear from spent fuel. The fact that more 96% of nuclear spent fuel could be reprocessed and recycle for use, leaving 0.33% of high level waste to be management has securely stored until its has decay into a manageable level. Malaysia should to invest in spent fuel processing technology available in market, to manage its nuclear spent fuel effectively. This may require careful study before being implemented.

3.0 Conclusion

The nuclear power programme for electricity generation may no longer be an option but a necessity for Malaysia, if she were to be economically competitive in future due to shortage of fossil fuel. The impending use of nuclear energy for electricity generation to drive the economic development of Malaysia in the future requires carefully planning and implementation that conform to nuclear standards and practices. Due to its high technological investment and also high return, the implementation of nuclear energy for electricity generation would not only provide affordable electricity tariff, but would also bring the Malaysian socio-economic development to the next level.

Acknowledgements

MNS would like to thank all parties involved in making drafting this paper, especially all the participants who attended the 'WORKSHOP ON NUCLEAR ENERGY' 2, 8 January 2011, Residence Hotel, Universiti Tenaga Nasional, 43000 Kajang, Selangor.

References

[1] http://en.wikipedia.org/wiki/Uranium_mining
[2] Malaysian Nuclear Society 2008 *White Paper on Nuclear Energy*
[3] International Atomic Energy Agency 2011 *Evaluation of Human Resource Needs for a New Nuclear Power Plant: Armenian Case Study* IAEA-TECDOC-1656
[4] Undang-Undang Malaysia, *Akta 304 Akta Perlesenan Tenaga Atom 1984*

4th International Conference on Energy and Environment 2013 (ICEE 2013) IOP Publishing
IOP Conf. Series: Earth and Environmental Science **16** (2013) 012137 doi:10.1088/1755-1315/16/1/012137

Characterization of a pulsed mode high voltage power supply for nuclear detectors

A B Ghazali[1], T S Ahmad[1] and N A Abdullah[2]
[1]Universiti Tenaga Nasional (UNITEN); [2]Malaysian Nuclear Agency, MOSTI

Email: Abakar@uniten.edu.my

Abstract. This paper discusses the characterization of a pulsed mode high voltage power supply (HVPS) using LT1073 chip. The pulsed modulated signal generated from this chip is amplified using a step-up ferrite core transformer of 1:20 turn ratio and then further multiplied and converted into DC high voltage output using a diode-capacitor arrangement. The circuit is powered by a 9V alkaline battery but regulated at 5V supply. It was found that the output for this setup is 520V, 87 μA with 10% load regulation. This output is suitable to operate a pancake-type GM detector, typically model LND 7317 where the plateau is from 475V to 675V. It was also found that when a β-source with intensity of 120 cps is used, the power consumption of the circuit is 5V, 10.1 mA only. When the battery was left 'on' for 40 hours continuously, the battery's voltage has dropped to 6.9V, meaning that the 5V supply as well as 520V output is still maintained. It is noted that the minimum output voltage of 475V has reached when the regulated supply has reduced to 4.6V and consequently the 9V battery dropped to 6.5V, and this had happened after approximately 3 days of continuous operation. The power efficiency for this circuitry was found to be 89.5%. This result has far better in performance since the commercial portable equipment of this type has normally specified that not less than 8 hours continuous operation only. On the circuit design for this power supply, it was found that the enveloped frequency is 133 Hz with approximately 50% duty cycle. The modulated frequency during 'on' state was found to be 256 KHz in which the majority of power consumption is required.

1. Introduction

The challenge to reduce energy consumption in instrumentation design, especially on portable equipment has never stopped. This includes selection and reduction in number of hardware components, such as the semiconductor chips and to improve the control strategy during software development. A power supply (PS) to drive electronic circuitry in the equipment plays a very important role in order to make the equipment functioning as required. This PS should be designed with high efficiency so that unwanted power lost in the components used for this power supply is minimal. A well-known design is to use a switched mode PS where the output is regulated using duty cycle control at optimal frequency to switch the transistors to fully on and off, so that very little resistive losses between input and the load [1]. Other method in the design is called pulsed mode PS where the duty cycle control is used to drive a power transistor that connected to an ignition coil so that a high transient that due to the release of coil's energy during switching off the transistor is created. This transient is then amplified and rectified by a step up transformer and the diode-capacitor circuit to obtain a DC output and this approach has been used for example by Bicron scintillation counter [2]. The later design is applied in this article where a micropower DC/DC converter, LT1073 chip is used to function as an oscillator as well as a regulator. In this report, a 9V battery is converted into 520V PS for operating a nuclear detector and a characterization of this PS is discussed here.

2. Circuit design

The circuit diagram for this HVPS is shown in Figure 1 where a LT1073 chip (U1) [3] as in Figure 2 is used. The transistor Q1 collector of U1 (pin 3) is connected to the primary winding of 1:20 step-up ferrite core transformer to produce a modulated signal of 256 KHz for about 50% duty cycle with enveloped frequency of 133 Hz. This duty cycle as well as the HV output are controlled by the feedback voltage at pin 8 of U1 obtained from a potential divider of R2, 100KΩ and R3, 40MΩ resistors. Pin numbers 6 and 7 are used for weak battery indicator and they do not consider in this circuit. After 20x amplification of signal from the transformer, it is then rectified and converted to direct current (dc) output at point D. The diodes D2,.., D4 and capacitors, C2,.., C4 arrangement as shown will get the HV output of $2V_{p(+)} + V_{p(-)}$ where V_p is the peak voltage in either positive or negative cycle of the waveform [4]. The output ripple due to the envelope frequency is reduced by a low pass filter of R4 and C5 before connected to a GM detector. The load regulation for this circuit was also performed by connecting the HV output to a number of resistor's values as the load and this will determine the power efficiency.

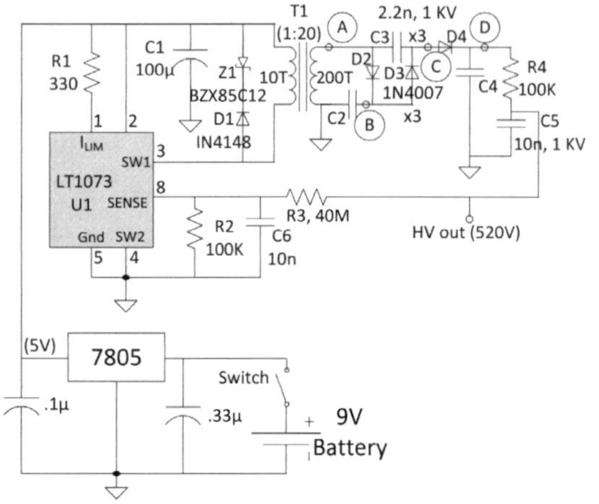

Figure 1. HV circuitry using LT1073 chip

Typical values:

Point A = 90.2V RMS

Point B = 187V d.c.

Point C = 320V RMS

Point D = 520V d.c.

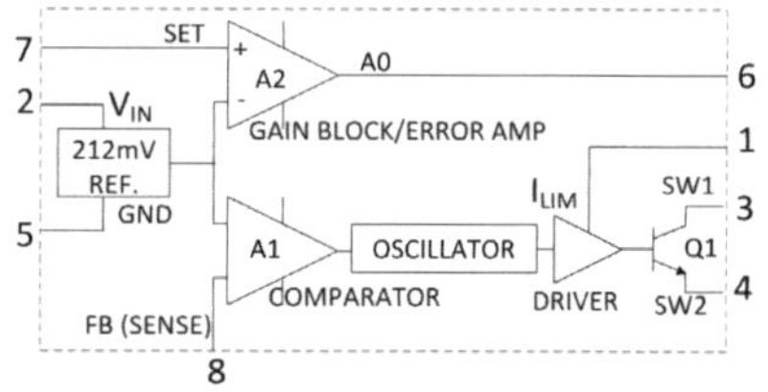

Figure 2. Block diagram of LT1073 chip

3. Result and discussion

A typical result with the prescribed feedback voltage to LT1073 chip is shown in Figure 3 where the modulated signal at point A is obtained from the chip that connected to the primary winding of the step-up transformer. Capacitor C3 in Figure 1 acts as an ac coupling and this signal is rectified by diode D3 to get 320V rms. The ac voltage that sitting of dc voltage at point C is further rectified by diode D4 and capacitor C4 to obtain the HV output of 520V dc at point D with ripple of $15V_{pp}$, 133 Hz. This ripple is then reduced to $5V_{pp}$ by low pass filter of R4 and C5.

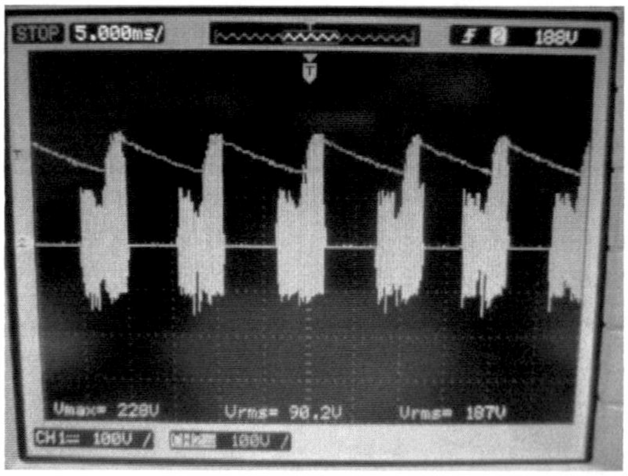

(a) Points A (yellow-packet) and B (green-rectified)

IOP Conf. Series: Earth and Environmental Science **16** (2013) 012137 doi:10.1088/1755-1315/16/1/012137

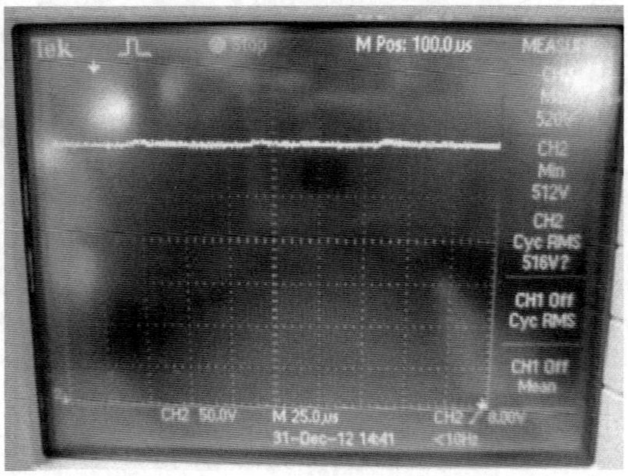

(b) Point D

Figures 3. HV output at various points

A load regulation for this portable HVPS was conducted by connecting a number of resistor values as the load. The result is shown in Figure 4 where the output of 520V (no load) has dropped to 468V (10% regulation) at about 87 µA of load current. This amount of current is more than adequate since an average current required to produce a pulse is in the range of ρA only [5]. The current flow from 5V regulator to this HVPS circuit was also measured and found that to be 10.1 mA. It means that the circuit efficiency in term of power delivery is (520V x 87 µA)/(5V x 10.1 mA) = 89.5%.

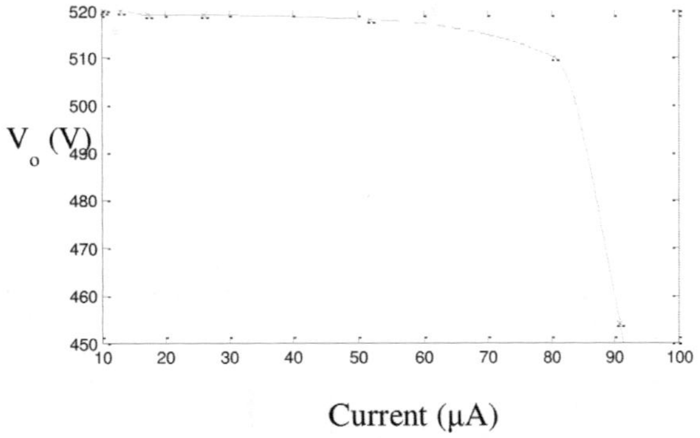

Figure 4. Load regulation responses for discrete load values from 5 MΩ to 50 MΩ

The 9V battery's life span was also monitored, initially for 40 hours continuous operation. A LND 7317 GM detector is connected to this HVPS output as shown in Figure 5 (a) where a β-source radionuclide is placed near to the detector. The pulses as shown in Figures 5 (b) and (c) were obtained from the detector and a pulse shaper respectively. For radiation intensity of 120 pulses per second (cps), it was found that the battery's voltage has dropped to 6.9V, meaning that the 5V supply as well as 520V output is still maintained. Note that the minimum plateau of this detector (475V) has reached when the regulated supply had reduced to 4.6V and consequently the 9V battery dropped to 6.5V, and this has happened after approximately 3 days of continuous operation.

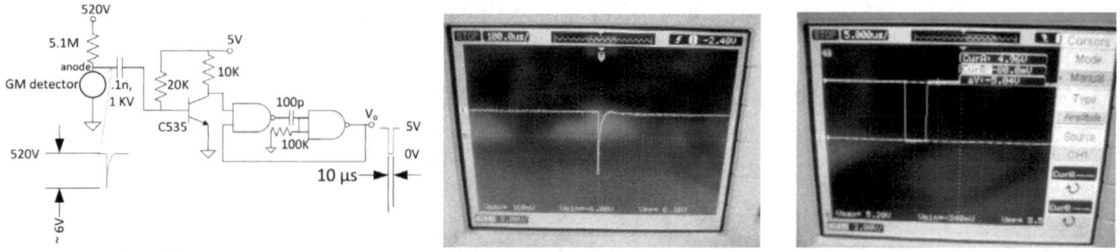

(a) Pulse shaper (b) A nuclear signal (c) Signal after pulse shaper

Figure 5. Nuclear signal from a Geiger Muller (GM) detector

4. Conclusion

This article has shown that a LT1073 IC is suitable for HVPS design to operate a nuclear detector used for portable equipment, such as a survey meter. It is also shown that a pulsed mode power supply has very good power efficiency and this is in comparable to a good performance of a switching mode power supply.

5. References

[1] 2002 *SWITCHMODE Power Supplies: Reference Manual and Design Guide* Rev. 3B (On Semiconductor) July.

[2] Gifford A Z 2000 Service Manual of Bicron Scintillation Counter *Model 2000 Saint-Gobain Ceramics and Plastics, Inc.*

[3] LT1073 Micropower DC/DC Converter Datasheet *Linear Technology Corporation* 1073fa LT/TP 0301 2K REV A- Printed in USA.

[4] Godse A P and Bakshi U A 2008 D.C. power supplies in elements of electronics engineering *Pune: Technical Publications* 64.

[5] Knoll G F 2010 *Radiation Detection and Measurement* 4th ed. Hoboken, N.J (John Wiley and Sons) 295-296.

IOP Conf. Series: Earth and Environmental Science **16** (2013) 012138 doi:10.1088/1755-1315/16/1/012138

Modeling the effect in of criticality from changes in key parameters for small High Temperature Nuclear Reactor (U-Battery^{TM}) using MCNP4C.

A M Pauzi

Department of Mechanical Engineering, College of Engineering, Universiti Tenaga Nasional (UNITEN), Jalan UNITEN-IKRAM, 43000 Kajang, Selangor, Malaysia

E-mail: anas@uniten.edu.my

Abstract. The neutron transport code, Monte Carlo N-Particle (MCNP) which was wellkown as the gold standard in predicting nuclear reaction was used to model the small nuclear reactor core called "U-battery^{TM}", which was develop by the University of Manchester and Delft Institute of Technology. The paper introduces on the concept of modeling the small reactor core, a high temperature reactor (HTR) type with small coated TRISO fuel particle in graphite matrix using the MCNPv4C software. The criticality of the core were calculated using the software and analysed by changing key parameters such coolant type, fuel type and enrichment levels, cladding materials, and control rod type. The criticality results from the simulation were validated using the SCALE 5.1 software by [1] M Ding and J L Kloosterman, 2010. The data produced from these analyses would be used as part of the process of proposing initial core layout and a provisional list of materials for newly design reactor core. In the future, the criticality study would be continued with different core configurations and geometries.

1. Introduction

In order to meet the energy needs without harming our planet, the application of nuclear power would be highly important. One of the challenges faced in applying nuclear energy comes from the high 'capital cost involved in setting up large power reactors. Hence the development small nuclear reactor would be a solution. A nuclear reactor produces its energy through nuclear processes that involves an understanding of the neutron interaction at microscopic level. A model of the nuclear reactor from the perspective of reactor physics is created by using MCNP4C software that could precisely and accurately predict the neutronic behaviour, specifically, the criticality of the reactor.

U-Battery

The term U- Battery is used to refer to a small High Temperature Reactor (HTR) and emphasizes on its special characteristic of long – life core, transportability and inherent safety [1]. The basic parameters of the U – Battery are suggested as below:

Parameters	Value
Reactor type	Block Type HTR
Energy conversion system	Rankine steam cycle
Fuel type	UO_2, PuO_2, TRISO particle
Fuel enrichment	< 30% wt% U – 235
Thermal Power	~20 MW
Inlet Helium Temperature	490 °C
Outlet Helium temperature	850 °C
Core lifetime	5 – 10 years
Number of control rod	7
Geometric parameter:	
- Height	2.70 m
- Diameter	3.24m
Coolant	Helium gas, CO_2 gas, FLiBe molten salt

Table 1. Basic parameters of the U-Battery. [1]

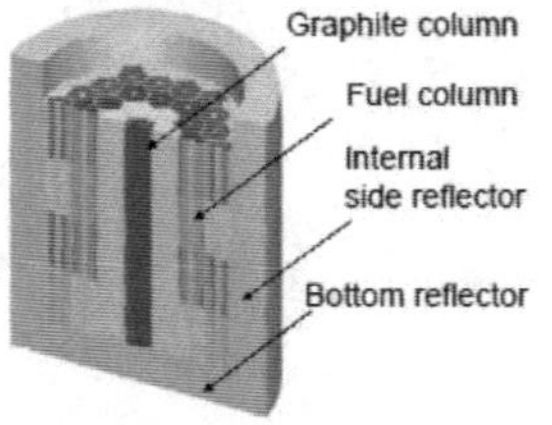

Figure 1. Rough layout of the U-Battery reactor core [1]

IOP Conf. Series: Earth and Environmental Science **16** (2013) 012138 doi:10.1088/1755-1315/16/1/012138

2. Review on reactor core material.

The criticality of a reactor core is closely related to the material type and composition of the core. To model the reactor inside MCNP requires research on detail materials of each section inside the core including the percentage and composition of all molecules for each material.

Parameters variation for the test

The U-Battery was simulated for different enrichment level of uranium oxide fuel, from natural uranium of 0.7204% enrichment to weapon grade of 90% enrichment. Altogether, 72 criticality calculations had been made.

Table 2 Models of U-Battery

Parameters	1	2	3	4
TRISO coating	SiC	ZrC		
Coolant	CO_2	He	F-Li-Be	
Control Rod	Void	Ag-In-Cd	B_4C	HfC_2

Neutron capture cross section

The criticality of a reactor core would highly depend on the value of neutron cross section of its materials. It is define as the capability to capture neutron by a nuclei. The figure below shows some materials that have very high thermal neutron capture and mostly used as material in a reactor core control rod, and wall.

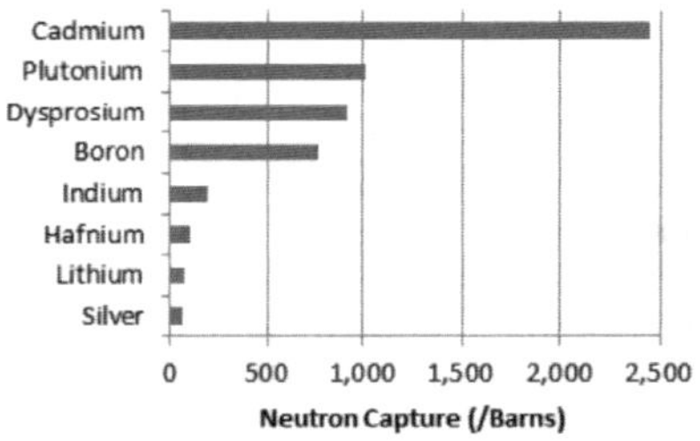

Figure 3. Thermal neutron capture of some material in reactor core [4]

3. Modelling the U-Battery using MCNPv4C software

The MCNP input files describes the problem geometry, specifies the materials and source, and defines the results you desire from the calculation. The geometry is constructed by defining cells that are bounded by one or more surfaces. Cells can be filled with a material or be void. Figure 4 shows the input file structure.

Geometry modelling - Cell card and surface card

The cell card defines the geometrical regions from all first and second degree surfaces and fields, and then combines them with Boolean operators. Next, the surface card defines the surface by with coefficients and constants of the first and second degree equations. Usually, to ease the modelling process, the first step is to write the surface cards, as this would build the outline structure of the model. Only then would be to complete the cell cards, where volume of spaces bounded by surfaces. Lastly would be to specify the materials and calculating the criticality.

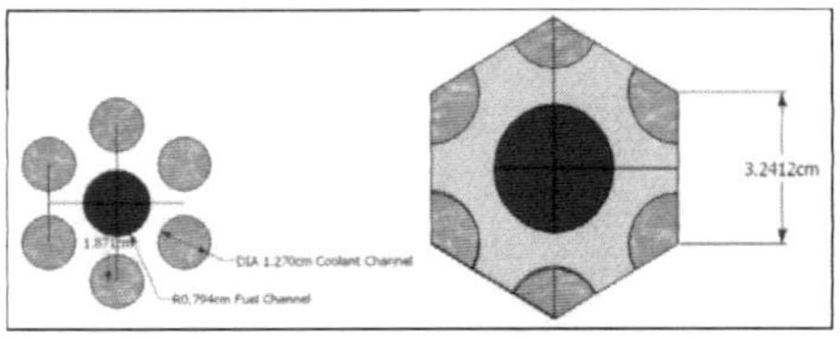

Figure 4: Hexagonal assembly lattice.

Assume all channel form a hexagonal lattice with one coolant in centre and six fuel channels.

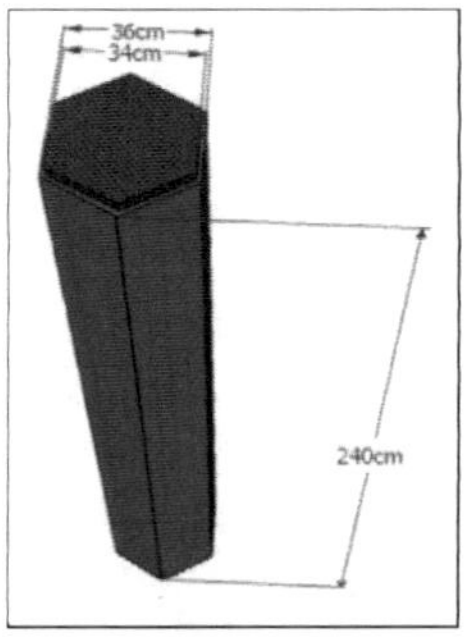

Figure 5: Assembly

Six 40cm height are stacked together, where in MCNP, it is considered as one stack of height 240cm

IOP Conf. Series: Earth and Environmental Science **16** (2013) 012138 doi:10.1088/1755-1315/16/1/012138

3.1.1. Assembly. Consist of hexagonal graphite block which have multiple holes where coolant channels, fuel rods and poison channels are placed.The assembly would then be repeated to form the reactor core and limited by cylindrical surface. Graphite reflectors are placed at assemblies at the edge that were cut by the cylindrical surface and the upper and bottom part.

3.1.2. TRISO particles. The TRISO particle contains uranium dioxide surrounded by different layers of graphite. It is placed in the fuel rod. In real situation, TRISO particles are random among the graphite matrix. Since it would be really difficult to place randomly in MCNP, the TRISO are placed in regular lattice where each coated particle take up one cell.

3.1.3. Burnable poison channel. Consume neutron to maintain the reactivity constant above one. At the beginning life of the reactor, the reactivity is needed to reduce. Hence, the capacity to consume neutron are reduce and eventually, near the end on reactor life, they do not absorb neutron and allow the remaining reactivity in the fuel to drive the reactor.

3.1.1. Control rod. Consists of boron or neutron absorber that would control the reactivity during shutdown, or to increase in reactivity when overpower. Control rod is located in the middle of the fuel assembly; hence, a new type of assembly with control rod is defined. The fuel rod and coolant channel that were cut by the control rod were replaced with graphite.

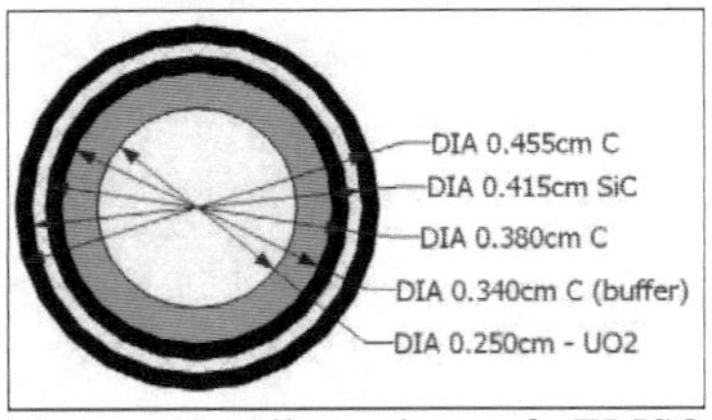

Figure 6: Illustration of TRISO particle with dimension.

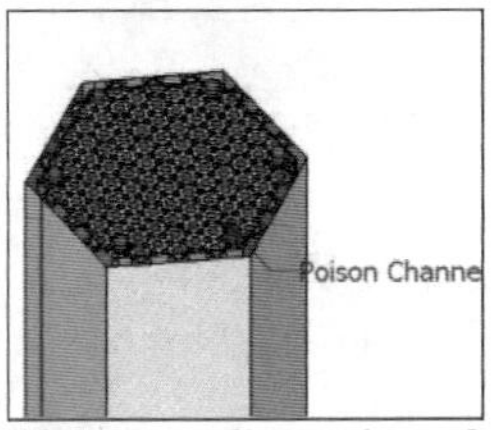

Figure 7: Illustration of assembly with poison channels

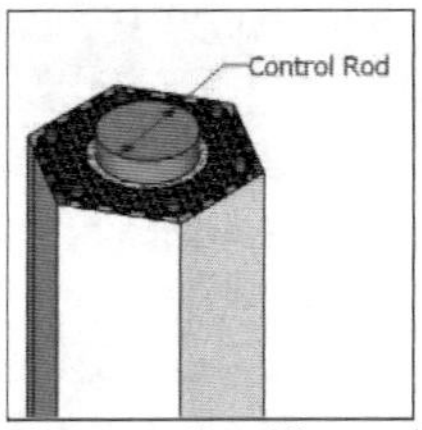

Figure 8: illustration of assembly with control rod

Data card

Would usually the lengthiest is the source card which consists of the source descriptions, tally instructions, cross sections and material composition.

3.1.2. Material. Based on the geometry defined, the materials were filled and research on the criticality of the U-Battery is done by varying the materials. In order to generate model of the materials in MCNP, the three basic information needed to be calculated are:

a) All the fraction atoms available in the material including different isotopes and fragment nuclide. This need to be calculated
b) ZAID – Atomic number followed by atomic mass and the extension based on ENDF/B-VI data library of MCNPv4C
c) Nuclide atom fraction (atom/b-cm)

Table 3: Description of the MCNP input code

MCNP code	m1	92238.61c	1.692-4	92235.61c	2.30-02	8016.60c	4.64-02	$ 0.7204%
Description	Material number	ZAID U238	Atomic Density	ZAID U235	Atomic Density	ZAID O	Atomic Density	Percentage Enrichment

The procedure to produce material card shown above is applied for all other material of the core

3.1.3. Criticality. The data is submitted to MCNP using the "kcode" where information the Number of cycles and number of particles per cycle to be started in eigen-mode calculations. Then the "ksrc" would specify the initial spatial fission distribution. A fission source point will be placed at each point with coordinates, X, Y, and Z.

```
C  Criticality Control Cards
kcode 5000 1.0 50 250
ksrc   1.6206  0.935650 1.6206 -0.93565 0 0  1.87130
       -1.6206 -0.935650 -1.6206  0.93565 0 0 -1.8713 0
```

Figure 9. The criticality control card.

4. Results and discussions

A few criticality output from the simulation were validated using the SCALE 5.1 software by [1] M Ding and J.L. Kloosterman, 2010. The results for the criticality on U-Batteri with the same core configuration with same enrichment level were compared. Since SCALE produces results of the criticality of the full life cycle of the burnable fuel, the result from the MCNPv4C are compared with the results of year 0 of the effective full power year (EFPY).

Table 4: Comparison result from SCALE5.1 and MCNPv4C

U235 Enrichment	20%
K_{eff} MCNPv4C	1.1162
K_{eff} SCALE 5.1	1.162
Dev (%)	3.774

From the results above, the deviation are acceptably high. The results of k_{eff} from varying material are shown in figure 10 – 12.

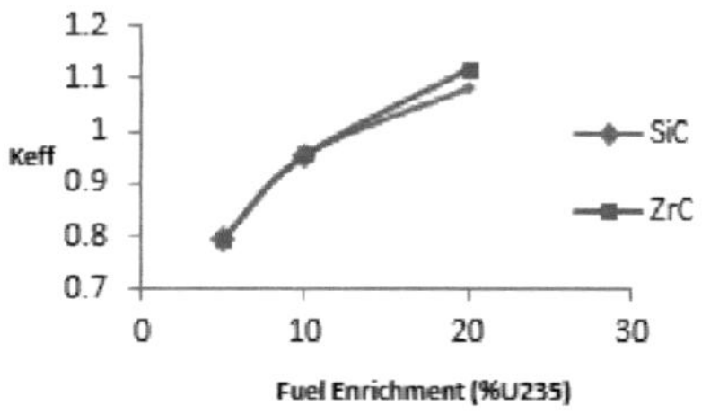

Figure 10. Graph of k_{eff} against enrichment level for different coating material

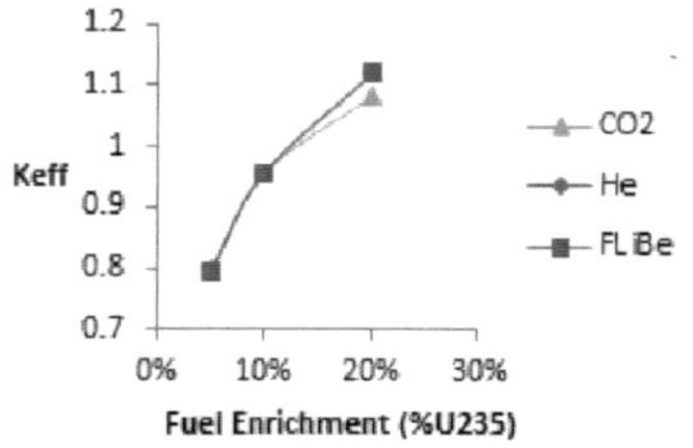

Figure 11. Graph of k_{eff} against enrichment level for different coolant

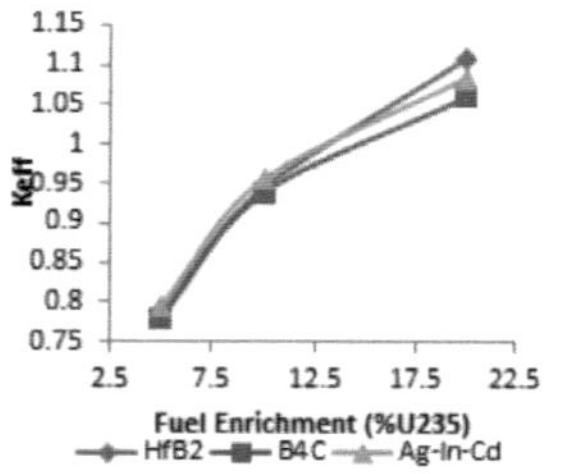

Figure 12. Graph of K_{eff} against U235 enrichment for different control rod material at closer view

Conclusion

TRISO coating material should have the characteristic of low thermal neutron cross section in order not to waste the neutron produce hence reduces the efficiency. The test results that shows little difference when varying the commonly SiC to ZrC show a positive results for the criticality, besides its physical advantages.

All three coolants would be good options in terms of neutronic properties where there are insignificant effects on criticality. Both gas type and even liquid salt have little effect on criticality, hence, in order to optimise the design of the reactor, only parameters regarding the heat transfer and other physical properties needed to be considered.

Finally, by applying control rods to the reactor core would significantly lower the criticality. With different type of control rods, the magnitudes of neutron control are very similar. A good control rod material should have the ability to decrease as much k_{eff} value as possible. From the model, B4C control rod is the best neutron absorber as proven by [4]

References

[1] A Ming D. and J. L. Kloosterman, 2010, Parametric Neutronics Design of a Small and Long-Life HTR, Proceedings of HTR 2010, p113

[2] Richard S, 2011, Reactor Core Design, MACE30632 Nuclear System, University of Manchester, unpublished

[3] Charles D H and et al, 2009, Criticality Calculations with MCNP 5: A Primer, Second Edition, Los Alamos National Laboratory (LANL).

[4] V D Risovany, E. E. Varlashova, D. N. Suslov, 2000, Dysprosium Titanate as an absorber material for control rod, Journal of Nuclear Material 2000, p84-89

Design of sample carrier for neutron irradiation facility at TRIGA MARK II nuclear reactor

Y Abdullah[1], N A Hamid[2], M A Mansor[2], M H A R M Ahmad[1], M R Yusof[1], H Yazid[1] and A A Mohamed[2]
[1]Malaysian Nuclear Agency, Bangi, Kajang 43000 Selangor, Malaysia
[2]Centre for Nuclear Energy, College of Engineering, Universiti Tenaga Nasional, 43000 Kajang, Selangor, Malaysia

E-mail: yusofabd@nuclearmalaysia.gov.my

Abstract. The objective of this work is to design a sample carrier for neutron irradiation experiment at beam ports of research nuclear reactor, the Reaktor TRIGA PUSPATI (RTP). The sample carrier was designed so that irradiation experiment can be performed safely by researchers. This development will resolve the transferring of sample issues faced by the researchers at the facility when performing neutron irradiation studies. The function of sample carrier is to ensure the sample for the irradiation process can be transferred into and out from the beam port of the reactor safely and effectively. The design model used was House of Quality Method (HOQ) which is usually used for developing specifications for product and develop numerical target to work towards and determining how well we can meet up to the needs. The chosen sample carrier (product) consists of cylindrical casing shape with hydraulic cylinders transportation method. The sample placing can be done manually, locomotion was by wheel while shielding used was made of boron materials. The sample carrier design can shield thermal neutron during irradiation of sample so that only low fluencies fast neutron irradiates the sample.

1. Introduction

The major infrastructure set up by Nuklear Malaysia includes a nuclear research reactor known as Reactor TRIGA PUSPATI (RTP) which is the only research nuclear reactor in Malaysia. The RTP is a 1MW TRIGA Mark II nuclear reactor that was established in 1983. It is a pool type reactor and has four horizontal beam ports that are currently used for neutron imaging and scattering [1] wherein the neutron irradiation at reactor core incorporates activities for thermal neutron and γ-radiation studies and applications of these include Neutron Activation Analysis (NAA), radioisotope production for medical, industrial and agricultural purposes. Irradiation is a process where an object or sample is exposed to radiation and is usually carried out in the irradiation chamber of the RTP. Presently, there is no such sample carrier or transport system to transfer the sample for irradiation experiments at beam ports. In general, the reactor core in RTP produces neutron with total flux at around 10^{13} n/cm^2/s. Neutron carries no charge and can penetrate materials easily without being much affected by electromagnetic force. The types of neutron produced fast neutron (energy : >~ 1 MeV), intermediate neutrons (in between fast and thermal) and thermal neutron (~ 0.025 eV) [2]. Thermal neutron is usually used to study material physical properties whereas fast neutron can be used to induce physical defect in materials and this is important in research of electronic materials. Therefore, fast neutron

IOP Conf. Series: Earth and Environmental Science **16** (2013) 012139 doi:10.1088/1755-1315/16/1/012139

irradiation to material can change the electronic properties but high fluencies of fast neutron can cause irreparable damage to the electronic materials. The beam port is more suitable place to irradiate the material as fast neutron fluencies is considerably low. The irradiation experiment thus requires special sample carrier that would only allow fast but exclude other neutrons irradiating the sample and will be carried out at beam port instead of reactor core. Here we discuss a conceptual design of the carrier.

2. Methodology

The concept generation of this work was done with the aid of building morphology [3, 4]. The first step of building morphology is decomposing the function. The functions in this work was identified as casing shape, transportation method, input source, locomotive, material and shielding. Next for each function that is identified concepts are developed. After doing these two steps then a combination of the individual concepts into complete conceptual designs can be done. The sample carrier design was based on a hydraulic cylinder transportation method by means of electrically driven and secured with a cylindrical shaped casing equipped with wheels. The advantages of this design are ease of manufacturing and fast delivery system with shorter travel time. The sample holder material considered in this design is aluminium. The shielded material considered is a layer of boron materials whereas sliding cylinder materials considered are stainless steel to carry the sample load. The sliding cylinders can extend to 3 meters. The hydraulic pumps are needed to operate the hydraulic cylinder. The sample is mounted in the holder at the far end of the sliding cylinders and this holder can be accessed in the shaft centre tube [5] with the cylinders retracted. The design of the sample carrier system was done using Pro Engineer and simple stress analysis was carried out using Pro Mechanica software. The sample carrier with proper dimensions was designed using the Pro Engineer Wildfire [6].

3. Results and Discussion

Figure 1 and Figure 2 show the Pro Engineer design drawings for the sample carrier system with images shown not to scale. The modification made to sample carrier design consists of BOSCH REXROTH ISO Cylinders of 3 meters long, cylinder casing was extended to 1.5 meters and legs (wheels) are added to stabilize the extension of cylinder casing. The sample carrier designed made by stainless steel and coated with boronated layer to add shielding to the system as neutron shielding during irradiation exposure. Composites of boron or boronated carbide (B_4C), is considered to be used for this shielding material. The sample carrier consisted of six steps to be operated such as follow: sample placing into the sample carrier, setting timer for the duration of sample to be irradiated, timer signaling the pneumatic pump to extend the pneumatic sliding cylinder, the irradiation starting of irradiation process and the timer switch signals the pneumatic pump to return the pneumatic cylinder and finally irradiated sample can be collected from the sample carrier.

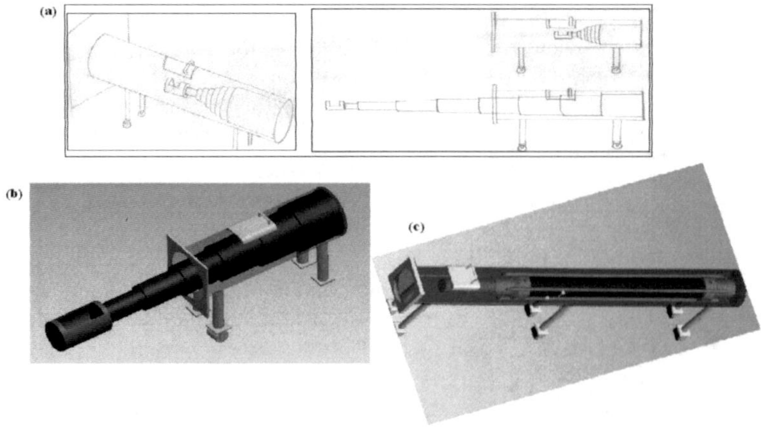

IOP Conf. Series: Earth and Environmental Science **16** (2013) 012139 doi:10.1088/1755-1315/16/1/012139

Figure 1. (a) The concept design of an automatic sample carrier (b) CAD Model Concept (c) Showing BOSCH REXROTH ISO Cylinder

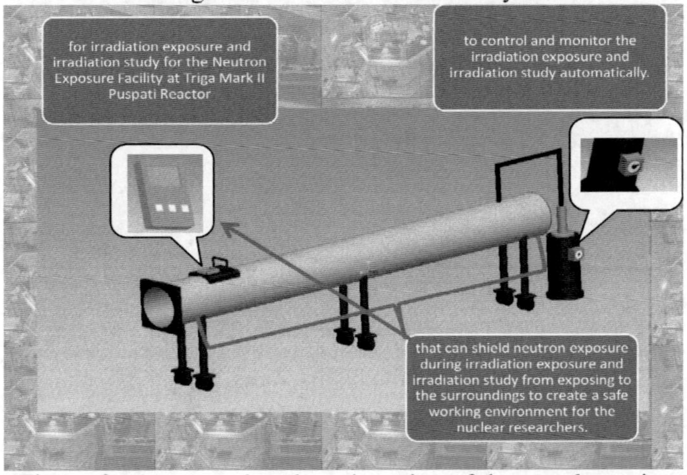

Figure 2. Automated cycle and monitor of the sample carrier

A simple stress analysis was done to preliminary determine the stress and strain in the materials or structure to forces of load both in static or dynamic situations. Figure 3 shows the results of the stress analysis and stress analysis graph indicated the outcome of the stress analysis. This simple stress analysis can be used gauge whether the structure may behave desirably under the prescribed loading.

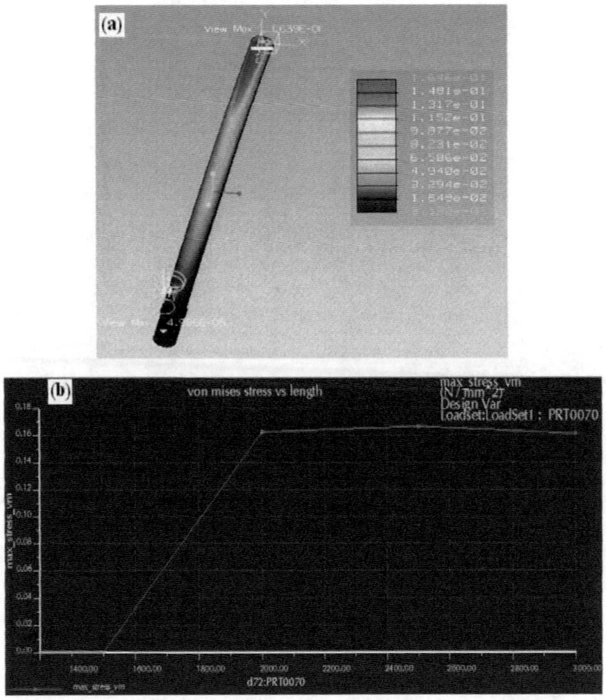

Figure 3. (a) Stress analysis (b) Stress analysis graph

This was done on the extending rod of the system where the load will be placed. The sliding cylinders are assumed as one long cylinder to simplify the stress/strain analysis. The desired loading weights needed for this design are roughly around the weight of 10 grams but during the stress analysis, it was run with a load of 1 kg to test the system to a possible limit and it was found to withstand the 1kg of load. The stress analysis results obtained indicates that the system does not fail after the given force was exerted. The model however does not include the whole conceptual design as this is more complicated to modelled and analysed.

4. Conclusion

The design developed here is a sample carrier for neutron irradiation study, automatic control and safe operation from the neutron radiation and it can be said that this conceptual design facility can be very useful to neutron irradiation process experiment. This facility can contribute to the safe arrays of others apparatus that can be used by researchers at RTP and promote the growth of neutron irradiation studies in Malaysia. Further works may include better modeled of strain and stress that may also include the load of sliding cylinders on the steps shielding inside the beam port.

Acknowledgement

This work was supported by the International Atomic Energy Agency Coordinated Research Project (CRP) under IAEA-CRP No. 17087, Collaboration with Universiti Tenaga National (UNITEN) and PQRD funds from Malaysian Nuclear Agency Project No. NM-R&D-11-20.

References

[1] Nuclear Malaysia 2008 Radiation Safety Kajang: MOSTI.
[2] Lamarsh J R 1975 *Introduction to Nuclear Engineering* **3** London: Addison-Wesley.
[3] Shigley J E, Mischke C R and Budynas R G 2004 *Mechanical Engineering Design* **7** New York: Mc Graw Hill.
[4] Mansor M A 2013 Design of sample carrier for neutron exposure facility at TRIGA Mark II PUSPATI Reactor, Degree of Bachelor of Mechanical Engineering, Universiti Tenaga Nasional, Kajang, Malaysia.
[5] Profio A E 1979 Radiation Shielding and Dosimetry *Toronto: John Wiley & Son, Inc.*
[6] Ullman D G 2009 *The Mechanical Design Process* **4** New York: Mc Graw Hill.

IOP Conf. Series: Earth and Environmental Science **16** (2013) 012140 doi:10.1088/1755-1315/16/1/012140

Impact of radiation exposure on mechanical and superconducting properties of Bi-2212 superconductor ceramics

A A Rahman, N A Hamid[1] and M S N Asbullah
Center for Nuclear Energy, College of Engineering, Universiti Tenaga Nasional, 43000 Kajang, Selangor, MALAYSIA

E-mail: Nasri@uniten.edu.my

Abstract. In the last few years, rapid improvements have been made to improve the quality of high-temperature superconductors. Amongst the high temperature superconductors, the Bi-based (BSCCO) consists of interest for various applications. $Bi_2Sr_2CaCu_2O_8$ (Bi-2212) have been used to make superconducting tapes and wires. Unlike conventional compound superconductors, the critical current, I_c of oxide superconducting tapes in the elastic strain is generally almost constant and degrades suddenly when it is subject to mechanical force by a strain beyond the limit. In this research, the Bi-2212 samples were prepared by solid state reaction method. Precursors oxide powders were pressed to pallets under hydrostatic pressure around 7 tons or 70 000 psi and then sintered at temperature of 850°C for 24 hours. The effect of radiation before and after irradiation on mechanical and superconducting properties of the samples was studied. Irradiation was carried out with a beam of 3 MeV, current of 10 mA and radiation dose of 100 and 200 KGray. The x-ray diffraction analysis is used to verify Bi-2212 phase. The samples were also characterized through electrical properties by using the four-point probe method. The microstructure of the samples was studied by using the scanning electron microscopy (SEM), and compression test was also conducted using the stress-strain relationship. The phase structure and electrical properties of the samples degrade slightly with irradiation exposure. Nevertheless the microstructure showed that when initial electron radiation dose was increased up to 100 kGray, the grain growth, texture and core density improved slightly but the grain growth, size and core density begin to deteriorate after the electron radiation dose is increased to 200 kGray. This may be due to the formation of larger size defects within the microstructure of the Bi-2212 phase as the radiation dose increases.

1. Introduction

Several research groups have been using irradiation techniques such as neutron irradiation to introduce larger defects in superconducting material and consequently enhanced the flux pinning properties of the material [1-2]. In another report, it indicates that the electron irradiation increased the normal state resistance and decreased the critical temperature (T_c) of bismuth-based superconductors [3]. The report shows that high-energy ($E \approx 3$ MeV) electrons at fluencies up to 1.8×10^{16} cm^{-2} increase the critical currents (I_c) at low temperatures and this phenomena was attributed from the changes in the concentration of charge carriers and the pinning centers in the lattice structure, especially in the CuO_2 planes. According to Krutzler et al. [4] in their study on MgB_2 doped with carbon, the defect structure

[1] To whom any correspondence should be addressed.

introduced by neutron irradiation provides a more efficient pinning structure than defects created by carbon doping. Zehetmayer et al. [2] in their study on Hg-based high temperature superconductor discovered that the effect of neutron irradiation is more pronounced at elevated temperature due to large defects created by the irradiation. Furthermore, Aoki et al. [1] found that the superconducting characteristics of superconducting tapes that they studied did not deteriorate under irradiation.

In general, the properties of Bi-2212 superconductor can be studied by analyzing the phase purity, crystal structure, composition, and defect structure of the materials. In view of these, the aim of this paper is to investigate the impact of electron irradiation exposure on the mechanical and superconducting properties of Bi-2212 superconductor ceramics. From literature reviews, very few studies on mechanical properties of superconductor under electron irradiation exposure have been conducted. Investigation also includes the study on the effect of irradiation dose on the mechanical and superconducting properties of the superconductor ceramics. Improvement in the microstructure is usually associated with enhancing the connectivity of the adjacent grains within the superconducting grains and as such should improve the mechanical strength and superconducting characteristics of the ceramics. Nevertheless, there is also possibility that the superconducting properties remained intact but the mechanical property of the ceramics deteriorated with irradiation exposure.

1. Experimental procedures

$Bi_2Sr_2CaCu_2O_8$ (Bi-2212) superconductor ceramics in powder form used in this work was the product of Sigma-Aldrich, USA. Preparation of bulk Bi-2212 samples was carried out using the conventional solid-state reaction method where the precursor's oxide powder was palletized and then sintered to achieve the required Bi-2212 phase. The powder was thoroughly mixed and grounded in an agate mortar. After the initial grounding process, the compound went through annealing process at temperature of 800 ^{o}C for 12 hours. The powder was then pressed into pallets under hydrostatic pressure around 7 tons or 70 000 psi, and then sintered at temperature of 855 ^{o}C for 24 hours. Electron irradiation on the samples was carried out using a beam of 3 MeV, current measurement of 10 mA and radiation doses of 100 KGray and 200 KGray, respectively. The irradiation was conducted using the EPS-3000 electron-beam accelerator at the Malaysian Nuclear Agency. Structural investigations of non-irradiated and irradiated samples were conducted using the Bruker D8 Advanced X-Ray Diffractometer (XRD) to confirm their phase formation. Microstructure investigation was carried out using the Hitachi S3400N Scanning Electron Microscope (SEM). The resistivity measurements at cryogenic temperatures were measured using the four-point probe method. Furthermore compression test on the samples at room temperature was also performed to study the plot of stress-strain curve and the mechanical property of the superconductor ceramics was investigated.

2. Results and analysis

Figure 1 shows the XRD patterns of the superconducting specimens of non-irradiated and irradiated specimens. Sharp peaks of (002), (008) and (115) that correspond to the Bi-2212 phase appear in the XRD patterns. The XRD patterns for both the non-irradiated and irradiated of Bi- 2212 superconductor ceramics show a well-defined peak all of which could be indexed on the basis of a Bi-2212 structure. The (002) peak which is the characteristic of the Bi-2212 phase, can be observed clearly in all the samples. A few peaks of unknown phases and other impurities are also observed. The intensity of the (002) peak is highest and sharpest for sample with dosage of 100 kGray. This indicates that small dosage of electron irradiation increased the volume fraction of Bi-2212 phase and enhanced the degree of crystallinity of the superconductor ceramics. But as the dosage is increased to 200 kGray, the (002) peak starts to deteriorate. Nevertheless the phase structure of Bi-2212 is still maintained. The presence of unidentified peaks is most probably due to inappropriate heat treatment during the sintering process which resulted in unidentified secondary phases.

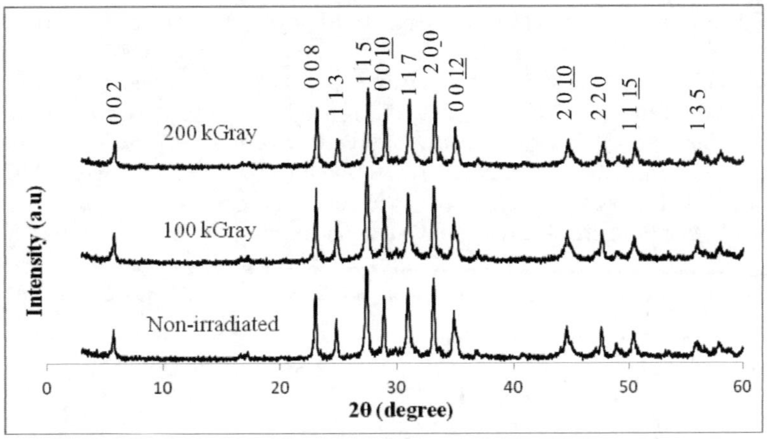

Figure 1. The XRD patterns of non-irradiated and irradiated Bi-2212 superconductor ceramics.

The SEM micrographs in figure 2(a) and figure 2(b) show the microstructure of Bi-2212 phase of non-irradiated and irradiated samples. The size of the Bi-2212 platelets in both samples remained about the same. Nevertheless micrograph of sample exposed to electron irradiation shows more random grain orientation. In addition, the SEM microstructure shows higher porosity and the orientation of the grains is much more perpendicular to the c-axis. With such microstructure, both the mechanical and superconducting properties of the sample are expected to be affected.

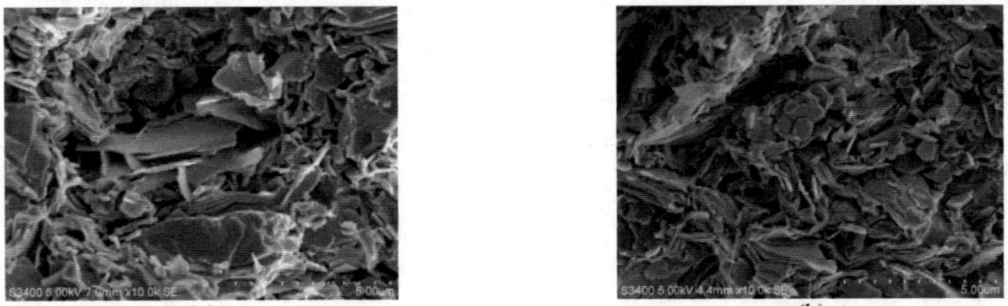

(a) (b)

Figure 2. SEM micrographs of (a) non-irradiated (b) irradiated Bi-2212 superconductor ceramics.

The resistivity measurements were used to determine the transition temperature (T_c) of the Bi-2212 superconductor ceramics. Table 1 shows the values of T_c for both non-irradiated and irradiated samples. As the irradiation dose increases, the value of T_c decreases. The decreasing trend is attributed to the change in oxygen contents as described by Bódi et al. [5]. The change in oxygen contents is the result of interaction between the electron particles and the oxides within the Bi-2212 superconductor ceramics.

Table 1. Transition temperature (T_c) of non-irradiated and irradiated Bi-2212 samples.

Sample	T_c (K)
Non-irradiated	107
100 kGray irradiation	92
200 kGray irradiation	87

The relationship between the microstructure and the strength of the Bi-2212 superconductor ceramics is useful to determine the effect of irradiation on the mechanical property of the samples. Table 2 shows the maximum strength of both non-irradiated and irradiated samples at room temperature. The results show that the irradiated samples have similar trend of deformability. In such case, fractured starts to occur at about 17.83% for the 100 kGray irradiated sample, and 17.24% for the 200 kGray irradiated sample. The high porosity microstructure as shown in the SEM micrographs of the irradiated samples resulted in weaker interface bonding between the Bi-2212 superconducting grains. This information is very useful since the mechanical properties such as stiffness, strength and toughness at room temperature are able to predict such behaviors at cryogenic temperature [6].

Table 2. Maximum strength of non-irradiated and irradiated Bi-2212 samples.

Sample	Maximum strength (MPa)
Non-irradiated	120
100 kGray irradiation	87
200 kGray irradiation	96

3. Conclusion

In this study, it was found that the mechanical property of Bi-2212 superconductor ceramics is affected by electron irradiation doses. The higher grain orientation and porous nature of the microstructure of irradiated samples contributed to the weak interface bonding between the superconducting grains. The strength of irradiated samples decreases when exposed to electron irradiation and eventually deformed. On the other hand, electron irradiation does not have significant effect on the superconducting properties of the samples. The Bi-2212 phase structure is maintained even though the T_c has slightly lowered due to decrease in oxygen contents. This study indicates that the electron irradiation deteriorate the mechanical strength of Bi-2212 superconductor ceramics.

Acknowledgements
This project is completed under collaboration with the Malaysian Nuclear Agency.

References

[1] Aoki T, Ueda H, Ishiyama A, Miyahara N, Kashima N and Nagaya S 2011*IEEE Trans. Appl. Supercond.* **21** 3200
[2] Zehetmeyer M, Eisterer M, Sponar S, Weber H W, Wisniewski A, Puzniak A, Panta P, Kazakov S M and Karpinski J 2004 *Physica C* **408-410** 30
[3] Ogun S E, Goktas H, Ozkan H and Gasanly N M 2005 *Surface & Coating Technology* **196** 118
[4] Krutzler C, Zehetmeyer M, Eisterer M, Weber H W, Zhigadlo N D and Karpinski J 2007 *Physica C* **460-462** 555
[5] Bódi A C, Csikai J, Pepelnik R, Halász I, Kirschner I, Kovács G, Porjesz T and Träger T 1990 *Physica C* **165** 83
[6] Hamid N A, Shamsudin N F and See K W 2009 *Mater. Res. Innov.* **13** 379

IOP Conf. Series: Earth and Environmental Science **16** (2013) 012141 doi:10.1088/1755-1315/16/1/012141

Microstructural investigations of materials for low temperature co-fired ceramic (LTCC) based fuel cell using small angle neutron scattering

A A Mohamed[1*], M H Ahmad[2], A Ibrahim[3], A Azman[2], R Alias[3], Z Ambak[3], S Shapee[3], E G Putra[4], A Patriati[4], M A Sharom[5], H Yazid[2], M R Mamat[2], J A Karim[2], F M Idris[2], K Yazid[1] and M R Zin[2]

[1]Centre for Nuclear Energy, Universiti Tenaga Nasional, 43000 Kajang, Selangor, Malaysia
[2]Malaysian Nuclear Agency, 43000 Bangi, Selangor, Malaysia
[3]Telekom Malaysia R&D, 63000 CyberJaya, Malaysia
[4]Neutron Scattering Laboratory, BATAN, Serpong, Indonesia
[5]Faculty of Science, Universiti Malaya, 50603 Kuala Lumpur, Malaysia

*E-mail: azizm@uniten.edu.my

Abstract. The concept and the realization fuel cell based on LTCC technology require the investigations of fired LTCC microstructures. The majority of the works involved using small angle neutron scattering studies on the microstructural of LTCC ceramic tape and development of neutron tomography for future tool to visualize channels inside the fired tape. Most SANS characterization were carried out at Smarter SANS instrument at BATAN, Indonesia. Standard sample for resolving tens of micron of object size were measured using simple neutron tomography setup utilizing monochromatic SANS beam at Malaysian Nuclear Agency. The initial microstructural findings indicates that organic additives shape the final microstructural of LTCC after firing with the glassy material possibly fill the space left by the burned organic additives. The tomography results showed that 40 micron size object can be differentiated. The conductor deposited on LTCC is preliminary investigated which will later be used as support for catalyst.

1. Introduction

Low temperature co-fired ceramics (LTCC) technology is a ceramic based electronics packaging technology. The technology allow circuits to be constructed within a ceramic block substrate that enables three-dimensional integration of passive elements such as resistors, inductors and capacitors for a complete electronic circuit [1]. The materials for LTCC microelectronic packaging need to meet stringent requirements in term of electrical, thermal and mechanical properties. Commercial available LTCC systems are glass+ceramic and glass-ceramic.

In the glass-ceramic system, crystallizable glasses were used. The degree of crystallization is crucial to be controlled because it will affect the final properties of tape (mechanical and electrical). Most crystallizable glass type LTCC systems are based on cordierite glass-ceramics. On the other hence, glass+ceramic is a mixture of high melting ceramic fillers and low softening glass and fired at elevated temperature. The properties of this type LTCC were controlled by the ration of glass to the ceramics and the individual properties of the mixtures. Typical systems are borosilicate + alumina and lead borosilicate + alumina. Apart from electronics, because of the modular structure of LTCC, it is possible that LTCC can be used as a fuel cell unit. The first part of the investigation is to look onto the microstructural of LTCC using small angle neutron scattering (SANS) technique and developing our own tomography system for fuel imaging.

2. Experiment

2.1. Sample preparations

LTCC tape sample was prepared by combination alumina, Al_2O_3 (Inframat) as filler and commercial borosilicate glass (Asahi Glass Co., LTD). The physical characteristics of the glass and alumina are shown in Table 1 as supplied by the manufacturer. The weight percentage of filler and the addition of titanium isopropoxide were varies in order to studied their effect to the physical properties of the develop tape. Table 2 shows the composition of each sample.

Oil was used as dispersant and an organic acid was used stabilizer in the mixture of ceramic particles and organic additives. Details of preparation and composition cannot be revealed because of proprietary reason. Slurries of ceramic particles with organic solvents were prepared and casted on a tape casting machine (KEKO equipment). The thickness of the tape was set to 0.5 mm. The green body of tape was then dried at 30 °C before it was cut for 5 to 8 layers and laminated at 21 MPa and 70 °C respectively. The laminated samples were then fired at temperature in the range of 850 to 1000°C for 10 minutes in a temperature-controlled tube furnace (Carbollite). This prepared tape was coded as batch S-45.

2.2. Characterizations

All samples were characterized in terms of their physical properties such as density, crystalline phase, surface morphology, dielectric and thermal properties. The thickness of the tape was measured using a vernier caliper. The laminated and fired density was determined using the standard relation of mass & volume. The crystalline phase of the tape was identified using a Bragg-Brantano XRD technique (PANalytical) with Cu anode. The instrument was operated at a power rating of 40 kV and 40 mA, scanning parameters of 2θ ranging from 20 to 80° with typical step size of 0.02° and at a measuring time of 25s per step. The microstructure and surface morphology was analyzed from field enhanced scanning electron microscope (FE-SEM) image (FEI Nova NanoSEM). Commercial silver paste (Ferro) was printed on surface of and it was fired at temperature in the range of 850 to 1000°C. The compatibility of paste is evaluated trough observation and conductivity test.

Measurements were carried out using the 36 m SANS instrument (SMARTer) at Neutron Scattering Lab (NSL) at BATAN (Serpong, Indonesia). A small circular sample area of 10 mm diameter was exposed to the neutron beam for 0.5 or 1 hour. The scattered elastic part, due to inhomogeneities in the sample microstructures, was registered on a two-dimensional (2D) position sensitive detector (128 x 128 pixels). The efficiency of the detector was calibrated using water as scattering standard. The scattering data were corrected for empty beam, background counts and detector efficiency, and then radially summed. No neutron tomography measurement was carried out on the samples but a neutron tomography simple system was evaluated to see whether such system can differentiate micron size structures for later use on the samples.

3. Results and discussion

Table 1 shows the physical properties of the sintered tape at 900°C. The tape shrank in the z-direction (thickness) up to 29% giving a final thickness of about 70 micron. The shrinkage in x- and y-direction is quite significant but did not cause the tape to bend. The fired tape remained flat. The material is mostly crystalline as concluded from the XRD pattern obtained [3].

Figure 1 a) shows the SEM image on the surface of the tape. Most pores seemed to be covered by glass material that melted during sintering process. SANS scattering profile indicates that the existence of oriented structure shown by the non-isotropic scattering [2]. The intensity versus scattering vector (Figure 2) shows small bump at 0.03 $Å^{-1}$. The formation of this structure was believed to be caused by melting glass that replaced the space left behind when organic additives with oriented structure burned during sintering.

Table 1. The physical properties of the sintered tape fired at 900°C.

Specification	**S-45 (2011)**
Thickness (green), mm	0.1
Shrinkage x, y (%) ±0.3*	17
Shrinkage z (%) ±0.5*	*29*
Dielectric constant	5.47
Dissipation Factor @ 1MHz (Max)	2.6 x 10^{-3}
Density (g/cm^3)	2.2

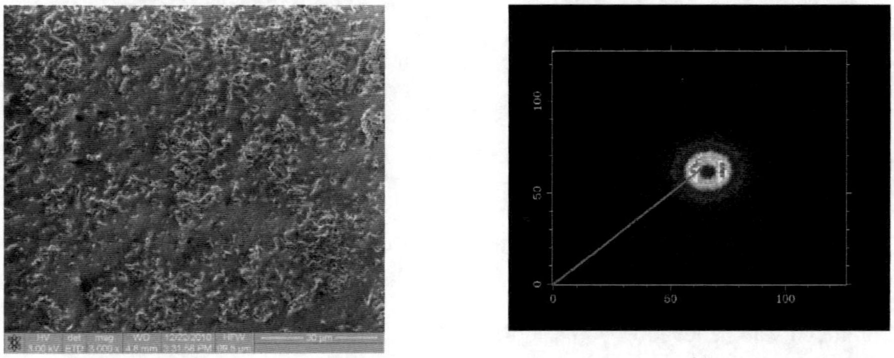

a) b)

Figure 1. a) SEM image of tape sintered at 900°C and b) scattering profile at 400cm sample to detector position (wavelength = 3.9Å)

It is not believed that this non-isotropic scattering was caused by artifacts of Smarter instrument. This oriented structure findings needs to be compared with other SANS instrument for confirmation and the Malaysian Nuclear Agency's SANS instrument (mySANS) can be used. The mySANS is now almost fully operational after a lot of repaired and refurbished done. It is now has a new data acquisition system based on IgorPro [4] and direct beam shows homogenous distribution (Figure 3). Initial investigation on the sample are hampered by reactor operation time as the sample need to be irradiated for a long time to match the experiment at Smarter which has a much higher flux.

In the case of the electrode material, the silver paste when applied on the surface of the tape did not evaporate after firing. The paste was found to be well deposited on the surface and silver conductivity not affected.

Neutron tomography was not used to study the sample yet. Most works involved on imaging standard sample to evaluate the performance of the tomography setup. CCD camera was used for the imaging (Photonic Science). This involved resolving object with size in micron. Standard samples were provided by Paul Scherrer Institute (PSI).

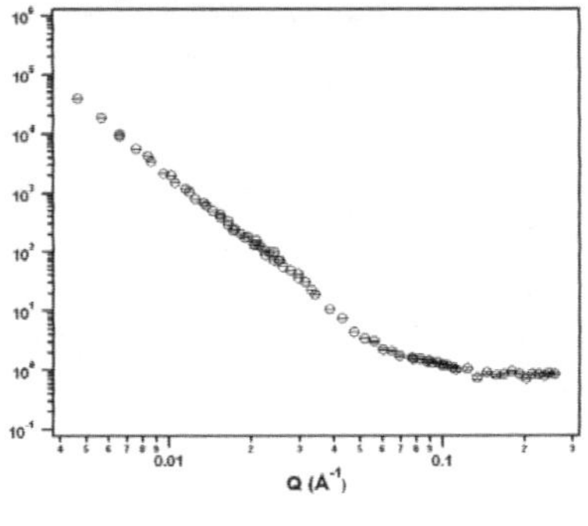

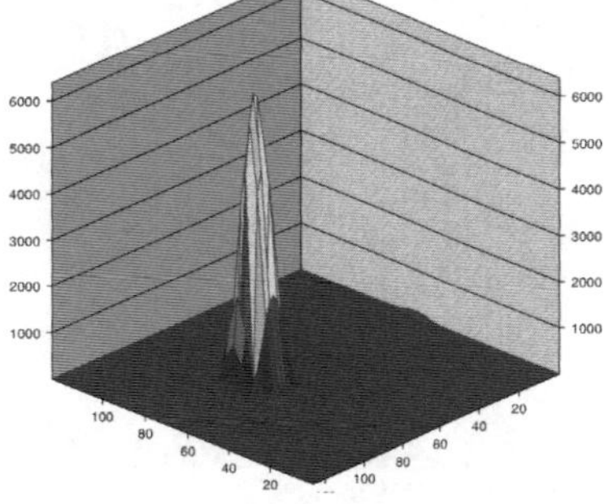

Figure 2. Intensity vs. scattering vector profile for tape sample sintered at 900°C. Oriented structure possibly at 0.03 Å^{-1}.

Figure 3. Direct beam profile at 2-dimensional position sensitive detector of mySANS.

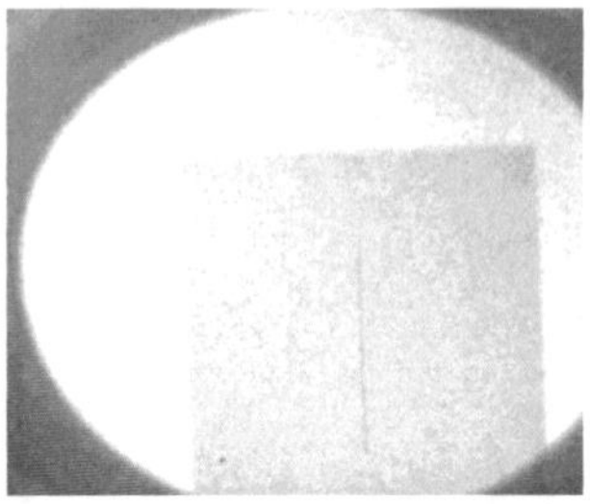

Figure 4. Radiographic image of thin iron solid layer (40 micron) sandwiched between aluminium blocks.

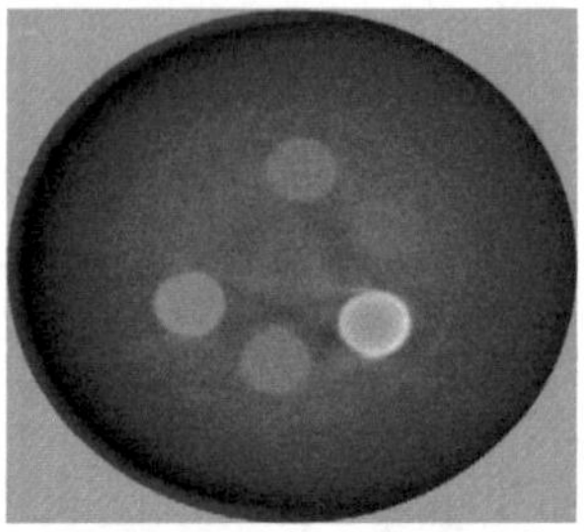

Figure 5. Tomographic image of aluminum cylinder with smaller cylinder of different materials inside.

Figure 4 shows the radiographic image of thin iron solid layer (40 micron) sandwiched between aluminum blocks. The layer is the fine vertical dark line in the middle of the block. Figure 5 shows tomographic image of aluminum cylinder with smaller cylinder of different materials inside it. This tomography utilize monochromatic neutron of wavelength about 5Å.

4. Conclusion

The LTCC material was believed to contain oriented structures that originated from organic additives used. This structure was formed by glassy material that replaced the organic additives after firing. This can be utilized to generate lines for fuel movement if porosity structure can be controlled. The possible setup of fuel cell LTCC may utilize beta-aluminate as solid electrolyte and silver catalyst electrode. Oriented structure and porosity analysis will be carried based on the scattering data obtained and the oriented structure will be confirmed with mySANS. Tomography characterization will be carried out on the tape sample when the system is established.

References

[1] Dernovsek O, Eberstein M and Schiller W A 2001 *J. Eur. Ceram. Soc.* **21** 1693
[2] Richards R W 1995 *Ellis Horwood Series of Polymer Science and Technology*
[3] Zhang Y F, Bai S L, Miao_M and Jin Y F 2009 2001 *J. Eur. Ceram. Soc.* **29** 1077
[4] Asraf Sharom M A, Ibrahim Z A, Wan Abdullah W A T, Ahmad M H A M, Idris F M and Mohamed A A 2011 *Chinese J. Phys* **50(2)** 229

IOP Conf. Series: Earth and Environmental Science **16** (2013) 012142 doi:10.1088/1755-1315/16/1/012142

Renewable Energy Supply for Power Dominated, Energy Intense Production Processes – A Systematic Conversion Approach for the Anodizing Process

D Stollenwerk[1], T Kuvarakul[1], I Kuperjans[1]

[1] Institut NOWUM-Energy, FH Aachen - University of Applied Science, Heinrich-Mussmannstr. 1, 52428 Jülich, DE

E-Mail: info@nowum-energy.com

Abstract. European countries are highly dependent on energy imports. To lower this import dependency effectively, renewable energies will take a major role in future energy supply systems. To assist the national and inter-European efforts, extensive changes towards a renewable energy supply, especially on the company level, will be unavoidable. To conduct this conversion in the most effective way, the methodology developed in this paper can support the planning procedure. It is applied to the energy intense anodizing production process, where the electrical demand is the governing factor for the energy system layout. The differences between the classical system layout based on the current energy procurement and an approach with a detailed load-time-curve analysis, using process decomposition besides thermodynamic optimization, are discussed. The technical effects on the resulting energy systems are shown besides the resulting energy supply costs which will be determined by hourly discrete simulation.

1. Introduction

According to [1], in Germany the energy consumption in the industrial sector is about 4.2 EJ in 2011. The energy consumption per 1000 € turnover in the different industrial sectors ranges between 0.2 GJ/1000€ in the tobacco processing industry and 32.3 GJ/1000€ in the coal mining industry. The third largest energy intensity per turnover is located in the fabrication of metal products and metal processing with 8 GJ/1000€ turnover in the year 2011. Thus the developed methodology introduced in this paper is applied to the anodizing process which is an important metal processing process for aluminium products.

The German energy sector is mainly driven by conventional fuels for space-, process-heating and electricity production. Referring to [2], the import dependency in the German energy sector changed from 60.74 % in the year 2006 to 59.78 % in the year 2010. With regard to [3] the share of renewables in the energy consumption in Germany increased from 8 % to 10 % in the same period. Therefore implementation of renewable energy technologies on the national, company and household level will be the key for lowering the import dependency in the future.

2. The Classic Layout Approach

The classical layout approach, depicted in **Figure 1**, for retrofitting renewables in an existing energy system, is taking the "Current Process and Energy System" and the energy procurement as basis for the system layout. Due to the fact that detailed knowledge about the process itself and its internal energy distribution is normally only existing in the process design company, not in the retrofitting company, the process is regularly regarded as black box. The energy flows in the company and the energy converter technologies are known, hence the "Supply", quality levels and in the best case "Supply-Time Curves" for the different energy carriers are known and can be used. However in most cases Supply-Time Curves are not available and only production data is existent for modelling.

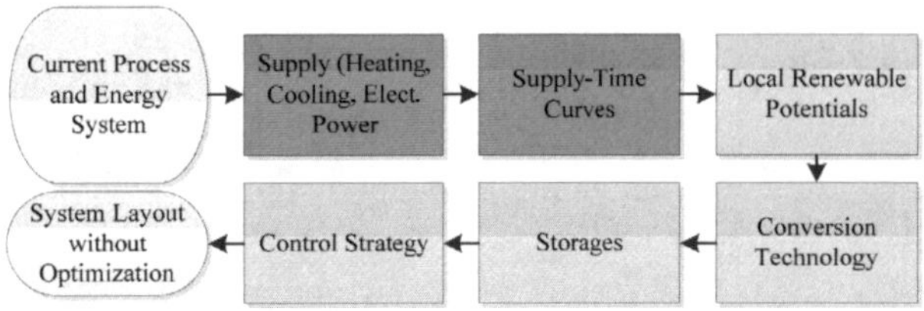

Figure 1. Classical layout process for renewable energy supply retrofitting

The "Local Renewable Potentials" depend strongly on the location of the investigated facility. In this study, the weather data for a test facility in Germany are used. The annual solar irradiation is 950 kWh/m². The mean wind speed in 10 m height is 4.5 m/s.

For the electrical sector, wind energy, solar irradiation, and a bio-diesel driven combined heat and power (CHP) unit are considered. Due to the fact that the demand cannot be satisfied with local wind and solar potential, the bio-diesel CHP is used for filling the gap. For the heating sector, solar irradiation and the CHP are considered. Geothermal potential is also interesting, but only for low temperature operation, which was negligible for the considered anodizing process.

In the considered case, the energy demand was supplied through a 130 °C thermal oil grid and a 230/400 V AC-network. Therefore the "Conversion Technologies" must achieve these energy quality levels. Hence technologies like near surface ground source heat pumps (GSHP), flat plate (FPC) and conventional evacuated tube collectors (ETC) could be excluded from this approach, because they are not able to supply heat on this temperature level with reasonable efficiencies [4–6]. The resulting schematic, following the aforementioned rules for the classical approach, is depicted in **Figure 2**.

The energy supplied to the thermal oil grid, is in this case supplied by a µ-gas-turbine CHP unit. All thermal energy is concentrated in the exhaust gas stream and can be extracted using a gas/thermal oil heat exchanger. Additionally, if the thermal demand is already satisfied and the unit runs to satisfy the electrical demand, the waste heat can be ejected through the exhaust without any further action. Thus the system has two main advantages compared to a conventional motor driven CHP unit: 1. All thermal energy concentrated in one stream with T_{ex}>200 °C; 2. No cooling of the unit necessary.

Energy Storages are under investigation, they are mainly used to bridge gaps between demand and supply. In this study heat storage systems and electrical storages are considered. Cold storage systems are not important for the investigated process but can be implemented in further studies.

The "Control Strategy" decision in this study is taking all locally available renewable energies first. If these sources cannot satisfy the demand, the CHP unit and the storage will cover the gap.

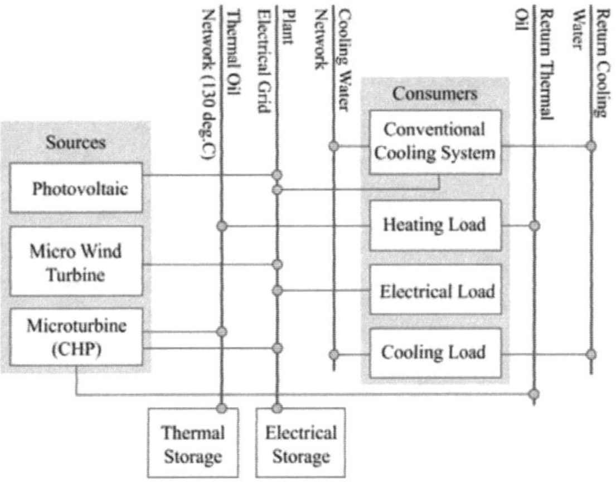

Figure 2 Resulting Energy Structure for the Classical Planning Approach

The electricity is supplied by small wind turbines, photovoltaic cells and the CHP-unit. The heat is supplied by the CHP-unit only, because the solar thermal potential for the needed energy quality (130°C) cannot be supplied by solar thermal collectors with reasonable efficiencies under the given weather conditions.

3. The Systematic Layout Approach

The systematical layout approach, depicted in **Figure 3**, differs strongly from the aforementioned classical approach. In this approach the process is investigated in detail, in contrast to consider it as a black box. First of all the process is decomposed to its part processes, the minimal energy demand of each part process is determined by the laws of thermodynamic and compared to reference processes and benchmarks (I). The minimal energy flows and energy quality level, e.g. pressures, temperatures are determined (II). Internal heat recovery and other process optimizations like lowering process temperatures or introducing newer processing technologies are applied, by using thermodynamic principles like pinch-analysis (III). The load time curves are investigated with regard to load shifting and smoothing possibilities (IV). The production site is also taken into consideration, because for the retrofitting case, normally there are big potentials due to the progress in building materials over the last 20 years (V). The results of these detailed studies are the “Energy-Demand Curves” (VI) of the specific process. Blocks VII-X corresponds to Blocks described in the Classical Approach, but the valid technologies may differ with regard to the newly derived demands which will then result in a different system layout.

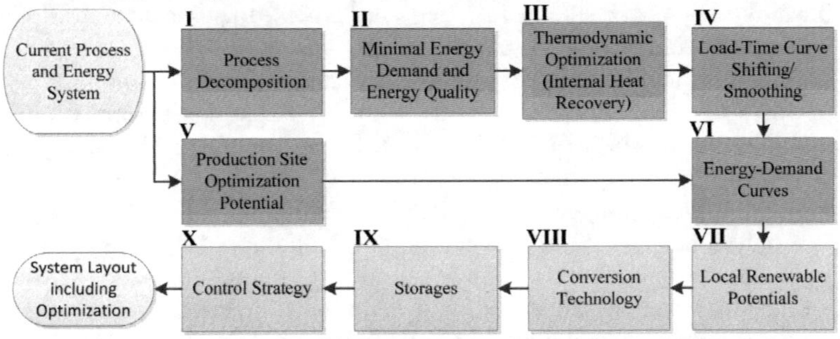

Figure 3 Systematic Layout Approach for Energy System Layout with Renewable Energy Sources

In **Figure 4** the new structure for the systematic layout approach is depicted. The heat supply is now divided into a high temperature (100 °C) and a low temperature (60 °C) network. Therefore other energy converter technologies become valid for satisfying the demand. A heat pump and a free cooling system can be introduced. The depicted cooling load was supplied with the conventional compression chiller system with electrical energy in the conventional approach and its energy supply was integrated in the electrical load. In this approach it is used as backup only, because the process depends vital on constant temperature levels.

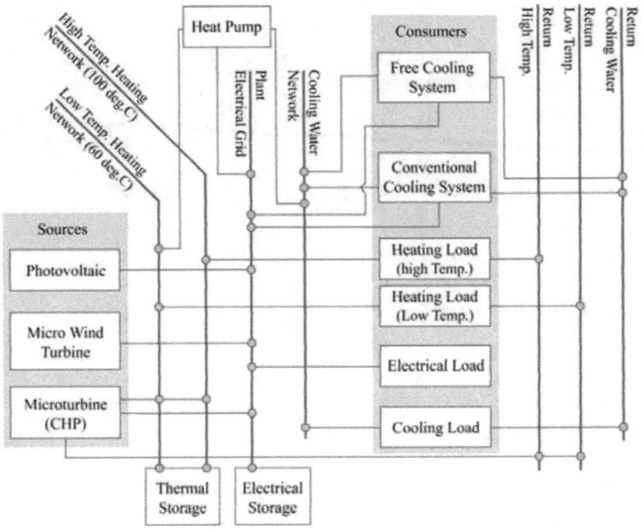

Figure 4 Resulting Energy Structure for the Systematic Planning Approach

The electricity is supplied by small wind turbines, photovoltaic cells and the CHP-unit. The heat is supplied by the CHP-unit and a heat pump which is also feeding the cooling load.

4. Simulation Results and Discussion

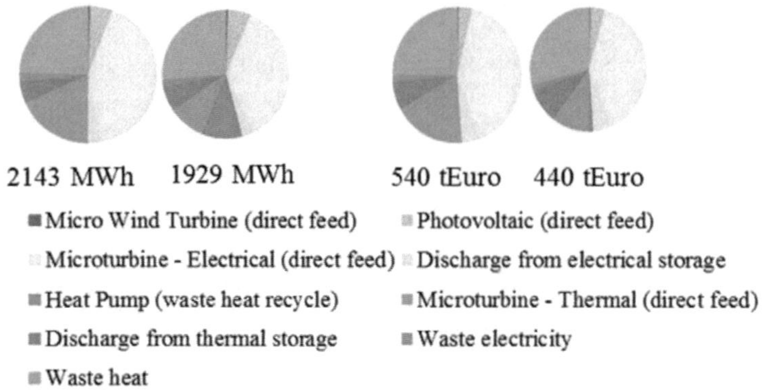

Figure 5 Annual Energy Generation and Energy Costs for the Classical (left) and the Systematic (right) Approach

Both approaches are applied to restructure an existing energy system to a 100 % renewable supplied system, the cost structure includes operation, investment into new equipment, maintenance and fuel costs. The simulation results, depicted in **Figure 5** show that the hereby introduced systematic layout approach can help, applied to the energy intense anodizing process, to reduce energy consumption up to 10 % and energy costs up to 18 % respectively. The amount of energy fed in the system by the micro wind turbine and the photovoltaic cells stay constant, due to the applied strategy of taking all locally available renewable sources first. Due to internal heat recovery with the usage of a heat pump, the energy directly fed into the heat grid by the CHP could be reduced to 45 %, relating to the conventional planning approach. Approximately 25% of the thermal energy produced by the system is supplied by internal heat recovery which also substantially lowers the electricity demand for the cooling system. The discharge energy from the electrical storage decreases, while the discharge of the thermal storage increases. Due to different cost structures this effect isn't reflected in the cost pie charts. The amount and costs of wasted energy could be reduced by 5% using the systematic layout approach. In total, the systematic layout approach gets better result, regarding energy savings and costs, even if the higher investment costs for the heat pump are considered.

5. Conclusion

The presented systematic layout approach will significantly reduce future costs for the energy supply for producing companies. The additional expenditure of time for the planning and layout of the system, including detailed process decomposition, benchmarking and thermodynamic analysis for the planning company, will turn to account regarding the typical durability of an energy system with approximately 20 years. In this case, the systematic layout approach could have saved 2 million Euro regarding a lifespan of 20 years for the system, compared to the classical approach.

Acknowledgments
The research is partially financed by the Federal Ministry of Education and Research (BMBF). Contract No. 17N1011

References

[1] Statistisches Bundesamt 2012 *Energieverwendung. Energieverbrauch des Verarbeitenden Gewerbes nach ausgewählten Wirtschaftszweigen 2011.* Available at https://www.destatis.de/.
[2] Eurostat 2013 *Energy dependence. Indicator: tsdcc310.* Share of dependency on energy imports by Member State (%). Available at http://epp.eurostat.ec.europa.eu.
[3] Federal Ministry for the Environment, Nature Conservation and Nuclear Safety 2013 *Zeitreihen zur Entwicklung der erneuerbaren Energien in Deutschland.* Available at http://www.erneuerbare-energien.de/.
[4] Atkins M J, Walmsley M Rand Morrison A S 2010 *Energy* **35** 1867.
[5] Chua K, Chou Sand Yang W 2010 *Applied Energy* **87** 3611.
[6] Chen Z, Furbo S, Perers B, Fan Jand Andersen E 2012 *Energy Procedia* **30** 65.

4th International Conference on Energy and Environment 2013 (ICEE 2013) IOP Publishing
IOP Conf. Series: Earth and Environmental Science **16** (2013) 012143 doi:10.1088/1755-1315/16/1/012143

Simulation of Hybrid Solar Dryer

Y M Yunus, H H Al-Kayiem[1]
Mechanical Engineering Department, Universiti Teknologi PETRONAS
[1]E-mail: hussain_kayiem@petronas.com.my

Abstract. The efficient performance of a solar dryer is mainly depending on the good distribution of the thermal and flow field inside the dryer body. This paper presents simulation results of a solar dryer with a biomass burner as backup heater. The flow and thermal fields were simulated by CFD tools under different operational modes. GAMBIT software was used for the model and grid generation while FLUENT software was used to simulate the velocity and temperature distribution inside the dryer body. The CFD simulation procedure was validated by comparing the simulation results with experimental measurement. The simulation results show acceptable agreement with the experimental measurements. The simulations have shown high temperature spot with very low velocity underneath the solar absorber and this is an indication for the poor design. Many other observations have been visualized from the temperature and flow distribution which cannot be captured by experimental measurements.

1. Introduction

Computational fluid dynamics (CFD) software could be used in solar drying field to predict air velocities and temperature distribution which is in manually will require a lot of sensors. Besides, the CFD may be used as a drying optimization tool to improve the design and to predict the drying time if connected to the thin-film equation [1]. [1] simulated a dryer which was designed to dry fruits and vegetables, consist of heat exchanger to enhance the heated air supplied from LPG gas burner. Centrifugal fan was located inside the dryer chamber to force the air movement. They found two areas of low air velocity under different trays location.

[2] have used CFD method to analyze the airflow and heat transfer by varying positions of a fan in a solar dryer. They had identified that the second position produced more uniform temperature and air flow. The maximum airflow was observed at the inlet and reduced as it approaches outlet. The simulation model described the real condition up to 89%.

[3] studied the heat and mass transfer in a solar dryer with biomass backup burner to obtain the optimum operating temperature using CFD software, STAR-CD. The modeling was performed on an empty chamber without the pepper berries. The simulation had been conducted under natural and forced convection. They had identified that heat and mass transfer by natural convection is more suitable for drying pepper berries with solar radiation. The simulation results indicate that the thermal heat distribution in the drying chamber was uniformly distributed.

There are no precise references to the literature concerning numerical simulation of air flows and heat transfer in the interior of dryers. However, there were shortages in the literature for the attempts of dryer simulation. The studies that had been conducted by previous researchers have shown that the CFD tools are very flexible to generate result close to the real processes.

The main objective of the present work is to investigate the drying performance of hybrid solar dryer using thermal backup by CFD simulation. The results are to be presented in form of velocity and temperature contours to visualize the flow and thermal fields' distributions.

2. Computational implementations

A 2-D model of solar dryer geometry was created using GAMBIT version 2.2.30. The inlet source for solar and biomass modes were different according to the type of the operational mode. The solar drying mode was design with air inlet located near the absorber plate at the side wall. In order to maximize the heat received from the back up burner and to minimize the hot air escape to the outside, the air inlet was closed in biomass mode. Label of the model is shown in figure 1.

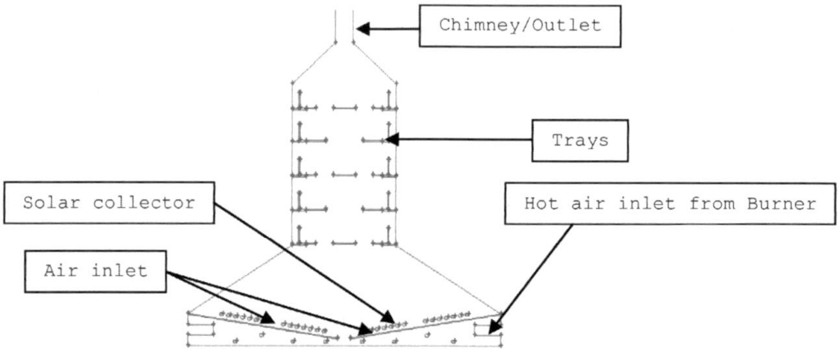

Figure 1. Labeled GAMBIT dryer model.

2.1 Computational Grid

The domain was discretized into a finite set of control cells. It was meshed with tri mesh elements and the face meshing type was pave, meaning an unstructured grid. The bottom part of the model was meshed with smaller interval size of 1.5 in order to get more accurate results surrounding the solar collector and biomass burner inlet. The middle and upper parts of the dryer were meshed with larger interval size of 2 and 3 to reduce the simulation time.

2.2 Governing Equations of Thermo Fluid Process

The governing equations for CFD are based on conservation of mass, momentum, and energy. FLUENT uses a finite volume method (FVM) to solve the governing equations. The FVM involves discretization and integration of the governing equation over the control volume. Several assumptions have been made for the solar dryer model. The design model is in two dimensional 2-D, the flow is in x and y-direction, steady, and compressible flow. Since the working fluid is gas, the body forces are very small and could be neglected.

2.3 Simulation procedure

Radiation Model: P-1 radiation model had been used. The radiation flux, q_r is given by (1).

$$q_r = \frac{1}{3(\alpha + \sigma_s) - C\sigma_s} \nabla G \tag{1}$$

where, α is the absorption coefficient, σ_s is the scattering coefficient, G is incident radiation and C is the linear-anisotropic phase function coefficient.

Turbulence Modeling: In this study, a turbulence model of two equations, standard k-ε model [7] had been used for the case of biomass burner mode. The turbulence model for mode that involves radiation was modeled using Renormalization Group (RNG). It is based on the two equation models of the Reynolds-Averaged Navier Stokes (RANS) equations. This RNG k-epsilon model is able to produce more accurate results than the standard k-epsilon model since it has additional term in its ε equation.

Boundary Conditions: The boundary conditions of the model were divided into two categories, which were solar mode and biomass mode. The boundary conditions of both modes were consisted of six zones. In solar mode only, the burner was disconnected and left as the air inlet. The thermal condition is depending on the type of model. The thermal conditions with and without P1 model were set as mixed and convection respectively. Further details are available in [8].

3. Results and discussion

3.1 Numerical Solution

Within this CFD simulation, the equations for: Flow, Turbulence, Energy and Radiation were solved. The discretization and solution schemes used for these equations are: Semi Empirical Pressure (SIMPLE), Standard and First Order Upwind scheme for Momentum, Turbulence Kinetic Energy, Turbulence Dissipation Rate and Energy. Convergence was monitored by checking the information from computed residuals of all equations. Scaled residuals were used. Computations were continued until the values of residuals were progressively reducing by typically five or six orders of magnitude, although in a few cases this could not be achieved. The calculations were considered convergent if the scaled residual for the continuity equation, the energy equation, P-1 radiation was less than 10^{-3}, 10^{-8} and10^{-6} respectively.

3.2 Analysis of Results

The results were focusing on the airflow and temperature distribution inside the dryer. It was presented according to the mode of drying from the unloading experimental results. The validation of the result was done by simulating the initial operating conditions of experimental measurement. The percentage of error was calculated by finding the difference between the experimental and simulated result.

The velocity vectors of air flow inside the solar dryer under solar mode are shown in figure 2. Ambient air flows into the dryer through the side opening holes at the base of the solar dryer and the front holes near the collector. The maximum velocity that can be achieved is 0.26 m/s and it is found to be concentrated at small path in the middle of collector. The air flows to the drying chamber and when it reaches small path of first tray, some of the air flow hits the bottom face of the trays causes a backflow to occur. This is similar to the results reported by [1]. The velocity increases as it flows through the alternate holes gap of the trays across the drying chamber and finally out at chimney. From simulation results, mean velocity of air escaping from chimney is shown to be 0.17 m/s. The value is approximately same as the velocity obtained through experiments which was 0.19 m/s. Observation of the velocity distribution indicated that the flow velocity in the solar dryer is not uniform. The flow velocity is high mainly at the gaps of the trays.

The temperature distribution inside the solar dryer is shown in figure 3. It could be seen that the highest temperature is concentrated surrounding the solar absorber plate which is around 330K to 341K. The heat was transferred from bottom to the top part of the dryer by natural convection. The first and second tray recorded the highest temperature around 318K since it is exposed to the high temperature air flow. However, as the location of the tray become farther, the temperature decreases due to the heat losses to the surroundings and to heat up the lower trays. The lowest temperature was spotted below the absorber plate and it represents the ambient air temperature which flow from the side walls of dryer. The simulation results indicated that the heat convection and heat conduction received from the absorber plate and walls respectively maintains the drying chamber at high temperature.

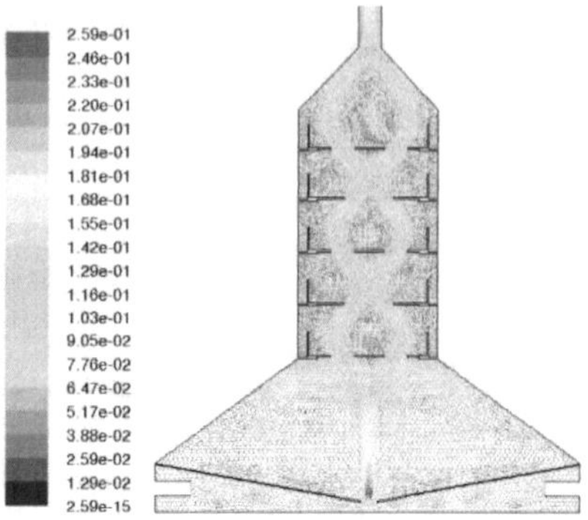

Figure 2. Velocity vector of fluid flow under solar mode.

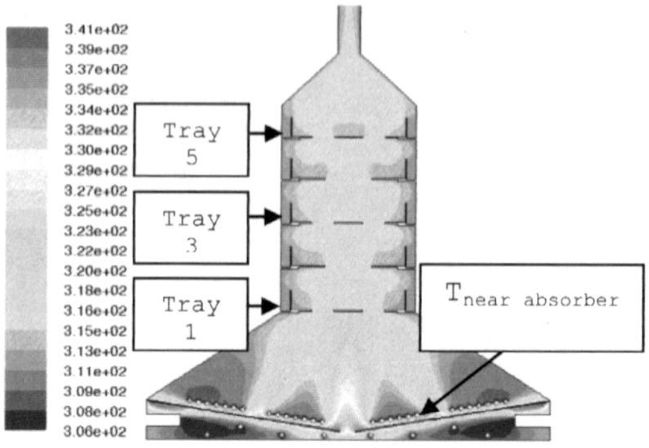

Figure 3. Temperature contours under solar mode.

The comparison between the simulation and experimental results is shown in Table 4. The results show that temperature near the absorber plate is in good agreement since the difference is around 11%. The measured values of the temperatures are higher than the simulated value due to the presence of solar radiation. The highest difference is at tray 5 since it is located at the top of drying chamber; hence it has the largest area exposed to the sun. In addition, the setting of P1 radiation heat transfer at the wall of dryer in is simulation is inadequate as compared to real condition.

Table 1. Percentage difference between the experimental and simulation result under solar mode

	$T_{\text{near collector}}$ (°C)	Tray 1 (°C)	Tray 3 (°C)	Tray 5 (°C)
Experimental	66.5	59.8	53.8	65.8
Simulation	59.0	45.1	43.4	45.0
Difference (%)	11.3	24.6	19.3	31.6

Conclusions

The simulations gave a reasonable result to real conditions. The simulation results were validated by comparing with the results obtained from experiment measurements. The results indicate that the calculated percentage of errors in thermal backup and hybrid modes were in good agreement. Hybrid mode is the most appropriate mode for drying application since it provides the highest temperature and falls within the required drying temperature range. The low temperature spot was identified and will be assisted in the improvement of dryer design. . The simulation results provide practical information by identifying the weakness of the system and consequently assist in design improvement.

[1] Mathioulakis E, Karathanos V T, and Belessiotis V G 1998 *J. of Food Eng* **36(2)** 183.
[2] Dyah W, Nelwan L O, Kamaruddin A, and Indra S A 2003 *IPB* (Bogor Agricultural University) **17(1)**.
[3] Rigit A R and Low P T 2010 *WASET* 115-118.
[4] Bird R, Stewart W, and Lightfoot E *Transport Phenomena* New York: John Wiley & Sons Inc.
[5] Incropera F P, Dewitt D P, Bergman T L, and Lavine A S 2007 5 Ed. John Wiley & Sons.
[6] Vest C M 1979 *Holographic interferometry* New York: John William & Sons.
[7] Launder B and Spalding D, 1974 *Computer Methods in App. Mech. and Eng* **3** 269-289
[8] Md Yunus Y 2011 MSc thesis Universiti Teknologi PETRONAS. http://utpedia.utp.edu.my/3060/

IOP Conf. Series: Earth and Environmental Science **16** (2013) 012144 doi:10.1088/1755-1315/16/1/012144

Cofiring biomass with coal: Opportunities for Malaysia

A A Rahman*, A H Shamsuddin
Centre for Renewable Energy, Universiti Tenaga Nasional, Putrajaya Campus, Jalan IKRAM-UNITEN, 43000 Kajang, Selangor, Malaysia

Email: adlansyah@uniten.edu.my

Abstract. Malaysia generated 108,175 GWh of electricity in 2010 where 39.51 % was sourced from coal. Coal power generation is also planned to overtake natural gas as the main fuel for electricity generation within the next two decades. Malaysia also has a vast biomass resource that is currently under-utilised for electricity generation. This paper studies the option of cofiring biomass in existing Malaysian coal power plants to increase the nation's renewable energy mix as well as to reduce its power sector carbon dioxide emission. Benefits of cofiring to the nation were discussed and agricultural residues from palm oil and paddy was identified as a potential source of biomass for cofiring. It was also found that there is a willingness for cofiring by stakeholders but barriers existed in the form of technical issues and lack of clear direction and mechanism.

Keywords: *cofiring; agricultural residues; Malaysia coal power stations*

1. Introduction

Malaysia's currently has a total of 7680 MW_e of installed generation capacity from six coal power stations. These power stations accounted for 39.51 % of the energy mix and are as listed in Table 1 [1]. Additional capacity of 1000 MW_e each at Manjung and Tanjung Bin has been approved for construction, and another plant up for 2000 MW_e is going through the request for quotations process. It is estimated that the share of electricity sourced from coal is going to surpass natural gas to become the largest contributor to the nation's energy mix by the year 2030. The nation's high dependence on coal had resulted in high carbon dioxide (CO_2) emission of 47.63 million tonnes in 2010 [2]. This accounted to 52.29 % of the total CO_2 emission from electricity and heat production [3].

Table1. Malaysia's coal power generation.

Power Station	Location	Coal Generation Capacity
Sultan Azlan Shah Power Station	Manjung, Perak	2100 MW_e (+ 1000 MW_e)
Tanjung Bin Power Station	Tanjung Bin, Johor	2100 MW_e (+ 1000 MW_e)
Sultan Salahuddin Abdul Aziz Power Station	Kapar, Selangor	1600 MW_e
Jimah Power Station	Jimah, Negeri Sembilan	1400 MW_e
Mukah Power Station	Mukah, Sarawak	270 MW_e
Sejingkat Power Station	Sejingkat, Sarawak	210 MW_e

Cofiring is defined as combustion of two or more different fuels in a single furnace although through previous experience, the term has been specifically used for describing partial substitution of coal as a main fuel in a utility scale boiler [4]. Cofiring is very popular in Europe and Australia where it is commercially practiced at almost all of the coal fired power plants there. Most of these plants are cofiring less than 20 $\%_{th}$ substitution although some plants are capable of 100 $\%_{th}$ fuel switching from coal to biomass. Cofiring is currently going through various trials and demonstration stages in the United States of America and Japan. It is also garnering interests of legislators and operators in Korea, China and other coal power producing countries as a mean to increase utilization of biomass in their renewables portfolio.

2. Rationale for Cofiring in Malaysia

Generally, the main benefit of cofiring is the reduction of fossil fuel CO_2 emission from electricity generation as coal is the most carbon dense fossil fuel. CO_2 emission reduction can be determined from the level of biomass substitution as it directly replaces the amount of coal to be combusted for a specific energy output. Reduction in sulphur dioxide (SO_2) emission is also expected as sulphur content in biomass is very small when compared to coal [5].

Cofiring is a cost effective way of increasing biomass utilization. The cost for a Malaysian biomass power plant firing empty fruit bunch recently costs RM 120M for 12 MW_e of installed capacity [6]. The costs for cofiring can be substantially lower for a similar sized of biomass substitution, for example 2 $\%_{th}$ substitution in a 600 MW_e furnace. This is because at low substitution levels, only minor modification is required to facilitate fuel handling and storage [7]. Furthermore, there is no additional cost downstream of the power plant required, for example for grid connection and grid reinforcement, as these are already in place at the existing power stations.

Another important advantage of cofiring is that coal fired furnaces operate at higher combustion temperatures compared to a dedicated biomass power plant resulting in higher conversion efficiencies. The Energy Commission reported that in 2010 Malaysian coal power stations achieved an average thermal efficiency of 33.1 %. In contrast, modern dedicated biomass power plant operates at thermal efficiencies of in the range of 25 % – 28 %. Furthermore, future planned coal fired furnaces are of the supercritical-steam type that would have thermal efficiency of over 40 %. This would ensure better utilization of the nation's available biomass resources.

Cofiring biomass residues diverts it from being landfilled and hence would reduce the volume of waste material to deal with. More importantly, it would directly reduce the amount of methane being released into the atmosphere from biomass decomposition. Methane is a greenhouse gas that is 23 times more potent than CO_2 in trapping heat in the atmosphere.

Several motivations exist at the national level for cofiring biomass in existing Malaysian coal fired power plants. Firstly, cofiring could be used to meet the current target of 330 MW_e of installed biomass capacity by 2015 where currently, grid connected biomass capacity is still below 50 MW_e. Cofiring only 5 $\%_{th}$ substitution at the four largest existing coal power stations would have surpassed this target at 360 MW_e. Secondly, replacing coal with biomass when cofiring would assist in meeting Malaysia's voluntary target of 40% reduction of CO_2 emission compared to 2005 levels as announced at the 15th Conference of Parties (COP15) in 2009. Cofiring could also be used to cater for more aspiring targets on renewables and greenhouse gas emission that will be necessary in the future. Finally, cofiring would also improve fuel security and supply as Malaysia currently imports most of its coal. Only Mukah Power Station utilizes coal from its local Mukah-Balingan reserve, which accounted for 9.92 % of total coal consumption in 2010.

3. Cofiring Potential in Malaysia

3.1. Biomass Resources and Availability

Malaysia is abundance in biomass resources especially from its main agricultural industries that can be used for cofiring [8]. The current study identified residues from two of the main agricultural activities in Malaysia namely palm oil and paddy. Biomass availability was estimated using previous work done by the Malaysian Palm Oil Board (MPOB) and Malaysian Agricultural Research and Development Institute (MARDI) for oil palm and paddy residues respectively [9, 10]. Calorific values from biomass characterisation, and reported average equivalent availability factor for the coal power stations were then used to determine thermal power potential of each biomass for that particular year. This value was then adjusted for combustion at the average thermal efficiency of coal power stations as a measure of potential power generation available when cofiring.

Palm oil mill residues consisted of empty fruit bunch (EFB), mesocarp fibre, palm kernel shell (PKS) and palm oil mill effluent (POME). The last is a liquid residue and hence is not suitable for cofiring. In 2010, fresh fruit bunch (FFB) yield in Malaysia was 18.03 tonne/hectare, planted on a total area of 4,853,766 hectares [10]. This gives a value of 87.51 million tonnes of FFB have been processed at the mills. Rice husk is the major residue from a rice mill and in 2010, paddy production was 2,464,831 tonnes giving a value of 542,263 tonnes of rice husk [12]. The potential energy available from palm oil mills and rice mills residues in 2010 is shown in Table 2. The results showed a generation potential of 7147 MW_e when cofiring all the palm oil mill and rice mill residues. Notably, this value is the theoretical limit and it is impractical to utilize all of the available residues for cofiring. Even so, it can be showed that even if only 5 % of all this residues is captured would give a value of 357.35 MW_e which surpassed the nation's biomass target for 2015.

Furthermore, all six coal power stations in Malaysia are located in rural areas close to the shore as sea water is used as a cooling medium. This type of land is also flat lying and fertile making it suitable for agriculture. Conveniently, the power stations are surrounded by palm plantations and in the case of Sultan Salahuddin Abdul Aziz Power Station, rice fields, which would indicate that palm oil mills and rice mills are in relatively close proximity. This would facilitate in lowering the cost of transportation of biomass to the power stations.

Table 2. Potential cofiring power generation from selected agricultural residues in 2010.

Residues	**Amount /tonnes**	**Calorific Value /MJ/kg**	**Thermal Potential /MW_{th}**	**Generation Potential /MW_e**
Empty fruit bunch	20,128,082	14.6	11,054	3,659
Mesocarp fibre	11,376,742	14.8	6,334	2,096
Palm kernel shell	5,425,831	19.0	3,878	1,284
Rice husk	542,263	16.0	326	108

3.2. Issues for Cofiring

Discussions with cofiring stakeholders have highlighted several concerns for cofiring to be practised commercially in Malaysia which can be classified as technical and non-technical issues. Feedback was received from major private and independent palm oil millers as well as all coal power station operators with the exception of Sejingkat Power Station.

Technical issues concerning powers station operators were mainly on logistics and boiler performance when introducing a new fuel to their existing setup and procedures. Awareness exists of the additional costs and installations to cater for transport, storage and handling of biomass. Most of these issues are actually manageable and lessons can be learned from international experiences of cofiring. Generally, some form of biomass upgrading, such as densification and torrefaction, was deemed to be necessary to facilitate in biomass logistics. Several studies in this area are currently being carried out with interests from both power station operators and biomass millers [13, 14, 15]. Boiler performance issues raised were on temperature profile, emission, ash deposition and corrosion. These issues are dependent on fuel being fired and combustion conditions. Hence, further studies are required in this respect especially when dealing with local biomass that is currently not being used for cofiring. However, it was generally accepted that cofiring low substitution levels of up to 5 $\%_{th}$ would not significantly alter boiler performance.

The non-technical issues are mainly associated with the costs and effort for cofiring. All parties agreed that any additional work would not directly add value to their respective core businesses. The lack of clear direction and policy on cofiring in Malaysia also meant that current work is on voluntary basis, purely from an environmental perspective. All parties also agreed that cofiring should be possible and are expecting for some form of mechanism from the authorities to facilitate implementation.

4. Conclusion

Coal will be the major fuel source for Malaysia resulting in high fossil fuel CO_2 emission. Cofiring biomass at existing coal power station has been identified to be the most suited method to efficiently use available biomass, reduce greenhouse gas emission and reduce energy import. Residues from palm oil and paddy are a good source of biomass feedstock and coal power stations is in close proximity to these resources. National motivations and stakeholders' interest do exist, but lack direction and mechanism. Cofiring is essential in not only meeting set commitments but also for future oncoming targets. There are several other technical issues that are manageable and warrant the need to further research to suit cofiring for local biomass and power stations operation.

References

[1] Energy Commision, 2012, Electricity Supply Industry in Malaysia: Performance and Statistical Information 2010, Energy Commision Malaysia, Putrajaya.

[2] U. S. Energy Information Administration, "International Energy Statistics," [Online]. Available: http://www.eia.gov/cfapps/ipdbproject/IEDIndex3.cfm. [Accessed 1 3 2013].

[3] International Energy Agency, 2012, CO2 Emissions from Fuel Combustion Highlights, IEA Publications, Paris.

[4] IEA Bioenergy Task 32, "Biomass Combustion and Cofiring," , [Online]. Available: http://www.ieabcc.nl/. [Accessed 1 3 2013].

[5] Sami M, Annamalai K and Wooldridge M, 2001, Co-Firing of Coal and Biomass Fuel Blends, *Progress in Energy and Combustion Science,* **27**, 171-214, 2001.

[6] Bahari B, 2012, Palm Oil a Good Source of Renewable Energy, Business Times, The New Straits Times Press (Malaysia) Berhad, [Online]. Available: http://www.btimes.com.my/Current_News/BTIMES/articles/palmo/Article/print_html. [Accessed 1 3 2013].

[7] Baxter L, 2005, Biomass-Coal Co-Combustion: Opportunity for Affordable Renewable Energy, *Fuel,* **84**, 1295-1302.

[8] Agensi Innovasi Malaysia, 2011, National Biomass Strategy 2020: New Wealth Creation for Malaysia's Palm Oil Industry, *Agensi Innovasi Malaysia*, Kuala Lumpur.

[9] Global Green Synergy Sdn Bhd, The Palm Biomass Industry, [Online]. Available: http://www.ggs.my/index.php/main-services/palm-biomass. [Accessed 1 3 2013].

[10] Economics & Industry Development Division, Malaysian Palm Oil Board, 2012, *Overview of the Malaysian Oil Palm Industry 2011.*

[11] Muhammad M D A, 2003, BERNAS Management of Rice Husk Disposal, *Business Opportunities in Agricultural Waste.*

[12] Department of Statistics, Malaysia, 2012, Malaysia Economic Statistics - Time Series 2011.

[13] Nasrin A B, Ma A N, Choo Y M, Mohamad S, Rohaya M H, Azali A, Zainal Z, Oil Palm Biomass as Potential Substitution Raw Materials for Commercial Biomass Briquettes Production, *American J. of Applied Sciences,* **5(3)**, 179-183, 2008.

[14] Uemura Y, Omar W N, Tsutsui T and Yusup S, 2011, Torrefaction of Oil Palm Wastes, *Fuel,* **90(8)**, 2585-2591.

[15] Daud D, Rahman A A, Shamsuddin A H, Abidin N, Zulkepli I I and Ja'afar M N, 2012, Malaysian Biomass Upgrading by Torrefaction *International Conference on Environmental Research and Technology.*

IOP Conf. Series: Earth and Environmental Science **16** (2013) 012145 doi:10.1088/1755-1315/16/1/012145

TNB Experience in Developing Solar Hybrid Station at RPS Kemar, Gerik, Perak Darul Ridzuan

K A Aziz[1], K N Shamsudin[2],
TNB Energy Services Sdn Bhd, Malaysia

Email: [1]khairulaa@tnb.com.my; [2]khiruln@tnb.com.my

Abstract. This paper will discuss on TNB experience in developing Solar Hybrid Station at RPS Kemar, Gerik, Perak. TNB has been approached by KKLW to submit proposal to provide electricity in the rural area namely RPS Kemar. Looking at area and source available, Solar Hybrid System was the best method in order to provide electricity at this area. This area is far from national grid sources. Solar Hybrid System is the best method to produce electrical power using the renewable energy from Solar PV, Battery and Diesel Generator Set. Nowadays, price of petroleum is slightly high due to higher demand from industry. Solar energy is good alternative in this country to practice in order to reduce cost for produce of electrical energy. Generally, Solar will produce energy during daytime and when become cloudy and dark, automatically battery and diesel generator set will recover the system through the hybrid controller system.

1. Introduction

Solar Hybrid System is operates automatically to supply the load using combination of Solar PV, Battery, Diesel Generator Set and Inverter Controller System. In this scheme, the priority of energy source shall be solar energy. If the solar energy insufficient to supply the load, the energy stored in battery will be used by discharging the battery. After certain period, the battery will reach its depth of discharge, and then the Inverter Controller shall instruct the diesel generator set to run, serve the load and charge the battery. Once the battery is fully charged, the diesel generator set will shut off to allow other energy source i.e Solar and Battery to supply the load. Solar Hybrid Station is located in RPS Kemar, Gerik, Perak and coordinate N 05° 12.038' E 101° 23.765. The site is accessible via speed boat ride from Jetty Banding or Jetty Trojen Tasik Temenggor, Gerik which takes 1 hour journey to reach Jetty Kemar. From Jetty Kemar, 4WD can be rented for transportation to the administration centre and the distance is about 6km from the jetty is approximately 15 minutes.

2. Load Profile

Based on finding and discussion with JAKOA, the tabulated number of consumer as Table 1.

Besides of this number of houses, there also have a few government building such as school, clinic, police station, shop and Surau. Based on KKLW need statement in this project, energy consumed per house shall be 8kWh and maximum demand (MD) shall be 1kW. Load profile was forecast based on previous similar experience and number & type of consumer.There are two peak period stipulated on the profile which comes from operation of government building on the daytime and night activity on the night time. Based on the above profile the max demand is about 365kW and the energy consume per day is about 3013kWh/day. The high demand during the night time was contributed by Hostel in the school.

Besides of this number of houses, there also have a few government building such as school, clinic, police station, shop and Surau. Based on KKLW need statement in this project, energy consumed per house shall be 8kWh and maximum demand (MD) shall be 1kW. Load profile was forecast based on previous similar experience and number & type of consumer.There are two peak period stipulated on the profile which comes from operation of government building on the daytime and night activity on the night time. Based on the above profile the max demand is about 365kW and the energy consume per day is about 3013kWh/day. The high demand during the night time was contributed by Hostel in the school.

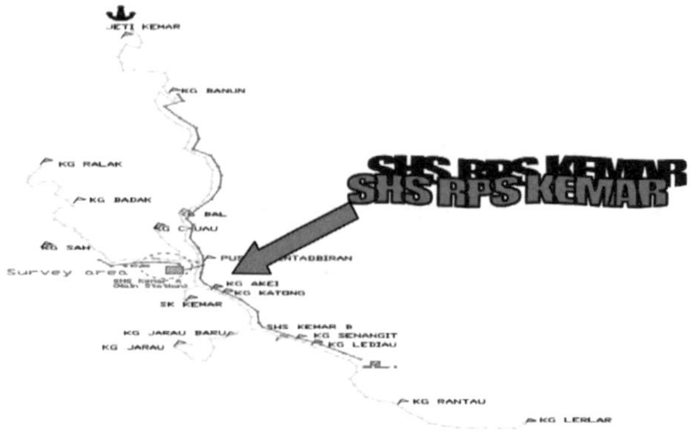

Figure 1. Location of Solar Hybrid RPS Kemar, Gerik, Perak.

Table 1. Number of houses

BIL	VILLAGES	VILLAGERS	HOUSES
1	KAMPUNG LERLAR	208	17
2	KG RANTAU	448	19
3	KG SENANGIT	186	27
4	KG LEDIOU	153	12
5	KG JARAU LAMA	237	22
6	KG JARAU BARU	108	26
7	KG SHAH	222	20
8	KG RALAK	212	12
9	KG CHUWAU	186	23
10	KG BADAK	158	17
11	KG BANUN	73	23
12	KG PENDERAS	112	3
13	KG BAL	577	45
14	PUSAT PENTADBIRAN	-	10
15	KG KATONG	295	30
16	KG AKEI	135	24
	TOTAL	**3093**	**276**

3. Design Configuration

Based on the load profile and to suite with KKLW design requirement which need 70% energy generates will come from Solar Energy and battery will only allow one (1) complete cycle charge and discharge per day, the design configuration as follows.

Plant installation data are shown in figure 2 for Solar PV, figure 3 for diesel generator set, figure 4 for Battery and figure 5 for Inverter.

Single line diagram for Solar Hybrid System at RPS Kemar was shown in Figure 5.

This power from Solar Hybrid Station have been distributed to consumer using the 11kV network and 2 x 1MVA step up transformer vector group YND11 before step down to 415V. It happened because the end point of consumer located far from the Solar Hybrid Station and will cause a voltage drop. This system is design based on two couple system which AC + DC couple. The reason why it was design in such away because to get the best efficient energy from Solar PV during the daytime (AC Couple) and the Solar energy stored in the battery to be use in night time (DC Couple).

4. Station Performance

Since commissioning on 30 November 2012, the performance of the station as follows:

IOP Conf. Series: Earth and Environmental Science **16** (2013) 012145 doi:10.1088/1755-1315/16/1/012145

Table 2. Component and configuration

Component	Configuration
Solar PV	• 850kWp
Diesel Set Generator	• 2 x 350kW • 2 x 450kW
Bi-Directional Inverter c/w controller	• 600kVA
PV Inverter	• 250kVA
Battery	• 4800kWh

Figure 2. Solar PV

Figure 3. Diesel Generator Set

Table 2. Solar specification

Solar Type	: Poly Crystalline
Solar Manufacturer	: AUO Corporation
Country of Origin	: Taiwan
Solar Model	: ECO DUO PM 200POO
Module Power Rated	: 220Wp
Module Efficiency	: 13.8%

Table 3. Generator specification

Generator Type	: Synchronous
Generator Speed	: 1500 rpm
Engine Maker	: Volvo Penta
Alternator Maker	: Sincro
Voltage	: Y 415/240
Frequency	:50 Hz
Power Factor	: 0.8

Figure 4. Battery

Figure 5. Inverter

Table 4. Battery specification

Battery Manufacturer	:System Sunlight
Battery Type	:Flooded Lead Acid
Country of Origin	: Greece
Battery Model	: 20 OPzS 2500
Capacity	: 2500Ah
Nominal Voltage	: 2VPC

Table 5. Inverter specification

Inverter Manufacturer	: Leonics Co. Ltd
Country of Origin	: Thailand
Bi-Directional Inverter	: Apollo MTP-6113H-P
PV Inverter	: Apollo GTP-512
Nominal Output	: 415V
Frequency	: 50Hz

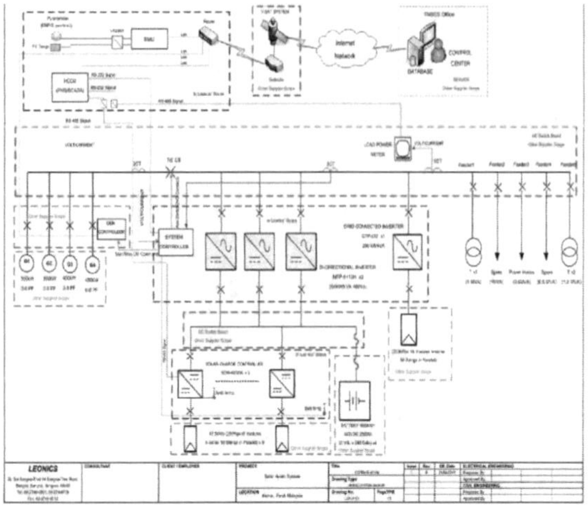

Figure 5. Single line diagram

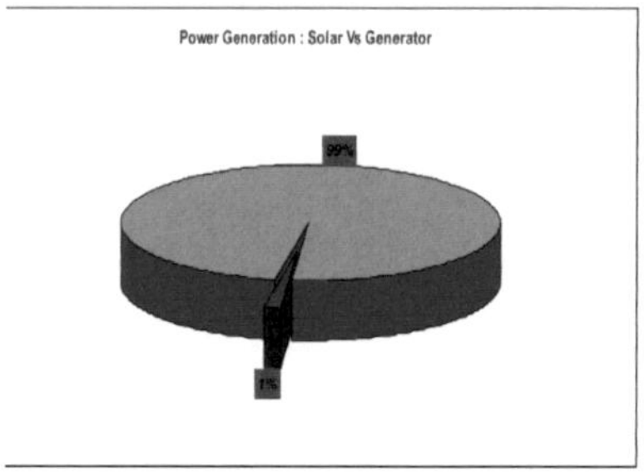

Figure 6. Energy Generated

Based on above pie chart, it shown that RE fraction of the system were 99% and the remaining 1% come from the diesel Generator Set. RE fraction value was calculated based on the following formula :

$$= \frac{\text{Total RE generation based on kWh meter}}{\text{Total RE + Diesel generation based on kWh meter}}$$

RE fraction is higher compare to design due to low load. Based on previous experience, load will increase from time to time due to economic growth at that place.

5. Analysis of Saving on Diesel Consumption

Solar has generated 89.35MWh energy to supply the load. Based on that figure, it already saves fuel consumption 26,715 liters after 2 months in operation. From the information available, diesel consumption to generate 1kWh energy is 0.299 liters and price for a liter diesel is RM2.60 (price 2012).

6. Conclusion

System Solar Hybrid is the best solution for Rural Electrification to reduce the diesel consumption. With the current station performance, the station was operating successfully. The current total estimated saving on diesel consumption at Solar Hybrid Station RPS Kemar is the great achievement.

7. Acknowledgement

The authors are thankful to all members for their direct and indirect contributions, and our beloved family members for their continuous support.

References

[1] HOMER- The Micropower Optimization model software, v2.81
[2] Fundementals of Solar Photovoltaic Technology, SEDA Malaysia

IOP Conf. Series: Earth and Environmental Science **16** (2013) 012146 doi:10.1088/1755-1315/16/1/012146

Smart Grid: Network simulator for smart grid test-bed

L C Lai [1], H S Ong[1], Y X Che[1], N Q Do[1] and X J Ong[1].
[1]College of Engineering, University Tenaga Nasional, Jalan IKRAM-UNITEN, 43000 Kajang, Selangor, Malaysia.

E-mail: laileechee@gmail.com

Abstract. Smart Grid become more popular, a smaller scale of smart grid test-bed is set up at UNITEN to investigate the performance and to find out future enhancement of smart grid in Malaysia. The fundamental requirement in this project is design a network with low delay, no packet drop and with high data rate. Different type of traffic has its own characteristic and is suitable for different type of network and requirement. However no one understands the natural of traffic in smart grid. This paper presents the comparison between different types of traffic to find out the most suitable traffic for the optimal network performance.

1. Introduction

Based on the definition from U.S. Department of Energy (DOE), "A smart grid uses digital technology to modernize the electric system from large generation, through the delivery systems to electricity consumption. A strong communication system is required in order to run Smart grid application [1]. The choice of a communication architecture and infrastructure represents a major issue that needs to be addressed [2]. A number of technologies were deployed in various smart grid initiatives. This ranges from wired technologies such as ADSL to POTS (plain old-fashion telephone) carried by incumbent telephone company to WIMAX and GSM provided by broadband access service providers. If a power utility wish to operate its own infrastructure, wired technology such as PLC and short-range wireless technology such as WiFI, ZigBee, and some proprietary wireless link will be the natural selection. Choice of a technology is depended on cost, accessibility, geographic location, terrain, population and building density, etc. In most instances, hybrid of technologies will be the best solution [3.4].

In order to investigate the performance and scalability of the network, network simulation is needed. Network simulator is a software program that imitates the working of a computer network. In simulators, the computer network is modeled with applications, devices, traffic and the performance is analyzed. Users can then customize the simulator to fulfill their specific analysis needs. In this simulation, Network simulator 2 is being chosen to perform the investigation because it is an open source software and more suitable for this network topology.

The main objective in this paper is to determine the type of traffic that suitable for the designed network by using different traffic generator. Three traffic generators will be tested in this project which is Constant Bit Rate (CBR) traffic generator, Pareto traffic generator, and Exponential traffic generator. CBR traffic generator is a fundamental traffic generator which sends data in a constant rate. On/ Off Pareto traffic generator is an advance traffic generator which embedded in oTCL application. Packets are sent at a fixed rate which specified by user during on periods, and no packets are sent during off periods [5]. An exponential on/off traffic generator acts as a CBR traffic generator during an ON interval and does not generate any payload during an OFF interval. On and OFF periods are both exponentially distributed [5]. This paper will discuss on the comparison of the result on different type of traffic. Thus, find out the suitable traffic type for the designed network.

2. Validation of Network Simulation

The main concern in this project is on getting the lowest delay and packet drop for different scenario in this designed network. This network represents smaller scale of current smart grid network with the combination of different topologies. The designed network consists of two technologies which are wired and wireless and three different types of network topology, mesh topology, star topology and point to point topology.

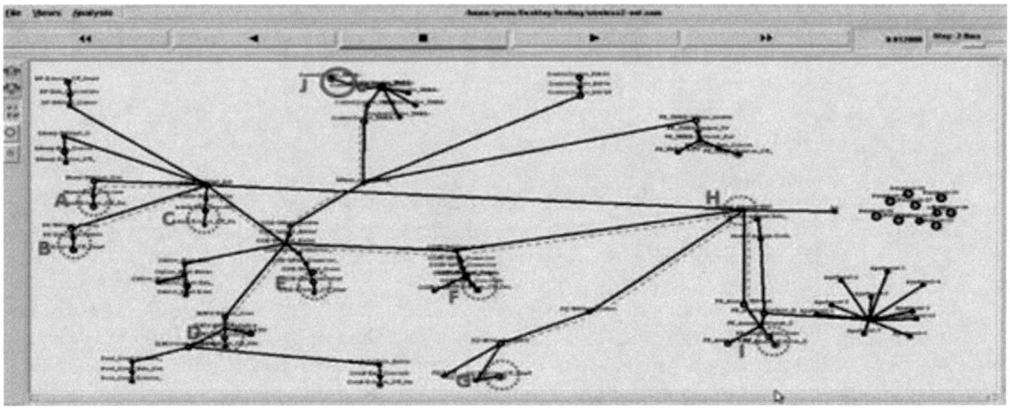

Figure 1. Network Design for Smart Grid Test-bed.

In the testing, nine substations which label as A to I will ping to J to check on the delay. The testing results are shown in Table 1. The result will then verify by "Student Test T", to prove that the null hypothesis is true. The null hypothesis in this testing is that there are no different between the result for real life testing and simulation. The t-test gives the probability that the difference between the two means is caused by chance. It is customary to say that if this probability is less than 0.05, that the difference is 'significant', the difference is not caused by chance [6]. The probability shows the percentage of null hypothesis for being true.

Table 1. Real Time Testing versus Simulation for each link.

Link	*Average Latency (ms)*	
	Real time testing	*Simulation*
A - J	*340*	*321.506*
B - J	*312*	*307.51*
C - J	*300*	*301.36*
D - J	*299*	*300.812*
E - J	*267*	*284.148*
F - J	*304*	*303.639*
G - J	*326*	*325.617*
H - J	*322*	*321.112*
I - J	*362*	*361.811*

"Paired student t-test" has been performed to compare the similarity of the network simulator with the real testing. The probability of the result is 0.873, which is 87.3 percent. Therefore, it can be conclude that result for real time testing and result for simulation are coming from the same family.

3. Result on simulation- Comparison on Traffic Generator

In the simulation, traffic generator will be placed in nine substations which label as A, B, C, D, F, G, H and I in the figure 1. All the traffic generators will then generate traffic to J, control center in Figure 1. The simulation will be conducted for 20 minutes to check on the capability of the link to handle traffic during burst time. Traffic settings for the simulation is depend on the project's background and its requirement. There are total 104 nodes in this simulation and all nodes use UDP transport protocol. The packet size and interval time defined as 1000 bytes and 0.01 second in the simulation.

In this section, we simulate the network by using different type of traffic generator and with different sending rate. About the simulation scenario, first we build the simulation network based on the project's requirement which include wired and wireless technologies and different network topology. Nine traffic generators are place in A, B, C, D, E, F, G, H and I in Figure 1. All traffic generators will generate traffic to a control center which label as J in Figure 1. Based on the designed network we test on the performance and reliability of 3 different type of traffic generators, Constant Bit Rate, Exponential and Pareto traffic generator which available in NS2. The performance test and reliability test include testing on maximum delay and drop packet over the simulation. The total simulation time is 1200 seconds for all scenarios.

First, we compare maximum delay for three different traffic generators. Traffic generators will generate packet based on the sending rate to the control center. The sending rate that tested in the simulation varies from 1 kbit/s to 1000kbit/s. The maximum delay increases until it reaches the saturation point where the delay is about 0.85 second. In Figure 2, we compare the maximum delay among three different traffic generator with various sending rate. There is no significant different on the maximum delay for this three traffic. CBR traffic generator has a little higher delay when the sending rate is 200kbit/s. The maximum delay for CBR traffic generator varies from 0.32 second to 0.8 second when it reaches the maximum sending rate. While Pareto traffic generator and Exponential traffic generator generates a constant delay when they reach the maximum sending rate for the network. As describe from Figure 5, CBR traffic shows a slightly better result on performance compare to Pareto and Exponential traffic generator.

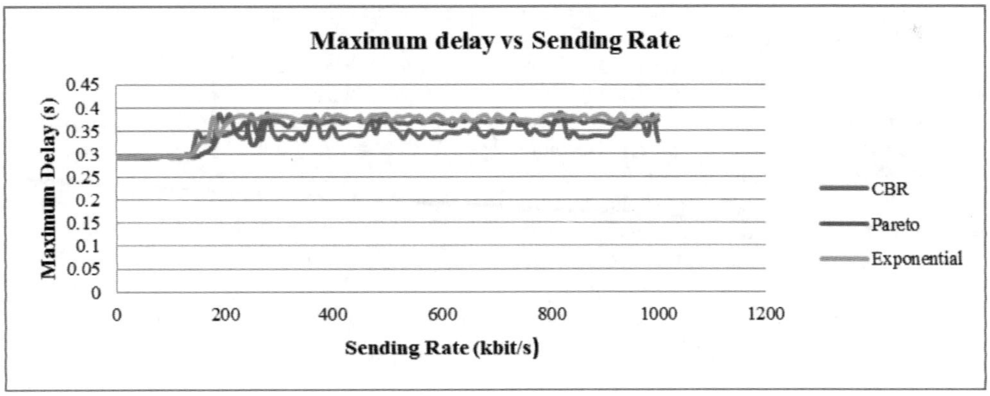

Figure 2. Compare Maximum delay among Different Traffic Generators.

Second, packet drop for different traffic generator based on different sending rate is shown in Figure 6. Similarly to the first test, the packet size for each data is 1000 bytes, but the sending rate that tested in this simulation varies from 1 kbit/s to 2000kbit/s. Figure 3 shows the result on comparing the packet drop for three traffic generators. The percentage packet drop increases until it reaches the saturation point where the sending rate is about 1200kbit/s. The higher packet drop implies the reliability for the network is lower. As observe from Figure 6, Pareto traffic generator and Exponential traffic generator are more suitable for this network because of their capability in delivering packets in acceptable packet lost range.

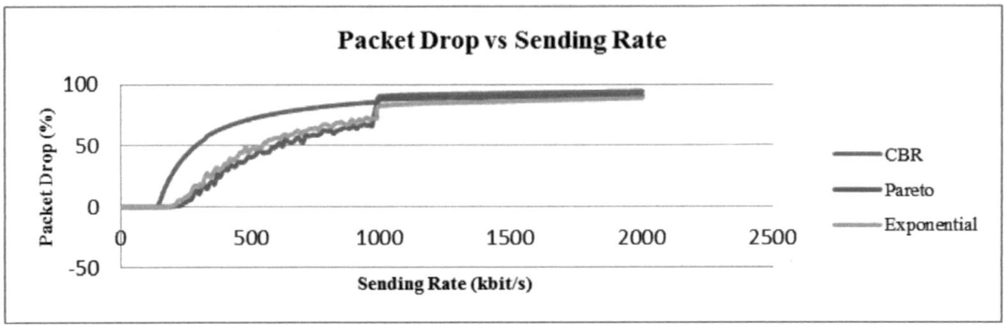

Figure 3. Compare Percentage of Packet Drop among Different Traffic Generators.

4. Conclusion

Three different types of traffics that generate by CBR, Pareto and Exponential traffic generator have been tested in the simulation. The reliability and the performance have been analyzed for UDP transport protocol. Results achieved by the analysis for both simulations suggest that the designed network is suitable for Pareto and Exponential traffic. The designed network has a better performance on handling delay for CBR traffic, but the reliability of the network is lower. The network has the ability to handle the delay and packet drop when the Pareto traffic is saturated. Exponential traffic is also suitable for this network. The performance is only slightly lower on handling the Exponential traffic compare to Pareto traffic.

5. References

[1] Austin T X 2012-06-12 $109M Malaysian Smart Grid Martket By 2016.
[2] Lin Y F, Zhong J and Felix W 2009-12-004 Discussion on Smart Grid Supporting Technologies, Power System Technology (Hong Kong, China).
[3] Qiu B 2002 Next Generation Information Communication Infrastructure and Case Studies for Future Power Systems, Ph. D Dissertation, Department of Electrical and Computer Engineering, Virginia Polytechnic Institute and State University.
[4] 2010 *Smart Grid Maturity Model* **1.1** Technology maturity, Carnegie Mellon Software Engineering Institute.
[5] Jean W 2012 NS Simulator for Beginner, Morgan & Claypool, chapter 2 29
[6] 2013-01-15 Simple Interactive Statistical Analysis (http://www.quantitativeskills.com/sisa/statistics/t-thlp.htm)

IOP Conf. Series: Earth and Environmental Science **16** (2013) 012147 doi:10.1088/1755-1315/16/1/012147

Performance testing framework for smart grid communication network.

D N Quang, O H See, L L Chee, C Y Xuen, S Karuppiah
College of engineering, Universiti Tenaga Nasional

Email: milkydove@yahoo.com

Abstract. Smart grid communication network is comprised of different communication mediums and technologies. Performance evaluation is one of the main concerns in smart grid communication system. In any smart grid communication implementation, to determine the performance factor of the network, a testing of an end-to-end process flow is required. Therefore, an effective and coordinated testing procedure plays a crucial role in evaluating the performance of smart grid communications. In this paper, a testing framework is proposed as a guideline to analyze and assess the performance of smart grid communication network.

1. Introduction

Multiple communication technologies in heterogeneous smart grid creates major challenges for conventional testing approaches, as legacy testing schemes are designated only for a single communication technology or standard in the network. There is a lack of a holistic testing scheme for heterogeneous network that comprises of diverse information and communication technologies. Currently, there is no standard way to evaluate and benchmark the performance of heterogeneous and hybrid smart grid communication system.

2. Objective and Scope

This paper proposes a performance testing framework for a heterogeneous and hybrid smart grid communication network. The testing scheme is carried out for TNBR-Uniten smart grid testbed.

3. Background

3.1 Smart Grid Testbed

Wired and wireless are proposed as communication mediums for smart grid. The communication technologies that support the network include Power Line Communication (PLC), unlicensed Radio Frequency (RF), and WiMax/4G. PLC is one of the common technologies used to transmit data over wired network. RF mesh is suitable to use in smart metering applications [1], and WiMax/4G is used for inter-substations and backbone.

3.2 Related Work

PLC is one the communication technologies for wired network in smart grid. As suggested in [2], the performance of PLC modems is evaluated by placing them in various locations within the power distribution network and carrying out a variety of tests to measure the average throughput. Besides wired network, wireless is another part of smart grid communication system. In order to analyze and investigate the performance of an ad hoc wireless network, Ping was proposed in [3] as an effective network testing tool. In wireless network, WiMax is one of the wel-known technologies. Iperf, a common network utility, can be used to generate traffic to analyze the performance of a WiMax testbed [4]. In previous researches, dedicated environment was set up in order to evaluate the performace of smart grid communication system. Using network emulation, a small-scale smart grid communication network is built in [5] to evaluate the network delay performance. Similarly, an emulated smart grid testbed is implemented in [6] to provide a framework for testing network requirements. The criteria used to evaluate the emulated communication network include throughput and latency. Iperf utility is used to perform throughput testing while Ping is utilized to troubleshoot network latency. The testing procedure for smart grid communication network is based on RFC 2544, a methodology to benchmark network devices. RFC 2544 describes various tests, including throughput, latency, and frame loss [7].

4. Methodology

The communication network in this Smart Grid testbed is supported by three different technologies, including WiMax, Mesh RF and PLC. Based on RFC 2544 standard, the same testing procedure was carried out to evaluate the reliability, latency and throughput of each technology. Reliability consists of availability and packet loss. In smart grid communication network, the availability of numerous devices is monitored by Nagios, a network management software. Each deived is assigned a fixed IP address and is identified as a host on Nagios. Each host's availability is monitored over a month and the obtained results are recorded. Ping, a common network testing tool, is used to measure packet loss and latency. The script runs on Windows for 24 housrs and the ping statistics are recorded in a text file, including packet loss and the average round-trip time. Throughput is calculated depending on the available technologies. For instance, Bing is used in order to measure the throughput of WiMax links. In addition, Iperf is utilized as a bandwidth measuring tool for Mesh RF, and a customized software is used to calculate the throughput for PLC modems. Last but not least, a series of stress tests are performed to test outside the boundary of what the network can handle. A single board computer is used to generate traffic at the potential bottlenecks in the network. The packet loss and latency are then measured to analyze and evalute the network performance under stressed conditions.

5. Results and Discussion

5.1 Availability

Table 1. Reliability (1): Availability.

No	Technology	Availability (% per month)		Requirement
		Under stable condition	Worst case scenario	
1	WiMax	100%	55.49%	99% - 99.99%
2	Mesh RF	99.69%	35.22%	99% - 99.99%
3	PLC	99.41%	16.81%	99% - 99.99%

Table 1 shows the availability of several devices supported by three different communication technologies in smart grid testbed. Under stable condition, the availability meets the standard requirement for smart grid [8]. However, under worst case scenarios, the availability is out of the requirement range. This happens due to a number of factors such as WiMax service problems, change of IP address, device failure or power trip at the substations.

5.2 Packet loss

The percentage of packet loss for different communication technologies is displayed in table 2. PLC shows the lowest level of packet loss with the average percentage of 0.69%. WiMax appears to be the least reliable technology among the three (1.97%). However, the average percentage of packet loss for all technologies is still within the acceptable range (0-1%) [9].

Table 2. Reliability (2): Packet loss.

No	Technology	Packet loss (% per day)		Range of packet loss
		Range	Average	
1	WiMax	0.45% – 3.45%	1.97%	0-1% Good
2	Mesh RF	0.56% – 2.29%	1.38%	1-2.5% Acceptable 2.5-5% Poor
3	PLC	0.15% – 1.65%	0.69%	5-12% Very poor >12% Bad

5.3 Latency

Table 3 shows the communication latency of three different technologies in smart grid. WiMax shows the highest latency, while Mesh RF shows the lowest. In general, the average latency of all technologies meets the requirement for smart grid [10], and is acceptable for SCADA applications (1-3s) [11].

Table 3. Latency.

No	Technology	Delay (ms)		Requirement
		Range	Average	
1	WiMax	178.2 – 186.8	181.8	< 5s
2	Mesh RF	0.60 – 1.32	6.84	< 5s
3	PLC	11.60 – 29.63	20.23	< 5s

5.4 Throughput

The practical throughput of three technologies is displayed in table 4. WiMax throughput appears to be the one closest to the theoretical value, while PLC shows half of the maximum specified bandwidth, and Mesh RF occupies only 39.29% of the theoretical amount.

Table 4: Throughput.

No	Technology	Throughput		Theoretical bandwidth	Compared to maximum specified bandwidth
		Range	Average		
1	WiMax	517.25 – 916.52 Kbps	700.71Kbps	1Mbps	70.07%
2	Mesh RF	9.86 – 27.04 Mbps	21.22Mbps	54Mbps	39.29%
3	PLC	35.08 – 96.50 Mbps	56.56Mbps	100Mbps	56.56%

5.5 Stress Test

The obtained results of stress tests are summarized in table 5. At 28 Mbps, traffic does not have significant impact on the network if the duration of traffic flow is less than 450ms. A significant increase is shown in communication latency when the traffic duration is within the range from 450ms to 1s. With the traffic busrt lasting more than 1s, the network experiences 100% of packet loss.

Table 5. Stress test.

No	Traffic	Duration	Packet loss	Delay
1	28 Mbps	< 450 ms	0%	0.7 ms – 23.5 ms
2	28 Mbps	450 ms – 1 s	0%	22.56 ms – 42.07 ms
3	28 Mbps	> 1s	100%	_

6. Conclusion

A performance testing framework is presented in this paper in order to unify different testing methods and software tools, as well as to provide a common platform for network performance evaluation based on measurement parameters. The performance of three different technologies in smart grid, including WiMax, Mesh RF and PLC, is analyzed and evaluated based on several performance metrics such as reliability, latency and throughput. Stress tests are also carried out to determine the operating limit and capacity of traffic in the communication network.

References

[1] Lichtensteiger B, Bjelajac B, Muller C and Wietfeild C 4-6 October 2010 RF mesh systems for smart metering: system architecture and performance *1st IEEE Int. Conf. on Smart Grid Communications* Gaithersburg 379-384

[2] Drosopoulos A, Xatziprokopiou M, Kotsifas A and Tagios C 21-23 November 2007 Performance evaluation of commertially available PLC modems *7th WSEAS Int. Conf. On Electric Power Systems, High Voltages, Electric Machines* Venice 192-197

[3] Mansor M, and Abdullah J 25-26 July 2009 Evaluating the communication performance of an ad hoc wireless network using mesh connectivity layer (MCL) protocol *Conf. on Innovative Technologies in Intelligent Systems and Industrial Applications* Bandar Sunway 192-197

[4] Yousaf F Z, Daniel K, and Wietfield C 17-19 October 2007 Performance evaluation of IEEE 802.16 WiMax link with respect to higher layer protocols *4th Int. Sump. on Wireless Communication Systems* Trondheim 180-184

[5] Wang W, Xu Y, and Khanna M 27 October 2011 *A survey on the communication architectures in smart grid Computer Networks* **55** Raleigh: Elsevier 3604-3629

[6] Kush N, Clark A, Foo E and Ahmed E 11 November 2010 Smart grid test bed design and implementation Queensland

[7] Bradner S and McQuaid J March 1999 Benchmarking methodology for network interconnect devices United States: RFC Editor 1-31

[8] U. S. Department of *Energy Communication requirements of Smart Grid Technologies* retrieved on 20 December 2012, available at : http://energy.gov/gc/downloads/communications-requirements-smart-grid- technologies

[9] The PingER Project, retrieved on 3 January 2013, available at http://sdu/ictp.it/pinger.html

[10] Kansal P and Bose A September 2012 Bandwidth and latency requirements for smart transmission grid applications IEEE Trans. on Smart Grid *IEEE J. M.* **3** 1344-42.

[11] Schmidt R A and Zamzow K 14 April 2010 Bandwidth planning for development of smart grid applications Power System Engineering, Inc.

IOP Conf. Series: Earth and Environmental Science **16** (2013) 012148 doi:10.1088/1755-1315/16/1/012148

Performance Comparison of BPL, EtherLoop and SHDSL technology performance on existing pilot cable circuits under the presence of induced voltage

Y X Che[1], H S Ong[1], L C Lai[1], S Karuppiah[2], X J Ong[1] and N Q Do[1]
[1] College of Engineering, Universiti Tenaga Nasional, Selangor, Malaysia
[2] Department of Communication, Tenaga Nasional Berhad Research, Selangor, Malaysia

E-mail: yxche828@gmail.com

Abstract. Pilot cable is originally used for utility protection. Then, pilot cable is further utilized for SCADA communication with low frequency PSK modem in the early 1990. However, the quality of pilot cable communication drops recently. Pilot cable starts to deteriorate due to aging and other unknown factors. It is also believed that the presence of induced voltage causes interference to existing modem communication which operates at low frequency channel. Therefore, BPL (Broadband Power Line), EtherLoop and SHDSL (Symmetrical High-speed Digital Subscriber Line) modem technology are proposed as alternative communication solutions for pilot cable communication. The performance of the 3 selected technologies on existing pilot cable circuits under the presence of induced voltage are measured and compared. Total of 11 pilot circuits with different length and level of induced voltage influence are selected for modem testing. The performance of BPL, EtherLoop and SHDSL modem technology are measured by the delay, bandwidth, packet loss and the long term usability SCADA (Supervisory Control and Data Acquisition) application. The testing results are presented and discussed in this paper. The results show that the 3 selected technologies are dependent on distance and independent on the level of induced voltage.

1. Introduction

Pilot cable circuit is a continuous copper wire being used in TNB (Tenaga Nasional Berhad) to provide services such as power system protection and as the communication medium for the low bandwidth SCADA application. Pilot cables were usually laid along with power supply cable between 2 substations. Pilot cable consists of 37 twisted pairs of PE insulated tinned copper wire 0.914mm diameter. The 74 wires are covered by dual insulation with an inner core of cellular polythene and outer skin of solid polythene with hard aluminum wire in between as the wire armor[1].

The pilot cables are subjected to severe induced voltage as it follows the same route as power cable that causes the communication to be intermittent/lost. The easy solutions are to hire lease line service from telecommunication companies or lay fiber optic line to support the SCADA communication. However, these are subject to recurring charges and high cost solutions. There is an immediate need to solve the communication issues of pilot cable for current SCADA applications. Therefore, BPL, EtherLoop and SHDSL modem technologies are proposed to overcome the communication issues. Unlike the existing PSK modem, the 3 selected modem technologies using OFDM to encode the data signal. Theoretically, the 3 selected modems are not susceptible to induced voltage and able to qualify the requirement of

SCADA communication with 9.6kbps bandwidth and 2s delay. The performance of each chosen modem technology are measured and verified in the field test.

2. Methodology

The selected 3 modem technologies are required to qualified performance requirement of SCADA communication. These requirements include low delay time, acceptable bandwidth and high reliability. Therefore, several testing are required to verify the capability of the modems in establishing SCADA communication on existing pilot cable circuits.

The figure 1 below shows the configuration setup of modem testing:

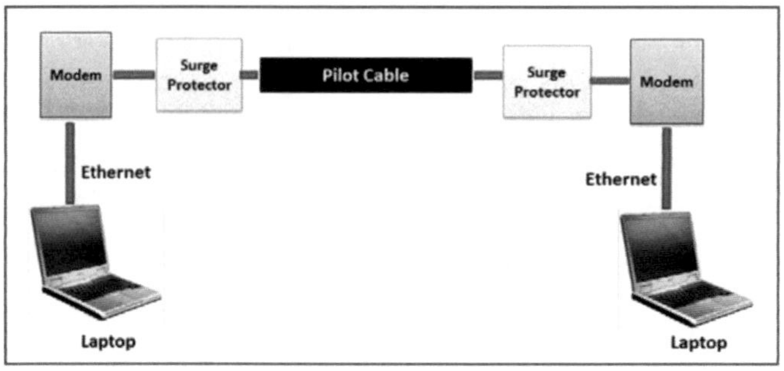

Figure 1. Configuration of Modem Testing

Each of the pilot cables are tested based on the configuration in the Figure 1 above. The testing was carried out at both ends of the circuits. Laptops and surge protectors were required to perform the testing at both ends. In order to test delay, bandwidth and packet loss, Ethernet connection is required to connect both laptops to modems. Then, the output of the modem will be connected to a surge protector before connecting to the respecting cores of pilot cables to protect the modem from any voltage spikes.

The 3 tests below will be carried out to determine the performance of pilot cable under difference induced voltage levels and pilot cable lengths (distances):

- Delay Test
- Bandwidth Test
- Reliability Test

Ping application is used for delay and reliability test while Bing application is selected to perform bandwidth test. The ping test is run for around 1 minute with 1 second interval before it is aborted. The average delay of the 1 minute test is calculated for analysis. Bing test is also run continuously for around 1 minute before it is aborted to capture the average bandwidth. For reliability test, Ping application is executed continuously overnight (for around 20 hours) to capture the packet loss. The number of packet loss is recorded and the reliability of the modem technology is calculated.

3. Result

3.1. Delay and Bandwidth Test

The results of delay and bandwidth test of the modems are as the table 1 below:

Table 1. Delay and Bandwidth Test

Station	Distance (km)	Induced Voltage (Vac)	Average Bandwidth (Mbps)			Average Delay (ms)		
			BPL	EtherLoop	SHDSL	BPL	EtherLoop	SHDSL
1	0.4	0.28	7.780	2.926	2.469	3.06	31.36	3.145
2	0.8	2.26	7.273	0.661	2.087	2.667	50.66	3.200
3	2.3	4.78	2.027	2.564	2.255	3.333	33.00	3.474
4	2.9	0.80	3.663	1.842	2.229	3.308	24.65	3.100
5	>3.0	0.568	n/a*	4.714	2.197	n/a*	56.00	3.210
6	3.4	9.54	n/a*	4.715	2.196	n/a*	37.19	3.907
7	4.8	2.51	n/a*	0.352	2.139	n/a*	52.67	3.454
8	5.7	21.18	n/a*	0.648	2.157	n/a*	45.50	3.031
9	6.4	3.87	n/a*	0.759	2.074	n/a*	55.28	3.174
10	9.3	14.41	n/a*	n/a*	2.029	n/a*	n/a*	3.350
11	>10	41.51	n/a*	n/a*	n/a*	n/a*	n/a*	n/a*

*Unable to communicate

Based on the observation in table 1, each of the modem is able to communicate up to certain distance (BPL: 2.9km, EtherLoop: 6.4km and SHDSL: 9.3km). The average delay of EtherLoop modem technology varies from 24ms to 56ms. The average delay obtained for both BPL and SHDSL technology is below 4ms. The delay measured for each technology is within the acceptable range for SCADA communication. The bandwidth of the EtherLoop technology measured has a range from 0.352 Mbps to 4.715 Mbps. BPL modem achieves bandwidth range from 2Mbps to 7.7Mbps. Meanwhile, SHDSL technology has around 2Mbps bandwidth on all of the circuits. The huge bandwidth difference between the substations is believed to be caused by the condition of pilot cable (multiple joints in between). The bandwidths obtained from all pilot cable circuits are qualified for SCADA communication which required only 9.6 kbps. None of the delay or bandwidth test done on the pilot cable circuits with the length >10km because all modems are unable to communicate at this distance range. There is no relationship found between the induced voltage and the delay/bandwidth of the communication for each modem technology.

3.2. Reliability Test

In order the study the reliability of each modem technology, overnight testing were conducted on 10 circuits to investigate the packet loss. However, the reliability of BPL modem is not tested because it is only able to communicate on 4 pilot cable circuits. The result of the overnight testing for EtherLoop modem is shown in the table 2 below:

Table 2. EtherLoop Reliability Test

Station	Packet		
	Packet Sent	Packet Loss	% Packet Loss
2	4452	1	0.0225
4	4624	0	0
7	4136	0	0
8	560	0	0
9	4060	0	0

The table 2 above shows that EtherLoop modem has no or extremely low packet loss communicating over the 24 hour testing. The highest packet loss observed is 0.22%. On the other hand, the modem technology has 0% packet loss on the remaining pilot cable circuits.

IOP Conf. Series: Earth and Environmental Science **16** (2013) 012148 doi:10.1088/1755-1315/16/1/012148

There is no packet loss on the pilot cable circuit which has 21.18Vac induced voltage. Based on the result obtained, EtherLoop modem is capable to perform reliable communication regardless of the induced voltage influence.

The reliability of SHDSL technology is tested on the remaining 5 pilot cable circuits. The result is as the table 3 below:

Table 3. SHDSL Reliability Test

	Packet		
Station	Packet Sent	Packet Loss	% Packet Loss
1	3676	0	0
3	4812	3	0.0623
5	11344	0	0
6	4256	0	0
10	4064	105	2.5836

The table 3 above shows that SHDSL modem has no or extremely low packet loss communicating over the 24 hour testing. The highest packet loss observed is 2.583%. The second highest packet loss obtained is 0.062% and no packet loss found for the 3 remaining circuits. There is no packet loss found on pilot cable circuits which has 41.51 Vac induced voltage. Therefore, SHDSL modem is capable to perform reliable communication regardless of the induced voltage influence.

4. Conclusion

The objective of this project is to investigate the performance of BPL, EtherLoop and SHDSL communication on existing pilot cable circuits under the influence of induced voltage. Based on the results, the proposed technologies are qualified competent to be an alternative solution for pilot cable communication. The lowest bandwidth measured for BPL (2.027Mbps), EtherLoop (0.352Mbps) and SHDSL (2.029Mbps) modem technologies qualified for the 9.6kbps SCADA bandwidth requirement. The maximum delay obtain for BPL (3.333ms), EtherLoop (56ms) and SHDSL (3.907ms) modem technology also fulfilled the 2s SCADA delay requirement. Besides that, all three of the selected modem technologies are able to communicate under the presence of induced voltage. However, the performance of each modem technology is dependent by the length/distance of pilot cable circuits. The distance limits for each technology are 2900m, 6400m, and 9300m respectively for BPL, EtherLoop and SHDSL modem technology.

References

[1] Mohd Rodzi M Z, Saparon A and Hamzah N R 11-14 April 2010 Detection of faulty pilot cable using ALTERA Cyclone II board with time-domain reflection (TDR) technique *Electronic Devices, Systems and Applications (ICEDSA), 2010 Intl Conf on* 125-130.

[2] Goralski W 2009 *The Illustrated Network* (Burlington: Morgan Kaufmann)

[3] Roebuck K 2011 *Broadband Over Power Lines: High-Impact Emerging Technology - What You Need to Know: Definitions, Adoptions, Impact, Benefits, Maturity, Vendors* (Quensland: Emereo Pty Limited)

[4] Chen W Y 2003 *Home Networking Basis* (New Jersey: Prentice Hall PTR)

IOP Conf. Series: Earth and Environmental Science **16** (2013) 012149 doi:10.1088/1755-1315/16/1/012149

A numerical study of laminar forced convection flow of Al_2O_3-water nanofluid in triangular-corrugated channel

M A Ahmed [1,2], M Z Yusoff [2], N H Shuaib [3], K C Ng [2]
[1] Department of Mechanical Engineering, College of Engineering, University of Anbar Ramadi, Anbar, Iraq
[2] Department of Mechanical Engineering, College of Engineering, Universiti Tenaga Nasional (UNITEN) Jalan IKRAM-UNITEN, 43009 Kajang, Selangor, Malaysia
[3] TNB Research Sdn. Bhd., Jalan Ayer-Hitam, 43000 Kajang, Selangor, Malaysia

E-mail: moh0891@yahoo.com, zamri@uniten.edu.my, norshahhafeez@tnbr.com.my, nkching@uniten.edu.my

Abstract. In this paper, laminar forced convection flow of Al_2O_3-water nanofluid in triangular-corrugated channel is numerically studied. The governing mass, momentum and energy equations in body-fitted coordinates are solved using finite volume method. Reynolds number and nanoparticle volume fractions are in the ranges of 100- 800 and 0- 5%, respectively. The effect of Reynolds number, nanoparticles volume fraction and nanoparticles diameter on the flow and thermal characteristics are examined. The results indicate that the Nusselt number increased as Reynolds number and nanopartiles volume fraction increased but the pressure drop increased as well. Also, Nusselt number increased as the nanoparticle diameter decreased, while there is no effect of nanoparticle diameter on the pressure drop.

1. Introduction

Heat transfer enhancement in many thermal devices such as heat exchangers is required to reduce size, weight and cost of these devices. For this purpose, the corrugated surfaces which are one of the passive methods have been used to improve the thermal performance of heat exchangers in several applications. For further enhancement in heat transfer, the addition of naonparticles to the working fluids can improve thermal conductivity and consequently enhancement in heat transfer rate. Several numerical and experimental studies on the forced convection flow of conventional fluids or nanofluids in channels exist in the literature [1-8]. Laminar forced convection of copper-water nanofluid in straight channel was numerically studied by Santra et al. [4]. Manca et al. [5] and Mohammed et al. [6] numerically studied of the convective heat transfer of nanofluid in a ribbed channel. Heidary and Kermani [7] and Ahmed et al. [8] studied on the forced convection flow of copper-water nanofluid in a wavy channel. Results displayed that the heat transfer rate increased with Reynolds number, amplitude of wavy wall, and nanoparticle volume fraction.

In this article, forced convection of nanofluid in triangular-corrugated channel is numerically studied using finite volume method. The effects of Reynolds number, nanoparticles volume fraction(ϕ) and nanoparticle diameter(dp) on flow and thermal chaerctersicts are presented and analyzed.

[1] To whom any correspondence should be addressed.

2. Geometry and boundary conditions

The physical model of present study is shown in Figure 1. The model consists of upper and lower walls with average spacing between these walls of (H). The length of corrugated channel is six cycles (pitch) and the amplitude of corrugated wall is (a= 0.2H). The axial length of each pitch is (λ=2H). The boundary conditions for the computational domain were specified. Dirichlet boundary conditions were assigned at the channel entrance (i.e. $u= u_{in}$, v= 0 and $\theta= 0$) and Neumann conditions adopted in the channel outlet. A long the walls of channel, a no-slip boundary condition were prescribed with a constant wall temperature condition (i.e. u= 0, v= 0 and $\theta= 1$).

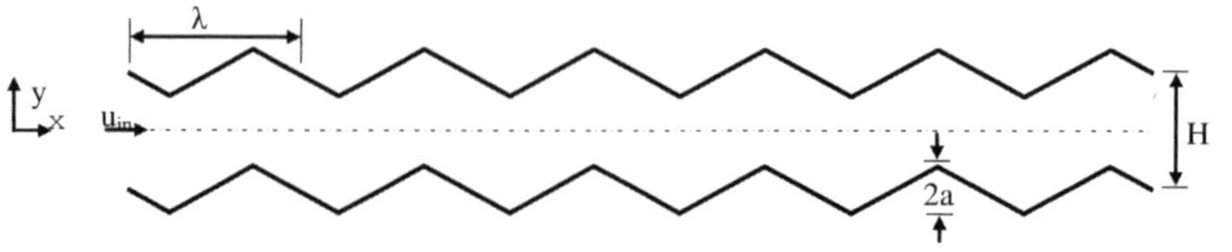

Figure 1. Physical model of the present study.

3. Governing equations and Thermophysical properties of nanofluid

Governing equation in non-dimensional form and in term of body-fitted coordinates for steady, laminar, incompressible and two-dimensional flow are:

$$\frac{\partial}{\partial \zeta}(\rho\varphi U^c)+\frac{\partial}{\partial \eta}(\rho\varphi V^c)=\Gamma\left[\frac{\partial}{\partial \zeta}(\beta_{11}\frac{\partial\varphi}{\partial \zeta})+\frac{\partial}{\partial \eta}(\beta_{22}\frac{\partial\varphi}{\partial \eta})+\frac{\partial}{\partial \zeta}(\beta_{12}\frac{\partial\varphi}{\partial \eta})+\frac{\partial}{\partial \eta}(\beta_{21}\frac{\partial\varphi}{\partial \zeta})\right]+S_{\varphi} \quad (1)$$

The thermophysical properties of the nanofluid used in present study are the ones used by Mohammed et al.[6].

4. Numerical method

The non-dimensional governing equations are discretized using finite volume method and solved iteratively using SIMPLE algorithm [9]. The convection term is discretized using power-law scheme and the collocated grid arrangement is used in this study. The under-relaxation of velocities and pressure is applied to achieve the convergence of numerical simulation and the convergence criterion is set to 10^{-5}.

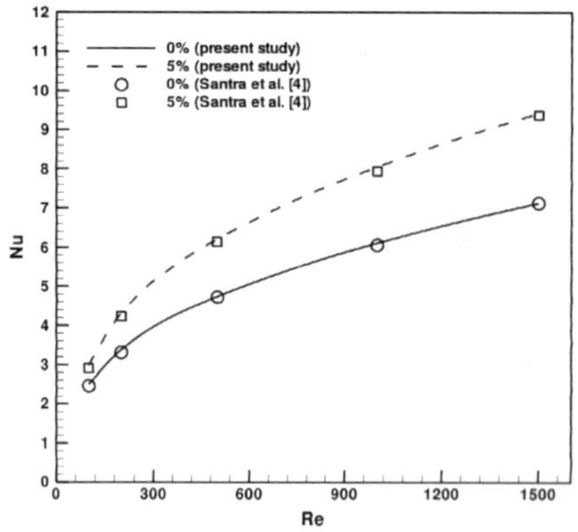

Figure 2. Average Nusselt number vs. Reynolds number.

5. Code validation and grid independence test

In order to validate the CFD code which is developed in this study, Nu for copper-water nanofluid flow in a straight channel is obtained and compared with previous numerical results of Santra et al.

[4], see Figure 2. The results are in good agreement. To test the grid independence, the grid sizes of 201x41, 301x51, 401x51, 401x61 and 501x61 are tested. It is found that the grid size of 401x61 can gives grid independent solution and is adopted in present study.

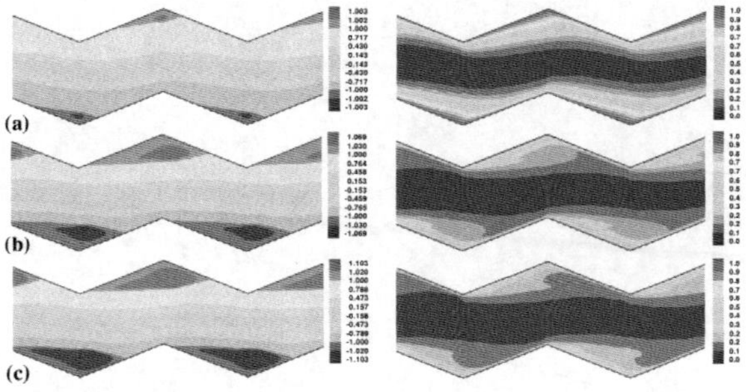

Figure 3. Streamlines (left) and isotherms contours (right) for different Reynolds numbers (a) Re=100, (b) Re=500 and (C) Re=800 at ϕ=5% and dp=30nm.

6. Results and discussion

Figure 3 shows that the streamlines and isotherms contours. It is found that from the streamlines contours the secondary flow regions appear in cavities near the upper and lower walls of channel. The size of these regions increase as Re increased and consequently the flow becomes more disturbed. Looking at isotherms contours, the temperature gradient near the walls of channel is increased with Re due to improve the mixing of the cold fluid in core with hot fluid near the walls. Figure 4 display that the non-dimensional pressure drop for different values of ϕ and dp. From Figure 4(a), it is observed that the pressure drop increase as ϕ increases due to increase the density and viscosity of nanofluid. Moreover, the nanoparticles diameters have no effects on the pressure drop, see Figure 4(b).

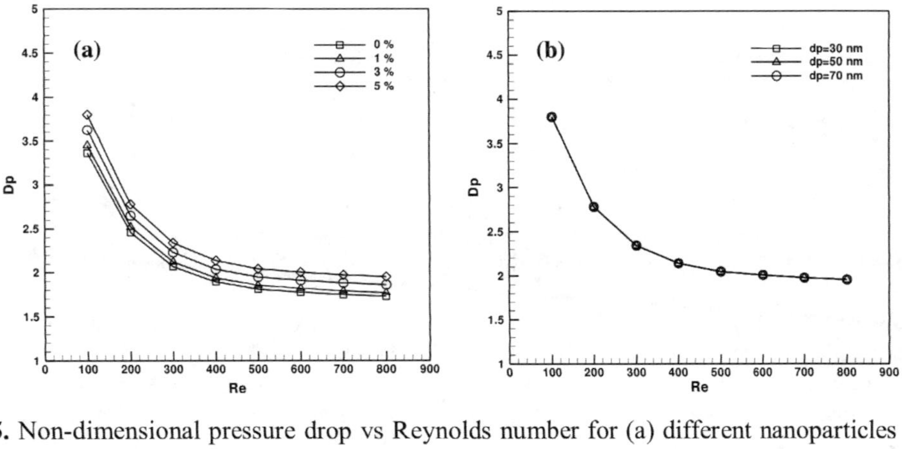

Figure 5. Non-dimensional pressure drop vs Reynolds number for (a) different nanoparticles volume fraction at dp=30 nm; (b) different nanoparticles diameters at ϕ= 5%.

The effect of nanoparticles volume fractions on Nu at dp= 30 nm is displayed in Figure 6(a). As expected, Nu increases with ϕ due to improve thermal conductivity of nanofluid. From Figure 6(b), it is found that Nu increases with a decreases in dp. Because of as dp decreases, the surface area of nanoparticle per unit volume and the Brownian motion of nanoparticles are increased.

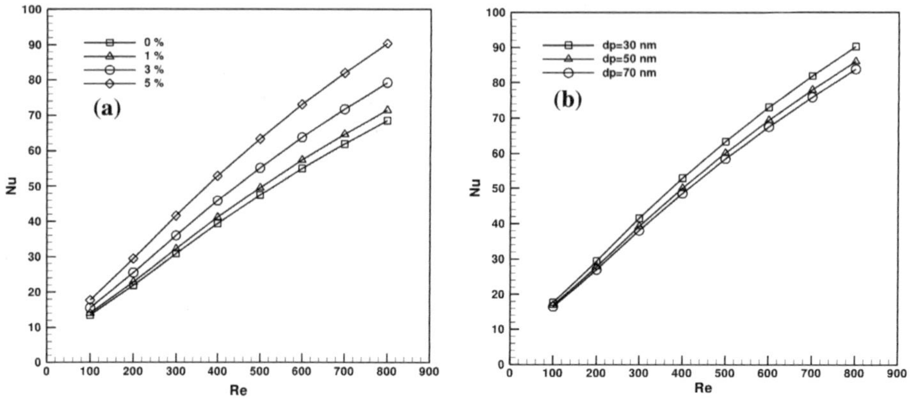

Figure 6. Nusselt number vs Reynolds number for (a) different nanoparticles volume fraction at dp= 30 nm; (b) different nanoparticles diameters at $\phi = 5\%$.

7. Conclusion

The results of numerical solution showed that Nu increases with increasing the Re, and nanoparticles volume fraction but the pressure drop penalty also increased. Further, as the nanoparticle size decreases Nu increased, while the nanoparticle size has no effect on the pressure drop. Based on the above results, the use of nanofluid in triangular-corrugated channel is a suitable method to achieve a good enhancement in the performance of many thermal devices.

Acknowledgements
The authors would like to sincerely thank the Ministry of Higher Education (MOHE) of Malaysia for the provision of a grant with code no. 20110106FRGS to support this work.

References

[1] Rush T A, Newell T A and Jacobi A M 1999 An experimental study of flow and heat transfer in sinusoidal wavy passages *Int. J. Heat and Mass Transfer* **42** 1541-1553.

[2] Wang C C and Chen C K 2002 Forced convection in a wavy-wall channel, *Int. J. Heat and Mass Transfer* **45** 2587–2595.

[3] Zhang L and Che D 2011 Influence of corrugation profile on the thermalhydraulic performance of cross-corrugated plates *Numerical Heat Transfer, Part A,* **59** 267-296.

[4] Santra A K, Sen S and Charaborty N 2009 Study of heat transfer due to laminar flow of copper-water nanofluid through two isothermally heated parallel plates *Int J.Thermal Science* **48** 391–400.

[5] Manca O, Nardini S and Ricci D 2012 A numerical study of nanofluid forced convection in ribbed channels *Applied Thermal Engineering* **37** 280-292.

[6] Mohammed H A, Al-Shamani A N and Sheriff J M 2012 Thermal and hydraulic characteristics of turbulent nanofluids flow in a rib-groove channel *Int Comm. Heat and Mass Transfer* **39** 1584-1594.

[7] Heidary H and Kermani M J 2010 Effect of nano-particles on forced convection in sinusoidal-wall channel, *Int Comm. Heat Mass Transfer* **37** 1520-1527.

[8] Ahmed M A, Shuaib N H and Yusoff M Z 2012 Numerical investigations on the heat transfer enhancement in a wavy channel using nanofluid *Int J. Heat and Mass Transfer* **55** 5891-5898.

[9] Versteeg H K and Malalasekera W 2007 An introduction to computational fluid dynamics the finite volume method 2nd edition. Longman Scientific and Technical, England.

IOP Conf. Series: Earth and Environmental Science **16** (2013) 012150 doi:10.1088/1755-1315/16/1/012150

One-dimensional analysis of unsteady flows due to supercritical heat addition in high speed condensing steam

N A Malek[1], H. Hasini[1] and M Z Yusoff[1]
[1]Centre for Advanced Computational Engineering, College of Engineering, Universiti Tenaga Nasional, Jalan IKRAM-UNITEN, 43000 Kajang, Selangor, Malaysia

E-mail: norhazwani@uniten.edu.my

Abstract. Unsteadiness in supersonic flow in nozzles can be generated by the release of heat due to spontaneous condensation. The heat released is termed "supercritical" and may be responsible for turbine blades failure in turbine cascade as it causes a supersonic flow to decelerate. When the Mach number is reduced to unity, the flow can no longer sustain the additional heat and becomes unstable. This paper aims to numerically investigate the unsteadiness caused by supercritical heat addition in one-dimensional condensing flows. The governing equations for mass, momentum and energy, coupled with the equations describing the wetness fraction and droplet growth are integrated and solved iteratively to reveal the final solution. Comparison is made with well-established experimental and numerical solution done by previous researchers that shows similar phenomena.

1. Introduction

The aim of every steam turbine design is an optimum efficiency operation. However, overall efficiency of a steam turbine power plant strongly depends on the turbine's performance. Design of the nozzle and the blades is important as it affects the performance of the turbine. Thus any slight improvement, can increase power availability, decrease costs, and generate savings in operation and development cost. While the High Pressure (HP) turbine and Intermediate Pressure (IP) turbine operate with superheated steam, Low Pressure (LP) turbine has to work with wet steam. This is due to the occurrence of supersaturation and spontaneous condensation in expanding the steam flow at this stage.

To understand the effect of spontaneous condensation on LP turbines, the fundamental process of nucleation must first be understood. When steam expands through a nozzle, droplets do not appear instantly as it crosses the saturation line, but it first becomes metastable or usually referred to as supersaturated or subcooled. Supersaturation is a necessary but not a sufficient precondition for spontaneous condensation to occur. Before it is initiated by either homogeneous or heterogenous nucleation, it must exceed the limiting condition, at which random kinetic motions of molecules create sufficient stable microclusters for equilibrium to be recovered through condensation. In contrast to heterogenous nucleation, homogeneous condensation occurs when there are no foreign nuclei, dust particles, ions, etc and when a rapid formation and growth of clusters from metastable to stable or equilibrium size starts. Condensation in a nozzle usually occurs in the divergent section just downstream of the throat with the consequence of the considerable release of latent heat, Q, to the flow. Apart from that, it also has the effect of moving the flow Mach number towards unity as the additional heat causes the flow to decelerate. The effect produced when Mach number equals to unity was described as "thermal choking" by Pouring [1]. This phenomenon results in flow instability and induces oscillations. Previously, flow oscillations induced by the supercritical heat addition during condensation of water vapour near the throat of converging-diverging nozzles have been observed in details by Pouring [1], Wegener [2], Barschdorff [3] and Skillings [4].

In the current study, a one-dimensional (1-D) time-marching compressible Euler solver has been developed to simulate the phenomenon in steam nozzle flows. It uses the second order cell-vertex finite volume spatial discretization and the standard fourth order Runge-Kutta temporal integration has been developed. As the flow is unsteady, the time accuracy of the scheme is retained. Inclusion of higher virial coefficient was made in order to increase the accuracy for the calculation of real gas properties. This model has been applied to the case of unsteady supercritical heat addition in Skilling's nozzle [4]. The shape of the nozzle is shown in Fig. 1. In this case, the inlet total temperature and inlet total pressure were set to be 347.9 K and 0.3514 bar respectively.

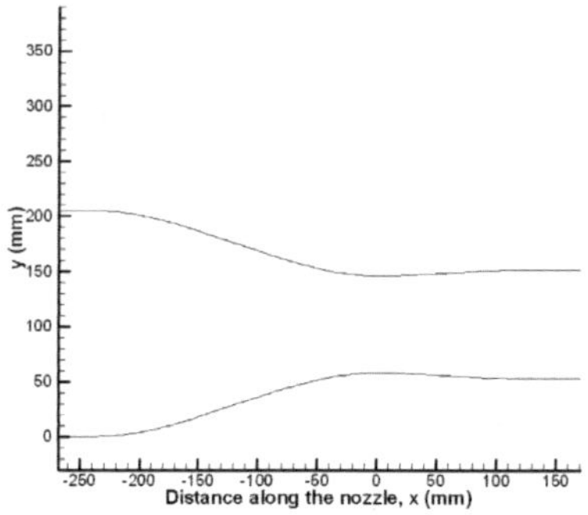

Figure 1. Skilling's nozzle [5].

2. The governing equations

The governing equations describing the conservation of mass, momentum and energy are written in the form of partial differential equations, as shown in Equations (1), (2) and (3) respectively:

Conservation of mass: $\frac{\partial(\rho A)}{\partial t}+\frac{\partial(\rho AV)}{\partial x}=0$ (1)

Conservation of momentum: $\frac{\partial}{\partial t}(\rho AV)+\frac{\partial}{\partial x}(\rho AV^2)=-A\frac{(\partial p)}{\partial x}$ (2)

Conservation of energy: $\frac{\partial}{\partial t}\left[\rho\left(e+\frac{V^2}{2}\right)A\right]+\frac{\partial}{\partial x}\left[\rho\left(e+\frac{V^2}{2}\right)AV\right]=-\frac{\partial}{\partial t}(\rho AV)$ (3)

The equation of state used for the calculation of real gas is truncated at second virial coefficient based on Keenan *et al.* [5]. It is shown in equation (4):

$$p=\rho RT_G(1+B\rho) \tag{4}$$

Where, R = 0.4615 kJ/(kg.K) is a gas constant and B is the second virial coefficient. The coupling between the vapour, G and droplet, L phase is by means of wetness fraction, which is denoted by ω:

$$\omega=\frac{m_L}{m_G+m_L} \tag{5}$$

The wetness fraction is introduced into the expression for mixture enthalpy h and density ρ.

3. Results and discussions

In order to investigate the unsteady effect, the frozen Mach number within the nozzle at different time intervals during one period of oscillations is plotted in Fig. 2. At time t = 0, it can be observed from the figure that the nucleation occurs some distance just downstream of the throat with the Mach number at the onset ranges from nearly above unity to below unity (1.02 – 0.94). The pattern is similar at time interval t = 0.2 T and t = 0.4 T, but the range of values is slightly lower with not less than 0.9. This is due to the supercritical heat addition that retards the flow and restores the sonic flow at the throat and condensation in the diverging section. As time progresses, the process then repeats itself giving rise to oscillations.

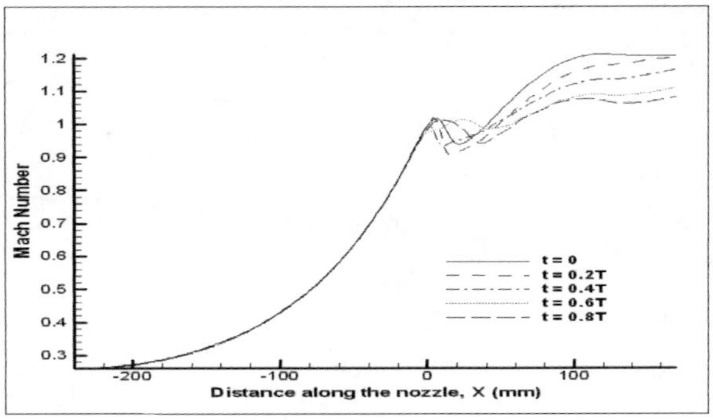

Figure 2. Frozen Mach number at different time intervals during one period of oscillations

The variations of static pressure at a point located just downstream of the nozzle's throat is plotted against the number of nucleation and time as in Fig. 3(a) and (b) respectively. As the flow becomes unstable in Fig. 3(a), the pressure decreases significantly and this pressure drops restores the sonic flow at the throat and gives rise to oscillations. In this case, regular oscillations with a frequency of 302 Hz can be clearly seen in Fig. 3(b). The amplitude of oscillations of about 17 mbar was also produced using the current method. The result is good compared to the experimental measurement made by Skillings [4], which is of frequency 380 Hz and amplitude of 18 mbar. He has also attempted to simulate similar one-dimensional model by using MacCormack's scheme. In the subsequent years, attempt was also been made to model the supercritical heat addition effect to one-dimensional flow using Denton's method by Guha [6]. Although the unsteady time-marching treatment for condensing flow of steam was adopted by Skillings [4] and Guha [6], however, the frequency was over-predicted, namely 540 Hz to 650 Hz respectively. This may be due to the use of different method in calculating the droplet growth compared to the present method. In recent decades, the study of supercritical heat addition in high speed condensing flow has been carried out by White and Young [7] and Yusoff [8]. The predicted frequency was calculated to be 420 Hz and 393 Hz respectively. Yusoff [8] has also predicted the amplitude of pressure of 20 mbar. The frequency and pressure difference of current results are compared with previous data obtained from other researchers and tabulated in Table 1.

The current model has shown good agreement with the experimental data of Skillings [4]. Reasonable agreement was also achieved when compared to numerical work carried out by other researchers. It is also noted that better agreement was obtained by two-dimensional models due to the significant velocity changes resulted from gradual changes in the nozzles' cross sectional area.

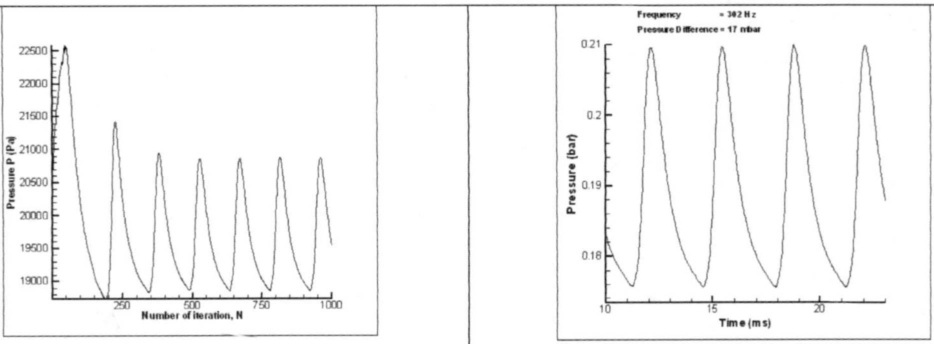

(a) (b)

Figure 3. Static pressure downstream of *Skilling's* nozzle's throat against (a) number of iterations and (b) time.

Table 1. Comparisons of frequency and pressure difference of current work with previous attempts.

Author	Case	Frequency, Hz	Pressure Difference, mbar
Skillings [4]	Experiment	380	20
Skillings [4]	1D program	640	-
Guha [6]	1D program	540	-
White and Young [7]	2D program	420	-
Yusoff [8]	2D program	393	18
Current work	1D program	302	17

4. Conclusions

This study reveals promising results indicating the capability of the model to calculate unsteady flow due to supercritical heat addition phenomenon which might cause instability in steam turbine blade channel. The heat addition causes the flow to retard and gives rise to oscillations. Comparison with published experiments and numerical simulations is generally good although the computed frequency and amplitude of oscillation is slightly lower. This might be due to fact that some of the nozzle geometries exhibit significant two-dimensional effects, and these are likely to be modified by boundary layer growth in two-dimensional case.

5. Acknowledgement

The author would like to acknowledge the financial support received from the Ministry of Higher Education Malaysia (MOHE) under the Fundamental Research Grant Scheme (2012122FRGS).

References

[1] Pouring A A 1965 *Phys. Fluids* **8(10)** 1802-1810
[2] Wegener P P 1969 *Non-Equilibrium Flows* (P.P. Wegener, ed.), Marcel Dekker, New York, **1** 163-243
[3] Barschdoff D 1970 *3rd Int. Conf., Rain Erosion and Allied Phenomena*, Farnborough
[4] Skillings S A 1987 *Ph.D Thesis*, University of Birmingham
[5] Keenan J H, Keyes F G, Hill P G 1978 *Steam Tables – Thermodynamic Properties Of Water Including Vapour, Liquids And Solid Phases* John Wiley and Sons
[6] Guha A 1990 *Ph.D. Thesis*, University of Cambridge
[7] White A J, Young J B 1993 *J. Propulsion and Power* **9(4)**.
[8] Yusoff M Z 1997 *Ph.D. Thesis*, University of Birmingham

IOP Conf. Series: Earth and Environmental Science **16** (2013) 012151 doi:10.1088/1755-1315/16/1/012151

Development of a Lagrangian Meshless Flow Solver based on the Moving Particle Semi-implicit (MPS) Method

K C Ng[1,2], Y L Ng[1] and M Z Yusoff[1]
[1]Centre for Advance Computational Engineering (CACE), Universiti Tenaga Nasional, Malaysia.

[2]Email: nkching@uniten.edu.my

Abstract. This paper discusses on the development of a meshless flow solver based on the Lagrangian particle method. The differential operators are discretised by using the particle interaction models proposed in the numerical framework of the Moving Particle Semi-implicit (MPS) technique. The MPS method is attractive from the viewpoint of no mesh is required and fluid is purely represented by points (the virtue of meshless algorithm). Some flow applications attempted by the current meshless solver will be shown and results are compared with the published experimental data.

1. Introduction

There are various numerical techniques available nowadays to simulate fluid flow. The mesh-based technique, particularly the Finite Volume Method (FVM), has gained immense popularity within the CFD community since the past few decades. In fact, most of the CFD commercial software has adopted FVM as the core numerical method in developing the fluid flow and heat transfer algorithms.

Although FVM has proven its versatility in many engineering problems, one should take note on the limitations of most of the mesh-based methods, for instances: inaccuracy due to inferior mesh quality, numerical diffusion due to convection discretization, smearing of highly rigorous free surface, interpolation error due to local remeshing when moving-boundary problem is encountered, and many others. The development of the Lagrangian particle method aims to circumvent the above issues. Fluids are treated as particles that are freely to move in space (Lagrangian); therefore, the diffusion due to convection discretization can be reasonably eliminated. Being meshless in nature, the information of mesh structure (such as normal area vector, volume of polygons, etc.) is not needed and the free surface can be traced easily and naturally; enabling one to capture the evolution (both space and time) of free surface effectively. MPS method has been used mainly to capture the rapidly changing flow interfaces such as those encountered in the dam-break problem [1], flow boiling [2] general two-phase flow [3] and even heat transfer problem such as Rayleigh-Benard convection [4].

The current paper intends to provide a brief introduction on the MPS method. In the following sections, the numerical methods of MPS will be highlighted, followed by its applications that have been attempted by the authors recently.

2. Numerical Method

The full incompressible Navier-Stokes equation is considered in the current work:

$$\frac{1}{\rho}\frac{D\rho}{Dt} + \nabla \bullet \vec{u} = 0, \quad (1)$$

$$\frac{D\vec{u}}{Dt} = -\frac{1}{\rho}\nabla P + \nu\nabla^2\vec{u} + \vec{g}. \quad (2)$$

Here, ν is the kinematic viscosity, ρ is the fluid density and $\vec{g}$ is the gravitational acceleration. The above equations are solved in the Lagrangian manner. The differential operators (gradient and Laplacian operators) appeared in the governing differential equations are treated in the following manner:

$$\nabla\phi_i = \frac{d}{n^0}\sum_{j\neq i}\frac{\phi_j - \phi_{\min}}{|\vec{r}_j - \vec{r}_i|^2}(\vec{r}_j - \vec{r}_i)w(|\vec{r}_j - \vec{r}_i|) \quad (3)$$

$$\nabla^2\phi_i = \frac{2d}{\lambda_i n^0}\sum_{j\neq i}(\phi_j - \phi_i)w(|\vec{r}_j - \vec{r}_i|) \quad (4)$$

where

$$\lambda_i = \frac{\sum_{j\neq i}|\vec{r}_j - \vec{r}_i|^2 w(|\vec{r}_j - \vec{r}_i|)}{\sum_{j\neq i} w(|\vec{r}_j - \vec{r}_i|)}. \quad (5)$$

$\vec{r}_i$ is the position vector of particle *i*. $\phi_{\min}$ is the minimum ϕ value of neighbouring particles within the radius of influence r_e. d is the number of dimensions (d=2 for 2D and d=3 for 3D). n^0 and $w(|\vec{r}_j - \vec{r}_i|)$ are the initial particle number density and kernel function respectively. The initial particle number of a particle i can be computed as the summation of kernel function $w(|\vec{r}_j - \vec{r}_i|)$ of its neighboring particles within its radius of influence.

The above equations are solved for each time step, following the fractional-step method for coupling between the continuity and momentum equations.

3. Result and Discussion

3.1. Sloshing in a Container

The MPS code is used to simulate the sloshing flow inside an oscillating tank. The boundary condition as well as the experimental data reporting on the time variation of the moving free-surface is available from [5].

During the simulation, the time step size and initial particle spacing is prescribed as 0.0005 sec. and 3.33mm, respectively. Figure 1 shows the comparison of instantaneous water level at t=1.652s between those tested and simulated by using the current method. As seen, the predicted data agree considerably well with those measured, showing that the current numerical procedure is suitable to predict liquid sloshing.

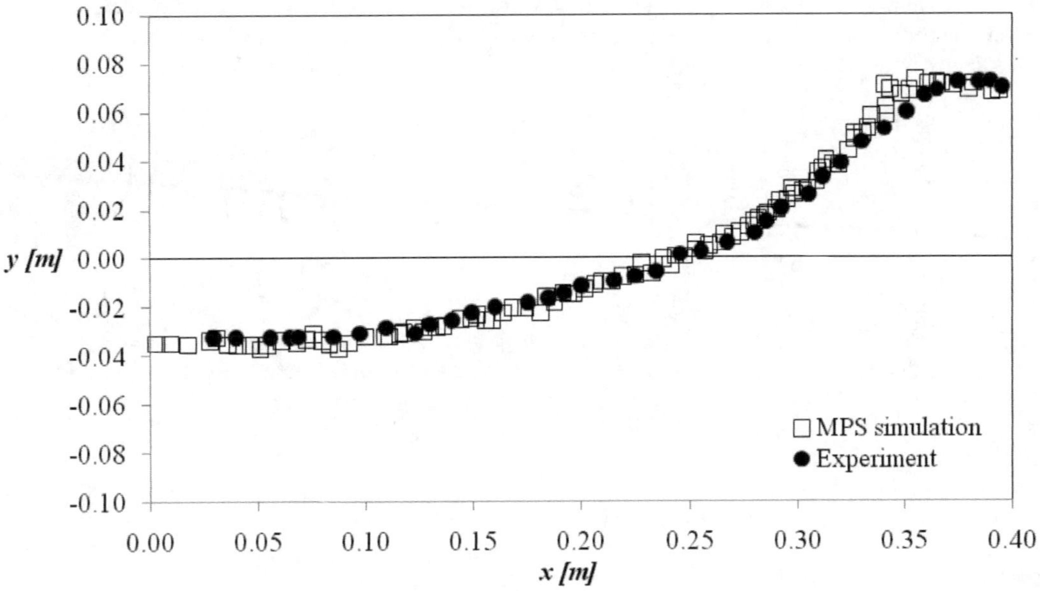

Figure 1. Comparison of free surfaces at t=1.652 sec.

3.2. Oscillating Impeller in a Cylindrical Tank

The test case considered here is the process of stirring of fluid in a cylindrical vessel agitated by a plate impeller with an infinitely long height. It has been anticipated by [6] that the reciprocating motion of the impeller may enhance the mixing performance. The sinusoidal motion of the impeller can be obtained from [6].

Figure 2 shows the instantaneous streakline at t~10s obtained from MPS and experimental observation [6]. The agreement is promising.

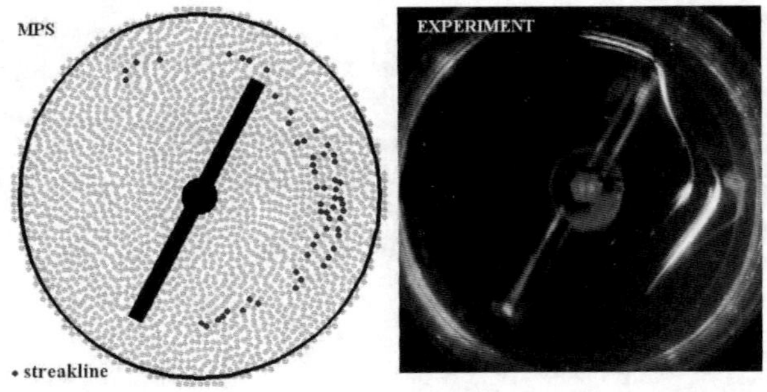

Figure 2. Streaklines obtained by MPS (left) and experimental method (right) at t~10s. Experimental data is obtained from [6].

3.3. Dam Break

Dam-break flow occurs when a column of water is subjected to collapse due to gravity after the sudden removal of a barrier. Figure 3 shows the predicted locations of water particles at t=1.0s. Indeed, by assuming laminar flow in the current work, rigorous movement of flow particles can be observed, and the proper description of eddy viscosity is effective, at least to certain extent, in suppressing these unphysical scatterings.

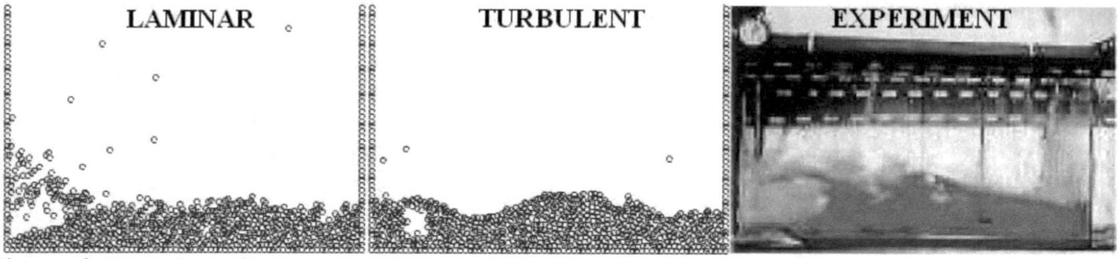

Figure 3. Location of water particles at t=1.0s. The experimental data is obtained from [7].

4. Conclusions

The basic algorithm of the meshless code based on Moving Particle Semi-implicit (MPS) method has been developed and verified against the experimental data. Future work would be the incorporation of more complex flow models in order to study other interesting flow physics such as granular flow, multiphase flow (solid-liquid, gas-liquid), etc. Meanwhile, we will be looking into some numerical aspects of MPS in order to enhance its numerical stability.

Acknowledgement

The first author is highly indebted to Professor Komoda and Professor Koshizuka for giving their written permissions in using the experimental data appearing in Figure 2 and Figure 3, respectively, in the current paper.

References

[1] Kondo M and Koshizuka S 2011 Improvement of Stability in Moving Particle Semi-Implicit Method *International Journal for Numerical Methods in Fluids* **65** 638-654.

[2] Koshizuka S, Ikeda H and Oka Y 1999 Numerical Analysis of Fragmentation Mechanisms in Vapor Explosions *Nuclear Engineering and Design* **189** 423-433.

[3] Shirakawa N, Yamamoto Y, Horie H and Tsunoyama S 2003 Analysis of Subcooled Boiling with the Two-Fluid Particle Interaction Method *Journal of Nuclear Science and Technology* **40 (3)** 125-135.

[4] Zhang S, Morita K, Fukuda K and Shirakawa N 2006 An Improved MPS Method for Numerical Simulations of Convective Heat Transfer Problems *International Journal for Numerical Methods in Fluids* **51** 31-47.

[5] Andrillon Y and Alessandrini B 2004 A 2D+T VOF Fully Coupled Formulation for the Calculation of Breaking Free-Surface Flow *Journal of Marine Science and Technology* **8** 159-168.

[6] Komoda Y, Senda S, Takeda H, Hirata Y and Suzuki H 2012 2D Fluid Deformation Induced by a Rotational Reciprocating Plate Impeller in a Cylindrical Vessel *14th European Conference on Mixing, Warszawa* **10-13** September 2012 211-216.

[7] Koshizuka S, Tamako H and Oka Y 1995 Particle Method for Incompressible Viscous Flow with Fluid Fragmentation *Computational Fluid Dynamics Journal* **4** 29-46.

The study of pressure source term in moving particle semi-implicit (MPS)

Y L Ng[1,2], K C Ng[1], and M Z Yusoff[1]
[1]Centre for Advance Computational Engineering (CACE), Universiti Tenaga Nasional, Malaysia.

Email: Yluon@uniten.edu.my

Abstract. In this paper pressure source terms proposed by different researcher are compared in term of pressure stability. In Moving Particle Semi-Implicit (MPS) method, the pressure error is compensated back into pressure Poisson equation as source term in order to enforce the incompressibility. However the non physical pressure fluctuation still remains an open problem in MPS method. In present study the hydrostatic model is used as example to study the pressure stability for each of the pressure source terms. The pressure at the fixed point over time is plotted and the fluctuation from the theoretical pressure is discussed.

1. Introduction

Most of the conventional computational fluid dynamics (CFD) employs finite volume or finite element method. The result generated is highly affected by the quality of the descretized domain, or so called grid. Recently, a new class of solving approach, called particle or meshless method had drawn much research interest. The particle methods are the Largrangian meshless method where particles are free to move in given space. This result in more feasible and effective performance in highly deformed, free surface flow, especially the problem involving complicated boundary and fluid/structure coupling compare with conventional gird based approach.

One of the particle methods, the Moving Particle Semi-Implicit (MPS) is first introduced by Koshizuka [1] for incompressible flow problems. The Navier-Stoke equation is solved by two different stages: the viscous part, which is solved explicitly, and the pressure gradient part, which require Poisson equation of pressure to be solved beforehand.

The density deviation error is compensating back as source term in Poisson pressure equation to take account of the density change after the particle moved. However the original pressure source term in [1] often results in high pressure fluctuation and even lead to solver failure.

In this paper a pressure source terms with different modification will be compared. In the following section, the pressure source terms will be introduced, followed by the investigation on pressure stability for different source terms.

2. Numerical Method

The continuity and Navier Stokes equation are:

$$\frac{1}{\rho}\frac{D\rho}{Dt} + \nabla \cdot \vec{u} = 0 \qquad (1)$$

$$\frac{D\vec{u}}{Dt} = -\frac{1}{\rho}\nabla P + v\nabla^2\vec{u} + \vec{g} \quad (2)$$

Where v is the kinematic viscosity, ρ is the fluid density, and $\vec{g}$ represents the gravitational acceleration. The equation (2) is divided into viscous and pressure term, where

$$\left(\frac{D\vec{u}}{Dt}\right)^{viscous} = v\nabla^2\vec{u} + \vec{g} \quad (3)$$

$$\left(\frac{D\vec{u}}{Dt}\right)^{pressure} = -\frac{1}{\rho}\nabla P \quad (4)$$

The pressure term in equation (4) is differentiated to get the Poisson equation, and to be solved implicitly.

$$\nabla \cdot \vec{u} = 0$$
$$\frac{D}{Dt}(\nabla \cdot \vec{u}) = \nabla \cdot \frac{D\vec{u}}{Dt} = 0$$
$$\frac{1}{\rho_0}\nabla^2 P = \nabla \cdot \frac{D\vec{u}}{Dt} = 0 \quad (5)$$

In ideal case, the equation (5) is equated to zero when the fluid is incompressible by nature. That is, the volume remains unchanged over time. However as the particle is f4 to move, the local density may not satisfy the incompressibility all the time. Hence, the error of the density will need to compensate back to equation (5). Assuming the fluid density is proportional to particle number density (PND), the density error is compensated back to equation (5) as:

$$\frac{1}{\rho_0}\nabla^2 P = -\frac{\rho}{\Delta t^2}\frac{n_i^* - n_0}{n_0} \quad (6)$$

The left hand side is descretized by diffusion model. The right hand side of the Poisson equation is the deviation of the temporal particle number density (PND) from the standard value. The term is used to maintain the PND to enforce the incompressibility. However the source term also induces large pressure instability, and the pressure at f4 surface will always be negative hence need special treatment.

In order to obtain smoother result, Kondo and Koshizuka [2] had proposed a source term consists of main part and two error compensating parts which multiplied by coefficient to improve the high frequency numerical oscillation of original source term in equation (6).

$$\frac{1}{\rho_0}\nabla^2 P = \frac{-1}{n^0\Delta t^2}\left[\left(n_i^* - 2n_i^k + n_i^{k-1}\right) + \beta\left(n_i^k - n_i^{k-1}\right) + \gamma\left(n_i^k - n_0\right)\right] \quad (7)$$

Where the β and γ are the constant satisfying $0 \leq \gamma \leq \beta \leq 1$. When both β and γ equal to 1, the equation reduced to original source term in equation (6).

Instead of fixed constant throughout the entire calculation for all vertices, Khayyer and Gotoh [3] had proposed the β and γ that depended on instantaneous flow feature. The error compensating term will cancel out each other when the already expanded fluid is being compressed, or vice versa.

$$\beta = \left|\frac{n^k - n^0}{n^0}\right| \qquad \gamma = \left|\left(\frac{\Delta t}{n^0}\left(\frac{Dn}{Dt}\right)^k\right)\right| \tag{8}$$

3. Result and Discussion

A hydrostatic model is used as test case for the pressure stability calculation. The dimension of the model is follow as in [2]. The initial particle spacing is 0.012 and the time step size is 0.001 s. The weight kernel in present study is same as in [2]. In present study only the different source terms are tested with same condition. Other improvement steps for particular literature such as surface treatment and pressure gradient correction matrix in [3] are not implemented. Figure 1 shows the pressure history at point A with different pressure source term used. Theoretical pressure is plotted as red dashed line.

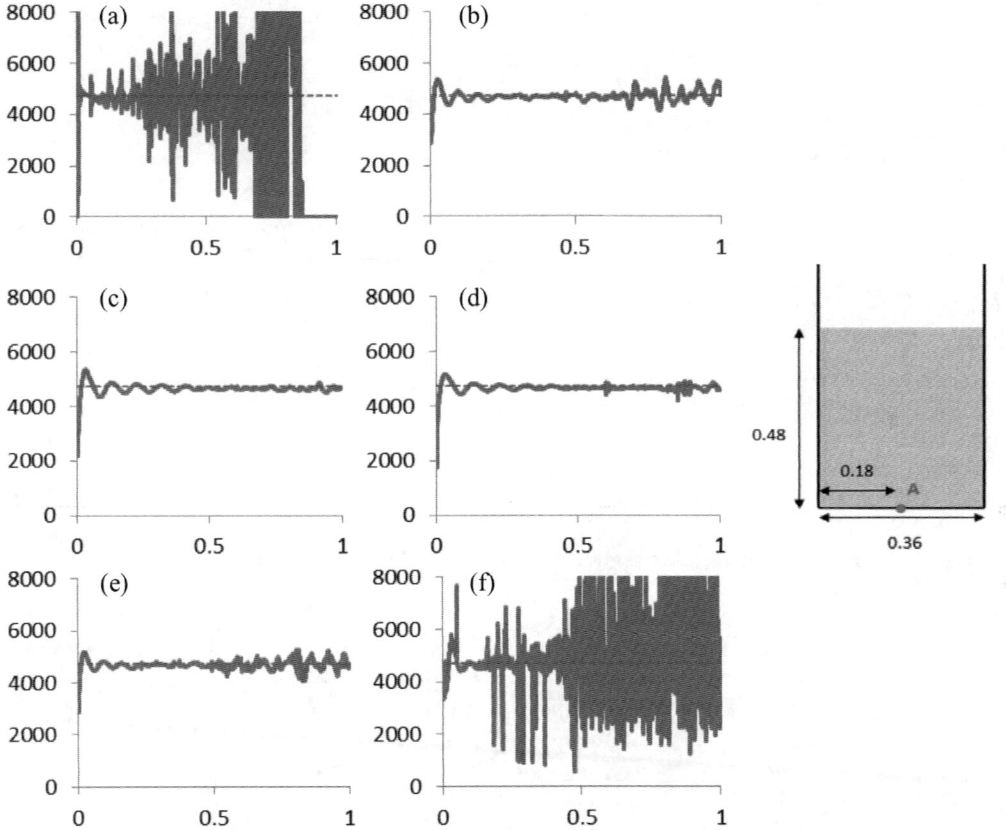

Figure 1. Right: Hydrostatic model used. Left: Pressure at point A over time. With source term from (a) equation (6), (b) equation (7), $(\beta, \gamma) = (0.5, 0.1)$, (c) equation (7), $(\beta, \gamma) = (0.5, 0.05)$, (d) equation (7), $(\beta, \gamma) = (1.0, 0.1)$, (e) equation (7), $(\beta, \gamma) = (1.0, 0.2)$, (f) (β, γ) through equation (8)

From the results obtained, the original source term without any coefficient (equation (6)) give large pressure fluctuation as shows in figure 1(a), and with magnitude of the fluctuation increased as the time progress. This pressure fluctuation eventually leads to solver failure after 0.9s. This is due to the nature of the source term with any small change in position will directly reflect in pressure change without any "buffer", and cause the rapid change of pressure as the particle moves.

From (b) to (e), Different β and γ are tested for the source term in equation (7). The pressure history obtained is much closer to theoretical value compare with case (a). It is found that in general the ratio of 1/10 of β and γ give better result, with (β, γ) of (0.5, 0.05) produced most stable result compared with other value. From the result obtained, the larger γ is found able to reduce the initial pressure fluctuation. However it will cause higher pressure noise as the time progress.

From the result in (f), the pressure history of Khayyer and Gotoh is surprisingly shows very high pressure fluctuation despite the claim of smoother pressure from the theory and literature. The β and γ obtained from equation (8) are much smaller than optimal value from Kondo and Koshizuka in [2] and current study as shows in figure 2 below. This may explain the large pressure fluctuation. Another reason of the high pressure fluctuation may be due to different weight function as used in [3], and lack of implementation of pressure gradient correction matrix.

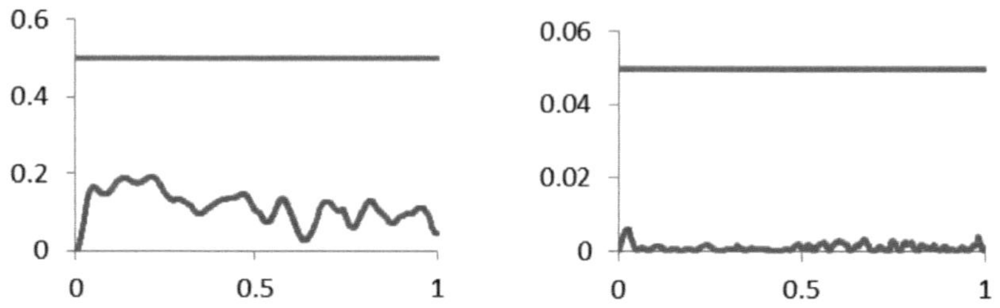

Figure 2. Compare of the β and γ from ideal case (blue line) with from equation (8) (red curve). Left: β versus time, Right: γ versus time

4. Conclusions

The performance of the different pressure source term in the MPS had been compared in term of pressure stability. It is found that the pressure stability can be improved more by fine-tuning the relevant constant in the pressure source term studied, compared with through instantaneous flow feature. Although this may be cause by the lack of implementation of some other improvement steps in the later source term. Even though the smooth pressure can be obtained by current parameter, however the best parameter in one case may not be ideal for other flow type. In future, more investigation will be done in determine the ideal constant for the more violence flow type.

References

[1] Koshizuka S, Oka Y, Tamako H 1995 A particle method for calculating splashing of incompressible viscous fluid. International Conference *Mathematics and Computations, Reactor Physics, and Environment Analyses* **2** 1514-1521.

[2] Kondo M, Koshizuka S 2010 Improvement of stability in moving particle semi-implicit method *Int. J. Numer. Methods Fluids* **10** 1002-2207.

[3] Khayyer A, Gotoh H 2011 Enhancement of stability and accuracy of the moving particle semi-inplicit method *Journal of Computational Physics* **230** 3093-3118.

[4] Tanaka M, Masuaga T 2010 Stabilization smoothing of pressure on MPS method by quasi-comp4sibility *Journal of Computational Physics* **229** 4279-4290.

[5] Kondo M, Koshizuka S December 3-6, 2006 Supressing pressure oscillation in MPS fluid analysis *Proceedings of the Third APCOM in Conjunction with 11th EPMESC, Kyoto.*

[6] Lee B H, Park J C, Kim M H, Hwang S C, 2011 Step-by-step improvement of MPS method in simulating violent f4-surface motions and impact-loads *Comput. Methods Appl. Mech. Engrg* **200** 1113-1125.

IOP Conf. Series: Earth and Environmental Science **16** (2013) 012153 doi:10.1088/1755-1315/16/1/012153

Power quality problems and solutions

A Atputharajah[1], **V K Ramachandaramurthy**[2], **J Pasupuleti**[2]
[1]*Department of Electrical and Electronic Engineering, University of Peradeniya, Peradeniya, 20400, Sri Lanka*
[2]*Power Quality Research Group, Department of Electrical Engineering, Universiti Tenaga Nasional, 43000, Kajang, Selangor, Malaysia*

E-mail: atpu@ee.pdn.ac.lk, vigna@uniten.edu.my

Abstract. This paper addresses the power quality problems and solutions related to (i) wind power generation and (ii) large industrial customers. Intermittent nature of wind power affects the quality of its electrical power output. The usage of low cost energy saving equipments also generates harmonics and voltage flicker thus affects the quality of supply voltage. A wide rang of solutions are being already proposed with the development on power electronic devices and electrical machines. This paper is organized to discuss the (i) power quality problems in brief, (ii) four major developments in wind generator technologies and (iii) solutions to large industrial customers using Dynamic Voltage Restorer (DVR), Active Power Filter (APF) and Static Compensator (STATCOM). Finally development in power electronic devices' control is briefed in each of the devices, which has utilizes these power electronic devices with integrated solutions to solve number of power quality problems.

1. Introduction to the power quality problems

The common power quality problems are voltage sag, voltage flicker, harmonics and switching transients. Mostly these are caused by intermittent nature of the non conventional power generations, large industrial non linear loads, power system faults due to thunder attack, transients due to switching of capacitor and large load feeders. Malfunction of devices and random failures of components are some results of power quality issues such as harmonic, voltage sags, flickering and interruptions [1, 2].

The conventional power generation technologies are well developed and they produced high quality of power. However, the popular renewable energy generation technologies, such as wind and solar, were in developing phase to maximise the power extraction while satisfying the grid code requirements. The research on wind generation technology and interconnection of solar plants using power electronics has brought the power quality at the grid connection point within the acceptable level [3]. The industrial loads are mostly driven by power electronic devices, which are mostly based on chopping techniques of supply voltages to achieve energy efficient operations. These kind of large industrial loads also severely affect the quality of power supply. However, the latest technologies on rectifiers and variable speed drives, which are used by modernised industries, reduce the harmonic injection. Some external devices such as APF (filter harmonics from load), DVR (compensate voltage sag from supply), STATCOM (applied to both load and supply sides) are also used to maintain power quality from both industries and utility [4].

IOP Conf. Series: Earth and Environmental Science **16** (2013) 012153 doi:10.1088/1755-1315/16/1/012153

2. Power quality improvement and development in wind generator technologies

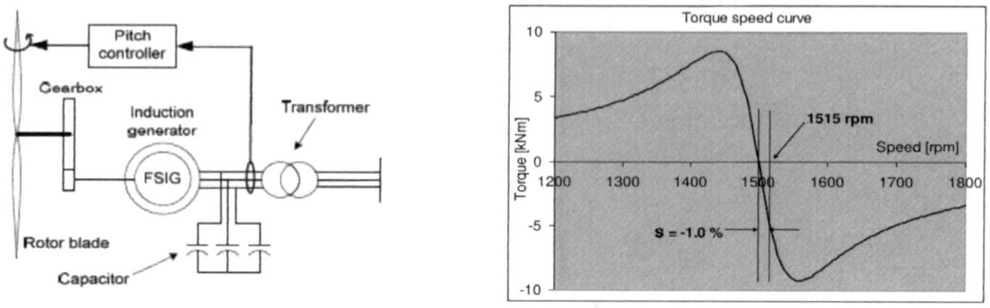

Figure 1. Fixed Speed Induction Generator Technology

Fixed Speed Induction Generator (FSIG), shown in Figure 1, is one of the oldest generator technology used in wind power development. Here, the generator speed is almost fixed with less than 5% of speed adjustment. Power Factor Correction capacitors are used at the generator terminal to compensate the reactive power absorbed by the generator. The speed control especially at high wind speeds is achieved by pitch control. Later this technology was improved with changing of number of poles in two steps, which allows the turbine to operate around two speeds.

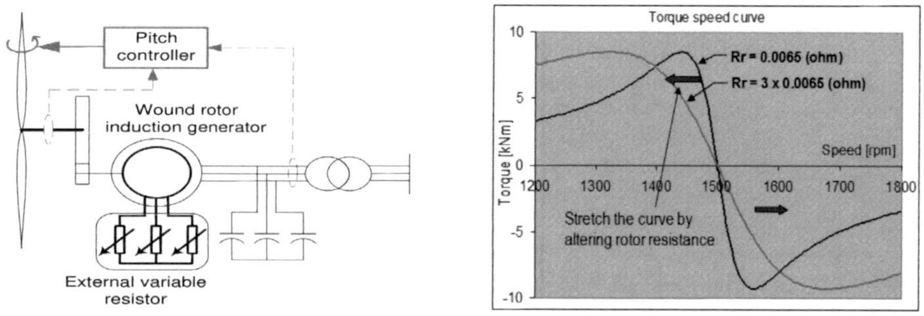

Figure 2. Opti-slip Induction Generator Technology

The second generation of the wind turbine was the opti-slip wind turbine, shown in Figure 2. The wound rotor induction machines were used with an external resistor bank at the rotor. The effective rotor resistance was varied through a power electronic to vary the operating speed by about 10%. Hence, it improves the power quality and reduces the mechanical loading on turbine components. However, the copper losses across the rotor resistors decrease the efficiency of the wind turbine.

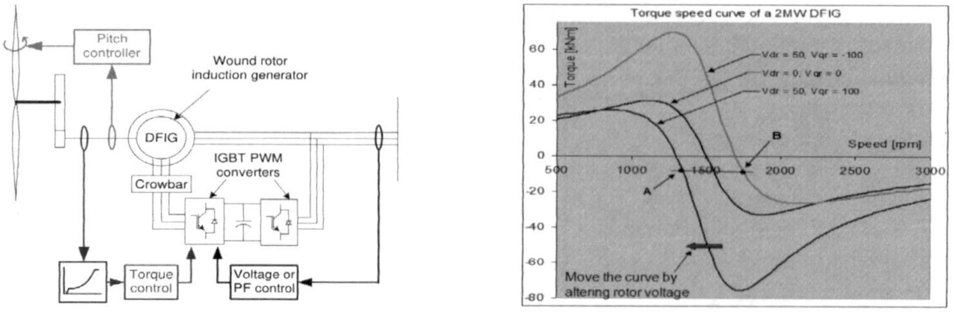

Figure 3: Doubly Fed Induction Generator Technology

The Doubly Fed Induction Generator (DFIG) was introduced to cover larger speed range, shown in Figure 3. The rotor can be operated at a different electrical frequency by absorbing or injecting active power by the rotor converter. Hence, the generator speed can be varied about 40% of its rated. Decouple operation of the DFIG results in the following advantages: (i) terminal voltage or power factor control, (ii) reduced mechanical stress, (iii) maximized power extraction and (iv) improved fault ride through performance. However, fault transient affects the generator since the stator is directly connected with the grid.

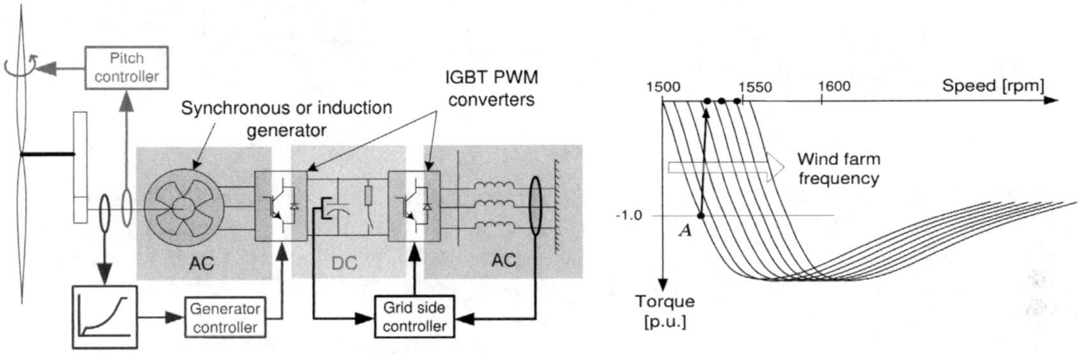

Figure 4. Full Range wind Generator Technology

The full range wind generation technology is shown in Figure 4. The generator is fully decoupled from the grid through a back to back voltage source converter. Since the generator has the full control of speed and decoupled from the grid, any network transients do not affect the generator dynamics and vice versa. The generator side converter to supplies the reactive power and controls the generator torque by shifting torque slip curve. The grid side converter maintains the dc link voltage at its reference value by controlling the active power flow to the grid. Wide speed operation, flicker elimination, direct control on real and reactive power, maximum power extraction, reduced mechanical stress and full controllability during a fault are some of the advantages exists in this wind turbines. The evolution of wind turbine technologies has resulted in power quality issues such as flicker and harmonics. These issues can be solved using a STATCOM or advance wind turbine controllers.

3. Power quality improvement at Industrial loads through power electronic devices

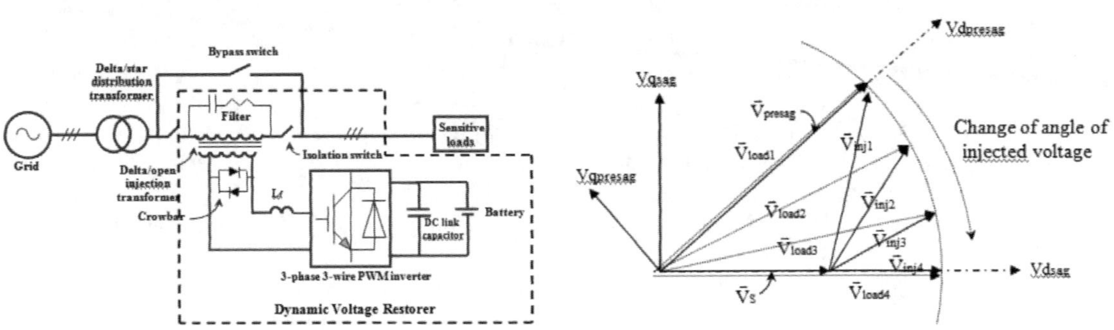

Figure 5. Dynamic Voltage Restorer for large industrial loads

The DVR, shown in Figure 5, is used to compensate voltage sags, swells and flicker from power grid and maintain the power quality at the load side. In DVR controls: pre-sag, in-phase and hybrid of both compensation techniques have been well reported. Further, minimising energy storage requirement is also addressed through reactive power injection to boost the voltage by injecting voltage 90 degree leading with respect to load current [4, 5].

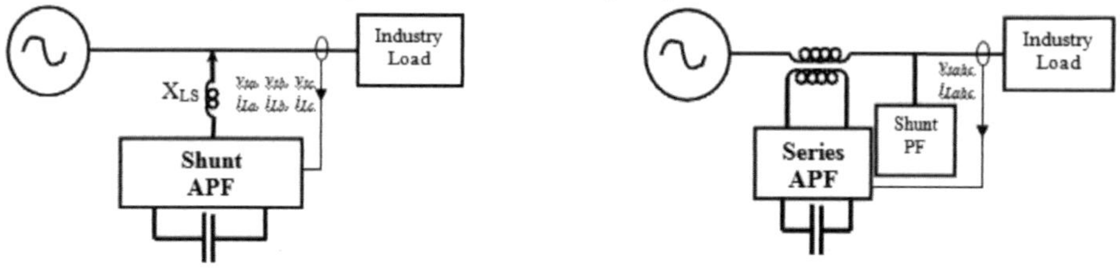

Figure 6. Active Power Filters (series and shunt)

The APFs are used to compensate or block the harmonic current injection from the industrial load centres [6]. Shunt APF control technique extracts the harmonic current components from the measured load currents and absorbs the harmonic current. Series APF block the harmonic current by injecting voltage, which represent the high impedance to the harmonic while low impedance to the fundamental. Therefore, it forces the harmonic current to be absorbed by the Passive Filter (PF). The STATCOM is also has similar configuration of shunt APF but with system level control to look after supply and load side problems [7]. All stable control technique must obtain the phase angle, where the vector based PLL [8] shows excellent performance in the power system control applications.

4. Conclusion

This paper has summarized the development of wind generator technologies, power quality improving devices for large industrial load customers with the brief on device integrated control concepts. This overall analysis on solutions for power quality problem will help to develop a premium city electricity supply concept.

References

[1] Stones J, Collinson A, 2001 Power quality. *IEE Power Engineering J.*, 58-64.

[2] Rodney H G, Ramachandaramurthy V K, 2012 Voltage Sag Acceptability Assessment Using Multiple Magnitude-Duration Function. *IEEE Transactions on Power Delivery*, **27(4)**, 1984 – 1990.

[3] Arulampalam A, Ramtharan G, Jenkins N, Ramachandaramurthy V K, Ekanayake J B, Strbac G, 2007 Trends in wind power technology and grid code requirements, *IEEE second International Conference on Industrial and Information Systems*, 129 - 134.

[4] Ramachandaramurthy V K, Fitzer C, Arulampalam A, Zhan C, Barnes M, Jenkins N, 2002. Control of a battery supported dynamic voltage restore. *IEE Proceedings on Generation Transmission and Distribution*, **149(5)**, 533-542.

[5] Changjiang Z, Ramachandaramurthy V K, Arulampalam A, Fitzer C, Kromlidis S, Barnes M, Jenkins N, 2001 Dynamic voltage restorer based on voltage-space-vector PWM control. *IEEE Transactions on Industry Applications*, **37(6)**, 1855-1863.

[6] Hirofumi A. Trends in Active Power Line Conditioners, 1994, IEEE Transaction on Power Electronics, **9(3)**, 263-268.

[7] Arulampalam A, Barnes M, Jenkins N, Ekanayake JB., 2006. Power quality and stability improvement of a wind farm using STATCOM supported with Battery Energy Storage. *IEE Proceedings on Generation Transmission and Distribution*, **153(6)**, 701 - 710.

[8] Changjiang Z., Fitzer C, Ramachandaramurthy V K, Arulampalam A, Barnes M, Jenkins N., 2001. Software phase-locked loop applied to dynamic voltage restorer (DVR). *IEEE Power Engineering Society Winter Meeting*, **3**, 1033-1038.

Power System Study for Renewable Energy Interconnection in Malaysia

O F Askar[1], **V K Ramachandaramurthy**[2]
[1]Engineering Executive, Energy Commission Malaysia
[2]Power Quality Research Group, Department of Electrical Engineering, UNITEN

E-mail: omarali@outlook.com, vigna@uniten.edu.my

Abstract. The renewable energy (RE) sector has grown exponentially in Malaysia with the introduction of the Feed-In-Tariff (FIT) by the Ministry of Energy, Green Technology and Water. Photovoltaic, biogas, biomass and mini hydro are among the renewable energy sources which offer a lucrative tariff to incite developers in taking the green technology route. In order to receive the FIT, a developer is required by the utility company to perform a power system analysis which will determine the technical feasibility of an RE interconnection to the utility company's existing grid system. There are a number of aspects which the analysis looks at, the most important being the load flow and fault levels in the network after the introduction of an RE source. The analysis is done by modelling the utility company's existing network and simulating the network with the interconnection of an RE source. The results are then compared to the values before an interconnection is made as well as ensuring the voltage rise or the increase in fault levels do not violate any pre-existing regulations set by the utility company. This paper will delve into the mechanics of performing a load flow analysis and examining the results obtained.

1. Introduction

The renewable energy industry in Malaysia has seen a steady uprising in the past year largely due to the introduction of the Feed-In-Tariff incentive by the ministry of Energy, Green Technology and Water. The FIT in an incentive handed out by the government to entice industry players to commit to long term renewable energy generation with a substantial profit gain for the industry. This incentive is within the government's effort to promote renewable energy plants that will reduce the amount of generation supplied from conventional power plants which simultaneously reduces the amount of CO2 released into the atmosphere by the burning of fossil fuel.

With the introduction of the FIT, an overseeing entity was set up to implement and ensure the fairness and transparency of the system. The Sustainable Energy Development Authority (SEDA) is the regulatory body which has been tasked by the government to handle the renewable energy industry while operating at all times within the guidelines of the Renewable Energy Act. [1] There are a number of pre-requisites one must have to be awarded with the FIT and most notably in the list of the pre-requisites is the Power System Study (PSS). This paper will look in to the technical aspect of the PSS to give readers a brief overview of the process involved in performing a PSS.

---

[1] To whom any correspondence should be addressed.

IOP Conf. Series: Earth and Environmental Science **16** (2013) 012154 doi:10.1088/1755-1315/16/1/012154

2. Power System Study Overview

In this section we discuss the overall aspect of the PSS and the parameters observed when performing a PSS. The PSS is valid for a number of renewable energy sources such as *Photovoltaic, Mini-Hydro, Biogas* and *Biomass*. There are four main aspects to the PSS. They include;

- Load Flow Analysis
    - Power Flow
    - Voltage Regulation
    - Network Losses
- Short Circuit Analysis
- Dynamic Analysis (Photovoltaic Plants)

The PSS is conducted by using DigSilent Power Factory, a power systems analysis program tool developed by **DIgSILENT GmbH,** a Consulting and Software Company providing highly specialized services in the field of electrical power systems for transmission, distribution, generation and industrial plants. These analyses are performed based on the loading in PSS Adept, obtained from the utility company in Malaysia.

2.1 Load Flow Analysis

2.1.1 Power Flow analysis is performed to determine the direction of power flow in the distribution network with the interconnection between TNB and the proposed power plant. This would determine the adequacy of the distribution network to absorb the active power from the proposed PV plant. The load flow analysis is performed under steady state conditions. It is important to note the utility's requirement that reverse power flow must not occur back to the transmission network.

2.1.2 Voltage regulation analysis is performed to determine if the steady state voltage levels at the busses are within the limits planned and declared by TNB at $\pm$ 5% variation during normal operating conditions. If and when these voltages are above the limits declared by TNB, the interconnection of an RE source becomes impossible unless certain measures are taken to reduce the voltage levels to acceptable values. One such measure is to control the power factor of the generator to induce a drop in the voltage levels.

2.1.3 Network Losses analysis is performed to determine the effect on network losses with the connection of an RE plant. The losses which are taken into consideration for the PSS is the interconnection cable losses between the RE plant and TNB's network as well as the transformer losses at the RE plant. The overall network losses is not discussed in the PSS. However, it has been shown that there is a reduction in losses for the overall network with the interconnection of an RE plant.

2.2 Short Circuit Analysis

Short circuit analysis is performed to indicate that all nodes are within the equipment short time rating as specified by TNB, hence complying TNB's circuit breaker duty capability. Fault level data for the 132 kV buses at the source is obtained from TNB Transmission. These values are used in DigSilent to calculate the source impedance as well as the downstream distribution buses short circuit values. Based on the results of short circuit analysis, the operational safety of the network configuration for interconnection between the proposed power plant and TNB substation can be determined.

2.3 Dynamic Analysis

For PV plants, a dynamic analysis is important since the power generation is constantly varying due to the changing solar radiation throughout the day. The varying power generation injected into the utility network is bound to create voltage fluctuations at the interconnection point and other buses upstream on the grid. The objective of the dynamic analysis is to determine the level of voltage fluctuations. At present, there is no guideline in TNB regulations, the Malaysian Distribution Code, or Malaysian Standards on the maximum percentage of voltage fluctuation allowed due to intermittent generation like solar PV, although the Malaysian Distribution Code allows voltage fluctuation of ± 3 % for load switching.The following limits for intermittent generation are provided by some of the international codes;

Table 1. Summary of International Codes for Intermittent Voltage Fluctuation [2]

Standards	*Lower limit*	*Upper limit*
IEEE 1547	*-8.3%*	*+5.8%*
German Grid Code	*±2% - Once in 3 minutes*	*±5% - Generator disconnections*
Spanish Grid Code	*-3 to 5%*	*+3 to 5%*

It appears that the maximum voltage fluctuation range that can be safely allowed is 6 % (±3). Beyond that, there is a danger of utility and consumer equipment getting heated up. The dynamic analysis is carried out in DigSilent using the load data from PSS Adept and solar radiation data which is recorded using a pyrometer. The simulation is typically carried out for the period 7am to 7pm, with 3 minute interval readings of the solar radiation.

3. Example of Load Flow Analysis

In this section, an example of the results obtained from a load flow analysis will be shown. An arbitrary study is taken and the results are displayed below;

- Voltage

  Load flow studies are performed on the system to assess the impact of on the voltage at the sub-stations involved, with and without the RE plant.

Table 2. Summary of Voltage Impact at TNB Substations

At point of connection PE ABC			*At PMU DEF (33kV)*		
Without RE	*With 1MW$_{PK}$ PV*	*Diff [%]*	*Without RE*	*With 1MW$_{PK}$ PV*	*Diff [%]*
11.028	*11.104*	*0.7*	*33.902*	*33.902*	*0.0*

Findings: Slight change of voltage at PE ABC. No significant steady-state voltage variation as a result of Solar PV system connection. Change in voltage is within ±5% limits.

- Summary of Losses

  Losses level is compared based on overall simulated network with and without the PV connection.

IOP Conf. Series: Earth and Environmental Science **16** (2013) 012154 doi:10.1088/1755-1315/16/1/012154

Table 3. Summary of Losses

Losses	*TROUGH LOAD*			
	Without interconnection		*With PV Plant (1 MW_{PK})*	
	P (kW)	*Q (kVAr)*	*P (kW)*	*Q (kVAr)*
Losses in 1 km ABC 240mm^2 cable (11kV to PE)	-	-	*0.05*	*-1.03*
Transformer Losses	-	-	*1.00*	*23.88*

Findings: There is a slight increase in losses with introduction of Solar PV generation. However, the overall losses in the distribution network is reduced with the introduction of PV generation.

4. Conclusion

A power system study is done to determine the technical feasibility of an RE interconnection to the utility company's existing grid system. As discussed, there are a number of aspects which the analysis looks at, the most important being the load flow and fault levels in the network after the introduction of an RE source. The PSS will thus determine if it is at all possible for an RE interconnection to occur. Once the PSS is done, the developer is required to obtain the report and use it to apply for the FIT therefor making the PSS a first step in developing an RE plant.

References

[1] Sustainable Energy Development Authority Malaysia (www.seda.gov.my)
[2] Grid Code, High and extra high voltage, EON Netz, German TSO

AUTHOR INDEX

AUTHOR INDEX

AUTHOR INDEX

AUTHOR INDEX